AF581904

Radar and Sonar Imaging and Processing

Radar and Sonar Imaging and Processing

Editors

Andrzej Stateczny
Krzysztof Kulpa
Witold Kazimierski

MDPI • Basel • Beijing • Wuhan • Barcelona • Belgrade • Manchester • Tokyo • Cluj • Tianjin

Editors

Andrzej Stateczny
Gdansk Technical University
Poland

Krzysztof Kulpa Warsaw
University of Technology
Poland

Witold Kazimierski
Maritime University of Szczecin
Poland

Editorial Office
MDPI
St. Alban-Anlage 66
4052 Basel, Switzerland

This is a reprint of articles from the Special Issue published online in the open access journal *Remote Sensing* (ISSN 2072-4292) (available at: https://www.mdpi.com/journal/remotesensing/special_issues/radar_sonar_imageprocessing).

For citation purposes, cite each article independently as indicated on the article page online and as indicated below:

LastName, A.A.; LastName, B.B.; LastName, C.C. Article Title. *Journal Name* **Year**, *Volume Number*, Page Range.

ISBN 978-3-03943-971-3 (Hbk)
ISBN 978-3-03943-972-0 (PDF)

Contents

About the Editors

Andrzej Stateczny is Professor at Gdansk Technical University Poland and President of Marine Technology Ltd. His research interests are mainly centered on navigation, hydrography, and geoinformatics. Current RF research activities include radar navigation, comparative navigation, hydrography, artificial intelligence methods focused on image processing and multisensory data fusion. He has been the Principal Investigator or Co-Investigator in a wide range of research projects in both civil and defense fields. He has published or presented over 200 journal and conference papers in the above areas, including numerous books such as "Radar Navigation", "Comparative Navigation", "Methods of Comparative Navigation", and "Artificial Neural Networks for Marine Target Recognition". He has headed many research projects and supervised the completion of 16 doctoral theses.

Krzysztof Kulpa received his M.Sc., Ph.D. and D.Sc. degrees from the Department of Electronic Engineering, Warsaw University of Technology (WUT) in 1982, 1987 and 2009, respectively. From 1985 to 1988, he was employed at the Institute of Electronic Fundamentals, WUT. From 1988 to 1990, he was an associate professor at the Electrical Engineering Department of the Technical University of Bialystok. From 1990 to 2005, he worked as a scientific consultant at WZR RAWAR. Since 1990, he has been a professor at the Institute of Electronic Systems in WUT. In 2014, he was appointed as a full professor by the President of Poland. Currently, Prof. Kulpa is the head of the Radar Technology Research Group and the Scientific Director of the Defense and Security Research Center at WUT. Prof. Kulpa's research interests are in the areas of the digital signal processing and radar signal processing. Specifically, his research interests include 2D and 3D maneuvering target tracking, maritime patrol radar, low RCS target detection and tracking, noise and passive radars and synthetic aperture radar imaging. His most recent research interest is that of airborne passive radars. Prof. Kulpa's work has been implanted in several radars produced by the Polish radar industry, and he was involved in the creation of the first Polish SAR system. Presently, he is involved in several research projects related to PCL, ESA and Noise radars, as well as SAR and ISAR imaging.

Witold Kazimierski is Associate Professor at the Maritime University of Szczecin, Poland. He serves as Chair of Geoinformatics in the Faculty of Navigation. He used to work at sea as a navigational officer and as an offshore hydrographer. He graduated from the Research and Innovation Management Course at University of California, Berkeley. He leads a research team in various projects focused on spatial data processing and analysis. His main research activities cover anti-collision systems, radars, data fusion, sensor integration, hydrography, and the use of artificial intelligence in the aforementioned areas. He has published or presented around 100 journal and conference papers and acts as reviewer for numerous international journals and research agencies. He has been the Principal Investigator or Co-Investigator in a wide range of research projects which he has also chaired in some cases. He holds 2 patents as a co-author of anti-collision and decision support systems for vessels. He has also recently been focusing on autonomous surface and underwater vehicles.

Editorial

Radar and Sonar Imaging and Processing

Andrzej Stateczny [1,*], Witold Kazimierski [2] and Krzysztof Kulpa [3]

[1] Department of Geodesy, Gdansk University of Technology, 80-233 Gdansk, Poland
[2] Department of Geoinformatics, Maritime University of Szczecin, 70-500 Szczecin, Poland; w.kazimierski@am.szczecin.pl
[3] Institute of Electronic Systems, Warsaw University of Technology, 00-665 Warszawa, Poland; kkulpa@elka.pw.edu.pl
* Correspondence: andrzej.stateczny@pg.edu.pl; Tel.: +48-609-568-961

Received: 27 May 2020; Accepted: 2 June 2020; Published: 3 June 2020

Abstract: The 21 papers (from 61 submitted) published in the Special Issue "Radar and Sonar Imaging Processing" highlighted a variety of topics related to remote sensing with radar and sonar sensors. The sequence of articles included in the SI dealt with a broad profile of aspects of the use of radar and sonar images in line with the latest scientific trends. The latest developments in science, including artificial intelligence, were used.

Keywords: radar; sonar; data fusion; sensor design; target tracking; target imaging; image understanding; target recognition

1. Introduction

Over the last few years, radar and sonar technology has been at the center of several major developments in remote sensing in both civilian and defense applications. Although radar technology has existed for more than 100 years, it is still developing and it is now implemented in many maritime, air, satellite, and land applications. New technologies, such as sparse image reconstruction and multistatic active and passive SAR and ISAR imaging, are changing the quality of images and areas of application. The rapid development of automotive radars in 3D dimensions, able to recognize different objects and assign the risk of collision, is one example of the progress of this technology. In maritime radars, the application of FMCW technology is becoming more and more popular, aside from classical pulse radars. Simultaneously, sonar technology has also been used for dozens of decades, at the beginning only for military solutions but, today, using 3D versions, it is used for many underwater tasks, such as underwater surface imaging, target detections, and tracking, among others. The impact of sonar technologies has been growing, particularly at the beginning of the autonomous vehicle era. Recently, the influence of artificial intelligence on radar and sonar image processing and understanding has emerged. Radar and sonar systems are mounted onboard smart and flexible platforms and also on several types of unmanned vehicles. Both of these technologies focus on the remote detection of targets and both may encounter many common scientific challenges. Unfortunately, specialists from the radar and sonar fields do not interact much with each other, slowing down progress in both areas.

The Special Issue entitled "Radar and Sonar Imaging and Processing" was focused on the latest advances and trends in the field of remote sensing for radar and sonar image processing, addressing original developments, new applications, and practical solutions to open questions. The aim was to increase the data and knowledge exchange between these two communities and allow experts from other areas to understand the radar and sonar problems.

In this article we provide a brief overview of the published papers, in particular the use of advanced modern technologies and data fusion techniques. These two areas seem to be the right direction for the future development of radar and sonar imaging and processing.

2. Overview of Contributions

2.1. Radar Imaging and Processing

The radar research presented in the Special Issue included many application fields from satellite level observation via airplane levels and maritime navigation and safety for ground and underground investigation.

The new method of parallax correction for clouds observed by geostationary satellites is presented by Bielinski [1]. The parallax shift effect of clouds occurs in satellite imaging, especially in the case of the high angles of satellite observations. The developed methods were compared with a known analytical method, namely the Vicente et al./Koenig method. It approximates the position of the cloud by means of an ellipsoid with the half-axis increased by the height of the cloud with an error of up to 50 m. The next two methods proposed in the article allow for significant error reduction. The first method proposed by the author, being an extended version of the Vicente et al./Koenig method, allows researchers to reduce the error to centimeters. The second method, by adjusting the number of iterations, allows researchers to reduce the error to a value close to zero. The article presents an example procedure of a numerical solution using the Newton method and also describes a simulation experiment, verifying the proposed methods. Due to the fact that the resolution of a functioning geostationary earth observation (EO) satellite currently ranges from 0.5 km to 8 km and the pixel dimensions are much larger than 50 m, the proposed method will be applied when the resolution of geostationary EO satellites reaches the assumed 50 m.

New satellite computing capabilities and extended applications for SAR imaging products have resulted in research into real-time synthetic aperture radar imaging. The orbit determination data of the SAR platform in space is essential for the SAR imaging procedure. In the case of real-time SAR imaging, the orbital determination data on board cannot reach a level of accuracy equivalent to the orbital ephemeris in ground-based SAR processing, which requires long processing times using the commonly used ground-based SAR imaging procedures. It is important to investigate the impact of errors in real-time orbiting data on the quality of the SAR imaging. Yan et al. [2], instead of the commonly used numerical simulation method, proposed an analytical model of square phase error approximation (QPE) introduced by orbit determination errors. The model can provide approximation results at two granulations: approximation with the true anomaly of the satellite as an independent variable and approximation for all positions in the whole orbit of the satellite. The proposed analytical approximation model reduces the complexity of the simulation, the calculation range, and the processing time. Moreover, the model reveals the essence of the process in which errors are transferred to the QPE calculations. A detailed comparison of the proposed method with the numerical simulation method demonstrates the accuracy and reliability of the analytical approximation model.

Due to advantages such as its low power consumption and higher concealment, deceptive jamming against synthetic aperture radar (SAR) has received extensive attention during the last few decades. However, large-scene deceptive jamming is still a challenge because of the huge computing burden. Yang et al. [3] propose a new large-scene deceptive jamming algorithm. First, the time-delay and frequency-shift (TDFS) algorithm is introduced to improve the jamming processing speed. The system function of the jammer (JSF) for a fake scatter is simplified to the multiplication of the scattering coefficient, a time-delay term in the range dimension and a frequency-shift term in the azimuth dimension. Then, in order to solve the problem that the effective region of the TDFS algorithm is limited, the scene deceptive jamming template is divided into several blocks according to the SAR parameters and the imaging quality control factor. The JSF of each block is calculated by the TDFS algorithm and added together to achieve the large-scene jamming. Finally, the correction algorithm in squint mode is derived. The simplification and parallel-block processing could improve the calculation efficiency significantly. The simulation results verified the validity of the algorithm.

Another interesting approach to SAR data processing is presented by Chen et al. [4]. As a result of the method developed by the authors, image quality and depth of field have been significantly improved. The improved method enables the efficient processing of high resolution and low frequency SAR data in a wide range. It is commonly known that synthetic high resolution, low frequency aperture radar (SAR) has severe phase-to-immutaneous coupling due to its high bandwidth and long integration time. High resolution SAR processing methods are essential to concentrating the raw data of such radars. The generalized surgical scaling algorithm (GCSA) is widely accepted as an attractive solution to focus low frequency, high bandwidth, and wide beam SAR systems. However, as bandwidth and/or beam width increases, severe phase coupling reduces the performance of the current GCSA and degrades imaging quality. This degradation is mainly due to two main reasons: the residual high order phase coupling and the insignificant error introduced by linear fixed phase point zoom using the stationary phase principle (POSP). The authors first present the principle of determining the required range frequency sequence. After compensating for the independent feedback phase sequence above the third order, the GCSA's analytically improved GCSA statement based on the Lagrange inversion is derived. The Lagrange inversion allows for the accurate compensation of the coupling phase dependent on the high order range. The results of the imaging of the SAR data in the P and L bands indicate the excellent performance of the proposed algorithm compared to the existing GCSA.

The phenomenon of the periodical penetration of synthetic aperture radar (SAR), which is induced in various ways, creates challenges in concentrating raw SAR data. To deal with this problem, Qian and Zhu [5] propose a new method. Complex deconvolution is used to reconstruct the azimuthal spectrum of the complete data from the raw data acquired in the proposed method. In other words, the proposed method provides a new approach to dealing with periodically extracted raw SAR data using complex deconvolution. The proposed method provides a robust implementation of deconvolution to process raw data obtained from azimuth. The algorithm consists mainly of the phase compensation and recovery of the azimuth spectrum of raw data using complex deconvolution. The obtained data become less frequent in the Doppler domain after phase compensation. Then, it is possible to recover the azimuth spectrum of complete raw data by complex deconvolution in the Doppler domain. Then, the traditional SAR imaging algorithm is able to focus on the reconstructed raw data in this work. The effectiveness of the proposed method has been confirmed by simulating a point and surface target. Furthermore, actual SAR data was used to better demonstrate the validity of the proposed method.

Appreciating the great importance of synthetic aperture radar (SAR) image processing in the range of moving targets to be defocused due to unknown motion parameters, an effective algorithm to change the focus of SAR for moving targets is presented in [6]. For fast-moving targets, range cell migration (RCM), Doppler frequency migration, and Doppler ambiguity are complex problems. As a result, focusing on fast-moving targets is difficult. The algorithm proposed by Wan et al. [6] consists mainly of three stages. First, the RCM is corrected by reversing the sequence, multiplying the matrix complex and improving the second order RCM correction function. Secondly, a 1D scale Fourier transform is introduced to estimate the remaining chirp speed. Thirdly, a matched filter based on the estimated chirp speed is proposed to focus the maneuvering target in the azimuth time range. The method described in the paper is computationally effective as it can be implemented by a fast Fourier transform (FFT), reverse FFT, and uneven FFT. A new deramp function is proposed to further solve the serious Doppler ambiguity problem. A procedure for incorrect peak recognition based on cross-sectional analysis is proposed. Simulated and actual data processing results demonstrate the validity of the proposed targeting algorithm and false peak recognition procedure.

An interesting approach to imaging using interferometer radars with inverted synthetic aperture (InISAR) was presented by Zhang et al. [7]. A technique involving the strong scattering of fusion centers (SSCF) was proposed in order to estimate the parameters of the translational movement of the maneuvering target. Compared to previous InISAR image recording methods, the SSCF technique is beneficial due to its high computational efficiency, excellent anti-nose performance, high recording precision, and simple system structure. Thanks to InISAR's one-dimensional, three-output terahertz system, the parameters of translational motion in both the azimuth and height directions are precisely estimated. First of all, motion measurement curves are taken from the spatial spectra of independent strong dispersion centers, which allows researchers to avoid the adverse effects of noise and the "angular scintillation" phenomenon. Next, translational motion parameters are obtained by matching motion measurement curves to phase unwinding and intensity-weighted fusion processing. Finally, ISAR images are accurately captured by compensating for the estimated translational motion parameters, and high quality InISAR imaging results are obtained. The validity of the proposed method was proven by both simulation and experimental results.

The use of radar techniques to classify aircraft objects was undertaken by Wang et al. [8]. With conventional narrow-band radars, detectable target information is limited, and the radar has difficulty in accurately identifying the type of target. In particular, the probability of classification can be further reduced if some echo data are omitted. By extracting target characteristics in the time and frequency domains from the scarce echo data of multi-wave gateways, a classification algorithm in the conventional narrowband radar is presented to identify three different types of aircraft target, i.e., helicopter, propeller, and jet. The classical algorithm for the reconstruction of a weak echo of an object is used to reconstruct the frequency spectrum of single-wave gateways with weak echo data. The micro-Doppler effect caused by rotating parts of different targets is analyzed, and then features based on the reconstructed echo data are extracted, such as the amplitude deviation factor, wave entropy in the time domain, and wave entropy in the frequency domain, in order to identify targets. Finally, the target characteristics that were extracted from the multi-wave gateways of the reconstructed echo data are weighted and combined to improve classification accuracy. Finally, the vectors of the combined elements are fed into the support vector machine model (SVM) for classification. The presented algorithm can effectively process scarce echo data and achieve a higher classification probability by combining the characteristics of weighted multi-wave gateway echo data. The results of simulation tests confirming the correctness of the algorithm are presented.

The problem of protection against the common occurrence of small unmanned aerial vehicles (UAV) in recent years has been addressed by Nowak et al. [9]. UAV, popularly known as drones, are used to carry out many tasks, but they are mainly used for observation by both private individuals and professionals. Intrusions into the airspace of airports or other dangerous events involving drones have been observed. More and more attention is being paid to finding solutions to prevent such incidents. The cost analysis excludes in many cases the idea of building stationary UAV detection systems. It seems to be advisable to develop mobile anti-drone systems using continuous wave frequency modulated radars (FMCW). The common operation of the radar chain requires that the measurements be reduced to a common reference surface and that the direction of the radar is uniform in relation to the north. Adequate measurement of the constant corrections of the measured angles is a necessity in this case. The authors propose a method involving the quick, simultaneous calibration of a set of mobile FMCW operating in a network. The method has been tested by means of a numerical experiment consisting of 95,000 tests. Satisfactory results were obtained to confirm the assumptions made by improving the north orientation of the radar over the whole range of initial errors. The conducted experiments allow researchers to put forward a thesis about the advisability of practical use of the proposed method.

A major part of the Special Issue covered topics related to the maritime use of radar. In the article by Hessner et al. [10], the authors used X-band marine radar (MR) to obtain data on sea surface currents. The quality of the measurements was verified by the control system working in near real time. The obtained results were verified by appropriate measurements using a Doppler acoustic current measurement device (ADCP). Numerous experiments were carried out under various wave, current, and weather conditions. The obtained results confirmed the accuracy and reliability of marine surface currents MR measurements.

Another example of the use of marine navigation radar, this time in the task of collision prevention, can be found in the article by Lisowski and Mohamed–Seghir [11]. The authors present a method of optimizing collision prevention maneuvering in the navigator's decision support system. The decision-making process is presented as a multi-stage optimization in a fuzzy and game environment. In the decision-making process, objective and subjective navigation parameters are analyzed. An interesting experiment was conducted on the basis of the actual navigation situation of passing three encountered ships in the Skagerrak Strait, with good and limited visibility at sea. According to the authors, the presented solution can be practically implemented in the decision support system of the ship's navigator.

The next example utilizing automotive radar sensors in the 3D variant in the task of collision prevention can be found in the article by Stateczny et al. [12]. Measuring the missions of unmanned vehicles, especially in autonomous missions mode, requires the detection and identification of objects both on the water and in the shore zone. The authors present the empirical results of their research on 3D automotive radar's detection capabilities in water environments, which can be used in the future development of tracking and collision prevention systems for autonomous surface vehicles (ASV). The conducted experiments concerned the field of radar vision and determination of the detection range in terms of the detection of various objects, both floating and fixed on the shore. The obtained results confirm the usefulness of automotive radars for navigation tasks on bodies of water for small ASVs performing measurement missions, especially performing tasks in an autonomous mode.

Another application of the 3D sensor, this time for future oriented road signs that can display the speed limit autonomously in cases where the road situation requires it, is presented by Czyzewski et al. [13]. Future oriented road signs contain a number of types of sensors, among which the Doppler sensor and acoustic probe, improved by the authors, are presented in the article. The authors present the method of vehicle detection and tracking, as well as the determination of vehicle speed, on the basis of Doppler sensor signals working on continuous waves. The algorithm for counting vehicles and determining their direction of movement by means of an acoustic vector sensor was also tested experimentally with the use of an improved Doppler radar and a developed sound intensity probe. The authors also present the assumptions of the method using the spatial distribution of sound intensity as determined by means of an integrated (3D) sound intensity sensor.

After space, aeronautical, marine, and land-based applications, it is now the turn of the subsurface application. Kang et al. [14] proposed a three-dimensional underground cavity detection network (UcNet) to prevent the collapse of furrows in complex urban roads based on radar images (GPR). UcNet is being developed based on a convulsive neural network (CNN) integrated with the phase analysis of super-resolution GPR images. CNNs are popularly used for the automatic classification of GPR data, as the interpretation of GPR mass data from urban roads by experts is usually cumbersome and time consuming. However, conventional CNNs often provide erroneous classification results due to the similar characteristics of earth granules automatically taken from any underground objects such as cavities, wells, gravels, subsoil backgrounds, etc. In particular, properties unrelated to cavities are often wrongly classified as actual cavities, which reduces the performance and reliability of the neural network. UcNet improves the detection of underground cavities by generating SR GPR images of cavities taken from the neural network and analyzing their phase information. The proposed UcNet is experimentally verified using GPR data collected on site from complex urban roads in Seoul, South Korea. The results of the validation test reveal that the incorrect classification of underground cavities is significantly reduced compared to conventional CNN cavities.

2.2. Sonar Imaging and Processing

Sonar imaging and processing covers a wide set of methods and techniques aiming at better detection and interpretation of the data and information acquired with underwater acoustic systems. A relatively wide variety of topics is also presented in the papers published in this Special Issue, relating not only to the processing of raw measurements but also to sonar image analysis, up to fusion with multi-beam echosounders. The issues undertaken relate to side-scan sonars, multi-beam sounders, and synthetic aperture sonar, aiming at better formulation and understanding of the acquired information. Most of the proposed solutions were verified with real data and some in simulations.

In Zhang et al. [15], the authors described multi-receiver synthetic aperture sonar (SAS) and propose a new method for providing high resolution images in systems. The idea is to overcome the problem of the approximation of the point target reference spectrum (PTRS), azimuth modulation, and coupling term in signal processing, as it results in the degradation of the accuracy of the obtained images. In the proposed method, the PTRS, azimuth modulation, and coupling term are deduced based on the accurate time delay. They are further exploited to develop the imaging processor, which compensates the coupling phase based on the sub-block processing method. It is also important that the proposed imaging scheme can be easily extended to any other PTRS, as it does not require the series expansion of the PTRS with respect to the instantaneous frequency. Thus, a novel imaging algorithm for the multi-receiver SAS, based on the accurate time delay and numerical evaluation method, is composed. The proposed method was verified firstly in simulation and then with real data. The results showed that it achieves high performance results compared with traditional methods. Based on simulations, it has been shown that the effectiveness of the traditional method in focusing is significantly reduced, as indicated by the residual error. The new method overcomes this problem, resulting in more accurate images from the multi-receiver SAS.

Other papers are focused more on image processing than imaging itself. Ye et al. [16] proposed a modified Retinex algorithm (known for its image processing) for processing sonograms in order to perform gray scale correction. The original side-scan sonar image has uneven gray distribution, which affects the interpretation of the side-scan sonar image and the subsequent image processing. Various algorithms were proposed to overcome this problem, including Retinex. The authors propose the modification of it and the goal is to achieve comparable accuracy with less computational and time complexity. The idea is to apply sonar image characteristics in the algorithm, and thus an enhanced Retinex method is obtained. Compared with the commonly used gray scale correction methods for side-scan sonar images, this method avoids limitations such as the need to know the side-scan sonar parameters, the need to recalculate or reset the parameters for different side-scan sonar image processing, and the poor image enhancement effect. The method was verified with a large set of real data. The research showed that, compared with the latest image enhancement algorithms based on Retinex, the methods have similar image enhancement indexes, and our method is the fastest. When it is necessary to adjust the brightness of the corrected image, only the magnitude of constant coefficient A in the algorithm needs to be adjusted. Usage of the method provides a good basis for further image processing.

Interesting research on the processing of side-scan sonar images aiming at detection of targets is presented by Wang et al. [17]. Taking into account the fact that the denoising and detecting of underwater sonar images is crucial for the proper interpretation of the image, the authors proposed a new adaptive approach for this. Firstly, an adaptive non-local spatial information denoising method based on the golden ratio is proposed, and then, a new adaptive cultural algorithm (NACA) is proposed to accurately and quickly complete the underwater sonar image detection in this paper. For denoising, the method makes use of earlier developments found in the literature; however, the thresholds for an adaptive non-local spatial information denoising method are calculated based on the golden ratio. For detecting NACA, the study makes use of an adaptive initialization algorithm based on the data field (AIA-DF) and then modification of the quantum-inspired shuffled frog leaping algorithm (QSFLA) is proposed—a new update strategy is adopted to update cultural individuals. The experimental results, as presented in the paper, demonstrate that the proposed denoising method can effectively remove noise and reduce the difficulty of the following underwater sonar image recognition. The method is also faster and has more advantages in its search ability. Thus, it can be considered an effective and important method for underwater sonar image detection, resulting in feature extraction for effective seabed topography.

Another important issue in side-scan sonar image processing is bottom tracking, which is examined by Yan et al. [18]. The research aimed at proposing a new method for real-time bottom tracking based on artificial intelligence (Convolutional Neural Network—CNN) for the processing of an image. Bottom tracking can be effectively used for accurately obtaining the sonar height from the seabed by finding the first echo that reaches the seabed. This knowledge about sonar height is crucial for the proper interpretation of sonar images. The proposed approach consists of three steps for obtaining effective bottom tracking. First, according to the characteristics of the side-scan backscatter strength sequences, positive and negative samples are extracted, representing, respectively, the bottom sequences and water column and seabed sequences to establish the sample sets. Secondly, a one-dimensional CNN is designed and trained by using the sample set to recognize the bottom sequences. Thirdly, a complete processing procedure of the real-time bottom tracking method is established by traversing each side-scan ping datum and recognizing the bottom sequences. This approach introduces the use of a deep learning algorithm for solving the problem, while most of the methods which have been used up until now have been based on fixed thresholds and deterministic numerical filtering. The method is verified with real measured data. The experimental results described in the paper showed that the proposed method is highly robust to the effects of noise, rich seabed texture, and artificial targets and proved its accuracy and real-time performance. The average bottom tracking accuracy reached for the experimental data was 94.7% with a 4.5% miss-ping rate and 99.2% excluding the missing data, showing that the method provides an effective algorithm for bottom tracking.

Sonar data processing may also be an important issue for navigation. Stateczny et al. [19] indicate that underwater sonar data can be processed with big data methods. In this particular research, 3D sonar data were processed and the purpose was the near real-time processing for so-called comparative navigation. A new approach of acquiring and simultaneously processing a set of bathymetric observations is presented. It includes fragmentary data acquisition and fast reduction (the optimum dataset method—OptD) within the acquired measuring strips in almost real time and the generation of DTMs. The OptD method was modified for this purpose by introducing a loop (FOR instruction) for fragmentary data processing. All processes in this approach were carried out at the first stage of data acquisition, but during the measurement the entire data set was not obtained, but rather a fragment of the data set was obtained. The proposed approach was compared with the method that uses full sets of bathymetric data. The results showed that it quickly obtained, reduced, and generated DTMs in almost real time for comparative navigation. The most important step during the processing was reduction, because a reduced number of data allowed faster 3D bottom model generation, which can be compared with other types of data within terrain reference navigation. In this paper, the research was based on the 3D Sidescan 3DSS-DX-450 sonar system, which provides bottom and water column data.

Xu at al. [20] work not with bottom data but with water column data, showing a very interesting case of the use of multi-beam measurements. The goal of this research was to propose an effective method for detecting gas leaks from bottom pipelines based on an analysis of water column images (WCI). WCIs use the differences in acoustic characteristics, such as backscattering strength or target strength, to detect solid, liquid, or gas targets by distinguishing them from the background images. Gas leakages can be detected with the use of so-called motion-estimation techniques. A gas bubble is considered to move in the consecutive scans and based on this movement can be detected. The authors proposed to use the optical flow method for this purpose, as it had already been validated using suspended objects but for different sensors. The entire image processing chain is analyzed including side lobe suppression, coordinates transformation, and other factors, resulting in the modified optical flow algorithm adjusted for multi-beam WCI analysis. The method is based on the combination of motion, and the intensity information of WCI pixels was studied in this paper. The method has been verified in two experiments with real sensors in real environments (pool and lake) with simulated gas leakages. It can be seen that the velocities of the gas bubbles obtained based on two variants of the method had relatively good consistency. The great potential of the method was proved. Further research is planned in which bottom tracking technology will be introduced and the influence of sound velocity changes for the thresholds will be analyzed.

Underwater surveys nowadays are more and more often dealing with more than one data source. Joint analysis of the various sources can in many cases provide important added value in situational awareness. An example of this can be found in [21], where Shang et al. propose a new method for acquiring a high resolution seabed topography and surface details that are difficult to obtain using MBES or SSS alone. It makes use of the observation that MBES data are well positioned, while SSS data (especially towed) provides high resolution images but with inaccurate positions. The authors proposed a method to combine both sources of data. Through taking the image geographic coordinates as the constraint when using the Speeded-Up Robust Features (SURF) algorithm for initial image matching, the authors have obtained more correct initial matched points compared to those obtained without constraint. Then, the finer matching step is conducted by adopting a template matching strategy which uses the dense local self-similarity (DLSS) descriptor to reflect the shape properties of the area's centered feature points. The method was empirically verified with real data, showing that the proposed method can overcome the limitations of adopting a single MBES or SSS for seabed mapping. High resolution and high accuracy seabed topography and surface details can be represented together, which is meaningful for understanding and interpreting seabed topography. Meanwhile, this paper discusses the accuracy of the reckoned SSS positions and uses it as a reference threshold in the image matching process. In addition, this paper discusses the impact of sonar frequency on the sonar backscatter image and provides some useful suggestions when dealing with multi-frequency sonar image matching.

3. Conclusions

The Special Issue entitled "Radar and Sonar Imaging Processing" comprised 21 articles on many topics related to remote sensing with radar and sonar sensors. In this paper, we have presented short introductions of the published articles.

It can be said that both radar and sonar imaging and processing still remains a "hot topic" and a lot of work in this is being done worldwide. New techniques and methods for extracting information from radar and sonar sensors and data have been proposed and verified. Some of these will provoke further research; however, some are already mature and can be considered for industrial implementation and development.

Author Contributions: A.S. wrote the first draft, A.S. revised and rewrite radar section, W.K. revised and rewrite sonar section, K.K. read the final version. All authors have read and agreed to the published version of the manuscript.

Acknowledgments: We would like to thank all the authors who contributed to the special issue and the staff in the editorial office.

Conflicts of Interest: The authors declare no conflict of interest.

References

1. Bieliński, T. A Parallax Shift Effect Correction Based on Cloud Height for Geostationary Satellites and Radar Observations. *Remote Sens.* **2020**, *12*, 365. [CrossRef]
2. Yan, X.; Chen, J.; Nies, H.; Loffeld, O.D. Analytical Approximation Model for Quadratic Phase Error Introduced by Orbit Determination Errors in Real-Time Spaceborne SAR Imaging. *Remote Sens.* **2019**, *11*, 1663. [CrossRef]
3. Yang, K.; Ye, W.; Ma, F.; Li, G.; Tong, Q. A Large-Scene Deceptive Jamming Method for Space-Borne SAR Based on Time-Delay and Frequency-Shift with Template Segmentation. *Remote Sens.* **2019**, *12*, 53. [CrossRef]
4. Chen, X.; Yi, T.; He, F.; He, Z.; Dong, Z. An Improved Generalized Chirp Scaling Algorithm Based on Lagrange Inversion Theorem for High-Resolution Low Frequency Synthetic Aperture Radar Imaging. *Remote Sens.* **2019**, *11*, 1874. [CrossRef]

5. Qian, Y.; Zhu, D. Image Formation of Azimuth Periodically Gapped SAR Raw Data with Complex Deconvolution. *Remote Sens.* **2019**, *11*, 2698. [CrossRef]
6. Wan, J.; Zhou, Y.; Zhang, L.; Chen, Z.; Yu, H. Efficient Algorithm for SAR Refocusing of Ground Fast-Maneuvering Targets. *Remote Sens.* **2019**, *11*, 2214. [CrossRef]
7. Zhang, Y.; Yang, Q.; Deng, B.; Qin, Y.; Wang, H. Estimation of Translational Motion Parameters in Terahertz Interferometric Inverse Synthetic Aperture Radar (InISAR) Imaging Based on a Strong Scattering Centers Fusion Technique. *Remote Sens.* **2019**, *11*, 1221. [CrossRef]
8. Wang, W.; Tang, Z.; Chen, Y.; Zhang, Y.; Sun, Y. Aircraft Target Classification for Conventional Narrow-Band Radar with Multi-Wave Gates Sparse Echo Data. *Remote Sens.* **2019**, *11*, 2700. [CrossRef]
9. Nowak, A.; Naus, K.; Maksimiuk, D. A Method of Fast and Simultaneous Calibration of Many Mobile FMCW Radars Operating in a Network Anti-Drone System. *Remote Sens.* **2019**, *11*, 2617. [CrossRef]
10. Hessner, K.; Naggar, E.; Von Appen, W.; Strass, V.; El Naggar, S.; Von Appen, W.-J. On the Reliability of Surface Current Measurements by X-band Marine Radar. *Remote Sens.* **2019**, *11*, 1030. [CrossRef]
11. Lisowski, J.; Mohamed-Seghir, M. Comparison of Computational Intelligence Methods Based on Fuzzy Sets and Game Theory in the Synthesis of Safe Ship Control Based on Information from a Radar ARPA System. *Remote Sens.* **2019**, *11*, 82. [CrossRef]
12. Stateczny, A.; Kazimierski, W.; Gronska-Sledz, D.; Motyl, W. The Empirical Application of Automotive 3D Radar Sensor for Target Detection for an Autonomous Surface Vehicle's Navigation. *Remote Sens.* **2019**, *11*, 1156. [CrossRef]
13. Czyżewski, A.; Kotus, J.; Szwoch, G. Estimating Traffic Intensity Employing Passive Acoustic Radar and Enhanced Microwave Doppler Radar Sensor. *Remote Sens.* **2019**, *12*, 110. [CrossRef]
14. Kang, M.-S.; Kim, N.; Im, S.; Lee, J.-J.; An, Y.-K. 3D GPR Image-based UcNet for Enhancing Underground Cavity Detectability. *Remote Sens.* **2019**, *11*, 2545. [CrossRef]
15. Zhang, X.; Tan, C.; Ying, W. An Imaging Algorithm for Multireceiver Synthetic Aperture Sonar. *Remote Sens.* **2019**, *11*, 672. [CrossRef]
16. Ye, X.; Yang, H.; Li, C.; Jia, Y.; Li, P. A Gray Scale Correction Method for Side-Scan Sonar Images Based on Retinex. *Remote Sens.* **2019**, *11*, 1281. [CrossRef]
17. Wang, X.; Li, Q.; Yin, J.; Han, X.; Hao, W. An Adaptive Denoising and Detection Approach for Underwater Sonar Image. *Remote Sens.* **2019**, *11*, 396. [CrossRef]
18. Yan, J.; Meng, J.; Zhao, J. Real-Time Bottom Tracking Using Side Scan Sonar Data Through One-Dimensional Convolutional Neural Networks. *Remote Sens.* **2019**, *12*, 37. [CrossRef]
19. Stateczny, A.; Błaszczak-Bąk, W.; Sobieraj-Żłobińska, A.; Motyl, W.; Wisniewska, M. Methodology for Processing of 3D Multibeam Sonar Big Data for Comparative Navigation. *Remote Sens.* **2019**, *11*, 2245. [CrossRef]
20. Xu, C.; Wu, M.; Zhou, T.; Li, J.; Du, W.; Zhang, W.; White, P.R. Optical Flow-Based Detection of Gas Leaks from Pipelines Using Multibeam Water Column Images. *Remote Sens.* **2020**, *12*, 119. [CrossRef]
21. Shang, X.; Zhao, J.; Zhang, H. Obtaining High-Resolution Seabed Topography and Surface Details by Co-Registration of Side-Scan Sonar and Multibeam Echo Sounder Images. *Remote Sens.* **2019**, *11*, 1496. [CrossRef]

Article

A Parallax Shift Effect Correction Based on Cloud Height for Geostationary Satellites and Radar Observations

Tomasz Bieliński

Department of Geoinformatics, Faculty of Electronics, Telecommunications and Informatics, Gdańsk University of Technology, 11/12 Gabriela Narutowicza Street, 80-233 Gdańsk, Poland; tomasz.bielinski@pg.edu.pl

Received: 18 December 2019; Accepted: 19 January 2020; Published: 22 January 2020

Abstract: The effect of cloud parallax shift occurs in satellite imaging, particularly for high angles of satellite observations. This study demonstrates new methods of parallax effect correction for clouds observed by geostationary satellites. The analytical method that could be found in literature, namely the Vicente et al./Koenig method, is presented at the beginning. It approximates a cloud position using an ellipsoid with semi-axes increased by the cloud height. The error values of this method reach up to 50 meters. The second method, which is proposed by the author, is an augmented version of the Vicente et al./Koenig approach. With this augmentation, the error can be reduced to centimeters. The third method, also proposed by the author, incorporates geodetic coordinates. It is described as a set of equations that are solved with the numerical method, and its error can be driven to near zero by adjusting the count of iterations. A sample numerical solution procedure with application of the Newton method is presented. Also, a simulation experiment that evaluates the proposed methods is described in the paper. The results of an experiment are described and contrasted with current technology. Currently, operating geostationary Earth Observation (EO) satellite resolutions vary from 0.5 km up to 8 km. The pixel sizes of these satellites are much greater than for maximal error of the least precise method presented in this paper. Therefore, the chosen method will be important when the resolution of geostationary EO satellites reaches 50 m. To validate the parallax correction, procedure data from on-ground radars and the Meteosat Second Generation (MSG) satellite, which describes stormy events, was compared before and after correction. Comparison was performed by correlating the logarithm of the cloud optical thickness (COT) with radar reflectance in dBZ (radar reflectance – Z in logarithmic form).

Keywords: parallax; cloud; earth observation; geostationary satellite; meteorological radar; MSG; SEVIRI

---

1. Introduction

The precision of remote space observations is important when investigating and monitoring various components of global ecological systems, such as marine, forestry, and climate environments [1–4]. Satellite data integration with external marine and other datasets is crucial in various applications of remote sensing techniques [5,6]. For climate and meteorological investigations, observations of clouds and precipitation on a global scale are usually performed using ground-based radar data and observations from geostationary satellites, due to their high temporal and moderate spatial resolution [7–9]. However, during data comparison and integration from these sources, the problem of parallax shift occurs [7,10], which is particularly observable for mid and high latitudes, and also for longitudes far from the sub-satellite point. Parallax shift is also important for cloud shadow determination, which is a significant issue for solar farms [11] and for flood detection [12]. Parallax phenomena also have a significant impact on the comparison of data from low-orbit satellites from different sensors [13–16].

In terms of mathematical problem formulation, the parallax shift effect for the geostationary satellites is actually a special case amongst low-orbit satellites, and it is easier to investigate due to higher temporal resolution data acquisition and the fixed satellite position.

There have been several attempts to solve parallax shift for geostationary satellites. One of them was proposed by Roebeling et al. [7,17,18], and was based on liquid water path (LWP) value pattern matching. This approach was suitable for stormy events and other inhomogeneous cloud formations, however it usually failed to perform correction in the case of homogeneous spatial LWP distribution. Another attempt proposed by Greuell et al. and Roebeling [19,20] used a simplified geometric model, which assumes Earth to be locally flat, as well as a sort of a priori knowledge about cloud height above the Earth's surface. There were also attempts by Li, Sun, and Yu [12] to solve this problem using a spherical model. Finally, there is the Vicente et al./Koenig method [21,22] based on the geometric properties of parallax shift phenomena, incorporating an ellipsoid model of Earth. The Vicente et al./Koenig method will be presented further in this paper.

There are two methods proposed by the author in this paper, which are based on the same assumptions as the Vicente et al./Koenig method. The first is an augmented version of the mentioned method. This augmentation reduces the correction error to centimeters. The second method proposed by the author is an original work which is based as before on a priori knowledge of cloud height, a geodetic equation of an ellipsoid, and numerical methods for solving the equation set. This method allows the correction error to be reduced to almost zero (assuming Earth to have an ellipsoidal shape).

2. Nature of Parallax Shift Problem and Vicente et al./Koenig Method

2.1. Problem Description

A parallax shift error in satellite observations occurs when the apparent image of the object is placed in the wrong location on the ellipsoid, considering the ellipsoid's normal line passing through the observed point. This geometric phenomena is particularly observable in geostationary and polar satellite observations due to the high angles of observations, particularly for edge areas of image scenes. In Figure 1, the problem is presented considering the case of a geostationary satellite. As a result, this phenomena causes pixel drift from the original position towards the edge of the observation disk. Consequently, the higher the cloud top layer is, the bigger the shift that occurs.

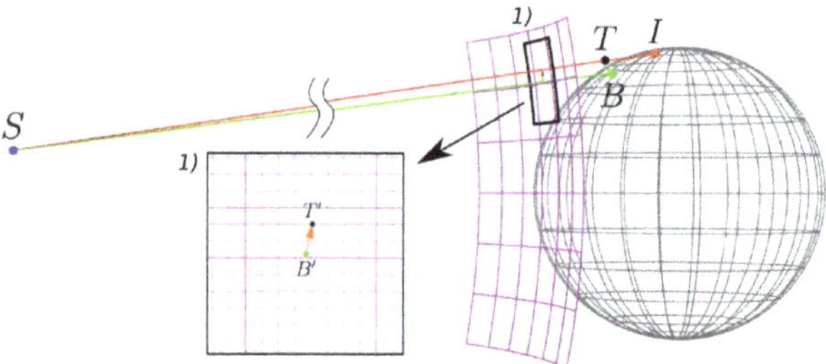

Figure 1. Parallax shift problem. The violet surface represents an image obtained from a geostationary satellite. The cloud top (T) is observed by the satellite as T′ (on the violet surface). The result of the reprojection of point T′ to ellipsoidal coordinates is I, which is not true the location of the cloud. The true location of the cloud is denoted as B, and from the perspective of the satellite sensor is observed as B′ on the violet surface. The square (marked as 1) shows how parallax shift affects the satellite image, where T′ is an image of the cloud top and B′ is where the cloud top should be placed according to its geodetic coordinates. The scale of the cloud height is not preserved.

The position of the cloud top (T) in Cartesian coordinates can be formulated as follows [23]:

$$\begin{cases} x = \left(N\left(\varphi_g\right) + h\right) \cos \varphi_g \cos\left(\lambda_g - \lambda_0\right) \\ y = \left(N\left(\varphi_g\right) + h\right) \cos \varphi_g \sin\left(\lambda_g - \lambda_0\right) \\ z = \left(N\left(\varphi_g\right)\left(1 - e^2\right) + h\right) \sin \varphi_g \end{cases} \tag{1}$$

where $N(\varphi_e) = \frac{a}{\sqrt{1-e^2 \sin^2 \varphi_g}}$ is the prime vertical radius of the curvature, $e^2 = \frac{a^2-b^2}{a^2}$ is the square of eccentricity, a is Earth's semi-major axis, b is Earth's semi-minor axis, h is the cloud top height, φ_g, λ_g is the geodetic latitude and longitude, and λ_0 is the longitude above which the geostationary satellite is floating. In this case, Equation (1) models the cloud position on a tangent line at coordinates φ_g, λ_g (see Figure 2). This is a more precise model than the flat-earth model or the spherical model. Note that: all longitudes (λ_g, λ_c and λ_p) are equal and the same. Subscripts are given to formally distinguish these values between corresponding latitudes that have different definitions (see Figure 2).

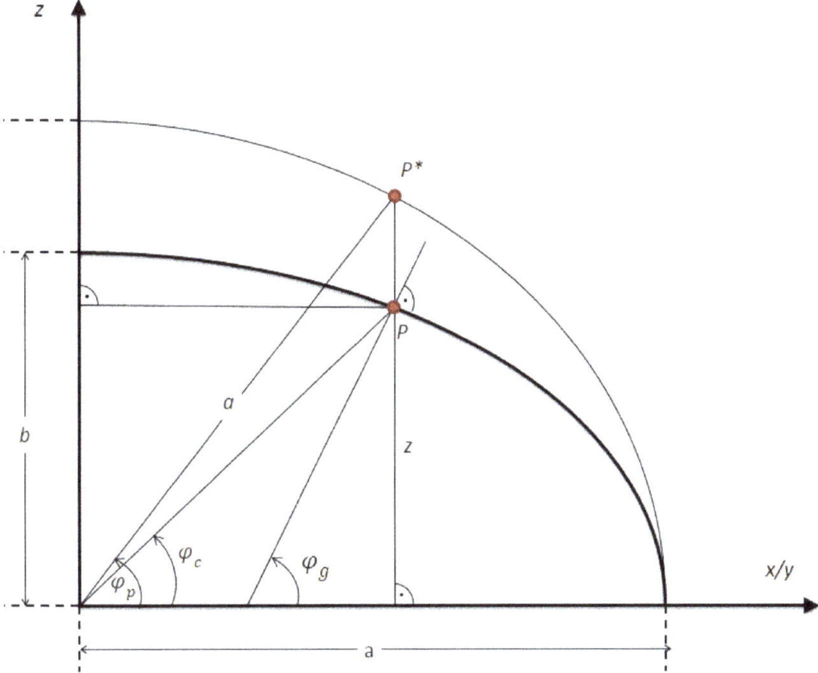

Figure 2. Three types of latitude: where φ_g is the geodetic latitude, φ_c is the geocentric latitude, φ_p is the parametric latitude, P is the point of interest on the ellipsoid, and P* is the image of the point of interest on a sphere. Based on figures from [23,24].

Pixel displacement in satellite view coordinates is defined as:

$$p_{disp}(h) = \sqrt{c_y^2(\varphi_s(h) - \varphi_s(0))^2 + c_x^2(\lambda_s(h) - \lambda_s(0))^2} \tag{2}$$

where c_x and c_y are constants that allow for sensor inclination angles to be converted to pixels or distance units in the satellite view space. Also, $\varphi_s(h)$ and $\lambda_s(h)$ are defined as:

$$\varphi_s(h) = \tan^{-1}\frac{z(h)}{\sqrt{(x(h)-l)^2 + y(h)^2}} \tag{3}$$

$$\lambda_s(h) = -\tan^{-1}\frac{y(h)}{x(h)-l} \tag{4}$$

where $x(h)$, $y(h)$, and $z(h)$ are cloud top coordinates from Equation (1) as functions of h; $l = a + h_s$– distance from center of Earth to satellite; a is the Earth's semi-major axis; h_s is the distance from the surface of Earth to the satellite. In order to illustrate $p_{disp}(h)$, the following analysis presented in Figure 3 was performed. Namely, depending on the geographical localization of the affected pixel and cloud top height, the absolute shift error in observations is expressed in Spinning Enhanced Visible Infra-Red Imager (SEVIRI) pixel units (In this case $c_x = c_y = \frac{h_s}{3\ km/px}$). It is worth noting that in many cases, especially for observations of clouds over 5000 m, this can cause pixel shift in the SEVIRI instruments used for the purpose of this study.

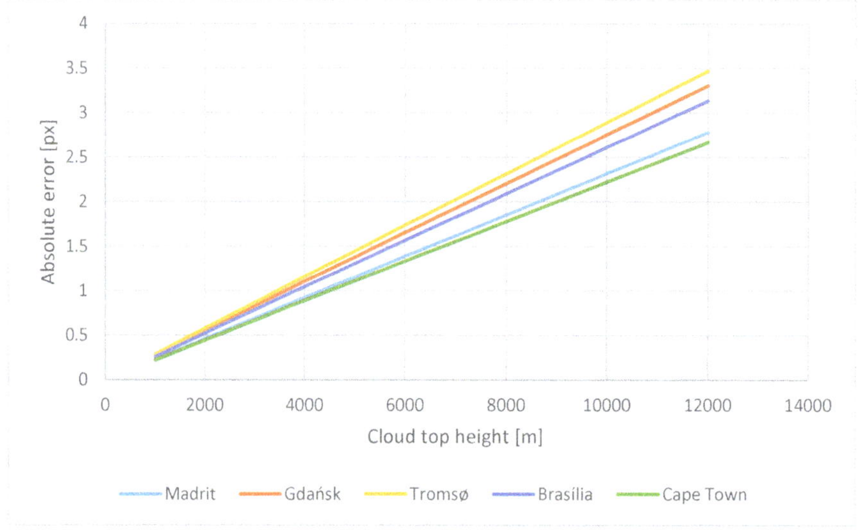

Figure 3. Error in pixels caused by cloud height parallax effect for 5 chosen cites, assuming the observation is acquired by SEVIRI instrument at longitude of 0°. Spatial resolution was assumed as 3 km/pixel.

As mentioned earlier, this effect hinders the comparison process between satellite and ground-based radar data [7]. An example is depicted in Figure 4.

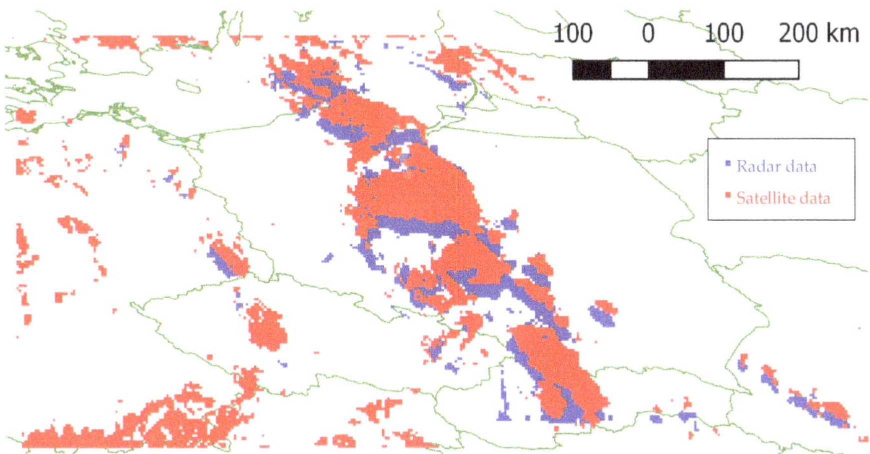

Figure 4. Comparison of detected precipitation mask based on ground-based radar data (blue) and data from Meteosat Second Generation (red). A parallax shift is particularly visible for small storm clouds in the bottom-right corner. The height of the cloud tops reaches 12 km. The stormy event is dated July 24, 2015, 13:00 UTC. EuroGeographics was used for the administrative boundaries.

2.2. Vicente et al./Koenig Method

The parallax shift problem is solved using a geometrical model, assuming that the surface of Earth is an ellipsoid, and with a priori knowledge of cloud top height. One of the approaches considered in this work is the method proposed by Vicente et al. [21] and implemented by Marianne Koenig [22]. This approach, similar to the rest of the methods presented in this paper, assumes a priori knowledge of the cloud top height, which can be calculated using the observed brightness temperature [7,25]. In this method, the Cartesian coordinates of the cloud image are described as:

$$\begin{cases} x = R_{loc}(\varphi_c)\cos\varphi_c\cos(\lambda_c - \lambda_0) \\ y = R_{loc}(\varphi_c)\cos\varphi_c\sin(\lambda_c - \lambda_0) \\ z = R_{loc}(\varphi_c)\sin\varphi_c \end{cases} \tag{5}$$

where a is Earth's semi-major axis; b is Earth's semi-minor axis; h is the cloud top height; φ_c and λ_c are the geocentric latitude and longitude (see Figure 2), respectively; λ_0 is the latitude of the geostationary satellite position; and $R_{loc}(\varphi_c)$ is the local radius of ellipsoid for the geocentric latitude model:

$$R_{loc}(\varphi_c) = \frac{a}{\sqrt{\cos^2\varphi_c + R_{ratio}^2\sin^2\varphi_c}} \tag{6}$$

where:

$$R_{ratio} = \frac{a}{b} \tag{7}$$

Satellite position (S) is defined as:

$$\begin{cases} x_s = a + h_s \\ y_s = 0 \\ z_s = 0 \end{cases} \tag{8}$$

where a is Earth's semi-major axis and h_s is the distance from the surface of Earth to the satellite.

The correction procedure is as follows:

1. Designate satellite position S in the Cartesian coordinates system;

2. Designate the position of cloud top image I in the Cartesian coordinates system using Equation (5);
3. Designate vector $|\vec{IS}|$;
4. Designate coefficient c, which allows Cartesian coordinates of the cloud top to be calculated using the following equations (see Figure 5):

$$|\vec{OT}| = |\vec{OI}| + c|\vec{IS}| \tag{9}$$

where $|\vec{OT}|$ is described by the ellipsoid parametric equation:

$$\begin{cases} x_{|\vec{OT}|} = (a+h)\cos\varphi_p \cos(\lambda_p - \lambda_0) \\ y_{|\vec{OT}|} = (a+h)\cos\varphi_p \sin(\lambda_p - \lambda_0) \\ z_{|\vec{OT}|} = (b+h)\sin\varphi_p \end{cases} \tag{10}$$

where φ_p and λ_p are the parametric latitude and longitude (see Figure 2). Therefore, Equation (9) can be presented as a set of equations:

$$\begin{cases} (a+h)\cos\varphi_p \cos(\lambda_p - \lambda_0) = x_{|\vec{OI}|} + cx_{|\vec{IS}|} \\ (a+h)\cos\varphi_p \sin(\lambda_p - \lambda_0) = y_{|\vec{OI}|} + cy_{|\vec{IS}|} \\ (b+h)\sin\varphi_p = z_{|\vec{OI}|} + cz_{|\vec{IS}|} \end{cases} \tag{11}$$

Squaring each equation and adding them according to their sides leads to a square equation, which can be solved with respect to c:

$$\frac{(x_I + cx_{|\vec{IS}|})^2 + (y_I + cy_{|\vec{IS}|})^2}{(a+h)^2} + \frac{(z_I + cz_{|\vec{IS}|})^2}{(b+h)^2} - 1 = 0 \tag{12}$$

5. Apply c to calculate the Cartesian coordinates of T - $x_{|\vec{OT}|}$, $y_{|\vec{OT}|}$, and $z_{|\vec{OT}|}$.
6. Calculate the geocentric ellipsoidal coordinates of T:

$$\begin{cases} \varphi_c = \tan^{-1} \frac{z_{|\vec{OT}|}}{\sqrt{x^2_{|\vec{OT}|} + y^2_{|\vec{OT}|}}} \\ \lambda_c = \tan^{-1} \frac{y_{|\vec{OT}|}}{x_{|\vec{OT}|}} + \lambda_0 \end{cases} \tag{13}$$

7. If required for further computation, a geodetic latitude can be calculated:

$$\varphi_g = \tan^{-1} \frac{a^2}{b^2} \frac{z_{|\vec{OT}|}}{\sqrt{x^2_{|\vec{OT}|} + y^2_{|\vec{OT}|}}} \tag{14}$$

Note that Equation (10) does not describe the cloud top position as it was defined in Equation (1) in Section 2.1. The coordinates of the point are shifted to height h above the ellipsoid along the normal vector. Instead, it describes the point on the ellipsoid with the semi-axes increased by h, therefore this method is burdened with error because of the inadequacy of the model.

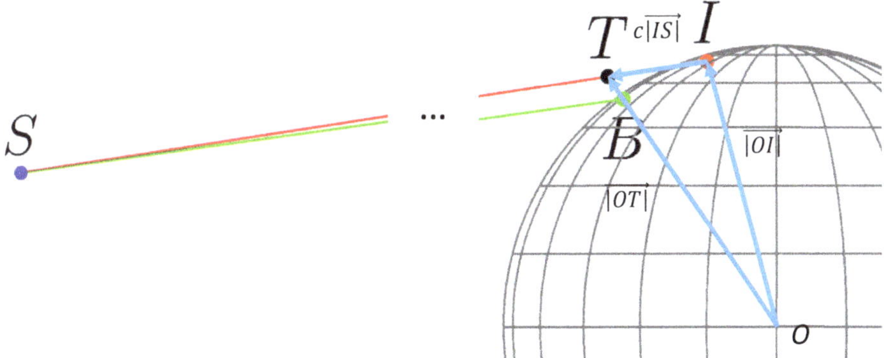

Figure 5. Vector notation in the Vicente et al./Koenig method.

3. Parallax Error Correction Methods with Lower Error

3.1. Vicente et al./Koenig Augmentation

The Vicente et al./Koenig method can be augmented in the final steps, where the latitude of the cloud bottom position is calculated. When using the Vicente et al./Koenig method, it is assumed that the cloud top is located on the ellipsoid with semi-axes increased by h, and therefore the geodetic latitude can be calculated taking into account the mentioned assumption:

$$\varphi_g = \tan^{-1} \frac{(a+h)^2}{(b+h)^2} \frac{z_{|\vec{OT}|}}{\sqrt{x^2_{|\vec{OT}|} + y^2_{|\vec{OT}|}}} \quad (15)$$

If further computation requires the geocentric latitude, this can be calculated using the following equation:

$$\varphi_c = \tan^{-1} \frac{b^2}{a^2} \tan \varphi_g \quad (16)$$

This modification allows the correction error to be reduced to centimeters. Details will be presented in the experimental section.

3.2. Ellipsoid Model with Geodetic Coordinates: Numeric Method

This method incorporates the cloud top position defined in Section 2.1 in Equation (1). With the described cloud top position, the geostationary satellite observation line should be defined as:

$$\begin{cases} x = -q \cos \varphi_s \cos \lambda_s + l \\ y = -q \cos \varphi_s \sin \lambda_s \\ z = q \sin \varphi_s \end{cases} \quad (17)$$

where $l = a + h_s$ is the distance from Earth's center to the satellite; a is Earth's semi-major axis; h_s is the distance from the surface of Earth to the satellite; φ_s and λ_s are satellite inclination angles; q is the distance from the satellite along the observation line. To solve this problem, an intersection point between the surface above the ellipsoid and the observation line needs to be calculated. Equations (1) and (17) should be merged, obtaining the following set of equations:

$$\begin{cases} \left(N(\varphi_g) + h\right) \cos \varphi_g \cos(\lambda_g - \lambda_0) = -q \cos \varphi_s \cos \lambda_s + l \\ \left(N(\varphi_g) + h\right) \cos \varphi_g \sin(\lambda_g - \lambda_0) = -q \cos \varphi_s \sin \lambda_s \\ \left(N(\varphi_g)(1 - e^2) + h\right) \sin \varphi_g = q \sin \varphi_s \end{cases} \quad (18)$$

The root of the above system of equations (φ_g and λ_g) is the geodetic coordinates of point ***B***. However, due to the entanglement of the φ_g variable in Equation (18), the root of the equations was designated using the numerical approach. The above method was implemented in C++ and Matlab. The Matlab implementation uses the fsolve function [26], which is part of the optimization toolbox. A detailed configuration of the fsolve function will be presented in the next section. The C++ implementation incorporates the Newton method, which is described below.

To solve the problem using the Newton method, the target function should be defined:

$$f(\varphi_g, \lambda_g, q) = \left[f_1(\varphi_g, \lambda_g, q), f_2(\varphi_g, \lambda_g, q), f_3(\varphi_g, \lambda_g, q)\right] \tag{19}$$

where:

$$\begin{aligned} f_1(\varphi_g, \lambda_g, q) &= \left(N(\varphi_g) + h\right)\cos\varphi_g\cos(\lambda_g - \lambda_0) + q\cos\varphi_s\cos\lambda_s - l \\ f_2(\varphi_g, \lambda_g, q) &= \left(N(\varphi_g) + h\right)\cos\varphi_g\sin(\lambda_g - \lambda_0) + q\cos\varphi_s\sin\lambda_s \\ f_3(\varphi_g, \lambda_g, q) &= \left(N(\varphi_g)(1 - e^2) + h\right)\sin\varphi_g - q\sin\varphi_s \end{aligned} \tag{20}$$

In Equation (19), $\left\|f(\varphi_g, \lambda_g, q)\right\|$ can be interpreted as the distance between the current solution and the optimal solution, which in the optimal case should be equal to zero. For such a defined cost function, calculation of the next iteration of the solution for the Newton method is defined as:

$$p_{n+1} = p_n - \left(\nabla f(p_n)\right)^{-1} f(p_n) \tag{21}$$

where:

$$p := \left[\varphi_g, \lambda_g, q\right] \tag{22}$$

and:

$$\nabla f(p_n) = \begin{bmatrix} \nabla f_1(p_n) \\ \nabla f_2(p_n) \\ \nabla f_3(p_n) \end{bmatrix} \tag{23}$$

and:

$$\nabla := \begin{bmatrix} \frac{\partial}{\partial\varphi_g} & \frac{\partial}{\partial\lambda_g} & \frac{\partial}{\partial q} \end{bmatrix} \tag{24}$$

The stopping condition is defined as:

$$\left\|f(p_n)\right\| < \varepsilon \tag{25}$$

However, the convergence of the above-presented approach is difficult to obtain for areas located near the edges of the observation disk. Therefore, an alternative target function is defined as the element-wise square of Equation (19):

$$g(p_n) = \begin{bmatrix} \left(f_1(p_n)\right)^2 \\ \left(f_2(p_n)\right)^2 \\ \left(f_3(p_n)\right)^2 \end{bmatrix} \tag{26}$$

with the gradient defined as:

$$\nabla g(p_n) = \begin{bmatrix} 2f_1(p_n)\nabla f_1(p_n) \\ 2f_2(p_n)\nabla f_2(p_n) \\ 2f_3(p_n)\nabla f_3(p_n) \end{bmatrix} \tag{27}$$

and the stop condition:

$$\left\|g(p_n)\right\| = \left\|f(p_n)\right\|^2 < \varepsilon^2 \tag{28}$$

Another issue that occurs in the numerical calculation problem is the big difference in scale between φ_g, λ_g, which are expressed in radians, and q, which is expressed in meters. To handle this problem, all distances (a, b, h, l, q) should be divided by a. This operation will bring q to a similar scale as φ_g, λ_g.

An example result of parallax correction using the numerical method via the Newton algorithm is presented in Figure 6. Note that the radar data is better aligned with the satellite data than in Figure 4.

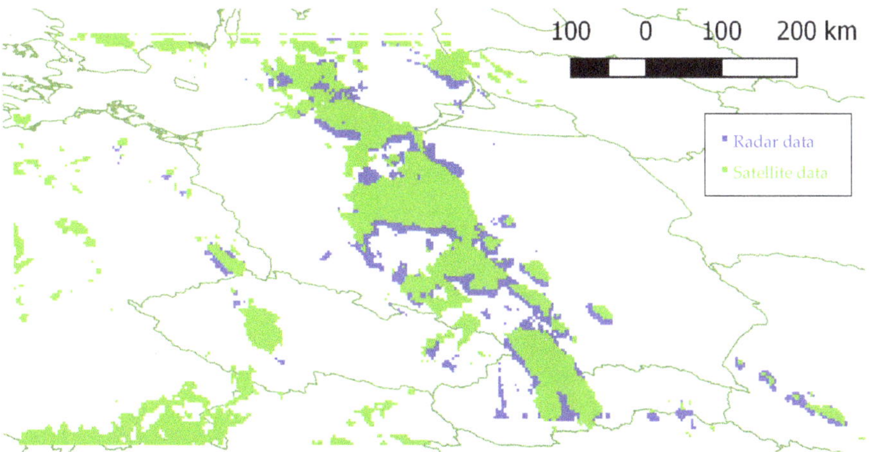

Figure 6. Comparison of detected precipitation mask based on ground-based radar data (blue) and data from Meteosat Second Generation with applied parallax correction using a numerical algorithm (green). Images of small storm clouds from satellite and meteorological radars in the bottom-right corner seem to overlap. The height of the cloud tops reaches 12 km. The stormy event is dated July 25, 2015, 13:00 UTC. EuroGeographics was used for the administrative boundaries.

4. Parallax Effect Correction Error Simulation

In order to compare the parallax effect correction obtained by the analyzed methods, a simulation experiment was performed. The main goal of the experiment was to generate several cloud top positions that simulate geostationary satellite observations, which result in φ_s and λ_s for simulated cloud top heights. With the φ_s and λ_s coordinates and a priori knowledge of the cloud height, correction methods were performed and their results were compared with the original (simulated) cloud position. The detailed procedure of the experiment is as follows:

1. Prepare a grid of geodetic coordinates: $\varphi_g \in \langle -90°; 90° \rangle$, $\lambda_g \in \langle -90°; 90° \rangle$, with 1° steps for each dimension;
2. Transform the grid coordinates to the geostationary view coordinates system, φ_s, λ_s (from now on called the base φ_s, λ_s) [27], and back to geodetic coordinates to specify which grid elements are out of scope; for out-of-scope elements, this operation will return Not a Number (NaN – floating point special value).
3. For each $h \in \{2\text{ km}, 4\text{ km}, 8\text{ km}, 12\text{ km}, 16\text{ km}\}$, the following steps are performed:

    a. For each φ_g and λ_g and with h, calculate the x, y, z coordinates using Equation (1);
    b. Using x, y, z, calculate the geostationary view coordinates φ_s and λ_s;
    c. With φ_s, λ_s, and h, run the correction algorithms: Vicente et al./Koenig, Vicente et al./Koenig augmented, and the numerical geodetic coordinates method;
    d. Each algorithm returns φ'_g, λ'_g, which should be transformed to φ'_s, λ'_s;

e. The distance between the simulated original base φ_s, λ_s and φ_s', λ_s' in the geostationary view space will be denoted as the correction error.

The correction error is calculated in the geostationary view coordinate space (violet surface on Figure 1), because it allows the impact of the correction error on a specific satellite sensor to be estimated. The coordinates in the above equation were expressed as an angle, however expressing them in radians and multiplying by h_s allows the result to be calculated in metric units (meters) as distances on a sphere of radius h_s around a geostationary satellite. This interpretation of geostationary coordinates is implemented in the PROJ software library [28].

In order to calculate the correction using the geodetic coordinates numerical method, the fsolve [26] function was applied. All distances were normalized with respect to the radius of the equator. The parameters of the fsolve function were as follows:

- Algorithm: Levenberg–Marquardt (instead of Newton);
- Function tolerance: $200m/a$;
- Specify objective gradient: yes;
- Input damping: 10^{-5}.

The results of the simulation using the Vicente et al./Koenig method and its augmented version are presented in Figures 7 and 8. The results using the geodetic coordinates numerical method are presented in Figures 9 and 10.

In Figure 7, the errors of the Vicente et al./Koenig method and its augmented version are depicted for certain cloud heights. Note that the error for the augmented version is 10^3 times smaller than for the unmodified version. Also, the median of error rises near linearly with the increase of the cloud height. Also note that the error rises as the distance from the equator and from the central meridian increases.

Figure 8 shows histograms of the errors presented in Figure 7. In the histograms, the error ratio between Vicente et al./Koenig and its augmented version can also be spotted, which can be estimated as 10^3. Another important piece of information is that for the assumed cloud heights, the maximal error of the Vicente et al./Koenig method can be estimated at 50 m, and for the augmented version, it can be estimated at 5 cm.

The errors of the geodetic coordinates numerical method for chosen cloud heights along with the number of iterations of the numerical method are shown in Figure 9. Note that the error is below 1 cm for almost the entire disc. The biggest errors appear near the edges in regions where the Vicente et al./Koenig method failed to compute a result (red NaN regions in Figure 7). The number of iterations increases as the height of the clouds and the distance from the center of the observation disc increase. However, during the performed experiments, the value for the majority of cases was less than or equal to five.

The histograms of errors for the geodetic coordinates numerical method and its number of iterations are presented in Figure 10. Based on the obtained results, the error histograms seem to be quite similar between the experiments—almost all values are classified as near zero. However, there are several occurrences of errors up to 3 meters, which are mainly caused by pixels in regions near the edge of the observation disc. The iteration histograms evolve along with the cloud height. As can be seen, the majority of occurrences fall below five iterations. Occurrences above this value refer to regions near the edge of the observation disc.

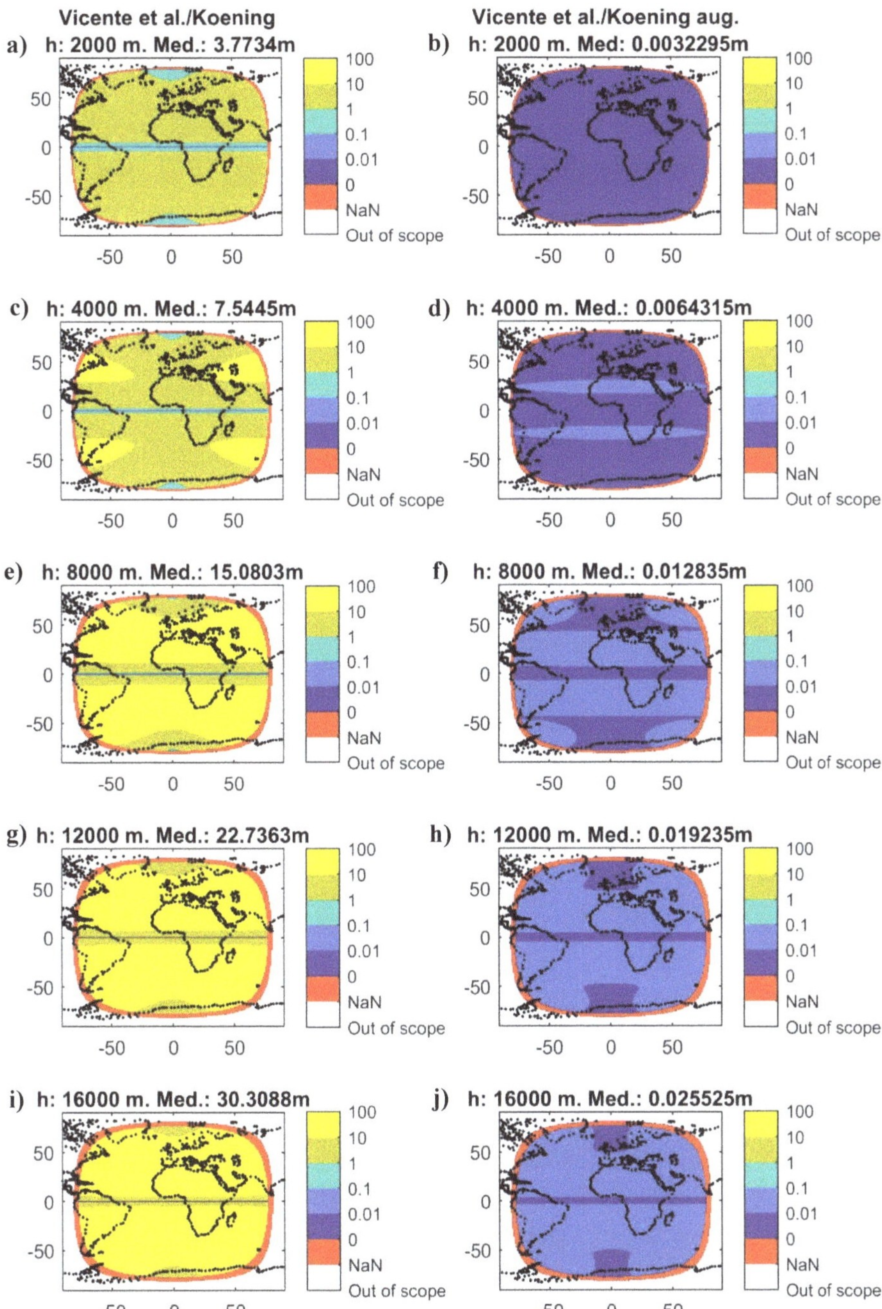

Figure 7. Maps showing error for the Vicente et al./Koenig (**a**,**c**,**e**,**g**,**i**) method and its augmented version (**b**,**d**,**f**,**h**,**j**) for several chosen cloud heights. Error are given in meters for the geostationary satellite coordinate system. NaN values for in-scope regions occur where the algorithm failed to calculate a solution. For each map, the median (med.) error was calculated.

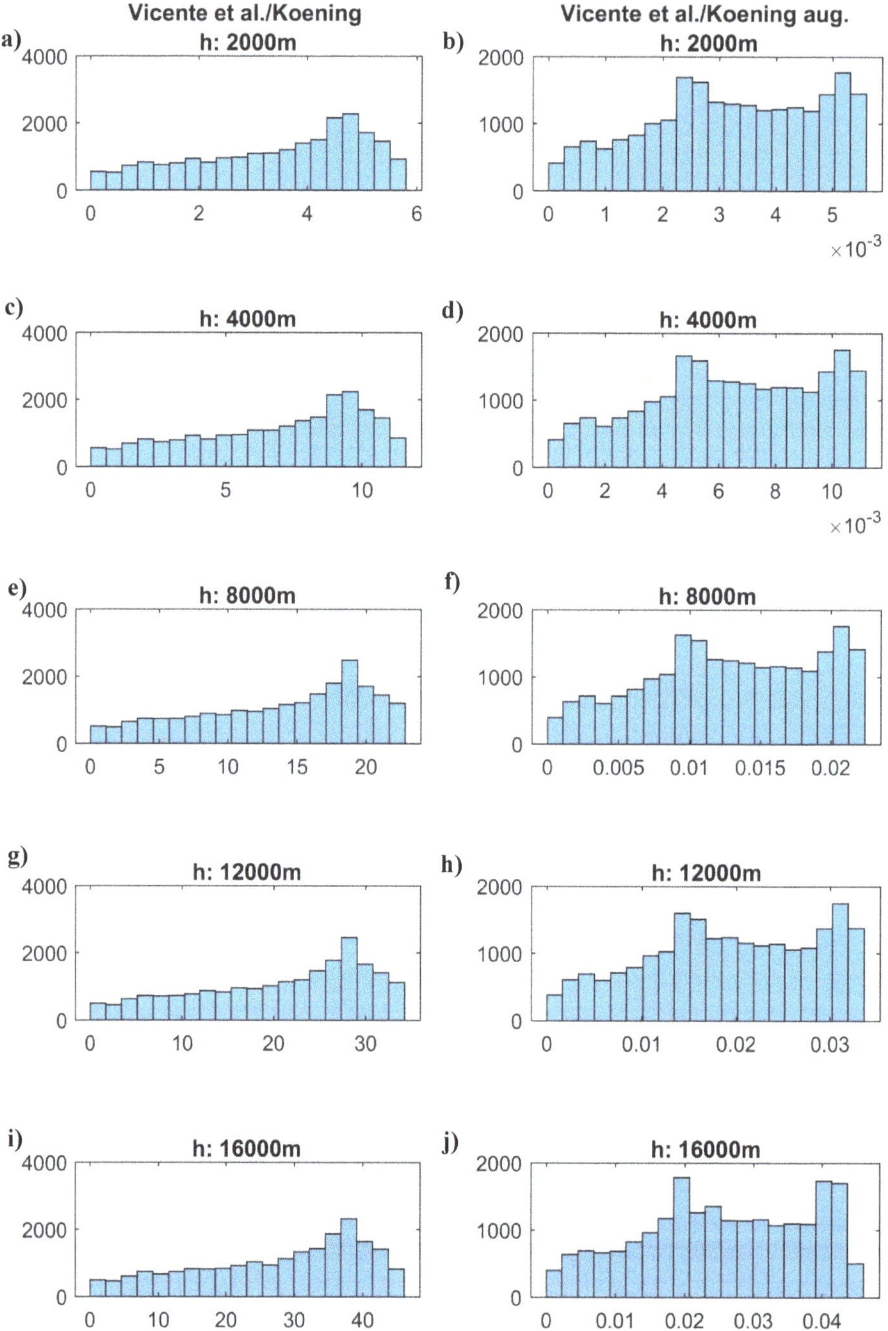

Figure 8. Error histograms for the Vicente et al./Koenig method (**a**,**c**,**e**,**g**,**i**) and its augmented version (**b**,**d**,**f**,**h**,**j**) for several chosen cloud heights. The Y-axis represents a count of 1 degree pixels, and the X-axis is the error in meters for the geostationary satellite coordinate system.

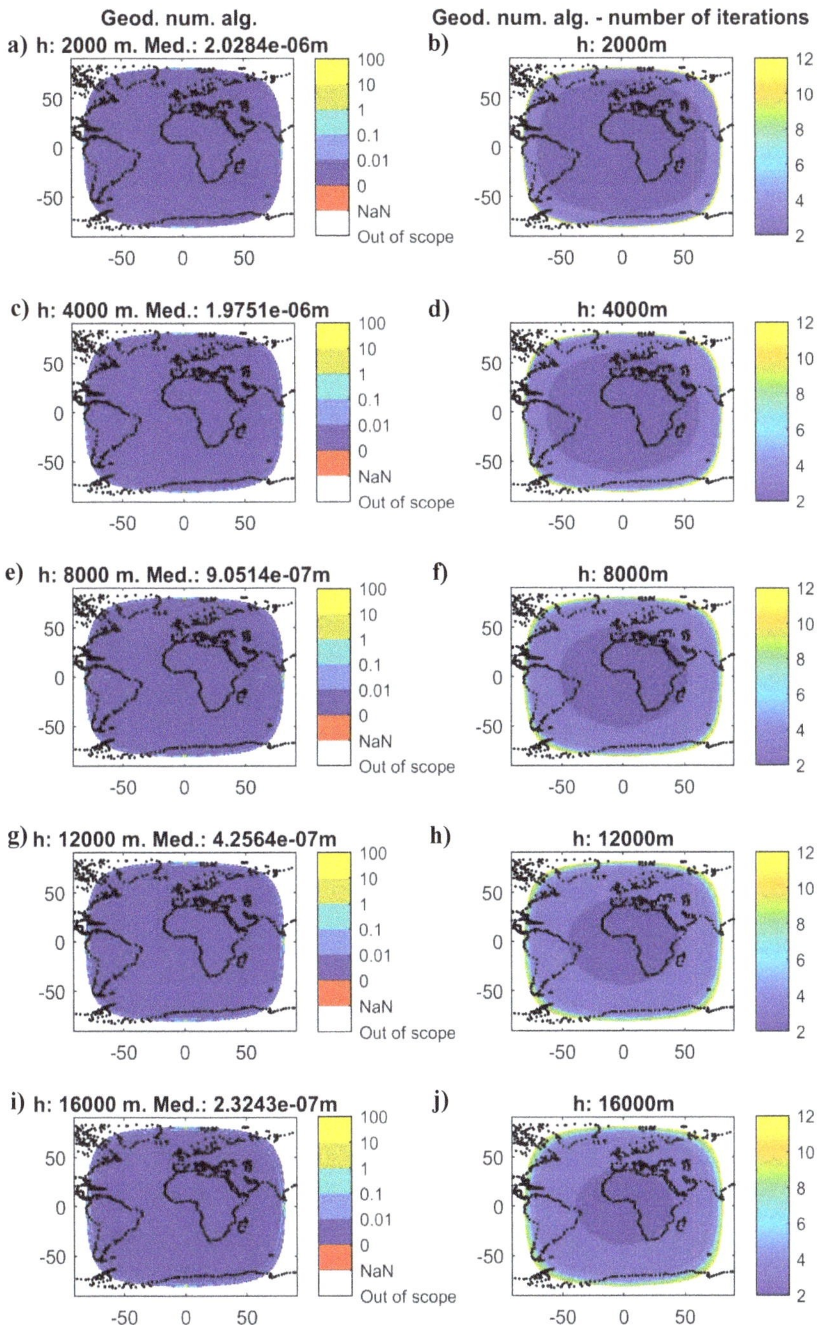

Figure 9. Maps with error (**a**,**c**,**e**,**g**,**i**) and number of iterations (**b**,**d**,**f**,**h**,**j**) for geodetic coordinates numerical algorithm (Geod. num. alg.) for several chosen cloud heights. Errors are given in meters for the geostationary satellite coordinate system. For each case, the median (med.) error was calculated.

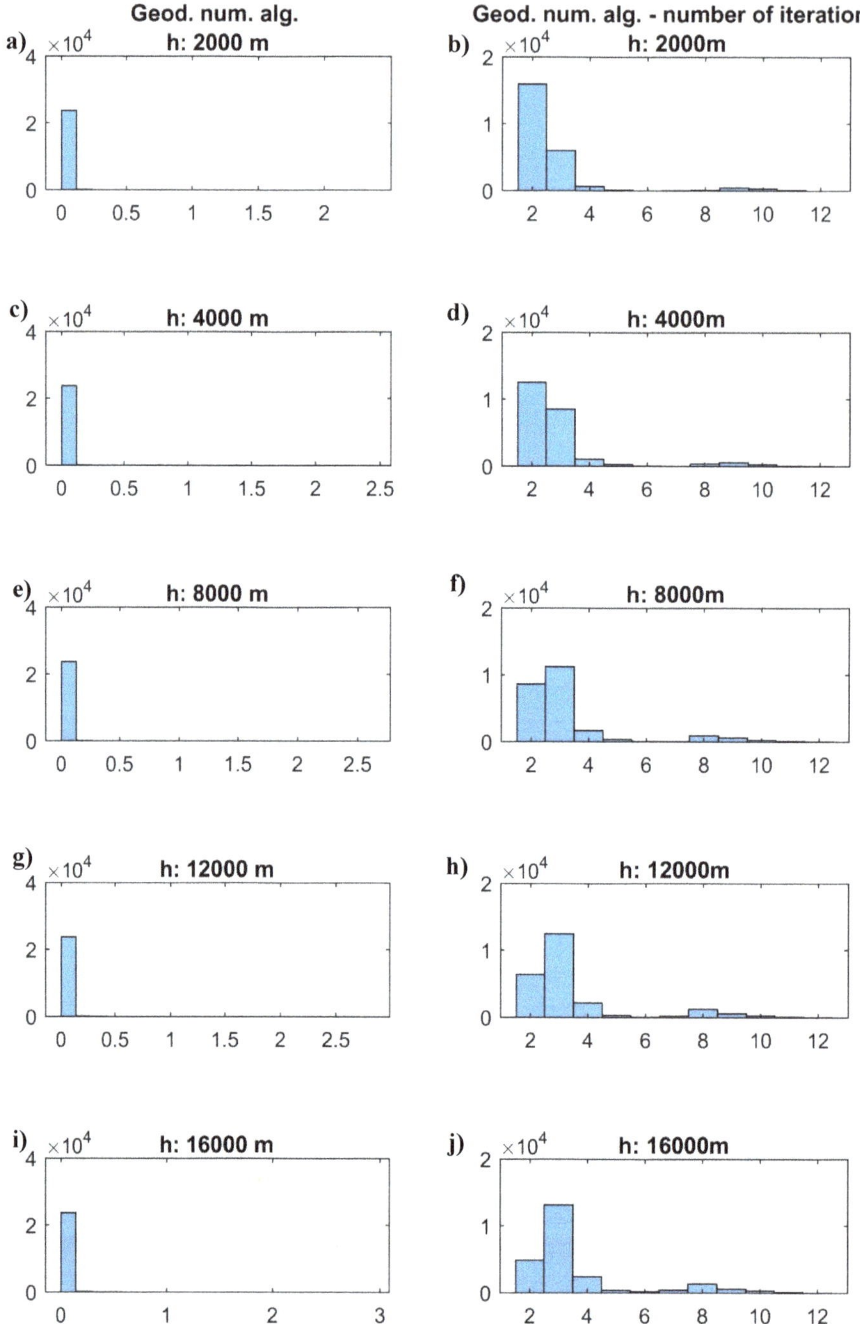

Figure 10. Histograms of error (**a**,**c**,**e**,**g**,**i**) and number of iterations (**b**,**d**,**f**,**h**,**j**) for geodetic coordinates algorithm (Geod. num. alg.) method for several chosen cloud heights. Error is given in meters for the geostationary satellite coordinate system.

5. Discussion

The results of the conducted experiments indicate that the Vicente et al./Koenig parallax effect correction method error is smaller than 50 meters in the geostationary satellite coordinates system for cloud heights of up to 16 km. The error for the augmented version of the Vicente et al./Koenig method proposed by the author allows the error values to be decreased to below 10 cm, which is negligible for current practical applications. As expected, the error of the geodetic coordinates numerical method is also negligible because it can be adjusted by the number of iterations. However, the advantage of the numerical approach is that it corrects the positions of pixels located near the edge of the observation disc (there is no NaN on Figure 9).

On the other hand, it must be noted that the proposed approach requires greater computational power than a method with a constant number of steps, such as the Vicente et al./Koenig method. However, experiments show that this is negligible, as the parallax correction problem was computed within minutes.

As was mentioned in the introduction, parallax effect correction is significant for the comparison and collocation of meteorological radar data and geostationary satellite data. This can be demonstrated by comparing radar reflectance in dBZ:

$$\mathrm{dBZ} = 10log_{10}Z \tag{29}$$

where Z is radar reflectance. Reflectance is described as the following empirical relation with precipitation rate R [mm/h] [29]:

$$Z = 200R^{1.6} \tag{30}$$

and cloud optical thickness (COT) in logarithmic form [30,31].

Figure 11 presents a scatterplot for radar reflectance and cloud optical thickness for satellite data without parallax effect correction, which should be mutually correlated in ideal cases. The Pearson's correlation value for that case is equal to 0.556. On the other hand, Figure 12 presents the same type of scatterplot but for the satellite data after parallax effect correction (by numerical method from Section 3.2). The correlation value for the corrected data is equal to 0.683. Note that a threshold effect occurs on top of both figures (presented as a horizontal set of points equal to 2.4), which is a consequence of Optimal Cloud Analysis (OCA) algorithm look-up table (LUT) limitations [31]. It is worth noticing that this effect is less significant for Figure 12, suggesting that data with parallax effect corrections are improved in terms of geometric accuracy.

Note that despite performed spatial correction, data presented in Figures 6 and 12 still differ. In this context, it is important to note that these differences are caused by other factors that influence data acquisition, namely:

- Different nature of the acquisition model, as on-ground radar and MSG satellite acquisition are registered with a slight temporal shift (less than 15 min);
- Both sensors utilize the different physical natures of acquisition. The on-ground radar is an active sensor which sends out an electromagnetic signal in the microwave spectrum and measurers the echo intensity scattered from precipitation particles. On the contrary, MSG SEVIRI is a passive sensor that measures radiation in a particular electromagnetic bandwidth (visible and near visible spectrum) coming from the sun and thermal radiance;
- Data acquired by MSG and on-ground radar is also characterized by different spatial resolutions. Therefore, in order to compare these datasets, additional resampling needs to be performed.

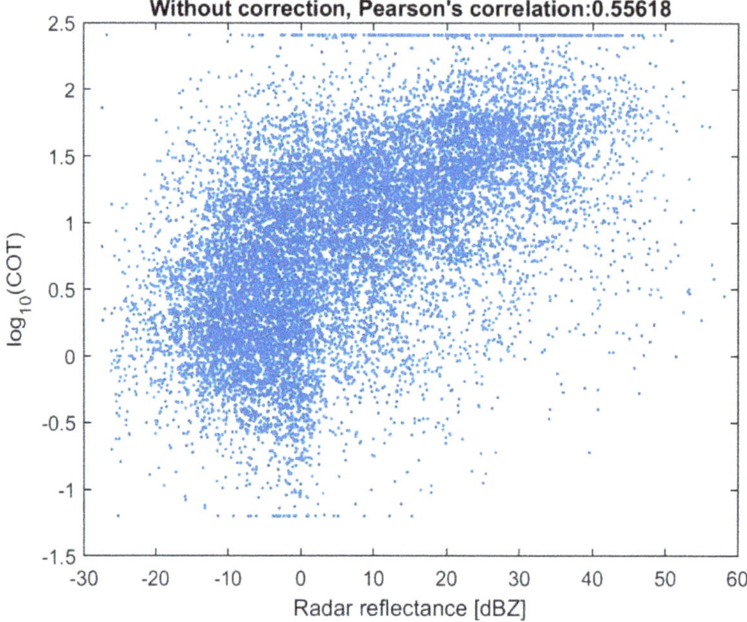

Figure 11. Scatterplot representing the dependence of radar reflectance and cloud optical thickness (logarithm) data for a stormy event on July 25, 2015, 13:00 UTC, without parallax effect correction (see Figure 4). The calculated Pearson's correlation coefficient is 0.556.

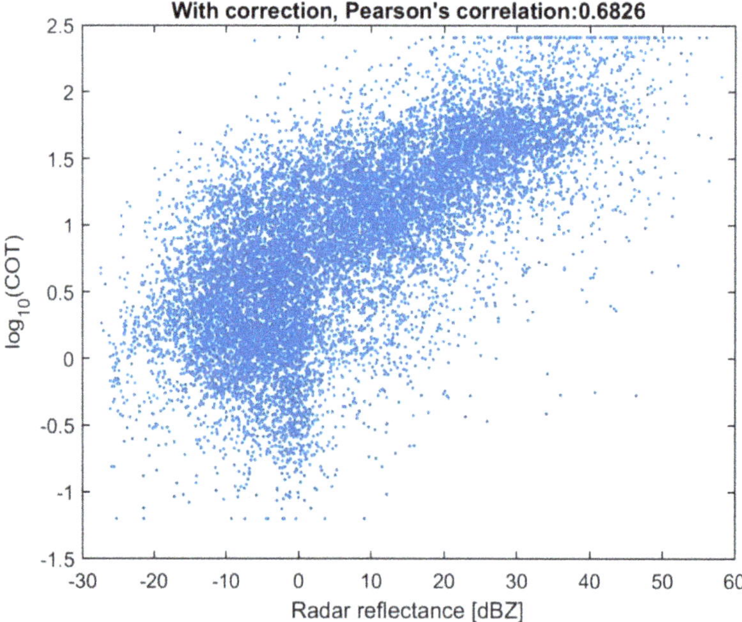

Figure 12. Scatterplot representing the dependence of radar reflectance and cloud optical thickness (logarithm) for a stormy event on July 25, 2015, 13:00 UTC, with parallax effect correction (see Figure 6). The calculated Pearson's correlation coefficient is 0.683.

Another aspect worth considering is algorithm sensitivity to the uncertainty of cloud top height. The easiest way to approximate this is to calculate the sensitivity of the parallax error itself due to changes of cloud top height. The sensitivity of the parallax error in satellite coordinates is defined as a derivative of pixel displacement (Equation (2)) in respect to h:

$$\frac{\partial p_{disp}(h)}{\partial h} \quad (31)$$

Because pixel displacement is nearly linear in respect to h (as can be noticed on Figure 3), the derivative (Equation (31)) should nearly be constant for assumed φ_g and λ_g. Therefore, it can be approximated as the mean slope $p_{disp}(h)$ in respect to h, for instance:

$$\frac{p_{disp}(12\text{ km})}{12\text{ km}} \quad (32)$$

where: $c_x = c_y = h_s$. The displacement sensitivity depends on φ_g and λ_g, therefore its value varies around the globe. Sensitivity values for cities from Figure 3 are presented in Table 1.

Table 1. The displacement function sensitivity in respect to h for five chosen cities for the geostationary sensor at longitude 0°. (N – North, S – South, E – East, W - West).

City	Geodetic Coordinates	Displacement Sensitivity as in Equation (32)
Cape Town	33.9253° S 18.4239° E	0.667
Madrid	40.4177° N 3.6947° W	0.696
Brasília	15.7839° S 47.9142° W	0.784
Gdańsk	54.3475° N 18.6453° E	0.827
Tromsø	69.6667° N 18.9333° E	0.868

Note that, the displacement sensitivity can be roughly approximated as less than 1. Therefore, SEVIRI instrument uncertainty of cloud height greater than or equal to 3 km may lead to one pixel size or greater error.

6. Conclusions

Data integration with data acquired from different sources requires developing additional methods that aim to reduce the discrepancies resulting from different physical aspects of observation. In this context, parallax shift correction for satellite data is a process that reduces geometric differences between observations, and in many cases can significantly improve the quality of corrected data in comparison with on-ground sources.

Regarding the scope of practical applications of the proposed approaches, it is important to note that the resolution of currently operating geostationary satellites varies between 1–3 km for a SEVIRI instrument [32] and 1–8 km for a Geostationary Operational Environmental Satellite (GOES) imager [33]. The upcoming series of Meteosat Third Generation (MTG) satellites will provide imagery with a spatial resolution between 0.5 and 2 km [34]. All of the above-presented parallax methods are effective enough for current and near future geostationary observation satellites, and the usefulness of the proposed methods is negligible. Selection of the proposed parallax effect correction method will be significant only when the spatial resolution of geostationary observations is comparable to 50 m. With high data resolution and precise parallax effect correction, the algorithm influence of precision on cloud height will become noticeable.

The parallax shift phenomena also affect the comparison between data collected from geostationary satellites and low-orbit satellites [14,35]. The parallax shift problem for geostationary satellites could be treated as a special case for low-orbit satellites. This problem for low-orbit satellites could be modeled with similar equations to those presented above, however taking into account the position and orientation of the satellite in the Cartesian coordinates space.

In this paper, the parallax effect was described using an ellipsoidal earth model. However, ellipsoidal model clearly does not fully reflect the real shape of Earth. Therefore, in situations where ellipsoidal model precision is insufficient, the numerical model of Earth's gravitational field and geoid values should be utilized. In this case, it would be necessary to describe the parallax effects using differential equations and solve them using a numerical approach. In this case, the most problematic issue would be to determine perpendicular paths to the equipotential boundaries of Earth's gravitational field.

Author Contributions: Conceptualization, T.B.; methodology, T.B.; software, T.B.; validation, T.B.; formal analysis, T.B.; investigation, T.B.; resources, T.B.; data curation, T.B.; writing—original draft preparation.; writing—review and editing, T.B.; visualization, T.B. All authors have read and agreed to the published version of the manuscript.

Funding: The research was supported under ministry subsidy for research for Gdansk University of Technology.

Acknowledgments: I would like to thank Andrzej Chybicki, PhD, and Tomasz Berezowski, PhD, for scientific and editorial support, as well as Marek Moszyński for supervising my work. Calculations were carried out thanks to the Academic Computer Centre in Gdańsk. Meteorological data from on-ground radars were provided by the Polish Institute of Meteorology and Water Management, National Research Institute.

Conflicts of Interest: The author declares no conflict of interest.

References

1. Kaminski, L.; Kulawiak, M.; Cizmowski, W.; Chybicki, A.; Stepnowski, A.; Orlowski, A. Web-based GIS dedicated for marine environment surveillance and monitoring. In Proceedings of the OCEANS 2009-EUROPE, Bremen, Germany, 11–14 May 2009; pp. 1–7.
2. Manzione, R.L.; Castrignano, A. A geostatistical approach for multi-source data fusion to predict water table depth. *Sci. Total Environ.* **2019**, *696*, UNSP 133763. [CrossRef] [PubMed]
3. Mishra, M.; Dugesar, V.; Prudhviraju, K.N.; Patel, S.B.; Mohan, K. Precision mapping of boundaries of flood plain river basins using high-resolution satellite imagery: A case study of the Varuna river basin in Uttar Pradesh, India. *J. Earth Syst. Sci.* **2019**, *128*, 105. [CrossRef]
4. Berezowski, T.; Wassen, M.; Szatylowicz, J.; Chormanski, J.; Ignar, S.; Batelaan, O.; Okruszko, T. Wetlands in flux: Looking for the drivers in a central European case. *Wetl. Ecol. Manag.* **2018**, *26*, 849–863. [CrossRef]
5. Stateczny, A.; Bodus-Olkowska, I. Sensor data fusion techniques for environment modelling. In Proceedings of the 2015 16th International Radar Symposium (IRS), Bonn, Germany, 24–26 June 2015; pp. 1123–1128.
6. Kazimierski, W.; Stateczny, A. Fusion of data from AIS and tracking radar for the needs of ECDIS. In Proceedings of the 2013 Signal Processing Symposium (SPS), Jachranka, Poland, 5–7 June 2013; pp. 1–6.
7. Roebeling, R.A.; Holleman, I. SEVIRI rainfall retrieval and validation using weather radar observations. *J. Geophys. Res. Atmos.* **2009**, *114*. [CrossRef]
8. Vicente, G.A.; Scofield, R.A.; Menzel, W.P. The Operational GOES Infrared Rainfall Estimation Technique. *Bull. Amer. Meteor. Soc.* **1998**, *79*, 1883–1898. [CrossRef]
9. Zhao, J.; Chen, X.; Zhang, J.; Zhao, H.; Song, Y. Higher temporal evapotranspiration estimation with improved SEBS model from geostationary meteorological satellite data. *Sci. Rep.* **2019**, *9*, 14981. [CrossRef] [PubMed]
10. Henken, C.C.; Schmeits, M.J.; Deneke, H.; Roebeling, R.A. Using MSG-SEVIRI Cloud Physical Properties and Weather Radar Observations for the Detection of Cb/TCu Clouds. *J. Appl. Meteor. Climatol.* **2011**, *50*, 1587–1600. [CrossRef]
11. Miller, S.D.; Rogers, M.A.; Haynes, J.M.; Sengupta, M.; Heidinger, A.K. Short-term solar irradiance forecasting via satellite/model coupling. *Solar Energy* **2018**, *168*, 102–117. [CrossRef]
12. Li, S.; Sun, D.; Yu, Y. Automatic cloud-shadow removal from flood/standing water maps using MSG/SEVIRI imagery. *Int. J. Remote Sens.* **2013**, *34*, 5487–5502. [CrossRef]

13. Wang, C.; Luo, Z.J.; Huang, X. Parallax correction in collocating CloudSat and Moderate Resolution Imaging Spectroradiometer (MODIS) observations: Method and application to convection study. *J. Geophys. Res. Atmos.* **2011**, *116*. [CrossRef]
14. Guo, Q.; Feng, X.; Yang, C.; Chen, B. Improved Spatial Collocation and Parallax Correction Approaches for Calibration Accuracy Validation of Thermal Emissive Band on Geostationary Platform. *IEEE Trans. Geosci. Remote Sens.* **2018**, *56*, 2647–2663. [CrossRef]
15. Chen, J.; Yang, J.-G.; An, W.; Chen, Z.-J. An Attitude Jitter Correction Method for Multispectral Parallax Imagery Based on Compressive Sensing. *IEEE Geosci. Remote Sens. Lett.* **2017**, *14*, 1903–1907. [CrossRef]
16. Frantz, D.; Haß, E.; Uhl, A.; Stoffels, J.; Hill, J. Improvement of the Fmask algorithm for Sentinel-2 images: Separating clouds from bright surfaces based on parallax effects. *Remote Sens. Environ.* **2018**, *215*, 471–481. [CrossRef]
17. Roebeling, R.A.; Feijt, A.J. Validation of cloud liquid water path retrievals from SEVIRI on METEOSAT-8 using CLOUDNET observations. In Proceedings of the EUMETSAT Meteorological Satellite Conference, Helsinki, Finland, 12–16 June 2006; EUMETSAT: Darmstadt, Germany, 2006; pp. 12–16.
18. Roebeling, R.A.; Deneke, H.M.; Feijt, A.J. Validation of Cloud Liquid Water Path Retrievals from SEVIRI Using One Year of CloudNET Observations. *J. Appl. Meteor. Climatol.* **2008**, *47*, 206–222. [CrossRef]
19. Greuell, W.; Roebeling, R.A. Toward a Standard Procedure for Validation of Satellite-Derived Cloud Liquid Water Path: A Study with SEVIRI Data. *J. Appl. Meteor. Climatol.* **2009**, *48*, 1575–1590. [CrossRef]
20. Schutgens, N.A.J.; Greuell, W.; Roebeling, R. Effect of inhomogeneity on the validation of SEVIRI LWP. In *Current Problems in Atmospheric Radiation (irs 2008)*; Nakajima, T., Yamasoe, M.A., Eds.; Amer Inst Physics: New York, NY, USA, 2009; Volume 1100, p. 424. ISBN 978-0-7354-0635-3.
21. Vicente, G.A.; Davenport, J.C.; Scofield, R.A. The role of orographic and parallax corrections on real time high resolution satellite rainfall rate distribution. *Int. J. Remote Sens.* **2002**, *23*, 221–230. [CrossRef]
22. Koenig, M. Description of the parallax correction functionality. Available online: https://cwg.eumetsat.int/parallax-corrections/ (accessed on 17 January 2020).
23. Wolfgang, T. *Geodesy, An Introduction*; De Gruyter: Berlin, Germany, 1980; ISBN 3-11-007232-7.
24. Czarnecki, K. *Geodezja współczesna*; Wyd. 3 (1 w WN PWN)-1 dodruk; Wydawnictwo Naukowe PWN: Warszawa, Poland, 2015; ISBN 978-83-01-18380-6.
25. Meteorological Products Extraction Facility Algorithm Specification Document. Available online: https://www.eumetsat.int/website/wcm/idc/idcplg?IdcService=GET_FILE&dDocName=PDF_TEN_SPE_04022_MSG_MPEF&RevisionSelectionMethod=LatestReleased&Rendition=Web (accessed on 28 November 2019).
26. Solve System of Nonlinear Equations - MATLAB Fsolve. Available online: https://www.mathworks.com/help/optim/ug/fsolve.html (accessed on 23 October 2019).
27. Wolf, R. Coordination Group for Meteorological Satellites LRIT/HRIT Global Specification. Available online: https://www.cgms-info.org/documents/pdf_cgms_03.pdf (accessed on 28 November 2019).
28. PROJ contributors. *PROJ Coordinate Transformation Software Library*; Open Source Geospatial Foundation: Beaverton, OR, USA, 2019.
29. Marshall, J.S.; Gunn, K.L.S. Measurement of snow parameters by radar. *J. Meteor.* **1952**, *9*, 322–327. [CrossRef]
30. Roebeling, R.A.; Feijt, A.J.; Stammes, P. Cloud property retrievals for climate monitoring: Implications of differences between Spinning Enhanced Visible and Infrared Imager (SEVIRI) on METEOSAT-8 and Advanced Very High Resolution Radiometer (AVHRR) on NOAA-17. *J. Geophys. Res. Atmos.* **2006**, *111*. [CrossRef]
31. Optimal Cloud Analysis: Product Guide. Available online: http://www.eumetsat.int/website/wcm/idc/idcplg?IdcService=GET_FILE&dDocName=PDF_DMT_770106&RevisionSelectionMethod=LatestReleased&Rendition=Web (accessed on 8 January 2020).
32. MSG Level 1. Available online: http://www.eumetsat.int/website/wcm/idc/idcplg?IdcService=GET_FILE&dDocName=PDF_TEN_05105_MSG_IMG_DATA&RevisionSelectionMethod=LatestReleased&Rendition=Web (accessed on 28 November 2019).
33. GOES N Databook. Available online: https://goes.gsfc.nasa.gov/text/GOES-N_Databook/databook.pdf (accessed on 29 November 2019).

34. MTG FCI L1 Product User Guide. Available online: http://www.eumetsat.int/website/wcm/idc/idcplg?IdcService=GET_FILE&dDocName=PDF_DMT_719113&RevisionSelectionMethod=LatestReleased&Rendition=Web (accessed on 28 November 2019).
35. Hewison, T.J. An Evaluation of the Uncertainty of the GSICS SEVIRI-IASI Intercalibration Products. *IEEE Trans. Geosci. Remote Sens.* **2013**, *51*, 1171–1181. [CrossRef]

Article

Optical Flow-Based Detection of Gas Leaks from Pipelines Using Multibeam Water Column Images

Chao Xu [1,2,3], Mingxing Wu [2], Tian Zhou [1,2,3,*], Jianghui Li [4], Weidong Du [1,2,3], Wanyuan Zhang [2] and Paul R. White [4]

1 Acoustic Science and Technology Laboratory, Harbin Engineering University, Harbin 150001, China; xuchao18@hrbeu.edu.cn (C.X.); duweidong@hrbeu.edu.cn (W.D.)
2 College of Underwater Acoustic Engineering, Harbin Engineering University, Harbin 150001, China; wmx0909@hrbeu.edu.cn (M.W.); zhangwanyuan@hrbeu.edu.cn (W.Z.)
3 Key Laboratory of Marine Information Acquisition and Security (Harbin Engineering University), Ministry of Industry and Information Technology, Harbin 150001, China
4 Institute of Sound and Vibration Research, University of Southampton, Southampton SO17 3AS, UK; J.Li@soton.ac.uk (J.L.); P.R.White@soton.ac.uk (P.R.W.)
* Correspondence: zhoutian@hrbeu.edu.cn; Tel.: +86-13895736718

Received: 11 November 2019; Accepted: 30 December 2019; Published: 1 January 2020

Abstract: In recent years, most multibeam echo sounders (MBESs) have been able to collect water column image (WCI) data while performing seabed topography measurements, providing effective data sources for gas-leakage detection. However, there can be systematic (e.g., sidelobe interference) or natural disturbances in the images, which may introduce challenges for automatic detection of gas leaks. In this paper, we design two data-processing schemes to estimate motion velocities based on the Farneback optical flow principle according to types of WCIs, including time-angle and depth-across track images. Moreover, by combining the estimated motion velocities with the amplitudes of the image pixels, several decision thresholds are used to eliminate interferences, such as the seabed, non-gas backscatters in the water column, etc. To verify the effectiveness of the proposed method, we simulated the scenarios of pipeline leakage in a pool and the Songhua Lake, Jilin Province, China, and used a HT300 PA MBES (it was developed by Harbin Engineering University and its operating frequency is 300 kHz) to collect acoustic data in static and dynamic conditions. The results show that the proposed method can automatically detect underwater leaking gases, and both data-processing schemes have similar detection performance.

Keywords: multibeam echo sounder; water column image; gas emissions; automatic detection; optical flow

---

1. Introduction

Multibeam echo sounders (MBESs) are important remote-sensing acoustical systems whose primary goal is mapping the seabed. They are also widely used to detect targets in water columns [1]. Many types of MBESs can collect water column image (WCI) data, which carry backscattering signals of scatters from the transducer to the seabed. The images can be used to detect artificial or natural structures in water columns, such as gas bubbles rising from seep sites [2–8] or gas pipelines [9], shipwrecks [10], fish schools [11], in addition to serving as a reference for the quality control of multibeam bathymetric data.

WCIs use the differences in acoustic characteristics, such as backscattering strength or target strength, to detect solid, liquid, or gas targets by distinguishing them from the background images. For the gas emissions discussed in this paper, their appearance in images is flare-like [12] and tends to rise from the source. In addition, the ascending gases and other scatterers may be deflected by water/sea currents [4]. Generally, these may be the result of leaks from man-made structures, such as underwater gas pipelines, or gases that were released from the seabed. Over the past two decades, research on gas emissions using image data has mainly focused on the aspects of detection (presence or absence) and positioning, as well as their quantification [2,12–18]. A typical method in the detection of gas emissions is to mark the suspected gas target by manually screening the image data collected by the MBES [1,5,12,16]. Therefore, it is necessary to study methods automatically detecting gas emissions and improving data processing efficiency [19].

In general, from the perspective of different input data sources, there are two ways to automatically detect leaking gases using MBES. One way is to detect targets in water columns by processing beam space data of MBES, such as the multi-detection technique proposed by Christoffersen et al. [20]. Compared to traditional multibeam seabed detection technologies (e.g., the weighted mean-time and the zero-crossing of the phase difference), multi-detection algorithms have been applied in some MBES applications and can be applied to detect seabed topography and targets suspended in water columns at the same time.

Another method is to detect targets using the WCIs as data sources. For instance, some automatic detection methods of bubble leakage via the images have been presented by Urban et al. and Zhao et al. [12,15], and were used to localize the bubble vent. However, it is often difficult to avoid the strong interference from sidelobes when the pixel intensity data in the images are used for target detection [12,21]. One solution to address this problem is to exclude WCI data outside a minimum slant range (MSR) [12], while the excluded area may contain part of the target image, if one is present. Furthermore, if a gas-detection algorithm relies only on image intensity information, without other auxiliary methods in data interpretation, it cannot confirm whether the detected targets are rising gases or if it is from other targets, such as underwater artificial structures with the same outline as gas flares. Therefore, coherent flow structures have been used as a feature of detection [22], and the motion of rising bubbles was considered [18] to detect leakage via cross-correlation of the WCIs in the depth-across orientation. This method not only discriminates the presence or absence of spilled gases, but also distinguishes them from fish, which is one of the major reasons for misdetections. However, in this method, the potential situation of false but large motion speed due to continuous frame changes of the strong backscatter pixels from the seafloor has not been considered.

The optical flow method is another motion-estimation technique. It has been validated using suspended objects (e.g., leaking gases [23] or sulfur dioxide flux) from infrared images or CCD images and has already been suggested as a promising method for detection of gas leaks moving patterns by von Deimling et al. [18]. In this paper, two types of information, amplitude and velocity, are comprehensively considered, and an automatic detection method for underwater gas leaks is designed to distinguish other strong scatterers in the water such as the seafloor. In addition, the above methods are mainly based on "depth-across track" (D-T) image, which needs the process of converting from the beam space data to the D-T image. This paper reduces the above process, and further proposes to design a more efficient parallel structure based on the beam data. In other words, we use the Farneback optical flow method to estimate the motion of underwater gas emissions. In our process, the potential interference of sidelobes on the leaking gas images is suppressed before WCI generation. Subsequently, we designed two processing schemes corresponding to the two types of images, and the intensities and velocity information of the images are combined for discrimination. Finally, we tested the effectiveness of the above two processing schemes by simulating pipeline leakage scenarios in water tanks and Songhua Lake in China.

2. Materials and Methods

2.1. Water Column Image (WCI) Generation and Sidelobe Suppression

After receiving underwater multi-channel echo signals, the MBES usually needs to perform array beamforming processing to obtain the time series of echo amplitudes at different beam angles.

Then, sonar equation [24] can be used to compute the intensity of the echo signals at each beam, the backscattering strength from the seabed, or the target strength from other targets in water. In general, these echo data at all beam directions can be visually displayed in two typical WCIs [17,25]. One is that the amplitude/intensity time series are directly arranged into a two-dimensional data matrix, which is displayed as a two-dimensional image with coordinate axes corresponding to time and beam angle, referred to as a "time-angle" (T-A) image. Figure 1a is a T-A image, in which the amplitude time sequences are obtained using the discrete Fourier transformation (DFT) of the multi-channel echo signals. Moreover, the T-A image is often used as a data source for seafloor detection methods. The vertical and horizontal coordinates of this image correspond to the beam angle and the two-way travel time (TWTT) of the echo signals. In the figure, multi-channel echo data was acquired on the lake by using the HT300 PA MBES mentioned later. The emission signal of sonar is a CW pulse signal with a pulse width of 0.1 ms. The other WCI is the amplitude/intensity time series in polar coordinates, converted into two-dimensional images in Cartesian coordinates, with its horizontal and vertical coordinates corresponding to the horizontal distance and vertical depth, respectively. This is referred to as a "depth-across track" (D-T) image (as shown in Figure 1b). The complexity of the coordinate transformation is embodied in the need to consider the influence of irregular beam space [25], and thus it may be necessary to interpolate pixels without data in the beam sector of the D-T image. Moreover, it is necessary to correct the spatial position of the amplitude/intensity data in Cartesian coordinates by ray tracking when the sound velocity in water varies significantly with the depth. The coordinate system corresponding to Figure 1b is more suitable for observation with the human eye; thus, it is often displayed in the display and control software of a sonar system. In addition, from the areas within the red boxes in Figure 1a, it can be seen that echo energy from the beam, perpendicular to the seabed, leaks into the main lobe direction of all other beams, such that obvious stripe patterns appear at and outside the MSR position. Correspondingly, the straight stripe patterns appear curved in Figure 1b.

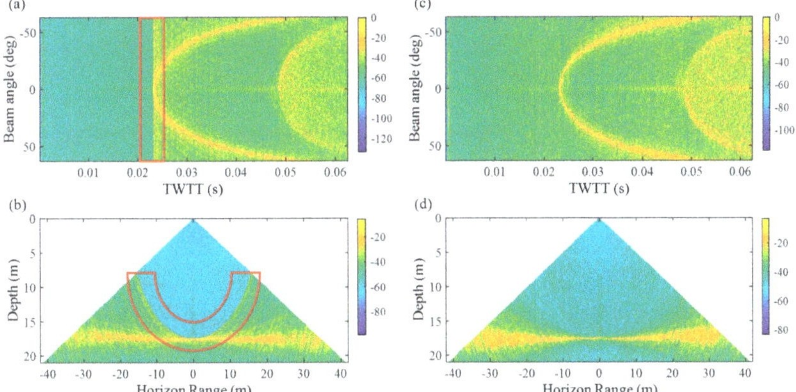

Figure 1. Examples of T-A and D-T images, and the results of sidelobe effect elimination based on the minimum variance distortionless response (MVDR) beamforming methods [26]. (**a**,**c**) are the T-A images, whose ordinates and horizontal coordinates correspond to the beam angle and two-way travel time (TWTT). (**b**,**d**) are the D-T images, whose ordinates and horizontal coordinate correspond to vertical depth and horizon range. (**a**,**b**) are obtained using a discrete Fourier transformation (DFT) beamforming method, and (**c**,**d**) are the processing results via a MVDR beamforming method.

Currently, in the above WCI generation process, most MBESs must use DFT or fast Fourier transformation (FFT) techniques to achieve beamforming tasks. These techniques offer simplicity and good real-time performance; however, they have energy leakage, which can cause sidelobe interference. Thus, the echo energy from the beam in the direction perpendicular to the seabed or the strong scattering in the water column leaks into the main lobe direction of all the other beams. For bathymetric measurements, a typical effect would be to mistake the flat seabed topography for a curved seabed topography with both sides upturned. The sidelobe effect here is also called a "tunnel effect" [27]. The sidelobe effect is a strong interference source in seabed topography survey and underwater target detection. Tangible and physical targets in water columns, gaseous or not, are real structures and can produce backscatters. However, the sidelobe effect can also produce a false target image and its intensity level cannot be ignored, which can hinder the MBES to detect real targets. Therefore, it is important to acquire WCI with suppressed sidelobe effects before the target detection methods are performed [12,28].

When the sidelobes are eliminated based on MBES, object data can be mainly processed in three types: channel signals, beam output sequences, and WCIs in different stages of data processing of the MBES. Firstly, the channel signal is processed using beamforming techniques, such as the minimum variance distortionless response (MVDR) algorithm [26], to suppress the sidelobe leakage. Secondly, adaptive filters can be used to dispose beam output sequences to offset the sidelobe effect [29]. Finally, the suppression of sidelobe effects can be achieved by excluding data outside the MSR. In this work, the MVDR beamforming method is used to process the channel signal data collected by the MBES to reduce sidelobes in the WCIs. The array output of this beamforming method can be expressed as [30]:

$$F(n) = \mathbf{w}^H \mathbf{x}(t_e) \quad (1)$$

where $\mathbf{x}(t_e)$ is the array data vector; t_e is discrete time of echo signal; and $\mathbf{w}$ is the array weight coefficient, which is defined as:

$$\mathbf{w} = \frac{\mathbf{R}_{xx}^{-1}\mathbf{a}(\theta)}{\mathbf{a}(\theta)^H \mathbf{R}_{xx}^{-1}\mathbf{a}(\theta)} \quad (2)$$

where $\mathbf{a}(\theta)$ is the array manifold vector, H denotes matrix Hermitian transpose, and $\mathbf{R}_{xx} = \mathrm{E}\left[\mathbf{x}(t_e)\mathbf{x}^H(t_e)\right]$ is the correlation matrix of the array data. Furthermore, for a spherically isotropic noise field, the directional gain of the array can be expressed by a directivity index (DI) written as [30]:

$$DI = 10\log\frac{\left|\mathbf{w}^H\mathbf{a}(\theta)\right|^2}{\mathbf{w}^H\mathbf{R}_{vv}\mathbf{w}} \quad (3)$$

where $\mathbf{R}_{vv}$ represents an isotropic noise correlation matrix whose mnth matrix element is defined as:

$$\{\mathbf{R}_{vv}\}_{mn} = \frac{\sin[(m-n)kd]}{(m-n)kd},\ m,n = 1,2,\ldots,M \quad (4)$$

where M is the number of array elements, d is the element spacing, and k is the wave number. Figure 1c shows a T-A image obtained by using the MVDR beamforming method with the same data as Figure 1a, and Figure 1d is the D-T image corresponding to Figure 1c. By comparing the four figures in Figure 1, it can be seen that the MVDR beamformer can successfully suppress the sidelobe leakage.

2.2. Motion Estimation of Gas Emissions Via Farneback Optical Flow

2.2.1. Motion Estimation Using D-T Images

Leaking gases exist in the form of bubble groups in the water column. They diffuse and rise to the surface. These gases result in changes in the intensity distribution in multi-frame WCIs, which can be exploited using the optical flow method to estimate the motion information of the gas emissions. The optical flow calculation [31] uses the variation and correlation of pixel values in two sequential image frames to determine the "motion" of each pixel location. As a basic condition for the application of the optical flow method, it is assumed that the pixel values in the two images satisfy $\mathrm{d}F(x,y,t_f)/\mathrm{d}t_f = 0$, that is:

$$F(x,y,t_f) = F(x+\Delta x, y+\Delta y, t_f+\Delta t_f) \quad (5)$$

The generation and extinction of the bubble group, their distribution, and changes in size as they rise will cause rapid changes in the image, which does not occur in sonar images of fixed-shape objects. Therefore, to satisfy the above assumptions, the time interval between frames cannot be too long. Thus, in this paper, when the experimental data is collected in pool and lake, the frame rates of images are 20 Hz and 16 Hz, respectively. Motion information of leaking gasses has been estimated using the Farneback optical flow method, which is a classical dense optical flow estimation algorithm for target motion using images, and its basic principle and implementation are described in detail in [31]. The main idea of this method is to approximate the neighborhood near each pixel of an image as a quadratic polynomial, and to estimate the displacement field of the two frames using a polynomial expansion. Specifically, the neighborhood of each pixel of two images (F_1, F_2) at different times can be approximated as:

$$\begin{cases} F_1(\mathbf{z}) = \mathbf{z}^T\mathbf{A}_1\mathbf{z} + \mathbf{b}_1^T\mathbf{z} + c_1 \\ F_2(\mathbf{z}) = \mathbf{z}^T\mathbf{A}_2\mathbf{z} + \mathbf{b}_2^T\mathbf{z} + c_2 \end{cases} \tag{6}$$

where $\mathbf{z}$ is coordinate vector of image pixel, parameter $\mathbf{A}$ is a symmetric matrix, $\mathbf{b}$ is a vector and c is a scalar. It is assumed that F_2 is equal to F_1 with a global displacement $\mathbf{d}$, that is,

$$F_2(\mathbf{z}) = F_1(\mathbf{z} - \mathbf{d}) \tag{7}$$

By using Equation (6) to expand the two functions at both ends of Equation (7) and making the coefficients of the items equal, the global displacement can be obtained. From a visual point of view, an image whose horizontal and vertical axes are represented by the distance is more conducive to directly show the shape of the target. This is also the reason why D-T images are commonly shown in the display and control software of the sonar system rather than T-A images. When the displacement is estimated from two D-T images and divided by the time interval between the two frames, the velocity of each pixel can be obtained. Scheme A of Figure 2 describes a processing flow from beamforming processes of multi-channel echo data to estimate velocity fields and detect targets. The multi-channel echo data are converted into beam space data or a T-A image through the beamforming processing, and then the D-T image can be produced using a coordinate transformation and interpolation.

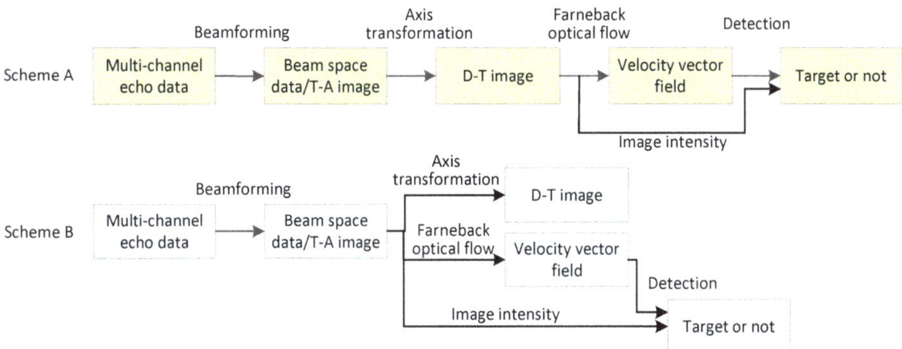

Figure 2. Two types of data processing schemes in motion estimation of gas emissions.

Some of the methods (e.g., [1,12,17]) to detect gas leaks based on MBESs only used the amplitude/intensity information of D-T images in Scheme A. However, if the detection is only based on amplitude/intensity, it cannot automatically distinguish between strong scatterers such as gas leaks and the seafloor. The motion of rising bubbles was considered [18] to detect leakage via cross-correlation; however, the potential situation of false but large motion speed due to continuous frame changes of the strong backscatter pixels from the seafloor has not been considered. In this paper, two types of information, intensities and velocity vectors, are comprehensively considered, and an automatic detection method for underwater gas leaks is designed to distinguish other strong scatterers in the water such as the seafloor (see Section 2.2.3). Specifically, the D-T images are processed using Farneback optical flow to obtain the velocity vector field.

Furthermore, in Scheme A, all feature information originates from the D-T images; thus, a process of converting from the beam space data to the D-T image is required. For a sonar system with automatic online detection requirements, the serial structure brings with it large hardware costs and affects real-time computing efficiency. For this reason, this paper further proposes the design of a more efficient parallel structure based on beam data. This new structure will affect velocity vector estimation; thus, it is necessary to further derive a corresponding conversion calculation method.

2.2.2. Motion Estimation Using T-A Images

Scheme A in Figure 2 is mainly a serial structure in which stringent requirements are often imposed on the real-time processing capability of the sonar hardware platform. This paper thus proposes Scheme B to directly estimate the motion information of the leaking gases using the T-A image, thereby reducing the processing steps from multi-channel data acquisition to target detection based on velocity. Moreover, the velocity field estimation can be processed in parallel with the D-T image and bathymetric estimation algorithms, instead of relying on serial processes as is done in Scheme A. In Scheme B, the Farneback optical flow method is used to process two frames of the T-A image to obtain relative displacement (t_0, θ_0) in the direction of echo time and beam angle axis. When the displacement is divided by the time interval, the velocity vector (u_t, v_θ) in the coordinate system with time/orientation as the coordinate axis can be obtained. During the hydrographic surveying, with the fluctuation of the vessel, the sonar array will produce changes in orientation, such as roll and pitch, which will bias the estimated target motion displacement. Fortunately, modern multibeam sonar products usually have roll- and pitch-compensation capabilities, and even medium and deep water MBES can compensate for the heading information, thus avoiding the above-mentioned errors.

The velocity vector (u_t, v_θ) estimated from the T-A images can be used to complete the task of leaking gas detection. There is often a requirement to display velocity information in conjunction with the D-T image. In such cases, it is necessary to convert (u_t, v_θ) into a motion velocity vector $\left(u_x, v_y\right)$ with horizontal and vertical displacement per unit time, that is, velocity, as coordinates. Under the T-A image with time-beam angle as the coordinate system, the coordinate transformation of (t_0, θ_0) can be expressed linearly as:

$$\begin{pmatrix} t' \\ \theta' \end{pmatrix} = \begin{pmatrix} t \\ \theta \end{pmatrix} + \begin{pmatrix} t_0 \\ \theta_0 \end{pmatrix} \tag{8}$$

Then, the displacement (x_0, y_0) of the point in the Cartesian coordinate system can be expressed as:

$$\begin{pmatrix} x_0 \\ y_0 \end{pmatrix} = \begin{pmatrix} \frac{c(t+t_0)\sin(\theta+\theta_0)-ct\sin(\theta)}{2} \\ \frac{c(t+t_0)\cos(\theta+\theta_0)-ct\cos(\theta)}{2} \end{pmatrix} \tag{9}$$

where c is the speed of sound at the specified depth of the pixel. Then, $\left(u_x, v_y\right)$ can be obtained by the division of (x_0, y_0) and time interval.

2.2.3. Detection of Gas Emissions

When using the MBES to detect gas emissions, the possible interference sources include sidelobe effects, backscatter contributions from non-gas targets in water (e.g., volume reverberation, seabed reverberation), and background ambient noise, among others [15]. An amplitude threshold can be used to suppress the interference of volume reverberation with low energy contributions. This threshold can be set to a dynamic value varying with the frames, or to a fixed value. The following threshold can be used:

$$|P(i,j) - P_{ref}| > k\sigma \tag{10}$$

where $P(i,j)$ is the value of each pixel in the WCI, k is the adjustable weight factor, and the values P_{ref} and σ are the median and standard deviation of all pixel values of the image, respectively. If the pixel in the frame satisfies Equation (10), then it is judged to belong to a part of a leaking gas image; otherwise, the pixel value is eliminated.

For a strong backscattering and relatively static target, the same decision criterion as in Equation (10) can be used, but the data involved in the calculation is replaced by the magnitude of the velocity estimated, that is, $V = \sqrt{u_x^2 + v_y^2}$. Furthermore, with regard to the relatively flat seabed topography, when a survey vessel is sailing, the image of the area corresponding to the seabed in WCIs of the adjacent pings will show the partial changes of the energy distribution in the horizontal direction, which is mainly due to the statistical fluctuation of the seabed echo energy caused by the fluctuation of the micro-topography. This phenomenon may also bring about large velocities corresponding to the seabed portion of the velocity field, and most of the velocity direction should be similar to the seabed terrain. In consideration of the above situation, a directional threshold is further set to exclude the target image in the horizontal direction or close to it, in the estimated velocity field. Specifically, in this paper, the threshold on the velocity direction relative to the horizontal is set at 40 degrees. The rising gases may be deflected by currents [4], and thus the range of the velocity direction threshold should be adjusted according to the direction of the ocean current in the sea test. In addition, schools of fish are also common underwater moving targets. Through this direction threshold, some fish school images shown in the horizontal direction can be eliminated.

3. Results

3.1. Pool Experiment

3.1.1. Experimental Design and Equipment

In this study, a scene of leaking gases from a pipeline was simulated in a pool, and the acoustic data was collected by an HT300 PA MBES developed by Harbin Engineering University, China. As shown in Figure 3, the MBES is mainly composed of an underwater subsystem (i.e., the sonar head), an interface unit, and a computer. The underwater subsystem is responsible for the emission, reception, and real-time processing of the underwater acoustic signal. The interface unit is used for real-time information collection of the auxiliary equipment and network data exchange, and the HTCS 3.4 display and control software of this MBES can be operated on a computer. The sonar system emits an acoustic wave with a frequency of 300 kHz, a CW pulse signal with a pulse width of 0.1 ms, and a ping rate of 20 Hz. In addition, its beam width is 2.5° × 1.5° and max swath width is 126°.

The distribution of the bubble size and number of bubbles changes continuously during their rising process because they are affected by factors such as water depth, nozzle size, pressure in the pipeline, flow rate, ocean current, and so on. Also, the backscattering strength (or sound attenuation) of the bubbles is closely related to the signal frequency, bubble size, and the number of bubbles [32–35]. For a nozzle, its size decides the sizes of bubbles. For example, the radius of the detached bubble could be expressed as a non-monotonic function of the nozzle diameter [36], and the radius showed an increasing trend during the change of the nozzle diameter (0.1 mm–2 mm). For a single bubble, the different sizes correspond to different resonance frequencies, and the backscattering strength reaches a peak at this frequency [35]. The relationship between the scattering cross-section σ and a bubble of a radius a can be expressed as [37]:

$$\sigma(a) = \frac{4\pi a^2}{\left[\left(\frac{f_0}{f}\right)^2 - 1\right]^2 + (ka)^2} \tag{11}$$

where f_0 is resonance frequency of the bubble and k is the wave number. This model was applied in [37] to analyze the relationship between backscattering strength and bubble radius at 40 kHz, 180 kHz and

300 kHz. In actual applications, MBESs operate at frequencies from 12 kHz to 400 kHz [3,37], and they have been applied in researching the detection of gas leaks and other targets in water. In general, in shallow water (<200 m), MBESs with a frequency range of 200–400 kHz are usually used for marine surveys, and MBESs with these frequencies are also suitable for the deployment of the unmanned underwater vehicles due to their size. Because the purpose of this paper is to detect underwater pipeline leaks in offshore waters (usually less than 200 m in depth), and comprehensive consideration of factors such as maximum sounding capability, resolution of the multibeam sonar, and frequency response characteristics of the target of interest, this type of MBESs was finally used for experimental research.

During the experiment, the sonar head of the MBES is fixed at 0.5 m below the water surface with a tilt angle of $\alpha = 30^{\circ}$, as shown in Figure 4. The maximum coverage angle of this MBES is from −63 degrees to +63 degrees and the depth of the pool is shallow; thus, in order to enable the sonar to accurately study the process wherein the bubbles rise to the surface, it is installed vertically with a 30-degree tilt. The leaking gases are generated by an air compressor and released through a long rubber tube, which is connected to the nozzle at the bottom of the pool (5 m depth). Figure 5 is a video capture of the gas leaking scenario during the experiment.

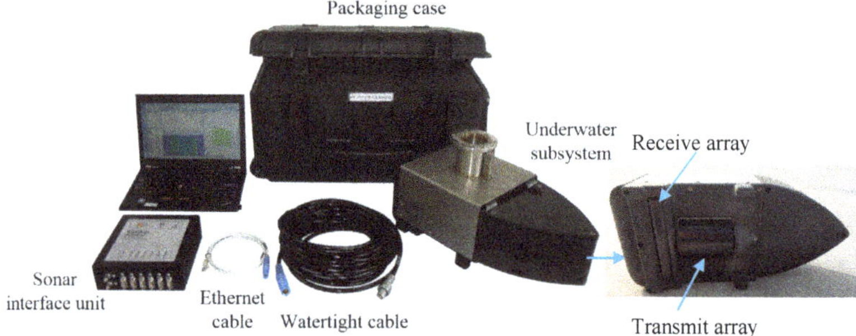

Figure 3. A photograph of HT300 PA MBES.

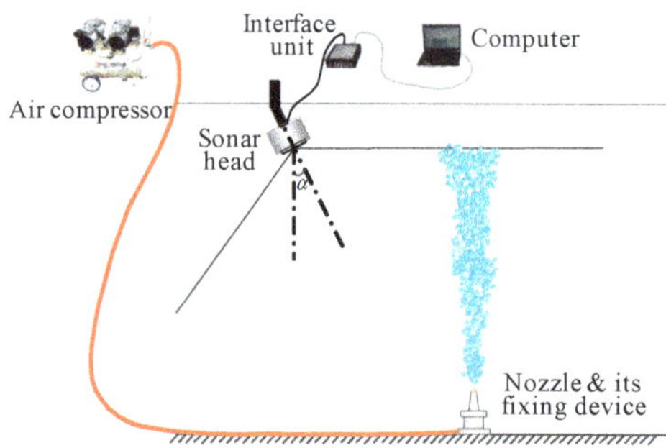

Figure 4. Schematic diagram of the pool experiment for simulation of leaking gases.

Figure 5. An underwater video screenshot of the bubbles being released.

3.1.2. Optical Flow Calculation of the Water Column Images (WCIs) Containing Leaking Gases

Figure 6a,b are D-T images taken via the conventional beamformer (CBF) and the MVDR beamformer, respectively, and both matrix data are normalized and displayed in decibels. The fixed nozzle diameter is $\phi = 0.6$ mm; the output intensity and the output flow rate of the air compressor are 0.1 MPa and 6.3 L/min, respectively; and the sonar head is stationary. In Figure 6a, fan-shaped bright strings are formed at the MSR, and their brightness is close to the intensity level of gas emissions. In this experiment, the gassy release area is designed outside the MSR and its purpose is to consider whether the detection of the gas emissions is still valid when they occur in an area that have significant interference from sidelobe contamination. Figure 6b shows that the image at the MSR (i.e., the region where the sidelobe effect energy is the largest) was well-suppressed. There also is sidelobe leakage outside of the MSR, but the energy level is usually lower than that of the leaking gases.

Using Figure 6b and its adjacent frame images, the motion field distribution of the obtained image region is an implementation of the Farneback optical flow algorithm in MATLAB with Scheme A. In Figure 7a, the estimated velocity field is displayed in conjunction with the WCI of Figure 6b. The direction of the arrow in the figure is indicated as the direction of velocity, and the length of the arrow is the magnitude of the velocity. Since an estimated velocity value can be obtained for each pixel of the D-T image, to avoid a large number of arrows being too dense to observe, the velocity matrix here is subjected to equal interval sampling processing to display. The image of the small frame on the left side in Figure 7a is the enlarged result of the image in the red frame in the gas release zone. Figure 7b shows the estimation results of the leaking gas motion obtained by processing the same data using Scheme B, and the parameters used by the Farneback algorithm are the same as those in Figure 7a. The estimated velocity is converted to the form of $\left(u_x, v_y\right)$ for display. Then, Figure 8 shows a set of WCIs and motion field estimation results after changing the experimental conditions, where the nozzle is replaced by one with a diameter of 1.2 mm, the output intensity of pressure is set to 0.5 MPa, and the output flow rate is 51 L/min.

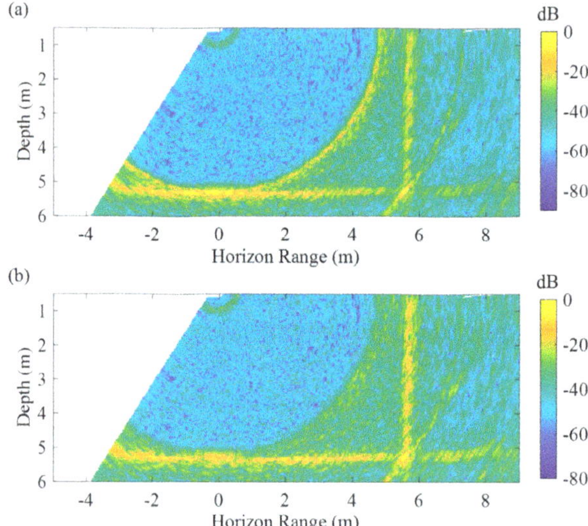

Figure 6. The WCIs containing gas emissions, obtained by the conventional beamformer (CBF) beamformer (**a**) and MVDR beamformer (**b**).

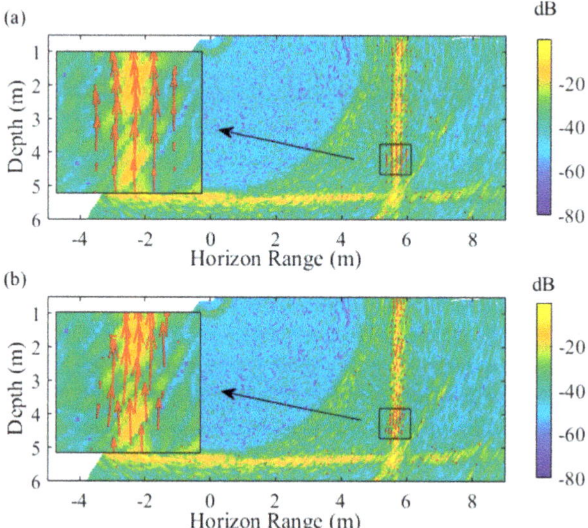

Figure 7. The joint display of the WCI and velocity field estimated by frame A (**a**) and frame B (**b**), respectively. The fixed nozzle aperture is $\phi = 0.6$ mm and the output intensity of pressure and the output flow rate was 0.1 MPa and 6.3 L/min, respectively.

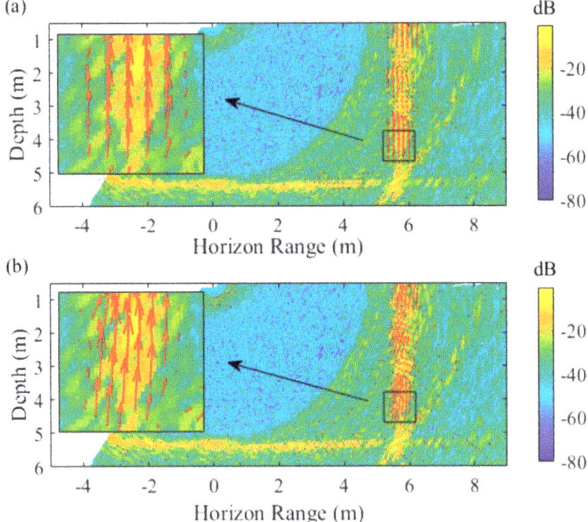

Figure 8. The joint display of the WCI and velocity field estimated by frame A (**a**) and frame B (**b**), respectively. The fixed nozzle aperture is ϕ= 1.2 mm and the output intensity of pressure and the output flow rate are 0.5 MPa and 51 L/min, respectively.

It can be seen from Figures 7 and 8 that the velocity field trends estimated by the two processing schemes are roughly similar, but the calculated velocity field matrices have different distributions in the D-T image. The velocity field data calculated by Scheme A and the individual pixels of the D-T image are sometimes in one-to-one correspondence to their spatial position. However, since the velocity field matrix calculated using Scheme B is equally spaced when using the coordinate system with the beam angle and the TWTT as the coordinate axes, the velocity field data near the sonar head is dense, but gradually becomes sparse as the distance increases. In addition, the velocity field of the region containing gas can be visually different from those of other regions. The direction of the velocity in the region containing gas is mostly in the quasi-vertical direction. Since the sonar head is stationary, the ensonified area on the bottom of the pool has not changed. Under this condition, the energy of the individual pixels corresponding to the bottom area in different ping images changes very little; thus, the estimated velocities in this area will be very small, which can be excluded by considering the magnitude of the velocity.

To quantitatively analyze the consistency of the estimated velocities of the two schemes, a square with a side of 5 cm is selected as an observation area to calculate magnitudes of the velocity at different times in this region. The center point coordinate of the observation area is (5.65 m, 3 m). Figure 9 shows the variation of the average of all the estimated velocities in the observation area, along with the sample time. The horizontal axis is the sample time or the image frame number, and the longitudinal axis is the average of the magnitudes of the velocities. For each pixel, the corresponding magnitude of the velocity can be calculated by $V = \sqrt{u_x^2 + v_y^2}$. The curves plotted in Figure 9a,b respectively correspond to the results of the two settings of the nozzle sizes and the output intensities of the pressure conditions, while the data processing is identical. It can be seen from the two figures that the estimated velocity values of the two frames are relatively similar in general, but there are also a small number of different cases. This may be because the number and spatial distribution of velocities estimated by the two frames in the observation area are different. When the difference in velocity values in this area is large, the two statistical averages will show a large deviation. However, this difference does not change the nature of the relatively high-speed movement of the gas emissions from the pipelines, and thus it will not affect the subsequent detection of gas emissions. For leaking gases from undersea pipelines, the velocities of bubbles in water columns were composed of two parts [38]. One is the free rising velocity of the gas bubbles, and the other is the initial velocity when they were ejected at the nozzle. Considering these contributions, the authors of [38] used Doppler shifts of ultrasonic scanning to estimate the leaking gases simulated in the laboratory; the estimated velocity is 3.2–9.7 m/s, which is similar to the results of Figure 9. The echo intensities at different flow rates are further measured and are shown in Figure 10. In the two data acquisitions, the nozzle diameter is always 0.6 mm; the flow rates of the air compressor are 21 L/min and 6.3 L/min, respectively; and the multibeam sonar is set with the same parameters, including the transmit power, fixed gain, and time varying gain. As shown in Figure 10, the large flow rate corresponds to stronger echo intensity as a whole, which is similar to the results found in [37]. This phenomenon implies that there is a certain correlation between the flow rate and the backscattering signal. Compared with results in [39,40], the derived rise velocities in Figure 9 appear on the lower scale. This may be due to the limited output pressure capability of the used air compressor and the absence of currents, which affects the intensity and direction of the motion velocities.

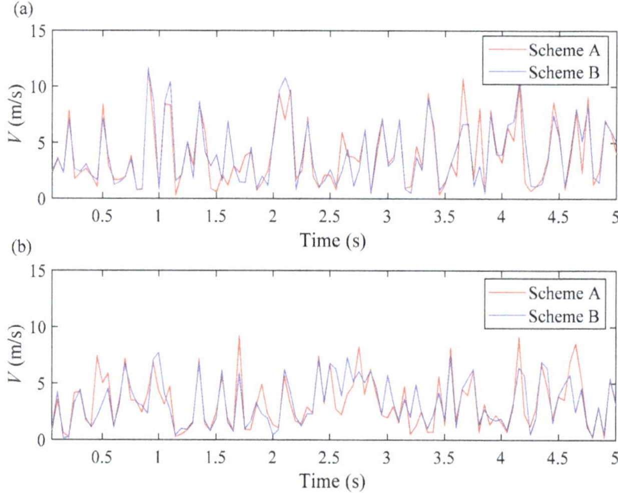

Figure 9. The magnitudes of the velocities estimated from Scheme A and Scheme B. (**a**) and (**b**) correspond to the two experimental conditions of Figures 7 and 8, respectively.

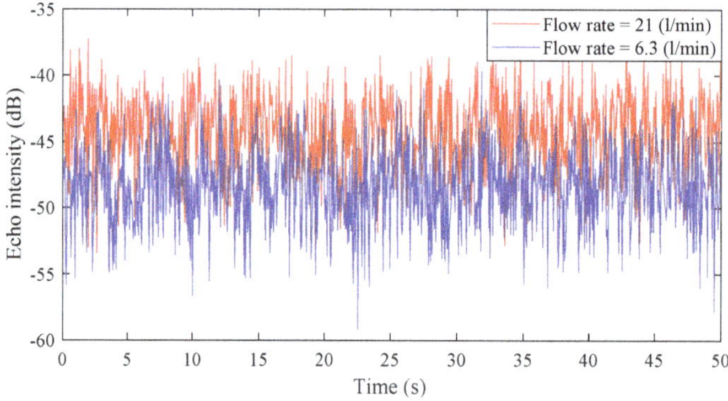

Figure 10. Comparison of echo intensities at two flow rate outputs.

3.1.3. Detection of Gas Emissions

Due to the WCIs and velocity field distributions in the pool experiment, the thresholds of the image amplitudes and velocity values were set according to Equation (10) to detect gas emissions. The weight coefficient of the amplitude threshold is 1, and the weight coefficient of the velocity intensity threshold is 1. By further eliminating the data below the thresholds in Figures 7 and 8, the results can be seen in Figures 11 and 12, respectively. The velocity field data used in Figure 11a, Figure 12a was estimated using Scheme A, while the velocity field data in Figures 11b and 12b were estimated using Scheme B. As can be seen in Figures 11 and 12, the image amplitude and velocity intensity threshold could be used to exclude images with small velocities but strong backscattering.

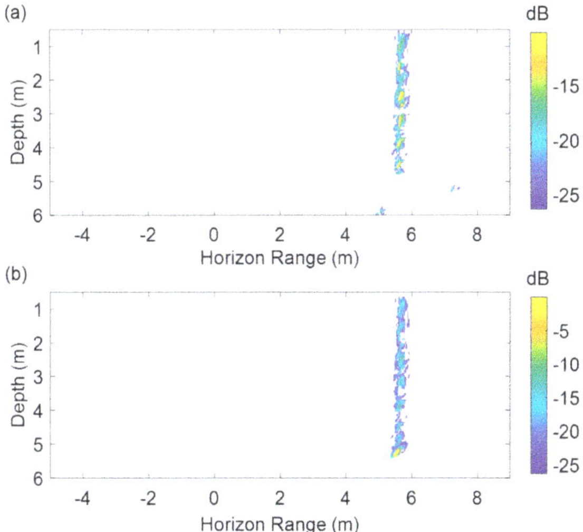

Figure 11. Results obtained by screening the data in Figure 7 using thresholds of image amplitude and velocity intensity. (**a**) and (**b**) correspond to the processing results of Scheme A and Scheme B, respectively.

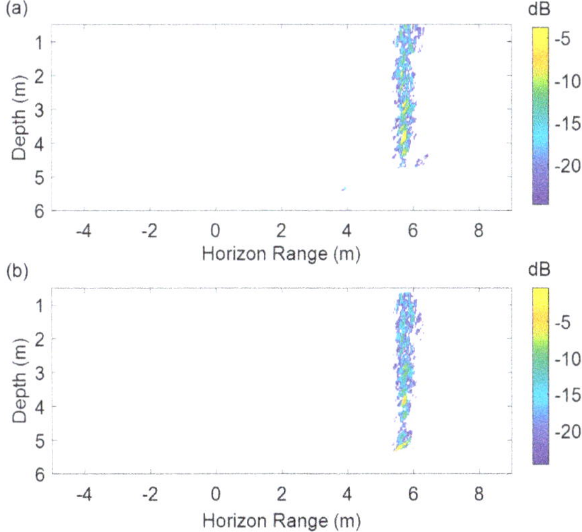

Figure 12. Results obtained by screening the data in Figure 8 using thresholds of image amplitude and velocity intensity. (**a**) and (**b**) correspond to the processing results of Scheme A and Scheme B, respectively.

3.2. Lake Experiment

3.2.1. Experimental Scenario

To quantitatively analyze the estimated velocities of the two schemes under the same conditions, the above-mentioned pool experiment is a static measurement where the relative position of the measuring platform and the gas-releasing device is constant. A dynamic experiment is further designed here to measure the simulated gas emissions from a pipeline using an MBES mounted on a survey vessel. The experimental site is in Songhua Lake, Jilin Province, China, and the MBES and air compressor used in the experiment are the same as in the pool experiment. During the experiment, one end of the rubber tube connecting the nozzles is fixed on a triangular holder seated at a flat bottom of water depth of about 13 m, while the other end is extended from the bottom to a barge on the coast and connected to the air compressor. The fixed nozzle aperture is $\phi = 1$ mm and the output intensity of pressure is $P_{out} = 0.68$ MPa. When the survey ship is traveling, the MBES collects the underwater target echo at a ping rate of 16 Hz.

3.2.2. Detection of Gas Emissions

The WCIs in Figure 13 are obtained when the survey ship travels above the leaking gases, and the velocity field distributions of Figure 13a,b are obtained using Schemes A and B, respectively. From the partial enlargement of the two figures, the corresponding velocity field of the regions containing gas is mainly on the upward trend, and the high magnitudes of velocity are also found in the seabed area. However, as analyzed in Section 2.2.3, the velocity field directions in the seabed area are mostly similar to the seabed terrain tendency. Correspondingly, the velocity direction threshold is set to 40 degrees. For instance, the data in Figure 13a are judged only by the amplitude threshold to obtain the detection result as shown in Figure 14a, and the detection result after additional estimation of the magnitude of the velocity and direction threshold is shown in Figure 14b.

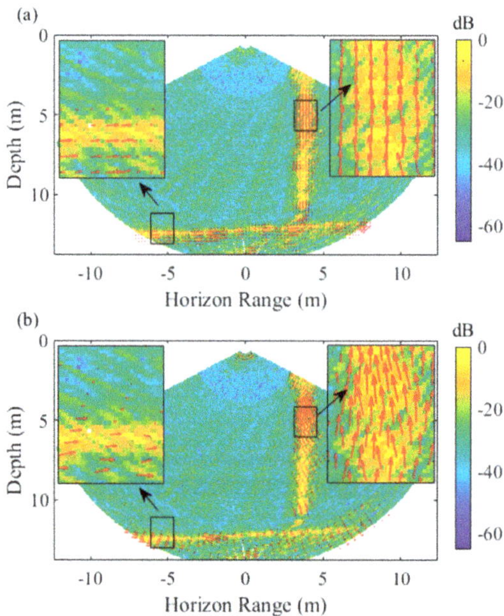

Figure 13. The joint display of the WCI and velocity field estimated by frame A (**a**) and frame B (**b**), respectively, from the experimental data of a measurement period in Songhua Lake.

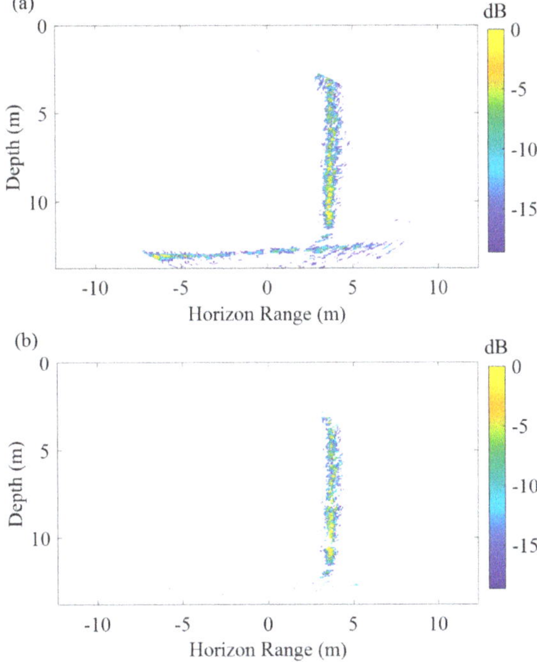

Figure 14. The result of the comprehensive detection of the data in Figure 12 using thresholds of pixel intensity, velocity magnitude, and velocity direction. (**a**) is the detection result using only the amplitude threshold, and (**b**) is the detection result of adding the magnitude of the velocity and direction threshold.

It is possible to eliminate pixel data with relatively low intensity, low speed, or high horizontal speeds by comparing Figures 13 and 14 after the detection threshold is set using both the energy and the displacement of the image pixels. Thus, an image with gas emissions can be detected from the complex WCI data.

4. Discussion

In this paper, the motion field in continuous multi-frame WCIs was estimated using an optical flow method. If only the intensity information of each pixel in the WCI is used to detect the leaking gases, the results could detrimentally include the seabed and other strong interference targets in the water column. Although the images of the seabed and sidelobe interference could simply exclude the portion outside the MSR as was done in [12,15], this strategy may also remove parts of the gassy image, and any strong targets suspended within the MSR region cannot be identified and ignored. As shown in Figure 15, only the amplitude threshold was used to detect the WCI data measured on a short survey line in the Songhua Lake. It can be seen from the figure that it is possible to detect other false targets using only the amplitude threshold, thereby affecting the autonomous and high-accuracy recognition of the gas emissions. For this reason, the estimated motion vector information was further used to eliminate the image pixels of relatively low-speed targets or targets whose backscatter is high but moving horizontally, such as a flat seabed or part of a school of fish. Inevitably, small spots of discrete distributions, as shown in Figure 14, may appear after the processing. These spots are mainly residues after the above threshold decision, and they have a low number of pixels with a discrete distribution. Here, these abnormal pixels can be eliminated by a small sliding window. As shown in Figure 16, the three-dimensional morphology of the leaking gases with a relatively clean background was obtained after multi-threshold detection, which achieves the automatic detection of gas emissions.

Considering that sidelobe contamination may adversely affect the amplitude and velocity threshold detection, the MVDR beamforming algorithm was used in the beamforming process to suppress the sidelobes.

In addition, two kinds of schemes were designed for data processing. Similar to [18], in Scheme A, the velocity of the movement is estimated from the D-T image, so that the dimension of the velocity vector matrix is consistent with the D-T image, which is more convenient for visual observation. However, considering that the data processing of Scheme A is mainly a serial structure, and the automatic and real-time detection of the leaking gases on the sonar hardware platform is also a potential demand in the future, this paper further proposes directly estimating the gas motion using T-A images and to judge whether the gas emissions are present or not. In this way, the velocity field estimation can be processed in parallel with the D-T image generation and the seabed terrain detection process from the perspective of the data processing flow, and it is not necessary to wait for the conversion of the image from T-A to D-T, which improves the feasibility of real-time realization in the hardware platform of the MBES.

To verify the effectiveness of the proposed method, experiments were carried out on pools and lakes with relatively shallow water depths. The underwater condition may be more complicated with the increase of water depth. For example, the backscattering strength of the bubbles is also related to water depth (or ambient hydrostatic pressure P_0). The resonance frequency in Equation (11) can be derived [37] as:

$$f_0 = \frac{1}{2\pi a}\sqrt{\frac{3\gamma P_0}{\rho}} \tag{12}$$

where ρ is the density of the water, and γ is the ratio of specific heats. It can be seen from the combination of Equations (11) and (12) that the scattering cross section is also a function of the hydrostatic pressure. This is especially the case in deep water conditions, where the effect of hydrostatic pressure is more obvious. Therefore, when the background noise level is isotropic and the leakage rates and sonar frequencies are the same, the signal-to-noise ratio corresponding to the echo of gases at different depths will also change. At this point, the amplitude threshold of Equation (10) should vary with the depth. Therefore, to fully test the detection performance and adjust its adaptability with changes in depth, in future research we will further carry out experimental verification studies in deep water environments. Of course, the rising gases may be deflected by ocean currents [4], and be affected by water density, salinity, temperature, and static pressure, which are important for seismic oceanography [41]. Therefore, the impact mechanism should be further analyzed and the velocity direction threshold adjusted according to the above conditions in sea experiments.

In a pool experiment, the MBES is stationary. When experimenting on the lake, the MBES is installed on the survey vessel and moves along the survey line. With fluctuation of the vessel, the sonar produces changes in orientation, such as roll and pitch, which will bias the estimated target motion displacement. Therefore, when processing the data on the lake, motion attitude information is compensated. In addition, in the pool experiment, the position of the illuminated bottom is unchanged; thus, the fluctuation of the echo from bottom in WCIs of the adjacent pings is very small. When a survey vessel is sailing, the fluctuation shows more significant changes, which is mainly due to statistical fluctuation of the seabed echo energy. Therefore, in terms of the bottom part of the velocity field in Figures 8 and 13, the estimated velocity in the lake experiment appears to be larger than that in the pool experiment.

This paper aims to detect moving targets through high and low-speed fields, without considering the rigorous verification of the accuracy of the estimated speed. This will be considered in subsequent studies, such as using video-in-situ observations [13] for comparison. We will also consider the combination of the backscattering strength of the bubble group to quantitatively estimate the flow rate of the leaking gases. In addition, it is assumed that the seabed terrain is relatively flat when the velocity direction threshold is used. If the terrain changes steeply, the appropriate values of this threshold may be difficult to obtain. Moreover, fish may be one of the major reasons for misdetections, although the direction threshold is set. Therefore, in future research, seabed detection technology will be applied to judge the seabed and eliminate seabed image interference. In addition, other features, such as target shapes, will be considered to eliminate interference from fish.

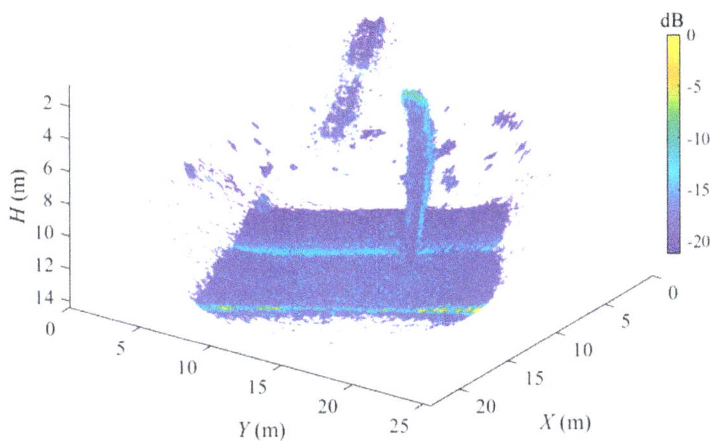

Figure 15. The result of the amplitude threshold detection of WCI data measured on a short survey line in the Songhua Lake.

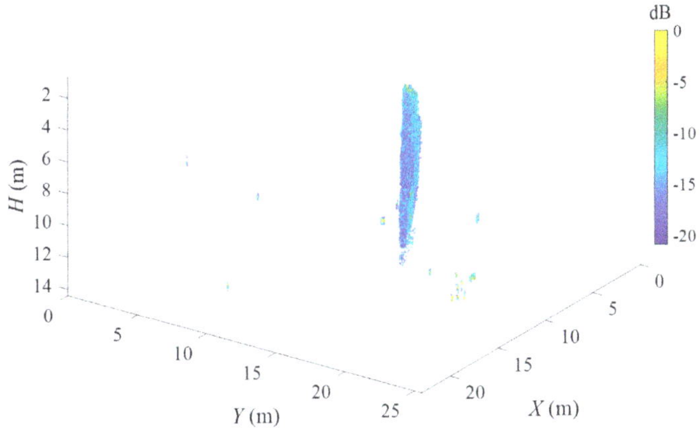

Figure 16. The result of the multi-threshold detection of WCI data measured on a short survey line in the Songhua Lake.

5. Conclusions

To achieve automatic detection of leaking gases from underwater pipelines, a detection method based on the combination of motion and intensity information of WCI pixels was studied in this paper. The motion of the image pixel was estimated using the Farneback optical flow principle, and two data processing schemes were designed according to two types of WCI data. Through the data analysis of two simulation experiments in a pool and a lake, it can be seen that the velocities obtained based on the above two schemes had relatively good consistency. In addition, after the comprehensive threshold selection using amplitude and velocity information, leaking gases could be detected more efficiently. From the perspective of our designed structure, Scheme B is more suitable for real-time implementation of sonar hardware platforms, while Scheme A is more suitable for execution in display and control software, and in post-processing.

In this paper, it is assumed that the trend of seabed topography is relatively flat in the design of the velocity direction threshold. In future research, bottom-tracking technology will be introduced to judge and eliminate the seabed image with more complex trends of topography, and morphological features will also be introduced to judge and eliminate the other high-velocity motion and strong-backscatter targets in water columns. Moreover, the influence of sound velocity changes for the thresholds will be analyzed.

Author Contributions: C.X., T.Z., and W.Z. developed and designed the experiments; C.X., W.Z., M.W., and W.D. performed the experiments; C.X. and W.Z. analyzed the data; C.X., T.Z., and M.W. wrote the paper, and J.L. and P.R.W. modified and polished the paper. All authors have read and agreed to the published version of the manuscript.

Funding: This research was supported by the National Natural Science Foundation of China (grant numbers U1709203, 41606115 and 41876100), the National Science and Technology Major Project of China (grant number 2016ZX05057005), the National Key R&D Program of China (grant number 2016YFC1402303), and financial assistance from the Postdoctoral Scientific Research Developmental Fund of Heilongjiang (grant number LBH-Q18042). The work of Jianghui Li and Paul R. White was partly supported by the European Union Horizon 2020 research and innovation programme under grant agreement number 654462 (STEMM-CCS).

Acknowledgments: The authors would like to thank Yongxiang Hu and Dongyang Li for their tireless work during the experiment on the lake.

Conflicts of Interest: The authors declare no conflict of interest.

References

1. Colbo, K.; Ross, T.; Brown, C.; Weber, T. A review of oceanographic applications of water column data from multibeam echosounders. *Estuar. Coast. Shelf Sci.* **2014**, *145*, 41–56. [CrossRef]
2. Merewether, R.; Olsson, M.S.; Lonsdale, P. Acoustically detected hydrocarbon plumes rising from 2-km depths in Guaymas Basin, Gulf of California. *J. Geophys. Res.* **1985**, *90*, 3075–3085. [CrossRef]
3. von Deimling, J.S.; Weinrebe, W. Beyond bathymetry: Water column imaging with multibeam echo sounder systems. *Hydrographische Nachrichten* **2014**, *31*, 6–10.
4. von Deimling, J.S.; Greinert, J.; Chapman, N.R.; Rabbel, W.; Linke, P. Acoustic imaging of natural gas seepage in the North Sea: Sensing bubbles controlled by variable currents. *Limnol. Oceanogr. Methods* **2010**, *8*, 155–171. [CrossRef]
5. Nikolovska, A.; Sahling, H.; Bohrmann, G. Hydroacoustic methodology for detection, localization, and quantification of gas bubbles rising from the seafloor at gas seeps from the eastern Black Sea. *Geochem. Geophys.* **2008**, *9*, 1–13. [CrossRef]
6. Conti, A.; D'Emidio, M.; Macelloni, L.; Lutken, C.; Asper, V.; Woolsey, M.; Jarnagin, R.; Diercks, A.; Highsmith, R.C. Morpho-acoustic characterization of natural seepage features near the Macondo Wellhead (ECOGIG site OC26, Gulf of Mexico). *Deep Sea Res. Pt II* **2016**, *129*, 53–65. [CrossRef]
7. Li, J.; White, P.R.; Bull, J.M.; Leighton, T.G. A noise impact assessment model for passive acoustic measurements of seabed gas fluxes. *Ocean Eng.* **2019**, *183*, 294–304. [CrossRef]
8. Li, J.; Roche, B.; Bull, J.M.; White, P.R.; Davis, J.W.; Deponte, M.; Gordini, E.; Cotterle, D. Passive acoustic monitoring of a natural CO_2 seep site—Implications for carbon capture and storage. *Int. J. Greenh. Gas Control* **2020**, *93*, 102899. [CrossRef]
9. Zhang, W.; Zhou, T.; Peng, D.; Shen, J. Underwater pipeline leakage detection via multibeam sonar imagery. *J. Acoust. Soc. Am.* **2017**, *141*, 1. [CrossRef]
10. Hughes Clarke, J.E.; Lamplugh, M.; Czotter, K. Multibeam water column imaging: Improved wreck least-depth determination. In Proceedings of the Canadian Hydrographic Conference, Toronto, ON, Canada, 2–4 June 2006.

11. Innangi, S.; Bonanno, A.; Tonielli, R.; Gerlotto, F.; Innangi, M.; Mazzola, S. High resolution 3-D shapes of fish schools: A new method to use the water column backscatter from hydrographic MultiBeam Echo Sounders. *Appl. Acoust.* **2016**, *111*, 148–160. [CrossRef]
12. Urban, P.; Koser, K.; Greinert, J. Processing of multibeam water column image data for automated bubble/seep detection and repeated mapping. *Limnol. Oceanogr. Methods* **2017**, *15*, 1–21. [CrossRef]
13. Leblond, I.; Scalabrin, C.; Berger, L. Acoustic monitoring of gas emissions from the seafloor. Part I: Quantifying the volumetric flow of bubbles. *Mar. Geophys. Res.* **2014**, *35*, 191–210. [CrossRef]
14. Bayrakci, G.; Scalabrin, C.; Dupre, S.; Leblond, I.; Tary, J.B.; Lanteri, N.; Augustin, J.M.; Berger, L.; Cros, E.; Ogor, A.; et al. Acoustic monitoring of gas emissions from the seafloor. Part II: A case study from the Sea of Marmara. *Mar. Geophys. Res.* **2014**, *35*, 211–229. [CrossRef]
15. Zhao, J.H.; Meng, J.X.; Zhang, H.M.; Wang, S.Q. Comprehensive detection of gas plumes from multibeam water column images with minimisation of noise interferences. *Sensors* **2017**, *17*, 2755. [CrossRef] [PubMed]
16. von Deimling, J.S.; Brockhoff, J.; Greinert, J. Flare imaging with multibeam systems: Data processing for bubble detection at seeps. *Geochem. Geophys. Geosyst.* **2007**, *8*, 7. [CrossRef]
17. Veloso, M.; Greinert, J.; Mienert, J.; De Batist, M. A new methodology for quantifying bubble flow rates in deep water using splitbeam echosounders: Examples from the Arctic offshore NW-Svalbard. *Limnol. Oceanogr. Methods* **2015**, *13*, 267–287. [CrossRef]
18. von Deimling, J.S.; Papenberg, C. Technical note: Detection of gas bubble leakage via correlation of water column multibeam images. *Ocean Sci.* **2012**, *8*, 175–181. [CrossRef]
19. Blomberg, A.E.A.; Saebo, T.O.; Hansen, R.E.; Pedersen, R.B.; Austeng, A. Automatic detection of marine gas seeps using an interferometric sidescan sonar. *IEEE J. Ocean. Eng.* **2017**, *42*, 590–602. [CrossRef]
20. Christoffersen, J.T.M. Multi-detect algorithm for multibeam sonar data. In Proceedings of the OCEANS'13 MTS/IEEE, San Diego, CA, USA, 23–27 September 2013.
21. Wilson, D.S.; Leifer, I.; Maillard, E. Megaplume bubble process visualization by 3D multibeam sonar mapping. *Mar. Petrol. Geol.* **2015**, *68*, 753–765. [CrossRef]
22. Best, J.; Simmons, S.; Parsons, D.; Oberg, K.; Czuba, J.; Malzone, C. A new methodology for the quantitative visualization of coherent flow structures in alluvial channels using multibeam echo-sounding (MBES). *Geophys. Res. Lett.* **2010**, *37*, 1–6. [CrossRef]
23. Sandsten, J.; Andersson, M. Volume flow calculations on gas leaks imaged with infrared gas-correlation. *Opt. Express.* **2012**, *20*, 20318–20329. [CrossRef] [PubMed]
24. Lurton, X. *An Introduction to Underwater Acoustics. Principles and Applications*, 1st ed.; Springer Praxis Books & Praxis Publishing: London, UK, 2002; pp. 206–209.
25. Clarke, J.E.H. Applications of multibeam water column imaging for hydrographic survey. *Hydrogr. J.* **2006**, *120*, 3.
26. Li, H.S.; Gao, J.; Du, W.D.; Zhou, T.; Xu, C.; Chen, B.W. Object representation for multi-beam sonar image using local higher-order statistics. *EURASIP J. Adv. Signal Process.* **2017**, *2017*, 7.
27. Du, W.D.; Zhou, T.; Li, H.S.; Chen, B.W.; Wei, B. ADOS-CFAR Algorithm for Multibeam Seafloor Terrain Detection. *Int. J. Distrib. Sens. Netw.* **2016**, *12*, 1719237. [CrossRef]
28. de Moustier, C. OS-CFAR detection of targets in the water column and on the seafloor with a multibeam echosounder. In Proceedings of the OCEANS'13 MTS/IEEE, San Diego, CA, USA, 23–27 September 2013.
29. Weng, N.; Li, H.; Yao, B.; Zhou, T.; Wei, Y.; Chen, B.; Liu, X. Tunnel effect in multi-beam bathymetry sonar and its canceling with error feedback lattice recursive least square algorithm. In Proceedings of the MTS/IEEE Kobe Techno-Ocean, Kobe, Japan, 8–11 April 2008.
30. Bai, M.R.; Ih, J.-G.; Benesty, J. *Acoustic Array Systems: Theory, Implementation, and Application*; Wiley/IEEE Press: Singapore, 2013; pp. 68–76.
31. Farneback, G. Two-frame motion estimation based on polynomial expansion. In Proceedings of the SCIA 2003 13th Scandinavian Conference, Halmstad, Sweden, 2–29 June 2003.
32. Ainslie, M.A.; Leighton, T.G. Review of scattering and extinction cross-sections, damping factors, and resonance frequencies of a spherical gas bubble. *J. Acoust. Soc. Am.* **2011**, *130*, 3184–3208. [CrossRef]
33. Ainslie, M.A.; Leighton, T.G. Near resonant bubble acoustic cross-section corrections, including examples from oceanography, volcanology, and biomedical ultrasound. *J. Acoust. Soc. Am.* **2009**, *126*, 2163–2175. [CrossRef]
34. Fan, Y.; Li, H.; Xu, C.; Zhou, T. Influence of bubble distributions on the propagation of linear waves in polydisperse bubbly liquids. *J. Acoust. Soc. Am.* **2019**, *145*, 16–25. [CrossRef]

35. Fan, Y.; Li, H.; Xu, C.; Chen, B.; Du, W. Acoustic scattering from bubble clouds near the sea surface based on linear oscillations. *Acta Acust.* **2019**, *44*, 312–320.
36. Longuet, M.S.; Kerman, B.R.; Lunde, K. The release of air bubbles from an underwater nozzle. *J. Fluid Mech.* **1991**, *230*, 365–390. [CrossRef]
37. Greinert, J.; Nutzel, B. Hydroacoustic experiments to establish a method for the determination of methane bubble fluxes at cold seeps. *Geo-Mar. Lett.* **2004**, *24*, 75–85. [CrossRef]
38. Wang, J.; Stewart, R.; Dyaur, N. Underwater gas-leak detection and imaging using ultrasonic scanning: Laboratory results. In Proceedings of the 2017 SEG International Exposition and Annual Meeting, Houston, TX, USA, 24–29 September 2017.
39. Clift, R.; Grace, J.R.; Weber, M.E. *Bubbles, Drops, Particles*; Academic Press: New York, NY, USA, 1978; pp. 169–199.
40. Leifer, I.; Luyendyk, B.P.; Boles, J.; Clark, J.F. Natural marine seepage blowout: Contribution to atmospheric methane. *Glob. Biogeochem. Cycles* **2006**, *20*, 1–9. [CrossRef]
41. Dickinson, A.; White, N.J.; Caulfield, C.P. Spatial variation of diapycnal diffusivity estimated from seismic imaging of internal wave field, Gulf of Mexico. *J. Geophys. Res. Oceans* **2017**, *122*, 9827–9854. [CrossRef]

Article

Estimating Traffic Intensity Employing Passive Acoustic Radar and Enhanced Microwave Doppler Radar Sensor

Andrzej Czyżewski, Józef Kotus and Grzegorz Szwoch *

Telecommunication and Informatics, Multimedia Systems Department, Faculty of Electronics, Gdansk University of Technology, Narutowicza 11/12, 80-233 Gdańsk, Poland; ac@pg.edu.pl (A.C.); jozef.kotus@pg.edu.pl (J.K.)
* Correspondence: grzszwoc@pg.edu.pl

Received: 15 November 2019; Accepted: 25 December 2019; Published: 29 December 2019

Abstract: Innovative road signs that can autonomously display the speed limit in cases where the traffic situation requires it are under development. The autonomous road sign contains many types of sensors, of which the subject of interest in this article is the Doppler sensor that we have improved and the constructed and calibrated acoustic probe. An algorithm for performing vehicle detection and tracking, as well as vehicle speed measurement, in a signal acquired with a continuous wave Doppler sensor, is discussed. A method is also experimentally presented and studied for counting vehicles and for determining their movement direction by means of acoustic vector sensor application. The assumptions of the method employing spatial distribution of sound intensity determined with the help of an integrated three-dimensional (3D) sound intensity probe are discussed. The enhanced Doppler radar and the developed sound intensity probe were used for the experiments that are described and analyzed in the paper.

Keywords: Doppler sensor; acoustic vector sensor; road traffic monitoring

---

1. Introduction

We develop innovative road signs that can autonomously determine and communicate (visually and over V2X, vehicle-to-everything radio messaging) the speed limit in cases where the traffic situation requires it in connection with the project that was carried out in our department. The project entitled "Intelligent Road Signs with V2X Interface for Adaptive Traffic Controlling" is carried out in response to the demand for improving road safety and traffic efficiency. The developed system of autonomous road signs will enable the prevention of the most common collisions on highways, resulting from the rapid stacking of vehicles that results most often from accidental heavy braking [1]. Figure 1 shows an example of a dangerous road situation, together with a system of autonomous road signs that display and wirelessly transmit (in the V2X standard) decreasing permissible speed as drivers approach the place with traffic obstruction.

The new design demands for solving various research and construction problems, such as effective and independent of weather conditions traffic monitoring based on simultaneous analysis of several types of data representation [2]. The engineering part of the project, as well as previous research results on this topic, as described in our earlier papers [3,4], were preceded by a series of experimental studies. Their results showed that measuring speed and traffic density causes a number of problems in practical conditions. For example, the use of visual analysis for this purpose encounters limitations that are associated with restrictions on the visibility of vehicles in both RGB and thermal cameras. The currently popular lidars also have some limitations and they are also relatively expensive. In this case, an estimation of the traffic exploits optical opacity of cars and laser beam reflections as the physical principle of working is often accompanied by advanced data processing [5]. Setting the sensor perpendicular to the axis of the road makes it impossible to count vehicles in the case of occlusion (vehicles present on both lanes simultaneously), which causes many missed detections. Inductive loops or pneumatic cables for counting vehicles are used as a source of reliable data. However, their use requires installation in the road pavement, which is cumbersome and is only suitable for permanent installations, rather than for temporary installation of road signs in hazardous locations.

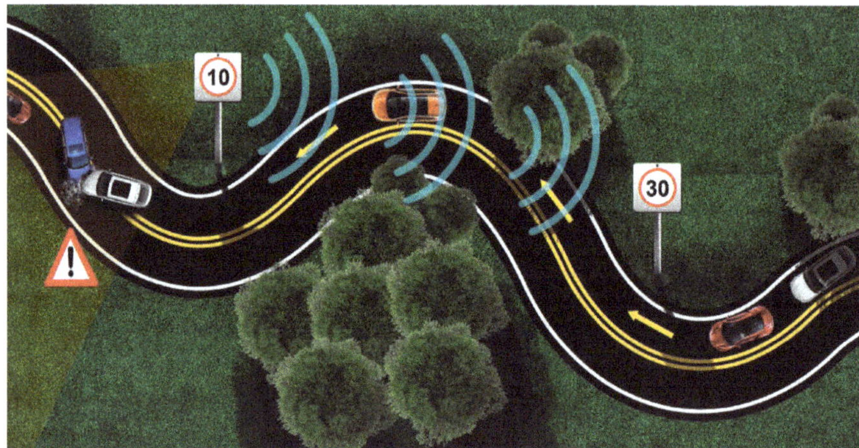

Figure 1. Illustration of the principle of limiting the speed when approaching a dangerous place on the road using autonomous road signs connected wirelessly to each other. Road signs communicate speed in a traditional way and also via radio channels using the V2X standard.

Another approach is based on recording anonymous Bluetooth MAC addresses of devices together with a timestamp as they pass by each detector, and then perform matching the addresses as vehicles pass through the next detector [6]. Specialized hardware was developed for related research (not covered in this paper), namely a radio module with software implementing the baseband controller and the firmware link manager layers. We constructed a practical vehicle counter in this way and then performed field tests showing that this technology makes a promising method of collecting real-time statistical traffic data and actual journey times from measurements on long distances, e.g., 1 km, which is not possible or difficult with other modalities [4]. However, counting vehicles that are based on the MAC addresses of Bluetooth devices can be unreliable, because these addresses are associated with both vehicles and mobile phones that are used by pedestrians. There can be several Bluetooth devices in one vehicle, e.g., audio and communication devices, as well as radio modules that are embedded in-vehicle diagnostic systems. Gupta et al. presented a different and interesting idea. They proposed the system for vehicles (i.e., bicycles, cars, trucks) counting by means of variations in the Wi-Fi signals strength [7]. Another approach that was related to the monitoring of traffic flow exploited the magnetic sensors to measure the vehicle's magnetic signature (VMS) evoked by moving vehicles [8].

Still, measurements of traffic intensity and vehicle speed while using Doppler microwave sensors find technical and economic justification. For example, a system that is based on high range resolution based on microwave radar sensor has been previously proposed by other authors to estimate the traffic flow rate and the flow rate of certain types of the vehicle [9,10]. However, microwave sensors are exposed to interference due to noise in the radio channels and reaching the receiver by parasitic reflections and microwave interferences. For this reason, we worked on this issue and presented an experimentally verified approach in this paper that allows for improving the results that were obtained while using a microwave radar.

Acoustic methods of road traffic estimation are, in practice, quite rarely used, although experimental research in this field is being conducted [11–13]. Our department has also been conducting such research for some time, which resulted in publications [14,15] and a recently defended doctoral dissertation [16].

A method is presented and experimentally studied for counting vehicles and determining their movement direction by means of the acoustic vector sensor application and the enhanced Doppler microwave sensor. The assumptions of the method employing spatial distribution of sound intensity determined with the help of an integrated three-dimensional (3D) intensity probe are discussed. The developed intensity probe was used for the experiments that brought the results discussed in the paper.

2. Materials and Methods

2.1. Vehicle Counting and Speed Measurement with Doppler Sensor

2.1.1. Doppler Sensor

The sensor that was used in the presented research emits a continuous wave with constant frequency within the K band (24.125 GHz) and it provides a dual-channel (I/Q) signal with frequencies below ca. 8 kHz, being proportional to the object's velocity, according to the Doppler effect. The sensor is characterized by a wide horizontal beam, which allows for capturing a vehicle's movement within a sufficiently long road segment (at least 50 m). This is different from the majority of radar sensors used in practical measurement systems, which measure the vehicle speed within a narrow zone.

The sensor transmits an electromagnetic wave with a constant frequency f_0. The frequency f_r of waves reflected from moving objects and received by the sensor differs from f_0, according to the Doppler effect [17,18]. An I/Q mixer produces a difference signal with frequency f_d, in two channels: in-phase (I) and quadrature (Q), which allows for the detection of the object's direction of movement (phase difference between I-Q channels is either 90° or −90°). The frequency f_d is related to the object's velocity v_r by an equation:

$$f_d = \left|f_r - f_0\right| = \frac{2}{\lambda}v_r = \frac{2f_0}{c}v_r = Sv_r, \tag{1}$$

where c is the speed of light. For a K-band sensor, f_0 = 24.125 GHz, and the scaling factor $S \approx 160.94$ (v_r in m/s). $f_d \leq 8.94$ kHz for road vehicles moving with speed up to 200 km/h (55.5 m/s). Therefore, the difference signal fits in the audio band (it is indeed audible), so standard audio signal processing algorithms may be applied for vehicle detection and speed measurement.

For practical reasons, it is not possible to directly mount the sensor on the vehicle's path of movement, therefore the sensor is usually placed alongside the road (Figure 2). As a result, the sensor only measures the radial component v_r of the velocity vector. As a vehicle moves through the detection zone of a sensor, v_r decreases when the vehicle approaches the sensor, and then increases as it moves away. This is called a 'cosine effect' [19], as the actual velocity is multiplied by a cosine of the angle to the object (Figure 2). Additionally, the angle difference between the front and the rear of the vehicle becomes larger as the vehicle moves closer to the sensor, which results in v_r values spanning a wider range. Figure 3 illustrates this, showing a spectrogram of a signal reflected from a road vehicle and recorded by a Doppler sensor. Signal frequency spans a range ($f_{\min}, f_{\max}$), given by:

$$f_{\min} = Sv\cos\alpha_{\max} = Sv\frac{r}{\sqrt{r^2+y^2}} \tag{2}$$

$$f_{\max} = Sv\cos\alpha_{\min} = Sv\frac{r+d}{\sqrt{(r+d)^2+y^2}} \tag{3}$$

where Figure 2 explains r, d, y, and α.

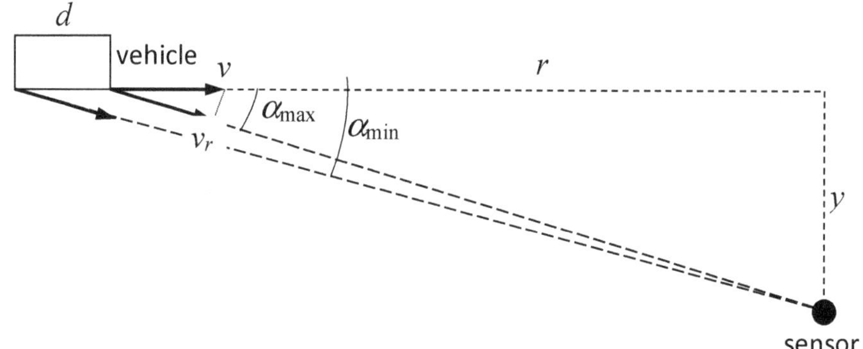

Figure 2. Measurement of the radial velocity of a vehicle with a Doppler sensor.

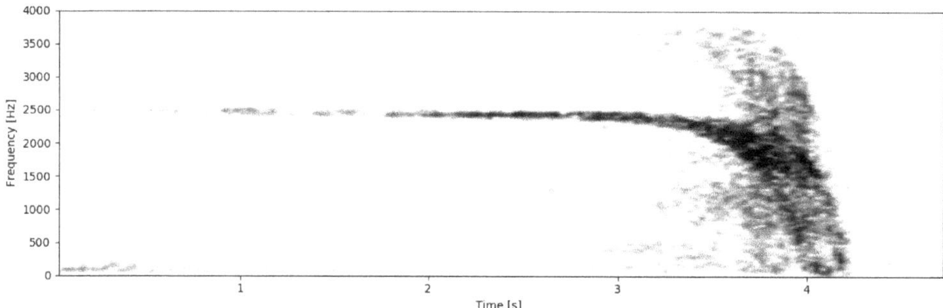

Figure 3. Spectrogram of a Doppler sensor signal recorded when a single vehicle was moving towards the sensor. The cosine effect is visible in the final phase (time 2.5–4.5 s).

Compensation of the cosine effect is problematic. Therefore, frequency is usually taken from the signal part captured at a large distance between the sensor and the object for speed measurement, so that the angle α is small (cos α close to 1), the difference between α_{min} and α_{max} is also small, and therefore the cosine effect is negligible. This approach was used in the proposed algorithm.

2.1.2. Algorithm for Processing of Doppler Sensor Signals

The task of the algorithm is to perform vehicle detection and tracking, as well as vehicle speed measurement, in a signal that was acquired with a continuous wave Doppler sensor. Figure 4 shows an overview of the processing algorithm. A dual-channel signal is received from an I/Q Doppler sensor. The first stage is the signal preprocessing, which suppresses noise and interference in the signal and then decomposes the signal into two components that represent opposite directions of movement. In the next stages, signal components that are reflected by moving vehicles are detected, and tracking of individual vehicles is performed. Finally, a velocity estimate is calculated from each identified vehicle track. The details of the algorithm are presented in the following Subsections.

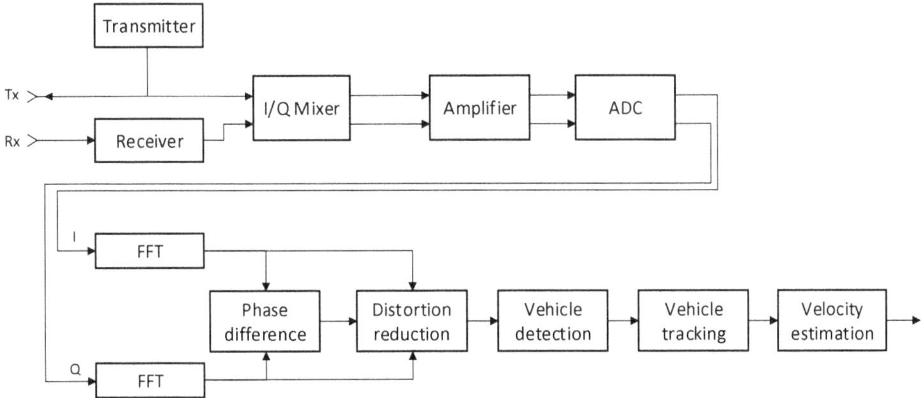

Figure 4. Block diagram of the Doppler sensor and the processing algorithm.

2.1.3. Suppression of Interference and Noise

The signals obtained from a Doppler sensor are often contaminated with distortions that make the detection and tracking processes more troublesome (Figure 5a). There are two main types of distortions that are observed in Doppler sensors. Noise is present in the signal as wide-band spectral components with random amplitude and phase, as a result of sensor imperfections, the nature of wave reflection, and environmental factors, such as wind. Electromagnetic interference (EMI) usually manifests as narrowband spectral components with a constant frequency. They may be induced by air (e.g., from nearby radio frequency transmitters, such as mobile network stations or airport radars) and by power lines. The amount of signal distortion depends on the sensor class, its positioning, and orientation, the quality of power supply, etc. As a result, a procedure for the suppression of distortions is necessary before the detection and tracking are performed.

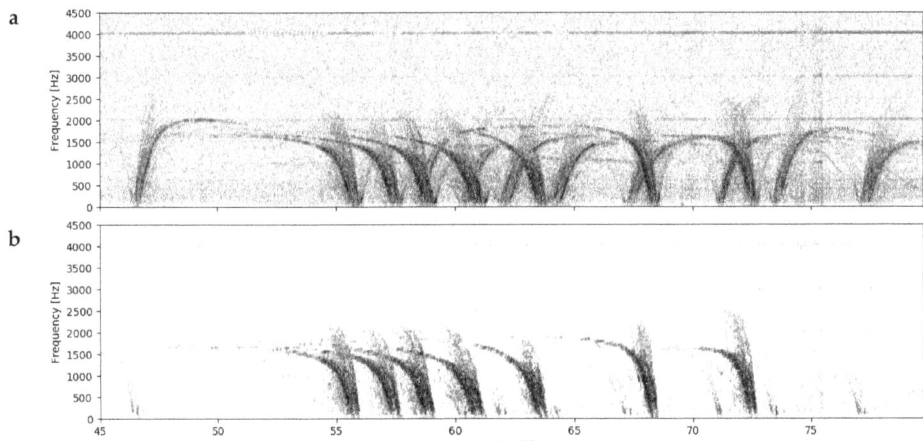

Figure 5. Spectrograms of road traffic signals: (**a**) signal recorded by the Doppler sensor, with electromagnetic interference (EMI) at multiplies of 1 kHz, and a wideband noise, (**b**) processed signal—suppression of noise, interference, and signals reflected from vehicles moving away from the sensor.

Noise reduction is usually performed by computing a noise profile and subtracting it from the signal. Such an approach requires the detection of signal parts containing only noise and constantly updating the profile. It is also not efficient for EMI removal. Therefore, a novel approach, which is based on the phase relationship between I/Q channels, is proposed. An additional benefit of this algorithm is the separation of the opposite directions of movement [20]. The algorithm is based on the phase difference $\Delta\phi$ between the I/Q channels of the sensor signal:

$$\Delta\phi = \arg Q(\omega) - \arg I(\omega) \tag{4}$$

where Q and I are the spectra of the respective signal channels.

In theory, $\Delta\phi$ should be equal to ±90° for signals that are reflected from moving objects. In practice, $\Delta\phi$ varies within a range, depending on the sensor class. The following observations were made from the analysis of phase in I/Q sensor signals [20]:

- signal components reflected from objects approaching the sensor or moving away from it have $\Delta\phi$ following a normal distribution with mean equal to 90° or −90°, respectively;
- for the noise, $\Delta\phi$ has a normal distribution with the mean value close to 0° and it might overlap the signal parts, depending on the sensor class; and,
- EMI is concentrated around $\Delta\phi = 0°$, as it influences both I/Q channels in an identical way.

Therefore, the algorithm for suppression of noise and EMI is based on the concept of a 'phase filter', as illustrated in Figure 6. The signal spectrum is multiplied by a weighting function w given by:

$$w(\omega) = \frac{1}{1 + e^{-\gamma(u(\omega)-0.5)}} \tag{5}$$

where u is given by:

$$u(\omega) = 1 - \left|\max\left(\frac{\Delta\phi(\omega)}{90}, 0\right) - 1\right| \tag{6}$$

for objects moving towards the sensor ('oncoming'), and

$$u(\omega) = 1 - \left|\max\left(\frac{-\Delta\phi(\omega)}{90}, 0\right) - 1\right| \tag{7}$$

for objects moving away from the sensor ('outgoing'), where $\Delta\phi$ is expressed in degrees [20]. The γ parameter controls the shape of the weighting function. As shown in Figure 6, part of the signal energy is lost if u is used as the weighting function (the 'no γ' case), and the overlap with the noise distribution is larger than in the case of additional shaping of the function. In the experiments, the authors found that $\gamma = 20$ is optimal, providing a proper balance between signal preservation and the suppression of distortions.

Figure 5b shows the example results of the preprocessing while using the proposed algorithm. As can be seen, EMI that occurred at multiplies of 1 kHz (most prominent at 4 kHz) is almost completely removed and wideband noise is significantly suppressed. The remaining speckle-noise results from the partial overlapping of signal and noise distributions, and from 'reflected' signals, captured when a vehicle has already passed the sensor. Such remaining noise components are discarded with amplitude thresholding in the later stage of processing. As can be seen, two opposite directions of the movement were separated, so that they may be individually analyzed. The detection and tracking phases are now significantly easier to perform due to the removal of occlusion by objects that are moving in opposite directions.

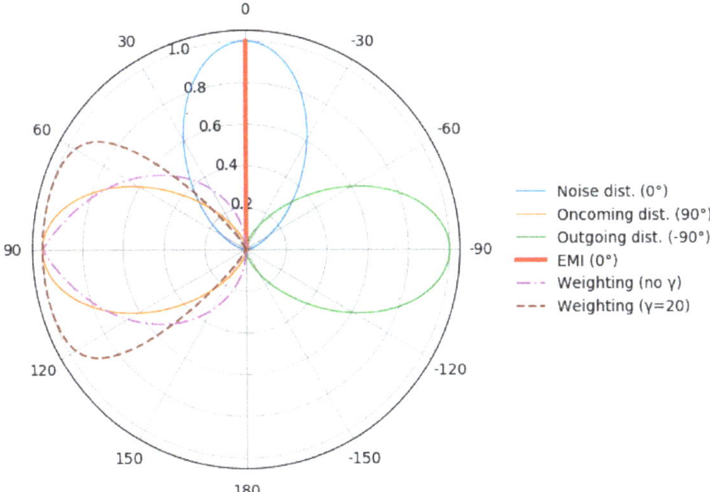

Figure 6. Polar plot illustrating the concept of phase filtering. Filled areas show distributions of signal and noise ($\sigma = 30°$), dashed lines show the shape of the weighting functions. The outer axis is scaled in the phase difference between I/Q channels (in degrees), the radial axis shows relative amplitude or gain.

2.1.4. Vehicle Detection, Tracking, and Velocity Estimation

The automatic measurement of the velocity of vehicles in road traffic, being performed by an unsupervised algorithm, requires performing three stages. First, the detection of spectral components that represent moving vehicles is performed in the preprocessed signal, in blocks of signal samples (short term analysis). In the second stage, the detection results have to be assigned to vehicles, so that changes in the signal frequency (caused by the object movement and the cosine effect) are tracked. The preprocessing algorithm that is presented in the previous Subsection, by decreasing the level of noise and interference, allows for easier vehicle detection, and eliminating occlusion from vehicles moving in the opposite direction allows for easier tracking. Occlusion from vehicles moving in the same direction still occurs and it is the main problem in the tracking. In the final stage of the analysis, the estimated velocity is extracted from each vehicle track. It should be noted that the velocity measurement with an automatic algorithm is much more problematic than where a human operator is able to relate the measurements to the observed vehicles.

The detection algorithm works on signal spectra that were computed in short windows (e.g., 2048 samples, 42.6 ms for 48 kHz sampling), after multiplication by the weighting function, as described earlier. The detection works by finding sequences of spectral bins with an amplitude above a threshold (that should be set according to the signal level and the remaining noise level). The threshold should be sufficiently low in order to detect weak signals when a vehicle is far from the sensor. Groups of spectral bins containing the reflected signal become larger when a vehicle approaches the sensor (Figure 3). In practice, gaps occur in such groups (due to weaker signal components at some frequencies), and such gaps result in the segmentation of signal parts. These fragments have to be merged in the detection phase. It is also inevitable that some strong noise components are incorrectly detected as signals.

The tracking algorithm merges the detection results from consecutive time windows. Each stored track is extrapolated to find the expected frequency in the current window. Linear interpolation is used in the initial phase and cubic interpolation in the later stage for objects approaching the sensor. Next, the detected groups of spectral bins are searched for the one closest to the expected value. If such a group is found, this group is appended to the track, its centroid frequency is computed, and the lowest and the highest frequency is stored in the track. The track is not updated when no matching group is found. Additionally, dead tracks are removed from the analysis, and finished tracks (i.e., with a sufficient length, and with a minimum frequency below a threshold of ca. 500 Hz, below which the vehicle is very close to the sensor) are passed to the velocity estimation phase.

The velocity of a vehicle should be measured within an initial ('thin' and flat) section of a track. Therefore, a finished track is analyzed in short sections. Frequency computed in each window is converted to velocity, and then the mean and the standard deviation are computed for each section. The algorithm selects a section with the highest mean, provided that the standard deviation is below the threshold. The latter condition allows for the elimination of tracking errors (track following a group of noise components) from velocity estimation. The computed standard deviation is used as a representation of the 'quality' of the result. If the value is obtained from an initial, flat section of a track, and the standard deviation is low (<0.1 km/h). The value is obtained from the later track part (with decreasing frequency), which will be reflected by the higher standard deviation, if there is no flat segment, e.g., due to occlusion by another vehicle ahead. The results with a high standard deviation (>1.0 km/h) should be discarded.

Figure 7 shows an example of a successful analysis in a signal recorded during the experiments. Only vehicles moving on the lane closer to the sensor (the oncoming traffic) were analyzed. The results of the detection stage are shown as gray points. Some of the detected components are assigned to the tracks of individual vehicles, color dots mark the computed centroids of the detected spectral bin groups, thus forming continuous tracks. The obtained velocity estimates are also shown in the plot. All of the estimates had a standard deviation not exceeding 0.25 km/h.

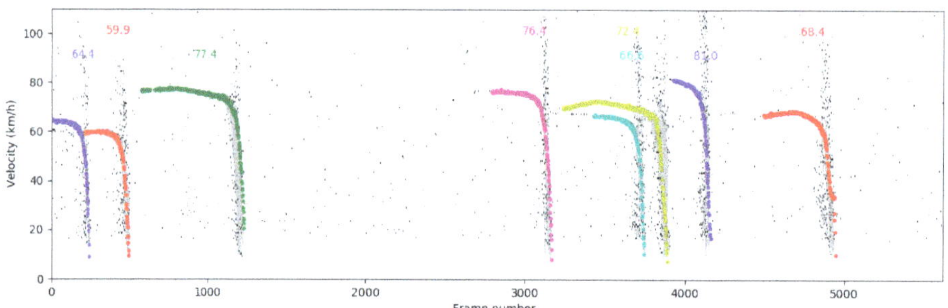

Figure 7. Example of vehicle detection and velocity measurement with the proposed algorithm. Gray points indicate the detection results, color dots mark the detection results assigned to the tracks of individual vehicles (distinguished by a color), values show the obtained velocity estimates (km/h).

2.2. Vehicle Counting and Speed Measurement with Acoustic Vector Sensor

2.2.1. Acoustic Vector Sensor

An intensity-based AVS (Acoustic Vector Sensor) described in this paper consists of three pairs of pressure sensors (microphones) positioned on three orthogonal axes, in equal distances from the center point. Each pair of microphones forms a *p-p* sensor, which is used for measuring the particle velocity in a given direction. The averaged pressure at the center point is computed as a mean of all six pressure signals. The intensity on each axis is proportional to the product of the particle velocity and the averaged pressure. Figure 8 shows the AVS used during experiments. Computation of intensity and angles are performed by the algorithm, as outlined in Figure 9. The developed algorithm for source localization and tracking has three main sections. The first part is related to the correction section. Correction of the pressure signals, as obtained by the microphones, should be applied for the proper determination of the sound intensity. This correction is realized in two steps. First, the frequency responses of all microphones are equalized, and the phase response in the microphone pair is also equalized (block denoted as *A&P.Corr* in Figure 9). Next, the particle velocity on each axis (ux, uy, uz) and the average acoustic pressure p are calculated. Then the second step of the correction is performed in which phase differences between the particle velocity and the average acoustic pressure signals are equalized (*P.Corr*). The authors' previous publication described the detailed description of the calibration and correction process [21].

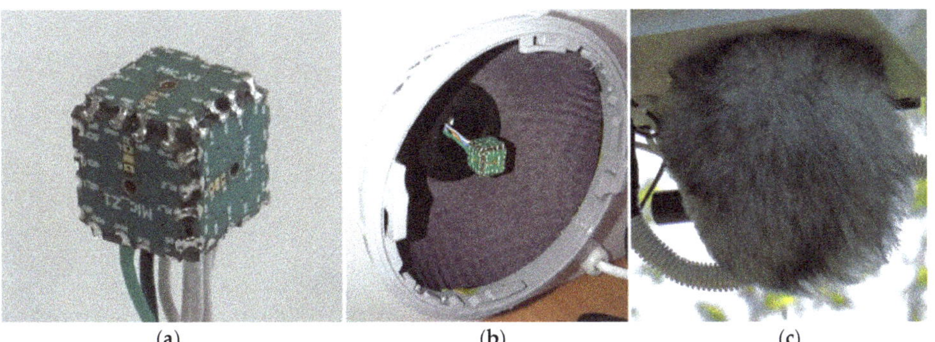

Figure 8. The acoustical vector sensor (AVS) designed by the authors (**a**), the AVS inside the windscreen (**b**), and the AVS during the measurements (**c**).

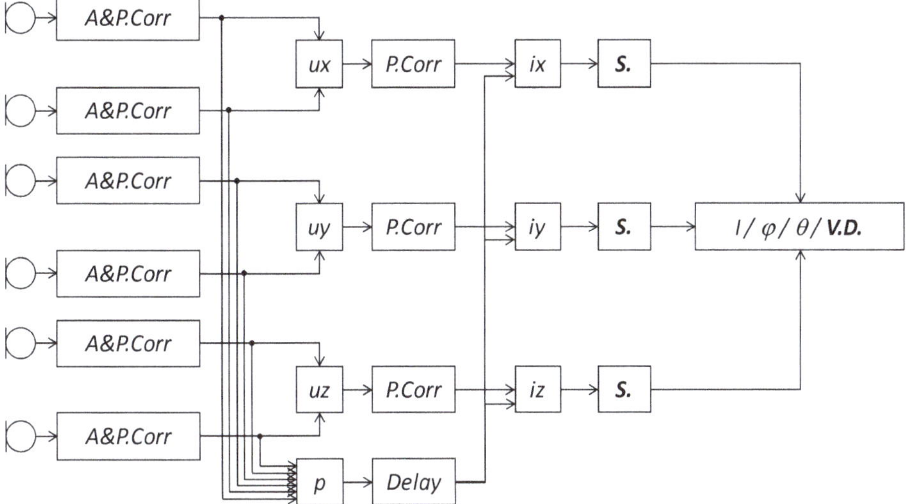

Figure 9. Block diagram of the developed algorithm for detection, localization, and tracking of moving sound sources.

After these steps, the sound intensity components (*ix*, *iy*, *iz*) can be determined by computing a product of each intensity component with the averaged pressure signal and by integrating the result [22, 23]. The second part of the algorithm contains smoothing blocks, labeled as *S*. Noise suppression procedure has to be applied to each intensity signal in order to make the vehicle detection possible. In the experiments described in this paper, a Savitzky–Golay filter was used in order to suppress noise and to obtain smoothed intensity functions [24]. The optimal values of the filter length (51) and the polynomial order (3) were experimentally found, as a good balance between the details and noise present in the processed signals. At the end of the second part, the three smoothed components I_X, I_Y, and I_Z are known.

The last section, labeled as *V.D.*—vehicle detection block—uses the intensity components and values of azimuth and elevation angle for making a decision regarding the presence of sound source (a vehicle in the considered scenario).

2.2.2. Intensity Computation

A three-dimensional (3D) AVS applied during our research is able to measure the acoustic particle velocity (referred to as "velocity" further in the text) in three orthogonal directions, as well as pressure in the central point of the sensor. Sound intensity vectors in three orthogonal directions may be obtained based on the velocity and pressure measurement results.

The acoustic pressure has to be measured at six points located on three orthogonal axes, at identical distances d from the origin. These points are denoted as $x1$, $x2$, $y1$, $y2$, $z1$, and $z2$, describing their location in the coordinate system, e.g., point $y2$ is located at $(0, d, 0)$ and $y1$ at $(0, -d, 0)$. Omnidirectional microphones of the same type are used to measure pressure $p_l(t)$ at six locations l. According to the Euler's formula [22], velocity vectors $\mathbf{u}_i(t)$ alongside axes X, Y, Z may be computed as:

$$\begin{bmatrix} \mathbf{u}_x(t) \\ \mathbf{u}_y(t) \\ \mathbf{u}_z(t) \end{bmatrix} = \begin{bmatrix} a_x & 0 & 0 \\ 0 & a_y & 0 \\ 0 & 0 & a_z \end{bmatrix} \cdot \begin{bmatrix} p_{x2}(t) - p_{x1}(t) \\ p_{y2}(t) - p_{y1}(t) \\ p_{z2}(t) - p_{z1}(t) \end{bmatrix} \tag{8}$$

where a_i are the scaling factors (determined during calibration procedure). The magnitude of the $\mathbf{u}_i(t)$ vector will be denoted as $u_i(t)$. Pressure $p(t)$ measured at the origin is averaged from two points at the given axis and it has to be equal on all three axes. In practice, the pressure is averaged from all six microphones (as in Equation (9)):

$$p(t) = \frac{p_{x1}(t) + p_{x2}(t) + p_{y1}(t) + p_{y2}(t) + p_{z1}(t) + p_{z2}(t)}{6} \tag{9}$$

The sound intensity I at a given axis can be then computed, as [22]:

$$I = \frac{1}{T} \int_{T} p(t) u(t) \, \mathrm{d}t \tag{10}$$

where T is the integration period.

If a single, omnidirectional sound source is put into the system at polar coordinates (r, ϕ, θ), the angles of the sound received by the AVS may then be computed as:

$$\phi = \arctan\left(\frac{I_y}{I_x}\right) \tag{11}$$

$$\theta = \arctan\left(\frac{I_z}{\sqrt{I_x^2 + I_y^2}}\right) \tag{12}$$

where I_x, I_y, I_z are the intensity signals measured along the axes of the coordinate system, being oriented as shown in Figure 10.

In general, the sound intensity is determined by means of the algorithm described above in the time domain while using broadband signals of acoustic pressure and particle velocity. For purposes that are considered in this article, it is important to perform sound intensity calculation in the frequency range related to the acoustic events produced by vehicles moving near the sensor. The frequency analysis of the background noise and pass-by vehicle sounds were performed to avoid the unwanted and disturbing sounds emitted by other sound sources during vehicle movement. Figure 11 shows the results of this analysis. The dotted line indicates the background noise. The solid line depicts the spectrum of pass-by vehicle event. It can be noticed that for frequency greater than 6 kHz the background noise is close to the noise that is emitted by the vehicle. The grasshoppers generated this high background noise. Low-frequency noise can be produced by the wind and other sound sources placed far away from the measurement point. For this purpose, the sound intensity analysis was limited to the frequency range: 400 Hz–4 kHz. It was shown in Figure 11 while using two vertical dotted lines. In this way, the essential part of the acoustic energy emitted by the moving vehicle was taken into consideration during the calculation of sound intensity and direction of arrival.

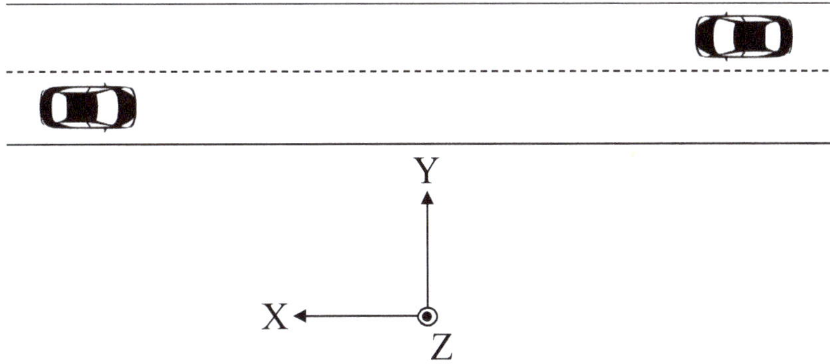

Figure 10. Orientation of the acoustic vector sensor relative to the road.

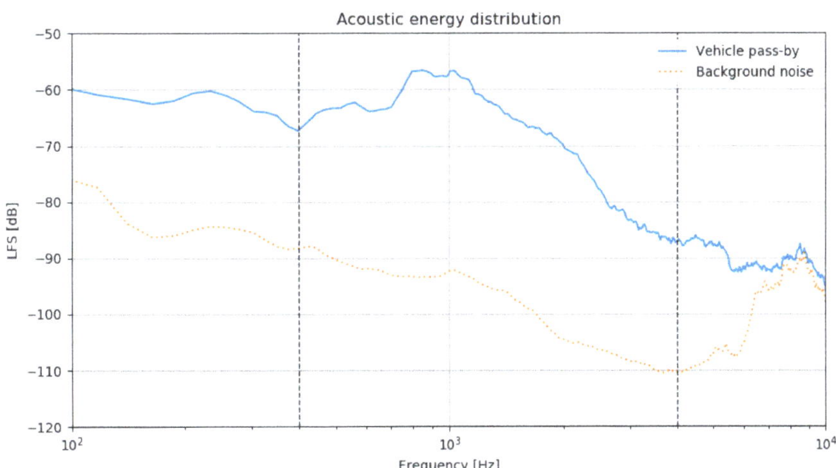

Figure 11. Acoustic energy distribution in the frequency domain for background noise and for the pass-by vehicle.

Figure 12 shows an example of the output of the algorithm described above for an acoustic event evoked by a single pass-by vehicle. The left chart depicted the sound intensity components. We can indicate the highest values that were obtained for the direction perpendicular to the road (Iy). Other components have relatively lover values. This observation will be an essential fact during the development of the vehicle detection module. The right chart includes the direction of arrival components, being expressed by azimuth and elevation angles. In Figure 13, an example of 120 s of sound intensity continuous analysis was shown. The acoustic events that are evoked by the vehicles are clearly visible. An event typical for a group of vehicles occurred around 80 s. The rapid changes of the azimuth angle can be noticed for this event. It is important to emphasize that no other acoustic events than passing vehicles can be observed. It confirms that the frequency range of sound intensity calculation was correctly selected.

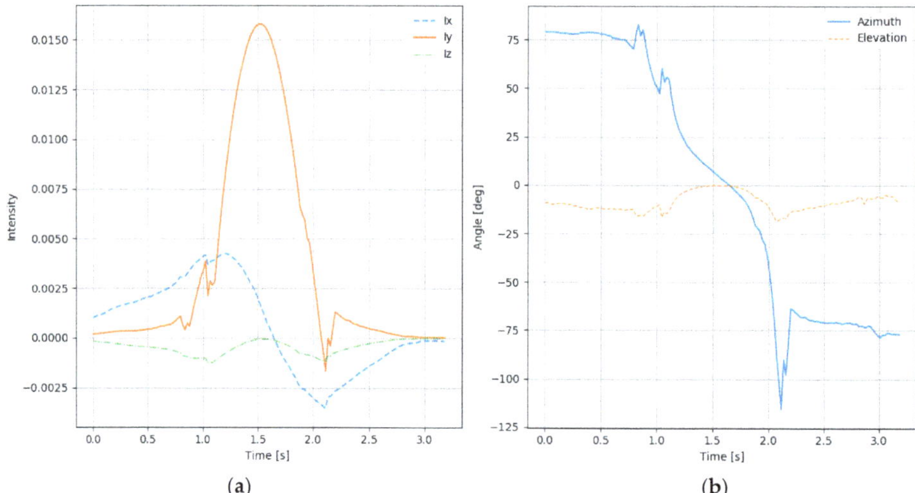

Figure 12. The results obtained from the AVS signal recorded for a single-vehicle: (**a**) intensity components, (**b**) azimuth and elevation.2.2.3. Vehicle Detection.

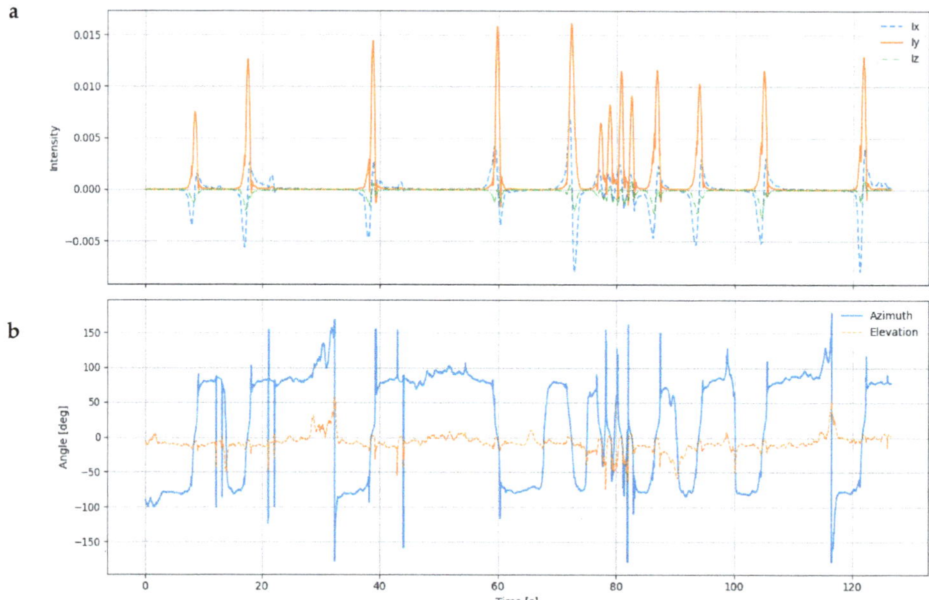

Figure 13. The results obtained from the AVS signal recorded for several vehicles: (**a**) intensity components, (**b**) azimuth and elevation.

The algorithm works in two stages. The first stage is based on the analysis of sound intensity signals and it detects acoustic events. The second stage analyses a detection function, based on the normalized source position, its task is to determine whether the acoustic event represents a vehicle passing the sensor and detecting its direction of movement. The detected and verified acoustic events may be analyzed further, e.g., for the velocity estimation.

In the first stage, the intensity signal is analyzed with a sliding window **w** of length N, moving with a step equal to one sample. In the experiments, the window span was about 640 ms. It was found that using the intensity signal that was measured on the axis perpendicular to the road (I_Y) provides better results than the total intensity. An acoustic event is detected if:

$$\mathbf{w}_{N/2} = \max_i \mathbf{w}_i \text{ and } \frac{1}{N}\sum_{i=1}^{N} I_{Y(i)} \geq T_{\text{int}} \tag{13}$$

where T_{int} is the minimum average intensity within the window and I is the sample index within the window. The value of T_{int} must be set according to the amplitude of the analyzed signal, so that both low-energy and short-term (impulse) acoustic events are discarded.

The second stage of the algorithm analyzes the changes in the normalized position of the sound source. Position x of the source moving along the trajectory parallel to the X-axis of the system, at a normalized distance of $y = 1$ m, is equal to:

$$x = \tan(\phi) = \frac{I_Y}{I_X} \tag{14}$$

where ϕ is the azimuth of the source.

If a vehicle moves along this trajectory with an approximately constant velocity, then smooth changes in x will be observed, and the direction of these changes will indicate the direction of the object's movement. For acoustic events that are not related to moving vehicles, much larger changes in the source position can be expected. Therefore, the detection metric d, as computed within the same window as in the first stage, is:

$$d = \frac{2}{N}\sum_{i=1}^{N/2}\left(\frac{I_{Y(i+1)}}{I_{X(i+1)}} - \frac{I_{Y(i)}}{I_{X(i)}}\right) \tag{15}$$

Only the first half of the window is used. In the case of isolated vehicles, the second half is redundant and, if several vehicles move close to each other, their measured intensities overlap, which usually causes more distortion in the second part of the window. The sign of d indicates the direction of movement: vehicles moving towards positive x values have $d < 0$, vehicles moving in the opposite direction have $d > 0$. Additionally, standard deviation within the window might be computed, similarly to d. It is expected that the standard deviation will be small for vehicles moving past the sensor, and high for unrelated acoustic events, which might be discarded with a maximum standard deviation threshold.

Figure 14 presents an example of detection. The maxima of the intensity function is detected as acoustic events (Figure 14a). The detection function (normalized x position) smoothly changes within the acoustic events and oscillates randomly when no events are present (Figure 14b). The value of d computed for each event indicates the direction of a vehicle's movement; this is marked with bars pointing upwards or downwards for $d < 0$ and $d > 0$, respectively.

Detections from the reference data are marked with dots, with the direction being indicated in the same way. It can be observed that most of the vehicles were correctly detected and identified. For isolated vehicles (frame 28429), the detection function changes smoothly within the whole detection window, and the analysis is straightforward. When multiple vehicles are moving close to each other, their detection functions overlap. In some events (e.g., frames 31804 & 31898), the results are correct. In the case of a heavy occlusion (multiple vehicles on both lanes), the probability of errors increases. In the presented example, the vehicle in frame 29835 is detected, but its direction is incorrect, due to the overlap of detection functions from many vehicles, and one vehicle (frame 29742) was missed, as two intensity maxima from two vehicles merged into a single one.

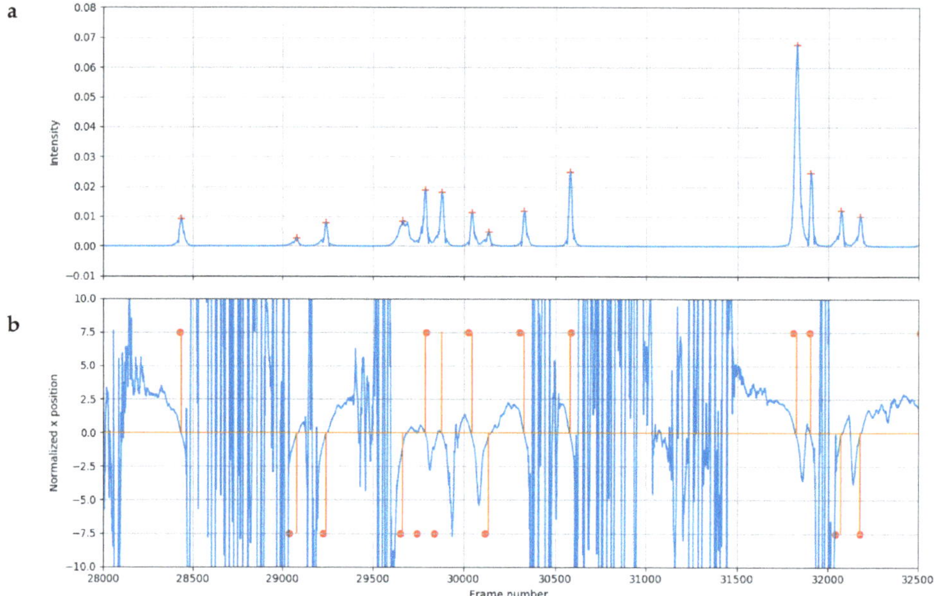

Figure 14. Example of the detection results: (**a**) sound intensity (line) and detected acoustic events (+); (**b**) detection function (line), detected vehicles (vertical bars, the direction of bars indicates the direction of movement), reference data (dots, vertical position indicates the direction of movement).

3. Results

3.1. Test Setup

A test setup was constructed and experiments were performed in a real-world scenario in order to verify the proposed algorithms. A low-cost RSM2650 Doppler sensor by B+B Sensors [25] was used. The sensor was connected through a custom-built amplifier and an analog-to-digital converter (48 kHz sampling rate) to a Raspberry Pi 3B microcomputer. All of the elements, together with an LTE router and a power supply, were placed in an enclosure (Figure 15). In the first stage of the research, signals from the sensor were recorded on the microcomputer and then downloaded for offline analysis. The analysis of the Doppler sensor signal was performed online on the microcomputer in the experiments described here, and the results (timestamp, velocity, and standard deviation) were available via the MQTT network protocol. The signal was analyzed in windows of 2048 samples (42.67 ms) with 75% overlap while using the Blackman window before FFT was computed. The processing algorithm was implemented in the Python programming language.

The AVS was constructed from six omnidirectional digital MEMS microphones, IvenSense INMP441 [26], operating at 48 kHz sampling rate with 24-bit resolution. Each microphone was mounted on a board of ca. 10×10 mm size that was connected through an I2S USB digital interface to a USB port on a computer (Figure 8). The sensor was mounted in a windshield at the bottom side of the enclosure. The six-channel signal was recorded on the microcomputer and then stored on a hard drive. The recordings were analyzed offline. Additionally, environmental sensors (temperature, pressure, precipitation, air quality), as well as a LiDAR sensor and a video camera, were mounted on the enclosure; these were not used in the experiments described here.

The test system was mounted on the outskirts of Gdańsk, Poland (near Leźno village), geographic coordinates: 54.344555, 18.443811. The monitored road section had one lane in each direction and the speed limit was 90 km/h. The measurements were performed on a straight and flat section of the road (Figure 16), where the typical speed of vehicles is 60 to 80 km/h. The enclosure was mounted 4 m away from the road edge and the bottom side of the box was 2.9 m above the ground. The Doppler sensor was positioned at 3.2 m above the ground, oriented at 45° azimuth, and −18° elevation relative to the road axis. The algorithm only analyzed the closer lane (eastbound traffic).

Figure 15. The test system mounted on the site. The Doppler sensor is located inside the box marked with the rectangle, the AVS in a windshield is visible at the bottom right, below the enclosure.

Figure 16. The test road section—a view from the camera mounted in the test system. The first measurement tubes are mounted at the trees visible in the back, the second pair is positioned near the bottom right corner of the photo.

A system based on pneumatic tubes (Metrocount MC5600 Vehicle Counter System) was mounted on the road in order to obtain reference data for the experiments. Two pairs of tubes were used. One pair was positioned near the test system, for comparison with the AVS results. The other pair of tubes were mounted ca. 100 m away, within the zone of Doppler sensor detection. Recordings that spanned a continuous period of 24 h (July 1st, 14:00 to 2nd July 2019) were obtained, and timestamps and velocity measured for each vehicle (from both lanes) were used for comparison with the Doppler and AVS sensors results. The temperature during the recording was 15 °C to 26 °C, average pressure was 997 hPa, the wind was up to 7 m/s from the West, and there were occasional periods of rainfall (about 15% of the total time).

3.2. Analysis of Vehicle Counting

Aggregating the detection results in 30-min. periods was undertaken to analyze the results of vehicle counting on the lane closer to the test setup. Data from the tubes were used as the reference and it is assumed that there are no detection errors in the recorded data (this was partially confirmed by reviewing selected sections of the recorded video). Table 1 presents the results that were obtained for both sensors within the measurement period. For the AVS, a total of 30 min. was missing from the recorded material due to technical difficulties. Figure 17 shows the vehicle count aggregated in 30 min. intervals, for the data from the Doppler sensor, analyzed by the proposed algorithm, and data from the reference device. Figure 18 presents a similar plot calculated for the AVS. The Pearson correlation coefficient between the sets of the calculated and the reference vehicle counts is 0.994 for the radar and 0.995 for the AVS.

Table 1. Summary of the vehicle detection results.

Sensor	Doppler	AVS
Analyzed time	24 h	23 h 30 min
Total number of vehicles	2998	2953
True detections	2742	2583
False negatives	256	370
False positives	44	189
Recall	91.46%	87.47%
Precision	98.42%	93.18%
Accuracy	90.14%	82.21%

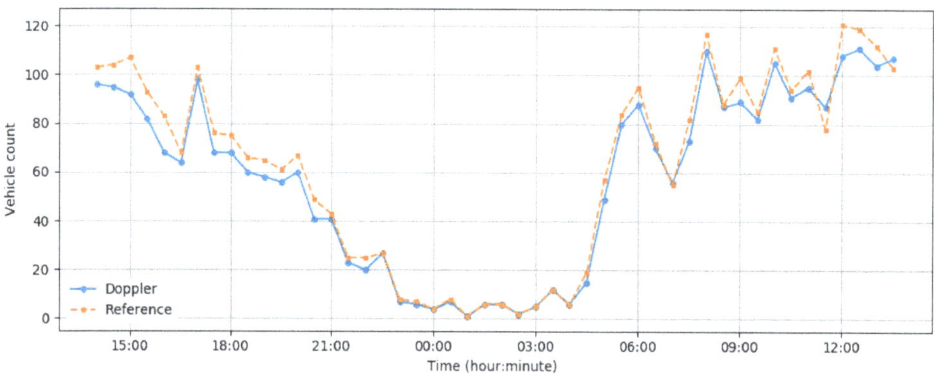

Figure 17. Vehicle count in 30-min. intervals—measured with the Doppler sensor and the proposed algorithm (solid line), and the reference data from the tube detector (dashed line).

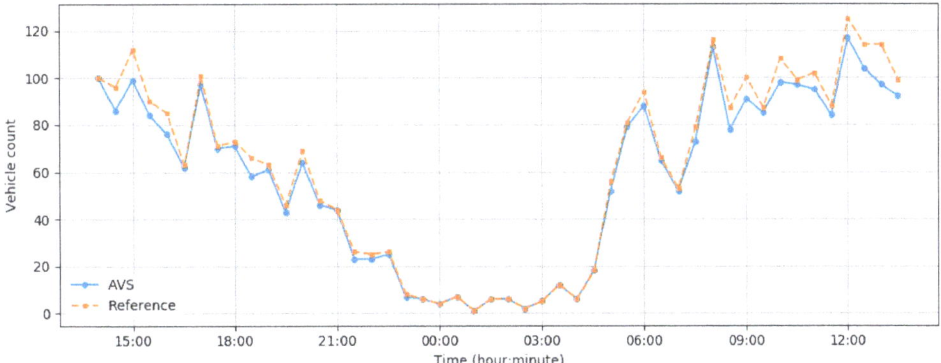

Figure 18. Vehicle count in 30-min. intervals—measured with the AVS and the proposed algorithm (solid line), and the reference data from the tube detector (dashed line).

'Ghost tracks' when the algorithm followed noisy components of the signal and produced duplicated detections mainly caused the observed false-positive results. To conclude, improving the algorithm accuracy requires its modifications that will make it more robust to the observed errors, and also tuning the algorithm parameters and, if possible, repositioning the sensor.

The results of vehicle counting that were obtained with the AVS and the proposed algorithm (Figure 18) were consistent with the results from the Doppler sensor. The number of false-positive and false-negative results is slightly larger for the AVS, the overall accuracy is lower (82% vs. 90%). The main source of incorrect detection results is the problem with determining the direction of movement in the case of occlusion (vehicles on both lanes, moving in opposite directions) and when several vehicles move close to each other. In such cases, the detection function does not allow for accurate direction analysis, because the signals from multiple vehicles overlap. As a result, some vehicles were detected, but their direction was incorrect, as shown in Table 2. However, this problem occurs on both lanes. In the presented experiment, the number of vehicles on each lane was similar, so that false positive and false negative results on each lane were mostly balanced. Some vehicles could not be detected at all, which was mostly due to high occlusion. As shown in Table 2, the number of vehicles in the closer lane that was not detected is almost equal to the number of vehicles detected with incorrect direction. The number of vehicles that were not detected is similar on both lanes, while the incorrect detection of the direction happens more often on the closer lane. The number of false detections is higher in the further lane, which results from the occlusion. In total, statistics that were obtained in 30-min. intervals were very similar to those from the Doppler sensor, they also slightly underestimate the real vehicle count in the case of high traffic. The trend also highly correlates with the reference data.

Table 2. Detailed analysis of vehicle detection in the AVS signal, on both lanes.

Lane	Closer Lane	Further Lane	Both Lanes
Number of vehicles	2953	2940	5893
Detected, correct lane	2583	2691	5274
Detected, wrong lane	191	80	271
Not detected	179	169	348
False detections	109	190	299

3.3. Analysis of Velocity Measurement Using Doppler Sensor

Comparison of velocity measured by the proposed algorithm analyzing signals from the Doppler sensor and by the reference device cannot be accurately performed due to the fact that the tube-based system measured the velocity at one point, about 100 m from the sensor, while the Doppler sensor measured velocity at different points within the zone approximately 50 to 100 m from the sensor. Therefore, any observed differences may be caused both by measurement errors and by vehicles that accelerate or brake within the detection zone. The mean squared difference (MSD) between the measurements of individual vehicles by both sensors is 19.45, with a standard deviation that is equal to 176.47, and the root of the MSD (RSMD) is 4.41 ± 13.28 km/h. The accuracy of measuring the velocity of individual vehicles with the proposed algorithm is satisfactory while taking the condition mentioned before into account.

Figure 19 shows the results of averaging the velocity in 30-min. intervals for both data sources. It can be observed that the results that were obtained from the evaluated algorithm are slightly lower than for the reference device (MSD 1.93 ± 1.68, RMSD 1.39 km/h). Both datasets follow a similar trend and the Pearson correlation coefficient is 0.92. This confirms that the evaluated algorithm provides velocity measurements with accuracy that is sufficient for collecting traffic statistics.

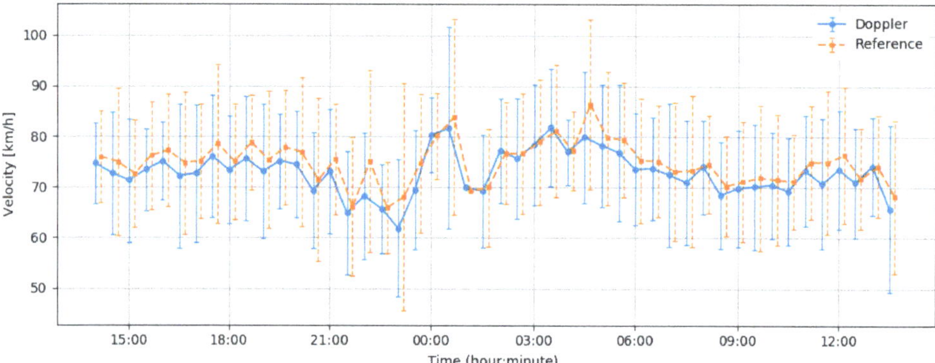

Figure 19. Vehicle velocity in 30-min. intervals—measured with the Doppler sensor and the proposed algorithm (solid line) and reference data from the tube detector (dashed line, shifted on the time axis for improved readability). Points represent mean velocity, error bars—standard deviation.

3.4. Analysis of Velocity Measurement Using Acoustic Vector Sensor

The vehicle velocity measurements using the Acoustic Vector Sensor are not so obvious as for the Doppler sensor. In the beginning, we need to mention that the AVS gives as the possibility to localize and tracking of moving sound sources. Let us consider a single point sound source moving through a linear trajectory, parallel to the X-axis of the sensor, with a constant velocity. The position of the source is $P(t) = (x(t), y)$, where y is constant. The velocity of the source might be expressed as:

$$v = \frac{\Delta x}{\Delta t} = \frac{\Delta(y \cdot \tan\phi)}{\Delta t} = \frac{y}{\Delta t}\Delta\left(\frac{I_X}{I_Y}\right) \tag{16}$$

where ϕ is the azimuth measured by the sensor.

Real vehicles are not single sound sources, but rather a setup of several sources (tires, engine, exhaust, etc.). Distances between sources, differences in the source power, and in the directivity of each source all contribute to the obtained results because the vehicles move close to the sensor. It was confirmed by a computer simulation, in which the azimuth that was computed for a single source and a setup of four sources, moving with the same velocity, was calculated. The obtained results (Figure 20), which are consistent with the measurements from the real AVS, indicate that the azimuth that was obtained for multiple sources moving together (solid line) deviates from the tangential line of the single-source case (dotted line). This is caused by an additional velocity component, representing the movement of a virtual sound source within a vehicle, which is nonlinear, with the opposite direction to the velocity vector of the moving vehicle. The AVS measures the position of a virtual sound source, so both velocity components are included. As a consequence, the velocity estimate computed by the algorithm (which assumes a single source) is lower than the real velocity. The difference depends on a number of factors, such as the width and the length of a vehicle, distance from the sensor, relative power of each source, and directivity of the sources. These variables are unknown, so it is not possible to correct the obtained velocity estimates.

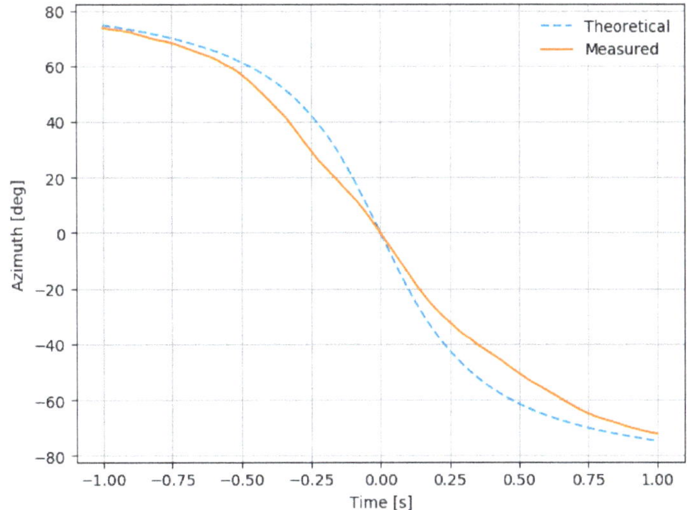

Figure 20. Computer simulation of the azimuth for a single source (dotted line) and an azimuth measured by the sensor (solid line).

4. Discussion

The accuracy of the proposed algorithm for vehicle counting in the Doppler sensor signals is approximately 90%, which is low for systems that are intended to detect each vehicle (e.g., traffic law enforcement systems), but it is sufficient for the intended application of collecting traffic statistics. Two main reasons for false-negative results were identified during the analysis of the obtained results. The first problem is common for all radar-based detection systems: it is difficult to distinguish each vehicle driving in a 'train', where several vehicles move close to each other. The most common error in the obtained result was skipping one vehicle in a sequence of four or more vehicles moving almost 'bumper to bumper'. The problem is that since all of the vehicles move with similar velocity, tracks of these vehicles overlap, and, in some cases, the tracking algorithm merges one track with another. It can be observed in Figure 17 that false-negative results mainly occur during the rush hours, which confirms these conclusions. One possible solution to this problem, left for future research, is implementing an additional algorithm that analyzes low-frequency components (final sections of the tracks) and detects vehicles moving past the sensor. Such an algorithm is not trivial, as noise and slow objects distort low frequencies (bicycles, pedestrians). The second problem was related to cars moving with high velocity, above 100 km/h. In the test system, the sensor was positioned too far from the road (due to practical constraints), which resulted in a low signal-to-noise ratio, causing a loss of the initial section of the tracks. For vehicles moving with high velocity, some tracks were too short to provide valid measurements. Moving the sensor closer to the road, and by decreasing the minimal track length, which in turn may increase the number of false-positive results, can mitigate this problem.

As discussed earlier, it is not possible to measure the exact velocity of individual cars with the AVS, as the obtained values underestimate the real velocity. Nevertheless, an attempt was made to calculate the averaged velocity within the time slots. For this purpose, the velocity of each vehicle was estimated within sections of the AVS signal, as determined by the vehicle detector. The distance between the sensor and the vehicles was chosen as 5.5 m for the closer lane and 8.5 m for the further one—a midpoint of each lane was selected as an average distance to the sound source. In some cases, due to occlusion by multiple vehicles present in the detection zone at the same time, the velocity estimation was not possible in some cases. In fact, as much as 11% of the detected vehicles were discarded from the analysis. The remaining speed estimates were averaged within 30-min. periods. The results are shown in Figure 21. Just like the results obtained for the individual vehicles, the averaged values are significantly lower than the real velocity from the reference device. However, the obtained pseudo-velocity function correlates with the reference data and Pearson's coefficient is 0.78. Therefore, the obtained values may be used as an indicator of relative velocity changes, allowing e.g., for detection of periods in which the averaged velocity falls below a selected limit, even if the actual velocity cannot be measured with the AVS. Moreover, the ratio of the obtained estimates to the reference data, as computed for each time slot, is 0.78 ± 0.03 (mean ± standard deviation), so it is stable within the whole observation period. A more accurate estimate of the real velocity is obtained if the obtained estimates are multiplied by the scaling factor equal to the mean ratio. The results of the experiments indicate that this scaling factor mainly depends on the distance between the sensor and the road, as other factors, related to individual vehicles (width, height, and velocity of a vehicle, position on the lane) are averaged and they do not influence the result. Therefore, the scaling factor might be potentially determined by measuring the distance between the sensor and the road. This aspect requires further research.

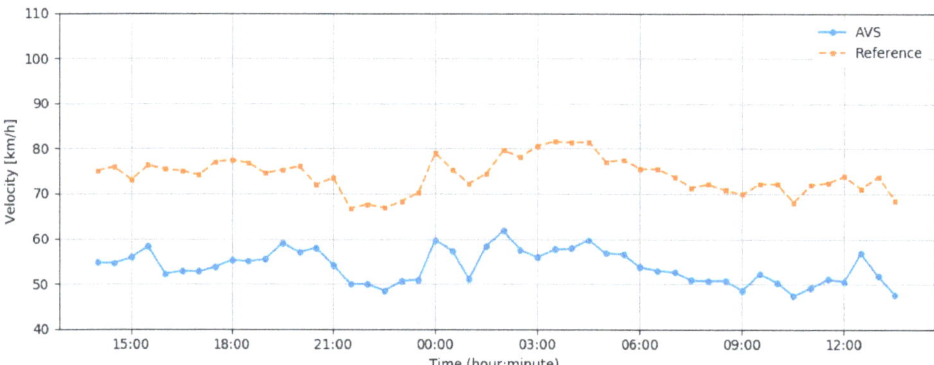

Figure 21. Velocity estimation from the AVS signal, averaged in 30-min. periods, as compared with the reference data.

The main advantage of the Doppler sensor is that it is capable of accurate measurement of the velocity of both individual vehicles and time-averaged statistics. The AVS is not able to achieve that, because the position of an apparent acoustic source within a vehicle is not constant. The correction of this effect is a separate problem that is outside of the scope of this paper and it is left for future research. Nevertheless, after applying an empirically found scaling coefficient, the AVS is able to provide traffic statistics with sufficient accuracy for the intended purpose, i.e., determining the trends in traffic speed variations and detection of periods in which the averaged velocity falls below a selected limit.

The vehicle counting accuracy relative to the traffic intensity is similar for both methods; they underestimate the vehicle count in high traffic scenarios. Both methods achieve accuracy close to 100% if the traffic intensity is below 100 vehicles per hour.

Although the acoustical vector sensor (AVS) has a lower accuracy than Doppler in vehicle counting and it is not able to measure the vehicle speed accurately, it has some advantages over the Doppler sensor. Namely, it does not emit any signals, it is not susceptible to electromagnetic interferences, and it allows for further analysis of audio signals, such as the assessment of the road surface state (e.g., wet/dry).

5. Conclusions

The authors developed and examined two methods for estimating traffic intensity in real traffic conditions. The first method is based on a microwave Doppler sensor. This method requires that the sensor emits a microwave signal, so it is an active method. This sensor is susceptible to electromagnetic interference from nearby sources, such as power lines, radar systems, cellular network base stations, etc. Additionally, waves that are emitted by the sensor may interfere with other devices operating in the same frequency range. The second method is based on an analysis of sound intensity vectors by means of the acoustic vector sensor. This method passively analyses acoustic signals that are emitted by moving vehicles. The sensor is not affected by electromagnetic interference, but acoustic noise might influence its operation. Both of the methods were evaluated in a real-world scenario while using a reference system based on pneumatic tubes. The conclusions are, as follows.

The vehicle counting accuracy of the proposed algorithm using the Doppler sensor signals is approximately 90%, which is sufficient for the intended application of collecting traffic statistics. The observed errors result from factors that are common for all radar-based measurement systems, namely the occlusion of multiple objects, inability to distinguish vehicles moving close to each other with similar speed, and duplicated detections for some vehicles. Modifications of the proposed algorithm and tuning its parameters to improve the detection accuracy in difficult situations, such as those observed in the experiments, is a topic of future research.

For the acoustic vector sensor, the proposed algorithm for vehicle counting provided results that are consistent with these from the Doppler sensor, with a slightly larger number of false-positive and false-negative results, and lower overall accuracy (82% vs. 90%). The main problem is related to determining the direction of movement in the case of occlusion (vehicles on both lanes, moving in opposite directions) and when several vehicles move close to each other. In such cases, an accurate direction analysis is problematic because the signals from multiple vehicles overlap. As a result, vehicles are detected, but their direction of movement is not correctly recognized. The algorithm also slightly underestimates the real vehicle count in case of high traffic. These problems will be addressed in future research.

Our observations that were collected in the course of experimental studies show that microwave sensors and acoustic sensors have application prospects for measuring traffic in order to discover traffic congestions by autonomous road signs or in other traffic measuring systems. The Doppler radar that we have improved and the constructed and the calibrated acoustic probe are applicable to perform vehicle detection and tracking, as well as vehicle speed measurement.

Both methods may be used for statistical analysis of traffic intensity and speed, providing valuable data for automated traffic management systems. The main advantage of the acoustic sensor over the microwave sensor is that it does not require sending any signals that could interfere with nearby devices, and it is not affected by any sources of electromagnetic signals in the vicinity.

The future work will focus on the optimizations of the proposed algorithms, which will lead to the increased accuracy of vehicle counting and velocity measurement. In the case of the AVS, the main topic of the research will be related to improving the algorithm for the detection of vehicle direction, and on developing a method for correction of the moving apparent sound source, which will allow for velocity measurement with the AVS.

Author Contributions: Conceptualization, A.C.; methodology, G.S., and J.K.; software, G.S. and J.K.; validation, G.S. and J.K.; formal analysis, A.C.; J.K.; G.S.; investigation, G.S. and J.K.; writing—original draft preparation, G.S. and J.K.; writing—review and editing A.C.; visualization, G.S.; supervision, A.C.; project administration, A.C.; funding acquisition, A.C. All authors have read and agreed to the published version of the manuscript.

Funding: Project financed by the Polish National Centre for Research and Development from under the EU Operational Programme Innovative Economy No. POIR.04.01.04-0089/16 entitled: INZNAK—"Intelligent road signs with V2X interface for adaptive traffic controlling".

Conflicts of Interest: The authors declare no conflict of interest.

References

1. Farmer, C.M.; Lund, A.K.; Trempel, R.E.; Braver, E.R. Fatal crashes of passenger vehicles before and after adding antilock braking systems. *Accid. Anal. Prev.* **1997**, *29*, 745–757. [CrossRef]
2. Ershadi, N.Y.; Menéndez, J.M.; Jiménez, D. Robust vehicle detection in different weather conditions: Using MIPM. *PLoS ONE* **2018**, *13*, e0191355. [CrossRef]
3. Czyżewski, A.; Sroczyński, A.; Śmiałkowski, T.; Hoffmann, P. Development of Intelligent Road Signs with V2X Interface for Adaptive Traffic Controlling. In Proceedings of the 6th International Conference on Models and Technologies for Intelligent Transportation Systems (MT-ITS), Kraków, Poland, 5–7 June 2019. [CrossRef]
4. Czyżewski, A.; Cygert, S.; Szwoch, G.; Kotus, J.; Weber, D.; Szczodrak, M.; Koszewski, D.; Jamroz, K.; Kustra, W.; Sroczyński, A.; et al. Comparative study on the effectiveness of various types of road traffic intensity detectors. In Proceedings of the 6th International Conference on Models and Technologies for Intelligent Transportation Systems (MT-ITS), Kraków, Poland, 5–7 June 2019. [CrossRef]
5. Sun, Y.; Xu, H.; Wu, J.; Zheng, J.; Dietrich, K.M. 3-D data processing to extract vehicle trajectories from roadside LiDAR data. *Transp. Res. Rec. J. Transp. Res. Board* **2018**, *2672*, 14–22. [CrossRef]
6. Tsubota, T.; Yoshii, T. An analysis of the detection probability of MAC address from a moving Bluetooth device. *Transp. Res. Procedia* **2017**, *21*, 251–256. [CrossRef]
7. Gupta, S.; Hamzin, A.; Degbelo, A. A Low-cost open hardware system for collecting traffic data using Wi-Fi signal strength. *Sensors* **2018**, *18*, 3623. [CrossRef] [PubMed]
8. Balid, W.; Tafish, H.; Refai, H. Intelligent vehicle counting and classification sensor for real-time traffic surveillance. *IEEE Trans. Intell. Transp. Syst.* **2017**, *19*, 1–11. [CrossRef]
9. Xuan, Y.; Meng, H.; Wang, X.; Zhang, H. A high-range-resolution microwave radar system for traffic flow rate measurement. In Proceedings of the 2005 IEEE Intelligent Transportation Systems, Vienna, Austria, 16 September 2005. [CrossRef]
10. Fang, J.; Meng, H.; Zhang, H.; Wang, X. A low-cost vehicle detection and classification system based on unmodulated continuous-wave radar. In Proceedings of the IEEE Intelligent Transportation Systems Conference, Seattle, WA, USA, 30 September–3 October 2007; pp. 715–720. [CrossRef]
11. Zhang, X.; Huang, J.C.; Song, E.L.; Liu, H.W.; Li, B.Q.; Yuan, X.B. Design of small MEMS microphone array systems for direction finding of outdoors moving vehicles. *Sensors* **2014**, *14*, 4384–4398. [CrossRef] [PubMed]
12. Na, Y.; Guo, Y.; Fu, Q.; Yan, Y. An acoustic traffic monitoring system: Design and implementation. In Proceedings of the 2015 IEEE 12th Intl Conf on Ubiquitous Intelligence and Computing and 2015 IEEE 12th Intl Conf on Autonomic and Trusted Computing and 2015 IEEE 15th Intl Conf on Scalable Computing and Communications and Its Associated Workshops (UIC-ATC-ScalCom), Beijing, China, 10–14 August 2015. [CrossRef]
13. Zu, X.; Zhang, S.; Guo, F.; Zhao, Q.; Zhang, X.; You, X.; Liu, H.; Li, B.; Yuan, X. Vehicle counting and moving direction identification based on small-aperture microphone array. *Sensors* **2017**, *17*, 1089. [CrossRef] [PubMed]
14. Marciniuk, K.; Kostek, B.; Czyżewski, A. Traffic noise analysis applied to automatic vehicle counting and classification. Multimedia Communications. *Serv. Secur. MCSS* **2017**, 110–123. [CrossRef]
15. Marciniuk, K.; Kostek, B.; Czyżewski, A. Classifying type of vehicles on the basis of data extracted from audio signal characteristics. *J. Acoust. Soc. Am.* **2017**, *141*, 3883. [CrossRef]
16. Marciniuk, K. Acoustic Road Traffic Monitoring Using Noise Information and Machine Learning. Ph.D. Thesis, Gdańsk University of Technology, Gdansk, Poland, 2019. (In Polish).
17. Richards, M.A. *Fundamentals of Radar Signal Processing*, 2nd ed.; McGraw-Hill Education: New York, NY, USA, 2014.
18. Brooker, G. Doppler measurement. In *Sensors and Signals*; Australian Centre for Field Robotics, University of Sydney: Sydney, Australia, 2006.
19. Sawicki, D.S. The cosine effect. In *Police Radar Handbook: A Comprehensive Guide to Speed Measuring Systems*; CreateSpace Independent Publishing Platform: Scotts Valley, CA, USA, 2013.
20. Szwoch, G. Suppression of distortions in signals received from Doppler sensor for vehicle speed measurement. In Proceedings of the IEEE Signal Processing: Algorithms, Architectures, Arrangements, and Applications (SPA), Poznań, Poland, 19–21 September 2018. [CrossRef]

21. Kotus, J.; Szwoch, G. Calibration of acoustic vector sensor based on MEMS microphones for DOA estimation. *Appl. Acoust.* **2018**, *141*, 307–321. [CrossRef]
22. Jacobsen, F. Sound intensity and its measurements and applications. *Curr. Top. Acoust. Res.* **2003**, *3*, 87–91.
23. Fahy, F. *Sound Intensity*, 2nd ed.; E & FN Spon: London, UK, 1995.
24. Savitzky, A.; Golay, M.J.E. Smoothing and differentiation of data by simplified least squares procedures. *Anal. Chem.* **1964**, *36*, 1627–1639. [CrossRef]
25. B+B Sensors. RSM2650 Radar Movement Alarm Unit—Data Sheet. Available online: http://www.produktinfo.conrad.com/datenblaetter/500000-524999/506343-da-01-en-RADARBEWEGUNGSM__MOD__STEREO_4_75__5_25V.pdf (accessed on 15 September 2019).
26. IvenSense. INMP441. Available online: https://www.invensense.com/products/digital/inmp441/ (accessed on 15 September 2019).

Article

A Large-Scene Deceptive Jamming Method for Space-Borne SAR Based on Time-Delay and Frequency-Shift with Template Segmentation

Kaizhi Yang [1,2], Wei Ye [1], Fangfang Ma [3,*], Guojing Li [1] and Qian Tong [2]

[1] Space Engineering University, Beijing 101416, China; yangkaizhi@aliyun.com (K.Y.); yeyuhan@sina.com (W.Y.); leeguojing1014@mail.dlut.edu.cn (G.L.)
[2] The Unit 94657 of PLA, Jiujiang 332104, China; qianx_tong@sina.cn
[3] Logistics Science and Technology Institute, Institute of Systems Engineering, Academy of Military Science, Beijing 100071, China
* Correspondence: fangf_ma@aliyun.com; Tel.: +86-181-9191-8806

Received: 2 December 2019; Accepted: 20 December 2019; Published: 21 December 2019

Abstract: Due to advantages such as its low power consumption and higher concealment, deceptive jamming against synthetic aperture radar (SAR) has received extensive attention during the past decades. However, large-scene deception jamming is still a challenge because of the huge computing burden. In this paper, we propose a new large-scene deceptive jamming algorithm. First, the time-delay and frequency-shift (TDFS) algorithm is introduced to improve the jamming processing speed. The system function of jammer (JSF) for a fake scatter is simplified to the multiplication of the scattering coefficient, a time-delay term in the range dimension and a frequency-shift term in the azimuth dimension. Then, in order to solve the problem that the effective region of the TDFS algorithm is limited, the scene deceptive jamming template is divided into several blocks according to the SAR parameters and imaging quality control factor. The JSF of each block is calculated by the TDFS algorithm and added together to achieve the large-scene jamming. Finally, the correction algorithm in squint mode is derived. The simplification and parallel-block processing could improve the calculation efficiency significantly. The simulation results verified the validity of the algorithm.

Keywords: synthetic aperture radar (SAR); space-borne SAR; deceptive jamming

1. Introduction

Synthetic aperture radar (SAR) is an effective system that uses electromagnetic waves for high-resolution imaging. Due to the unique advantages of its all-day, all-weather operation, and ability to penetrate camouflage compared with traditional optical remote sensing methods, SAR has become a major means of remote sensing and has been widely used, especially in the military field. At the same time, for the purpose of protecting sensitive targets and regions, electronic countermeasures against SAR have received intensive attention [1–5].

In general, active electronic interference against SAR is divided into two types: barrage jamming and deception jamming. The former uses high-power noise to cover the echo signal from the region of interest (ROI) and makes it impossible to form a clear and distinguishable image [6,7]. The latter emits an echo signal of a false target by the direct generation or modulation–retransmission method, which is mixed with the echo of the real target, affecting the image interpretation process and achieving the purpose of "hidden truth in falsehood" [8–19]. Compared with barrage jamming, deception jamming belongs to a type of smart jamming method which has lower power consumption, higher concealment, and more flexible application scenarios; thus, it is more attractive and does not arouse the awareness of the enemy.

At present, almost all SAR deceptive jamming methods are based on the modulation and retransmission mechanism. In each pulse repetition interval (PRI), according to a series of parameters of the SAR which should be jammed, including kinematic parameters, antenna parameters, and signal parameters, and combining the jamming scene template, the jammer modulates and retransmits the intercepted radar pulse to generate a jamming signal, which will form a false image after range and azimuth compression by the receiver. The deceptive jammer can be regarded as a linear time-invariant (LTI) system in a single PRI. The problem of obtaining the system function of jammer (JSF) is a focus in the field of SAR deceptive jamming. A straightforward method is to calculate the signal propagation delay difference between each scatter in the jamming scene template and the jammer during each PRI [8]. However, this method is computationally intensive and can hardly guarantee real-time processing. Subsequent research has mainly focused on reducing the computational complexity and increasing the processing speed. Usually, parts of the processing are performed in advance to reduce the computational burden during the implementation of jamming. In the specific implementation, this is divided into two categories: azimuth time-domain processing and azimuth frequency-domain processing. The former reduces the computational complexity by approximating the distance equation and is suitable for the broadside or low squint angle mode, including the inverse range-Doppler algorithm [9], phase pre-modulation [10], segmented modulation [11,12], and approach of multiple receivers [13,14]. The latter, including frequency-domain pre-modulation [15], the frequency-domain three-stage algorithm [16], the inverse Omega-K algorithm [17], etc., needs to perform 2-D Fourier transform and Stolt interpolation on the jamming scene template [20,21], which can work under a large squint angle but requires additional information such as the azimuth bandwidth.

Although the methods above improve the computational efficiency of the jamming process to varying degrees, the large computational burden is still the bottleneck of large-scene deceptive jamming for SAR. Zhou etc. proposed a large-scene deceptive jamming method by dividing the jamming scene template into sub-templates according to the depth of focus in the range dimension to simplify the JSF and decomposing the JSF into the slow-time independent terms generated off-line and slow-time dependent terms calculated in real-time [11]. However, this algorithm only works for space-borne SAR operating at the broadside mode, and the computational efficiency is still insufficient. Inspired by this, we propose a new large-scene deceptive jamming algorithm called time-delay and frequency-shift with template segmentation (TDFS-TS). First, the complex modulation process is simplified into the time-delay and frequency-shift operation to increase the computational efficiency. Second, the jamming scene template is divided both in the range dimension and azimuth dimension according to the imaging quality control factor. The correction algorithm in the squint situation is derived as well. Compared with other available deceptive jamming techniques, the proposed method can produce well-focused large deceptive scenes more efficiently.

This paper is organized as follows. Section 2 provides a detailed description of the TDFS-TS algorithm. We begin with the analysis of the basic principles of deceptive jamming against SAR; based on these, we propose the time-delay and frequency-shift (TDFS) jamming algorithm to simplify the process. Then the template segmentation (TS) method is used to achieve large-scene jamming, and the correction algorithm in squint mode is described. In Section 3, the TDFS-TS algorithm is verified by simulation and the computation complexity is analyzed. Section 4 discusses the results and Section 5 concludes this paper.

2. Large-Scene Deceptive Jamming Method Based on TDFS-TS

This section will derive the TDFS-TS deceptive jamming algorithm step by step. First, the principles of deceptive jamming against SAR are introduced. Then the TDFS algorithm is proposed which can significantly improve computational efficiency. The analysis of the jamming signal generated by TDFS shows that the effective region is limited. To solve this problem, the TS method is introduced, and the squint correction algorithm is derived to extend the application scope of the jamming algorithm. Finally, the TDFS-TS algorithm procedure is clarified.

2.1. Principles of Deceptive Jamming against SAR

The principle of SAR deceptive jamming based on modulation–retransmission is presented in Figure 1 [17]. The jammer performs a serious of operations including the amplification, down-conversion, analog-to-digital conversion (A/D), and fast Fourier transform (FFT) on the intercepted radar radio frequency (RF) signal to obtain the frequency domain representation of the baseband, while the JSF is calculated based on the template and the SAR parameters including kinematic parameters (platform position, velocity, etc.), antenna parameters (antenna direction, beam pattern, etc.), and signal parameters (carrier frequency, PRI, etc.). Then, we multiply the two and perform an inverse fast Fourier transform (IFFT) to obtain the baseband of the jamming signal and finally perform the digital-to-analog conversion (D/A), up-conversion, gain control, and retransmission. By repeating the above steps for each pulse, a false image can be generated by the receiver. The template is an array of false scatters that depicts the electromagnetic characteristic of the fake scene artificially fabricated by the jammer.

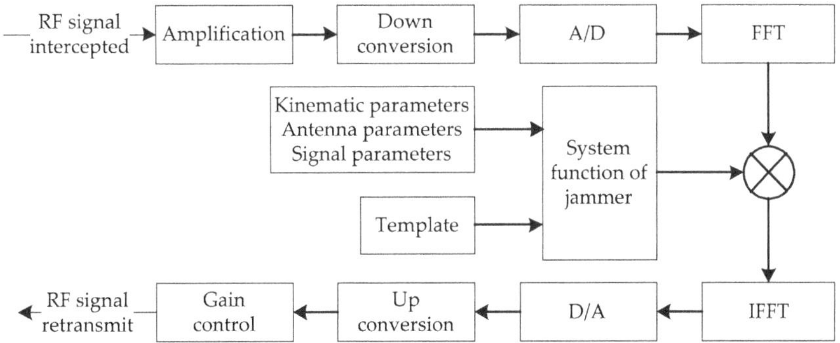

Figure 1. Principle of synthetic aperture radar (SAR) deceptive jamming.

The JSF is the key to the generation of jamming signals. The following illustrates the basic idea for JSF in combination with the geometric model of SAR jamming. As shown in Figure 2, we assume that the SAR platform moves at a constant velocity v and the azimuth time $t_a = 0$ when the plane of zero Doppler passes through the jammer. A Cartesian coordinate system with the location of the jammer as the origin is established in a two-dimensional slant range plane. The x-axis points to the range direction and the y-axis points to the azimuth direction. The shortest slant distance between the jammer and the SAR is R_{J0}, and the instantaneous slant distance is $R_J(t_a)$ at azimuth time t_a. An arbitrary false point scatter P is generated by the jammer, the scattering coefficient of P is σ_P and the location is $(x,\ y)$ in the coordinate above. $R_P(t_a)$ denotes the instantaneous slant distance between P and the jammer.

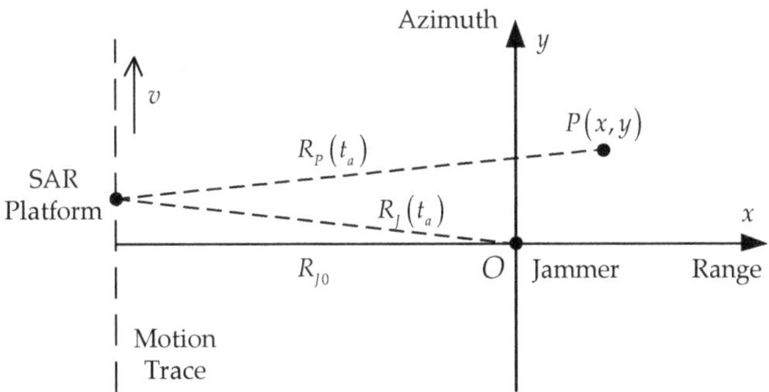

Figure 2. The geometric model of SAR deceptive jamming.

In order to generate the fake point target P in the SAR image, the JSF at azimuth time t_a is [17]

$$H_P(f_r,\ t_a) = \sigma_P \exp\left[-j2\pi(f_r+f_0)\frac{2\Delta R_P(t_a)}{c}\right], \tag{1}$$

where f_r is the range frequency, c is the velocity of light, $\Delta R_P(t_a)$ is the difference between $R_P(t_a)$ and $R_J(t_a)$:

$$\Delta R_P(t_a) = R_P(t_a) - R_J(t_a) = \sqrt{\left(x+R_{J0}\right)^2 + (y-vt_a)^2} - \sqrt{R_{J0}^2 + (vt_a)^2}. \tag{2}$$

By calculating the JSF for each scatter in the template T, the JSF for the deception scene can be derived as follows:

$$H(f_r,\ t_a) = \sum_{P\in T} H_P(f_r, t_a) = \sum_{P\in T} \sigma_P \exp\left[-j2\pi(f_r+f_0)\frac{2\Delta R_P(t_a)}{c}\right]. \tag{3}$$

In the implementation of jamming, the jammer must calculate JSF $H(f_r,\ t_a)$ and modulate the intercepted signal real-time in each PRI. The calculation of Equation (2) is time-consuming, therefore the method represented by Equation (3) cannot be used directly for large-scene deceptive jamming unless some improvements are made.

2.2. TDFS-TS Algorithm

2.2.1. Deceptive Jamming Based on TDFS

For space-borne SAR, we can assert that $R_{J0} \gg x$, $R_{J0} \gg y$, and $R_{J0} \gg vt_a$ throughout the synthetic aperture time. After the Taylor series expansion, Equation (2) can be approximated as follows [11]:

$$\begin{aligned} \Delta R_P(t_a) &\approx \left[x + R_{J0} + \tfrac{y^2}{2(x+R_{J0})} - \tfrac{yvt_a}{x+R_{J0}} + \tfrac{(vt_a)^2}{2(x+R_{J0})}\right] - \left[R_{J0} + \tfrac{(vt_a)^2}{2R_{J0}}\right] \\ &\approx x + \tfrac{y^2}{2R_{J0}} - \tfrac{yvt_a}{R_{J0}}. \end{aligned} \tag{4}$$

With the approximation above, the JSF—i.e., Equation (1)—can be rewritten as follows:

$$\begin{aligned} H_P(f_r,\ t_a) &= \sigma_P \ \exp\left[-j2\pi(f_r+f_0)\left(\tfrac{2x}{c} + \tfrac{y^2}{cR_{J0}} - \tfrac{2yvt_a}{cR_{J0}}\right)\right] \\ &= \sigma_P \ \exp\left[-j2\pi(f_r+f_0)\tfrac{2x}{c}\right]\exp\left(j2\pi f_0 \tfrac{2yvt_a}{cR_{J0}}\right) \\ &\quad \exp\left(-j2\pi f_0 \tfrac{y^2}{cR_{J0}}\right)\exp\left(j2\pi f_r \tfrac{2yvt_a - y^2}{cR_{J0}}\right). \end{aligned} \tag{5}$$

where the third exponential term is independent of f_r and t_a and is equivalent to introducing a fixed phase that has no effect on the imaging; thus, it can be ignored. The fourth exponential term can also be omitted when y is small enough (details will be analyzed in the next subsection). Thus, Equation (5) can be simplified as follows:

$$H_P(f_r,\ t_a) = \sigma_P \exp\left[-j2\pi(f_r+f_0)\frac{2x}{c}\right]\exp\left(j2\pi f_0 \frac{2yvt_a}{cR_{J0}}\right). \tag{6}$$

Actually, at the broadside mode, the azimuth frequency modulation rate of the SAR echo signal from the location of the jammer is [22]

$$K_a = -\frac{2f_0 v^2}{cR_{J0}}. \tag{7}$$

Combined with Equations (6) and (7), we can derive

$$H_P(f_r,t_a) = \sigma_P \exp\left[-j2\pi(f_r+f_0)\frac{2x}{c}\right]\exp\left(-j2\pi K_a\frac{y}{v}t_a\right). \tag{8}$$

This is equivalent to delaying the original echo signal $2x/c$ in fast-time to achieve position deception in the range dimension and a shifting frequency $-K_a y/v$ in slow-time. According to the frequency shifting property of the linear frequency modulation (LFM) signal, the frequency shifting is equivalent to time-delaying y/v in slow-time, which can implement position deception in the azimuth dimension as well.

If the scattering coefficient of the false point target which locates (x,y) in the template is $\sigma(x,y)$, the JSF for the deception scene can be derived as follows:

$$\begin{aligned} H(f_r,t_a) &= \sum_x\sum_y \sigma(x,y)\exp\left[-j2\pi(f_r+f_0)\tfrac{2x}{c}\right]\exp\left(j2\pi f_0\tfrac{2yvt_a}{cR_{J0}}\right) \\ &= \sum_y \exp\left(j4\pi f_0\tfrac{yvt_a}{cR_{J0}}\right)\sum_x \sigma(x,y)\exp\left[-j4\pi(f_r+f_0)\tfrac{x}{c}\right]. \end{aligned} \tag{9}$$

The second summation term in Equation (9) is independent of azimuth time t_a and only related to the relative position of each point in the false scene, which can be calculated off-line to reduce the real-time computational burden. This is the TDFS jamming algorithm, which has the advantages of simplicity and high computational efficiency.

2.2.2. Jamming Signal Analysis

Due to the approximation and simplification, the TDFS algorithm has high computational efficiency. However, the approximation and simplification will cause a decline in the image quality of the deceptive target at the same time. In this subsection, we will analyze the impact of the simplified operations above on the imaging results by comparing the difference between the jamming signal and real point target echo in the range-Doppler domain [23].

The echo signal of a real scatter point $P(x,y)$ is represented as follows [22]:

$$s_P(t_r,t_a) = \sigma_P w_a(t_a) w_r\left[t_r - \frac{2R_P(t_a)}{c}\right]\exp\left[-j4\pi f_0\frac{R_P(t_a)}{c}\right]\exp\left\{j\pi K_r\left[t_r-\frac{2R_P(t_a)}{c}\right]^2\right\}, \tag{10}$$

where $w_a(t_a)$ represents the azimuth amplitude, $w_r(t_r)$ is the SAR pulse complex envelope, and K_r is the frequency modulation rate of the SAR pulse.

The parabolic approximation of the instantaneous slant range $R_P(t_a)$ by Taylor series is [22]

$$R_P(t_a) = \sqrt{\left(x+R_{J0}\right)^2 + (y-vt_a)^2} = x + R_{J0} + \frac{(y-vt_a)^2}{2\left(x+R_{J0}\right)}. \tag{11}$$

We can derive the echo signal expression (12) of the real target point $P(x,y)$ in the range-Doppler domain by bringing Equation (11) into Equation (10) and using the principle of stationary phase (POSP) [22]:

$$\begin{aligned} S_P(t_r,f_a) = \sigma_P W_a(f_a) w_r\left[t_r - \tfrac{2R_P(f_a)}{c}\right] \\ \exp\left\{j\pi K_r\left[t_r-\tfrac{2R_P(f_a)}{c}\right]^2 - j\pi\tfrac{f_a^2}{K_a} - j2\pi f_a\tfrac{y}{v} - j4\pi f_0\tfrac{x+R_{J0}}{c}\right\}, \end{aligned} \tag{12}$$

where $R_P(f_a)$ is the range cell migration (RCM) curve of the target in the range-Doppler domain:

$$R_P(f_a) = x + R_{J0} + \frac{\left(x+R_{J0}\right)c^2}{8v^2 f_0^2} f_a^2, \tag{13}$$

and the azimuth frequency modulation rate is

$$K_a = -\frac{2v^2 f_0}{c\left(x + R_{J0}\right)}. \tag{14}$$

The echo signal will focus on the coordinate (x, y) after range compression, RCM correction (RCMC), and azimuth compression.

For comparison, when the JSF is Equation (8), the jamming signal is expressed as follows:

$$\begin{aligned} s_J(t_r, t_a) &= \sigma_P w_a(t_a) w_r\left[t_r - \tfrac{2R_J(t_a)}{c} - \tfrac{2x}{c}\right] \\ &\exp\left\{-j4\pi f_0\left[\tfrac{R_J(t_a)}{c} + \tfrac{x}{c} - \tfrac{y v t_a}{c R_{J0}}\right]\right\}\exp\left\{j\pi K_r\left[t_r - \tfrac{2R_J(t_a)}{c} - \tfrac{2x}{c}\right]^2\right\}. \end{aligned} \tag{15}$$

Similarly, the parabolic approximation and POSP is used to obtain the expression of the jamming signal in the range-Doppler domain:

$$\begin{aligned} S_J(t_r, f_a) &= \sigma_P W_a(f_a) w_r\left[t_r - \tfrac{2R_{JP}(f_a)}{c}\right] \\ &\exp\left\{j\pi K_r\left[t_r - \tfrac{2R_{JP}(f_a)}{c}\right]^2 - j\pi \tfrac{f_a^2}{K_{Ja}} - j2\pi f_a \tfrac{y}{v} - j4\pi f_0 \tfrac{x + R_{J0}}{c} + j2\pi f_0 \tfrac{y^2}{c R_{J0}}\right\}, \end{aligned} \tag{16}$$

where the RCM of fake target is

$$R_{JP}(f_a) = x + R_{J0} + \frac{R_{J0} c^2}{8 v^2 f_0^2} f_a^2 - \frac{y c}{2 v f_0} f_a + \frac{y^2}{2 R_{J0}}, \tag{17}$$

and the azimuth frequency modulate rate of jamming signal is

$$K_{Ja} = -\frac{2v^2 f_0}{c R_{J0}}. \tag{18}$$

It can be found that there are differences between the real target echo signal and the jamming signal in the RCM curve and azimuth frequency modulation rate, ignoring the phase terms unrelated to pulse compression. The effects of these differences on the imaging results and the corresponding effective region of deceptive jamming are analyzed in detail below.

First, it is obvious that the jamming signal introduces the azimuth frequency modulation rate error, which will cause a mismatch of the azimuth matched filter and finally lead to the main lobe broadening of the azimuth pulse compression result. The azimuth frequency modulation rate error is

$$\Delta K_a = K_{Ja} - K_a = -\frac{2v^2 f_0 x}{c R_{J0}\left(x + R_{J0}\right)}. \tag{19}$$

The effect of ΔK_a on the main lobe broadening can be measured by the quadratic phase error (QPE); the expression of QPE is as follows [22]:

$$\mathrm{QPE} = \pi \Delta K_a \left(\frac{T}{2}\right)^2 = -\frac{2\pi v^2 f_0 x}{c R_{J0}\left(x + R_{J0}\right)}\left(\frac{L}{2v}\right)^2, \tag{20}$$

where $T = L/v$ is the synthetic aperture time and L represents the synthetic aperture length.

For a typical Kaiser window with $\beta = 2.5$, if the broadening is required to be less than 2%, 5%, and 10%, the corresponding QPE absolute value should be less than 0.27π, 0.41π, and 0.55π [22]. Here, we define the azimuth QPE factor ε; when the condition $|\mathrm{QPE}| \leq \varepsilon\pi$ is required, the range coordinate x should satisfy

$$|x| \leq \frac{2\varepsilon c R_{J0}\left(x + R_{J0}\right)}{f_0 L^2} \approx 2\varepsilon \frac{c}{f_0}\left(\frac{R_{J0}}{L}\right)^2. \tag{21}$$

Second, in the range-Doppler domain, the RCM error of the jamming signal is

$$\Delta R_P(f_a) = R_{JP}(f_a) - R_P(f_a) = -\frac{xc^2}{8v^2 f_0^2} f_a^2 - \frac{yc}{2v f_0} f_a + \frac{y^2}{2R_{J0}}. \tag{22}$$

The last term in Equation (22) can be omitted because $R_{J0} \gg y$. The residual RCM introduced by $\Delta R_P(f_a)$ in broadside mode is represented as follows:

$$\mathrm{RCM}_{\mathrm{res}} = \left|\Delta R_P\left(\frac{B_a}{2}\right) - \Delta R_P\left(-\frac{B_a}{2}\right)\right| = \frac{B_a c}{2v f_0}|y|, \tag{23}$$

where B_a is the azimuth Doppler bandwidth and can be expressed as [22]

$$B_a = \frac{L}{v}\left|K_{Ja}\right| = \frac{2v f_0 L}{c R_{J0}}. \tag{24}$$

Combined with Equations (23) and (24), we can simplify the expression of residual RCM:

$$\mathrm{RCM}_{\mathrm{res}} = \frac{L}{R_{J0}}|y|. \tag{25}$$

The residual RCM will result in the main lobe broadening in both range and azimuth dimensions and the extent of broadening can be measured by the ratio of the residual RCM to the range resolution. The range resolution $\rho_r \approx c/2B$, where B is the signal bandwidth. Then, we define the residual RCM factor η; if we require $\mathrm{RCM}_{\mathrm{res}} \leq \eta\rho_r$, the azimuth coordinate y should satisfy

$$|y| \leq \eta \frac{\rho_r R_{J0}}{L} = \eta \frac{c R_{J0}}{2BL}. \tag{26}$$

According to [22], the residual RCM should be no more than 0.5 of the range cell; therefore, η should be no more than 0.5.

In addition, due to the difference between the false and real point targets in the instantaneous slant range history, the Doppler center frequency of the jamming signal shifts, and the azimuth main lobe broadening and ghost targets are introduced. This phenomenon limits the effective azimuth scale as well; according to [11], the azimuth coordinate y should satisfy

$$|y| \leq \frac{c R_{J0}}{2v f_0}\left(\mathrm{PRF} - \frac{v}{D}\right) - \frac{L}{2}, \tag{27}$$

where PRF is the pulse repetition frequency and D is the antenna aperture in the azimuth direction.

Equations (21) and (26) describe the effective region of the range and azimuth directions of the TDFS algorithm with the specified azimuth QPE factor ε and residual RCM factor η. The jamming signals representing the false target located beyond the region will not achieve the desired deception. Equation (27) described the inherent limitations of SAR deceptive jamming in the azimuth direction, which is beyond the scope of this article. In the following discussion, we suppose that the size of the template in the azimuth dimension meets the requirement of Equation (27). When the typical C-band space-borne SAR parameters (see Table 1 in Section 3) are set as an example, we can calculate the effective region: $|x| \leq 1.25$ km and $|y| \leq 0.75$ km with $\varepsilon = 0.25$ and $\eta = 0.5$; i.e., a rectangular area of 2.5 km × 1.5 km.

Table 1. The setting of SAR parameters in simulations.

Carrier Frequency	Chirp Rate	PRI	Pulse Width	Platform Velocity	Shortest Slant Range	Antenna Aperture
5.30 GHz	0.72 MHz/μs	0.79 ms	41.74 μs	7.06 km/s	989 km	10 m

2.2.3. Template Segmentation

The analysis in the previous subsection shows that the effective region of the TDFS algorithm is limited. In order to achieve deceptive jamming in a larger scene, we divide the jamming scene template into several blocks and apply a time-delay and frequency-shift in each block to calculate the partial JSF, which will be summed to get the JSF on the whole template. As shown in Figure 3, the template consisting of $m \times n$ point scatters is divided into $M \times N$ blocks; each block contains $U \times V$ point scatters, namely $U = m/M$ and $V = n/N$. If the range interval between each point is Δx and the azimuth interval is Δy, U and V should satisfy the following conditions according to the limitation of the effective region Equations (21) and (26) with the required ε and η:

$$U \leq \eta \frac{cR_{min}}{BL\Delta y}, \quad V \leq 4\varepsilon \frac{c}{f_0 \Delta x}\left(\frac{R_{min}}{L}\right)^2, \tag{28}$$

where R_{min} is the minimum value of the shortest slant range of all point scatters, which can be approximated by R_{J0}.

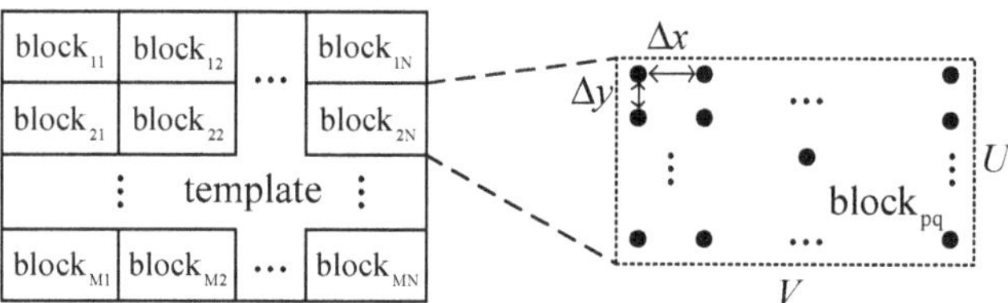

Figure 3. Jamming scene template segmentation diagram.

The scattering coefficients of point scatters in block_{pq} can be expressed as a matrix $\mathbf{T}_{pq}$:

$$\mathbf{T}_{pq} = \begin{bmatrix} \sigma_{11} & \sigma_{12} & \cdots & \sigma_{1V} \\ \sigma_{21} & \sigma_{22} & \cdots & \sigma_{2V} \\ \vdots & \vdots & \ddots & \vdots \\ \sigma_{U1} & \sigma_{U2} & \cdots & \sigma_{UV} \end{bmatrix}. \tag{29}$$

Suppose the coordinate of the block_{pq} geometric center is (x_q, y_p), where $p = 1, 2, \cdots, M$ and $q = 1, 2, \cdots, N$. According to Equations (1) and (2), the JSF on the block_{pq} geometry center $Hc_{pq}(f_r, t_a)$ is represented as follows:

$$Hc_{pq}(f_r, t_a) = \exp\left\{-\frac{j4\pi(f_r + f_0)}{c}\left[\sqrt{\left(x_q + R_{J0}\right)^2 + \left(y_p - vt_a\right)^2} - \sqrt{R_{J0}^2 + (vt_a)^2}\right]\right\}. \tag{30}$$

The jamming signal generated by $Hc_{pq}(f_r, t_a)$ can generate a well-focused point target in the center of block_{pq} after imaging, which is equivalent to moving the jammer to the geometric center of block_{pq}.

Then, we use the TDFS algorithm to generate the JSF on block_{pq}:

$$\begin{aligned} H_{pq}(f_r, t_a) = Hc_{pq}(f_r, t_a) \sum_{k=1}^{U} \exp\left[\tfrac{j4\pi f_0 v t_a \Delta y}{c(R_{J0}+x_q)}\left(k - \tfrac{U+1}{2}\right)\right] \\ \sum_{l=1}^{V} \sigma_{kl} \exp\left[-\tfrac{j4\pi(f_0+f_r)\Delta x}{c}\left(l - \tfrac{V+1}{2}\right)\right], \end{aligned} \tag{31}$$

where σ_{kl} is the element of row k and column l in the matrix $\mathbf{T}_{pq}$. Finally, the JSF on the whole template can be derived by summing JSF on all blocks:

$$H(f_r, t_a) = \sum_{p=1}^{M} \sum_{q=1}^{N} H_{pq}(f_r, t_a). \tag{32}$$

Since the number of blocks is very small compared to the total number of scatters in the template, Equation (30) has a limited effect on the computing load. In addition, the second summation term of Equation (31) is independent of slow time t_a and can be calculated offline. The template segmentation method can solve the problem that the effective region of the TDFS algorithm is limited and achieve the purpose of the rapid generation of large scene deceptive jamming signals.

2.2.4. Correction Algorithm in Squint Mode

The analyses above are based on the broadside mode with the squint angle $\theta = 0$. In order to extend the broadside jamming algorithm to squint mode, this subsection will discuss the effect of the squint angle on the jamming result and the corresponding correction method. It should be pointed out that the parabolic approximation of the instantaneous slant range in Equation (4) will no longer be applicable under the condition of a large squint angle, so the jamming method of this article is limited to the small squint angle and the medium aperture length SAR.

First, the Doppler center frequency of the echo signal $f_{ac} = 2vf_0 \sin\theta / c$, the presence of the squint angle will result in the non-zero Doppler center frequency. At this time, the azimuth signal can be regarded as the non-baseband signal, and the frequency modulation rate error will cause the position offset besides main lobe broadening in pulse compression [22]. According to the pulse compression principle [22], the azimuth main lobe position offset of scatter with coordinate (x, y) is

$$y_{ofs}(x, y) = -\frac{\Delta K_a}{K_a} t_{ac} v = x \tan\theta, \tag{33}$$

where $t_{ac} = -R_{J0} \tan\theta / v$ is the pulse center time of the jamming signal in the azimuth dimension.

On the other hand, when $f_{ac} \neq 0$, the RCM error will cause the RCM curve to shift along the range dimension in addition to introducing the residual RCM, causing main lobe broadening. According to Equation (22), the offset in the range dimension of fake point $P(x, y)$ is

$$x_{ofs}(x, y) = \Delta R_P(f_{ac}) \approx -\frac{x}{2} \sin^2\theta - y \sin\theta. \tag{34}$$

The residual RCM will increase at the same time:

$$\text{RCM}_{\text{res}} = \left|\Delta R_P\left(f_{ac} + \frac{B_a}{2}\right) - \Delta R_P\left(f_{ac} - \frac{B_a}{2}\right)\right| = \left|\frac{B_a c}{2 v f_0} y + \frac{B_a c \sin\theta}{4 v f_0} x\right|. \tag{35}$$

In order to ensure the image quality of the fake scene, the size of the blocks should be reduced. However, the RCM_{res} increment is not obvious when the squint angle is small. Thus, it can be ignored in this paper.

According to the analysis above, the main effect of the squint angle is the location offset of fake targets in the deceptive image, and the offset depends on the coordinates in the template. This effect will cause the distortion of the jamming image. Therefore, the coordinates of each scatter in the template should be corrected. For the false scatter with coordinate (x, y), the corrected coordinate (x_c, y_c) can be represented as follows:

$$\begin{cases} x_c = x - x_{ofs}(x, y) = x + \frac{x}{2}\sin^2\theta + y\sin\theta, \\ y_c = y - y_{ofs}(x, y) = y - x\tan\theta. \end{cases} \tag{36}$$

Correspondingly, Equation (31) will be modified as follows:

$$\begin{aligned} H_{pq}(f_r, t_a) = Hc_{pq}(f_r, t_a) \sum_{k=1}^{U}\sum_{l=1}^{V} \sigma_{kl} \exp\Big\{-\tfrac{j4\pi(f_r+f_0)\Delta x}{c}\Big[\Big(1+\tfrac{\sin^2\theta}{2}\Big)\Big(l-\tfrac{V+1}{2}\Big) \\ +\Delta y\sin\theta\Big(k-\tfrac{U+1}{2}\Big)\Big]\Big\} \exp\Big\{\tfrac{j4\pi f_0 v t_a}{c(R_{J0}+x_q)}\Big[-\Delta x\tan\theta\Big(l-\tfrac{V+1}{2}\Big)+\Delta y\Big(k-\tfrac{U+1}{2}\Big)\Big]\Big\}. \end{aligned} \tag{37}$$

The range and azimuth-related terms in Equation (37) are separable, so Equation (37) can be rewritten as matrix operations to increase the calculation speed. Here, we define the time-delay matrixes $\mathbf{Hr}_1$, $\mathbf{Hr}_2$ and frequency-shift matrixes $\mathbf{Ha}_{q1}$, $\mathbf{Ha}_{q2}$:

$$\mathbf{Hr}_1 = \begin{bmatrix} \exp\Big[-\frac{j4\pi(f_r+f_0)\Delta x}{c}\Big(1+\frac{\sin^2\theta}{2}\Big)\Big(1-\frac{V+1}{2}\Big)\Big] \\ \exp\Big[-\frac{j4\pi(f_r+f_0)\Delta x}{c}\Big(1+\frac{\sin^2\theta}{2}\Big)\Big(2-\frac{V+1}{2}\Big)\Big] \\ \vdots \\ \exp\Big[-\frac{j4\pi(f_r+f_0)\Delta x}{c}\Big(1+\frac{\sin^2\theta}{2}\Big)\Big(V-\frac{V+1}{2}\Big)\Big] \end{bmatrix}, \tag{38}$$

$$\mathbf{Hr}_2 = \begin{bmatrix} \exp\Big[-\frac{j4\pi(f_r+f_0)\Delta y\sin\theta}{c}\Big(1-\frac{U+1}{2}\Big)\Big] \\ \exp\Big[-\frac{j4\pi(f_r+f_0)\Delta y\sin\theta}{c}\Big(2-\frac{U+1}{2}\Big)\Big] \\ \vdots \\ \exp\Big[-\frac{j4\pi(f_r+f_0)\Delta y\sin\theta}{c}\Big(U-\frac{U+1}{2}\Big)\Big] \end{bmatrix}, \tag{39}$$

$$\mathbf{Ha}_{q1} = \begin{bmatrix} \exp\Big[-\frac{j4\pi f_0 v t_a\Delta x\tan\theta}{c(R_{J0}+x_q)}\Big(1-\frac{V+1}{2}\Big)\Big] \\ \exp\Big[-\frac{j4\pi f_0 v t_a\Delta x\tan\theta}{c(R_{J0}+x_q)}\Big(2-\frac{V+1}{2}\Big)\Big] \\ \vdots \\ \exp\Big[-\frac{j4\pi f_0 v t_a\Delta x\tan\theta}{c(R_{J0}+x_q)}\Big(V-\frac{V+1}{2}\Big)\Big] \end{bmatrix}, \tag{40}$$

$$\mathbf{Ha}_{q2} = \begin{bmatrix} \exp\Big[\frac{j4\pi f_0 v t_a\Delta y}{c(R_{J0}+x_q)}\Big(1-\frac{U+1}{2}\Big)\Big] \\ \exp\Big[\frac{j4\pi f_0 v t_a\Delta y}{c(R_{J0}+x_q)}\Big(2-\frac{U+1}{2}\Big)\Big] \\ \vdots \\ \exp\Big[\frac{j4\pi f_0 v t_a\Delta y}{c(R_{J0}+x_q)}\Big(U-\frac{U+1}{2}\Big)\Big] \end{bmatrix}. \tag{41}$$

Equation (37) is rewritten as follows:

$$H_{pq}(f_r, t_a) = Hc_{pq}(f_r, t_a)\Big[\big(\mathbf{Hr}_2 \circ \mathbf{Ha}_{q2}\big)^{\mathrm{T}}\mathbf{T}_{pq}\big(\mathbf{Hr}_1 \circ \mathbf{Ha}_{q1}\big)\Big], \tag{42}$$

where $(\cdot)^{\mathrm{T}}$ represents the matrix transposition and the operator $\circ$ represents the Hadamard product of the matrix.

We can get the JSF on the entire scene by superimposing the JSF on all blocks, which is the same as Equation (32). The time-delay matrixes $\mathbf{Hr}_1$ and $\mathbf{Hr}_2$ are independent of the azimuth time t_a, which can be calculated offline in advance to improve the real-time processing speed.

2.3. TDFS-TS Algorithm Procedure

The prerequisite for the successful implementation of SAR deceptive jamming is to obtain relevant intelligence on the jamming object, which mainly includes the follows aspects.

- Kinematic parameters of the SAR platform, including motion trajectory, motion velocity v, etc. The motion trajectory information is used to establish the jamming coordinate system and determine R_{J0}, the shortest distance between the jammer and SAR;
- Antenna parameters, including the squint angle θ, synthetic aperture length L, etc.;
- Signal parameters, including the carrier frequency f_0, bandwidth B, PRI, etc.

The specific detection methods of the parameters above will not be discussed in this paper; we simply suppose that the parameters have been obtained in advance.

As shown in Figure 4, the entire procedure includes two parts: preprocessing and real-time calculation. The first step in the preprocessing stage is template segmentation: according to the parameters including synthetic aperture length L, signal bandwidth B, shortest slant range of the jammer R_{J0} and the given factors ε and η, we divide the template into several blocks based on the limitation of Equation (28); then, we perform offline calculation and calculate the time-delay matrixes $\mathbf{Hr}_1$ and $\mathbf{Hr}_2$ according to Equations (38) and (39). In the real-time calculation stage, we calculate frequency-shift matrixes $\mathbf{Ha}_{q1}$, $\mathbf{Ha}_{q2}$ according to Equations (40) and (41); then, we calculate the JSF on each block based on Equations (30) and (42); finally, we add the JSF on all blocks to obtain the JSF on the entire scene.

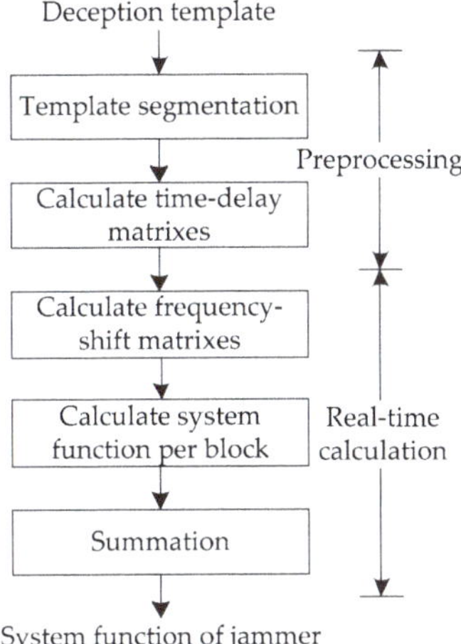

Figure 4. Time-delay and frequency-shift with template segmentation (TDFS-TS) algorithm procedure.

3. Simulation Results

In this section, the effectiveness of the TDFS-TS algorithm is verified by simulating the imaging results of false point targets and fake scenes, and the computational complexity is analyzed. The simulation results of the range dimension segmentation (RDS) algorithm proposed by Zhou et al. [11] are used as a comparison. The main parameters of the radar, which reference the satellite RADARSAT-1, are listed in Table 1.

3.1. Fake Point Scatters Case

In order to analyze the imaging result of fake point scatter at different positions after imaging, a deceptive scene template containing only four-point scatters is set as shown in Figure 5. The four points $P_0 \sim P_3$ are arranged in a rectangular shape with a distance of 6 km in the range dimension and 2 km in the azimuth dimension. According to the calculation results in Section 2.2.2, we set the length of blocks to 2.5 km in the range dimension and 1.5 km in the azimuth dimension. Since the imaging quality of fake scatters is only related to the position in the block, for the purpose of analyzing the jamming effect of the algorithm comprehensively, the template segmentation scheme is shown by the dashed line in Figure 5, meaning that P_0 is located at the center of block 1, P_1 is at the edge of block 3 in the range dimension, P_2 is at the azimuth edge of block 4, and P_3 is at the edge of block 6 in both range and azimuth dimensions. In addition, the position of the jammer (i.e., the origin position) is set at the point P_0; actually, the position of the jammer has little effect on the imaging result. In the simulation of the RDS algorithm, the template is divided into three segments with the same segmentation length (2.5 km) in the range dimension and is no longer segmented in the azimuth dimension.

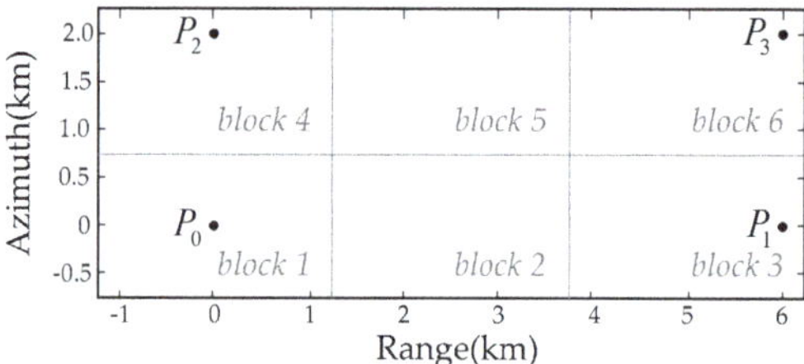

Figure 5. The deceptive scene containing four-point scatters.

Figures 6–9 show the imaging results of the four scatters by the RDS and TDFS-TS algorithm in the broadside mode and squint mode with the squint angle $\theta = 5^{\circ}$, including the close-up image, range profile, and azimuth profile. Tables 2 and 3 list the imaging quality parameters of range and azimuth dimensions in the two modes, including 3 dB impulse response width (IRW), main lobe position offset (MLPO), peak sidelobe ratio (PSLR), and integrated sidelobe ratio (ISLR). It can be seen that, compared with the RDS algorithm, the performance of the TDFS-TS algorithm is basically equivalent, the RDS algorithm is more advantageous on IRW, while TDFS-TS is dominant over MLPO. The IRW of TDFS-TS algorithm is increased especially for the azimuth dimension in squint mode; however, the maximum broadening does not exceed 3.6%. In the squint mode, the MLPO of the RDS algorithm can reach up to −80.32 m in the azimuth dimension, which can be eliminated basically by the TDFS-TS algorithm due to the corresponding correction. Because of the influence of the residual RCM, the MLPO in the range dimension cannot be completely corrected, but the overall image is affected very little. In short, the simulation of fake point scatters shows that the TDFS-TS algorithm is effective and has certain advantages in several areas.

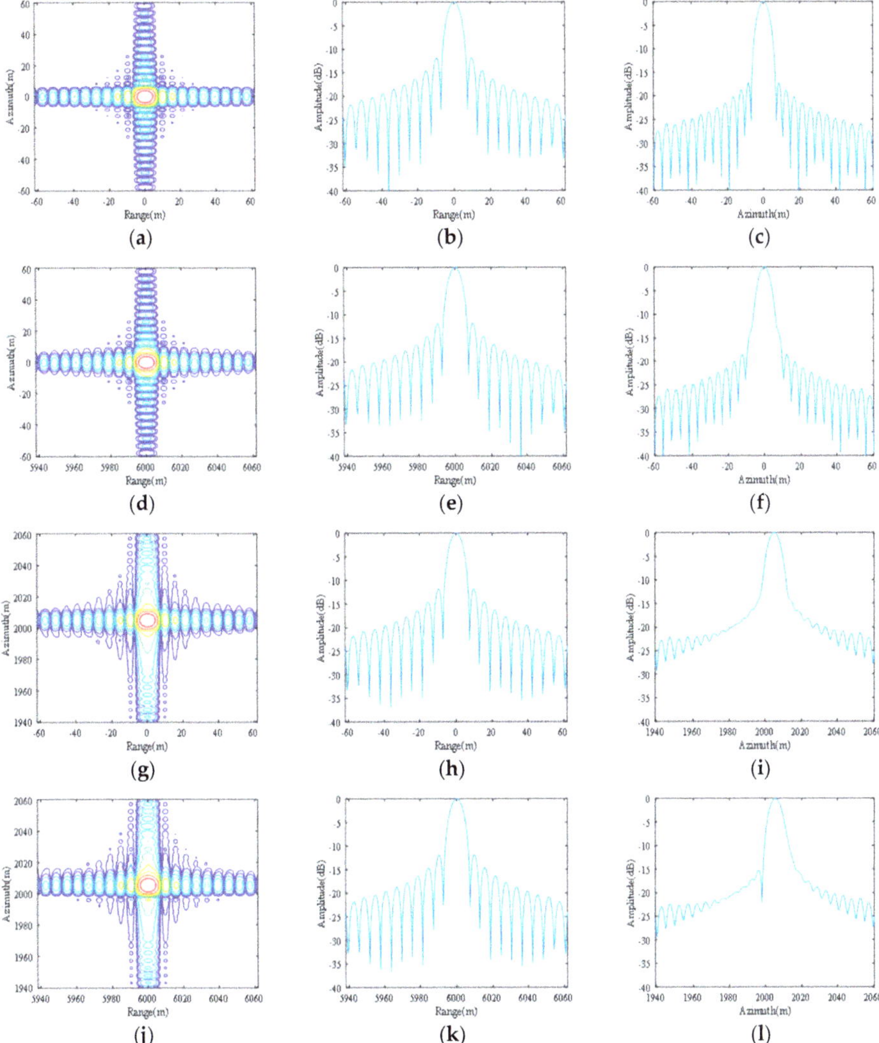

Figure 6. Simulation results of the RDS algorithm for false point scatters in broadside mode. (**a**–**c**) are the close-up image, range profile, and azimuth profile of P_0, respectively; (d–f) are the close-up image, range profile, and azimuth profile of P_1 respectively; (g–i) are the close-up image, range profile, and azimuth profile of P_2, respectively; (j–l) are the close-up image, range profile, and azimuth profile of P_3, respectively.

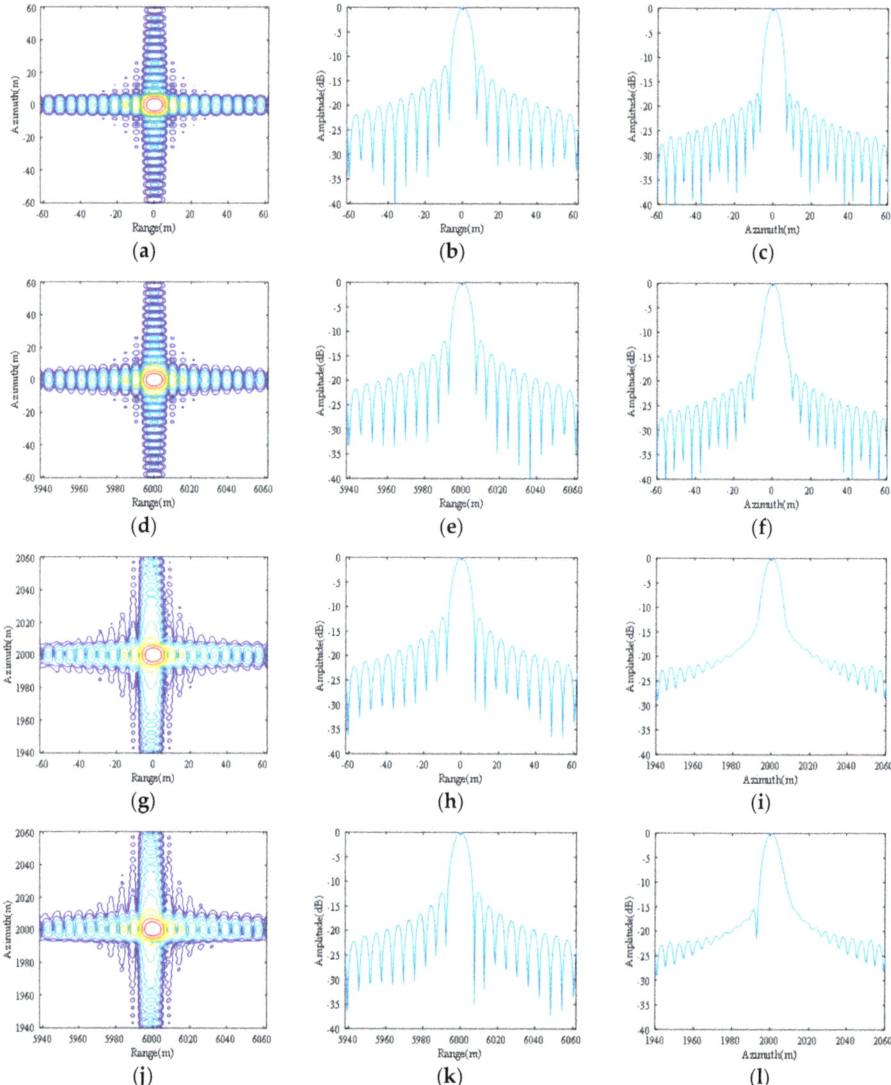

Figure 7. Simulation results of the TDFS-TS algorithm for false point scatters in broadside mode. (**a**–**c**) are the close-up image, range profile, and azimuth profile of P_0, respectively; (d–f) are the close-up image, range profile, and azimuth profile of P_1, respectively; (g–i) are the close-up image, range profile, and azimuth profile of P_2, respectively; (j–l) are the close-up image, range profile, and azimuth profile of P_3, respectively.

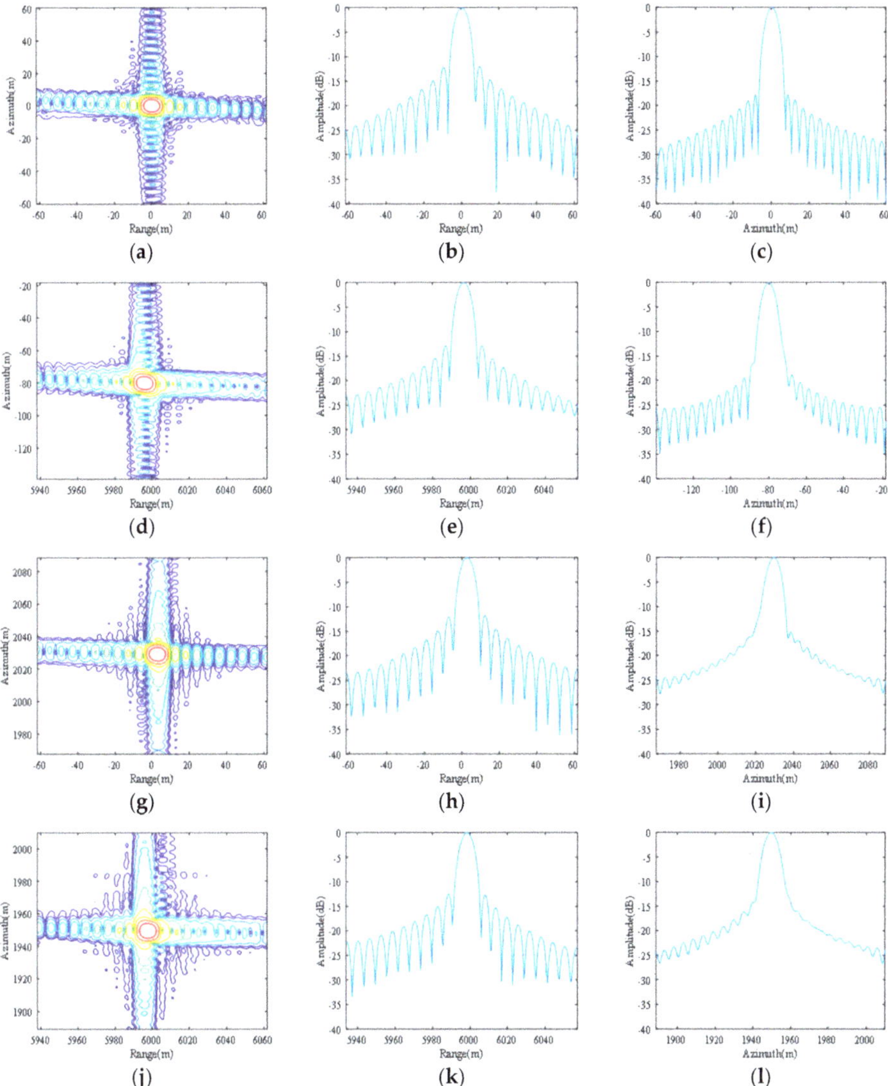

Figure 8. Simulation results of the RDS algorithm for false point scatters in squint mode with a squint angle $\theta = 5^{\circ}$. (a–c) are the close-up image, range profile, and azimuth profile of P_0, respectively; (d–f) are the close-up image, range profile, and azimuth profile of P_1, respectively; (g–i) are the close-up image, range profile, and azimuth profile of P_2, respectively; (j–l) are the close-up image, range profile, and azimuth profile of P_3, respectively.

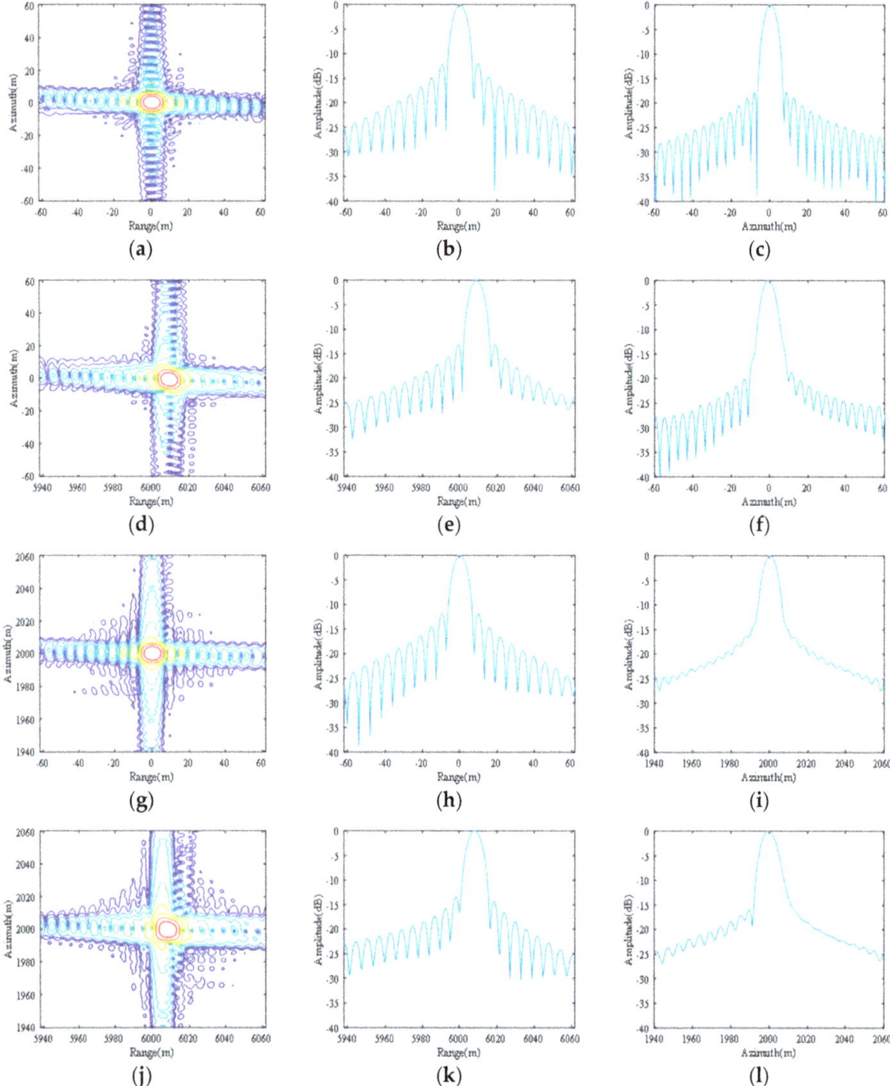

Figure 9. Simulation results of the TDFS-TS algorithm for false point scatters in squint mode with a squint angle $\theta = 5^{\circ}$. (a–c) are the close-up image, range profile, and azimuth profile of P_0, respectively; (d–f) are the close-up image, range profile, and azimuth profile of P_1, respectively; (g–i) are the close-up image, range profile, and azimuth profile of P_2, respectively; (j–l) are the close-up image, range profile, and azimuth profile of P_3, respectively.

Table 2. Comparison of imaging quality parameters between the RDS algorithm and TDFS-TS algorithm in broadside mode.

		Range Dimension				Azimuth Dimension			
		IRW (m)	**MLPO (m)**	**PSLR (dB)**	**ISLR (dB)**	**IRW (m)**	**MLPO (m)**	**PSLR (dB)**	**ISLR (dB)**
P_0	RDS	7.73	0	−11.78	−12.57	6.77	0	−17.24	−19.50
	TDFS-TS	7.73	0	−11.78	−12.57	6.77	0	−17.24	−19.50
P_1	RDS	7.73	0	−11.78	−12.57	6.88	0	−13.47	−15.05
	TDFS-TS	7.73	0	−11.78	−12.57	6.88	0	−13.39	−14.98
P_2	RDS	7.73	0	−11.78	−12.57	7.39	5.04	−14.46	−14.90
	TDFS-TS	7.78	−0.21	−12.03	−12.96	7.47	0	−13.95	−13.67
P_3	RDS	7.73	0	−11.77	−12.57	7.53	5.52	−15.03	−14.24
	TDFS-TS	7.78	−0.21	−12.04	−12.96	7.60	0.48	−15.12	−12.55

Table 3. Comparison of imaging quality parameters between the range dimension segmentation (RDS) algorithm and TDFS-TS algorithm in squint mode with squint angle $\theta = 5^{\circ}$.

		Range Dimension				Azimuth Dimension			
		IRW (m)	**MLPO (m)**	**PSLR (dB)**	**ISLR (dB)**	**IRW (m)**	**MLPO (m)**	**PSLR (dB)**	**ISLR (dB)**
P_0	RDS	7.73	0	−12.07	−12.79	6.77	0	−17.79	−20.30
	TDFS-TS	7.73	0	−12.07	−12.79	6.77	0	−17.79	−20.30
P_1	RDS	7.88	−3.48	−12.79	−13.87	7.07	−80.32	−16.56	−17.91
	TDFS-TS	7.88	7.69	−12.94	−13.92	7.15	−0.98	−18.76	−17.83
P_2	RDS	7.77	2.46	−12.02	−12.60	7.39	29.01	−15.03	−15.45
	TDFS-TS	7.74	0.01	−11.68	−12.48	7.34	−0.01	−14.59	−14.96
P_3	RDS	7.95	−1.17	−12.36	−13.47	8.05	−50.70	−14.05	−14.08
	TDFS-TS	8.16	7.73	−13.29	−14.87	8.34	−0.57	−15.95	−14.28

3.2. General Deceptive Scene Case

In this subsection, the TDFS-TS algorithm is applied to yield a fake scene. The jamming object is RADARSAT-1, whose parameters are listed in Table 1, and the squint angle $\theta \approx -1.58^{\circ}$. The raw data of the radar are obtained from the appendix in [22]. The fake scene template is another SAR image, as shown in Figure 10, whose length in the range dimension is 12.5 km and 4.5 km in the azimuth dimension. We divide the template according to the same block size calculated in Section 2.2.2, as shown by the yellow line in Figure 10.

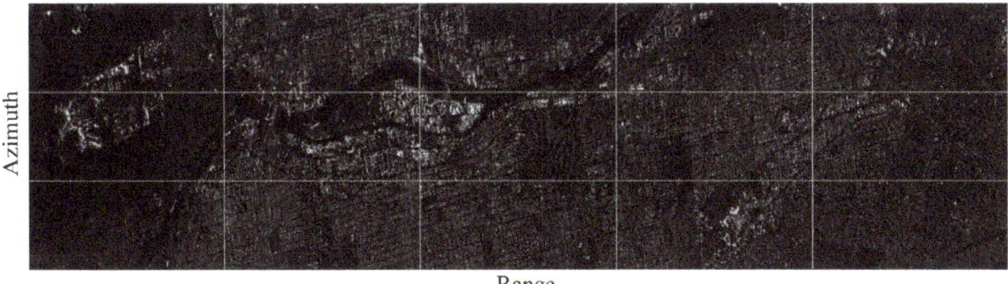

Figure 10. Deceptive jamming template with 12.5 km in the range dimension and 4.5 km in the azimuth dimension, which is divided into blocks by the yellow line. The block size is 2.5 km × 1.5 km.

First, the signals generated by the two algorithms are processed to get the images which are shown in Figure 11 after being amplitude-normalized, and partially enlarged images are shown in Figure 12. Due to the existence of the squint angle, each segment of the image generated by the RDS algorithm has geometric distortion, which is especially obvious at the splicing of each segment. Moreover, since the positions of adjacent scatters in the template are shifted after the imaging process, the image is blurred, and the ghost targets generated by the Doppler center frequency shifting in the azimuth dimension will be more obvious after the amplitude normalization. Actually, for the same reason, the brightness of the image is weakened as well, this phenomenon can be seen in Figures 13 and 14. These problems are solved by the TDFS-TS algorithm which corrected the geometric distortion caused by the squint angle.

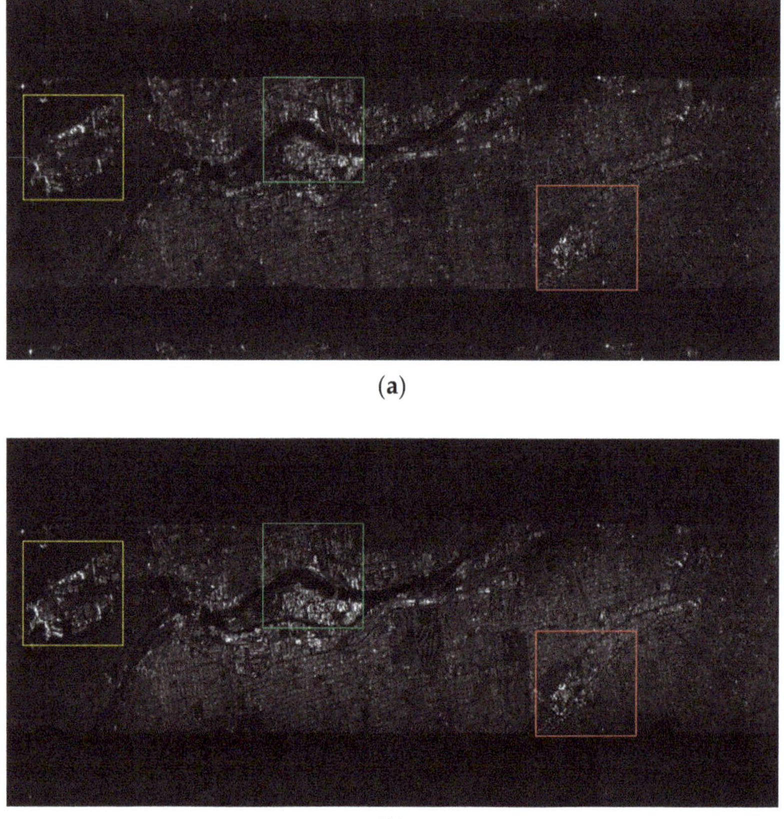

Figure 11. Imaging results of the fake scene in Figure 10 using (**a**) the RDS algorithm and (**b**) the TDFS-TS algorithm. Parts of the images marked by rectangular boxes are enlarged and shown in Figure 12.

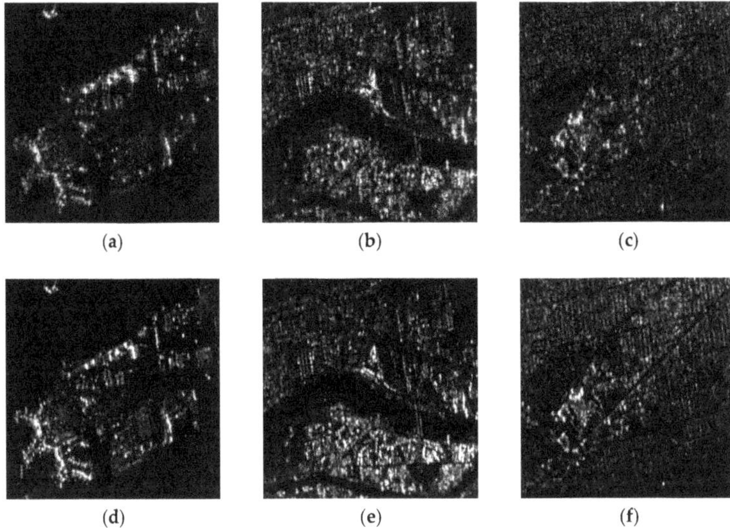

(a) (b) (c)

(d) (e) (f)

Figure 12. Partially enlarged image of Figure 11. (**a–c**) are the partial enlargement of the image generated by the RDS algorithm in the yellow, green, and red boxes, respectively; (**d–f**) are the partial enlargement of the image generated by the TDFS-TS algorithm in the yellow, green, and red boxes.

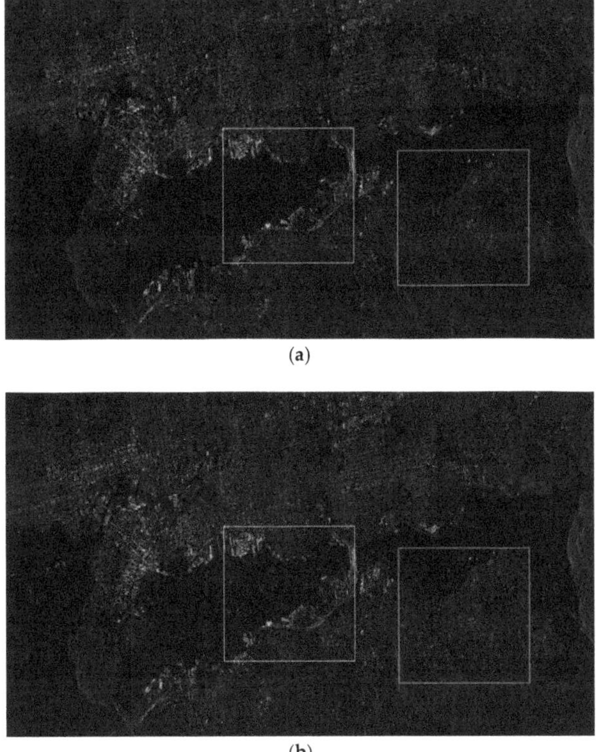

(a)

(b)

Figure 13. *Cont.*

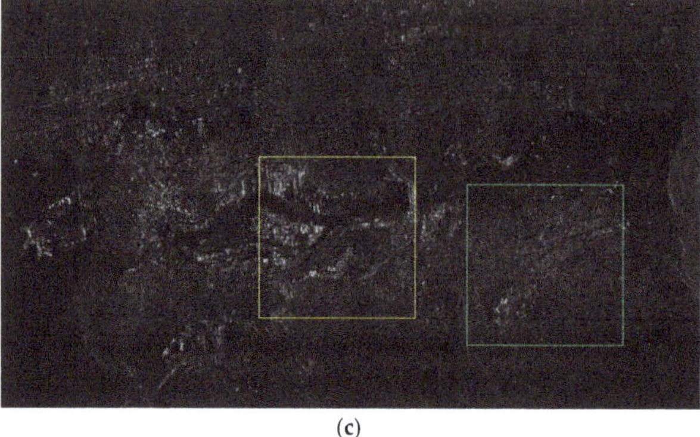

(c)

Figure 13. Comparison of jamming results. (**a**) is the image formed by the original signal, (**b**) is the image formed by the superposition of the original signal and jamming signal generated by the RDS algorithm, and (**c**) is the image formed by the superposition of the original signal and jamming signal generated by the TDFS-TS algorithm. Parts of the images marked by rectangular boxes are enlarged and shown in Figure 14.

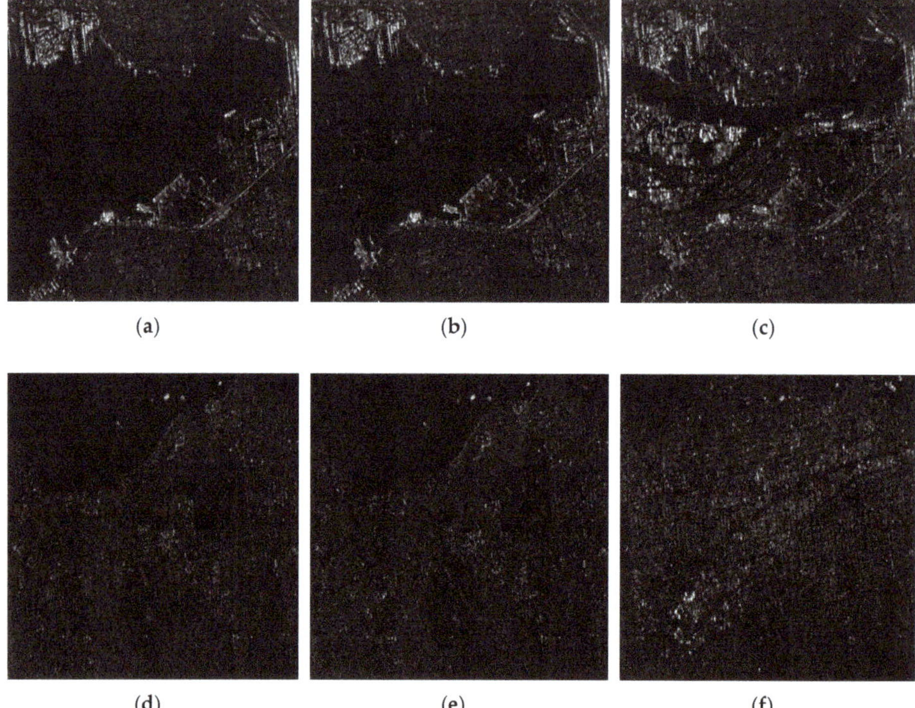

(**a**) (**b**) (**c**)

(**d**) (**e**) (**f**)

Figure 14. Partially enlarged image of Figure 13. (**a–c**) are the partial enlargement of the original image, the jamming image of the RDS algorithm, and the TDFS-TS algorithm, respectively in the yellow rectangular boxes; (**d–f**) are the partial enlargement of the original image, the jamming image of the RDS algorithm, and the TDFS-TS algorithm, respectively in the green boxes.

The imaging results of the superposition of the jamming signal and original signal are shown in Figure 13, and partially enlarged images are shown in Figure 14. It can be seen that the TDFS-TS algorithm achieves a good deception effect, generating clear false targets such as land and buildings and changing the topographic structure of the protected area. However, the fake scene generated by the RDS algorithm is not obvious, mainly because of the decrease in brightness caused by geometric distortion. The jamming power has to increase in order to achieve a satisfactory jamming purpose, but the ghost targets will be strengthened at the same time. In conclusion, the TDFS-TS algorithm has certain advantages compared with the RDS algorithm in large-scene jamming.

3.3. Computational Complexity Analysis

The effectiveness of the TDFS-TS algorithm is verified by simulation in the previous subsection. In this subsection, we will estimate the computational complexity of the TDFS-TS algorithm to assess its practical value. Assume that the jamming scene template consists of $m \times n$ point scatters and is divided into $M \times N$ blocks, each block contains $U \times V$ point scatters. Here, for ease of analysis, the addition, multiplication and power operations are all considered to be a basic operation. The following will analyze the number of basic operations to obtain the JSF $H(f_r, t_a)$ at a specific frequency f_r and azimuth time t_a.

In the preprocessing stage, calculating an element of matrix $\mathbf{Hr}_1$ requires 15 basic operations, and the operation amount of all elements is $15V$; however, because the elements in $\mathbf{Hr}_1$ are actually a geometric series, the other elements can be generated by multiplying the first element by the common ratio, meaning that the amount of calculation is reduced to $15 + V$. According to the same analysis, the operation amount of $\mathbf{Hr}_2$ is $12 + U$, so the total computational complexity in the preprocessing stage is

$$C_{pre} = 15 + V + 12 + U = U + V + 27. \tag{43}$$

In the real-time calculation stage, calculating the JSF of a block center requires 20 basic operations, and a total of $20MN$ operations are required for all blocks. The operation amount for calculating matrix $\mathbf{Ha}_{q1}$ is $15 + V$, which needs to be repeated N times. The calculation amount of matrix $\mathbf{Ha}_{q2}$ is $14 + U$ and needs to be repeated N times as well. The computational complexity of matrix multiplication in Equation (42) is $2UV + U + 2V - 1$, so the operation amount of Equation (42) is $2UV + U + 2V$ added to the multiplication with $Hc_{pq}(f_r, t_a)$, and the calculation of Equation (42) needs to be repeated MN times. Finally, the operation amount of the summation—i.e., Equation (32)—is $MN - 1$. Based on the analysis above, the total amount of computation in the real-time calculation stage is

$$\begin{aligned} C_{rt} &= 20MN + N(15 + V) + N(14 + U) + MN(2UV + U + 2V) + MN - 1 \\ &= mn\left(2 + \tfrac{2}{U} + \tfrac{1}{V} + \tfrac{21}{UV}\right) + n\left(1 + \tfrac{U}{V} + \tfrac{29}{V}\right) - 1. \end{aligned} \tag{44}$$

According to the same analysis method, the computational complexity of different jamming algorithms including TDFS-TS, RDS, TDFS-TS without squint correction (TDFS-TS-WSC) and the basic algorithm (BA) shown by Equation (3) is derived and shown in Table 4.

Table 4. Computational complexity comparison of different algorithms.

	TDFS-TS	RDS [1]	TDFS-TS-WSC	BA
Preprocessing	$U + V + 27$	$mn\left(14 - \frac{1}{V}\right)$	$2UV - U + V + 11$	-
Real-time calculation	$mn\left(2 + \frac{2}{U} + \frac{1}{V} + \frac{21}{UV}\right)$ $+n\left(1 + \frac{U}{V} + \frac{29}{V}\right) - 1$	$\frac{22mn}{V} - 1$	$\frac{21mn}{UV} + \frac{n}{V}(3U + 13) - 13)\ V$	$21mn - 1$

[1] the segment length in the range dimension of the RDS algorithm is the same as the block size of the TDFS-TS algorithm.

The block size U and V are determined by the parameters of the jamming object, the scatter density of the template, and the two imaging quality control factors ε and η. We fix U and V and observe the relationship between the computational complexity and the template size m and n. According to the simulation parameters in Section 4.1, we set $U = 267$ and $V = 539$, draw the relationship between the computational complexity and the total number of scatters in the template shown in Figure 15. Here, for the convenience of analysis, we set $m = n$ and take the logarithm of computational amount for display.

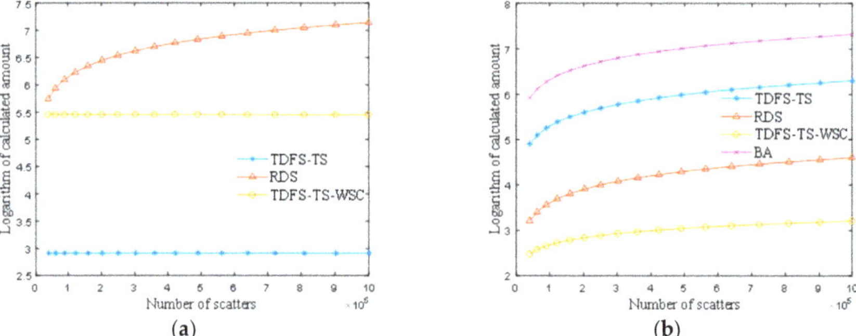

Figure 15. The relationship between the computational complexity and the total number of scatters in the template. (**a**) is the computational complexity in the preprocessing stage and (**b**) is the computational complexity in the real-time calculation stage.

It can be seen that, no matter in the preprocessing stage or in the real-time calculation stage, the computational complexity of the TDFS-TS-WSC algorithm is less than that of the RDS algorithm. In particular, in the real-time calculation stage, the computational complexity can be reduced by more than one order of magnitude because the TDFS algorithm in Equation (31) simplifies the calculation procedure significantly. When the squint correction is added, the calculation amount in the preprocessing stage is reduced, but increased in the real-time calculation stage.

4. Discussion

4.1. Imaging Quality

According to the analysis of the TDFS-TS algorithm in Sections 2.2.2 and 2.2.4, the simplification in the range dimension leads to a distortion of the azimuth profile including the main lobe broadening and position offset (in the squint mode), and the residual RCM introduced by the simplification in the azimuth dimension affects both the range and the azimuth profile. In addition, the inherent limitations of jamming signal, namely the Doppler bandwidth loss of the fake scatter far away from the jammer in the azimuth dimension, will broaden the azimuth main lobe. In summary, we can draw the following conclusions on the TDFS-TS algorithm:

1. The imaging quality of the jamming signal in the range dimension depends on the azimuth position of the false scatter ;
2. The distortion of the azimuth profile depends on both the range and the azimuth position of the false scatter;
3. These distortions become severe as the distance between the fake scatter and the block center increases.

These phenomena can be seen clearly in Table 2: P_0 and P_1 are both at the azimuth center of the block, thus they have the same imaging quality parameters in the range dimension; the imaging quality parameters of P_2 and P_3 in the range dimension are the same, and worse than that of P_0 and P_1 because the two points are both in the azimuth edge of the block; the situation in the azimuth dimension is more complicated: affected by both the range and azimuth position, the four points have different degrees of azimuth distortion, where P_3 is the worse and P_0 is the best; the impact of residual RCM in the azimuth dimension is greater than that of azimuth chirp rate error by comparing the imaging quality parameters of P_1 and P_2.

From Table 2 it can be seen that, in the broadside mode, the main lobe positions of P_2 and P_3 in the range dimension are slightly shifted. This is caused by the residual RCM as well. Due to the existence of residual RCM, the curve of jamming signal after RCM correction in the range-Doppler domain is not a straight line but a parabola which is expressed as Equation (22), thus the main lobe in the range dimension will shift towards the negative direction besides broadening. In the squint mode, the effect of residual RCM is more complicated, which can be seen in Table 3. Due to the nonlinear nature of the RCM error shown in Equation (22), the offset in the range dimension expressed in Equation (34) is just an approximation. In fact, it can hardly be eliminated completely, thus the images of P_1, P_2, and P_3 shift to varying degrees in the range dimension. In addition, by comparing the range offset of P_1 and P_2, we can consider that the range coordinate has a greater effect on the range offset in the squint mode. Although the TDFS-TS algorithm cannot correct the range offset accurately, the maximum residual offset shown in Table 3 is approximately equal to the range resolution, and has little impact on the deceptive image shown in Figures 11–13.

In the squint mode, the offset in the azimuth dimension caused by the frequency modulation rate error can be effectively corrected by the TDFS-TS algorithm; the results are shown in Table 3. As a comparison, the RDS algorithm divides the template in the range dimension as well, thus the problem of frequency modulation rate error still exists. Without a correction algorithm, the main lobe position of the RDS algorithm shift severely in the azimuth dimension and the image is distorted. In short, the imaging quality of the TDFS-TS algorithm can be guaranteed both in the broadside mode and in the squint mode with a small squint angle.

4.2. Computational Efficiency

According to the results in Section 3.3, in the broadside mode, the TDFS-TS-WSC algorithm has a quite high computational efficiency, and the squint correction is time-consuming. This is because the range coordinate correction is related to the azimuth coordinate and the azimuth coordinate correction is also related to the range coordinate. Therefore, the range and azimuth coupling terms are added, resulting in a large number of calculations needing to be completed in the real-time calculation stage, and the computational efficiency becomes worse. However, due to the calculation of each block being completely independent in the TDFS-TS algorithm and the introduction of matrix operations, it is convenient to apply parallel computing technology to greatly increase the calculation speed. Therefore real-time jamming is feasible. In summary, compared with the RDS algorithm, the TDFS-TS algorithm is more efficient in the broadside mode, and the application scenario can be extended to the squint mode.

5. Conclusions

In this paper, the large-scene electromagnetic deception of SAR is studied. The primary focus is to reduce the computational burden during the jamming process. For this purpose, the TDFS algorithm is proposed, which can improve the computational efficiency significantly. In addition, the focus capability of the jamming signal must be considered. In order to ensure the deceptive image quality of the TDFS algorithm in a large scene, the template is divided into several blocks according to the SAR parameters and imaging quality control factor. The correction algorithm in squint mode is introduced so that the TDFS-TS algorithm can be used to the SAR with a low squint angle and medium aperture length. Finally, simulation results and computational complexity analyses show that, compared to other jamming algorithms, the TDFS-TS algorithm has higher computational efficiency with less image quality decline in the broadside mode. Furthermore, the application of parallel computation can partially compensate for the computational performance decline in the squint mode.

The TDFS-TS algorithm is applicable to space-borne SAR operating at broadside mode or a low squint angle mode. In the future, we will investigate the rapid jamming method against the SAR with a significant squint angle and long synthetic aperture. Additionally, other problems such as intelligence gathering and gain control will also be studied.

Author Contributions: Conceptualization and supervision, W.Y.; formal analysis, F.M.; methodology, K.Y. and F.M.; experiment, G.L. and Q.T.; writing, K.Y. All authors have read and agreed to the published version of the manuscript.

Funding: This research was funded by Science and Technology on Complex Electronic System Simulation Laboratory, grant number DXZF-JC-ZZ-2017-007.

Conflicts of Interest: The authors declare no conflict of interest.

References

1. Walter, W.G. *Synthetic Aperture Radar and Electronic Warfare*; Artech House: Boston, MA, USA, 1993; pp. 25–39.
2. Condley, C.J. Some system considerations for electronic countermeasures to synthetic aperture radar. In Proceedings of the IEE Colloquium on Electronic Warfare Systems, London, UK, 14 January 1991.
3. Dumper, K.; Cooper, P.S.; Wons, A.F.; Condley, C.J.; Tully, P. Spaceborne synthetic aperture radar and noise jamming. In Proceedings of the 1997 Radar Edinburgh International Conference, Edinburgh, UK, 14–16 October 1997; pp. 411–414.
4. Fouts, D.J.; Pace, P.E.; Karow, C.; Ekestorm, S.R.T. A single-chirp false target radar image generator for countering wideband imaging radars. *IEEE J. Solid State Circuits* **2002**, *37*, 751–759. [CrossRef]
5. Kristoffersen, S.; Thingsrud, O. The EKKO II synthetic target generator for imaging radar. In Proceedings of the 5th European Conference on Synthetic Aperture Radar, Ulm, Germany, 25–27 May 2004; pp. 871–874.
6. Wei, Y.; Hang, R.; Shuxian, Z.; Li, Y. Study of noise jamming based on convolution modulation to SAR. In Proceedings of the 2010 International Conference on Computer, Mechatronics, Control and Electronic Engineering, Changchun, China, 24–26 August 2010; pp. 169–172.
7. Garmatyuk, D.S.; Narayanan, R.M. ECCM capabilities of an ultrawideband bandlimited random noise imaging radar. *IEEE Trans. Aerosp. Electron. Syst.* **2002**, *28*, 1243–1255. [CrossRef]
8. Yan, Z.; Guoqing, Z.; Yu, Z. Research on SAR jamming technique based on man-made map. In Proceedings of the International Conference on Radar, Shanghai, China, 16–19 October 2006; pp. 1–4.
9. Lin, X.; Liu, P.; Xue, G. Fast generation of SAR deceptive jamming signal based on inverse range doppler algorithm. In Proceedings of the IET International Radar Conference, Xi'an, China, 14–16 April 2013; pp. 1–4.
10. Xiaodong, H.; Jun, Z.; Jun, W.; Dongping, D.; Bin, T. False target deception jamming for countering missile-borne SAR. In Proceedings of the IEEE 17th International Conference on Computational and Engineering, Chengdu, China, 19–21 December 2014; pp. 1974–1978.
11. Feng, Z.; Bo, Z.; Mingliang, T.; Xueru, B.; Bo, C.; Guangcai, S. A Large scene deceptive jamming method for space-borne SAR. *IEEE Trans. Geosci. Remote Sens.* **2013**, *51*, 4486–4495. [CrossRef]
12. Qingyang, S.; Ting, S.; Shicheng, Z.; Bin, T.; Wenxian, Y. A novel jamming signal generation method for deceptive SAR jammer. In Proceedings of the IEEE Radar Conference, Cincinnati, OH, USA, 19–23 May 2014; pp. 1174–1178.
13. Bo, Z.; Feng, Z.; Zheng, B. Deception jamming for squint SAR based on multiple receivers. *IEEE J. Sel. Top. Appl. Earth Obs. Remote Sens.* **2015**, *8*, 3988–3998. [CrossRef]
14. Bo, Z.; Lei, H.; Feng, Z.; Jihong, Z. Performance improvement of deception jamming against SAR based on minimum condition number. *IEEE J. Sel. Top. Appl. Earth Obs. Remote Sens.* **2017**, *10*, 1039–1055. [CrossRef]
15. Qingyang, S.; Ting, S.; Kaibor, Y.; Wenxian, Y. Efficient deceptive jamming method of static and moving targets against SAR. *IEEE Sens. J.* **2018**, *18*, 3601–3618. [CrossRef]
16. Yongcai, L.; Wei, W.; Xiaoyi, P.; Dahai, D.; Dejun, F. A Frequency-domain three-stage algorithm for active deception jamming against synthetic aperture radar. *IET Radar Sonar Navig.* **2014**, *8*, 639–646. [CrossRef]
17. Yongcai, L.; Wei, W.; Xiaoyi, P.; Qixiang, F.; Guoyu, W. Inverse omega-K algorithm for the electromagnetic deception of synthetic aperture radar. *IEEE J. Sel. Top. Appl. Earth Obs. Remote Sens.* **2016**, *9*, 3037–3049. [CrossRef]
18. Bo, Z.; Lei, H.; Jian, L.; Maliang, L.; Jinwei, W. Deceptive SAR jamming based on 1-bit sampling and time-varying thresholds. *IEEE J. Sel. Top. Appl. Earth Obs. Remote Sens.* **2018**, *11*, 939–950. [CrossRef]

19. Yongcai, L.; Wei, W.; Xiaoyi, P.; Letao, X.; Guoyu, W. Influence of estimate error of radar kinematic parameter on deception jamming against SAR. *IEEE Sens. J.* **2016**, *16*, 5904–5911. [CrossRef]
20. Franceschetti, G.; Guida, R.; Iodice, A.; Riccio, D.; Ruello, G. Efficient simulation of hybrid stripmap/spotlight SAR raw signals from extended scenes. *IEEE Trans. Geosci. Remote Sens.* **2004**, *42*, 2385–2396. [CrossRef]
21. Ahmed, S.K.; Laurent, F.F.; Eric, P. Efficient SAR raw data generation for anisotropic urban scenes based on inverse processing. *IEEE Geosci. Remote Sens. Lett.* **2009**, *6*, 757–761. [CrossRef]
22. Ian, G.C.; Frank, H.W. *Digital Processing of Synthetic Aperture Radar Data: Algorithms and Implementation*; Artech House: Boston, MA, USA, 2005; pp. 45–63.
23. Yongcai, L.; Wei, W.; Xiaoyi, P.; Dahai, D. Effective region of active decoy jamming to SAR based on time-delay doppler-shift method. *J. Radars* **2013**, *2*, 46–53.

Article

Real-Time Bottom Tracking Using Side Scan Sonar Data Through One-Dimensional Convolutional Neural Networks

Jun Yan [1,2], Junxia Meng [3,4,*] and Jianhu Zhao [5,6]

1 School of Resources and Environmental Engineering, Anhui University, Hefei 230601, China; jun.yan@ahu.edu.cn
2 Anhui Province Engineering Laboratory for Mine Ecological Remediation, Hefei 230601, China
3 College of Civil Engineering, Anhui Jianzhu University, Hefei 230601, China
4 Anhui Rural Revitalization Research Institute, Hefei 230601, China
5 School of Geodesy and Geomatics, Wuhan University, Wuhan 430079, China; jhzhao@sgg.whu.edu.cn
6 Institute of Marine Science and Technology, Wuhan University, Wuhan 430079, China
* Correspondence: junxia.meng@ahjzu.edu.cn; Tel.: +86-1996-520-1627

Received: 13 November 2019; Accepted: 18 December 2019; Published: 20 December 2019

Abstract: As one of the most commonly used acoustic systems in seabed surveys, the altitude of the side scan sonar from the seafloor is always difficult to determine, especially when raw signal levels and gain information are unavailable. The inaccurate sonar altitudes would limit the applications of sonar image geocoding, target detection, and sediment classification. The sonar altitude can be obtained by using bottom tracking methods, but traditional methods often require manual thresholds or complex post-processing procedures, which cannot ensure accurate and real-time bottom tracking. In this paper, a real-time bottom tracking method of side scan data is proposed based on a one-dimensional convolution neural network. First, according to the characteristics of side scan backscatter strength sequences, positive (bottom sequences) and negative (water column and seabed sequences) samples are extracted to establish the sample sets. Second, a one-dimensional convolution neural network is carefully designed and trained by using the sample set to recognize the bottom sequences. Third, a complete processing procedure of the real-time bottom tracking method is established by traversing each side scan ping data and recognizing the bottom sequences. The auxiliary methods for improving real-time performance and sample data augmentation are also explained in detail. The proposed method is implemented on the measured side scan data from the marine area in Meizhou Bay. The trained network model achieves a 100% recognition of the initial sample set as well as 100% bottom tracking accuracy of the training survey line. The average bottom tracking accuracy of the testing survey lines excluding missed pings reaches 99.2%. By comparison with multi-beam bathymetric data and the statistical analysis of real-time performance, the experimental results prove the validity and accuracy of the proposed real-time bottom tracking method.

Keywords: side scan sonar; bottom tracking; one-dimensional convolutional neural network; signal recognition; real-time processing

---

1. Introduction

A side scan sonar can rapidly obtain large-area seabed images, has been widely used in seabed investigation, and plays an important role in seabed target detection [1–4] and investigation as well as research of the seabed ecological environment [5,6] due to its low cost and simple installation. A side scan sonar is usually dragged by a towing line to get close to the bottom of the sea to obtain high-resolution seabed images. Although the depth of the side scan sonar can be obtained by using depth sensors, the height of the sonar from the seabed cannot be accurately obtained [7]. Inaccurate sonar heights will lead to inaccurate geocoding sonar images [8], confuse the water column information with the seabed information, and cause serious problems in applications of target recognition and segmentation [9–11], image interpretation [12,13], and seabed sediment classification [14–16]. The bottom tracking of side scan data can accurately obtain the sonar height from the seabed by finding the first echo that reaches the seabed. Meanwhile, real-time bottom tracking can quickly detect changes in sonar height and seabed terrain, and enhance the safety of sonar equipment and ship navigation.

Side scan sonars can be installed on the vessel or towed close to the seabed from the survey ship. These sonars acquire high-resolution images by emitting sound pulses and recording the backscatter strengths from the water column and seabed [17]. The depth of the side scan sonar can be determined by the depth sensor, whereas the sonar height cannot be easily determined [18]. The sound wave transmits through the water column, and then arrives at the seabed. Given that the backscatter strengths from the water column are much lower than those from the seabed, the backscatter strengths recorded near the bottom positions differ from those recorded in other positions, which makes bottom position tracking possible [7].

With best practices, i.e., the gains are logged in the recorded files (e.g., *.jsf file for EdgeTech sonars) and the gains are kept track in the processing chain, all useable information, including the raw signal levels and gains, are available. Then the bottom can usually be easily determined with a very high signal-to-noise ratio (SNR), which makes the signal level of the bottom tens of dB larger than in the water column [17]. However, when the gains are lost, detecting the bottom over the seafloor becomes much harder. Moreover, as the development of the oceanographic survey, more researchers are stepping into the relative fields of sonar imaging. In many cases, if the researchers are not there to record enough useable data during the survey, valuable information will be lost and these researchers can only study the recorded side scan data with very little information. In addition, old side scanside scan data often need to be reprocessed to find new results or to be compared with the current study. Given that the recorded side scan data are used for seabed imaging, the depths and gains are usually not recorded in the data (e.g., eXtended Triton Format *.xtf files). In these situations, when the original sound signal levels are unknown and the echoes have been compensated with unknown gains (e.g., time varied gains), the recorded side scan data only include the converted backscatter strength data in special fixed ranges. Thereby, the bottom tracking methods are necessary. Moreover, certain effect factors, including sonar self-noises, ambient noises, and other object disturbances, also introduce challenges in bottom tracking methods [19].

To process these types of side scan data, most bottom tracking works are completed by using the threshold method assisted by expensive commercial software, such as Chesapeake SonarWiz and EdgeTech Discover [19]. Given that the threshold is usually determined on the basis of the operator's experience, this method also requires extensive manual work. Moreover, given the complexity of the seabed environment, the threshold changes during the processing. Using inappropriate threshold parameters can lead to incorrect bottom tracking results. Accordingly, researchers are looking for automatic algorithms to achieve enhanced efficiency. Some researchers have used the filtering method to remove noise, studied the variation features of the backscatter strengths of the side scan sonar, and used these feature differences for bottom tracking of the side scan data [20]. Given the continuity of sonar heights and the symmetry of the port and starboard side scan data, other researchers have built general models and used dynamic data filtering algorithms, such as Kalman filtering and time series, to repair abnormal data and improve accuracy [19]. Given the existence of many types of effect factors, the variations in backscatter strengths typically show a feature of regularity and local randomness. Traditional methods require manual threshold parameters or time-consuming post-processes, which, thereby, cannot guarantee accurate and real-time bottom tracking results.

Deep learning algorithms have been widely applied in image recognition and classification [21–24]. The one-dimensional convolutional neural network (1D-CNN) is a deep learning algorithm for processing one-dimensional sequence data, and has been proven to be an effective recognition and classification method for one-dimensional sequence data [25,26]. After introducing the deep learning idea, algorithms can simulate the human brain, learn the variation feature of the local backscatter strength sequence, and fulfill the bottom tracking of side scan sonars. Therefore, on the basis of the recognition of side scan bottom data sequences through 1D-CNN, this paper presents a new real-time bottom tracking method for side scan sonar data. First, the operation theory of the side scan sonar and the characteristics of the side scan backscatter strength data are briefly introduced. Second, according to the variation features of backscatter strengths, the proposed 1D-CNN model is designed and then trained by using the established sample sets for recognizing bottom sequences. Third, the bottom tracking of side scan data is implemented by traversing each ping to use the trained model to detect the bottom data sequences. Lastly, the proposed method is validated in the experiment by using the measured side scan data.

2. Theory and Method

This chapter introduces the proposed real-time bottom tracking method using the 1D-CNN model. The basic theories of side scan operation and data characteristics, the recognition of bottom data sequences, and bottom tracking using the trained model will be explained successively.

2.1. Side Scan Sonar Operation and Data Characteristics

The operation of a side scan sonar is shown in Figure 1. The side scan sonar, which is usually a towfish, is towed by the survey vessel using a tow cable to get close to the seabed. The side scan sonar transducer projects a single wide sound beam (e.g., 50°, as shown in Figure 1) at the port and starboard sides. After the sound is projected from the side scan transducer, the transducer receives and records the backscatter strengths in the time sequence at the port and starboard sides, respectively, and these strengths are used to construct side scan sonar images. During the sound propagation, the backscatter strengths received from the water column are usually much lower than those received from the seabed, as shown in Figure 1. The special variations in signal levels (or backscatter strengths) when the sound arrives at the seabed serve as the basis for the bottom tracking of side scan sonar data. However, given the effects of sonar self-noise, suspended objects in the water column, and other instrumental and environmental factors, the many uncertainties in the sonar data introduce difficulties in bottom tracking.

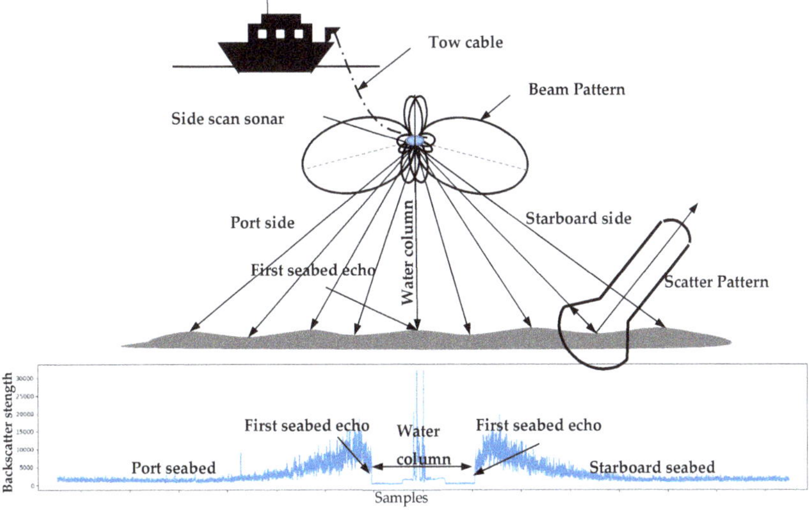

Figure 1. Operation of the side scan sonar and one-ping backscatter strength sequence.

As shown in Figure 1, the acoustic backscatter strengths of a ping have the following characteristics.

- The backscatter strengths in the water column area are much lower than those in the seabed area. The backscatter strengths in the water column area near the zero position are usually heavily affected by the self-noise of the side scan sonar.
- When the sound hits the seabed, the transducer receives higher echo signal intensities compared with those in the water column. Affected by the beam patterns and transmission losses [27], the strength sequences temporarily increase after the bottom position and, subsequently, decrease along with an increasing transmission length.
- The backscatter strength sequences of the port and starboard sides are almost symmetric at the zero position. However, the bottom tracking position may slightly differ due to the attitude of the side scan sonar and the seabed terrain slopes [19].

2.2. Recognition of Sonar Data Sequences Using 1D-CNN

When the sound hits the seabed, the local backscatter strength sequence shows a special variation feature, as illustrated in Figure 1. The proposed method uses the suitable 1D CNN model to recognize the bottom samples for bottom tracking. In this section, the positive and negative samples were first extracted to establish the sample sets. The sample sets were normalized and a 1D-CNN model was carefully designed and trained by using the sample sets.

2.2.1. Sampling

When using 1D-CNN to learn the variation features of the side scan data sequence, the one-ping backscatter data sequence should be divided into regional sub-sequences as samples, as illustrated in Figure 2. The sub-sequence/sample size should be properly selected to accurately reflect the variation characteristics, as discussed in Section 4.1. An improper sample size would cause the network to learn the wrong information and misjudge the results.

- A very large sample size cannot represent the special variation characteristics of backscatter strengths and
- A very small sample size can be easily affected by local noise.

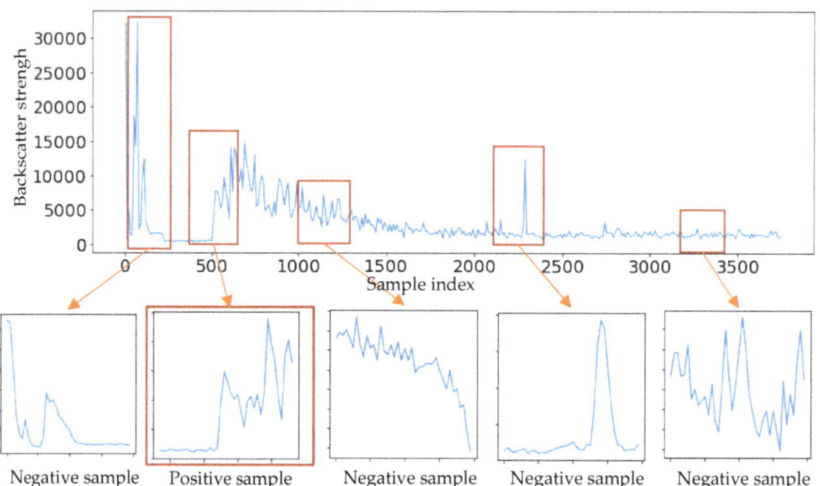

Figure 2. Data sequence sample of a ping data. The positive (bottom) and negative (noise, water column, and seabed) samples.

To establish the sample sets, the positive and negative samples need to be selected from raw sonar backscatter strength data sequences. The positive samples can be detected using the traditional method with manual intervention, whereas the negative samples should contain the samples in the water column area, those containing noise, and those in the seabed area, as shown in Figure 2.

2.2.2. Normalization of Sonar Data Sequences

As shown in Figure 2, the samples are in various strength ranges and need to be normalized into the same range for the network training. These samples can be normalized by using the z-score to ensure that they are in the same range [28]. Given that the side scan data are usually recorded in a fixed range (e.g., 0 to 2^{16}-1), the samples can be simply normalized by the equation below.

$$dB = \frac{dB}{Max_{dB}}\left(e.g.\frac{dB}{32767}\right). \tag{1}$$

After the normalization, the sample range should be normalized to (0~1), as shown in Figure 3.

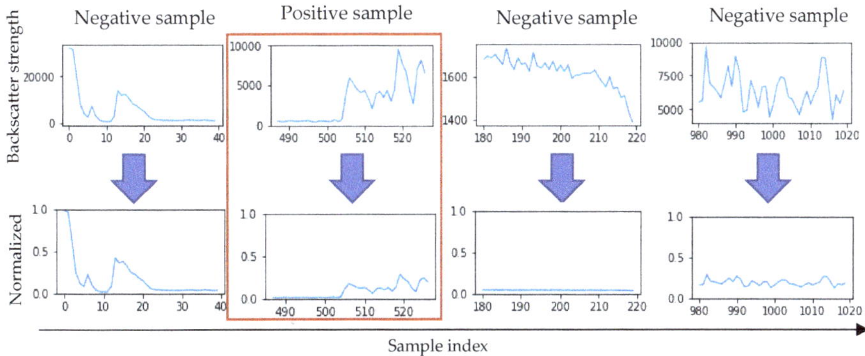

Figure 3. Normalization of the data sequence samples.

2.2.3. Network

The bottom tracking of the side scan sonar data aims to recognize special bottom backscatter strength sequences, and can be fulfilled via the 1D-CNN recognition of the normalized data sequences.

1D-CNN is the one-dimensional version of common CNNs, which also contain input layers, convolution layers, pooling layers, and the output layers. Given the characteristics of our problem, the input layer of the 1D-CNN contains backscatter strength sequences, whereas the output layer contains the positive (1) and negative (0) results, as shown in Figure 4.

The input layer contains the one-dimensional normalized backscatter strength samples, and the median layers are combinations of convolution and pooling layers. The one-dimensional convolution operation s of the data sequences in discrete form is shown below.

$$s(t) = (d * w)(t) = \sum_{a=-\infty}^{\infty} d(a)w(t-a) \tag{2}$$

where d is the input data sequence, w is the activation function, and t is the tth value of d.

The following rectified linear unit (ReLU) h is selected as the activation function for the convolution layers.

$$h_{w,b}(X) = \max(X{\cdot}w + b, 0) \tag{3}$$

where w and b are the trainable parameters, and X is the input data.

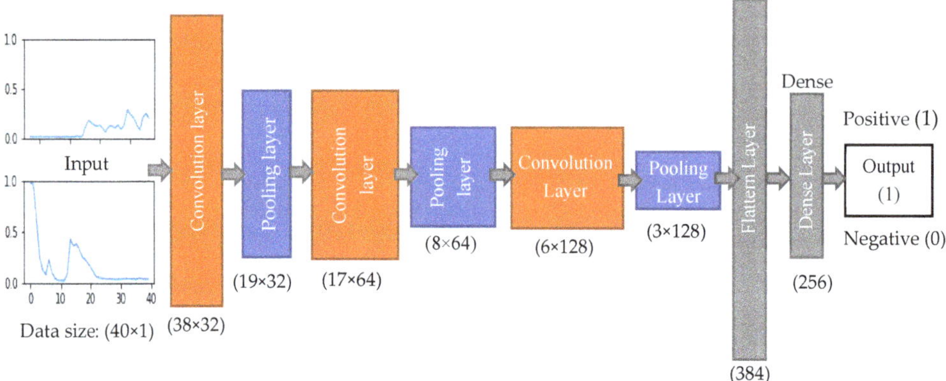

Figure 4. The structure of the one-dimensional convolution neural network (1D-CNN) with the positive and negative input samples and the corresponding output results.

After the convolution and pooling layers, the flattened layer reshapes the tensors into vectors, whereas the fully-connected layer usually uses the ReLU to connect the output layer.

The last layer is the output layer, with the following activation sigmoid function σ:

$$\sigma(x) = \frac{1}{1 + e^{-x}} \tag{4}$$

where x is the input data.

After each training loop, the loss function is used to calculate the difference between the predicted results and the ground truth. Given that the bottom tracking problem is a binary classification problem, the cross-entropy loss function is selected as the following loss function H.

$$H_{y_i}(y) = -\sum_i y_i' \log(y_i) \tag{5}$$

where y_i is the predicted result and $y_i{}'$ is the ground truth.

The root mean square propagation optimizer is chosen to update the parameters. After several loops, if the network learns the variation features of the samples properly, then the loss function would reach a stable low value, whereas the training and validation accuracies would reach stable high values. The well-trained 1D-CNN serves as the basis of the real-time bottom tracking method.

2.3. Bottom Tracking Using the Trained 1D-CNN

In this section, the trained network is used for the bottom tracking of the side scan data. The complete procedures of real-time bottom tracking are explained in detail, along with the auxiliary methods that can improve real-time performance and recognition accuracy.

2.3.1. Real-Time Bottom Tracking

By finding the first echo that reaches the seabed, the purpose of bottom tracking aims to accurately obtain the sonar height from the seabed, then geocode the sonar image, and fulfill other applications. Given that the trained network model can effectively understand the characteristics of the input samples, bottom tracking can be accomplished via the 1D-CNN recognition of local backscatter strength sequences. By traversing the ping data sequence in the propagation direction, the trained 1D-CNN recognizes the local backscatter strength sequence in each search window, and the predicted score of each local sequence can be obtained, as shown in Figure 5. The window size should be the same as the sample size.

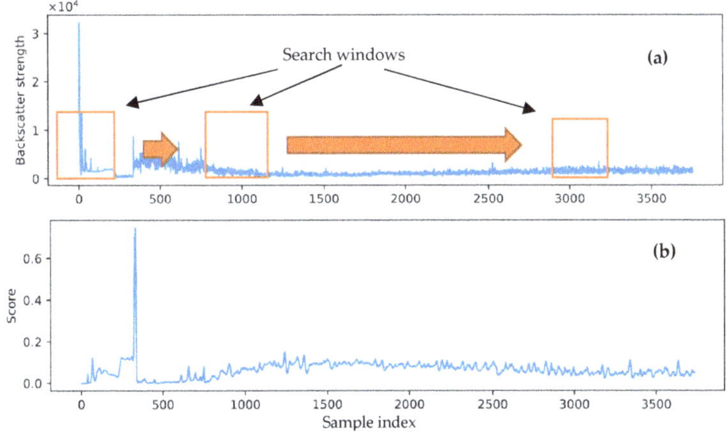

Figure 5. Prediction result for a one-side ping data obtained by using the trained network model with approximately 300 samples. (**a**) shows the one-side ping backscatter strength sequences, and (**b**) shows the corresponding prediction scores of each window.

The scores of the backscatter strength sequence in each search window are treated as the prediction results of the trained 1D-CNN. When the sound beam arrives at the seabed, a high score can be obtained by using the 1D-CNN prediction. Meanwhile, the scores of the data sequences in the other positions are far lower. Therefore, the maximum score position can be used to determine the bottom position of each ping.

Given the symmetry between the port and starboard data of each ping, a bottom tracking of the port and starboard data is carried out. The predicted port and starboard scores are then used to check the results and to achieve improved robustness to noise. The bottom tracking result of each ping is obtained on the basis of the port and starboard results. The complete bottom tracking procedures for each recorded ping is shown in Figure 6.

The bottom tracking accuracy of the survey line as obtained by 1D-CNN is calculated by using Equation (6) below.

$$acc = \frac{N_1}{N_0} \tag{6}$$

where N_1 is the number of successful-bottom-tracked pings and N_0 is the total number of pings.

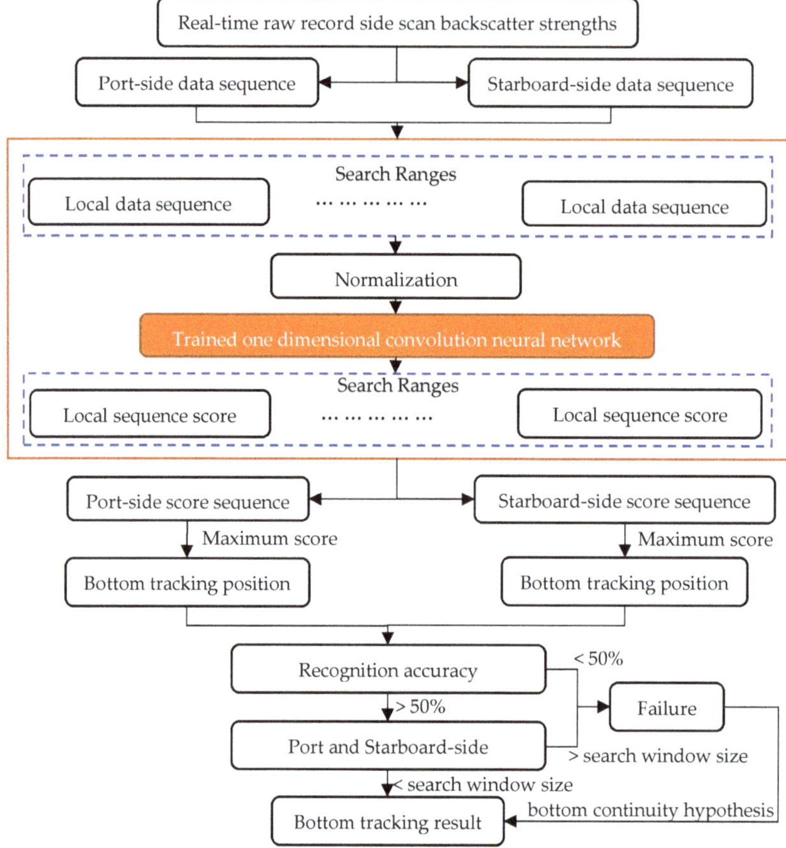

Figure 6. Flowchart of the real-time bottom tracking of side scan data.

2.3.2. Improving Speed: Narrow Search Range

The processing speed of the bottom tracking algorithm for each ping plays a key role in guaranteeing the real-time performance of the proposed method. Given the stringent hardware computing power requirements of the deep learning algorithm, the data sequences should be recognized in a limited search range instead of the whole ping. In the current common hardware platform (AMD R5-2600X CPU and GTX-2070 GPU), the relationship between different search scopes and corresponding times was analyzed and shown in Figure 7. The computing speed and search range are linearly dependent on the same hardware platform, which indicates that narrowing the search range can effectively improve the computing speed.

According to the continuity of the seabed terrain variation, the bottom tracking position (sonar height) of the former ping can be used as the initial search position, and the search range can be determined by the seabed terrain variation or bottom tracking position rate. The relationship between the bottom tracking position variation rate and search range is shown in Table 1. By combining the initial search position provided by the previous ping and the bottom tracking position variation rate between the previous pings, the search range of the proposed method can be adaptively controlled, to guarantee an excellent real-time performance.

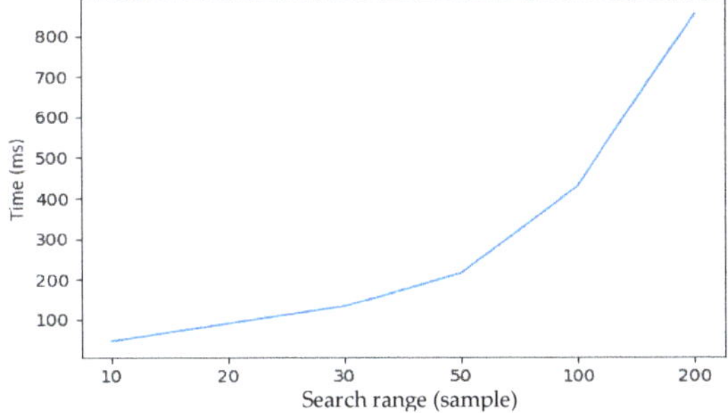

Figure 7. The relationship between consuming time and search range. The experiment was tested on the platform with AMD R5-2600X and GTX-2070.

Table 1. Auto-Adapted Search Ranges Depending on the Bottom Position Variation.

Bottom Position Variation Rate *r* (Number of Samples/Ping)	$r \leq 5$	$5 < r < 10$	$10 < r < 20$	$20 < r < 40$
Bottom position variation rate *r* (meter/ping)	$r \leq 0.2$	$0.2 < r < 0.4$	$0.4 < r < 0.8$	$0.8 < r < 1.6$
Search range (number of samples)	20	40	80	160
Search range (meter)	0.8	1.6	3.2	6.4

2.3.3. Improving Accuracy: Sample Data Augmentation

The accuracy and abundance of samples are key in ensuring an accurate bottom sequence recognition. However, traditional sampling methods are time consuming and require manual intervention to ensure enough accuracy. During its application, the trained network could process other types of side scan sonar data in other seabed environments. The network needs to learn the features of the new data by continuously increasing the number of samples to improve the recognition accuracy. In this paper, a continuous increase of the samples was realized by using the learning ability of the network and few manual assistances, as shown in Figure 8.

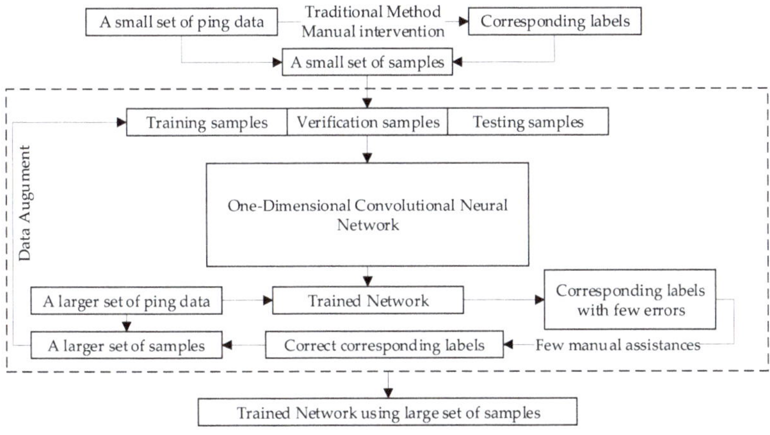

Figure 8. Flowchart of sample set establishment and augmentation.

As the number of samples increases, the recognition accuracy and robustness of the network can be further improved, and an accurate bottom tracking can be guaranteed.

3. Experiment and Results

To verify the validity and real-time performance of the proposed method, the side scan and multibeam data measured in Meizhou Bay, Fujian, China in 2012 were selected for the experiment, as shown in Figure 9. The coverage of the survey area was approximately 13.5 × 2.5 km^2, and the water depth ranged from 10 m to 40 m. The seabed sediments were mainly gravel and silt. According to the designed route, the measurements of the multibeam and side scan sonars were carried out successively in the same water area. In the multibeam measurement, Kongsberg EM 3002 with an operating frequency of 300 kHz, across-track beam aperture of 130° and maximum beam number of 131 was used. Meanwhile, in the side scan measurement, EdgeTech 4100P side scan sonar was towed at approximately 2 m underwater with an operating frequency of 500 kHz, maximum recorded slant range of 150 m (corresponding to 3751 sample numbers), vertical angular aperture of 50°, and horizontal angular aperture of 0.5°. The interval time between pings was 0.15 s on average. The side scan data were recorded in eXtended Triton Format (*.xtf) files, which only contained the backscatter strengths. The raw signal levels were unavailable, and the echoes were compensated with unknown gains.

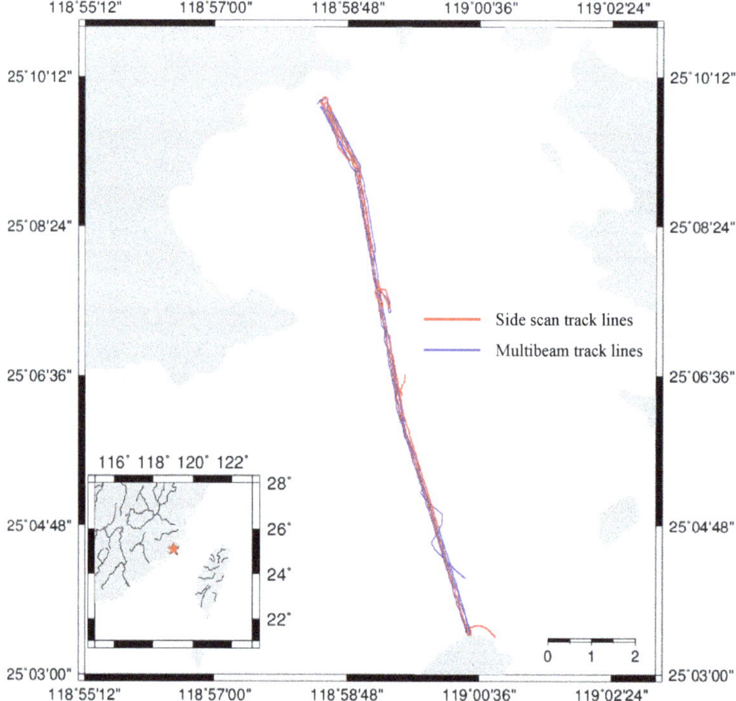

Figure 9. Survey track lines of both the side scan and multibeam sonars in Meizhou Bay, Fujian, China.

The experiment was divided into the following stages.

- The experiment began by sampling, training, and bottom tracking of a small side scan survey line. Small sets of samples were initially extracted from the side scan data. Then the proposed 1D-CNN network model was trained to learn the variation features of the sample set. The training network was eventually used for bottom tracking, and the predicted value was compared with the artificially selected truth value.

- The trained network was validated by using additional side scan data of other survey lines to obtain the tracking results of other lines.
    - The bottom tracking result obtained by the proposed method was compared with those obtained by the traditional method.
    - The water depth predicted by using the bottom tracking results were compared with the ground truth (depths measured by the multibeam sonar).
    - The real-time performance of the proposed method was analyzed and validated via statistical analysis.
- Additional experiments on the side scan data with heavy noise and rich texture were conducted.

3.1. Network Model Establishment and Bottom Tracking

The experiment started with a small side scan survey line of 121 pings. The raw recorded (*.xtf) data was decoded, and the corresponding waterfall sonar image was constructed, as shown in Figure 10a. The bottom tracking results were processed by manual recognition, as shown in Figure 10b.

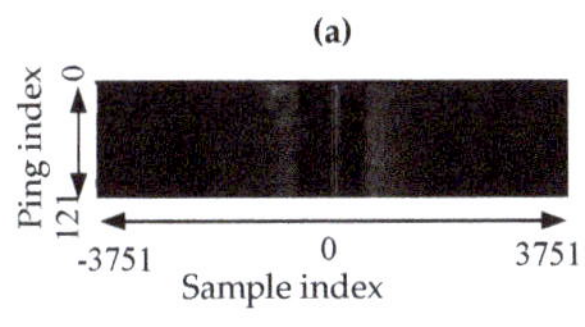

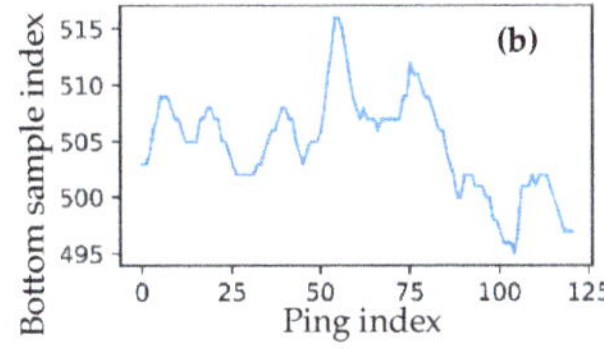

Figure 10. (**a**) Side scan waterfall sonar image and (**b**) manual bottom tracking result of the survey line.

The positive sample sequences were selected as the bottom backscatter strength sequences from the side scan data, according to the corresponding bottom tracking position. Meanwhile, the negative sample sequences were uniformly selected in the water column and seabed area. The positive and negative samples constituted the sample set for the model training. Given that the survey line had 121 pings, the sample set contained 242 positive and 2662 negative samples, respectively.

The sample set was normalized, according to Equation (1), and was imported into the network as the input layer. The corresponding label (1: positive, 0: negative) was imported as the output layer. During the training, the samples were randomly divided into training and validation sets in a 70–30% proportion. The 1D-CNN (Figure 4) was trained to learn the variation features of the data samples. The training and validation accuracies improved as the training epoch increased, as shown in Figure 11.

As shown in Figure 11, the training accuracy gradually improved as the training epoch increased, and eventually reached a stable value of 100% after approximately 40 training epochs. The validation accuracy fluctuated in 10 training epochs and reached a stable value after 20 epochs. The training and validation losses gradually decreased along with an increasing training epoch, and eventually decreased to 0. For the small sample set of the selected survey line, the network model proposed in this paper can effectively learn the features of the positive and negative samples, and accurately recognize them after training, which is the basis for the real-time bottom tracking of the survey line.

Based on the trained network model, each ping of the survey line was bottom tracked following the procedure illustrated in Figure 6. The bottom tracking results of the port and starboard side scan data were then processed (Figure 5) and compared with each other (Figure 12a). The corresponding bottom tracking result can be displayed in the side scan waterfall image (Figure 12b).

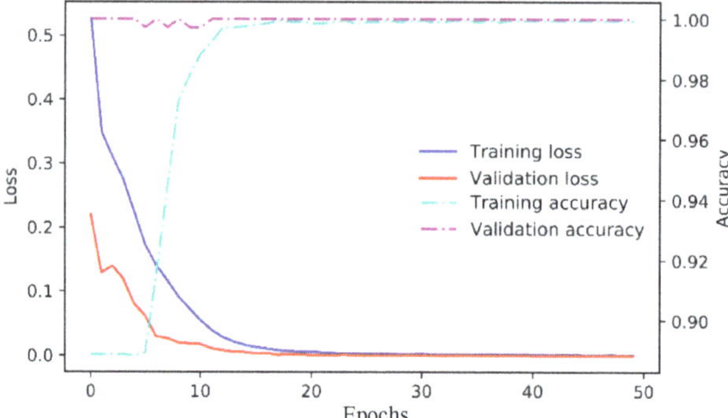

Figure 11. Training and validation accuracies and losses of the network in 50 epochs.

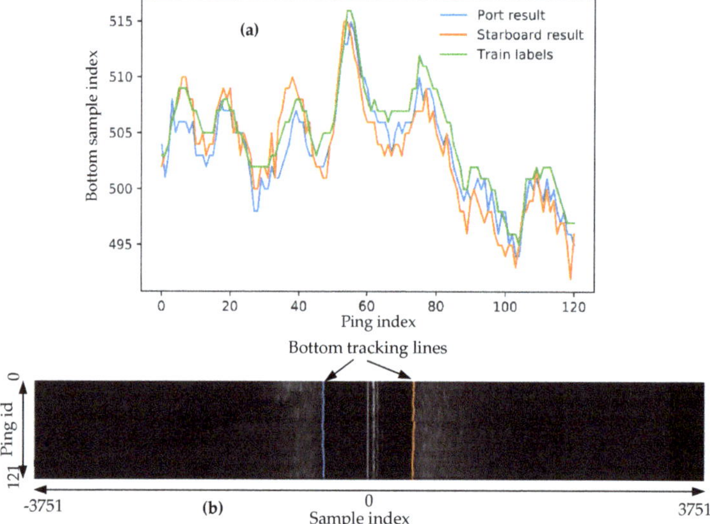

Figure 12. Bottom tracking results obtained by using the trained network. (**a**) This area shows the bottom positions (sample indexes) of the port and starboard sides, and (**b**) this area shows the waterfall image with the bottom tracking lines.

The bottom tracking results of the port and starboard data were highly consistent with each other, and all bottom position differences were less than four samples (0.16 m) because the seabed topography of the survey water area was relatively flat. Moreover, the tracking results in the waterfall diagram were highly intuitive to show the edges of the port and starboard seabed area, which agreed with the visual results. The comparison between the bottom tracking results and manual ones showed that the bottom tracking accuracy can reach 100% on the training survey line. These results prove the validity of the proposed bottom tracking method.

3.2. Method Validation and Comparison

To validate the generalization of the trained model and the effectiveness of the proposed method for the side scan data of other survey lines in the test area, the trained model was used to recognize unknown data for the bottom tracking of other survey lines.

3.2.1. Validation on a Larger Survey Line

The side scan data of a long survey line with 13,504 pings were used for the validation. The survey line spanned the seabed of two different sediments, which appeared as the clearly lighter and darker areas in the side scan image, as shown in Figure 13b. At the joint area of the two sediments, the seabed topography rapidly changed, as shown in Figure 13c. Based on the 1D-CNN model that was trained by using the sample set of a small survey line, the bottom tracking of the validation survey line was processed, and the results are shown in Figure 13. During the processing, the search ranges of each ping were self-adapted to improve the speed, according to Table 1.

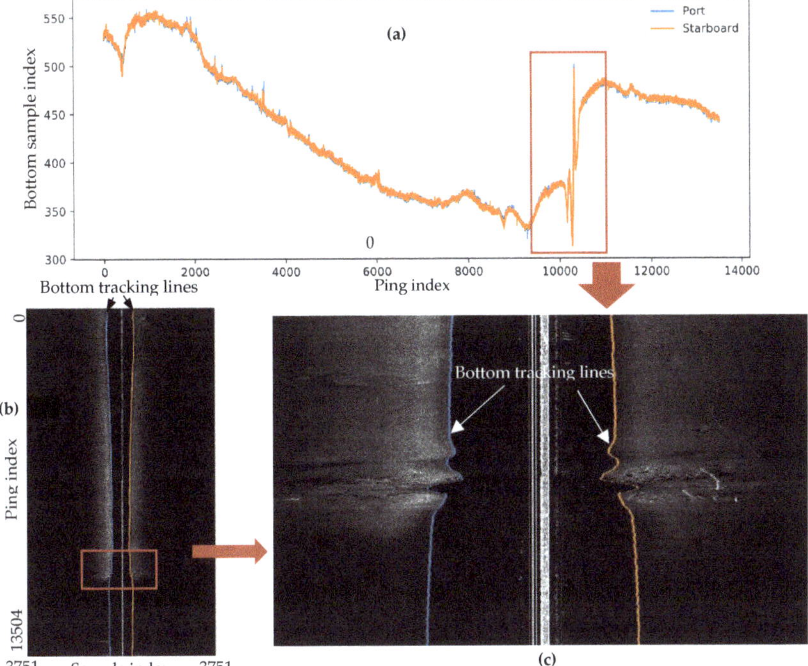

Figure 13. Bottom tracking of a larger survey line. (**a**) This area shows the port and starboard bottom tracking results, (**b**) this area shows the bottom tracking results represented in the side scan waterfall image, and (**c**) this part shows the seabed area where the terrain changes rapidly.

As shown in Figure 13a, the port and starboard results were consistent with each other, and the bottom tracking lines coincided with the edges of the port and starboard seabed in the waterfall image shown in Figure 13b. After removing the missing ping data of this survey line, the accuracy of the bottom tracking results reached 99.5%. In the area where the seabed terrain changed rapidly, the proposed bottom tracking method with auto-adapt search ranges still achieved good tracking results, which proved the validity of the proposed method in both flat and rugged seabed environments.

3.2.2. Comparison with Other Bottom Tracking Methods

To compare the proposed method with traditional methods, the survey line was processed by using the last peak method [19]. For real-time processing, the last peak method was used without post-processing, including Kalman filtering. The tracking results obtained by both methods are shown in Figure 14.

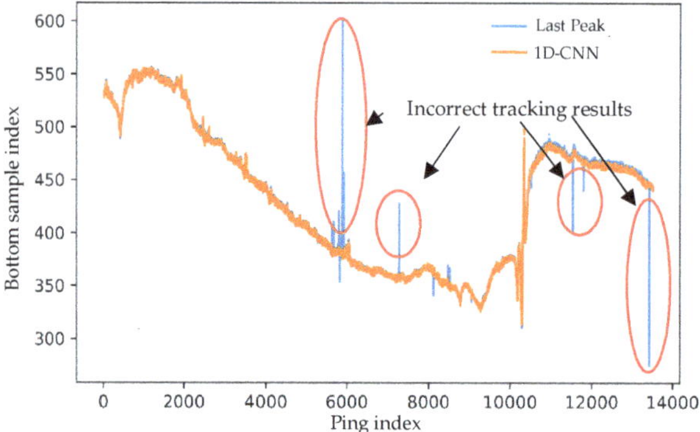

Figure 14. Bottom tracking results obtained by using the last peak method and 1D-CNN.

The bottom tracking results obtained by using these two methods were consistent in most positions. However, the results obtained by the traditional method based on the numeric features were sensitive to noise, such as water column noise and seabed objects., so the results could possibly be inaccurate without post-processing. As for the proposed method, by training the sample sets, the network could properly learn the variation feature of the backscatter strength sequences, and show better robustness to noise. As more samples were learned, the 1D-CNN could more accurately recognize the side scan data. The comparison proved the validity and performance of the proposed method.

3.2.3. Comparison Between the Bottom Tracking Depths and Ground Truth (Manual Annotations)

The manual annotations of bottom positions were used as the ground truth for the bottom tracking results of the side scan sonar data. Additionally, the bathymetric data measured by the multibeam sonar can be regarded as good references for the bottom tracking results. The depth of the side scan sonar sensor can be obtained by using its depth sensor, and the sonar height can be calculated by using the bottom tracking results. Therefore, the depth D of the corresponding seabed can be calculated using the equation below.

$$D = \frac{n \times t \times v}{2} + d \tag{7}$$

where n is the nth sample detected as bottom, t is the sampling interval time, v is the sound velocity, and d is the side scan sonar depth.

The digital elevation model was constructed by using the multibeam bathymetric data in the selected survey marine area (Figure 13c), as shown in Figure 15a. The track line of the multibeam data was extracted, and the corresponding water depths are shown in Figure 15d. The seabed depths of the side scan survey line that corresponded to the multibeam survey line were calculated by using the manual annotation and the predicted data, as shown in Figure 15b,c.

In the same water area, the seabed depths measured by the multibeam sonar (Figure 15d), those calculated by using manual annotations (Figure 15b), and the bottom tracking results (Figure 15c) were consistent with each other. The significant terrain fluctuations in the middle of the region coincided with the seabed variation shown in Figure 13c. Given that the multibeam and side scan data were measured at different times and that the multibeam data were not post-processed, the depths of the multibeam data had some errors and showed slight deviations from the depths calculated by using the side scan data. The terrain variation trends were consistent with each other, which proves the accuracy of the bottom tracking data.

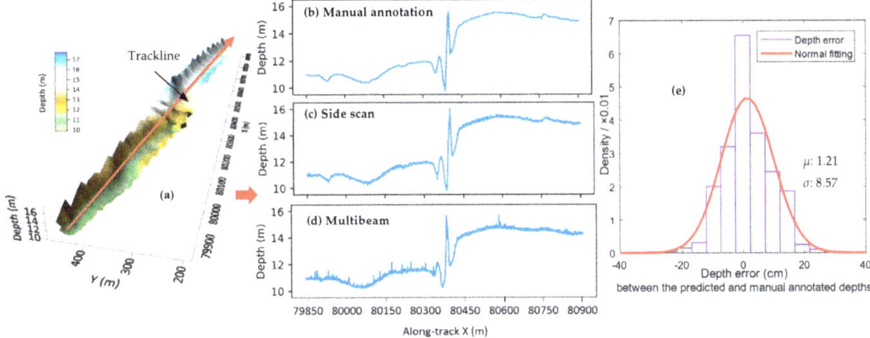

Figure 15. Depths comparison between the manual annotations, bottom tracking results, and multibeam bathymetric data. (**a**) This area shows the local seabed terrain, (**b–d**) show the depth sequences tracked by the manual annotations, predicted by the side scan data, and measured by using the multibeam sonar, respectively, and (**e**) shows the histogram and normal fitting (with the mean μ as 1.21 cm and standard deviation σ as 8.57 cm) of the depth errors between the predicted (**c**) and manual annotated (**b**) depths in centimeters.

The depth errors between the predicted and manual annotated depths were fitted using a normal curve with the mean μ equal to 1.21 cm and standard deviation σ equal to 8.57 cm, as shown in Figure 15e. Given that the errors of manual annotations were within ±3 samples (corresponding to ±12.0 cm), the depth errors were less than two times the error (i.e., 24.0 cm) can be acceptable. By statistical analysis, the depth errors (Figure 15e) within ±24.0 cm are in a 99.44% proportion. Thereby, the accuracy of the bottom tracking results compared with manual annotations is 99.44%.

3.2.4. Real-Time Experiment

To verify the real-time performance of the proposed method, the spend times of each ping were recorded during the bottom tracking experiment. The bottom tracking results and time sequences are shown in Figure 16a,b, respectively, whereas the spend times were statistically analyzed to evaluate the real-time performance, as shown in Figure 16c.

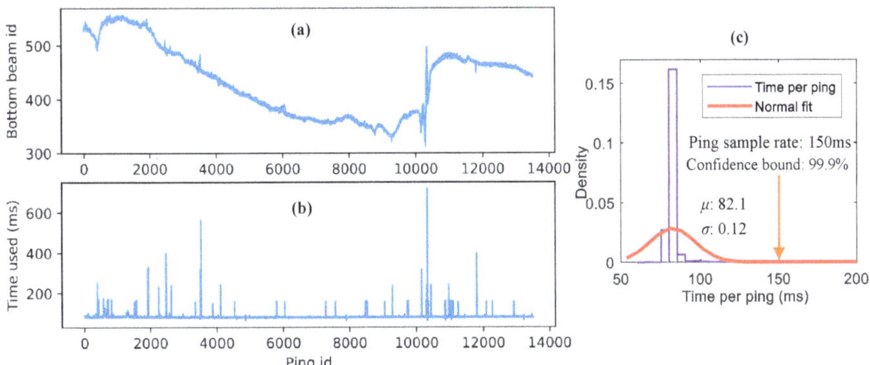

Figure 16. Real-time experimental results obtained by using AMD R5-2600X and GTX-2070. The necessary memory to run the algorithm should not be less that 2GB and the graphic memory should not be less than 8GB. (**a**) This area shows the bottom tracking results of the line, (**b**) this area shows the corresponding spend times of each ping, and (**c**) this area shows the normal fit of times and its 99.9% confidence bound at 150 ms.

Given the auto-adapt search ranges used in the bottom tracking experiment, the spend times of each ping changed along with the variation rate of the seabed terrain. The spend times of each ping were fitted by using the normal distribution curve with a mean μ of 82.1 cm and a variance σ of 0.12 cm. According to the statistical analysis results, the confidence bound of the side scan sampling interval time of each ping (150 ms) was 99.9%, which suggests a 99.9% possibility for the calculation time of each ping to be shorter than the sampling interval time. Moreover, it is guaranteed that, given the number of predicted sample sequences being less than 60, the calculation speed is always less than 150 ms, where 150 ms is the interval time between two pings. The statistical results proved the real-time feasibility of the proposed method.

Moreover, if the prior depth range is known, then the search range of each ping would be smaller. Moreover, with better hardware and multi-thread computing, the calculation speed would be improved, as discussed in Section 4.4.

3.3. Bottom Tracking of Side Scan Data with Noise and Rich Texture

To obtain the bottom tracking results of the other survey lines in the experimental area, data augmentation was applied on the sample sets as more survey lines were processed, as shown in Figure 8. The characteristic side scan data with large noise, rich seabed texture, and artificial targets were carefully processed and analyzed, as shown in Figure 17. The recorded side scan data contained missing pings, which had no backscatter strengths or very low backscatter strengths, as shown in yellow rectangles in Figure 17.

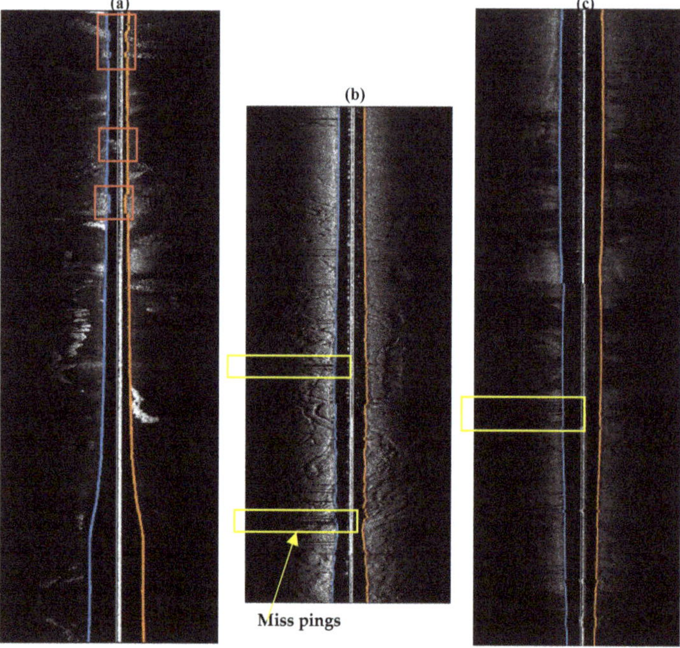

Figure 17. Bottom tracking of the characteristic side scan data with noise (**a**) and rich seabed texture (**b**) and artificial targets (**c**). The gaps shown in yellow rectangles between the pings are the missing data.

Figure 17a shows that the noises in the water column are relatively large. In the red rectangular area, the noises in the water column made the seabed and water column data indistinguishable, or made the edge variation of the seabed abnormal. The bottom tracking accuracy of this survey line as obtained by 1D-CNN was 97.3% with a 2.0% miss-ping rate. The accuracy excluding the missing pings was 99.3%.

Figure 17b shows that the seabed has rich textures and that some noise can be observed in the water column. The backscatter strength variation of the complex seabed texture would result in clear light and shade areas, which would interfere with bottom tracking. The bottom tracking accuracy of this survey line as achieved by 1D-CNN was 93.1% with a 6.1% miss-ping rate. The accuracy excluding the missing pings was 99.1%.

Figure 17c shows that the seabed contains artificial targets, such as submarine pipelines. These artificial targets can also cause light and shade areas in the side scan image, which would significantly affect bottom tracking. The bottom tracking accuracy of this survey line as achieved by 1D-CNN was 94.5% with a 4.9% miss-ping rate. The accuracy excluding the missing pings was 99.4%, as shown in Table 2.

Table 2. Bottom Tracking Accuracies of the Survey Lines Shown in Figure 17.

Survey Line (in Figure 17)	Tracking Accuracy Using 1D-CNN	Miss-Ping Rate	Tracking Accuracy Excluding the Missing Pings
a	97.3%	2.0%	99.3%
b	93.1%	6.1%	99.1%
c	94.5%	4.9%	99.4%

As shown in Table 2, by means of sample data augmentation, mutual inspection of the port and starboard results, and auto-adapt search ranges, the proposed method can guarantee the bottom tracking accuracy of the side scan data with large amounts of noise, a rich seabed texture, and artificial targets as well as simultaneously realize real-time calculation performance. The average bottom tracking accuracy of the overall testing survey lines as achieved by 1D-CNN was 94.7% with a 4.5% miss-ping rate. The tracking accuracy excluding the missing pings was 99.2%. The experiments proved that the proposed method has high robustness to noise, and can yield accurate results in complex seabed conditions.

4. Discussion

4.1. Determination of the Sample Size

Sample size is an important factor in accurately recognizing the bottom data samples and further realizing bottom tracking of the side scan data. If the sample size is too large, then the samples cannot represent the special variation characteristics of the bottom data sequences. However, if the sample size is too small, then the samples can be easily affected by local noise. For a better comparison, bottom tracking experiments were conducted with sample sizes of 10, 20, 40 (chosen in this paper), and 100, as shown in Table 3.

Table 3. Comparison of results obtained under different sample sizes after a 50-epoch training with 242 positive samples and 2662 negative samples.

Sample Size (Number of Recorded Samples)	Sample Size (m)	Training Accuracy (%)	Validation Accuracy (%)	Bottom Tracking Accuracy (%)
10	0.4	98.1	98.6	0.0
20	0.8	99.7	100	98.3
40	1.6	99.9	100	100
100	4.0	100	100	46.3

As shown in Table 3, when the sample size was as small as 10, although the training and validation accuracies were high enough, the bottom tracking accuracy was 0%, which suggests that the variation characteristics of the samples can be easily affected by noise. When the sample size was as large as 20, the training and validation accuracies were improved, and the bottom tracking accuracy reached as high as 98.3%. When the sample size was 40 (as used in this paper), the training and validation accuracies were further improved, and the bottom tracking accuracy increased to 100%, which suggests that the samples can accurately reflect the variation characteristics of backscatter strengths. However, when the sample size was 100, although the training and validation accuracies were 100%, the bottom tracking accuracy was only 46.3%, which suggests that the samples cannot properly reflect the variation characteristics of bottom backscatter strengths. The comparison results reveal that the proper sample size of the window should be 40 (as used in this paper) for the side scan sonar.

4.2. Net Comparison

To compare the performance of different networks, given the characteristics of the input sample data, the networks of different layers were established, trained, and used in bottom tracking experiments. The results are shown in Table 4.

Table 4. Comparison of Results Obtained by Using the Networks of Different Layers.

Network Layer	Training Accuracy (%)		Validation Accuracy (%)	Bottom Tracking Accuracy (%)	Time per Ping (s)
	10 Epochs	20 Epochs			
6	89.5	97.3	98.1	90.0	0.24
8	98.4	99.8	99.6	95.0	0.30
10	98.6	99.6	100	98.3	0.35

As seen from Table 4, with a sufficient number of samples, the deeper networks demonstrated better learning rates and higher training, validation, and bottom tracking accuracies, but required a longer calculation time. Meanwhile, each convolution operation would further reduce the data size. Therefore, the maximum number of network convolution layers was limited due to the limitations in the input data size. In this paper, a network of 10 layers was adopted (Figure 4).

4.3. Exceptional Situations

The validity of the proposed method was proven by conducting experiments using side scan data collected from Meizhou Bay. However, in some special cases when the backscatter strengths of sea bottom cannot be recognized, the proposed method may return invalid results. These possible exceptional situations are:

1. The sonar altitude to the seabed is too low (less than 5 m). When the sonar is too close to the seabed, the variation characteristics of the bottom sequences would be overridden by the sonar self-noises in the water column area, which would make bottom tracking impossible. This situation can be avoided by controlling the sonar altitude within the proper range.
2. The backscatter strengths of the seabed are too weak, and are no different from those of the water column area. This situation may be caused by the low-energy-level sonar emission, special sediment types, or maloperation of sonar instruments. In this situation, the backscatter strengths of the seabed are almost in the same range as those of the water column, cannot reflect the variation characteristics of the bottom echo sequences, and would make bottom tracking very difficult. This situation can be avoided by increasing the sound energy level, using the different-frequency sonars, and ensuring careful manual operation.

4.4. Other Methods for Improving Efficiency

The proposed method realizes bottom tracking of side scan data based on 1D-CNN recognition. In addition to the methods mentioned in this paper, some other ways to improve the efficiency of the proposed method include:

1. Define the depth range in advance. For the pre-surveyed water area, the previous bathymetric data can be used to define the depth range. The pre-known depth range can be used to control the detection ranges of the side scan data and to validate the bottom tracking results. Therefore, the pre-defined depth range can improve the calculation efficiency of the proposed bottom tracking method.
2. Improve the computing hardware. Given the high computing ability requirements of the deep learning algorithm, using better computing hardware can improve the calculation of the proposed method and reduce the bottom tracking time of each side scan ping. With the development of sonar technologies, given that sonars will have higher sample rates, a better computing hardware can improve the calculation efficiency of the proposed bottom tracking method.

4.5. Development of Modern Scanning Sonars

With the development of modern sonar technologies including interferometry, the newest scanning sonar could not only obtain sonar images but also bathymetric data [29] including the following.

1. Kongsberg GeoSwath sonars can simultaneously offer swath bathymetry and side scan seabed mapping with sufficient accuracy.
2. Teledyne Blueview's 3D multibeam scanning sonar can create high-resolution and laser-like imagery of underwater areas, structures, and objects of interest.

Although these sonars have many advantages, they are only used by a limited number of companies and research institutions because of their high cost.

The traditional side scan sonar remains one of the most widely used marine survey instruments because of its very low cost. Moreover, modern data process algorithms may provide new abilities for the traditional side scan sonar. In this paper, by using a real-time bottom tracking algorithm, the side scan sonar can measure the seabed depth. This enhances the potential applications of the side scan sonar.

4.6. Handling of Important Issues

The following important issues should be noted.

Low SNR. Our method processes side scan data that have been compensated and converted in fixed ranges when most information (e.g., the original signal level and time-varied gain) is unavailable. Under this situation, the echo intensities are almost in the same range. When the SNR is very small, the echo intensities of the bottom can be affected by noise, but the variation features remain. We believe that our method can process side scan data with very small SNR after training of the corresponding samples.

Obstacles in the water column. When obstacles (e.g., the fish school) exist above the seabed, the fishes can be easily distinguished by using the trained 1D-CNN with enough negative samples (i.e., fishes). By training with all types of obstacle samples, the network can distinguish the bottom, the fishes, and the other obstacle targets from one another. Moreover, when the bottom continuity hypothesis fails, our method will automatically search for the new bottom position.

Reproducibility. Each step of our method is described in detail, including how to create the samples from the recorded side scan data, how to design a suitable 1D-CNN, how to train the network, and how to use the proposed bottom tracking method. In the experiment, we demonstrate our complete processing procedure, including the sampling, training, and bottom tracking. We believe that the reader can easily reproduce our results by using their own side scan data.

5. Conclusions

Based on the 1D-CNN recognition of bottom backscatter strength sequences, this paper develops a high-accuracy and real-time bottom tracking method of side scan sonar data. This method was validated by using the measured side scan data from Meizhou Bay, and the validity of each step of this method was proven. The side scan sonar data from the experimental area were bottom tracked by using the proposed method, and the average bottom tracking accuracy reached 94.7% with a 4.5% miss-ping rate, and 99.2% excluding the missing data. The experimental results showed that the proposed method is highly robust to the effects of noise, rich seabed texture, and artificial targets and proved its accuracy and real-time performance. Our method can process side scan data in field measurements (i.e., when the operator has no control over the SNR, or when fishes or obstacles are present in the water column, or in analogue simulations by using the recorded data), and in post-processing (i.e., when the recorded data only contain compensated and converted backscatter strengths). The proposed method also demonstrates that the real-time sound is possible by using the side scan sonar, which may further expand the applications of side scan sonars.

Author Contributions: Conceptualization, J.Y. and J.M. Methodology, J.Y. Software, J.Y. Validation, J.Y., J.M., and J.Z. Formal analysis, J.Y. and J.M. Investigation, J.Y. and J.Z. Resources, J.Z. Data curation, J.Y. Writing—original draft preparation, J.Y. Writing—review and editing, J.Y., J.M., and J.Z. Visualization, J.Y. and J.M. Supervision, J.Y. and J.Z. Project administration, J.Y. and J.M. Funding acquisition, J.Y., J.M., and J.Z. All authors have read and agreed to the published version of the manuscript.

Funding: The National Natural Science Foundation of China (grant number 41906168, 41576107, and 51804001), Natural Science Foundation of Anhui Province (grant number 1908085QD161), and the University Science Research Key Project of Anhui Province (grant number KJ2019A0024) funded this research.

Acknowledgments: The Guangzhou Marine Geological Survey Bureau provided the data in this study. The authors are appreciative of their support. We gratefully thank the editor and the anonymous reviewers for their valuable comments and suggestions that greatly improve our manuscript.

Conflicts of Interest: The authors declare no conflict of interest.

References

1. Blondel, P. Automatic mine detection by textural analysis of COTS sidescan sonar imagery. *Int. J. Remote Sens.* **2000**, *21*, 3115–3128. [CrossRef]
2. Reed, S.; Petillot, Y.; Bell, J. An automatic approach to the detection and extraction of mine features in sidescan sonar. *IEEE J. Ocean. Eng.* **2003**, *28*, 90–105. [CrossRef]
3. Mishne, G.; Talmon, R.; Cohen, I. Graph-based supervised automatic target detection. *IEEE Trans. Geosci. Remote* **2015**, *53*, 2738–2754. [CrossRef]
4. Acosta, G.G.; Villar, S.A. Accumulated CA–CFAR process in 2-D for online object detection from sidescan sonar data. *IEEE J. Ocean. Eng.* **2015**, *40*, 558–569. [CrossRef]
5. Collier, J.S.; Humber, S.R. Time-lapse side-scan sonar imaging of bleached coral reefs: A case study from the Seychelles. *Remote Sens. Environ.* **2007**, *108*, 339–356. [CrossRef]
6. Degraer, S.; Moerkerke, G.; Rabaut, M.; Van Hoey, G.; Du Four, I.; Vincx, M.; Henriet, J.-P.; Van Lancker, V. Very-high resolution side-scan sonar mapping of biogenic reefs of the tube-worm *Lanice conchilega*. *Remote Sens. Environ.* **2008**, *112*, 3323–3328. [CrossRef]
7. Capus, C.G.; Banks, A.C.; Coiras, E.; Ruiz, I.T.; Smith, C.J.; Petillot, Y.R. Data correction for visualisation and classification of sidescan SONAR imagery. *IET Radar Sonar Nav.* **2008**, *2*, 155–169. [CrossRef]
8. Reed, S.; Ruiz, I.T.; Capus, C.; Petillot, Y. The fusion of large scale classified side-scan sonar image mosaics. *IEEE Trans. Image Process.* **2006**, *15*, 2049–2060. [CrossRef]
9. Huo, G.; Yang, S.X.; Li, Q.; Zhou, Y. A robust and fast method for sidescan sonar image segmentation using nonlocal despeckling and active contour model. *IEEE Trans. Cybern.* **2017**, *47*, 855–872. [CrossRef]
10. Song, Y.; He, B.; Zhao, Y.; Li, G.; Sha, Q.; Shen, Y.; Yan, T.; Nian, R.; Lendasse, A. Segmentation of sidescan sonar imagery using markov random fields and extreme learning machine. *IEEE J. Ocean. Eng.* **2019**, *44*, 502–513. [CrossRef]
11. Villar, S.A.; Paula, M.D.; Solari, F.J.; Acosta, G.G. A framework for acoustic segmentation using order statistic-constant false alarm rate in two dimensions from sidescan sonar data. *IEEE J. Ocean. Eng.* **2018**, *43*, 735–748. [CrossRef]

12. Le Bas, T.P.; Huvenne, V.A.I. Acquisition and processing of backscatter data for habitat mapping – Comparison of multibeam and sidescan systems. *Appl. Acoust.* **2009**, *70*, 1248–1257. [CrossRef]
13. Yan, J. Acquisition and superposition of the high-quality measurement information of multibeam echo sonar and side scan sonar. *Acta Geodaetica et Cartographica Sinica* **2019**, *48*, 400. [CrossRef]
14. Buscombe, D.; Grams, P.E.; Smith, S.M.C. Automated riverbed sediment classification using low-cost sidescan sonar. *J. Hydraul. Eng.* **2016**, *142*, 06015019. [CrossRef]
15. Berthold, T.; Leichter, A.; Rosenhahn, B.; Berkhahn, V.; Valerius, J. Seabed sediment classification of side-scan sonar data using convolutional neural networks. In Proceedings of the 2017 IEEE Symposium Series on Computational Intelligence (SSCI), Honolulu, HI, USA, 27 November–1 December 2017.
16. Yan, J.; Zhao, J.; Meng, J.; Zhang, H. A universal seabed classification method of multibeam and sidescan sonar images in consideration of radiometric distortion. *J. Harbin Inst. Technol.* **2019**, *51*, 178–184. [CrossRef]
17. Blondel, P. *The Handbook of Sidescan Sonar*; Praxis Publishing Ltd.: Chichester, UK, 2009; pp. 7–9, 35–37.
18. Capus, C.; Ruiz, I.T.; Petillot, Y. Compensation for changing beam pattern and residual TVG effects with sonar altitude variation for sidescan mosaicing and classification. In Proceedings of the 7th European Conference Underwater Acoustics, Delft, The Netherlands, 5–8 July 2004.
19. Zhao, J.; Wang, X.; Zhang, H.; Wang, A. A Comprehensive bottom-tracking method for sidescan sonar image influenced by complicated measuring environment. *IEEE J. Ocean. Eng.* **2017**, *42*, 619–631. [CrossRef]
20. Al-Rawi, M.; Elmgren, F.; Frasheri, M.; Cürüklü, B.; Yuan, X.; Martínez, J.; Bastos, J.; Rodriguez, J.; Pinto, M. Algorithms for the detection of first bottom returns and objects in the water column in sidescan sonar images. In Proceedings of the OCEANS 2017, Aberdeen, UK, 19–22 June 2017.
21. Ding, J.; Chen, B.; Liu, H.; Huang, M. Convolutional neural network with data augmentation for SAR target recognition. *IEEE Geosci. Remote Sens.* **2016**, *13*, 364–368. [CrossRef]
22. Bentes, C.; Velotto, D.; Tings, B. Ship Classification in TerraSAR-X images with convolutional neural networks. *IEEE J. Ocean. Eng.* **2018**, *43*, 258–266. [CrossRef]
23. Williams, D.P. Transfer learning with SAS-image convolutional neural networks for improved underwater target classification. In Proceedings of the 2019 IEEE International Geoscience and Remote Sensing Symposium (IGARSS 2019), Yokohama, Japan, 28 July–2 August 2019; pp. 78–81.
24. Xie, Y. Machine Learning for Inferring Depth from Side-Scan Sonar Images. Master's Thesis, KTH Royal Institute of Technology, Stockholm, Sweden, September 2019.
25. Sun, J.; Xu, G.; Ren, W.; Yan, Z. Radar emitter classification based on unidimensional convolutional neural network. *IET Radar Sonar Nav.* **2018**, *12*, 862–867. [CrossRef]
26. Abdeljaber, O.; Avci, O.; Kiranyaz, S.; Gabbouj, M.; Inman, D. Real-time vibration-based structural damage detection using one-dimensional convolutional neural networks. *J. Sound Vib.* **2017**, *388*, 154–170. [CrossRef]
27. Zhao, J.; Yan, J.; Zhang, H.; Meng, J. A new radiometric correction method for side-scan sonar images in consideration of seabed sediment variation. *Remote Sens.* **2017**, *9*, 575. [CrossRef]
28. Zhao, J.; Meng, J.; Zhang, H.; Yan, J. A new method for acquisition of high-resolution seabed topography by matching seabed classification images. *Remote Sens.* **2017**, *9*, 1214. [CrossRef]
29. Bates, C.R.; Oakley, D.J. Bathymetric sidescan investigation of sedimentary features in the Tay Estuary, Scotland. *Int. J. Remote Sens.* **2004**, *25*, 5089–5104. [CrossRef]

Article

Aircraft Target Classification for Conventional Narrow-Band Radar with Multi-Wave Gates Sparse Echo Data

Wantian Wang, Ziyue Tang, Yichang Chen *, Yuanpeng Zhang and Yongjian Sun

Air Force Early Warning Academy, Wuhan 430019, China; laodifang0120@126.com (W.W.); tang_zi_yue@163.com (Z.T.); zhangyuanpeng312@163.com (Y.Z.); bmdsun@126.com (Y.S.)

* Correspondence: cyc_2007@163.com; Tel.: +86-134-7620-2876

Received: 27 September 2019; Accepted: 15 November 2019; Published: 18 November 2019

Abstract: For a conventional narrow-band radar system, the detectable information of the target is limited, and it is difficult for the radar to accurately identify the target type. In particular, the classification probability will further decrease when part of the echo data is missed. By extracting the target features in time and frequency domains from multi-wave gates sparse echo data, this paper presents a classification algorithm in conventional narrow-band radar to identify three different types of aircraft target, i.e., helicopter, propeller and jet. Firstly, the classical sparse reconstruction algorithm is utilized to reconstruct the target frequency spectrum with single-wave gate sparse echo data. Then, the micro-Doppler effect caused by rotating parts of different targets is analyzed, and the micro-Doppler based features, such as amplitude deviation coefficient, time domain waveform entropy and frequency domain waveform entropy, are extracted from reconstructed echo data to identify targets. Thirdly, the target features extracted from multi-wave gates reconstructed echo data are weighted and fused to improve the accuracy of classification. Finally, the fused feature vectors are fed into a support vector machine (SVM) model for classification. By contrast with the conventional algorithm of aircraft target classification, the proposed algorithm can effectively process sparse echo data and achieve higher classification probability via weighted features fusion of multi-wave gates echo data. The experiments on synthetic data are carried out to validate the effectiveness of the proposed algorithm.

Keywords: narrow-band radar; target classification; signal reconstruction; features extraction; weighted features fusion

1. Introduction

Aircraft target classification is always a difficult problem for traditional narrow-band radar. Even for the three distinct targets, i.e., helicopter, propeller and jet aircraft, the recognition rate of traditional narrow-band radar is not high in practical applications. The main reasons for target recognition probability deteriorating in traditional narrow-band radar include the following three points: (1) it is limited of range resolution due to the narrow bandwidth of radar transmitting signal; (2) a traditional mechanical scanning radar has a short dwell time, which results in limited azimuth resolution; (3) special circumstances such as data missing increase the difficulty of target classification.

In order to overcome the aforementioned problems, many classification methods have been proposed to improve the aircraft recognition rate. An intuitive idea is to increase the signal bandwidth of the radar system. It is well known that the enlargement of signal bandwidth can improve the range resolution of radar, so wide-band radar can provide more information for target classification. The existing methods of aircraft classification in wide-band radar can be generally divided into three types: (1) methods based on image processing [1–5]. In Reference [6], the extracted image, instead of radar data, was fed into a three-layered feed forward artificial neural network for aircraft classification; (2) methods based on high-resolution range profile (HRRP) [7–11]. Liu et al. [12] proposed a multi-scale target classification method based on the scale-space theory through extracting features from HRRP; and (3) methods based on inverse synthetic aperture radar (ISAR) [13–17]. A shape extraction based aircraft target classification method using ISAR images is proposed in [18]. Although image processing, HRRP- and ISAR-based aircraft classification have achieved good simulation results in wide-band radar, the radar system is more complex and the detectable range is shorter than that in narrow-band radar. Therefore, it is still of great significance to study aircraft classification based on narrow-band radar.

In recent years, many micro-Doppler parameter estimation methods [19–21] were designed to tackle the aircraft target classification problem in narrow-band radar. Two techniques of cubic polynomial fitting and three-point models are developed to estimate the micro-Doppler parameters from a fraction of the period in real-world scenarios [22]. In Reference [23] the minimal mean-square error (MMSE)-based method was proposed for estimating the micro-Doppler parameter from a fraction of the period data. Li et al. [24] proposed the parametric sparse representation and pruned orthogonal matching pursuit to the micro-Doppler parameter estimation for target classification and recognition. There are also many methods used in micro-Doppler parameter estimation for aircraft classification, such as time-frequency transform [25,26], continuous wavelet transform [27], Hough transform [25] and so on. However, these methods have the same common problem of inaccurate parameter estimation while the micro-Doppler signal is weak and only single-wave gate echo data is utilized in the aforementioned methods. Moreover, with echo data missing in narrow-band radar, Wang et al. [28] proposed a complex Gaussian model [29] based and factor analysis model [30]-based signal reconstruction methods. On the basis of the complex Gaussian model, the method of directly reconstructing the time-frequency spectrum of the original is proposed in Reference [31]. Although these methods mentioned above can effectively reconstruct the echo signals of aircraft, they have not studied the problem of classification with missing samples.

In addition, for the limited information of narrow-band radar, deep learning and machine learning [32] based methods have also been introduced to aircraft target classification in recent years. One of the most frequently used methods is the convolutional neural network (CNN) [33,34]. A novel landmark and CNN based aircraft recognition method was proposed [35], in which it alleviates the work of human annotation and can be used for any type of aircraft not contained in the training data set without retraining, thus it is highly accurate and efficient. Zuo et al. [36] proposed a deep convolutional neural network (DCNN) [37,38] based aircraft type recognition framework. Additionally, conditional generative adversarial networks [39], a self-organizing neural network [40] and deep belief net [41] have been used in aircraft targets classification. The networks used in the aforementioned methods, as an intelligent technology developed recently, have achieved good performance on aircraft targets classification, but they should be trained with large-scale datasets, which is quite time-consuming to acquire and mark with label, and difficult to practice in radar equipment. Also there is still a certain gap in the actual application of equipment.

In this paper, an aircraft target classification method is proposed for conventional narrow-band radar with multi-wave gates sparse echo data. By contrast with the previous work where there are only extract features from single wave gate echo for aircraft classification [10], the proposed method in this paper uses multiple wave gates echo data for weight feature fusion, and combines sparse theory to improve the probability of target classification in the case of missing data. Firstly, smoothed l_0 norm (SL0) [42] and orthogonal matching pursuit (OMP) [43,44] algorithms are used to reconstruct the sparse echo data in order to solve the low classification probability. Then we analyse the echo data with three kinds of aircraft targets in time domain and frequency domain. According to the difference of micro-Doppler effects [45] of rotating parts due to the difference in structure and rotating speed, features of the amplitude deviation coefficient, time domain waveform entropy and frequency domain waveform entropy are extracted to classify targets. Finally, features extracted from multi-wave gates sparse echo data are weighted and fused to train and test the support vector machine (SVM) [46–48] model for classification. Experimental results show that the proposed algorithm can improve the classification probability, and four wave gates echo data in weighted features fusion used to extract features is the optimal wave gate number for target classification.

The rest of this paper is organized as follows. The echo model and reconstruction algorithms are reviewed in Section 2. The proposed algorithm based on multi-wave gates sparse echo data is summarized in Section 3. Section 4 verifies the effectiveness of the proposed algorithm by simulated experiments. Conclusions are presented in Section 5.

2. Reconstruction Algorithm of Sparse Echo Data

2.1. Echo Model

In narrow-band radar, the signal wavelength is much smaller than the target size, and the received signal by radar is composed of echoes reflected by multiple scattering points. For targets with rotors, such as a helicopter, propeller aircraft and jet aircraft, the echo can reflect not only the translation of scattering points of the fuselage, but also the fretting characteristics of the scattering points of the rotor blades.

Take a helicopter as an example; the special geometry between radar and a helicopter is shown in Figure 1a, in which the distance between the radar and rotor target center is denoted as R_C, and the angle of pitch is denoted as β. Considering a 2-D slant-range plane, the simplified geometry is shown in Figure 1b. A radar coordinate system XOY and target coordinate system $X'O'Y'$ are set up, in which the rotor center is denoted as O'. The rotation radius of scattering point P on the rotor blade is assumed to be r—i.e., the distance from P to O' is r—and the distance from P to the radar is denoted as R_P. The scattering point P rotates around the target coordinate system center O' with an angular velocity ω, and the rotation angle at the initial time is denoted as θ_0. Assume that the radial velocity of the helicopter's translational motion is v.

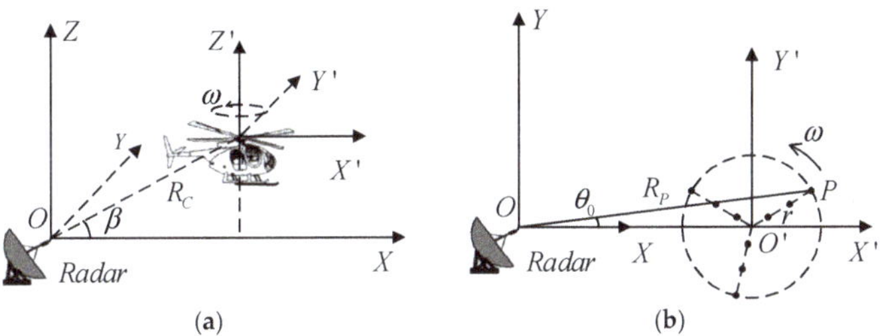

Figure 1. Geometry between radar and rotor target: (**a**) space geometry; (**b**) 2-D plane geometry.

In the case of the far field, the instantaneous distance between the scattering point P and the radar can be written as:

$$R_P(t_m) \approx R_C + vt_m + r\cos(\omega t_m + \theta_0), \tag{1}$$

where $t_m = mT_r$ is the slow time, m is the m-th echo pulse, and T_r is the pulse repetition period.

In this paper, we take the linear frequency modulation (LFM) as the transmitted signal, which can be expressed as:

$$s_t(\hat{t}, t_m) = \mathrm{rect}(\hat{t}/T_p)\exp\left(\mathrm{j}2\pi(f_c t + \mu\hat{t}^2/2)\right), \tag{2}$$

where $rect(\cdot)$ is the rectangular window, $\hat{t}$ is the fast time, T_p is the pulse width, f_c is the signal carrier frequency, μ is the chirp rate, t is the total time, and $t = \hat{t} + t_m$. There are two different time variables $\hat{t}$ and t in the transmitted signal described in Formula (2), the reason is that the signal carrier frequency

f_c exists on the whole pulse transmission time axis, while the chirp rate μ is used to adjust the change of Doppler frequency within a pulse. The echo signal of scattering point P can be expressed as:

$$s_r(\hat{t}, t_m) = \sigma \mathrm{rect}(t_m/T_a)\mathrm{rect}((\hat{t} - 2R_P(t_m)/c)/T_p) \exp\left(\mathrm{j}2\pi(f_c(t - 2R_P(t_m)/c) + \mu(\hat{t} - 2R_P(t_m)/c)^2/2)\right), \quad (3)$$

where σ is the scattering coefficient of the scattering point P, T_a is the observation time, and c is the speed of light. The target echo signal after pulse compression can be expressed as:

$$\begin{aligned} s_P(\hat{t}, t_m) = \; & \sigma T_p \sin \mathrm{c}[B(\hat{t} - 2R_P(t_m)/c)]\mathrm{rect}(t_m/T_a) \exp\left(-\mathrm{j}4\pi R_C/\lambda\right)\cdot \\ & \exp\left(-\mathrm{j}4\pi(vt_m + r\cos(\omega t_m + \theta_0))/\lambda\right) + w(\hat{t}, t_m) \end{aligned} \quad (4)$$

where B is the signal bandwidth, λ is the wavelength, and $w(\hat{t}, t_m)$ denotes the Gaussian white noise signal. By taking the derivative of the phase, the micro-Doppler frequency can be obtained as:

$$f_{d-P} = \frac{1}{2\pi}\frac{\mathrm{d}\phi}{\mathrm{d}t_m} = \frac{1}{2\pi}\frac{\mathrm{d}[-4\pi(vt_m + r\cos(\omega t_m + \theta_0))/\lambda]}{\mathrm{d}t_m} = -2(v - \omega r \sin(\omega t_m + \theta_0))/\lambda, \quad (5)$$

It can be seen from the above formula that the Doppler frequency of the scattering point echo on the target rotor blade is related not only to the radial velocity v of the scattering point echo on the target rotor blade is related not only to the radial velocity of the translational motion, but also to the angular velocity ω of the rotating component and the blade length r. Because the rotational motion of scattering points on the fuselage is negligible, it is equivalent to only translational motion. Therefore, the instantaneous distance between scattering point F on fuselage and the radar can be written as:

$$R_F(t_m) \approx R_C + vt_m, \quad (6)$$

The echo signal of scattering point F on fuselage after pulse compression can be expressed as:

$$s_F(\hat{t}, t_m) = \sigma T_F \sin \mathrm{c}[B(\hat{t} - 2R_F(t_m)/c)]\mathrm{rect}(t_m/T_a) \exp\left(-\mathrm{j}4\pi R_C/\lambda\right) \exp\left(-\mathrm{j}4\pi v t_m/\lambda\right) + w(\hat{t}, t_m), \quad (7)$$

Compared with the echo of blade scatterer in Formula (4), the fuselage echo lacks only the fretting term. In addition, the Doppler frequency of echo is only related to translational radial velocity, that is $f_{d-F} = -2v/\lambda$.

The frequency domain echo of helicopter, propeller aircraft and jet aircraft are simulated as shown in Figure 2. The simulation parameters of radar and transmitting signal are set as follows, the pulse repetition frequency F_r is 5000 Hz, the pulse-repetition period T_r is 0.2 ms, the pulse width T_p is 50 μs, the signal carrier frequency f_c of LFM signal is 1 GHz, the signal bandwidth B is 2 MHz, and the observation time T_a is 0.05 s. The parameters of three types of aircraft targets are shown in Table 1.

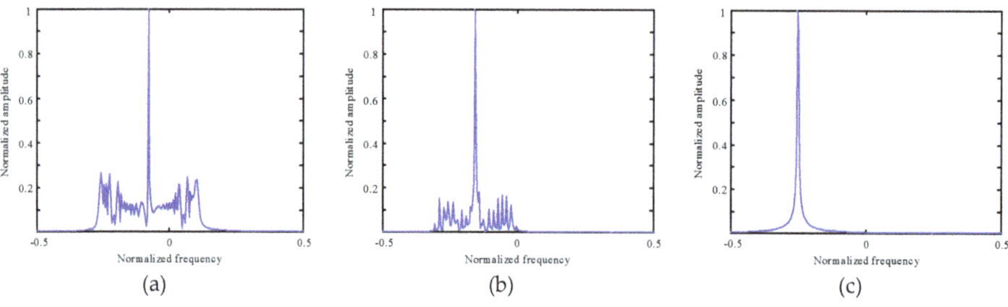

Figure 2. Target frequency domain echo: (**a**) helicopter; (**b**) propeller; (**c**) jet.

Table 1. Three kinds of aircraft targets simulation parameters.

Target Type	Rotor Length (m)	Blade Number	Rotation Velocity (rad/s)	Translational Velocity (km/h)
Helicopter	8	4	20	250
Propeller	2	4	130	500
Jet	1	27	380	800

The rotation plane of the helicopter main rotor is parallel to the ground, so the micro-Doppler effect produced can be observed easily by conventional ground radar. It can be seen from Figure 2a that in addition to the fuselage echo, there are strong echo components caused by micro-Doppler motion of rotor blades in the frequency domain echo of the helicopter, and the micro-Doppler spectrum width of the helicopter is higher than that of the propeller in Figure 2b, which is due to the fact that the length of the helicopter rotating parts is significantly longer than that of propeller. In addition, because the rotating plane of the propeller's engine blade is perpendicular to the flight direction of the aircraft, the blade is easily obscured by the fuselage, and its micro-Doppler effect is relatively weak. Because of the small size of jet engine blades and the particularity of its position, the micro-Doppler effect caused by blade rotation can hardly be observed by ground radar. It can be seen from Figure 2c that the echo of the jet aircraft only contains the fuselage component, but not the micro-Doppler component caused by blade rotation.

2.2. Reconstruction Algorithm

The multi-wave gates echo data can be obtained from the echo reflected by the transmitted signal after encountering the target during each resolution of the radar antenna in the continuous observation of the aircraft target by the radar. The definition of multi-wave gates echo data is shown in Figure 3.

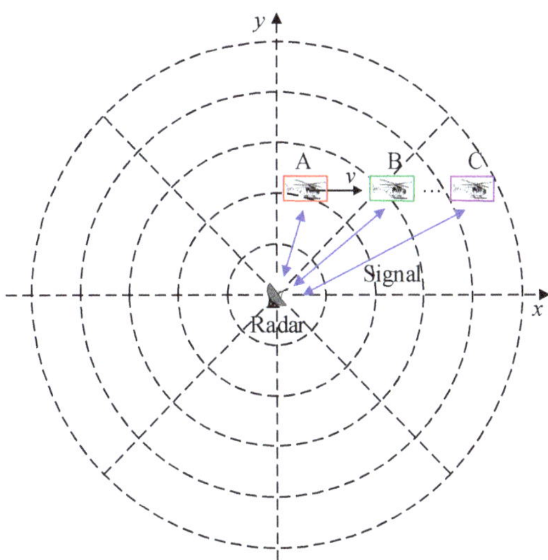

Figure 3. The definition of multi-wave gates echo data.

As we can see from Figure 3, it is a radar display that denotes the position relationship between the radar and the aircraft target from the perspective of top view. Let us assume that the aircraft target flies in a positive direction along the x-axis with a velocity of v, and the blue two-way arrow line in the Figure is the signal transmission route. In the rotation of radar antenna, when the aircraft target is seen for the first time, the aircraft target is located at position A, when the antenna comes to the aircraft target after a rotation cycle, the aircraft target is located at position B, the next position is C, and so on. While the radar irradiates the aircraft target, it will receive the echo data in the area, marked by red, green and purple square boxes in this Figure, where the aircraft target is located, and we count it as a wave gate echo. After multiple irradiations, we can obtain multi-wave gates echo data.

It is difficult for conventional narrow-band radar to obtain continuous observation of the same target for a long time, and it may lead to the loss of target echo pulse in one observation time which can be called sparse echo data. The description of complete and sparse echo data are shown in Figure 4.

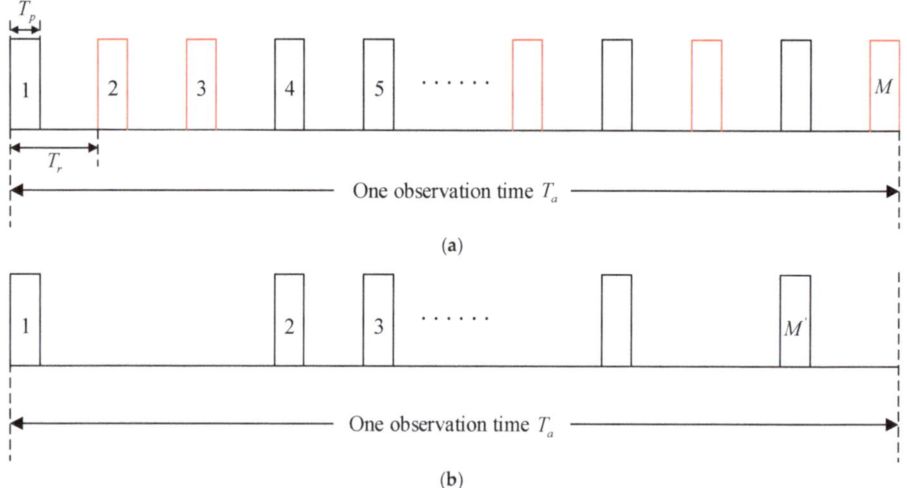

Figure 4. Description of complete and sparse echo data: (**a**) complete echo data; (**b**) sparse echo data.

As we can see from Figure 4a, there are M received echo pulses in one observation time which are called the single-wave gate echo data, among which the pulses marked in red randomly indicate the echo data that may be lost when the radar receives the echo. From Figure 4b, it can be seen that the number of pulses of sparse echo data is less than that of complete echo data in one observation time, we can also say that less echo information is available in sparse echo data, which is not conductive to aircraft target classification. Therefore, it is feasible and necessary to reconstruct sparse echo data by appropriate sparse signal recovery methods.

Since the emergence and development of compressed sensing (CS) [49,50] technology, sparse signal recovery algorithms have mainly been divided into greedy algorithms, non-convex function minimization algorithms, and Bayesian algorithms. The most typical and widely used greedy algorithm is orthogonal matching pursuit (OMP), which finds the best matching dictionary unit by solving the maximum inner product of the residual and the dictionary matrix, then obtains the approximate value of the sparse vector by using the least square method, and finally obtains the reconstruction signal by alternately updating the support set and solving the sparse vector. The smoothed l_0 norm (SL0) algorithm is one of the most famous non-convex function minimization algorithms, which transforms the l_0 norm minimization problem into an optimization problem by introducing a smooth Gaussian function to approach the l_0 norm. The reason for doing this is that we can avoid the non-deterministic polynomial (NP) time hard problem caused by the direct solution of l_0 norm minimization. Therefore, we use the SL0 and OMP algorithms to reconstruct the sparse echo signal, respectively in this paper.

Assuming that the number of pulses of sparse echo data is M', and the typical model of CS can be expressed as:

$$\mathbf{Y} = \mathbf{\Phi}\mathbf{X}, \tag{8}$$

where $\mathbf{Y}$ is an $M' \times 1$ measurement vector. Actually, $\mathbf{Y}$ is the superposition of sparse echo data of scattering point and fuselage, which can be expressed as:

$$\mathbf{Y} = s_P(t_m) + s_F(t_m) \quad m = 1, 2, \cdots, M', \tag{9}$$

$\mathbf{\Phi}$ is an $M' \times N$ dictionary matrix, and $\mathbf{X}$ is an $N \times 1$ sparse vector to be determined. According to CS theory, the sparse solution $\mathbf{X}$ can be obtained by:

$$\hat{\mathbf{X}} = \underset{\mathbf{X}}{\operatorname{argmin}} \|\mathbf{X}\|_0 \quad \text{s.t.} \quad \|\mathbf{Y} - \mathbf{\Phi} \cdot \mathbf{X}\|_2^2 \leq \varepsilon, \tag{10}$$

where $\|\cdot\|_0$ and $\|\cdot\|_2$ donate L_0 and L_2 norms respectively, ε is the error threshold in the sparse recovery processing. The solution for (10) can be obtained by the SL0 and OMP algorithms through iteration. The main steps of the two algorithms are summarized in Tables 2 and 3.

Table 2. Main steps of orthogonal matching pursuit (OMP) reconstruction algorithm.

Input: estimated signal $\mathbf{Y} \in \mathbb{C}^{M'}$, dictionary matrix $\mathbf{\Phi} \in \mathbb{C}^{M' \times N}$, error threshold ε_0.
Initialization: let the iterative counter $k = 1$, residual matrix $\gamma_0 = \mathbf{Y}$, the index set $\Lambda_0 = \varnothing$.
Iteration: at the k-th iteration
(1) Update the index set $\Lambda_k = \Lambda_{k-1} \cup \lambda_k$, where $\lambda_k = \underset{i=1,2,\cdots,N}{\operatorname{argmax}} |\langle \gamma_{k-1}, \varphi_i \rangle|$, φ_i is the i-th column of $\mathbf{\Phi}$.
(2) Update the support set $\mathbf{\Phi}_{\Lambda_k} = [\mathbf{\Phi}_{\Lambda_{k-1}}, \varphi_{\lambda_k}]$, and calculate the signal
$\hat{\mathbf{X}}_k = \operatorname{argmax} \|\mathbf{Y} - \mathbf{\Phi}_{\Lambda_k} \mathbf{X}_k\|_2 = (\mathbf{\Phi}_{\Lambda_k}{}^{\mathrm{H}} \mathbf{\Phi}_{\Lambda_k})^{-1} \mathbf{\Phi}_{\Lambda_k}{}^{\mathrm{H}} \mathbf{Y}$.
(3) Update the residual matrix $\gamma_k = \mathbf{Y} - \mathbf{\Phi}_{\Lambda_k} \hat{\mathbf{X}}_k$.
(4) Increment k, and return to Step (1) until the stopping criterion $\|\gamma_k\|_2 \leq \varepsilon_0$ is met. The selection of the error threshold ε_0 is related to the precision requirement.
Output: Reconstructed signal $\hat{\mathbf{X}} = \hat{\mathbf{X}}_k$.

Table 3. Main steps of smoothed l_0 norm (SL0) reconstruction algorithm.

Input: estimated signal $\mathbf{Y} \in \mathbb{C}^{M'}$, dictionary matrix $\mathbf{\Phi} \in \mathbb{C}^{M' \times N}$, the search step length α.
Initialization: Choose an appropriate standard deviation parameter decrement sequence $[\sigma_1, \sigma_2, \cdots, \sigma_I]$, the outer loop number is I, the inner loop number is J. The initial solution is the minimum L2 norm of $\mathbf{Y} = \mathbf{\Phi X}$, that is $\mathbf{X}_0 = (\mathbf{\Phi}^{\mathrm{H}} \mathbf{\Phi})^{-1} \mathbf{\Phi}^{\mathrm{H}} \mathbf{Y}$.
Iteration:
(1) The i-th outer iteration, $i = 1, 2, \cdots, I$, at this time $\sigma = \sigma_i$, $\mathbf{X} = \mathbf{X}_{i-1}$.
(2) The j-th inner iteration, $j = 1, 2, \cdots, J$
1. Update the signal with $\mathbf{X} = \mathbf{X} + \alpha \mathbf{d}$, where $\mathbf{d} = [-x_1 \exp(-x_1{}^2/2\sigma^2), \cdots, -x_n \exp(-x_n{}^2/2\sigma^2)]$.
2. Project $\mathbf{X}$ onto the feasible domain, that is $\mathbf{X} \leftarrow \mathbf{X} - \mathbf{\Phi}^{\mathrm{H}} (\mathbf{\Phi \Phi}^{\mathrm{H}})^{-1} (\mathbf{\Phi X} - \mathbf{Y})$.
(3) Update the reconstructed signal $\hat{\mathbf{X}}_i = \mathbf{X}$.
Output: Reconstructed signal $\hat{\mathbf{X}} = \hat{\mathbf{X}}_I$.

SL0 and OMP algorithms are used to reconstruct the sparse frequency domain echoes of three types of aircraft targets. In order to simulate the sparse echo data in real radar equipment, we randomly cut half of the pulses in the complete echo data as sparse echo data in this paper, that is the pulse number of sparse echo data equals $M' = M/2$. The reconstructed results of SL0 and OMP algorithms are shown in Figures 5 and 6, respectively, where the dictionary matrix is the Fourier transform matrix because the time domain echoes are reconstructed to obtain the frequency domain echoes in this paper and the error threshold is set as $\varepsilon_0 = 0.05\|\mathbf{Y}\|_2^2$, that is the iteration process is stopped when the residual energy is equal to or smaller than 5% of the received signal energy. Comparing the complete frequency domain echoes in Figure 2, we can see that SL0 and OMP algorithms can realize the reconstruction of sparse echo data. Compared with the reconstructed results of SL0 and OMP algorithms in Figures 5 and 6, it can be seen that the reconstructed result of SL0 algorithm is better than the OMP algorithm in the similarity to the complete frequency domain echo of Figure 2. The echo data reconstructed by these two algorithms are used to extract features, and the simulation experiment of classification probability will be given in Section 4.

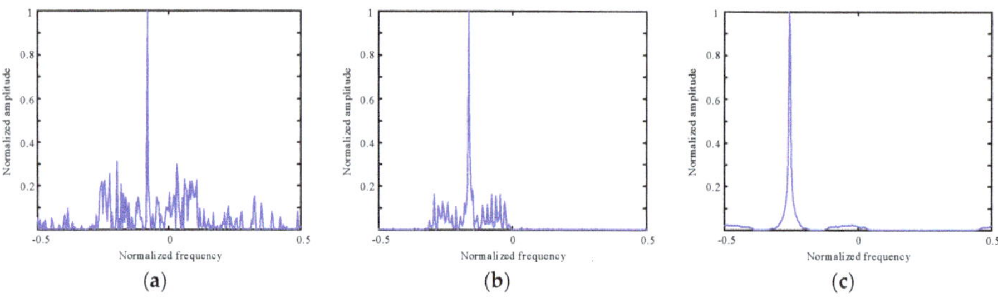

Figure 5. Reconstructed frequency domain echoes of SL0 algorithm: (**a**) Helicopter; (**b**) Propeller; (**c**) Jet.

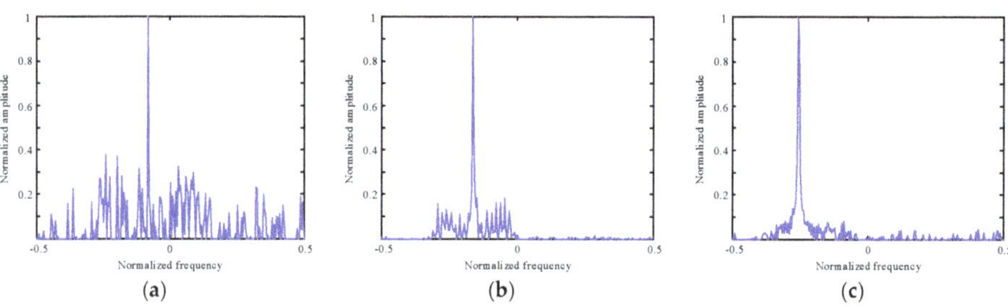

Figure 6. Reconstructed frequency domain echoes of OMP algorithm: (**a**) helicopter; (**b**) propeller; (**c**) jet.

3. Classification Algorithm Based on the Weighted Features Fusion of Multi-Wave Gates

By analyzing the characteristics of the helicopter, propeller aircraft and jet aircraft in time and frequency domains, we can classify three kinds of aircraft targets through the micro-Doppler effect caused by rotating parts due to the difference in structure and rotating speed.

3.1. Features Extraction

According to the difference of echoes in time domain and frequency domain, this paper classifies three types of aircraft targets by extracting amplitude deviation coefficient, time domain waveform entropy and frequency domain waveform entropy. Three feature extraction methods are described as follows:

1. Amplitude deviation coefficient

The amplitude deviation coefficient g_y of the discrete echo signal $\mathbf{Y} = \{y_i\}, i = 1, 2, \cdots, M'$ reflects the proportional relationship between the rotating parts of an airplane target and its fuselage, which can be defined as:

$$g_y = \sigma_y / \bar{\mathbf{Y}}, \tag{11}$$

where g_y denotes the amplitude deviation coefficient, $\sigma_y = \sum_{i=1}^{M'} (y_i - \bar{\mathbf{Y}})^2 / (M' - 1)$ is the variance of echo amplitude, $\bar{\mathbf{Y}} = \sum_{i=1}^{M'} y_i / M'$ is the mean of echo amplitude, M' is the length of the echo signal.

Generally speaking, the higher the complexity of the target structure, such as the helicopter and propeller aircraft, the larger the proportion of the micro-Doppler modulation component of the rotating parts to the radar echo, the greater the overall fluctuation of the echo amplitude and the larger the amplitude deviation coefficient of the echo.

2. Waveform entropy

Waveform entropy is usually used to describe the waveform characteristics of radar echo signals. From the analysis of Section 2, it can be seen that there are differences in the blade's number, length and rotating speed of the rotating parts in helicopter, propeller aircraft and jet aircraft, so the micro-Doppler effect of the rotating parts is different in the echo waveform. Therefore, we can distinguish the difference of waveform between different targets by extracting waveform entropy in time domain and frequency domain.

Time domain waveform entropy E_t and frequency domain waveform entropy E_f of echo signal are defined as follows:

$$E_t = -\sum_{i=1}^{M'} p_i \log_{10}(p_i), \tag{12}$$

$$E_f = -\sum_{i=1}^{M'} q_i \log_{10}(q_i) \tag{13}$$

where $p_i = y_i / \sum_{j=1}^{M'} y_j$ and $q_i = f_i / \sum_{j=1}^{M'} f_j$ are normalized signals in time domain and frequency domain respectively, $\mathbf{F} = \{f_i\}$, $i = 1, 2, \cdots M'$ is the result of fast Fourier transform (FFT) of echo signal $\mathbf{Y}$.

In this paper, three kinds of aircraft target echo models are established, and the time domain and frequency domain echoes of targets are simulated according to the parameters of the rotor in Table 1. Then, the amplitude deviation coefficient, time domain waveform entropy and frequency domain waveform entropy are extracted. We simulate 200 sparse echo signal samples of three types of aircraft targets respectively from different radar perspectives where one sparse echo signal sample corresponds to one observation angle which denotes the relationship between the aircraft target's flying direction and the radar's line of sight, and it changes uniformly from 0° to 360° at an interval of 1.8°. Therefore, the angle varies from different samples, and the dataset includes 600 samples in all. The results of features extracted from 600 sparse echo signal samples are shown in Figure 7, where the signal to noise ratio (SNR) of target echo before pulse compression is −13dB, which is defined as $\text{SNR} = \|\mathbf{Y}\|_2^2 / (M'\sigma^2)$, where σ^2 is the variance of noise.

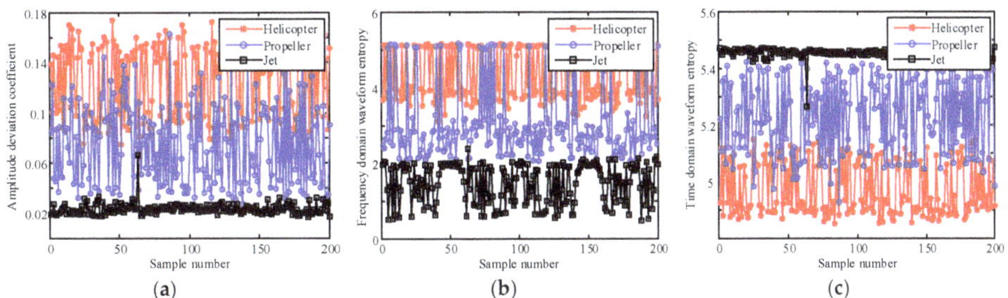

Figure 7. Results of extracted features: (**a**) amplitude deviation coefficient; (**b**) frequency domain waveform entropy; (**c**) time domain waveform entropy.

As can be seen from the above figure, in the case of SNR it is −13 dB, because of the difference among the rotating parts of three types of aircraft targets, the amplitude deviation coefficient, time-domain waveform entropy and frequency-domain waveform entropy are different among targets. Taking the amplitude deviation coefficient as an example, it can be seen from Figure 7a that the amplitude deviation coefficients extracted from the echoes of three kinds of targets have cross-values, which will inevitably lead to erroneous judgment in the process of target classification and reduce the classification probability. The reason may be the low SNR or the change of the angle of view between the aircraft and the radar, which results in small fluctuation of the extracted features. However, we can also see from the graph that the mean values of each feature differ greatly among the three kinds of aircraft targets and are more stable than those extracted from each sparse echo signal sample. Therefore, we need to adopt appropriate methods to make the features extracted from each sparse echo signal sample close to the mean value, so as to eliminate the impact of the target echo fluctuation model and improve the classification probability of the aircraft targets.

3.2. Weighted Features Fusion

On the basis of the above simulation analysis, we propose a target classification algorithm based on the weighted features fusion of multi-wave gates sparse echo data. This algorithm uses multi-wave gates echo data to extract features, which are fused by weighting to improve the correct classification probability. The fused features can be expressed as:

$$\tilde{F} = \sum_{i=1}^{K} \alpha_i F_i, \tag{14}$$

where $\tilde{F}$ is the fused feature, F_i is the feature extracted from the echo data of the i-th wave gate, K is the number of gates for feature fusion, α_i is the weight of the i-th wave gate feature.

In this paper, we consider that the features extracted from different gates have the same contribution to aircraft target type classification. Therefore, we adopt the same weighting value for feature fusion, that is to say, the weights $\alpha_i = 1/K$. In Section 3.1, we simulate 200 single-wave sparse echo signal samples of one aircraft target where one sparse signal sample corresponds to one radar observation angle. While in the simulation experiment of weighted features fusion, we collect four-wave gates sparse echo data at each radar observation angle which is set the same as that in Section 3.1 during the observation of the aircraft target. That is to say, in each observation angle, we reconstruct four-wave gates sparse echo signal samples, then extract the features from each reconstructed sample and fuse them as a fusion feature. Therefore, each type of feature consists of 200 fusion features for each aircraft target. Figure 8 shows the result of the fusion features extracted and fused from four wave gates echo data. Compared with Figure 7, the cross-value of extracted features between different targets is significantly reduced under the same SNR, we can also say that the fused features are more clustered near the mean of all samples.

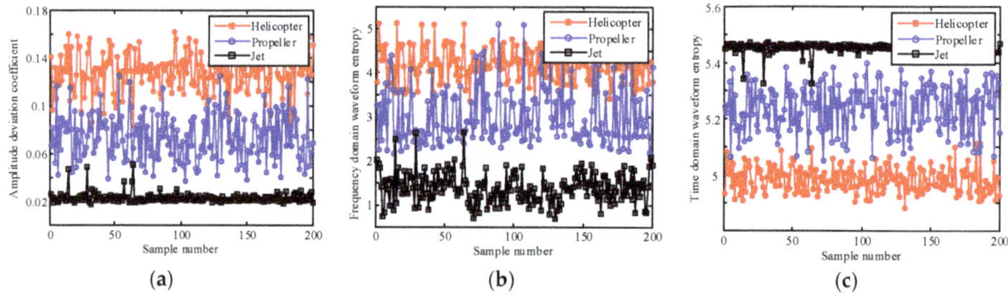

Figure 8. Result of fusing the features extracted from four wave gates echo data: (**a**) amplitude deviation coefficient; (**b**) frequency domain waveform entropy; (**c**) time domain waveform entropy.

We know that variance is a measure of the degree of dispersion of a set of data. In this paper, we calculate the variance of the fused features extracted from four wave gates sparse echo data with SNR is −13 dB, which is shown in Table 4. For comparison, we also compute the variance of the features that are not fused. We can see from Table 4 that the variance of the fused features is less than that of the features without fusion, no matter which kind of feature. In other words, fusion of extracted features is more conducive to distinguishing the three types of aircraft targets mentioned in this paper.

Table 4. Comparison of variance of extracted features whether to fuse or not.

Target Type	Amplitude Deviation Coefficient ($\times 10^{-4}$)		Frequency Domain Waveform Entropy		Time Domain Waveform Entropy ($\times 10^{-3}$)	
	No Fusion	Fusion	No Fusion	Fusion	No Fusion	Fusion
Helicopter	8.60	2.19	0.45	0.13	8.50	1.60
Propeller	11.00	2.70	0.98	0.31	18.60	4.80
Jet	0.21	0.14	0.27	0.09	0.31	0.27

3.3. *Classification Algorithm*

In this paper, support vector machine (SVM) method is used to classify the extracted fusion features of three types of aircraft targets. SVM was first proposed by Vapink for the classification of two types of liner separable data [51]. By finding the optimal hyperplane which makes the boundary distance between the two classes the maximum, the sample data was divided into two types. Later, it was extended to linear separable data. To solve the problem of three types of aircraft targets classification in this paper, we use the one-vs-one method to construct an SVM classifier between any two types, and construct three classifiers in total, and then obtain the final classification result by voting scheme. Three types of aircraft targets classification method based on SVM that we adopted in this paper are shown in Figure 9.

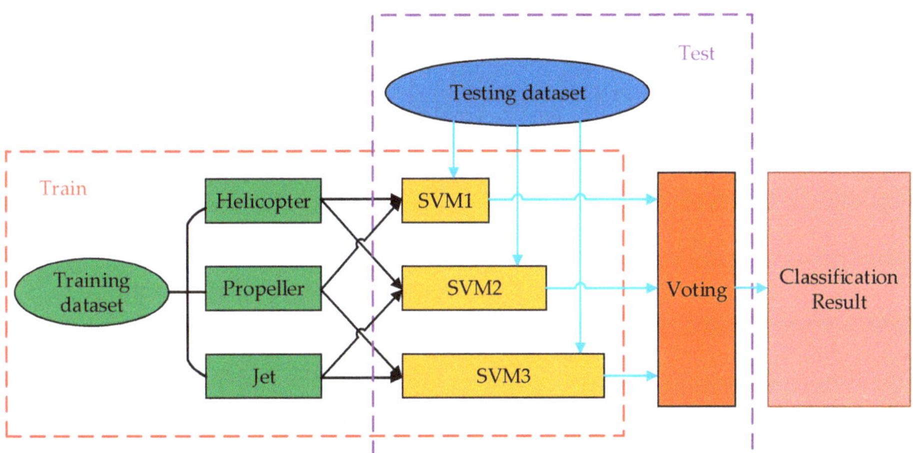

Figure 9. Three-class support vector machine (SVM) model.

In Figure 9, the flow of red dotted box marked is the training process, in which the training dataset labeled in advance are divided into three parts belonging to different aircraft targets, and the two parts of them are combined to train three SVM models, respectively. The flow of purple dotted box marked is the testing process, in which the testing dataset which are completely different from the training dataset are sent to three trained SVM models, and then vote on the results of the SVM model to get the final classification results.

To sum up, with the sparse echo data, the classification algorithm of aircraft targets based on the weighted features fusion of multiple wave gates is summarized in Figure 10. In the proposed algorithm, the multi-wave gates sparse echo data are obtained as described in Figure 3: K is the number of wave gates in weighted feature fusion, and the SL0 and OMP methods are used to reconstruct sparse echoes and by which three types of features are extracted: amplitude deviation coefficient, time domain waveform entropy and frequency domain waveform entropy. In addition, in order to improve the classification probability of three types of aircraft targets, a classification algorithm based on the weighted features fusion of multiple wave gates is proposed. Finally, the fused features are used to classify three aircraft targets by three class support vector machine model.

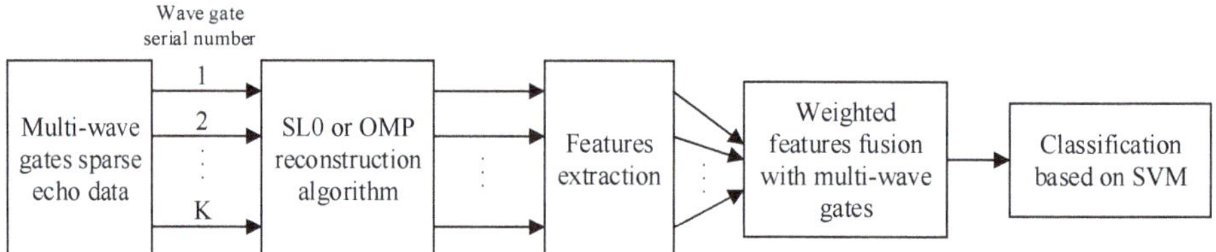

Figure 10. Flowchart of the proposed classification algorithm.

4. Experimental Results

4.1. Dataset Details

In experiments, the parameters of radar, transmitting signal and aircraft targets are the same as those in Section 2.1. In this paper, several groups of comparative experiments are constructed where the noise is all considered for classification probability. In the classification experiment of complete echo data, we simulate 200 single-wave gate complete echo signal samples for each aircraft target as the dataset where one complete signal sample corresponds to one radar observation angle. While the sparse echo data of multi-wave gates continuously are simulated in one angle in the classification experiment of sparse echo data, in other words, the dataset contains 200 multi-wave gates sparse echo signal samples for each aircraft target. What we want to emphasize is that the testing dataset are completely different from the training dataset with different aircraft target distance and view angle, and the aircraft target correct classification probability can be obtained by comparing the classification result of the SVM model with the real label of the aircraft target, which is equal to the number of correctly classified samples divided by the total number of samples.

4.2. Validity Experiment of Reconstruction Algorithm

With the aim of verifying the validity of SL0 and OMP reconstruction algorithm for the classification of aircraft targets, we conduct an experiment. In this experiment, in order to better simulate the actual work of radar target classification and recognition, the training dataset of 600 single-wave gate complete echo data samples are extracted features which are used to train the SVM model. Several comparative testing experiments are conducted with different testing dataset, two of which take the reconstructed echo data obtained by SL0 and OMP algorithms as the testing dataset, and the results are shown as the blue curve and the black curve, respectively, in Figure 11. As a comparison, we take another complete echo data samples as the testing dataset, and the result of average classification probability is shown as the red curve. In addition, we take the sparse echo data samples as the testing dataset to verify that the sparse echo data worsen aircraft targets classification due to the loss of several echo information elements.

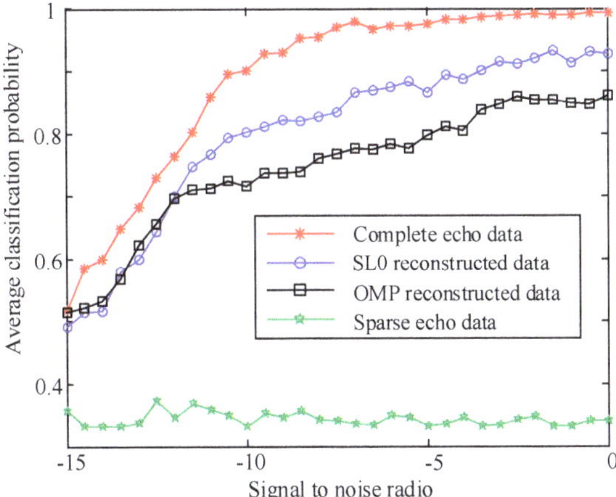

Figure 11. Contrast experiment of reconstruction algorithm.

From Figure 11, we can see that the classification probability of sparse echo data is the lowest, which distributes around 33%, and does not vary with the change of SNR. This shows that the sparse echo data loses the components reflecting the micro-Doppler effect of the rotating parts. We can also say that the helicopter, propeller aircraft and jet aircraft cannot be distinguished correctly by extracting the three kinds of features through the sparse echo data. The correct classification probability of the complete echo data is the highest, and with the increase of SNR, the correct classification probability increases gradually and stabilizes at 99.33%. Although the correct classification probability of reconstructed echo data obtained by the SL0 and OMP algorithms is lower than that of complete echo data, it is obviously higher than that of sparse echo data, which verifies the validity of the two kinds of reconstruction algorithms for reconstructing echo data. When the SNR is lower than −12 dB, the classification probability of the reconstructed echo data between the two algorithms is similar. When SNR is higher than −12 dB, the classification effect of the SL0 reconstruction algorithm is better than that of OMP reconstruction algorithm, which is consistent with the reconstructed results of frequency domain echoes of two kinds of algorithms in Section 2.2.

4.3. Selection of Wave Gate Number in Weighted Features Fusion

Although SL0 and OMP reconstruction algorithms can accurately reconstruct the sparse echo data, the correct classification probability of the extracted features still lags behind that of the complete echo data. However, as we analyzed in Section 3.2, after using the weighted features fusion method for the features extracted from the multi-wave gates reconstructed echo data, the fused features are more clustered near the mean of all samples. Therefore, a classification algorithm based on the weighted features fusion of multi-wave gates reconstructed echo data is proposed in this paper in order to improve the classification probability. In this experiment, we compare the influence of the wave gate number on the probability of target classification in order to get the best wave gate number in weighted features fusion. The training and testing dataset of features are all extracted from the reconstructed multi-wave gates echo data. The experimental results are shown in Figure 12.

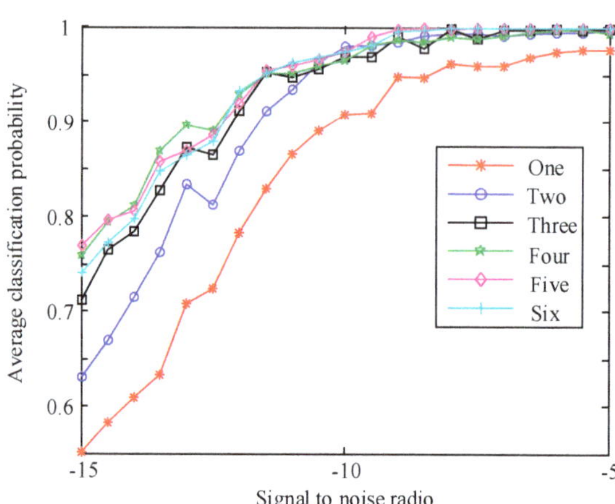

Figure 12. Selection experiment of wave gate number.

We can summarize from the above figure that the classification probability among the wave gate number from one to six in weighted features fusion increases with the raise of SNR. When only one wave gate reconstructed echo datum is used to extract features which is selected for classification, the probability is lower than that of multi-wave gates. When the SNR is low, the probability of choosing two fused wave gates features to classify aircraft targets is lower than that of choosing three to six wave gates fused features, but with increasing of the SNR, the probability of choosing two fused wave gates features is similar to that of using more. The experimental results also show that the classification probability curves of choosing three to six wave gates for weighted features fusion has the same change rule and is similar with each other under the same SNR. On the one hand, the experimental result shows that the classification effect of using multi-wave gates reconstructed echo data to classify three types of aircraft targets is better than that of using only one gate reconstructed echo datum. On the other hand, by fusing the features extracted from multi-wave gates reconstructed echo data, the fused features can be close to the mean value, and the number of cross-values of features extracted from different aircraft targets is reduced. However, if too many wave gates in weighted features fusion are selected, during this period, there are some differences between the extracted features due to the change of the aircraft's motion direction and flight attitude, and the probability of target classification does not increase with the increase of the number of wave gates. In summary, when the wave gate number in weighted features fusion is four, the classification probability of aircraft targets is the best.

4.4. Classification Experiment Based on Weighted Features Fusion with Four Wave Gates Sparse Echo Data

In this section, based on the target classification algorithm of multi-wave gates in weighted features fusion proposed in this paper, we conduct two comparative simulation experiments. One is that we train the SVM model with the dataset composed of single wave gate complete echo data, classify aircraft targets with the dataset consisting of four wave gates complete echo data and the dataset consisting of four wave gates reconstructed echo data respectively. The simulation results are shown in Figure 13. Another comparative simulation experiment is that we train the SVM model with the dataset composed of four wave gates complete echo data, while the testing datasets consist of four wave gates complete echo data and of four wave gates reconstructed echo data, respectively. The experimental results are shown in Figure 14.

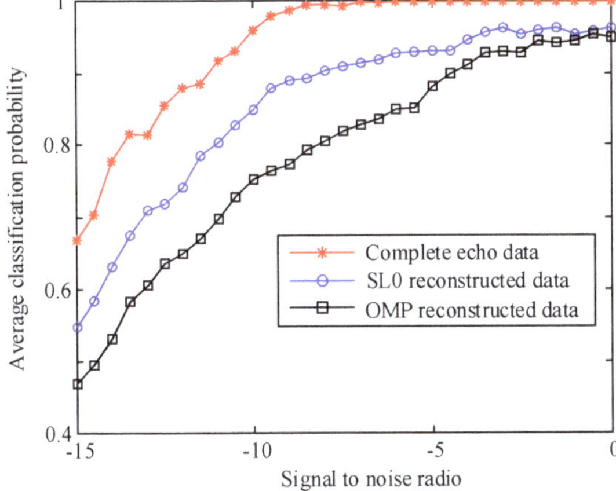

Figure 13. Classification results of single wave gate echo data for training and four wave gates echo data for testing.

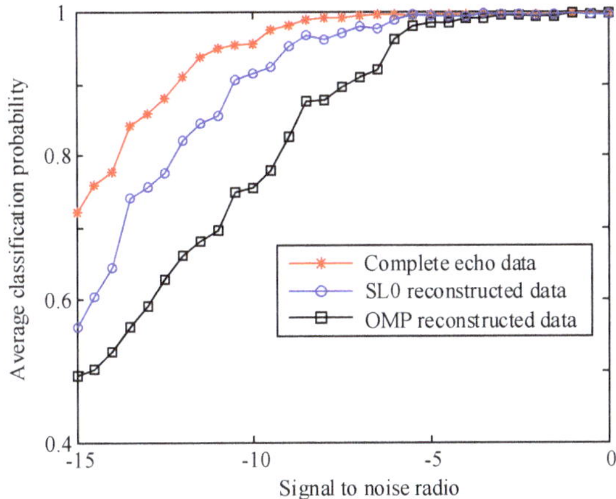

Figure 14. Classification results using four wave gates echo data for both training and testing.

Compared with the experimental result of single wave gate echo data as the testing dataset in Figure 11, we can conclude from the results of four wave gates echo data that the best way to classify targets, as testing dataset in Figure 13 that the classification probability of the complete echo data is obviously improved. The classification probability of features extracted from reconstructed echo data by SL0 and OMP algorithms after weighted features fusion is also higher than that of features without weighted features fusion. Therefore, we come to the conclusion that the echo data of multi-wave gates to classify the aircraft targets can improve the correct classification probability.

Comparing the experimental results in Figures 13 and 14, the classification probability of using four wave gates complete echo data for both training and testing is better than that of the single wave gate complete echo data for training and four wave gates complete echo data for testing, and the reason for this is that the SVM model can learn more target echo information by using four wave gates echo data. Moreover, the classification probability of the two kinds of reconstruction algorithms can reach 99.83% in Figure 14, which verifies the validity of both kinds of reconstruction algorithm and the effectiveness of the classification algorithm based on weighted features fusion of multi-wave gates reconstructed echo data. Therefore, it can be summed up that in the process of radar classification of three types of aircraft target, we can use four wave gates echo data as far as possible in weighted features fusion for training the SVM model, in which the parameters of the SVM model trained in this way are optimal. Also the classification probability of the target is highest when testing with four wave gates echo data.

5. Conclusions

In this paper, an aircraft target classification algorithm is proposed based on weighted features fusion of multi-wave gates sparse echo data. Not only are the SL0 and OMP algorithms utilized to reconstruct the sparse echo data to solve the problem of low classification probability in this case, but also the amplitude deviation coefficient, time domain waveform entropy and frequency domain waveform entropy are extracted to classify aircraft targets according to the analysis of the micro-Doppler effect of echo data. The proposed algorithm works on the multi-wave gates echo data in weighted features fusion, rather than single wave gate echo data, which is helpful for reducing the number of cross-value features of different targets. Experimental results show that the proposed algorithm can improve the classification probability of reconstructed echo data obtained by the SL0 and OMP methods, and four wave gates echo data in weighted features fusion used to extract and fuse features and to both train and test the SVM model is the optimal wave gate number for target classification.

Although our method is effective in aircraft target-type classification, it can still be improved further. In the future, we not only study the aircraft target classification algorithm within unmanned aerial vehicle (UAV) types, but also verify the effectiveness of the algorithm in the actual radar equipment by conducting an experiment with measured data.

Author Contributions: Writing—original draft preparation, W.W.; methodology, Z.T. and Y.C.; writing—review and editing, W.W. and Y.C.; data curation, Y.Z. and Y.S.

Funding: This work was supported in part by the National Natural Science Foundation of China under Grant 61901514, and in part by Young Talent Program of Air Force Early Warning Academy under Grant TJRC425311G11.

Acknowledgments: The authors would like to thanks to the editor and anonymous reviewers for processing our manuscript.

Conflicts of Interest: The authors declare no conflict of interest.

References

1. Pan, X.R.; Yang, F.; Gao, L.R.; Chen, Z.C.; Zhang, B.; Fan, H.R.; Ren, J.C. Building extraction from high-resolution aerial imagery using a generative adversarial network with spatial and channel attention mechanisms. *Remote Sens.* **2019**, *11*, 917. [CrossRef]
2. Deledalle, C.A.; Denis, L.; Poggi, G.; Tupin, F.; Verdoliva, L. Exploiting patch similarity for SAR image processing: The nonlocal paradigm. *IEEE Signal Process. Mag.* **2014**, *31*, 69–78. [CrossRef]
3. Zhao, C.H.; Wang, Y.; Masahide, K. An optimized method for image classification based on bag of words model. *J. Electron. Inf. Technol.* **2013**, *34*, 2064–2070. [CrossRef]
4. Wang, Y.P.; Hu, Y.H.; Lei, W.H. Aircraft target classification method based on texture feature of laser echo time-frequency image. *Acta Optica Sin.* **2017**, *37*, 1128004. [CrossRef]
5. Fu, K.; Dai, W.; Zhang, Y.; Wang, Z.R.; Yan, M.L.; Sun, X. Multiple class activation mapping for aircraft recognition in remote sensing images. *Remote Sens.* **2019**, *11*, 544. [CrossRef]
6. Karacor, A.G.; Torun, E.; Abay, R. Aircraft classification using image processing and artificial neural networks. *Int. J. Patt. Recogn. Artif. Intell.* **2011**, *25*, 1321–1335. [CrossRef]
7. Hwang, J.; Lin, K.; Chiu, Y. Automatic target recognition based on high-resolution range profiles with unknown circular range shift. In Proceedings of the IEEE International Symposium on Signal Processing & Informational Technology, Bilbao, Spain, 18–20 December 2017.
8. Jiang, Y.; Li, Y.; Cai, J.J.; Wang, Y.H.; Xu, J. Robust automatic target recognition via HRRP sequence based on scatter matching. *Sensors* **2018**, *18*, 593. [CrossRef] [PubMed]
9. Zhao, F.X.; Liu, Y.X.; Huo, K.; Zhang, S.H.; Zhang, Z.S. Radar HRRP target recognition based on stacked autoencoder and extreme learning machine. *Sensors* **2018**, *18*, 173. [CrossRef]

10. Suresh, P.; Thayaparan, T.; Obulesu, T. Extracting micro-Doppler radar signatures from signatures from rotating targets using Fourier-Bessel transform and time-frequency analysis. *IEEE Trans. Geosci. Remote Sens.* **2014**, *52*, 3204–3210. [CrossRef]
11. Smith, G.E.; Mobasseri, B.G. Robust through-the-wall radar image classification using a target-model alignment procedure. *IEEE Trans. Image Process.* **2012**, *21*, 754–767. [CrossRef]
12. Liu, J.; Fang, N.; Wang, B.F.; Xie, Y.J. Scale-space theory-based multi-scale features for aircraft classification using HRRP. *Electron. Lett.* **2016**, *52*, 475–477. [CrossRef]
13. Tang, N.; Gao, X.Z.; Li, X. Target classification of ISAR images based on feature space optimization of local non-negative matrix factorization. *IET Signal Process.* **2012**, *6*, 494–502. [CrossRef]
14. Lopez-Rodriguez, P.; Fernandez-Recio, R.; Bravo, I.; Gardel, A.; Lazaro, J.L.; Rufo, E. Computational burden resulting from image recognition of high resolution radar sensors. *Sensors* **2013**, *13*, 5381–5402. [CrossRef]
15. Wang, Y.; Zhu, P.K. Novel and comprehensive approach for the feature extraction and recognition method based on ISAR images of ship target. *J. Harbin Inst. Technol.* **2017**, *5*, 12–19.
16. Karine, A.; Toumi, A.; Khenchaf, A.; EI Hassouni, M. Radar target recognition using salient keypoint descriptors and multitask sparse representation. *Remote Sens.* **2018**, *10*, 843. [CrossRef]
17. Wang, F.; Sheng, W.X.; Ma, X.F.; Wang, H. ISAR image recognition with fusion of Gabor magnitude and phase feature. *J. Electron. Inf. Technol.* **2013**, *35*, 1813–1818. [CrossRef]
18. Saidi, M.N.; Daoudi, K.; Khenchaf, A.; Hoeltzener, B.; Aboutajdine, D. Automatic target recognition of aircraft models based on ISAR images. In Proceedings of the Geoscience & Remote Sensing Symposium IEEE, Honolulu, HI, USA, 25–30 July 2010.
19. Biondi, F.; Addabbo, P.; Orlando, D.; Clemente, C. Micro-motion estimation of maritime targets using pixel tracking in Cosmo-Skymed Synthetic Aperture Radar data-An operative assessment. *Remote Sens.* **2019**, *11*, 1637. [CrossRef]
20. Ji, J.Z.; Jiang, J.X.; Al-Armaghany, A.; Shu, C.Y.; Huang, P.L. Nutation and geometrical parameters estimation of cone-shaped target based on micro-Doppler effect. *Optik Int. J. Light Electron Opt.* **2017**, *150*, 1–10. [CrossRef]
21. Abdullah, R.S.A.R.; Alnaeb, A.; Salah, A.A.; Rashid, N.E.A.; Sali, A.; Pasya, I. Micro-Doppler estimation and analysis of slow moving objects in forward scattering radar system. *Remote Sens.* **2017**, *9*, 699. [CrossRef]
22. Thayaparan, T.; Stanković, L.J.; Daković, M.; Popović, V. Micro-Doppler parameter estimation from a fraction of the period. *IET Signal Process.* **2010**, *4*, 201–212. [CrossRef]
23. Zuo, L.; Li, M.; Zhang, X.W. Micro-Doppler parameter estimation from a fraction of the period data with the MMSE criterion. *J. Xidian Univ.* **2013**, *40*, 123–129.
24. Li, G.; Varshney, P.K. Micro-Doppler parameter estimation via parametric sparse representation and pruned orthogonal matching pursuit. *IEEE J. Sel. Top. Appl. Earth Observ. Remote Sens.* **2014**, *7*, 4937–4948. [CrossRef]
25. Liu, Y.X.; Li, X.; Zhuang, Z.W. Estimation of micro-motion parameters based on micro-Doppler. *IET Signal Process.* **2010**, *4*, 213–217. [CrossRef]
26. Song, C.; Wu, Y.R.; Zhou, L.J.; Li, R.M.; Yang, J.F.; Liang, W.; Ding, C.B. A multicomponent micro-Doppler signal decomposition and parameter estimation method for target recognition. *Sci. China Inf. Sci.* **2019**, *62*, 029304. [CrossRef]
27. Mujica, F.A.; Leduc, J.P.; Murenzi, R.; Smith, M.J.T. A new motion parameter estimation algorithm based on the continuous wavelet transform. *IEEE Trans. Image Process.* **2000**, *9*, 873–888. [CrossRef]
28. Wang, B.S.; Du, L.; He, H.; Liu, H.W. Reconstruction method for narrow-band radar returns with missing samples based on complex Gaussian model. *J. Electron. Inf. Technol.* **2015**, *37*, 1065–1070.
29. Yang, J.B.; Liao, X.J.; Yuan, X.; Llull, P.; Brady, D.J.; Sapiro, G.; Carin, L. Compressive sensing by learning a Gaussian mixture model from measurements. *IEEE Trans. Image Process.* **2015**, *24*, 106–119. [CrossRef]
30. Xue, H.; Zhang, S.J.; Su, Y.K.; Wu, Z.Z. Capital cost optimization for prefabrication: A factor analysis evaluation model. *Sustainability* **2018**, *10*, 159. [CrossRef]
31. Shi, H.R. Research on Feature Extraction of Micro Motion Target and Reconstruction of Incomplete Signal. Master' Thesis, XiDian University, Xi'an, China, May 2017.
32. Pedregosa, F.; Varoquaux, G.; Gramfort, A.; Michel, V.; Thirion, B.; Grisel, O.; Blondel, M.; Prettenhofer, P.; Weiss, R.; Dubourg, V.; et al. Scikit-learn: Machine learning in python. *J. Mach. Learn. Res.* **2011**, *12*, 2825–2830.

33. Li, Y.; Fu, K.; Sun, H.; Sun, X.W. An aircraft detection framework based on reinforcement learning and convolutional neural networks in remote sensing images. *Remote Sens.* **2018**, *10*, 243. [CrossRef]
34. Rikhtegar, A.; Pooyan, M.; Manzuri-Shalmani, M.T. Ga-optimized structure of CNN for face recognition applications. *IET Comput. Vis.* **2016**, *10*. [CrossRef]
35. Zhao, A.; Fu, K.; Wang, S.Y.; Zuo, J.W. Aircraft recognition based on landmark detection in remote sensing images. *IEEE Geosci. Remote Sens. Lett.* **2017**, *14*, 1413–1417. [CrossRef]
36. Zuo, J.W.; Xu, G.L.; Fu, K.; Sun, X.W. Aircraft type recognition based on segmentation with deep convolutional neural networks. *IEEE Geosci. Remote Sens. Lett.* **2018**, *15*, 282–286. [CrossRef]
37. Krizhevsky, A.; Sutskever, I.; Hinton, G. ImageNet classification with deep convolutional neural networks. In Proceedings of the 26th Annual Conference on Neural Information Processing Systems, Lake Tahoe, NV, USA, 3–6 December 2012.
38. Simonyan, K.; Zisserman, A. Very deep convolutional networks for large-scale image recognition. *arXiv* **2014**, arXiv:1409.1556.
39. Zhang, Y.H.; Sun, H.; Zuo, J.W.; Wang, H.Q.; Xu, G.L.; Sun, X. Aircraft type recognition in remote sensing images based on feature learning with conditional generative adversarial networks. *Remote Sens.* **2018**, *10*, 1123. [CrossRef]
40. Wang, D.; He, X.; Yu, H. A method of aircraft image target recognition based on modified PCA features and SVM. In Proceedings of the 9th International Conference on Electronic Measurement & Instruments, Beijing, China, 16–19 August 2009.
41. Diao, W.; Sun, X.; Dou, F.; Yan, M.; Wang, H.; Fu, K. Object recognition in remote sensing images using sparse deep belief networks. *Remote Sens. Lett.* **2015**, *6*, 745–754. [CrossRef]
42. Donoho, D.L. Compressed sensing. *IEEE Trans. Inf. Theory* **2006**, *52*, 1289–1306. [CrossRef]
43. Tropp, J.A.; Gilbert, A.C. Signal recovery from random measurements via orthogonal matching pursuit. *IEEE Trans. Inf. Theory* **2007**, *53*, 4655–4666. [CrossRef]
44. Li, S.D.; Pei, W.J.; Yang, J.; Hu, G.Q. OMP reconstruction algorithm via Bayesian model and its application. *Syst. Eng. Electron.* **2015**, *37*, 246–251.
45. Li, M.; Wu, J.J.; Zuo, L.; Song, W.J.; Liu, H.M. Aircraft target classification and recognition algorithm based on measured data. *J. Electron. Inf. Technol.* **2018**, *40*, 2606–2612.
46. Mario, G.; Thorsten, R.; Christoph, E.; Thomas, U. Remote sensing based binary classification of maize. Dealing with residual autocorrelation in sparse sample situations. *Remote Sens.* **2019**, *11*, 2172.
47. Man, Q.X.; Dong, P.L. Extraction of urban objects in cloud shadows on the basis of fusion of airborne LiDAR and hyperspectral data. *Remote Sens.* **2019**, *11*, 713. [CrossRef]
48. Gapper, J.J.; EI-Askary, H.; Linstead, E.; Piechota, T. Coral Reef change Detection in Remote Pacific islands using support vector machine classifiers. *Remote Sens.* **2019**, *11*, 1525. [CrossRef]
49. Kutyniok, G. *Compressed Sensing: Theory and Applications*; Cambridge University Press,: Cambridge, UK, 2012; pp. 1–20.
50. Rauhut, H.; Schnass, K.; Vandergheynst, P. Compressed sensing and redundant dictionaries. *IEEE Trans. Inf. Theory* **2008**, *54*, 2210–2219. [CrossRef]
51. Vapnik, V.N. *Statistical Learning Theory*; Wiley: New York, NY, USA, 1998.

Article

Image Formation of Azimuth Periodically Gapped SAR Raw Data with Complex Deconvolution

Yulei Qian * and Daiyin Zhu

Key Laboratory of Radar Imaging and Microwave Photonics, Ministry of Education, College of Electronic Information Engineering, Nanjing University of Aeronautics and Astronautics, 29 Jiangjun Avenue, Nanjing 211100, China; zhudy@nuaa.edu.cn

* Correspondence: qian_yulei@nuaa.edu.cn; Tel.: +86-025-8489-2410

Received: 3 October 2019; Accepted: 15 November 2019; Published: 18 November 2019

Abstract: The phenomenon of periodical gapping in Synthetic Aperture Radar (SAR), which is induced in various ways, creates challenges in focusing raw SAR data. To handle this problem, a novel method is proposed in this paper. Complex deconvolution is utilized to restore the azimuth spectrum of complete data from the gapped raw data in the proposed method. In other words, a new approach is provided by the proposed method to cope with periodically gapped raw SAR data via complex deconvolution. The proposed method provides a robust implementation of deconvolution for processing azimuth gapped raw data. The proposed method mainly consists of phase compensation and recovering the azimuth spectrum of raw data with complex deconvolution. The gapped data become sparser in the range of the Doppler domain after phase compensation. Then, it is feasible to recover the azimuth spectrum of the complete data from gapped raw data via complex deconvolution in the Doppler domain. Afterwards, the traditional SAR imaging algorithm is capable of focusing the reconstructed raw data in this paper. The effectiveness of the proposed method was validated via point target simulation and surface target simulation. Moreover, real SAR data were utilized to further demonstrate the validity of the proposed method.

Keywords: Synthetic Aperture Radar (SAR); focusing; periodically gapped data; complex deconvolution

1. Introduction

The development and advancement of Synthetic Aperture Radar (SAR) improve humans' abilities to obtain more abundant information for target surveillance with high resolution images and videos [1–9]. Along with the development process of SAR, the stability of missing or gapped raw data attracts research interest due to new mission requirements, sensor geometries, jamming, or interference that cause sparse and irregular sampling [10]. State-of-the-art radar systems usually operate with large bandwidth and produce huge amount of data, according to the Nyquist theorem [11]. To achieve a balance between the requirement of large bandwidth and a huge amount data, azimuth periodically gapped sampling is considered to contribute to this tradeoff. Azimuth periodically gapped sampling has the potential to reduce the pressure of the big mass of data caused by large bandwidth requirement. Figure 1 shows the sampling scheme of periodically gapped sampling in the azimuth direction, and Figure 2 demonstrates the sampling scheme of uniform sampling in the azimuth direction. As shown in Figure 1, radar transmits M_R pulses and gaps M_G pulses repeatedly in the scenario of periodically gapped sampling. In Figure 2, M_P equals the sum of M_R and M_G. Compared with aperiodic gapping, periodic gapping is more attractive. Periodic gapping is more practical than aperiodic gapping for SAR system design. It is more feasible to design a SAR system with transmitting periodically gapped pulses than a SAR system with transmitting randomly gapped pulses. Therefore, it is valuable to develop methods for focusing azimuth periodically gapped raw SAR data. This paper mainly presents a study on image formation from azimuth periodically gapped raw SAR data.

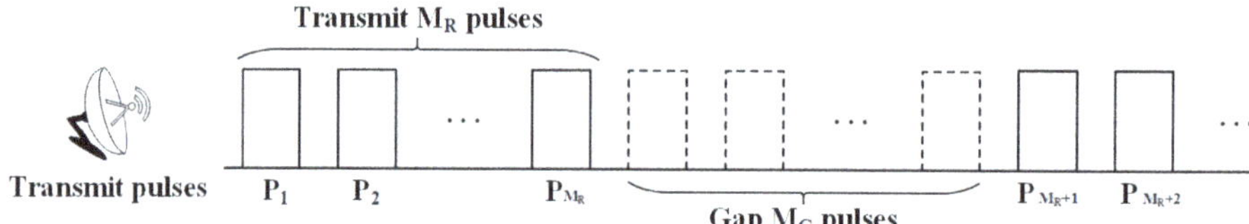

Figure 1. Sampling scheme of periodically gapped sampling in the azimuth direction.

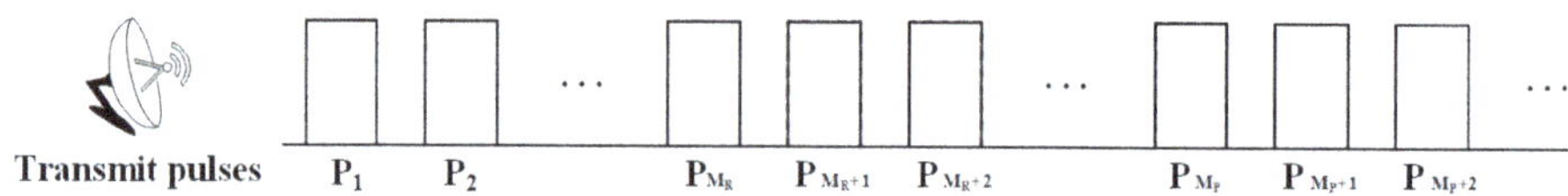

Figure 2. Sampling scheme of uniform sampling in the azimuth direction.

Nevertheless, conventional SAR imaging algorithms usually perform unsatisfactorily in focusing periodically gapped raw SAR data. Therefore, novel methods are expected to be developed to obtain focused images from periodically gapped raw SAR data.

Efforts have been invested into investigating the missing or gapped SAR data. In [12], the mitigating effects of missing data for coherent SAR images are presented. The authors of [12] showed that it is not necessary for coherent change detection to construct the information within the missing data samples. In [13], a novel geometric approach is proposed to reconstruct a coherent pair, which can be extended for coherent stacks. However, the data loss is less than 2% of the synthetic aperture length in [13]. In addition, the Burg algorithm, which is an auto-regressive linear prediction method, is applied to cope with gapped SAR raw data [14,15]. The Burg algorithm is an order recursive algorithm for obtaining the reflection coefficients that minimize the sum of the forward and backward prediction errors, which deduces auto-regressive parameters via the Levinson–Durbin recursion [16]. The Burg algorithm is utilized to interpolate the missing data in the gaps in order to improve the SAR focusing quality from the gapped data. In [14], the Burg algorithm is implemented to achieve gap-filling in gapped SAR raw data, which accommodates interrupted SAR data collections caused by the severe timeline requirements of the multi-mode functions of modern radars. In [15], an improved full-aperture imaging algorithm for scanning SAR is presented, which fills the data gaps between bursts with the Burg algorithm. The scanning SAR data can be regarded as periodically gapped data in the scenario of full-aperture processing. The proposed method in [15] can significantly improve the image quality of scanning SAR from periodically gapped data via aperture interpolation with the Burg algorithm.

Besides the auto-regressive linear prediction method, several spectral analysis methods are also proposed to cope with gapped data. The Amplitude and Phase Estimation (APES) approach is a successful filterbank-based estimator [17,18]. The Gapped-data Amplitude and Phase Estimation (GAPES) method was developed as an extension of the APES estimator for gapped data and is based on minimizing the least square criterion with respect to the gapped data [19]. The GAPES is mainly used to estimate the adaptive filter, obtaining the corresponding spectrum of the adaptive filter via APES and interpolating the gapped part of the data with a least square fit [20]. Subsequently, the Periodically-Gapped APES (PG-APES) method is devised in [21] to process periodically gapped data. The PG-APES method exploits the structure of the gapped data and is an interpolation-free extension of the APES method.

Additionally, as missing or gapped data can be regarded as sparse sampling data, the sparse optimization method can also be implemented on sparse SAR and Inverse SAR (ISAR) data [22–26]. In [22], a novel ISAR imaging algorithm with a relaxation technique is proposed to focus azimuth randomly sparse ISAR data. The proposed method in [22] is capable of obtaining high quality images from partially available ISAR data. In [23], a new method is proposed to deal with azimuth periodically gapped SAR raw data. The presented method in [23] mainly consists of phase compensation and reconstructing raw data in the range of the Doppler domain with the greed method [27,28].

The implemented greedy method in [23] is the Generalized Orthogonal Matching Pursuit (GOMP) [29] algorithm. In [24], a novel approach for sparse aperture bistatic ISAR is proposed to acquire a focused image of a highly mobile target with complex motion. The presented method in [24] effectively achieves fundamental motion compensation, range, and cross-range scaling, as well as simultaneously achieving bistatic distortion correction for bistatic ISAR. In [25], a new sparse SAR imaging method using the multiple measurement vectors model is proposed with a reduced computation cost and enhanced image quality. The presented method in [25] utilizes the structural information and matched filter processing to implement imaging under sub-Nyquist sampling. In [26], a novel method is proposed to focus the azimuth missing SAR raw data via segmented recovery. The method in [26] mainly consists of multiplication with a designed reference function in the time domain and subsequent segmented recovery in the two-dimensional frequency domain with Stagewise Orthogonal Matching Pursuit (StOMP) [30].

To obtain a focused SAR image from the azimuth periodically gapped raw data, a novel method based on complex deconvolution is proposed in this paper. The proposed method allows the robust implementation of deconvolution to handle azimuth gapped raw data. The proposed method has the potential to process azimuth periodically gapped data for a complicated scene with more frequent gapping. Complex deconvolution is utilized to recover the azimuth spectrum of complete data from the gapped raw data in the proposed method. Thus, a new approach is presented through the proposed method for processing periodically gapped SAR raw data via complex deconvolution. Azimuth periodically gapped raw data can be considered as the multiplication between complete raw data and the rectangular pulse sequence in the azimuth time domain after zero padding in gapped positions. As a result, the spectrum of the gapped signal is the convolution of the spectrum of the original signal and the spectrum of the gapping pattern [31]. In other words, the azimuth spectrum of the gapped data after zero padding can be expressed as the convolution between the azimuth spectrum of the complete data and the spectrum of the rectangular pulse sequence. Consequently, it is practicable to acquire the azimuth spectrum of the complete data by deconvoluting the azimuth spectrum of azimuth periodically gapped data with a spectrum of rectangular pulse sequences. Inspired by the Polar Format Algorithm (PFA) [32], a phase compensation function is designed to concentrate the energy of the azimuth spectrum of periodically gapped data in order to obtain a sparser azimuth spectrum. After multiplication of the phase compensation function in the range frequency domain, the azimuth spectrum of the complete data is able to be restored via complex deconvolution in the range of the Doppler domain. The expression of the spectrum of the rectangular pulse sequence is derived in this paper and treated as a point spread function in complex deconvolution. The Iterative Shrinkage-Thresholding Algorithm (ISTA) [33] is utilized to implement deconvolution in this paper. After restoration of the azimuth spectrum of complete data, the Range Migration Algorithm (RMA) [34] is implemented to focus the recovered SAR raw data. The RMA is a traditional SAR imaging algorithm.

In comparison with existing algorithms, the proposed method has the potential to process azimuth periodically gapped data for a complicated scene with more frequent gapping. Moreover, the date loss rate can achieve 50%. In [13], the proposed method performs well on real SAR data. However, the data loss rate is less than 2% of the synthetic length, as mentioned in [13]. A method is proposed in [15] for scanning SAR and filling gaps with the Burg algorithm. In [15], the data loss rate in the azimuth direction is no greater than 33.3%, and gapping is not frequent. A cycle period has 600 gapped pulses and more than 1200 available pulses in [15]. The Burg algorithm performs unsatisfactorily when a cycle period has 16 gapped pulses and 16 available pulses. The PG-APES is proposed in [21] for periodically gapped data. Nevertheless, the PG-APES is capable of achieving better results for the data of a simple scene. In [23], a method is also proposed for dealing with azimuth periodically gapped data. However, the method in [23] tends to obtain results with unexpected fake strong scatters when a cycle period has 16 gapped pulses and 16 available pulses. In contrast, the proposed method has the potential to accommodate azimuth periodically gapped data of a complex scene with more frequent gapping and a 50% data loss. In experiments, point target simulation is operated to verify the validity of the proposed method. In experiments, 50% of the data is periodically gapped in the azimuth direction. Real SAR data is also utilized to further validate the effectiveness of the proposed method. Vectors and matrices are in bold italics, while variables are in italics in this paper.

The remaining part of this paper is structured as follows. The proposed method for focusing the azimuth periodically gapped SAR raw data via complex deconvolution is derived and presented in Section 2. In Section 3, experiments are presented that verified the effectiveness of the proposed method. The presented experiments included point target simulations and real SAR data processing. Analyses and discussions of the experimental results are included in Section 4. Finally, Section 5 gives the conclusions.

2. Focusing the Azimuth Periodically Gapped SAR Raw Data via Deconvolution

The proposed method for obtaining a focused image with deconvolution is described in this section. The phase compensation function, the restoration via deconvolution, a brief introduction of the RMA, and the procedure of the proposed method are presented.

2.1. Phase Compensation Function

Deconvolution can be regarded as a linear inverse problem. It is beneficial to solve linear inverse problems in a sparser domain [33]. In this case, it is significant to process the azimuth periodically gapped raw data in a sparser domain. In Section 2.1, a phase compensation function is designed to match the azimuth periodically gapped raw data in the range frequency domain. The phase compensation is implemented on raw data to remove modulation in the range direction and reduce most modulation in the azimuth direction. The data are compressed along the range direction in the range time domain due to the removal of range modulation. Moreover, the data are coarsely compressed along the azimuth direction in the azimuth frequency domain because of the reduction of the azimuth modulation. The azimuth frequency domain is also known as the Doppler domain. Therefore, the data are coarsely compressed in the range of the Doppler domain along both the range and azimuth direction after phase compensation. In other words, the energy of the data is more concentrated in the range of the Doppler domain after phase compensation. As a result, the gapped data become sparser in the range of the Doppler domain. Consequently, it is proper to restore the gapped data in the range of the Doppler domain after phase compensation. The description of the phase compensation function is derived and presented in this part.

In this paper, SAR is assumed to emit a pulse chirp signal and move with a uniform linear motion. After demodulation to its baseband, the echo data of a single point can be described by the range time τ and azimuth time η as below:

$$\begin{aligned} s(\tau,\eta) = \; & \boldsymbol{w_r}\left[\tau - \frac{2\boldsymbol{R}(\eta)}{c}\right]\boldsymbol{w_a}(\eta)\exp\left[\frac{-j4\pi f_0 \boldsymbol{R}(\eta)}{c}\right] \\ & \times \exp\left\{j\pi K_r\left[\tau - \frac{2\boldsymbol{R}(\eta)}{c}\right]^2\right\}, \eta \in Q_D \end{aligned} \tag{1}$$

In Equation (1), the amplitude factors are neglected. $\boldsymbol{w_r}$ is the range envelope, and $\boldsymbol{w_a}$ is the azimuth envelope. The velocity of light is denoted as c. The range frequency modulation rate is denoted as K_r. $\boldsymbol{R}(\eta)$ is the slant range between the target and radar at η, f_0 is the carrier frequency, and j is the imaginary unit. The complete SAR raw dataset is defined as a matrix with the size of $N_r \times N_a$. The complete SAR raw data have N_r and N_a sampling points in the range and azimuth directions, respectively. Set D is defined as a set that includes indices of the available data in the azimuth periodically gapped raw data. The quantity of elements in set D is represented as N_D. In comparison with the complete SAR raw data, the set of available azimuth sampling time in the azimuth periodically gapped data is denoted as Q_D.

The slant range model can be regarded as a parabolic approximation of the Taylor expansion, as below:

$$\boldsymbol{R}(\eta)=\sqrt{R_0^2+(v\eta)^2}\approx R_0+\frac{v^2\eta^2}{2R_0} \tag{2}$$

In Equation (2), v represents the velocity of the SAR platform, and R_0 represents the closest slant range. By operating Fourier transform (FT) along the range direction, the signal can be approximated as below:

$$\boldsymbol{S}(f_\tau,\eta)=\boldsymbol{W_r}(f_\tau)\boldsymbol{w_a}(\eta)\exp\left(-\frac{j\pi f_\tau^2}{K_r}\right)\exp\left[-j4\pi\frac{(f_0+f_\tau)\boldsymbol{R}(\eta)}{c}\right] \tag{3}$$

Inspired by PFA [32], the phase compensation function is designed as below:

$$\boldsymbol{\theta}(f_\tau,\eta)=\exp\left(\frac{j\pi f_\tau^2}{K_r}\right)\exp\left[\frac{j4\pi(f_0+f_\tau)\boldsymbol{R}_{ref}(\eta)}{c}\right] \tag{4}$$

The reference slant range at the azimuth time is presented in the following formula:

$$\boldsymbol{R}_{ref}(\eta)=\sqrt{R_{0,ref}^2+(v\eta)^2}\approx R_{0,ref}+\frac{v^2\eta^2}{2R_{0,ref}} \tag{5}$$

In Equation (5), $R_{0,\mathrm{ref}}$ is the closest reference slant range. Afterwards, the signal after multiplication with the phase compensation function can be expressed as below:

$$\begin{aligned}\boldsymbol{S}_c(f_\tau,\eta)&=\boldsymbol{S}(f_\tau,\eta)\boldsymbol{\theta}(f_\tau,\eta)\\&=\boldsymbol{W_r}(f_\tau)\boldsymbol{w_a}(\eta)\exp\left\{\frac{-j4\pi(f_0+f_\tau)\left[\boldsymbol{R}(\eta)-\boldsymbol{R}_{ref}(\eta)\right]}{c}\right\}\end{aligned} \tag{6}$$

By implementing inverse FT to $\boldsymbol{S}_c(\tau,f_\eta)$ along the range direction, the result can be presented as below:

$$\boldsymbol{s}_c(\tau,\eta)=\boldsymbol{p_r}\left\{\tau-\frac{2\left[\boldsymbol{R}(\eta)-\boldsymbol{R}_{ref}(\eta)\right]}{c}\right\}\boldsymbol{w_a}(\eta)\exp\left\{-j4\pi f_0\frac{\left[\boldsymbol{R}(\eta)-\boldsymbol{R}_{ref}(\eta)\right]}{c}\right\} \tag{7}$$

In Equation (7), $\boldsymbol{p_r}$ denotes the sinc function in the range direction. Substituting Equations (2) and (5) into Equation (7), $\boldsymbol{s}_c(\tau,\eta)$ can be presented as follows:

$$\boldsymbol{s}_c(\tau,\eta)=\boldsymbol{p_r}\left\{\tau-\frac{2\left[\boldsymbol{R}(\eta)-\boldsymbol{R}_{ref}(\eta)\right]}{c}\right\}\boldsymbol{w_a}(\eta)\exp\left\{-j4\pi f_0\frac{(R_0-R_{0,ref})}{c}\right\}\exp\left\{\frac{-j2\pi f_0v^2}{c}\left(\frac{1}{R_0}-\frac{1}{R_{0,ref}}\right)\eta^2\right\} \tag{8}$$

As illustrated in Equation (8), the second exponential term shows that the bandwidth of the azimuth modulation is almost removed after phase compensation in the range frequency. In addition, the signal has been compressed in the range direction, as the range modulation has been removed by phase compensation in the range frequency. Consequently, $\boldsymbol{s}_c(\tau,\eta)$ is considered to be coarsely compressed in the range of the Doppler domain. In other words, the $\boldsymbol{S}_c(\tau,f_\eta)$ is coarsely compressed.

The coarse compression in the range of the Doppler domain after phase compensation is presented in Figures 3 and 4. Figures 3 and 4 demonstrate the coarse compression of point targets and real SAR data, respectively. In the depiction of the coarse compression of point targets, nine point targets are positioned to generate echo data. A distribution of the nine point targets is displayed in Figure 3a. The original echo data in the time domain are shown in Figure 3b. As shown in Figure 3b, the original echo data are dense in the time domain. Figure 3c,d illustrates the original echo data and the data with compensation in the range of the Doppler domain, respectively. Figure 3c shows that the original echo data are also dense in the range of the Doppler domain. Figure 3d indicates that the data become sparser around the Doppler domain after phase compensation in the range of the frequency domain. Figure 4a shows the original echo data for real SAR data in the time domain, and Figure 4b demonstrates the original echo data for real SAR data in the range of the Doppler domain. Moreover, the real SAR data in the range of the Doppler domain after phase compensation is presented in Figure 4c. Figure 4c illustrates the real SAR data in coarse compression in the range of the Doppler domain after phase compensation. The $s(\tau,\eta)$ denotes the original echo data in the time domain. The $s(\tau,f_\eta)$ denotes the original echo data in the range of the Doppler domain. The $S_c(\tau,f_\eta)$ denotes the data in the range of the Doppler domain after phase compensation.

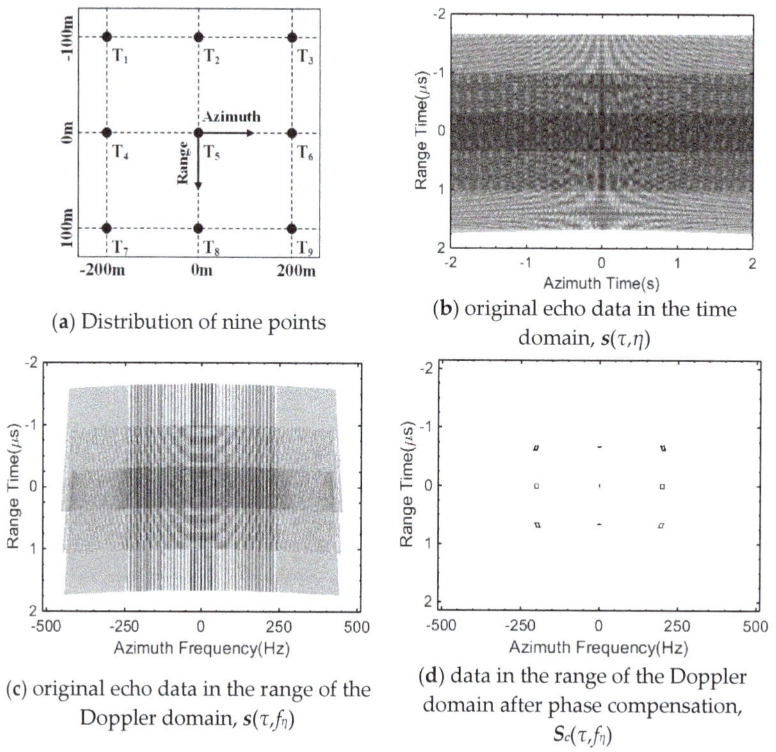

Figure 3. Coarse compression of point targets: (**a**) distribution of nine points; (**b**) original echo data in the time domain, $s(\tau,\eta)$; (**c**) original echo data in the range of the Doppler domain, $s(\tau,f_\eta)$; and (**d**) data in the range of the Doppler domain after phase compensation, $S_c(\tau,f_\eta)$.

To further verify the coarse compression in the range of the Doppler domain after phase compensation, the analysis of coarse compression is listed in Table 1. Image contrast (IC) [35] and Image Entropy (IE) [36] are chosen as the criteria to assess the sparsity of the original echo data in the time domain, the original echo data in the range of the Doppler domain, and the data in the range of the Doppler domain after phase compensation. The IE and IC in Table 1 are all obtained with whole images in Figure 3b–d, Figure 4a–c. The more concentrated the energy of the SAR image is, the sparser the SAR image is. Moreover, if the energy of the SAR image is more concentrated, the IC of the SAR image has an increasing trend, and the IE of SAR image has a decreasing trend. As demonstrated in Table 1, the data in the range of the Doppler Domain after phase compensation have the largest IC and smallest IE for point targets and real SAR data, in contrast with the original echo data in the time domain and the range of the Doppler domain. As demonstrated in Table 1, the effect of concentrating the energy of data is more apparent in point scatters than in real SAR data. Similarly, the presented transformation for making data sparser also has a limited effect on surface scatters. Consequently, it can be concluded that the data in the range of the Doppler Domain after phase compensation are sparser than the original echo data in the time domain and the Doppler domain. In summary, the SAR data become much sparser around the Doppler domain after phase compensation in the range frequency domain.

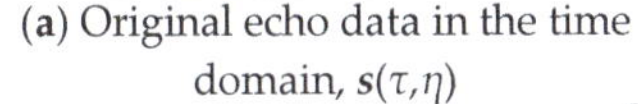

(**a**) Original echo data in the time domain, $s(\tau,\eta)$

(**b**) original echo data in the range of the Doppler domain, $s(\tau,f_\eta)$

(**c**) data in the range of the Doppler domain after phase compensation, $S_c(\tau,f_\eta)$

Figure 4. The coarse compression of real SAR data: (**a**) original echo data in the time domain, $s(\tau,\eta)$; (**b**) original echo data in the range of the Doppler domain, $s(\tau,f_\eta)$; and (**c**) data in the range of the Doppler domain after phase compensation, $S_c(\tau,f_\eta)$.

Table 1. Analysis of coarse compression.

	IE		IC	
	Point Target	**Real SAR Data**	**Point Target**	**Real SAR Data**
$s(\tau,\eta)$	15.6475	13.9677	1.3344	1.1625
$s(\tau,f_\eta)$	15.3237	13.8575	1.8670	1.4381
$S_c(\tau,f_\eta)$	8.5858	13.1141	344.5768	11.6098

Although the proposed phase compensation function is effective in the described scenario in Section 2.1, it also has limitations. The compensation occurs only for the reconstruction strategies because the interpolation error is data dependent. The proposed method performs satisfactorily over point scatters while the performance of the proposed method over surface scatters may decrease. If the platform of SAR has complicated motion rather than uniform linear motion, the approximation in Equation (5) may become invalid. A possible solution for the inaccuracy of Equation (5) is to develop a more accurate range model to accommodate for complicated motion. In addition, the proposed transformation does not consider high squint. Consequently, the proposed phase compensation requires extension in the scenario with high squint. Moreover, the proposed transformation may perform better in a wide-swath scenario with the implementation of a proper sub-block partition.

2.2. Restoration via Deconvolution

As presented in Section 2.1, the raw data become sparser in the range of the Doppler domain after phase compensation in the range of the frequency domain. Hence, it is feasible to recover the azimuth periodically gapped raw data in the Doppler domain after phase compensation. In this part, the description of restoring the azimuth periodically gapped raw data in the Doppler domain via complex deconvolution is presented.

Figure 5 describes the sampling of SAR azimuth periodically gapped data. The shadow square represents the received pulse and the white square represents the gapped pulse. The complete dataset has N_a sampling pulses, while the gapped data have N_D sampling pulses in the azimuth direction. The azimuth periodically gapped data has N_C clusters of samples, and each cluster contains M_R available pulses and M_G gapped pulses. M_P denotes the periodicity of the gapping and equals the sum of M_R and M_G. Therefore, N_a=(M_R+M_G)N_C=$M_P N_C$ and N_D=$M_R N_C$. In this paper, the proposed method is developed to cope with the SAR data, which is periodically gapped in the azimuth direction. As a result, the structure of the periodically gapped data is exploited to derive restoration with complex deconvolution in the following section.

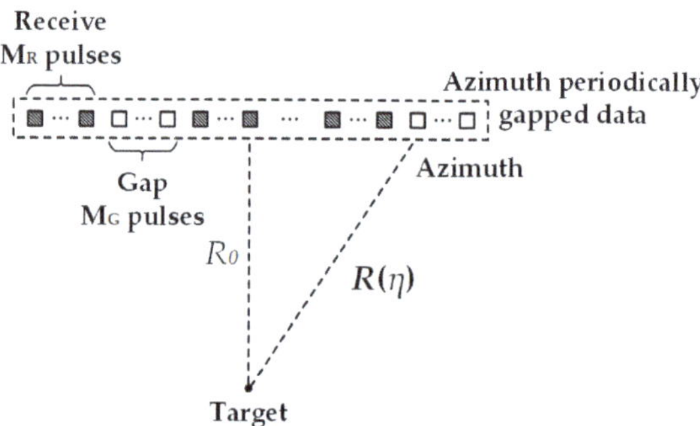

Figure 5. Sampling of the SAR azimuth periodically gapped data.

Figure 6 demonstrates filling gaps with zeros. As shown in Figure 6, the positions of gaps are filled with zero vectors. A zero vector has the same size as the pulse in the echo data. Therefore, the size of the data is N_r×N_a after the adding zeros operation.

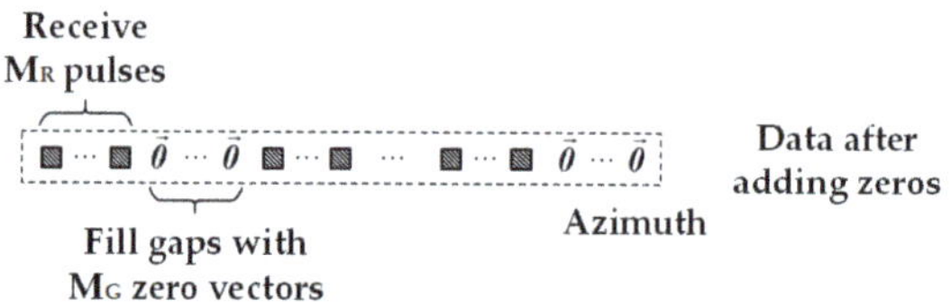

Figure 6. Demonstration of filling gaps with zeros.

After zero padding in the gapped positions of the azimuth gapped data, the azimuth periodically gapped raw data can be treated as the multiplication between the complete raw data and the rectangular pulse sequence. The rectangular pulse sequence can be expressed as follows:

$$y(m) = \begin{cases} 1, & 1 \le \mathrm{mod}(m, \mathrm{M_P}) \le \mathrm{M_R} \\ 0, & \mathrm{M_R} + 1 \le \mathrm{mod}(m, \mathrm{M_P}) \le \mathrm{M_P} \end{cases} \quad m = 1, 2, 3, \ldots, N_a \tag{9}$$

In Equation (9), the mod (m,$\mathrm{M_P}$) denotes modulo operation, where m is the dividend and $\mathrm{M_P}$ is the divisor. The modulo operation returns the remainder after the division of m by $\mathrm{M_P}$.

Figure 7 presents the demonstration of the rectangular pulse sequence. The rectangular pulse sequence is acquired according to Equation (9). In Figure 7, O denotes the origin of coordinates. The $y(m)$ is plotted on the vertical axis against m on the horizontal axis in Figure 7. The black solid lines in Figure 7 represent elements that equal "1" in the rectangular pulse sequence, and the red solid lines in Figure 7 represents elements that equal "0" in the rectangular pulse sequence. The $y(m)$ is an $\mathrm{N_a}\times 1$ vector.

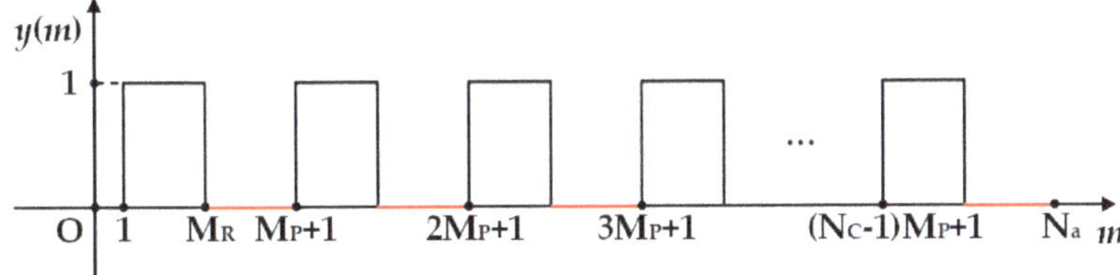

Figure 7. Demonstration of rectangular pulse sequence.

Azimuth periodically gapped raw data can be regarded as the multiplication between the complete raw data and rectangular pulse sequence in the azimuth time domain after zero padding. Hence, the azimuth spectrum of gapped data after zero padding can be represented as the convolution between the azimuth spectrum of complete data and the spectrum of the rectangular pulse sequence. Consequently, it is feasible to restore the azimuth spectrum of complete data by deconvoluting the azimuth spectrum of the azimuth periodically gapped data with the spectrum of the rectangular pulse sequence. In complex deconvolution, the spectrum of the rectangular pulse sequence acts as the point spread function. Therefore, it is necessary to derive the expression for the spectrum of the rectangular pulse sequence.

The expression for the spectrum of the rectangular pulse sequence in the discrete domain is derived with Equation (9) and the definition of Fast Fourier Transform (FFT), as follows:

$$\begin{aligned} \boldsymbol{Y}(k) &= \sum_{m=1}^{N_a} \boldsymbol{y}(m) W_{N_a}^{(\mathrm{m}-1)(\mathrm{k}-1)} \\ &= (W_{N_a}^{0\cdot(\mathrm{k}-1)} + W_{N_a}^{1\cdot(\mathrm{k}-1)} + W_{N_a}^{2\cdot(\mathrm{k}-1)} + \cdots + W_{N_a}^{(\mathrm{M_R}-1)\cdot(\mathrm{k}-1)})(1 + \boldsymbol{q} + \boldsymbol{q}^2 + \ldots + \boldsymbol{q}^{\mathrm{N_C}-1}) \\ k &= 1, 2, 3, \ldots, N_a \end{aligned} \tag{10}$$

The definitions of $\boldsymbol{W_{Na}}$ and $\boldsymbol{q}$ are expressed in Equations (11) and (12), as follows:

$$\mathbf{W}_{N_a} = \mathbf{e}^{-j\frac{2\pi}{N_a}} \tag{11}$$

$$\boldsymbol{q} = \mathbf{W}_{N_a}^{M_P(k-1)} \tag{12}$$

As shown in Equation (10), the spectrum of the rectangular pulse sequence can be represented as the multiplication of two geometric sequences. According to the summation formula of the geometric sequence, the spectrum of the rectangular pulse sequence in the discrete domain can be expressed as follows:

$$\mathbf{Y}(k) = \begin{cases} M_R N_C & ,k=1 \\ \frac{1-W_{N_a}^{M_R(k-1)}}{1-W_{N_a}^{k-1}} \cdot \frac{1-q^{N_C}}{1-q} & ,k=2,3,\ldots,N_a \end{cases} \tag{13}$$

Figure 8 presents the procedure for restoration via complex deconvolution. The azimuth periodically gapped raw data are processed in the Doppler domain after phase compensation. The $\boldsymbol{g}(\tau,\eta)$ denotes the azimuth periodically gapped data after phase compensation in the range frequency domain and a subsequent adding zeros operation in the time domain. The adding zeros operation fills the positions of gaps with zero vectors. The size of the $\boldsymbol{g}(\tau,\eta)$ is $N_r \times N_a$. The $\boldsymbol{s}_c(\tau,\eta)$ denotes the complete echo data in the time domain after phase compensation in the range frequency domain. The size of $\boldsymbol{s}_c(\tau,\eta)$ is $N_r \times N_a$.

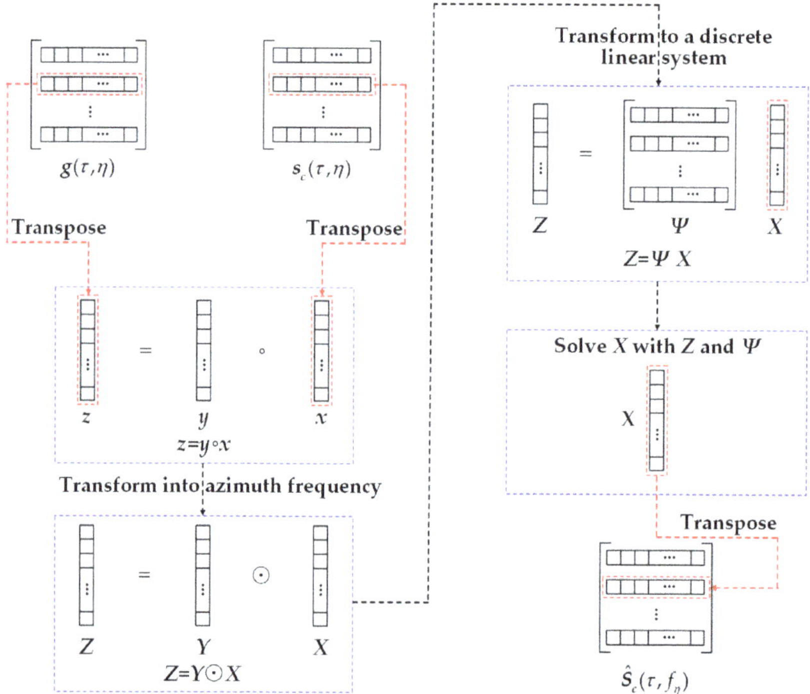

Figure 8. Procedure of restoration via complex deconvolution.

The $N_a \times 1$ vector $\boldsymbol{x}$ is defined as the transposed vector of any range time τ in the data $\boldsymbol{s}_c(\tau,\eta)$. The $N_a \times 1$ vector $\boldsymbol{z}$ is defined as the transposed vector of any range time τ in the data $\boldsymbol{g}(\tau,\eta)$. Thus, there exists a connection, as follows:

$$\boldsymbol{z} = \boldsymbol{y} \circ \boldsymbol{x} \tag{14}$$

In Equation (14), the $\circ$ denotes the pointwise multiplication operation between two vectors. On the basis of the properties of Fourier transform, the relationship between the spectrums of $\boldsymbol{x}$, $\boldsymbol{y}$, and $\boldsymbol{z}$ can be expressed as follows:

$$\mathbf{Z} = \mathbf{Y} \odot \mathbf{X} \tag{15}$$

In Equation (15), $\odot$ denotes the circular convolution operation between two vectors. As the lengths of $\boldsymbol{X}$, $\boldsymbol{Y}$, and $\boldsymbol{Z}$ are all N_a, the convolution in Equation (15) is circular convolution rather than linear convolution.

This is the key point to solving the unknown X with the known Y and Z in Equation (15). To solve this deconvolution problem more conveniently, it is feasible to transform Equation (15) into another discrete linear system, as follows:

$$\boldsymbol{Z} = \boldsymbol{\Psi X} \tag{16}$$

In Equation (16), the matrix $\boldsymbol{\Psi}$ is the circular convolution operator. Multiplication between the matrix $\boldsymbol{\Psi}$ and the vector $\boldsymbol{X}$ equals the circular convolution between vector $\boldsymbol{Y}$ and vector $\boldsymbol{X}$. The matrix $\boldsymbol{\Psi}$ is a circulant matrix, which is formed with vector $\boldsymbol{Y}$ according to the definition of circular convolution. The matrix $\boldsymbol{\Psi}$ can be expressed as follows:

$$\boldsymbol{\Psi} = \begin{bmatrix} Y(\mathrm{N}_a) & Y(\mathrm{N}_a-1) & Y(\mathrm{N}_a-2) & Y(\mathrm{N}_a-3) & \cdots & Y(1) \\ Y(1) & Y(\mathrm{N}_a) & Y(\mathrm{N}_a-1) & Y(\mathrm{N}_a-2) & \cdots & Y(2) \\ Y(2) & Y(1) & Y(\mathrm{N}_a) & Y(\mathrm{N}_a-1) & \cdots & Y(3) \\ \vdots & \vdots & \vdots & \vdots & \ddots & \vdots \\ Y(\mathrm{N}_a-2) & Y(\mathrm{N}_a-3) & Y(\mathrm{N}_a-4) & Y(\mathrm{N}_a-5) & \cdots & Y(\mathrm{N}_a-1) \\ Y(\mathrm{N}_a-1) & Y(\mathrm{N}_a-2) & Y(\mathrm{N}_a-3) & Y(\mathrm{N}_a-4) & \cdots & Y(\mathrm{N}_a) \end{bmatrix} \tag{17}$$

As shown in Equation (17), the matrix $\boldsymbol{\Psi}$ has a form where each row is a cyclic shift of the row above it. The size of the matrix $\boldsymbol{\Psi}$ is $N_a \times N_a$.

Consequently, the deconvolution problem is transformed by estimating the unknown $\boldsymbol{X}$ from the known $\boldsymbol{Z}$ and $\boldsymbol{\Psi}$. This problem can be cast as a linear inverse problem, and the ISTA is capable of dealing with this problem. The ISTA (Iterative Shrinkage-Thresholding Algorithm) [33] solves the linear inverse problem via the method of l_1 regularization. The deconvolution belongs to an ill-conditioned problem and has a very large norm. In this case, regularization is utilized to replace the large norm by approximants with a smaller norm, so numerically stable solutions can be defined and used as meaningful approximations of the true solution corresponding to the exact data [33]. The l_1 regularization seeks to find the solution of

$$\min_{\boldsymbol{X}}\left\{F(\boldsymbol{X}) \equiv \|\boldsymbol{\Psi X}-\boldsymbol{Z}\|^2+\beta\|\boldsymbol{X}\|_1\right\} \tag{18}$$

In Equation (18), the $\|\bullet\|_1$ stands for the sum of the absolute values of the components of $\boldsymbol{X}$. The general step of ISTA is

$$\boldsymbol{X}^{p+1} = \boldsymbol{\Gamma}_{\beta t}\left[\boldsymbol{X}^p - 2t\boldsymbol{\Psi}^T(\boldsymbol{\Psi X}^p - Z)\right] \tag{19}$$

In Equation (19), t is the proper stepsize, p denotes the iteration number, and Γ_α is the shrinkage operator defined as

$$\boldsymbol{\Gamma}_{\beta t}(\boldsymbol{B}) = \max(|\boldsymbol{B}| - \beta t, 0)\mathrm{sgn}(\boldsymbol{B}) \tag{20}$$

In Equation (20), the max(·) denotes finding the maximum value operation and the sgn(·) denotes the sign function. As the sign function is included in Equation (20), the shrinkage operator in Equation (20) can only be implemented in the field of real numbers. However, SAR data are complex data. To process complex data, the shrinkage operator needs modification. The modified shrinkage operator for dealing with complex data is given in [37], as follows:

$$\boldsymbol{\Gamma}_{\beta t}(\boldsymbol{B}) = \max(|\boldsymbol{B}| - \beta t, 0)\frac{\boldsymbol{B}}{|\boldsymbol{B}|} \tag{21}$$

With the implementation of ISTA via Equations (19) and (21), $\boldsymbol{X}$ can be restored from the known $\boldsymbol{Y}$ and $\boldsymbol{Z}$ via complex deconvolution. Consequently, the azimuth spectrum of complete data can be recovered by deconvoluting the azimuth spectrum of the azimuth periodically gapped data with the spectrum of the rectangular pulse sequence in the range of the Doppler domain. The estimation of $\boldsymbol{S}_c(\tau, f_\eta)$ is denoted as $\hat{\mathrm{S}}_c(\tau, f_\eta)$, which is restored from $\boldsymbol{g}(\tau, f_\eta)$ and $\boldsymbol{Y}$ via the operation of deconvolution. Then, the recovery of the two-dimensional (2D) spectrum of the raw data can be acquired as follows:

$$\hat{\mathrm{S}}_{2\mathrm{df}}(f_\tau, f_\eta) = \mathrm{FTa}\left\{\mathrm{FTr}\left\{\mathrm{IFTa}\left[\hat{\mathrm{S}}_c(\tau, f_\eta)\right]\right\}\cdot\mathrm{conj}[\boldsymbol{\theta}(f_\tau, \eta)]\right\} \tag{22}$$

The IFTa[·] denotes inverse Fourier Transform (FT) along azimuth direction. FTa[·] denotes the FT along the azimuth direction. FTr[·] denotes the FT along the range direction and conj[·] represents the conjugate operation. After the raw data have been recovered, traditional SAR imaging algorithms are capable of focusing the recovered SAR raw data.

2.3. Range Migration Algorithm

A brief introduction on the Range Migration Algorithm (RMA) is presented in this section. The RMA is proposed with the condition of uniform linear motion. By operating two-dimensional (2D) FT on Equation (1) along with the range model in Equation (2), the expression of the 2D spectrum of the echo signal can be given as follows:

$$S_{2df}(f_\tau, f_\eta; \mathrm{T}) = \boldsymbol{W}_r(f_\tau)\boldsymbol{W}_a(f_\eta)\exp\left[-j\frac{4\pi r}{c}\sqrt{(f_0 + f_\tau)^2 - \frac{c^2 f_\eta^2}{4v^2}} - j\pi\frac{f_\tau^2}{K_r}\right] \tag{23}$$

In Equation (23), f_τ and f_η are the range frequency and azimuth frequency. In addition, $\boldsymbol{W}_r(\bullet)$ is the envelope of the range frequency and $\boldsymbol{W}_a(\bullet)$ is the envelope of the azimuth frequency.

The first key step in RMA is the Reference Function Multiplication (RFM). The reference function in the 2D frequency domain can be selected as conj[$S_{2df}(f_\tau, f_\eta; \mathrm{T}_{\mathrm{ref}})$], which is the conjugated 2D spectrum of the reference target. $\mathrm{T}_{\mathrm{ref}}$ represents the reference point target, and conj[•] represents the conjugation operation.

The step of the RFM operation can be given as follows:

$$\begin{aligned} S_{RFM}(f_\tau, f_\eta; \mathrm{T}) &= S_{2df}(f_\tau, f_\eta; \mathrm{T})\mathrm{conj}\left[S_{2df}(f_\tau, f_\eta; \mathrm{T}_{\mathrm{ref}})\right] \\ &= \boldsymbol{W}_r(f_\tau)\boldsymbol{W}_a(f_\eta)\exp\left[-j\frac{4\pi(r - r_{ref})}{c}\sqrt{(f_0 + f_\tau)^2 - \frac{c^2 f_\eta^2}{4v^2}}\right] \end{aligned} \tag{24}$$

The RFM operation totally removes the range migration of every target in the reference range. However, the RFM only partly removes the range migration of each target in other ranges. Hence, a subsequent Stolt interpolation is used to cancel the residual quadratic and higher order phase modulation. In uniform linear motion, the Stolt interpolation is presented as follows:

$$\sqrt{(f_0+f_\tau)^2-\frac{c^2f_\eta^2}{4v^2}}=f_0+f_\tau' \tag{25}$$

By performing the Stolt interpolation, the SAR data are transformed into a new 2D frequency domain (f_τ', f_η), as follows:

$$\begin{aligned} &\mathbf{S}_{RFM}'(f_\tau', f_\eta; \mathrm{T}) \\ &= \mathbf{W}_r(f_\tau')\mathbf{W}_a(f_\eta)\exp\left[-j\frac{4\pi(r-r_{ref})}{c}(f_0+f_\tau')\right] \end{aligned} \tag{26}$$

By operating the 2D Inverse FT according to Equation (26), the echo data can be transformed into the time domain. Consequently, a well-focused SAR image is obtained as follows:

$$I_{RFM}'(\tau,\eta;\mathrm{T})=\sin c\left[\tau-\frac{2(r-r_{ref})}{c}\right]\sin c(\eta) \tag{27}$$

2.4. Procedure of Proposed Method

Figure 9 illustrates the flowchart of the proposed method. In the flowchart, IFFT denotes inverse FFT. The proposed method is generally comprised of three parts: phase compensation, restoration via complex deconvolution, and recovered data focusing. In phase compensation, the azimuth periodically gapped raw data are multiplied with $\theta(f_\tau,\eta)|\eta\in Q_D$ in the range frequency domain. $\Theta(f_\tau,\eta)|\eta\in Q_D$ is the phase compensation function for the azimuth periodically gapped data. After phase compensation in the range frequency domain, the gapped data are transformed back to the time domain and become sparser in the Doppler domain. The restoration via deconvolution starts with filling the azimuth gaps with zeros. Then, the estimation of $S_c(\tau,f_\eta)$ is recovered with $g(\tau,f_\eta)$ and Y via the operation of complex deconvolution. Complex deconvolution is implemented by seeking the solution for Equation (18). After estimating $S_c(\tau,f_\eta)$, the recovery of the 2D spectrum of the raw data can be obtained with Equation (22). As the 2D spectrum of raw data is recovered, the focused image can be acquired via recovered data focusing. The RMA is selected to focus the recovered data. The RFM and a subsequent Stolt interpolation are implemented to the recovered 2D spectrum of raw data. RFM and Stolt interpolation are two key steps in RMA. Afterwards, the focused image can be obtained by transforming the data into the time domain by 2D IFFT. The computational complexity of phase compensation is $O(2N_DN_r\log_2(N_r)+N_DN_r)$. The computation complexity of restoration via deconvolution is $O((N_a-N_D)N_r+3N_aN_r\log_2(N_a)+5PN_aN_r+N_aN_r\log_2(N_r)+N_aN_r)$. The computation complexity of recovered data focusing is $O(N_aN_r+N_aN_r\log_2(N_a)+N_aN_r\log_2(N_r)+8N_aN_r)$.

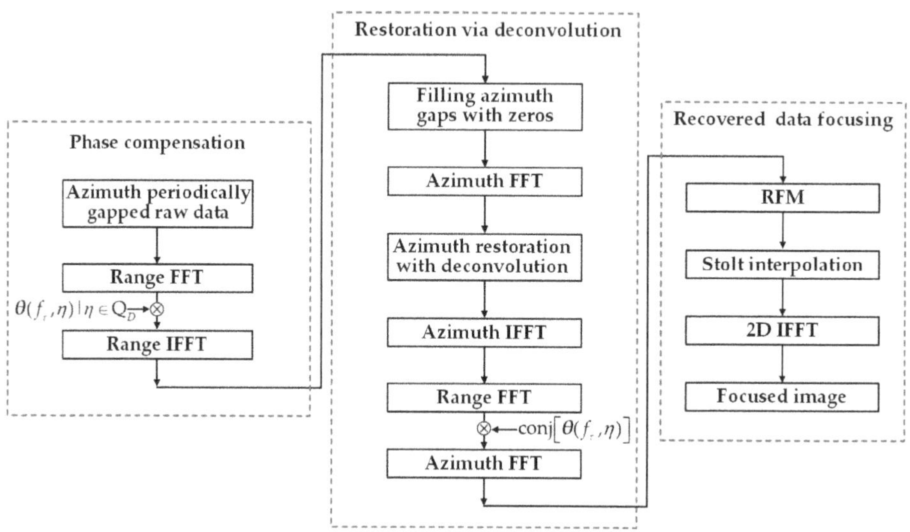

Figure 9. Flowchart of the proposed method.

3. Experimental Results

Experiments were implemented to assess the effectiveness of the proposed method. Firstly, the point target simulation was undertaken to demonstrate the validity of the proposed method. Then, the effectiveness of the proposed method was further validated with surface target simulation. Finally, real SAR data were utilized to demonstrate the performance of the proposed method.

3.1. Point Target Simulation

To appraise the performance of the proposed method, point target simulation was employed. Point target simulation was implemented under the condition of monostatic SAR by emitting a pulse linear frequency modulation signal. The system parameters for simulation are given in Table 2. In point target simulation, the resolution was expected to achieve 0.5 m in the range direction and the azimuth direction. The mode of SAR operation was chosen as the spotlight SAR in point target simulation due to the demand of resolution.

Table 2. System parameters for point target simulation.

Parameter Name	Value	Units
Slant range of scene center	8	km
Transmitted pulse duration	2	μs
Signal bandwidth	300	MHz
Range sampling rate	360	MHz
Pulse repetition frequency	1536	Hz
Radar center frequency	10	GHz
Radar velocity	120	m/s
Range resolution	0.5	m
Azimuth resolution	0.5	m

Figure 10 displays the distribution of point targets in simulation. Nine point targets were positioned for point target simulation; these targets were labeled from T_1 to T_9. The size of the complete SAR raw data in the point target simulation was 5120×3072. The complete data had 5120 and 3072 sampling points in the range and azimuth direction.

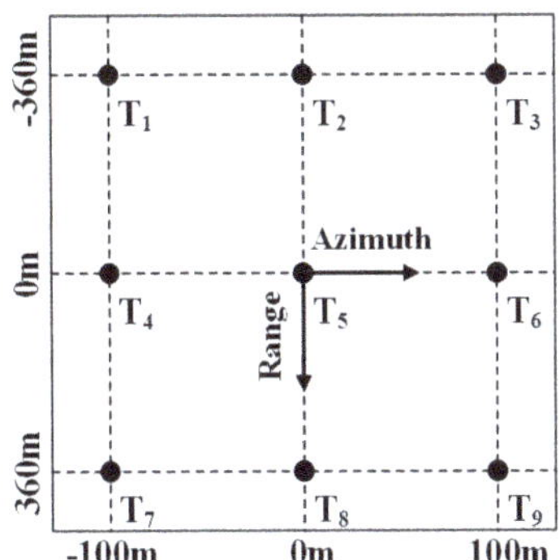

Figure 10. Distribution of point targets in simulation.

The azimuth periodically gapped SAR raw data for simulation was obtained by generating a new matrix with extracting columns from complete SAR raw data periodically. In simulation, the gapped SAR data were generated via gapping every 16 pulses, and the size of the gapped data was 5120×1536. The parameter M_R was selected to be 16, and the parameter M_G was chosen to be 16 in the point target simulation. The gapping rate for point target simulation was 50% in the azimuth direction. In the point target simulation, the parameter β was set as 0.25, and the iteration number was set as 1000. Stepsize t was set as the maximum singular value of the matrix $\boldsymbol{\Psi}$.

The focused results of the point target simulation are illustrated from Figures 11–13. The focused results of the placed nine point targets are depicted in Figures 11a, 12c and 13c. In Figures 11–13, the contour plots of the interpolated impulse response functions of the focused nine point targets are illustrated. The contour lines of -3 dB, -13 dB, -20 dB, and -30 dB were chosen to present the focused point targets. As displayed in Figures 11–13, all nine point targets were focused satisfactorily.

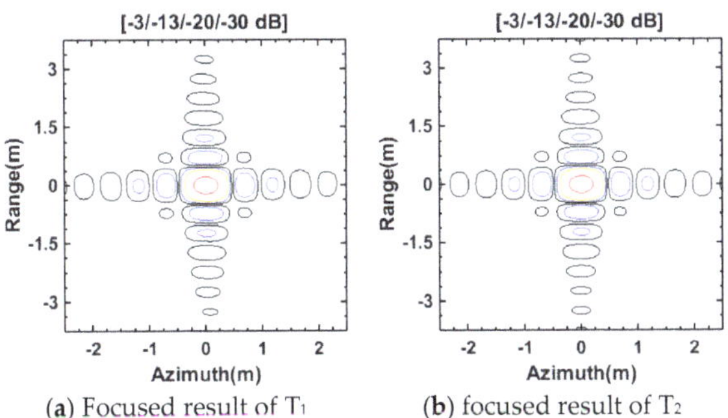

(**a**) Focused result of T_1 (**b**) focused result of T_2

Figure 11. *Cont.*

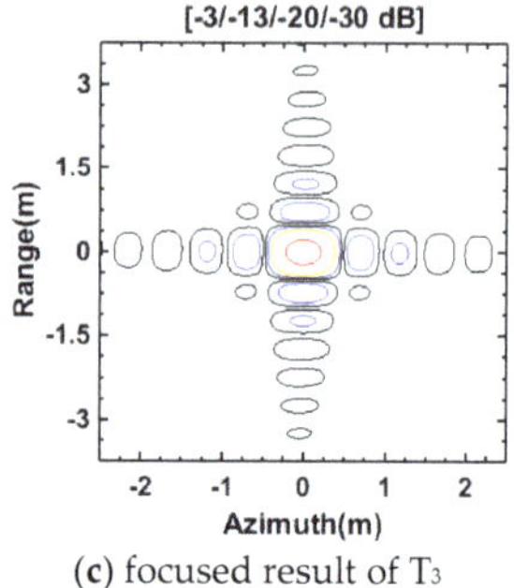

(c) focused result of T3

Figure 11. Focused point target results of T_1, T_2, and T_3: (a) focused result of T_1; (b) focused result of T_2; and (c) focused result of T_3.

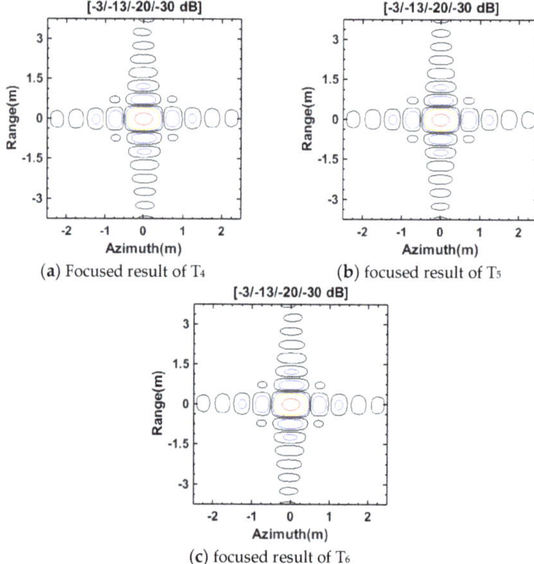

(a) Focused result of T4 (b) focused result of T5
(c) focused result of T6

Figure 12. Focused point target results of T_4, T_5, and T_6: (a) focused result of T_4; (b) focused result of T_5; and (c) focused result of T_6.

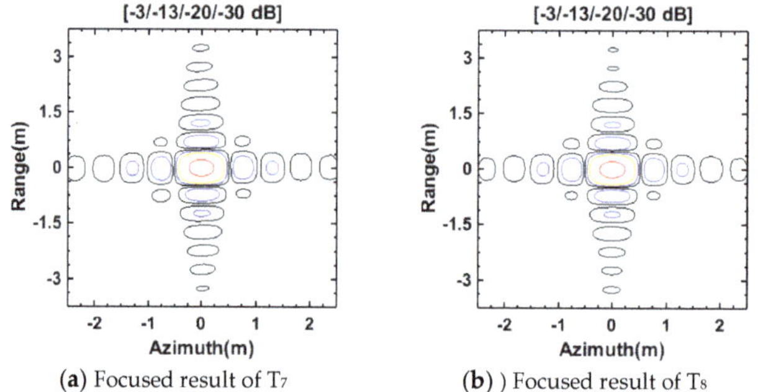

(a) Focused result of T7 (b)) Focused result of T8

Figure 13. *Cont.*

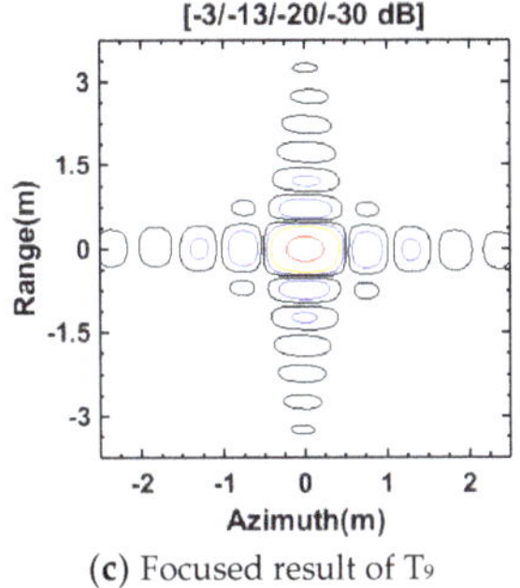

(**c**) Focused result of T_9

Figure 13. Focused point target results of T_7, T_8, and T_9: (**a**) focused result of T_7; (**b**) focused result of T_8; and (**c**) focused result of T_9.

Figures 14–16 demonstrate the one-dimensional (1D) azimuth profiles of the imaged nine point targets via the adding zeros operation and the proposed method. The adding zeros operation entailed filling the gapping positions in the azimuth direction with zeros and the subsequent implementation of RMA. The existence of fake targets in the azimuth direction was the main problem in focusing azimuth periodically gapped data, which was caused by periodical azimuth gapping. As shown in Figures 14–16, the fake targets were obviously suppressed via the implementation of the proposed method. Consequently, it can be concluded that the proposed method performed well in point target simulations. To assess the validity of the proposed method for point target simulation, the analysis of point target simulation is listed in Section 4.

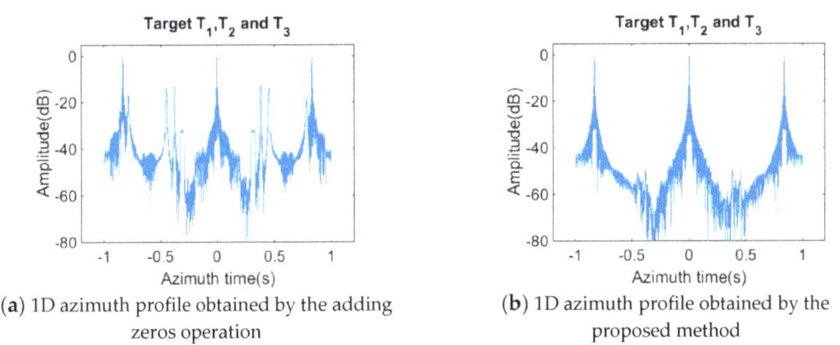

(**a**) 1D azimuth profile obtained by the adding zeros operation

(**b**) 1D azimuth profile obtained by the proposed method

Figure 14. 1D Azimuth profiles of T_1, T_2, and T_3: (**a**) 1D azimuth profile obtained by the adding zeros operation; and (**b**) 1D azimuth profile obtained by the proposed method.

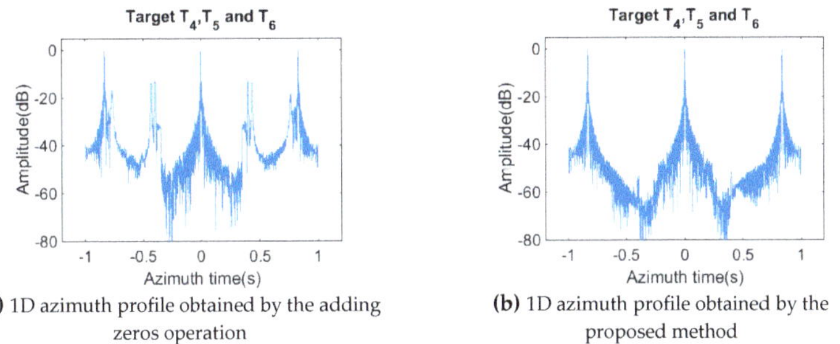

(**a**) 1D azimuth profile obtained by the adding zeros operation

(**b**) 1D azimuth profile obtained by the proposed method

Figure 15. 1D Azimuth profiles of T_4, T_5, and T_6: (**a**) 1D azimuth profile obtained by the adding zeros operation; and (**b**) 1D azimuth profile obtained by the proposed method.

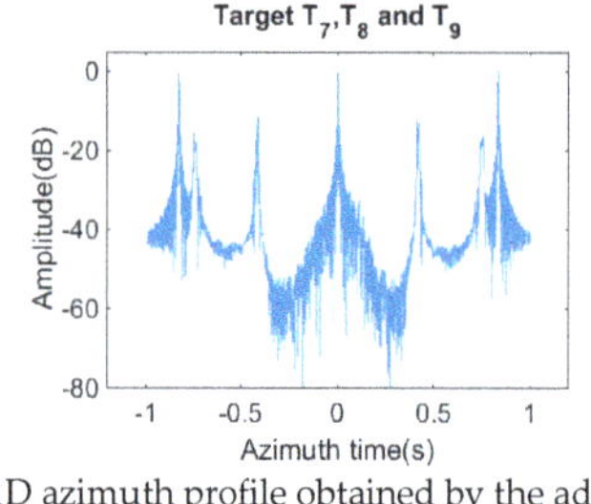

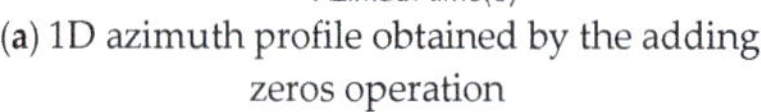
(a) 1D azimuth profile obtained by the adding zeros operation

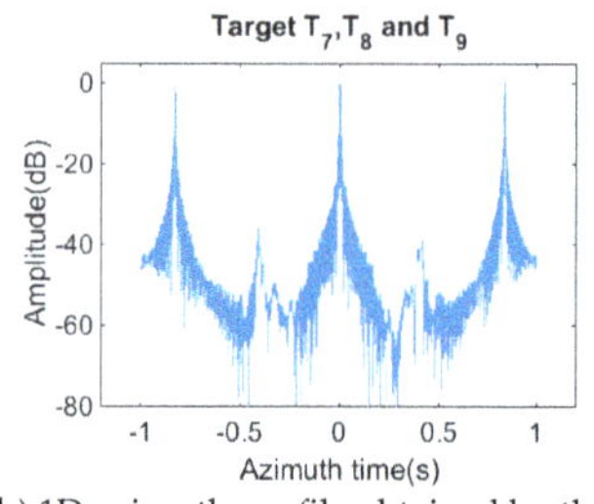

(b) 1D azimuth profile obtained by the proposed method

Figure 16. 1D Azimuth profiles of T_7, T_8, and T_9: (**a**) 1D azimuth profile obtained by the adding zeros operation; and (**b**) 1D azimuth profile obtained by the proposed method.

3.2. Surface Target Simulation

The surface target simulation is presented in this part that verified the effectiveness of the proposed method. Many point targets were placed to form a surface target with the shape of capital letter "T". The surface target simulation was carried out with the system parameters in Table 3. The complete raw SAR data in surface target simulation were a 3256×4096 matrix, which had range sampling points and azimuth sampling pulses. In surface target simulation, the gapped SAR data were generated by gapping every 16 pulses, and the size of the gapped data was 3256×2048. The parameter M_R was chosen to be 16, and the parameter M_G was selected to be 16 in the surface target simulation. The loss rate in surface target simulation was 50% in the azimuth direction. In the surface target simulation, the parameter β was set as 0.25, and the iteration number was set as 1000. Stepsize t was selected as the maximum singular value of the matrix $\boldsymbol{\Psi}$.

Table 3. System parameters for surface target simulation.

Parameter Name	Value	Units
Slant range of scene center	8	km
Transmitted pulse duration	2	μs
Signal bandwidth	600	MHz
Range sampling rate	720	MHz
Pulse repetition frequency	1024	Hz
Radar center frequency	10	GHz
Radar velocity	120	m/s
Range resolution	0.25	m
Azimuth resolution	0.25	m

Figure 17 demonstrates the results of surface target simulation with the adding zeros operation and the proposed method. Figures 18–20 illustrate the 1D azimuth profiles of results in surface target simulation via the adding zeros operation and the proposed method. The 1D azimuth profiles in Figures 18–20 are selected at -150 m, 0 m and 150 m in range direction, respectively. As shown in Figure 17a, many fake targets exist in the result obtained by the adding zeros operation. In contrast, Figure 17b demonstrates that fake targets were obviously suppressed with the implementation of the proposed method. Furthermore, the 1D azimuth profiles in Figures 18–20 present more detailed comparison between results obtained by the adding zeros operation and the proposed method.

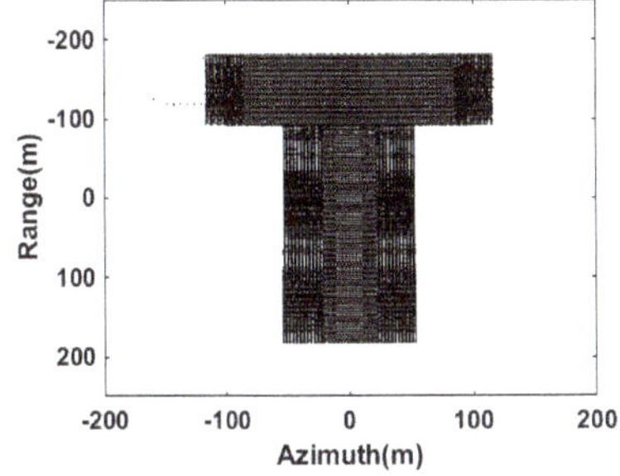

(a) Result obtained by the adding zeros operation

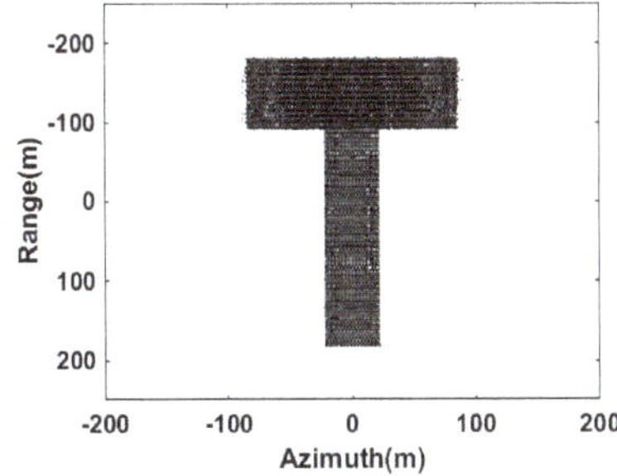

(b) result obtained by the proposed method

Figure 17. Imaged results of surface target simulation: (**a**) result obtained by the adding zeros operation; and (**b**) result obtained by the proposed method.

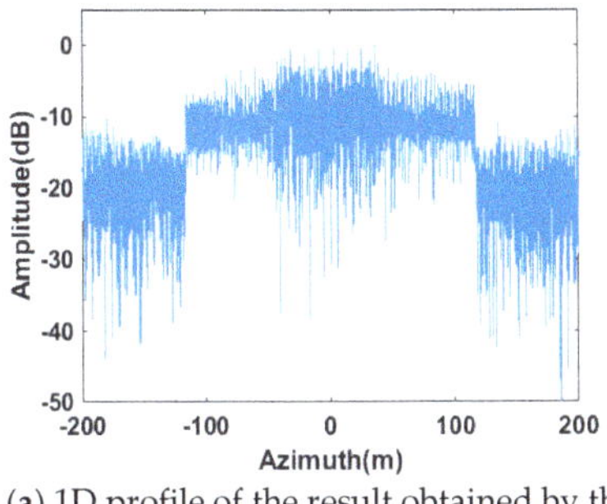

(a) 1D profile of the result obtained by the adding zeros operation

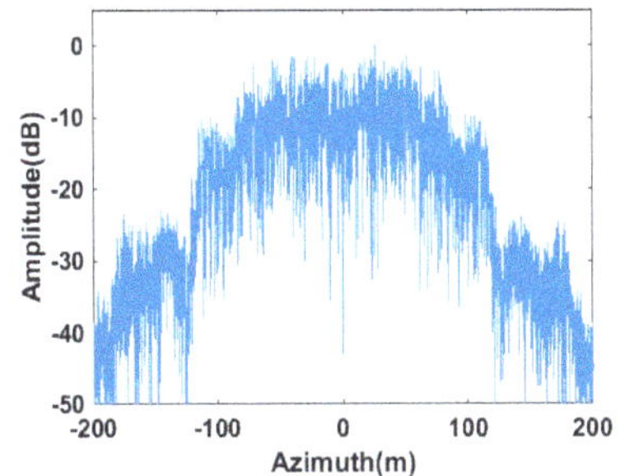

(b) 1D profile of the result obtained by the proposed method

Figure 18. 1D profiles of results at -150 m in range direction: (**a**) 1D profile of the result obtained by the adding zeros operation; and (**b**) 1D profile of the result obtained by the proposed method.

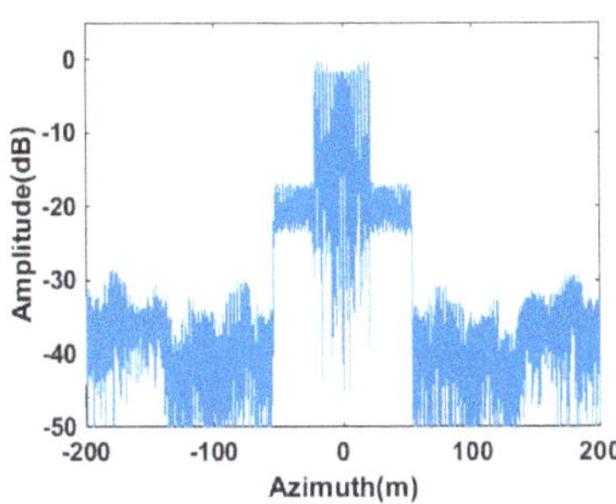

(a) 1D profile of the result obtained by the adding zeros operation

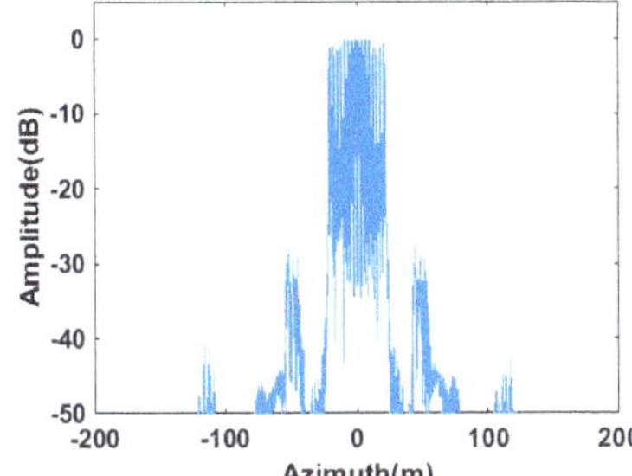

(b) 1D profile of the result obtained by the proposed method

Figure 19. 1D profiles of results at 0 m in range direction: (**a**) 1D profile of the result obtained by the adding zeros operation; and (**b**) 1D profile of the result obtained by the proposed method.

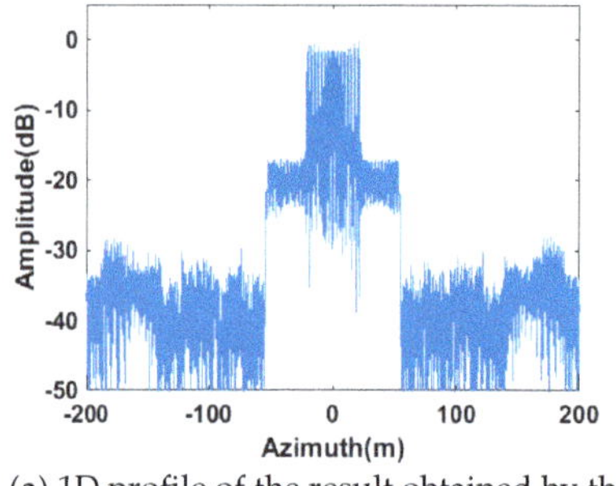

(a) 1D profile of the result obtained by the adding zeros operation

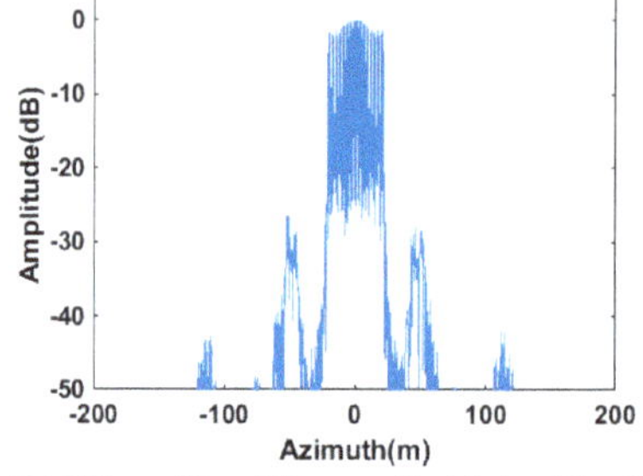

(b) 1D profile of the result obtained by the proposed method

Figure 20. 1D profiles of results at 150 m in range direction: (**a**) 1D profile of the result obtained by the adding zeros operation; and (**b**) 1D profile of the result obtained by the proposed method.

3.3. Real SAR Data Processing

Real SAR data were utilized to demonstrate the validity of the proposed method. The real SAR data for this experiment were collected by Sentinel-1A. The size of the complete real SAR data was 8192×8192. In other words, the complete real SAR data had 8192 sampling points in the range direction and 8192 sampling pulses in the azimuth direction.

The azimuth periodically gapped SAR raw data for the real SAR data processing were acquired by constructing a new matrix via extracting columns from complete real SAR data. In real SAR data processing, the gapped SAR data were obtained by gapping every 16 pulses, and the size of the gapped data was 8192×4096. The parameter M_R was chosen to be 16, and the parameter M_G was supposed to be 16 in real SAR data processing. The gapping rate for real data processing was 50% in the azimuth direction. In real SAR data processing, the parameter β was set as 0.25, and the iteration number was set as 800. The stepsize t was set as the maximum singular value of the matrix $\boldsymbol{\Psi}$.

Figure 21 depicts the results of real SAR data processing. Figure 21a shows the imaging result obtained from the complete raw data via RMA. Figure 21b demonstrates the imaging result acquired via the adding zeros operation from the gapped data. The result in Figure 21b was acquired via filling the gapping positions in the azimuth direction with zeros and the subsequent implementation of RMA. Figure 21c illustrates the imaging result acquired with PG-APES. The PG-APES used to obtain Figure 21c was implemented by recovering the gapped data via PG-APES in the range of the Doppler domain and the subsequent operation of the RMA. Figure 21d is the imaging result acquired with the Burg algorithm. The image result in Figure 21d was obtained by recovering the gapped data via the method in [15] and a subsequent implementation of RMA. Figure 21e displays the imaging result acquired with the method in reference [23]. Figure 21f shows the imaging result acquired with the proposed method. In real SAR data images, the range and azimuth directions were the vertical and horizontal directions, respectively.

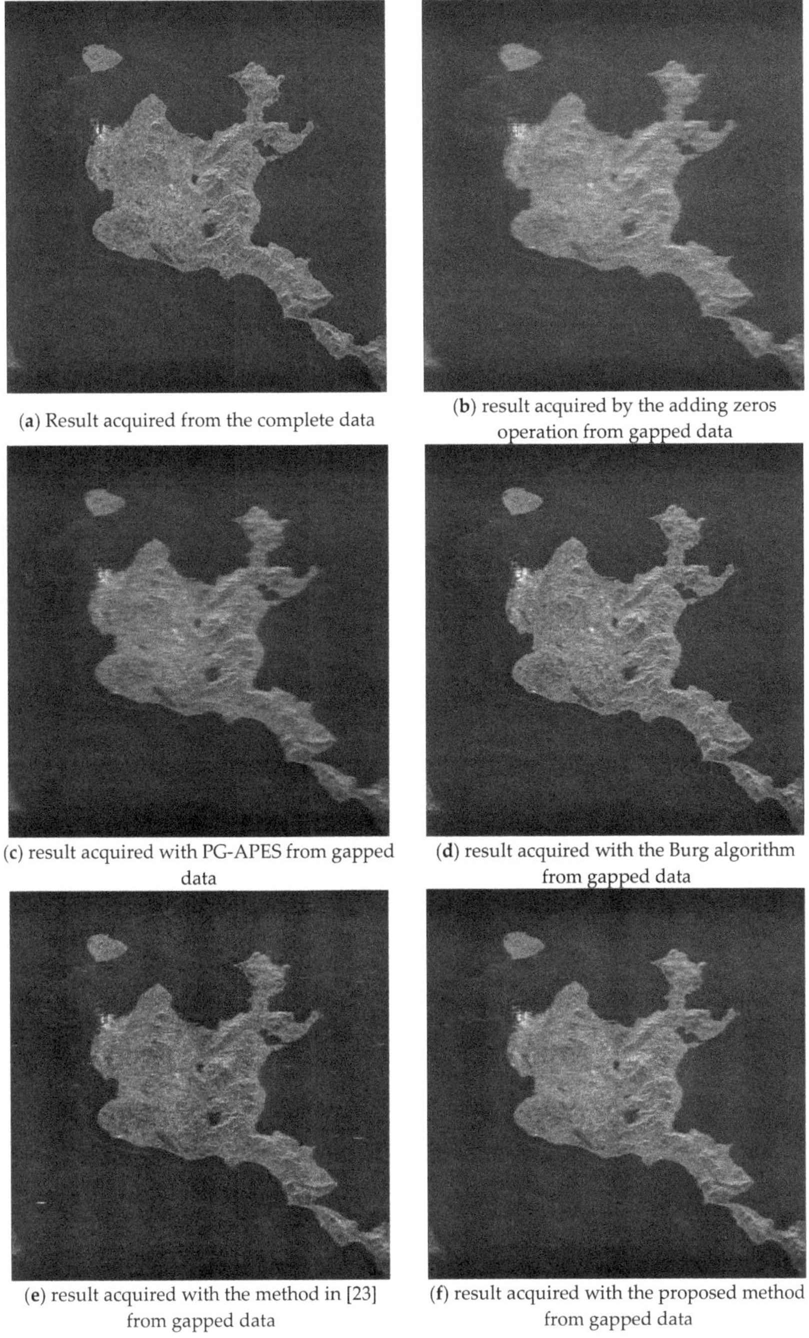

(a) Result acquired from the complete data

(b) result acquired by the adding zeros operation from gapped data

(c) result acquired with PG-APES from gapped data

(d) result acquired with the Burg algorithm from gapped data

(e) result acquired with the method in [23] from gapped data

(f) result acquired with the proposed method from gapped data

Figure 21. Results of real SAR data: (**a**) Result acquired from the complete data; (**b**) result acquired by the adding zeros operation from gapped data; (**c**) result acquired with PG-APES from gapped data; (**d**) result acquired with the Burg algorithm from gapped data; (**e**) result acquired with the method in [23] from gapped data; and (**f**) result acquired with the proposed method from gapped data.

As demonstrated in Figure 21b, the phenomenon of ghost targets exists in the azimuth direction due to azimuth periodical gapping. With the application of different methods, the phenomenon of ghost targets was suppressed in different degrees via the demonstration in Figure 21c–f. Figure 21c,d,f indicates that the proposed method performed better than the PG-APES and Burg algorithm in suppressing ghost targets in the azimuth direction. Moreover, Figure 17e illustrates that the method in [23] performed competently in suppressing ghost targets. However, as shown in Figure 17e, unexpected fake strong scatters were induced by the method in [23] when M_R and M_G were both set as 16. Consequently, it can be concluded that the proposed method performed better than the other four methods in suppressing ghost targets. The validity of the proposed method for real SAR data is demonstrated in Figure 21f.

To demonstrate the validity of the proposed method further, three local regions were chosen to show more details. The three chosen regions are displayed in Figure 22 and are identified with red boxes. The results of the enlarged local regions of real SAR data are demonstrated in Figures 23–25. The displayed enlarged local regions were acquired in six different ways. These six methods are as follows: obtaining the image from complete data via RMA, obtaining the image from the gapped data by adding zeros, obtaining the image from gapped data via PG-APES, obtaining the image from gapped data via the Burg algorithm, obtaining the image from the gapped data via the method in [23], and obtaining the image from the gapped data via the proposed method. The validity of the proposed method is verified further by Figures 23–25.

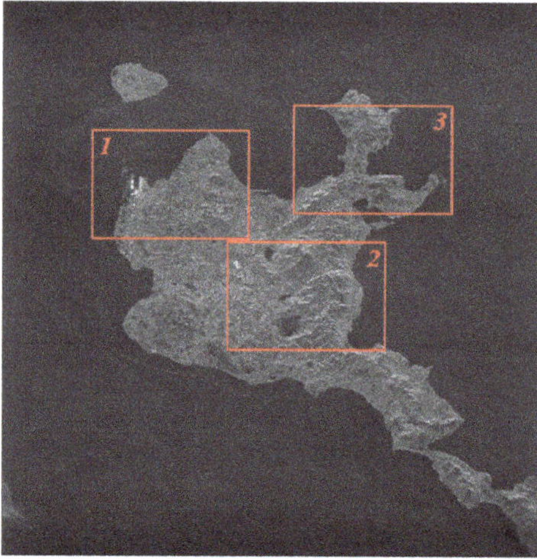

Figure 22. Chosen regions in real SAR data.

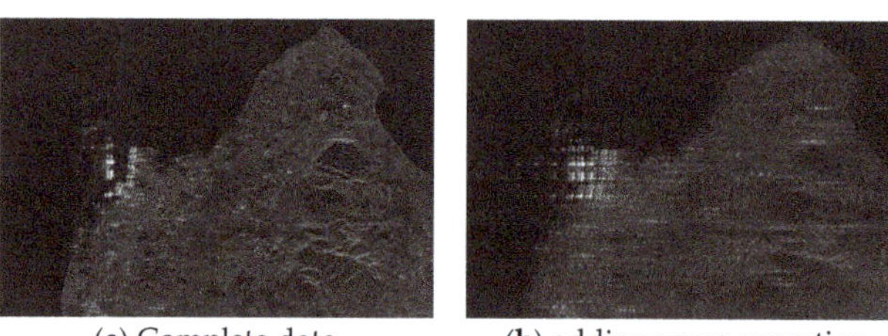

(**a**) Complete data (**b**) adding zeros operation

Figure 23. *Cont.*

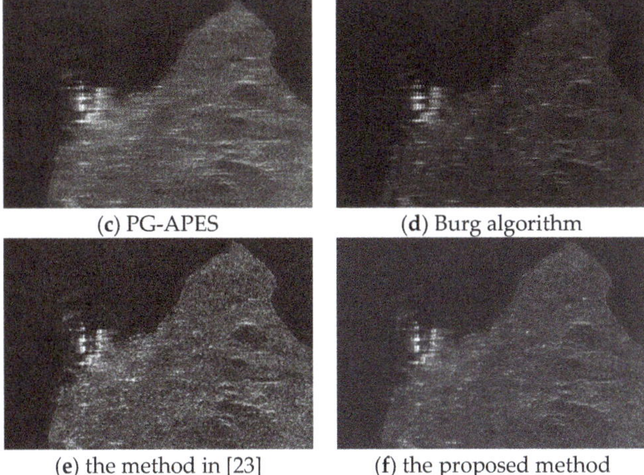

Figure 23. Results of Region 1: (**a**) complete data; (**b**) adding zeros operation; (**c**) PG-APES; (**d**) Burg algorithm; (**e**) the method in [23]; and (**f**) the proposed method.

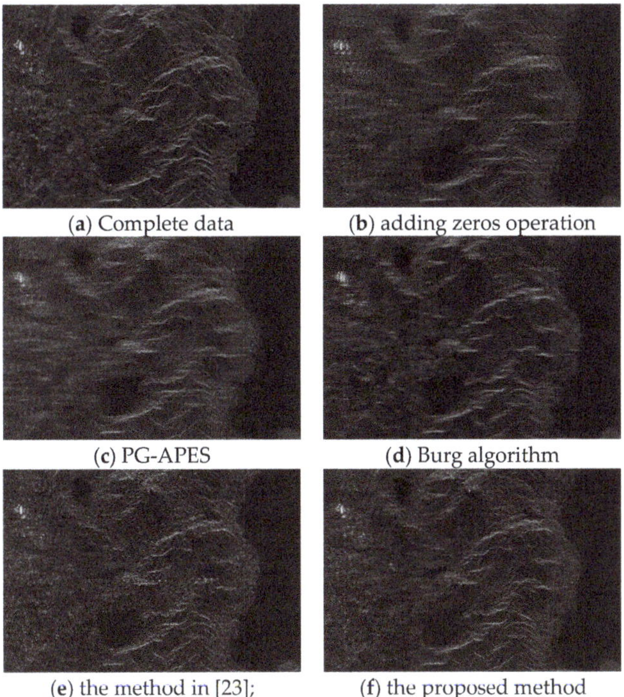

Figure 24. Results of Region 2: (**a**) complete data; (**b**) adding zeros operation; (**c**) PG-APES; (**d**) Burg algorithm; (**e**) the method in [23]; and (**f**) the proposed method.

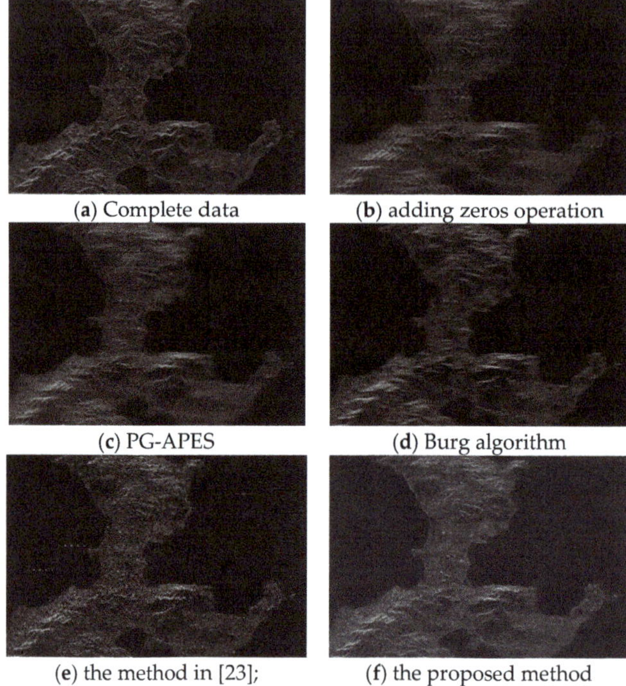

Figure 25. Results of Region 3: (**a**) complete data; (**b**) adding zeros operation; (**c**) PG-APES; (**d**) Burg algorithm; (**e**) the method in [23]; and (**f**) the proposed method.

4. Discussion

The focused results in Section 3 demonstrate the effectiveness of the proposed method. To evaluate the validity of the proposed method further, several criteria were selected to analyze the results of point target simulation and real SAR data processing.

To evaluate the performance of the proposed method on point target simulation, the impulse response width (IRW), peak sidelobe ratio (PSLR), and integrated sidelobe ratio (ISLR) were chosen as criteria for assessing the quality of focused point targets. To ensure the quality of focused point targets, the PSLR and ISLR are supposed to be less than -13 dB and -10.15 dB, respectively. In addition, because IRW is regarded as the image resolution in the scenario of SAR image, the IRW should not be inferior to 0.5 m when the resolution is expected to attain 0.5 m. In this paper, the main lobe width in ISLR is set to two times the IRW.

The analysis of the focused results in the point target simulation is illustrated in Table 4. As demonstrated in Table 4, all the focused point targets achieved the requirements of IRW, PSLR, and ISLR in both the range and azimuth direction. Table 5 demonstrates the analysis of fake targets in point target simulation. Row 1 contains 1D azimuth profiles of T_1, T_2, and T_3. Row 2 contains 1D azimuth profiles of T_4, T_5, and T_6. Row 3 contains 1D azimuth profiles of T_7, T_8, and T_9. The amplitude of the highest fake target in each row is listed in Table 5. As shown in Table 4, fake targets were obviously suppressed with the implementation of the proposed method. Therefore, it can be concluded that the proposed method performed satisfactorily in point target simulation.

Table 4. Analysis of point target simulation with the proposed method.

T	Range			Azimuth		
	IRW (m)	PSLR (dB)	ISLR (dB)	IRW (m)	PSLR (dB)	ISLR (dB)
1	0.4458	-13.52	-10.89	0.4275	-13.18	-10.61
2	0.4417	-13.42	-10.87	0.4250	-13.23	-10.63
3	0.4455	-13.51	-10.88	0.4263	-13.25	-10.60
4	0.4479	-13.35	-10.57	0.4487	-13.37	-10.88
5	0.4458	-13.22	-10.34	0.4438	-13.12	-10.56
6	0.4460	-13.35	-10.56	0.4500	-13.34	-10.87
7	0.4462	-13.54	-11.08	0.4663	-13.20	-10.62
8	0.4396	-13.38	-10.08	0.4630	-13.06	-10.49
9	0.4462	-13.48	-11.05	0.4665	-13.09	-10.61

Table 5. Analysis of fake targets in the point target simulation.

Row	Method	
	Adding Zeros Operation	The Proposed Method
1	-12.63 dB	-49.16 dB
2	-12.54 dB	-51.36 dB
3	-11.03 dB	-35.75 dB

The analysis of fake targets in the surface target simulation is demonstrated in the Table 6. In Table 5, Rows 1–3 represent 1D azimuth profiles at -150 m, 0 m and 150 m in the range direction, respectively. As illustrated in Table 6, fake targets were apparently suppressed with the operation of the proposed method. Comparing with the point target simulation, the performance of suppression on fake targets decreased in the surface target simulation.

Table 6. Analysis of fake targets in the surface target simulation.

Row	Method	
	Adding Zeros Operation	The Proposed Method
1	-5.18 dB	-10.76 dB
2	-16.63 dB	-27.21 dB
3	-16.83 dB	-26.31 dB

To analyze the results of real SAR data processing, IE and IC were utilized to assess the imaging quality of real SAR data results. In addition, the MSE (Mean Square Error) was also utilized to analyze the results of real SAR data processing. The focused result of complete real SAR data was defined as the ground truth image in this study. The MSE analysis of real SAR data processing was obtained with respect to the ground truth image. IE and IC analyses of the real SAR data results are given in Table 7. The results of the whole scene and local regions are analyzed in Table 7. Generally, better focusing leads to a larger IC and smaller IE [35,36]. As displayed in Table 7, the proposed method generally performed better than the adding zeros operation, PG-APES, and Burg algorithm with the IE and IC criteria. Because of the existence of some fake strong scattering points, the method in [23] had a larger IC in Region 3 than the proposed method, and the method in [23] had a smaller IE than the proposed method. The differences among the results acquired via the adding zeros operation, PG-APES, Burg algorithm, the method in [23], and the proposed method were not apparent in the

IE criterion. The IE and IC analyses of real SAR data processing were generally consistent with the results shown in Figures 21 and 23, Figures 24 and 25. The MSE analysis for the results of real SAR data processing is presented in Table 8. As displayed in Table 8, the proposed method had the lowest MSE in the whole scene and three regions. Consequently, it can be concluded that the proposed method performed better than the other methods for MSE in real SAR data processing.

Table 7. IE and IC analysis for the results of real SAR data processing.

	IE				IC			
	Whole	**Region 1**	**Region 2**	**Region 3**	**Whole**	**Region 1**	**Region 2**	**Region 3**
Complete data	17.02	13.68	14.38	14.30	7.99	12.71	2.54	2.68
Adding zeros	17.15	14.00	14.53	14.44	4.72	7.36	1.74	1.94
PG-APES	17.22	14.25	14.57	14.49	3.55	5.85	1.53	1.67
Burg	17.05	13.79	14.47	14.37	5.89	9.07	1.92	2.10
Method in [23]	16.98	13.65	14.24	14.28	7.60	12.55	2.64	2.67
Proposed method	17.04	13.72	14.41	14.39	9.13	14.29	2.69	2.61

Table 8. MSE analysis for the results of real SAR data processing.

	MSE			
	Whole	**Region 1**	**Region 2**	**Region 3**
Adding zeros	1.7382	6.5166	28.2920	15.7520
PG-APES	2.7807	5.1541	15.0794	10.9907
Burg	2.1794	5.4941	30.5636	20.1540
Method in [23]	1.6967	4.0241	22.7452	16.2771
Proposed method	1.0338	3.0572	14.3736	9.2486

5. Conclusions

A novel method for focusing azimuth periodically gapped SAR raw data is proposed and described in this paper. The proposed method introduces the complex deconvolution into solving the problem of imaging periodically gapped SAR raw data and offers a robust implementation of deconvolution for dealing with gapped SAR raw data. The proposed method mainly comprises phase compensation and recovering the azimuth spectrum of raw data with complex deconvolution. The gapped data become sparser in the range of the Doppler domain after phase compensation, and the deconvolution for recovering the azimuth spectrum of the raw data is also implemented in the Doppler domain. After the recovery of the azimuth spectrum, the RMA is selected to focus the recovered raw data. The effectiveness of the proposed method was validated via point target simulation. Moreover, real SAR data were utilized to verify the validity of the proposed method further. The proposed method is proposed under the consideration of uniform linear motion and side-looking. Future research can be focused on scenarios of complicated motion or extension towards situation of high squint. In addition, improvement on the performance of the proposed method over surface scatters also remains to be investigated in the future.

Author Contributions: Conceptualization, Y.Q. and D.Z.; methodology, Y.Q. and D.Z.; bibliography review, Y.Q.; acquisition, analysis and interpretation of data, Y.Q.; writing, review and editing, Y.Q.

Funding: This research was funded by the National Key R&D Program of China, grant number 2017YFB0502700; the Postgraduate Research and Practice Innovation Program of Jiangsu Province, grant number KYCX17_0266; and Aeronautical Science Foundation of China, grant number 20182052013.

Conflicts of Interest: The authors declare no conflict of interest.

References

1. Alexander, G.F.; Simon, H.Y.; Bryan, W.S.; Wenqing, T.; Akiko, K.H. On Extreme Winds at L-Band with the SMAP Synthetic Aperture Radar. *Remote Sens.* **2019**, *11*, 1093. [CrossRef]
2. Qian, Y.; Zhu, D. Focusing of Ultrahigh Resolution Spaceborne Spotlight SAR on Curved Orbit. *Electronics* **2019**, *8*, 628. [CrossRef]
3. Min-Ho, K.; Pavel, E.S.; Mikhail, I.B. A New Single-Pass SAR Interferometry Technique with a Single-Antenna for Terrain Height Measurements. *Remote Sens.* **2019**, *11*, 1070. [CrossRef]
4. Zhang, Y.; Zhu, D. Height Retrieval in Postprocessing-Based VideoSAR Image Sequence Using Shadow Information. *IEEE Sens. J.* **2018**, *18*, 8108–8116. [CrossRef]
5. Li, T.; Chen, K.; Jin, M. Analysis and Simulation on Imaging Performance of Backward and Forward Bistatic Synthetic Aperture Radar. *Remote Sens.* **2018**, *10*, 1676. [CrossRef]
6. Liu, Q.; Wang, Y. A Fast Eigenvector-Based Autofocus Method for Sparse Aperture ISAR Sensors Imaging of Moving Target. *IEEE Sens. J.* **2019**, *19*, 1307–1319. [CrossRef]
7. Qian, Y.; Zhu, D. High Resolution Spotlight Spaceborne SAR Focusing via Modified-SVDS and Deramping-based approach. *IET Radar Sonar Navig.* **2019**, *13*, 1826–1835. [CrossRef]
8. Wang, Y.; Rong, J.; Han, T. Novel Approach for High Resolution ISAR/InISAR Sensors Imaging of Maneuvering Target Based on Peak Extraction Technique. *IEEE Sens. J.* **2019**, *19*, 5541–5558. [CrossRef]
9. Xu, Z.; Chen, K. On Signal Modeling of Moon-Based Synthetic Aperture Radar (SAR) Imaging of Earth. *Remote Sens.* **2018**, *10*, 486. [CrossRef]
10. Cetin, M.; Stojanovic, I.; Onhon, N.O.; Varshney, K.R.; Samadi, S.; Karl, W.C.; Willksy, A.S. Sparsity-Driven Synthetic Aperture Radar Imaging: Reconstruction, autofocusing, moving targets, and compressed sensing. *IEEE Signal Process. Mag.* **2014**, *31*, 27–40. [CrossRef]
11. Ender, J.H.G. On Compressive Sensing Applied to Radar. *Signal Process.* **2010**, *90*, 1402–1414. [CrossRef]
12. Musgrove, C.H.; West, J.C. Mitigating Effects of Missing Data for SAR Coherent Images. *IEEE Trans. Aerosp. Electron. Syst.* **2017**, *53*, 716–721. [CrossRef]
13. Pinheiro, M.; Rodriguez-Cassola, M.; Prats-Iraola, P.; Reigber, A.; Krieger, G.; Moreira, A. Reconstruction of Coherent Pairs of Synthetic Aperture Radar Data Acquired in Interrupted Mode. *IEEE Trans. Geosci. Remote Sens.* **2015**, *53*, 1876–1893. [CrossRef]
14. Joseph, S.; Don, A.; Russell, L.; John, C.K. Interrupted Synthetic Aperture Radar (SAR). *IEEE Trans. Aerosp. Electron. Syst.* **2002**, *17*, 33–39. [CrossRef]
15. Li, N.; Wang, R.; Deng, Y.; Chen, J.; Zhang, Z.; Liu, Y.; Xu, Z.; Zhao, F. Improved Full-Aperture ScanSAR Imaging Algorithm Based on Aperture Interpolation. *IEEE Geosci. Remote Sens. Lett.* **2015**, *12*, 1101–1105. [CrossRef]
16. Francis, C. *Digital Spectral Analysis: Parametric, Non-Parametric and Advanced Methods*, 1st ed.; John Wiley & Sons: Hoboken, NJ, USA, 2011; pp. 145–168.
17. Li, J.; Stoica, P. An adaptive filtering approach to spectral estimation and SAR imaging. *IEEE Trans. Signal Process.* **1996**, *44*, 1469–1484. [CrossRef]
18. Li, H.; Li, J.; Stoica, P. Performance Analysis of Forward-Backward Matched-Filterbank Spectral Estimators. *IEEE Trans. Signal Process.* **1998**, *46*, 1954–1966. [CrossRef]
19. Larsson, E.G.; Stoica, P.; Li, J. Amplitude spectrum estimation for two-dimensional gapped data. *IEEE Trans. Signal Process.* **2002**, *50*, 1343–1354. [CrossRef]
20. Wang, Y.; Li, J.; Stoica, P. *Spectral Analysis of Signals: The Missing Data Case*, 1st ed.; Morgan & Claypool: San Rafael, CA, USA, 2005; pp. 13–30.
21. Larsson, E.G.; Li, J. Spectral Analysis of Periodically Gapped Data. *IEEE Trans. Aerosp. Electron. Syst.* **2003**, *39*, 1089–1097. [CrossRef]
22. Wang, Y.; Liu, Q. Super-Resolution Sparse Aperture ISAR Imaging of Maneuvering Target via the RELAX Algorithm. *IEEE Sens. J.* **2018**, *18*, 8726–8738. [CrossRef]
23. Qian, Y.; Zhu, D. High-resolution SAR imaging from azimuth periodically gapped raw data via generalised orthogonal matching pursuit. *Electron. Lett.* **2018**, *54*, 1235–1236. [CrossRef]
24. Kang, M.; Bae, J.; Lee, S.; Kim, K. Bistatic ISAR Imaging and Scaling of Highly Maneuvering Target with Complex Motion via Compressive Sensing. *IEEE Trans. Aerosp. Electron. Syst.* **2018**, *54*, 2809–2826. [CrossRef]

25. Ao, D.; Wang, R.; Hu, C.; Li, Y. SAR A Sparse SAR Imaging Method Based on Multiple Measurement Vectors Model. *Remote Sens.* **2017**, *9*, 297. [CrossRef]
26. Qian, Y.; Zhu, D. High Resolution Imaging from Azimuth Missing SAR Raw Data via Segmented Recovery. *Electronics* **2019**, *8*, 336. [CrossRef]
27. Tropp, J.A. Greed is good: Algorithmic results for sparse approximation. *IEEE Trans. Inf. Theory* **2004**, *50*, 2231–2242. [CrossRef]
28. Tropp, J.A.; Gilbert, A.C. Signal Recovery from Random Measurements via Orthogonal Matching Pursuit. *IEEE Trans. Inf. Theory* **2007**, *53*, 4655–4666. [CrossRef]
29. Wang, J.; Kwon, S.; Shim, B. Generalized Orthogonal Matching Pursuit. *IEEE Trans. Signal Process.* **2012**, *60*, 6202–6216. [CrossRef]
30. Donoho, D.L.; Tsaig, Y.; Drori, I.; Starck, J.L. Sparse Solution of Underdetermined Systems of Linear Equations by Stagewise Orthogonal Matching Pursuit. *IEEE Trans. Inf. Theory* **2012**, *58*, 1094–1121. [CrossRef]
31. Zhou, P.; Poularikas, A.D. Spectral estimation of segmented signal using deconvolution technique. In Proceedings of the IEEE Southeastcon 92, Birmingham, AL, USA, 12–15 April 1992.
32. Zhu, D.; Ye, S.; Zhu, Z. Polar Format Algorithm using Chirp Scaling for Spotlight SAR Image Formation. *IEEE Trans. Aerosp. Electron. Syst.* **2008**, *44*, 1433–1448. [CrossRef]
33. Daubechies, I.; Defrise, M.; De Mol, C. An Iterative Thresholding Algorithm for Linear Inverse Problems with a Sparsity Constraint. *Commun. Pure Appl. Math.* **2004**, *57*, 1413–1457. [CrossRef]
34. Mao, X.; He, X.; Li, D. Knowledge-Aided 2-D Autofocus for Spotlight SAR Range Migration Algorithm Imagery. *IEEE Trans. Geosci. Remote Sens.* **2018**, *56*, 5458–5470. [CrossRef]
35. Berizzi, F.; Corsini, G. Autofocusing of inverse synthetic aperture radar images using contrast optimization. *IEEE Trans. Aerosp. Electron. Syst.* **1996**, *32*, 1185–1191. [CrossRef]
36. Wang, J.; Liu, X. SAR minimum-entropy autofocus using an adaptive order polynomial model. *IEEE Geosci. Remote Sens. Lett.* **2006**, *3*, 512–516. [CrossRef]
37. Aberman, K.; Eldar, Y.C. Sub-Nyquist SAR via Fourier Domain Range-Doppler Processing. *IEEE Trans. Geosci. Remote Sens.* **2017**, *55*, 6228–6244. [CrossRef]

Article

A Method of Fast and Simultaneous Calibration of Many Mobile FMCW Radars Operating in a Network Anti-Drone System

Aleksander Nowak [1,*], **Krzysztof Naus** [2] **and Dariusz Maksimiuk** [3]

1 Faculty of Civil and Environmental Engineering, Department of Geodesy, Gdansk University of Technology, Narutowicza 11/12, 80-233 Gdansk, Poland

2 Faculty of Navigation and Naval Weapons, Polish Naval Academy, Smidowicza 69, 81-103 Gdynia, Poland; k.naus@amw.gdynia.pl

3 Faculty of Electrical Engineering, Mathematics & Computer Science, Department of Circuits and Systems, TU Delft, Mekelweg 4, 2628 CD Delft, The Netherlands; d.maksimiuk@tudelft.nl

* Correspondence: aleksander.nowak@pg.edu.pl

Received: 30 September 2019; Accepted: 7 November 2019; Published: 8 November 2019

Abstract: A market for small drones is developing very fast. They are used for leisure activities and exploited in commercial applications. However, there is a growing concern for accidental or even criminal misuses of these platforms. Dangerous incidents with drones are appearing more often, and have caused many institutions to start thinking about anti-drone solutions. There are many cases when building stationary systems seems to be aimless since the high cost does not correspond with, for example, threat frequency. In such cases, mobile drone countermeasure systems seem to perfectly meet demands. In modern mobile solutions, frequency modulated continuous wave (FMCW) radars are frequently used as detectors. Proper cooperation of many radars demands their measurements to be brought to a common coordinate system—azimuths must be measured in the same direction (preferably the north). It requires calibration, understood as determining constant corrections to measured angles. The article presents the author's method of fast, simultaneous calibration of many mobile FMCW radars operating in a network. It was validated using 95,000 numerical tests. The results show that the proposed method significantly improves the north orientation of the radars throughout the whole range of the initial errors. Therefore, it can be successfully used in practical applications.

Keywords: anti-drone systems; FMCW radars; drones detection; radars calibration

---

1. Introduction

A market of small unmanned aerial vehicles (UAVs)—commonly called drones—is developing very fast [1] and is stimulated by a very good quality to price ratio—for a relatively small amount a drone can be bought that has excellent performance regarding the possibility of flight and shooting a high-quality video from the air. Unfortunately, the drones could be used not only for commercial, research, surveillance, fun, or any other legal activities but also as a tool of crime. Sometimes intentionally, like in the case of drug smuggling across prison walls or dropping explosive materials [2]. Another time inappropriate and dangerous use of drones is caused by curiosity or lack of imagination. Camera-equipped UAVs can invade people's privacy or cause a threat to their life. For example, the report by Bard College's Center for the Study of the Drone [3] found that 327 incidents between December 2013 and September 2015 (USA only) posed a "proximity danger" where an unmanned aircraft got within 500 feet (152.4 m) of a plane, helicopter, or other manned aircraft (involved multiengine jet aircraft). The above mentioned, as well as similar accidents, have caused many institutions to start thinking about anti-drone solutions. In the case of objects permanently threatened by drones, for example, airports or buildings crucial for national security, stationary anti-drone systems can be justified. However, there are many cases when using a permanent solution seems to be aimless due to the high cost of the system not corresponding with threat frequency, for example, mass events or meetings with very important persons (VIP). For protecting people and objects in such cases, mobile systems (on vehicle or man-carried) perfectly meet the demands. As an example of such solutions, anti-drone systems Hawk by Hertz Systems Ltd (Poland) [4] and Radar Backpack Kit by SpotterRF (USA) [5], are presented in Figure 1.

Hawk by Hertz Systems Ltd
http://thehawksystem.com/pl/

Radar Backpack Kit by SpotterRF
https://spotterrf.com/products/mobile-solutions/

Figure 1. Examples of mobile anti-drone systems.

Many scientists and engineers involved in the development of anti-drone systems agree that a multi-sensor system could provide a higher probability of drone detection than a single sensor [6–12]. Experiments also suggest that one of the most efficient detectors of UAV is a frequency modulated continuous wave (FMCW) radar [6–12]. It has a big detection range, transmits low power [13], and is independent of electromagnetic signals and sounds emitted by drones (UAV operating in a fully autonomous mode and gliding can be detected, too).

Active radars do measurements in a local polar system, where distance and azimuth of the detected object are determined. Three-dimensional radars also provide information about the object elevation. It means that the accuracy of determining the drone localization in a global coordinate system depends not only on the radar performance but also on the accuracy of the radar coordinates and spatial orientation relative to the local horizon and the north. Admittedly, using precise Global Navigation Satellite System (GNSS) positioning, for example, Real Time Kinematic (RTK), can solve the problem of accurate localization. Orientation angles can also be relatively easily determined in the case of a stationary system. However, there would be some difficulties in mobile solutions, especially if the system changes location during the mission. Obviously, some sophisticated techniques, like Moving Base RTK, can provide precise orientation angles; however, using them is not always possible. They also do not ensure expected accuracy in every condition and require additional devices and often specialized knowledge. Although a magnetic compass could be a low-cost and convenient solution, it is not accurate enough if we consider that accuracy better than 1^0 is required. Due to magnetic variation, magnetic deviation, errors connected with radar and magnetic sensor installation on a vehicle, and method of measurement, accuracy better than 20^0 is difficult to achieve in a real application.

Long-range FMCW radars used in non-military anti-drone systems have detection ranges of micro-drones (DJI Phantom sizes) not exceeding 3 km, due to the small radar cross section (RCS) of such objects [14–17]. However, the classification range has a practical meaning, as anti-drone radars also detect birds and other small objects. Determining if the object is a drone allows relevant actions to be taken. The most common method of classification is the micro-Doppler signature (MDS) analysis. It allows investigating micromotions of the object detected by the radar. The contributions of the rotating turbine or propeller blades in radar backscattering in the form of micro-Doppler contents contain information that there are rotating parts on the object. On this basis, the method distinguishes a drone from a bird [18–23]. If the analysis of MDS is a classifier, the classification range is about 50% of the detection range, due to the RCS of the rotating turbine or propeller blades being smaller than the RCS of the drone body. The classification range is the range in which the classifier works accurately. Examples of detection and classification characteristics of the DJI Phantom 4 drone for one of the anti-drone radars are presented in Figure 2. It should be noted that it is currently one of the radars with the longest range of the micro-drone detection and classification available on the commercial market.

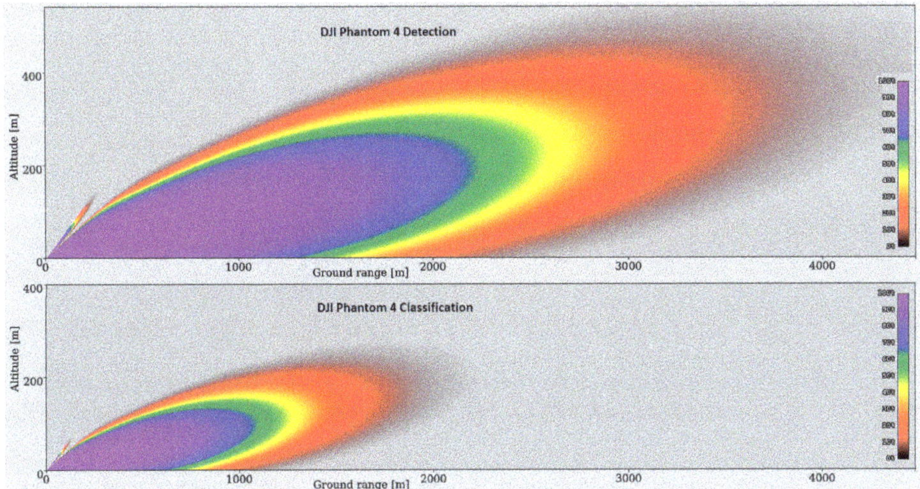

Figure 2. DJI Phantom 4 drone detection (upper) and classification (lower) ranges for one of FMCW radars available on a commercial market (the characteristics were obtained from the manufacturer, who asked not to disclose the name and model of the radar for obvious reasons).

Considering the real range of non-military FMCW radars and the fact that effective protection of the area or object requires detection and classification of the drone several hundred meters before the forbidden zone, it is often necessary to use several mobile systems operating in the network. One of the key conditions for their proper cooperation is the measurement of azimuths relative to the same direction (preferably geographical north, also called the true north, or just, the north). Otherwise, errors will occur in determining the actual location of the drone. In a single-radar system, a large error in determining the coordinates of the detected object may result in incorrect classification as an object "violating" or "not violating" the forbidden area (see Figure 3) and, consequently, an inadequate response. In network systems, where several radars operate in the same area and their coverages partly overlap, there will be additional uncertainty concerning a location of the detected drone—each of the radars will indicate different coordinates of the same object (see Figure 4). Moreover, one will not know how many objects are really in the area. A single drone can be classified as a swarm or vice versa. All described cases are very undesirable, and their elimination is necessary to ensure the proper functioning of the anti-drone system.

Due to the fact that non-military anti-drone systems are still developing and many of them are on a low Technology Readiness Level (TRL) - similarly as some of the radar solutions in this area [3,24], the market usually offers single radars or radar systems integrated on one platform or assembly set. It ensures that the measurements are in the same local coordinate system (not necessarily oriented to the north). Then, the problem of azimuths misalignment does not exist due to all the devices making the same constant error in azimuths measurements. However, if single radars are integrated into a bigger system, where sensors are distributed over a wide area, bringing all the measurements to a common coordinates system is required. It means that a misalignment reduction of azimuths is necessary. In other cases, after the transformation of local coordinates of the detected object to the global one, different results will be obtained, depending on the azimuth error of the particular radar (see Figure 4).

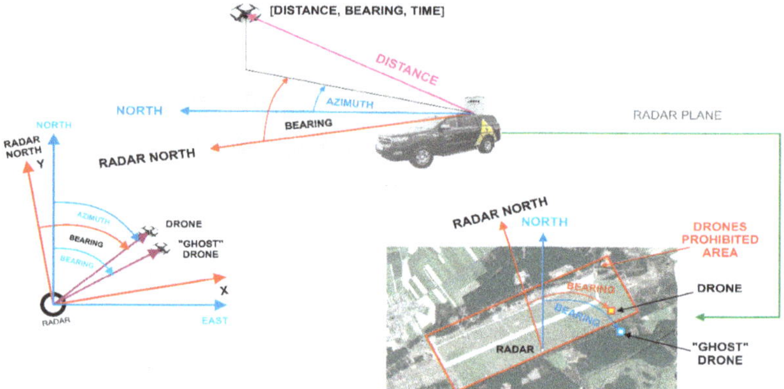

Figure 3. An example of drone misclassification due to a large azimuth error. The radar indicates "GHOST DRONE" (blue rectangle with a white background), and the system classifies it as an object that does not violate the airport zone. In fact, "DRONE" (red rectangle with a yellow background) violates the forbidden zone.

The co-author of the paper has participated in the research project granted by the Polish National Center for Research and Development since 2017. One of its aims is to develop an anti-drone system consisting of many mobile sensors (on the vehicles) spread over a wide area. Experiments already performed in a real environment show the problem of azimuth misalignment of particular sensors. It causes improper operation of the whole system and repeats every time the vehicles change location. As an example, the PrintScreen from the developing system console is shown in Figure 4.

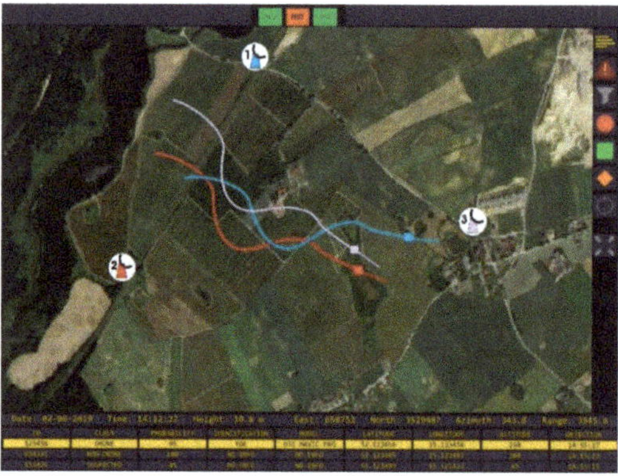

Figure 4. An example of improper operation of the system due to azimuths misalignment. The system detected and was tracking three objects instead of one.

In Figure 4, it can be seen that one object is identified as three by the system due to azimuth misalignment. The radars one, two, and three made constant azimuths errors of $+5.1^0$, -10.3^0, $+14.5^0$, respectively, relative to the north. The errors were determined using GNSS RTK on a drone and the radars.

The azimuth misalignment of 5^0 produces the drone coordinate error of about 87.5 m at a distance of 1 km (see Figure 5). It means that two radars will determine the coordinates of the same object differing by 87.5 m. As already mentioned, in mobile systems (especially on vehicles), it is difficult to obtain the radar azimuths alignment better than 20^0. Then, the coordinates error will be 364 m (see Figure 5) at a distance of 1 km. Doubtlessly, this will disturb the proper operation of the system. Moreover, variable errors of azimuth measurement of both radars will cause a further increase in the drone position uncertainty. This is a quite serious issue and must be taken into account.

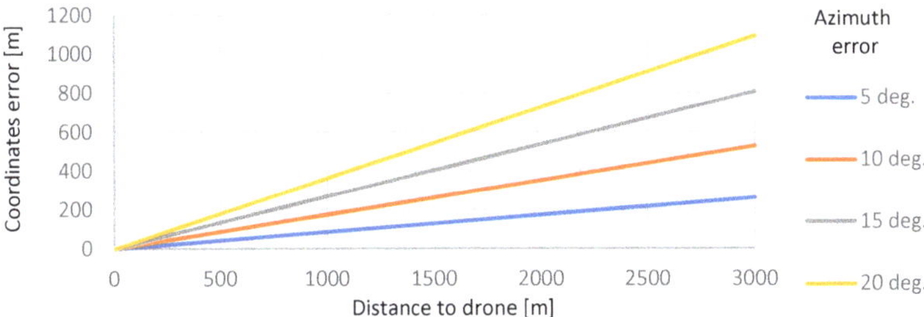

Figure 5. The additional error of drone coordinates caused by the azimuth misalignment.

As was mentioned earlier, regarding a permanent system, the azimuths misalignment can be reduced using, for example, geodetic methods. However, radars in a mobile system can change location during the mission. Therefore, in new places, they must be aligned to the rest of the system as quickly as possible. To the best knowledge of the authors, this issue has not been described in publications yet (probably due to the commercial multi-radars mobile systems are still on low TRL). This statement is based on an unsuccessful search for a suitable method description. 'Suitable' means giving the desired accuracy at an acceptable time. However, the authors are conscious that this statement may be a kind of abuse. Research done by the authors shows that, at the moment, in practical applications, each of the radars is only roughly set separately using magnetic sensors and more advanced devices, such as gyrocompasses, GNSS RTK, or satellite compasses, are not applied in non-military mobile systems. Such a method does not ensure the desired accuracy.

This work presents a proposal to solve the problem of azimuths misalignment, consisting of the initial/coarse orientation of the radar system to the north using a magnetic compass, and then carrying out a calibration procedure consisting of automatically calculating corrections to azimuths measured by individual radars. The calibration process uses the "friend" drone flying in autonomous mode along a fixed route. The position of the drone does not have to be known, and the drone does not need to communicate with the system. The proposed method is simple, automatic, and requires no specialized knowledge. It gives very good results and can be successfully applied in real applications.

2. Proposed Method of Reducing the Azimuths Misalignment

In an ideal situation (no measurement errors), the coordinates of the drone detected by the radar are expressed by the following system of equations:

$$\begin{cases} x_{D\_ENU} = x_{R\_ENU} + D_{2D} \cdot sin(A) \\ y_{D\_ENU} = y_{R\_ENU} + D_{2D} \cdot cos(A) \end{cases}, \tag{1}$$

where:

x_{D_ENU}, y_{D_ENU}	- true coordinates of the drone in the ENU system (East North Up),
x_{R_ENU}, y_{R_ENU}	- true coordinates of the radar in the ENU system,
D_{2D}	- horizontal distance (2D) to the drone,
A	- azimuth of the drone from the radar position.

However, in real conditions, all quantities occurring on the right side of equations in the system (1) are affected by errors. Therefore, (1) will take the form

$$\begin{cases} \hat{x}_{D\_ENU} = x_{R\_ENU} + \delta_{RP\_ENU} + (D_{2D} + \delta_{D\_2D})\cdot sin(A + \delta_{A\_ENU}) \\ \hat{y}_{D\_ENU} = y_{R\_ENU} + \delta_{RP\_ENU} + (D_{2D} + \delta_{D\_2D})\cdot cos(A + \delta_{A\_ENU}) \end{cases}, \tag{2}$$

where:

δ_{RP_ENU}	- radar coordinates determination error,
δ_{D_2D}	- horizontal distance measurement error,
δ_{A_ENU}	- azimuth measurement error in the ENU system.

Taking the practical experience in using FMCW radars for drone detection into consideration, it should be noted:

Note 1.: In practice, the radar position is fixed by GNSS before the operation and taken as a constant until the radar is moved to a new location. In such a case, δ_{RP_ENU} is a constant error. It can be minimized by using Differential GNSS or GNSS RTK measurement techniques and by mounting the satellite system antenna centrally above the center of the radar measuring system. As already mentioned, the GNSS RTK technique is not used in commercial systems. A convenient solution is a Satellite Based Augmentation System (SBAS)—GNSS augmented by the differential system based on satellites. This reduces the radar position error to a sub-meter level. It allows omitting it in further considerations due to a small impact on the final performance of the proposed method.

Note 2.: δ_{D_2D} is a random variable with some small constant component resulting from the leveling of the mobile platform or mounting system of the radar. The constant component can be reduced almost to zero by using a leveling system and then δ_{D_2D} can be modeled as an independent random variable with a normal standard distribution. Then, the highest probability density is concentrated around zero. Therefore, δ_{D_2D} is also omitted in the proposed method.

Note 3.: δ_{A_ENU} is a random variable with a small variable component and a very large constant component, resulting from magnetic declination and magnetic deviation, radar and compass assembly on the platform, and the physical possibilities of setting the mobile platform or mounting system of the radar in the direction indicated by the magnetic compass. In practical applications, the variable component varies within $1^0 - 3^0$ (depending on the class of anti-drone radar), while the constant component can reach 20^0. Such a large error in measuring the azimuth of the detected object will result in an error in determining its location (coordinates). According to Equation (2), it will be larger the greater the distance between the radar and the detected object is (see Figure 5). This will entail some complications, both in single- and multi-radar systems, which have already been mentioned in the Introduction.

Bearing in mind Notes 1 to 3, and that

$$A_m = A + \delta_{A\_ENU} \rightarrow A = A_m - \delta_{A\_ENU}, \tag{3}$$

the system of Equation (1) can be written as

$$\begin{cases} \hat{x}_{D\_ENU} = \hat{x}_{R\_ENU} + D_{m2D}\cdot cos(A_m - \delta_{A\_ENU}) \\ \hat{y}_{D\_ENU} = \hat{y}_{R\_ENU} + D_{m2D}\cdot sin(A_m - \delta_{A\_ENU}) \end{cases}. \tag{4}$$

This system has 3 unknowns: The estimated drone position in the ENU system $(\hat{x}_{D\_ENU}, \hat{y}_{D\_ENU})$ and the azimuth measurement error in the ENU system (δ_{A_ENU}). The radar measurement results are the horizontal distance (D_{m2D}) and the azimuth (A_m), where

$$D_{m2D} = D_{m3D} \cdot cos(E_m), \tag{5}$$

and:

D_{m3D}	- the spatial distance (3D) to the detected drone measured by the radar,
E_m	- the drone elevation measured by the radar or a priori assumed.

Estimated radar coordinates (fixed using DGNSS/SBAS measurement) in the ENU system $(\hat{x}_{R\_ENU}, \hat{y}_{R\_ENU})$ are also known. On the left side of the equations, instead of true coordinates, there are their estimators, because both the position of the radar and the measurements are burdened with the errors. The errors were omitted in the system (4); however, they have an impact and cause some error in determining the coordinates of the detected drone.

From (4), it follows that unambiguous determination of unknowns based on measurements from a single radar is not possible. The next radar measurement will introduce the next 2 unknowns (new 2D coordinates of the drone), and 2 independent equations. Therefore, regardless of the number of measurements done by a single radar, the number of independent equations will always be 1 less than the number of unknowns.

Let us now consider a situation where a few radars in the network are tracking the same object. Then, the following system of independent equations can be formed

$$\begin{cases} \hat{x}_{D\_ENU1} = \hat{x}_{R\_ENU1} + D_{m2D1} \cdot cos(A_{m1} - \delta_{A\_ENU1}) \\ \hat{y}_{D\_ENU1} = \hat{y}_{R\_ENU1} + D_{m2D1} \cdot sin(A_{m1} - \delta_{A\_ENU1}) \\ \vdots \\ \hat{x}_{D\_ENUn} = \hat{x}_{R\_ENUn} + D_{m2Dn} \cdot cos(A_{mn} - \delta_{A\_ENUn}) \\ \hat{y}_{D\_ENUn} = \hat{y}_{R\_ENUn} + D_{m2Dn} \cdot sin(A_{mn} - \delta_{A\_ENUn}) \end{cases}, \tag{6}$$

where:

$\hat{x}_{D\_ENUi}, \hat{y}_{D\_ENUi}$	- the estimators of the drone true coordinates calculated on the base of measurements done by i-th radar,
$\hat{x}_{R\_ENUi}, \hat{y}_{R\_ENUi}$	- the estimators of the i-th radar true coordinates,
D_{m2Di}, A_{mi}	- measurements results of horizontal distance and azimuth done by i-th radar,
n	- number of radars in the network.

Let us mark the position of the detected drone in the ENU system as $\hat{P}_{D\_ENUi}$. This is the position estimated on the base of measurements done by i-th radar. Since both estimators of the radar coordinates and radar measurements are affected by errors, then

$$\hat{P}_{D\_ENU1} \neq \hat{P}_{D\_ENU2} \neq \ldots \neq \hat{P}_{D\_ENUn}. \tag{7}$$

In this case, system (6) also has no unambiguous solution, because the number of unknowns will be $3n$, and the number of independent equations will be $2n$. However, due to all radars are observing the same object, we can assume a simplification:

$$\hat{P}_{D\_ENU1} = \hat{P}_{D\_ENU2} = \ldots = \hat{P}_{D\_ENUn} = \hat{P}_{D\_ENU}(x_{D\_ENU}, y_{D\_ENU}), \tag{8}$$

and then, system (6) will take the following form:

$$\begin{cases} x_{D\_ENU} = \hat{x}_{R\_ENU1} + D_{m2D1} \cdot cos(A_{m1} - \delta_{A\_ENU1}) \\ y_{D\_ENU} = \hat{y}_{R\_ENU1} + D_{m2D1} \cdot sin(A_{m1} - \delta_{A\_ENU1}) \\ \vdots \\ x_{D\_ENU} = \hat{x}_{R\_ENUn} + D_{m2Dn} \cdot cos(A_{mn} - \delta_{A\_ENUn}) \\ y_{D\_ENU} = \hat{y}_{R\_ENUn} + D_{m2Dn} \cdot sin(A_{mn} - \delta_{A\_ENUn}) \end{cases} . \quad (9)$$

In this situation, the number of unknowns will be $2 + n$, and the number of independent equations $2n$, and already for $n = 2$, all 3 unknowns can be fixed. However, for $n > 2$, redundant equations will appear. In such a case, we have $\binom{2+n}{2n}$ numbers of possible equations combinations, which give a solution in the form of $S\left(\hat{P}_{D\_ENU}, \hat{\delta}_{A\_ENU1}, \ldots, \hat{\delta}_{A\_ENUn}\right)$. Unfortunately, due to errors both in the estimated coordinates of the radar position and in the radar measurements, each of these combinations will give a slightly different solution. These differences will depend not only on the radar coordinates errors and the azimuth, elevation, and distance measurement errors but also on the relative location of the radars and the detected object (the system geometry described by the dilution of precision coefficient). It should be noted that the geometry of the system will change during the drone's flight. Therefore δ_{A_ENUi} calculated for the moment t, based on observations from n radars, can be expressed by the following equation:

$$\delta_{A\_ENUi}(t) = \delta^{const}_{A\_ENUi} + \delta^{var}_{A\_ENUi} = \delta^{const}_{A\_ENUi} + f\left(\delta_{A\_ENU1}(t), \ldots, \delta_{A\_ENUn}(t), \delta^{DOP}_{A}(t)\right), \quad (10)$$

where:

$\delta^{const}_{A\_ENUi}$	- constant component of $\delta_{A\_ENUi}(t)$,
$\delta^{var}_{A\_ENUi}$	- variable in time component of $\delta_{A\_ENUi}(t)$,
$\delta^{DOP}_{A}(t)$	- variable in time component of $\delta_{A\_ENUi}(t)$ resulting from the changing geometry of the radars-detected object arrangement due to drone flight.

$\delta^{const}_{A\_ENUi}$ is the sought difference between the radar north and the true north:

$$North_{real} + \delta^{const}_{A_{ENUi}} = North_{RADARi} \rightarrow \delta^{const}_{A\_ENUi} = North_{RADARi} - North_{real}. \quad (11)$$

Due to $\delta^{const}_{A\_ENUi} \gg \delta^{var}_{A\_ENUi}$, the determination of $North_{RADARi}$ would significantly improve both the accuracy of determining the coordinates of the detected object and minimize the likelihood of classifying a single drone as a swarm and vice versa. Both factors are crucial for the anti-drone system. The calculated $\delta^{const}_{A\_ENUi}$ value with the opposite sign would be a correction to the measured azimuth by the radar and could be automatically taken into account by the master system responsible for the multi-sensors data fusion. In the proposed method, determining the difference $North_{RADARi} - North_{real}$, and hence the correction for measuring azimuths consists of eliminating the $\delta^{var}_{A\_ENUi}$ component from Equation (10). According to the proposed idea, this is done by multi-radars observations of a flying drone with unknown coordinates. Azimuth measurement error δ_{A_ENU} is estimated in every measurement epoch (t_e), and then the constant component of the azimuth measurement error is calculated as:

$$\delta^{const}_{A\_ENU} = \frac{1}{m} \sum_{j=1}^{m} \delta_{A\_ENU}\left(t_{ej}\right), \quad (12)$$

where:

m	- number of measurement epochs common to all radars,
$\delta_{A\_ENU}\left(t_{ej}\right)$	- azimuth measurement error at the moment $\left(t_{ej}\right)$.

Therefore, the key issue is to estimate the azimuth measurement error δ_{A_ENU} in every measurement epoch. In the proposed method, instead of the classical form of system of equations describing the coordinates of the detected object (9), the following form is used:

$$\begin{cases} D_{m2D1} = \sqrt{(x_{D\_ENU} - \hat{x}_{R\_ENU1})^2 + (y_{D\_ENU} - \hat{y}_{R\_ENU1})^2} \\ A_{m1} = arctan\left(\frac{x_{D\_ENU1} - \hat{x}_{R\_ENU1}}{y_{D\_ENU1} - \hat{y}_{R\_ENU1}}\right) + \delta_{A\_ENU1} \\ \vdots \\ D_{m2Dn} = \sqrt{(x_{D\_ENU} - \hat{x}_{R\_ENUn})^2 + (y_{D\_ENU} - \hat{y}_{R\_ENUn})^2} \\ A_{mn} = arctan\left(\frac{x_{D\_ENU} - \hat{x}_{R\_ENUn}}{y_{D\_ENU} - \hat{y}_{R\_ENUn}}\right) + \delta_{A\_ENUn} \end{cases}. \tag{13}$$

Let us assume the approximate drone position and mark it as $P^0_{D\_ENU}\left(x^0_{D\_ENU}, y^0_{D\_ENU}\right)$. Let it be a position calculated on the basis of observations from any one radar from the network:

$$\begin{cases} x^0_{D\_ENU} = \hat{x}_{R\_ENUi} + D_{m2Di} \cdot cos(A_{mi}) \\ y^0_{D\_ENU} = \hat{y}_{R\_ENUi} + D_{m2Di} \cdot sin(A_{mi}) \end{cases}. \tag{14}$$

Now let us expand the right sides of the equations of the system (13) in the Taylor series relative to $P^0_{D\_ENU}$, limiting to the second order components:

$$\begin{cases} D_{m2D1} = D^0_{m2D1} + \frac{\partial}{\partial x^0_{D\_ENU}}\left(D^0_{m2D1}\right) \cdot \Delta x_{D\_ENU} + \frac{\partial}{\partial y^0_{D\_ENU}}\left(D^0_{m2D1}\right) \cdot \Delta y_{D\_ENU} \\ A_{m1} - \delta_{A\_ENU1} = A^0_{m1} + \frac{\partial}{\partial x^0_{D\_ENU}}\left(A^0_{m1}\right) \cdot \Delta x_{D\_ENU} + \frac{\partial}{\partial y^0_{D\_ENU}}\left(A^0_{m1}\right) \cdot \Delta y_{D\_ENU} \\ \vdots \\ D_{m2Dn} = D^0_{m2Dn} + \frac{\partial}{\partial x^0_{D\_ENU}}\left(D^0_{m2Dn}\right) \cdot \Delta x_{D\_ENU} + \frac{\partial}{\partial y^0_{D\_ENU}}\left(D^0_{m2Dn}\right) \cdot \Delta y_{D\_ENU} \\ A_{mn} - \delta_{A\_ENUn} = A^0_{mn} + \frac{\partial}{\partial x^0_{D\_ENU}}\left(A^0_{mn}\right) \cdot \Delta x_{D\_ENU} + \frac{\partial}{\partial y^0_{D\_ENU}}\left(A^0_{mn}\right) \cdot \Delta y_{D\_ENU} \end{cases}, \tag{15}$$

wherein:

$$D^0_{m2Di} = \sqrt{\left(x^0_{D\_ENU} - \hat{x}_{R\_ENUi}\right)^2 + \left(y^0_{D\_ENU} - \hat{y}_{R\_ENUi}\right)^2}, \tag{16}$$

$$A^0_{mi} = arctan\left(\frac{x^0_{D\_ENU} - \hat{x}_{R\_ENUi}}{y^0_{D\_ENU} - \hat{y}_{R\_ENUi}}\right). \tag{17}$$

Let's mark the individual derivatives as follows:

$$\frac{\partial}{\partial x^0_{D\_ENU}}\left(D^0_{m2Di}\right) = m_{Dxi}, \qquad \frac{\partial}{\partial y^0_{D\_ENU}}\left(D^0_{m2Di}\right) = m_{Dyi},$$

$$\frac{\partial}{\partial x^0_{D\_ENU}}\left(A^0_{mi}\right) = m_{Axi}, \qquad \frac{\partial}{\partial y^0_{D\_ENU}}\left(A^0_{mi}\right) = m_{Ayi}.$$

Then, the system of Equation (15) will take the form

$$\begin{cases} D_{m2D1} - D^0_{m2D1} = m_{Dx1} \cdot \Delta x_{D\_ENU} + m_{Dy1} \cdot \Delta y_{D\_ENU} \\ A_{m1} - A^0_{m1} = m_{Ax1} \cdot \Delta x_{D\_ENU} + m_{Ax1} \cdot \Delta y_{D\_ENU} + \delta_{A\_ENU1} \\ \vdots \\ D_{m2Dn} - D^0_{m2Dn} = m_{Dxn} \cdot \Delta x_{D\_ENU} + m_{Dyn} \cdot \Delta y_{D\_ENU} \\ A_{mn} - A^0_{mn} = m_{Axn} \cdot \Delta x_{D\_ENU} + m_{Axn} \cdot \Delta y_{D\_ENU} + \delta_{A\_ENUn} \end{cases}. \tag{18}$$

The system of Equation (18) can be written in matrix form as

$$\underbrace{\begin{bmatrix} D_{m2D1}-D^{0}_{m2D1} \\ A_{m1}-A^{0}_{m1} \\ D_{m2D2}-D^{0}_{m2D2} \\ A_{m2}-A^{0}_{m2} \\ \vdots \\ D_{m2Dn}-D^{0}_{m2Dn} \\ A_{mn}-A^{0}_{mn} \end{bmatrix}_{2n\times 1}}_{\mathbf{d}} = \underbrace{\begin{bmatrix} m_{Dx1} & m_{Dy1} & 0 & 0 & \cdots & 0 \\ m_{Ax1} & m_{Ay1} & 1 & 0 & \cdots & 0 \\ m_{Dx2} & m_{Dy2} & 0 & 0 & \cdots & 0 \\ m_{Ax2} & m_{Ay2} & 0 & 1 & \cdots & 0 \\ & & \vdots & & & \\ m_{Dxn} & m_{Dyn} & 0 & 0 & \cdots & 0 \\ m_{Axn} & m_{Ayn} & 0 & 0 & \cdots & 1 \end{bmatrix}_{2n\times(n+2)}}_{\mathbf{A}} \cdot \underbrace{\begin{bmatrix} \Delta x_{D_{ENU}} \\ \Delta y_{D_{ENU}} \\ \delta_{A_{ENU1}} \\ \delta_{A_{ENU2}} \\ \vdots \\ \delta_{A_{ENUn}} \end{bmatrix}_{(n+2)\times 1}}_{\mathbf{X}} . \quad (19)$$

X is a vector of unknowns. Estimated value of **X** using least square estimator is:

$$\hat{\mathbf{X}} = \left(\mathbf{A}^{T}\mathbf{A}\right)^{-1}\mathbf{A}^{T}\mathbf{d}. \quad (20)$$

The solution of (20) is:

$$\hat{\mathbf{X}} = \begin{bmatrix} \Delta\hat{x}_{D\_ENU} \\ \Delta\hat{y}_{D\_ENU} \\ \hat{\delta}_{A\_ENU1} \\ \hat{\delta}_{A\_ENU2} \\ \vdots \\ \hat{\delta}_{A\_ENUn} \end{bmatrix}_{(n+2)\times 1} . \quad (21)$$

Using $\Delta\hat{x}_{D\_ENU}, \Delta\hat{y}_{D\_ENU}$, the estimated position of the drone can be calculated as:

$$\hat{x}_{D\_ENU} = x^{0}_{D\_ENU} + \Delta\hat{x}_{D\_ENU}, \hat{y}_{D\_ENU} = y^{0}_{D\_ENU} + \Delta\hat{y}_{D\_ENU}. \quad (22)$$

If $\Delta\hat{x}_{D\_ENU} > 1m$ or $\Delta\hat{y}_{D\_ENU} > 1m$, the next iteration is necessary, taking the estimated drone position calculated from (22) as the new approximate position. The final value of the azimuth measurement error for each of the radars is the value of $\hat{\delta}_{A\_ENUi}$ from the last iteration. Then, the $\hat{\delta}^{const}_{A\_ENU}$ value is calculated based on (12). According to the presented idea the correction to measuring azimuths (the difference between the true north and the radar north) for a given radar is:

$$c_{Ai} = -\hat{\delta}^{const}_{A\_ENUi}. \quad (23)$$

Then, the azimuth estimator of the detected object for a given radar is:

$$\hat{A}_{i} = A_{mi} + c_{Ai}. \quad (24)$$

The practical implementation of the proposed method can be presented in the form of the following algorithm:

```
BEGIN      //drone started a calibration flight and all radars detected a drone
    j := 0;  //first measurements epoch
    n := number of radars;
    REPEAT
            FOR i := 1 TO n DO
                    Get the radars measurements M_i = {A_mi, D_m2Di};
            Bring M_i to a common moment (t_ej);  //explanation below the algorithm
        Compute δ_A_ENU(t_ej) for all radars;     //using Equation (20)
            j := j + 1;                           //next measurement epoch
    UNTIL drone ended a calibration flight
    m := j;                                       //number of measurements epoch
    Computation of δ^const_A_ENU;                 //using Equation (12)
END.
```

Since observations of the drone by each of the radars are generally not done at the same time, but in some time window (approximately equal to the period of scanning of a particular radar), the values of measured azimuths and distances must be brought to a common moment (t_e). The proposed method uses linear interpolation between the measurements immediately preceding and following (t_e), according to the formulas:

$$\begin{aligned} \hat{A}_m(t_e) &= A_m(t_0) + \tfrac{t_e-t_0}{t_1-t_0}(A_m(t_1) - A_m(t_0)), \\ \hat{D}_{m2D}(t_e) &= D_{m2D}(t_0) + \tfrac{t_e-t_0}{t_1-t_0}(D_{m2D}(t_1) - D_{m2D}(t_0)), \end{aligned} \tag{25}$$

where:

t_0	- the time immediately preceding (t_e), when the measurement was done,
t_1	- the time immediately following (t_e), when the measurement was done,
$\hat{A}_m(t_e)$, $\hat{D}_{m2D}(t_e)$	- estimated values of azimuth and distance at moment (t_e),
$A_m(t_0)$, $A_m(t_1)$	- measured azimuths at moments (t_0) i (t_1),
$D_{m2D}(t_0)$, $D_{m2D}(t_1)$	- measured distances at moments (t_0) i (t_1).

3. Method of Verification

Intensive numerical simulations were performed to verify the correctness of the developed method. There were simulated drone flights along a fixed route and indications of FMCW radars tracking them—measurements of azimuth and distance. The simulation software SimWizardADS developed by the Polish company Basic Solution was used in the research. The software was developed as part of a project granted by the Polish National Center for Research and Development (NCRD) in 2017 under the name SimWizardSSAD. The software was accepted by NCRD experts as a reliable tool for simulation of anti-drone systems in the scope of detection and is being further developing under the name SimWizardADS. An example of the simulation wizard window is shown in Figure 6.

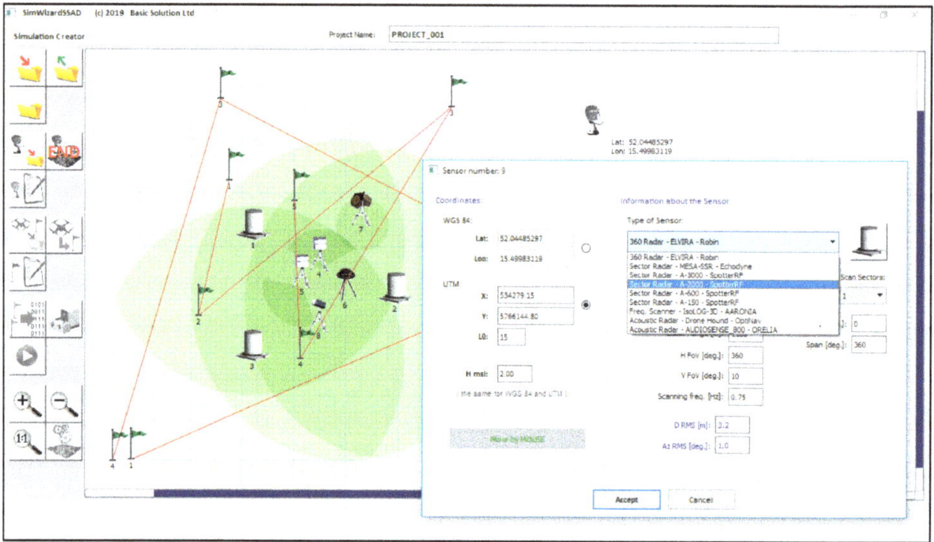

Figure 6. Example window of the SimWizzardADS application (creator of the simulation) of the Basic Solution company (Poland) [www.basicsolution.eu].

The numerical simulations were performed with the following assumptions:

1. Number of the simulations: 95,000
2. In each simulation, an anti-drone system consisting of 3 and 4 omnidirectional long-range radars will be simulated. The radars will measure distances and azimuths without elevations (2D radars):
    a. In the variant with 3 radars, they will be located in the nodes of an equilateral triangle. The side length will be 2 km (see Figure 7-left),
    b. In the variant with 4 radars, they will be located in the nodes of the rectangle. The side length will be 2 km (see Figure 7-right),
    c. Detection range of the radars will be 2.5 km,
    d. The beam width of the radars in the elevation will be 10^0. Due to the fact that the radars do not measure the height of the detected object, it is assumed that it equals half the beam width. As mentioned earlier, this is a common assumption in 2D radars,
    e. Range measurement errors will contain only the variable component. It will be generated randomly according to the normal distribution. In each simulation, each radar will be assigned a standard deviation of the range measurement error (RMS range). It will be selected randomly (for each radar independently) from a set of 0.6, 0.8, 1.0, or 1.2 m. These are typical values of the RMS range for anti-drone FMCW radars available on the commercial (open) market. The values will be constant for one simulation. In the next one, the new values will be drawn on the same rules,
    f. Azimuth measurement errors will contain a variable and constant component. The variable component will be generated randomly according to the normal distribution. In each simulation, a standard deviation of azimuth measurement error (RMS azimuth) will be assigned to each radar. It will be selected randomly (for each radar independently) from a set of 0.8, 1.0, 1.2, or 1.4 m. These are typical RMS azimuth values for anti-drone FMCW radars available on the commercial (open) market. The values will be constant for one simulation. The next one will draw new values on the same principles,

g. The constant component of the azimuth measurement error will be randomly generated according to a uniform distribution in the range of +/− 15^0. The values will be generated independently for each radar and will be constant for one simulation. In the next one, new values will be drawn on the same rules,
h. The radar scanning frequency will be 2, 1, 1.5, or 0.5 Hz. These are typical values for anti-drone FMCW radars available on the commercial (open) market. The values will be chosen randomly and will be constant for one simulation. In the next one, new values will be drawn on the same principles.

3. The calibration flight (using "friend" drone) will take place along the routes shown in Figure 8. The height will be 20 m above the ground. The justification for the route selection can be found below. The drone speed will be constant and equal to 10 m/s. This is a typical value for commercial drones. This means that the calibration flight will take 11 min 10 s for the variant with 3-radars and 10 min 40 s for the 4-radars option.
4. The analysis of the simulation results will be carried out in the context of:

    a. Calibration accuracy. The measure will be the difference between the radar north and the true north after calibration,
    b. The radar north increasing. A measure will be a value of which a difference between the radar north and the true north has changed,
    c. For each of the value listed in part 4, the following parameters will be computed:
    d. Mean value,
    e. Standard deviation (RMS),
    f. Absolute minimum and maximum values and parameters at which they occurred (radar configuration, actual radar north deviation from the true north, and a standard deviation of azimuth and distance measurement.

Experiences of the authors with commercial FMCW radars used in anti-drone systems indicate that when choosing a calibration flight route, certain factors should be considered. They result from the characteristics of commercial FMCW radars used in anti-drone systems. The idea is to minimize the likelihood of losing drone tracking, as this will adversely affect the effectiveness of the proposed method. At the same time, the likelihood of various random measurement errors associated with the mutual geometry between the drone and the radars should be maximized. Practical experiences show that:

1. The route, if possible, should be in a common coverage area for all radars. This will have a positive effect on the accuracy of the proposed method. In such a case, the system of observational equations will contain more redundant equations. It will increase the accuracy of the estimator of the azimuth measurement error for individual radars.
2. Commercial FMCW radars have difficulties with detecting and tracking low radial-speed objects (slowly approaching or moving away from the radar). Therefore, the calibration route should not include sections with a low radial-speed relative to any radar.
3. The accuracy of measuring horizontal distances by 2D radars decreases with increasing distance to the object, as the difference between the assumed and actual height of the drone increases (see Figure 5). Therefore, the flight should be at a minimum safe altitude. Then, the 3D and 2D distances will be close to each other, which will minimize the error associated with the uncertainty of the drone elevation. Practice shows that 20 m height is the optimal solution.
4. Measurement experiments with FMCW radars indicate that the accuracy of the determined drone parameters also depends on the flight direction relative to the radar (among others, due to the different position of the drone's hull relative to the radar, and thus different RCS). Therefore, the

calibration flight route should include sections where the drone is approaching and moving away from the radar.

5. The route should not be too long, as this will make the calibration process time consuming, and at the same time, should consider points 1–4 as much as possible.

Taking the above into account, several possible variants of calibration flight routes were analyzed. It was finally found that the optimal ones (for selected locations of the radars) will be as shown in Figure 7.

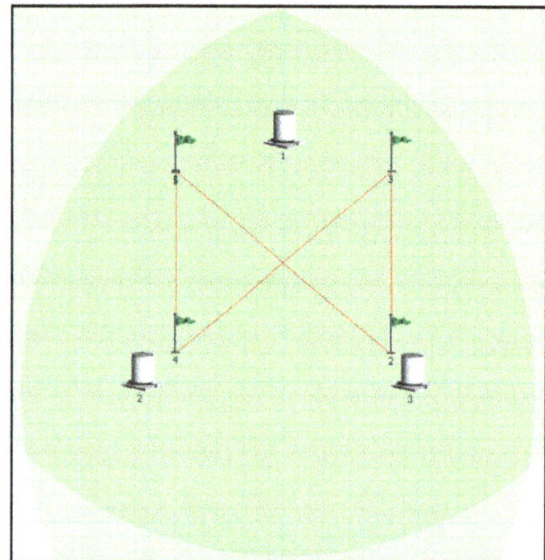

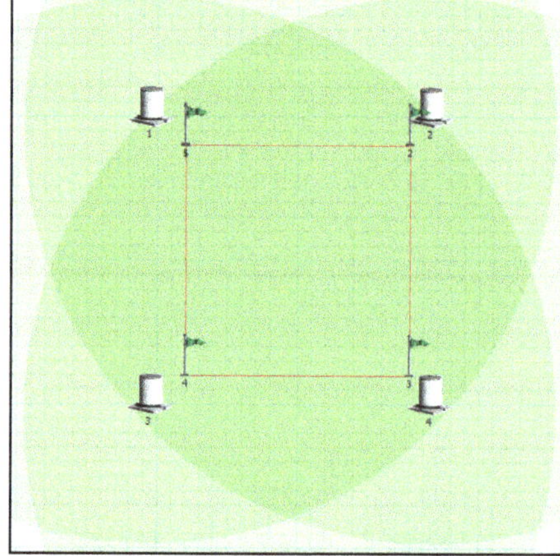

Figure 7. The optimal routes of the calibration flight for the selected location of 3 radars (left) and 4 radars (right).

4. Results

According to the assumptions, 95,000 simulations were performed for each of the two variants of the radar location (the 3- and 4-radars variants). Tables 1 and 2 show the calibration accuracy obtained due to the proposed method. The measure used is the difference between the radar north and the true north after calibration.

Table 1. Calibration accuracy in the network system consisting of 3 radars.

Measure	Radar 1	Radar 1	Radar 3
Mean Value	-0.24^0	0.23^0	0^0
Standard Deviation	3.42^0	3.36^0	3.31^0
Min Value (the best)	0^0	0^0	0^0
Max Value (the worst)	9.49^0	8.99^0	8.75^0

Table 2. Calibration accuracy in the network system consisting of 4 radars.

Measure	Radar 1	Radar 1	Radar 3	Radar 4
Mean Value	-0.09^0	-0.04^0	0.02^0	-0.08^0
Standard Deviation	3.40^0	3.79^0	3.12^0	3.32^0
Max Value (the best)	0^0	0^0	0^0	0^0
Min Value (the worst)	9.73^0	10.86^0	8.60^0	9.46^0

As it results from the data presented in Tables 1 and 2, the average value of the constant component of the azimuth error after calibration is close to 0 for each of the radars in the network, both in the case of the system consisting of 3 and 4 radars. In the best cases, it has been eliminated. However, in the worst cases, after calibration, it was still around $9 - 10^0$. Therefore, to better analyze the effectiveness of the proposed method, Figure 8 presents histograms of the azimuth error constant component for radars R1, R2, and R3 in the 3-radars network (see Figure 7 (left)) before and Figure 9 after calibration. On the horizontal axis of the charts, there are ranges of the azimuth error constant component in degrees, and the numbers in the blue rectangles are the numbers of individual ranges.

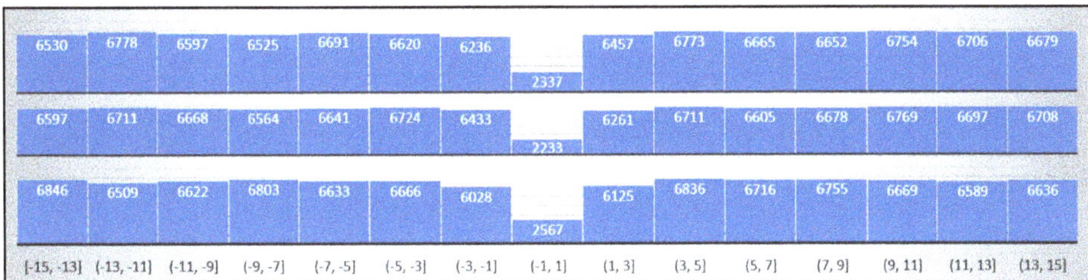

Figure 8. Histogram of the azimuth error constant component for radars R1(upper), R2(middle), and R3(lower) in the 3-radars system before calibration.

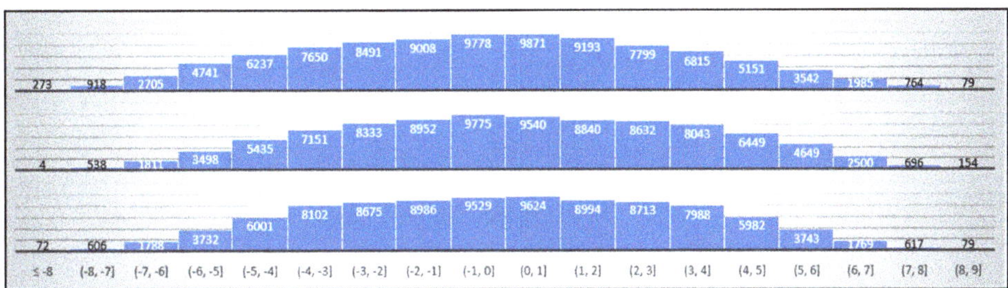

Figure 9. Histogram of the azimuth error constant component for radars R1(upper), R2(middle), and R3(lower) in the 3-radars system after calibration.

Figures 8 and 9 show that for each of the radars, there was a clear improvement—a decrease in the value of the constant component of the azimuth error. Moreover, before calibration, the errors had a uniform distribution (according to the simulation assumptions), while after calibration, it was similar to the Gaussian distribution, with the highest probability density centered around 0. It is worth noting that only about 5% percent of errors after calibration exceeded the value of 6^0, which should be considered a very good result, given that before calibration, such errors were about 65%. It is also apparent from Figures 8 and 9 that the method impacts on a constant azimuth measurement error for each of the radars in a similar way. The same is also true in the 4-radars network (shown in Figure 7 (right)). Histograms of the azimuth error constant component for radars R1, R2, R3, and R4 before calibration are shown in Figure 10 and after calibration in Figure 11.

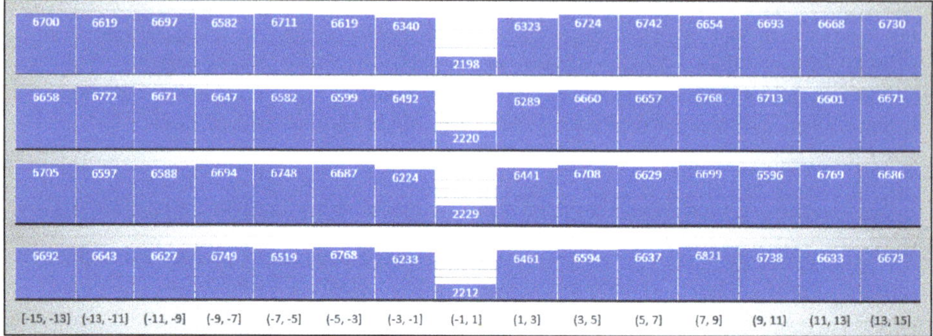

Figure 10. Histogram of the azimuth error constant component for radars R1(upper), R2(upper-middle), R3(lower-middle), and R4(lower) in the 4-radars system before calibration.

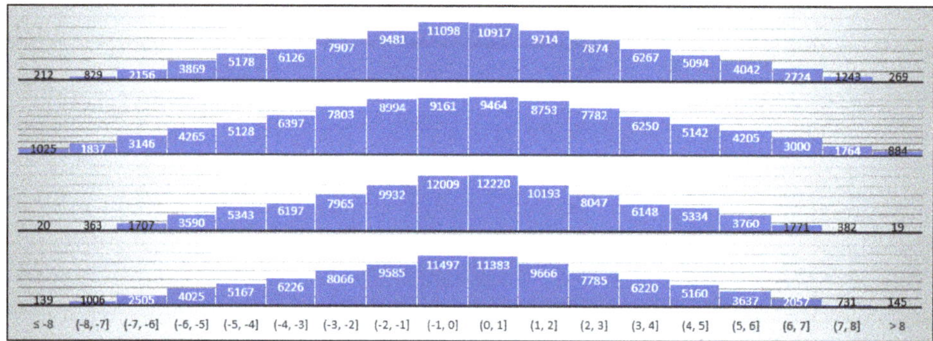

Figure 11. Histogram of the azimuth error constant component for radars R1(upper), R2(upper-middle), R3(lower-middle), and R4(lower) in the 4-radars system after calibration.

It can be seen in Figures 8–11 that the azimuth error constant component before and after calibration for both 3- and 4-radars networks have similar distributions. Therefore, the results of further analyses are presented based on one selected radar from both tested cases. The solution was applied to limit the volume of the paper. To maintain proper reliability, the results of the analyses are presented for the worst case (least improvement and greatest deterioration after calibration). This was the case of R1 radar from the 4-radars network (see Figure 7 (right)).

The graphs in Figures 8–11 show that the proposed method significantly improves the accuracy of the azimuth measurement by reducing azimuths misalignment (decreasing the constant error of the azimuths measurements). However, the distribution of the azimuth error constant component after calibration does not yet give a complete picture of the method's effectiveness. There are no answers to the following important questions:

a) Is there radar orientation improvement in each of the cases, or are there cases where the value of the azimuth error constant component increases after calibration?
b) If there are cases of radar orientation deterioration relative to the north after calibration, how large are they, and do they preclude the practical application of the proposed method?
c) What is the percentage of the radar orientation relative to the north improvement, in comparison to the initial value of the azimuth error constant component?

To answer the above questions, an analysis of the radar north improvement was performed. The measure used was the value by which the difference between the radar north and the true north changed due to calibration. To better show the effect of the proposed method, the radar north improvement was expressed as a percentage of the constant azimuth error before calibration. The results of the analysis for the selected radar (the worst case) are presented in Figures 12–14.

Figure 12 shows the percentage histogram of improvement and deterioration of the selected radar orientation relative to the north. The horizontal axis presents the percentages, and the numbers in the blue rectangles are the number of individual ranges. A negative percentage value (≤ 0) means that after calibration, the radar orientation has deteriorated.

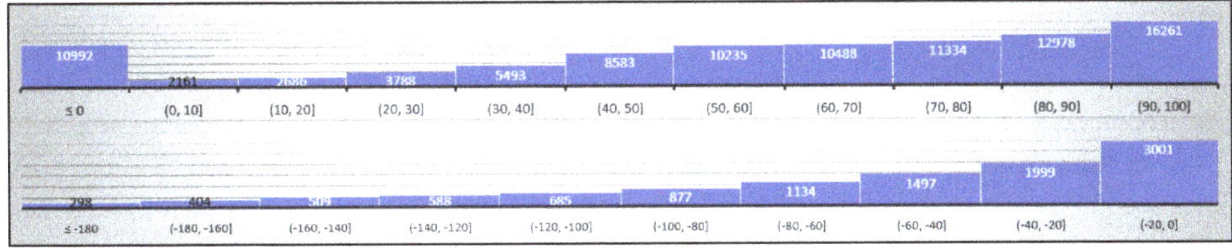

Figure 12. Histogram of the percentage improvement (upper) and deterioration (lower) of the R1 radar north in the 4-radars system after calibration.

Figure 12 shows that up to 11.6% of cases, after calibration by the proposed method, the radar orientation deteriorated relative to the true north. Most values of degradation did not exceed 100% of the pre-calibration value, but there were also cases where they reached 200%. At first glance, this seemed very worrying, because even a 20% deterioration may be unacceptable if it concerns large initial values (before calibration). Therefore, it seems to be important to answer the following questions:

a) How big is the deterioration in numerical values?
b) Does it concern small or large values of initial orientation errors?
c) How this might affect the practical application of the proposed method?

To answer them, deterioration cases were thoroughly analyzed. Figure 13 shows the histogram of the initial (before calibration) constant azimuth errors for the R1 radar in the 4-radars system, where the orientation was deteriorated after the calibration.

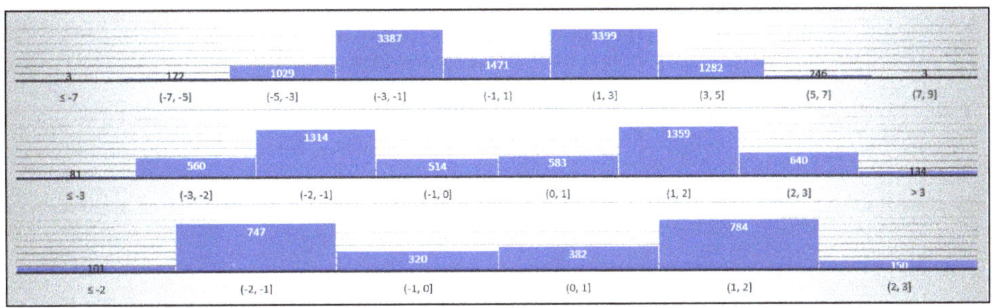

Figure 13. Histogram of initial (before calibration) errors of the R1 radar orientation in the 4-radars system, for which the azimuth misalignment increased after calibration: Above 0%(upper), above 50%(middle), and above 100%(lower).

Figure 13 shows that the large percentage deterioration of radar orientation relative to the north was related to small initial values (before calibration). The deterioration by more than 50% basically concerned angles not greater than 3^0, and 100 not greater than 2^0. This means that even in the worst case of deterioration, the azimuth misalignment after calibration did not exceed 4^0, which is completely acceptable from a practical point of view. Moreover, it should be emphasized that in all analyzed cases, the deterioration of radar orientation relative to true north after calibration, concerned only one radar in the network, while significantly improving the orientation of the other radars. The deterioration effect occurred when one radar had a small initial error and the rest had large ones.

To show the advantages of the proposed method for the same radar, the results of an analogous analysis regarding the improvement of orientation relative to true north are presented below. Figure 14 shows the distribution of the initial (before calibration) azimuth constant errors for which the R1 radar orientation improved.

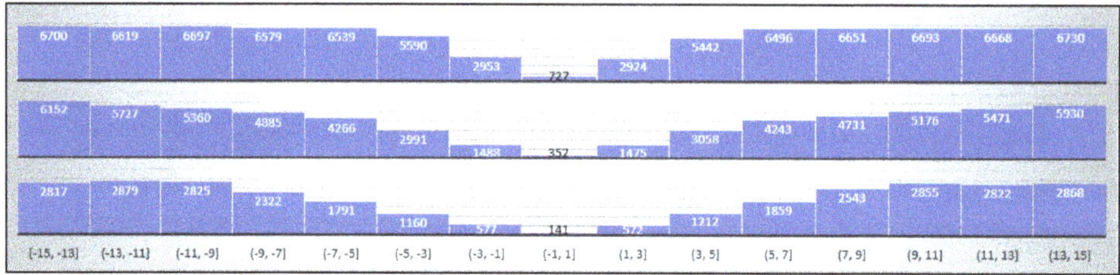

Figure 14. Histogram of initial (before calibration) errors of the R1 radar orientation in the 4-radars system, for which the azimuth misalignment decreased after calibration: Above 0% (upper), above 50% (middle), and above 80% (lower).

The graphs shown in Figure 14 show the advantages of the proposed calibration method. It can be seen that the method improves the initial orientation of the radar relative to the north over the entire range of initial errors. It is the most effective for errors above 3^0.

5. Discussion

Based on the obtained results, one can say:

1. The method improves the orientation of radars relative to the true north over the entire range of initial errors (before calibration). It is especially effective if they are above 3^0. Moreover, the method reduces the initial value of the error, the higher it is. This is of great practical importance, as it greatly reduces the error in determining the coordinates of the detected object by the radar. What is more, the method only requires a very coarse orientation of the radars to the north (e.g., using magnetic compass indications), which is also important in real applications, as it does not require the use of specialized equipment and significantly reduces the time needed to prepare the system for operation. The smallest improvement in the radar orientation occurs for angles below 3^0, which is not a practical problem.
2. Not always does the proposed method improve the initial value of the radar orientation. In about 10% of cases, deterioration occurs. However, it concerns small initial errors (below 3^0). As a result, after calibration, the azimuth misalignment did not exceed 4^0. This is perfectly acceptable from a practical point of view, because in all registered cases, the deterioration concerned only one radar in the network, while significantly improving the orientation of the others.
3. Further work will focus on improving the method's effectiveness. It is planned to use another method of calculating the initial position of the drone for the first iteration purposes. It was noted that the unfavorable results might be due to the fact that the approximate position was too far from the true one. Moreover, it should be checked whether weighing the measurements (depending on the azimuth and distance measurement accuracy) will significantly improve the calibration accuracy, or whether the increase will be insignificant and only the computational complexity will increase. Further research is also planned to look for the optimal calibration flight route. It would be desirable to achieve maximum calibration accuracy while minimizing flight length.

6. Conclusions

The proposed method of calibration improves the initial orientation of the radars relative to the north over the entire range of the initial errors. Moreover, it reduces the azimuth misalignment the bigger its initial value is. This is an undoubted advantage of the method and also has a great practical meaning, as it leads to a significant reduction of errors in determining the coordinates of the detected object (see Figure 5). Although there are cases when the azimuths misalignment increased after calibration, it has no practical meaning as it only concerned small initial errors. Therefore, after calibration, the orientation errors were still acceptable. The disadvantage of the method may be the need to perform a drone calibration flight along a fixed route. However, this appears to be a minor nuisance compared to other calibration methods.

Author Contributions: Conceptualization, A.N.; methodology, A.N., K.N. and D.M.; software, A.N.; validation, A.N., K.N. and D.M.; formal analysis, K.N.; writing—original draft preparation, A.N.; writing—review and editing, K.N. and D.M.; visualization, A.N. and D.M.

Funding: Own funds for the research activity of the Department of Geodesy, Faculty of Civil and Environmental Engineering of the Gdansk University of Technology.

Acknowledgments: The article was possible thanks to research on the use of remote sensing methods for the detection and classification of drones conducted in cooperation with the Gdansk University of Technology and the Naval Academy in Gdynia.

Conflicts of Interest: The authors declare no conflict of interest.

References

1. Goldman Sachs. Available online: https://www.goldmansachs.com/our-thinking/technology-driving-innovation/drones/ (accessed on 11 August 2019).
2. Interesting Engineering. Available online: https://interestingengineering.com/top-5-drone-intercepting-methods-you-should-know-about (accessed on 11 August 2019).
3. Michel, A.H. *Counter-Drone Syst*; The report by Bard College's Center for the Study of the Drone; Bard College's Center for the Study of the Drone: Dutchess Country, NY, USA, 2018.
4. Hertz Systems. Available online: http://thehawksystem.com/pl/ (accessed on 1 September 2019).
5. SpotterRF. Available online: https://spotterrf.com/products/mobile-solutions/ (accessed on 1 September 2019).
6. Farlik, J.; Kratky, M.; Casar, J.; Stary, V. Multispectral Detection of Commercial Unmanned Aerial Vehicles. *Sensors* **2019**, *19*, 1517. [CrossRef] [PubMed]
7. Laurenzis, M.; Hengy, S.; Hommes, A.; Kloeppel, F.; Shoykhetbrod, A.; Geibig, T.; Johannes, W.; Naz, P.; Christnacher, F. Multi-sensor field trials for detection and tracking of multiple small unmanned aerial vehicles flying at low altitude. In Proceedings of the Signal Processing, Sensor/Information Fusion, and Target Recognition XXVI, Anaheim, CA, USA, 10–12 April 2017.
8. Nassi, B.; Shabtai, A.; Masuoka, R.; Elovici, Y. SoK—Security and Privacy in the Age of Drones: Threats, Challenges, Solution Mechanisms, and Scientific Gaps. *arXiv* **2019**, arXiv:1903.05155.
9. Doroftei, D.; De Cubber, G. Qualitative and quantitative validation of drone detection systems. In Proceedings of the International Symposium on Measurement and Control in Robotics, Mons, Belgium, 26–28 September 2018. [CrossRef]
10. Zhahir, A.; Razali, A.; Mohd Ajir, M.R. Current development of UAV sense and avoid system. *IOP Conf. Ser. Mater. Sci. Eng.* **2016**, *152*, 012035. [CrossRef]
11. Eriksson, N. Conceptual Study of a Future Drone Detection System. Master's Thesis, Product Development, Chalmers University of Technology, Gothenburg, Sweden, 2018.
12. Hommes, A.; Shoykhetbrod, A.; Noetel, D.; Stanko, S.; Laurenzis, M.; Hengy, S.; Christnacher, F. Detection of acoustic, electro-optical and radar signatures of small unmanned aerial vehicles. In Proceedings of the Target and Background Signatures II, Edinburgh, UK, 26–27 September 2016; Volume 9997, p. 999701.
13. Stateczny, A.; Lubczonek, J. FMCW Radar Implementation in River Information Services in Poland. In Proceedings of the 16th International Radar Symposium (IRS), Dresden, Germany, 24–26 June 2015; pp. 852–857.
14. Farlik, J. Radar cross section and detection of small unmanned aerial vehicles. In Proceedings of the 17th International Conference on Mechatronics—Mechatronika (ME), Prague, Czech Republic, 7–9 December 2016; pp. 1–7.

15. Schroder, A. Numerical and Experimental Radar Cross Section Analysis of the Quadrocopter DJI Phantom 2. In Proceedings of the 2015 IEEE Radar Conference, Johannesburg, South Africa, 27–30 October 2015; pp. 463–468.
16. Ritchie, M. Micro-drone RCS Analysis. In Proceedings of the 2015 IEEE Radar Conference, Johannesburg, South Africa, 27–30 October 2015.
17. Li, C.J.; Ling, H. An Investigation on the Radar Signatures of Small Consumer Drones. *IEEE Antennas Wirel. Propag. Lett.* **2017**, *16*, 649–652. [CrossRef]
18. Guay, R.; Drolet, G.; Bray, J.R. Measurement and modelling of the dynamic radar cross-section of an unmanned aerial vehicle. *IET Radar, Sonar Navig.* **2017**, *11*, 1155–1160. [CrossRef]
19. Herschfelt, A.; Birtcher, R.C.; Gutierrez, R.M.; Rong, Y.; Yu, H.; Balanis, C.A.; Bliss, D.W. Consumer-Grade Drone Radar Cross-Section and Micro-Doppler Phenomenology. In Proceedings of the 2017 IEEE Radar Conference, Seattle, WA, USA, 8–12 May 2017; pp. 0981–0985.
20. Kim, B.K.; Kang, H.-S.; Park, S.-O. Experimental Analysis of Small Drone Polarimetry Based on Micro-Doppler Signature. *IEEE Geosci. Remote. Sens. Lett.* **2017**, *14*, 1670–1674. [CrossRef]
21. de Wit, J.J.M.; Harmanny, R.I.A.; Molchanov, P. Radar micro-Doppler feature extraction using the Singular Value Decomposition. In Proceedings of the 2014 International Radar Conference, Lille, France, 13–17 October 2014; pp. 1–6.
22. Molchanov, P.; Egiazarian, K.; Astola, J.; Harmanny, R.I.A.; de Wit, J.J.M. Classification of small UAVs and birds by micro-Doppler signatures. In Proceedings of the 2013 European Radar Conference (EuRAD), Nuremberg, Germany, 9–11 October 2013; pp. 172–175.
23. Harmanny, R.I.A.; de Wit, J.J.M.; Cabic, G.P. Radar micro-Doppler feature extraction using the spectrogram and the cepstrogram. In Proceedings of the 11th European Radar Conference (EuRAD), Rome, Italy, 8–10 October 2014; pp. 165–168.
24. Counter-Unmanned Aerial Systems Market Survey Report. Department of Homeland Security, 2017. Available online: https://www.dhs.gov/sites/default/files/SAVER_Counter-Unmanned-Aerial-Systems-MSR_0917-508.pdf (accessed on 11 August 2019).

Article

3D GPR Image-based UcNet for Enhancing Underground Cavity Detectability

Man-Sung Kang [1], Namgyu Kim [2], Seok Been Im [3], Jong-Jae Lee [2] and Yun-Kyu An [1,*]

1. Department of Architectural Engineering, Sejong University, Seoul 05006, Korea; kms102353@sju.ac.kr
2. Department of Civil and Environmental Engineering, Sejong University, Seoul 05006, Korea; namgyu.kim@sejong.ac.kr (N.K.); jongjae@sejong.ac.kr (J.-J.L.)
3. Research Institute for Infrastructure Performance, KISTEC, Jinju 52856, Korea; sbeeni@kistec.or.kr

* Correspondence: yunkyuan@sejong.ac.kr; Tel.: +82-2-6935-2426

Received: 1 October 2019; Accepted: 26 October 2019; Published: 29 October 2019

Abstract: This paper proposes a 3D ground penetrating radar (GPR) image-based underground cavity detection network (UcNet) for preventing sinkholes in complex urban roads. UcNet is developed based on convolutional neural network (CNN) incorporated with phase analysis of super-resolution (SR) GPR images. CNNs have been popularly used for automated GPR data classification, because expert-dependent data interpretation of massive GPR data obtained from urban roads is typically cumbersome and time consuming. However, the conventional CNNs often provide misclassification results due to similar GPR features automatically extracted from arbitrary underground objects such as cavities, manholes, gravels, subsoil backgrounds and so on. In particular, non-cavity features are often misclassified as real cavities, which degrades the CNNs' performance and reliability. UcNet improves underground cavity detectability by generating SR GPR images of the cavities extracted from CNN and analyzing their phase information. The proposed UcNet is experimentally validated using in-situ GPR data collected from complex urban roads in Seoul, South Korea. The validation test results reveal that the underground cavity misclassification is remarkably decreased compared to the conventional CNN ones.

Keywords: ground penetrating radar; underground cavity detection network; deep convolutional neural network; automated underground object classification; phase analysis; super-resolution

1. Introduction

A series of sudden sinkhole collapses continuously occur on complex urban areas over the world. Recently, several sinkhole activities have been reported globally, such as the United States [1–3], China [4,5], Japan [6], and South Korea [7]. These sinkholes have resulted in a major disruption of traffic flow and utility services causing significant economic losses as well as critical human injuries and fatalities [8,9]. Since these sinkholes have shown up without any forewarning, there is an increasing demand for their early detection especially in complex urban areas.

Recently, ground penetrating radar (GPR) has been widely employed for early detection of underground cavities, which are most likely propagating to sinkholes, thanks to its fast scanning speed, nondestructive inspection, and 3D imaging capabilities [10–12]. GPR transmitters emit electromagnetic waves into the underground at several spatial positions along the scanning direction, and GPR receivers measure the reflected signals to establish 2D GPR image called a radargram, also known as a B-scan image. If the multi-channel GPR transmitters and receivers parallel to the scanning direction are equipped, 3D GPR images including B- and C-scan images can be obtained at once. Since the electromagnetic waves propagating along the underground medium are dominantly reflected from the abrupt change of electromagnetic permittivity, the reflection signal features appear in the B- and C-scan images. To enhance visibility and detectability of the reflection signal features, a number of data processing techniques such as time-varying gain [13], subtraction [14,15], and basis pursuit-based background filtering [16] have been proposed. However, these techniques are highly susceptible to measurement noises especially in complex urban roads and sometimes unreliable due to decision making based on experts' experiences. Moreover, expert-dependent data interpretation becomes time consuming and cumbersome, as the amount of 3D GPR data increases.

To overcome the technical limitations, several researchers have made efforts to automate the GPR data classification process. Simi et al. proposed a Hough transform-based automatic hyperbola detection algorithm to reduce GPR data analysis time [17]. Li et al. utilized a randomized Hough transform to effectively find the parabola parameters [18]. In addition, histograms of oriented gradient features-based GPR data classification were proposed for automatically detecting underground objects [19,20]. More recently, neural networks and convolutional neural networks (CNNs) were widely used as a promising tool for automated GPR image recognition and classification. Mazurkiewicz et al. tried to identify underground objects using neural networks while reducing the processing time and human intervention [21–23]. Then, Al-Nuaimy et al. utilized both neural network and pattern recognition methods to automatically detect the buried objects [24]. Lameri et al. also applied CNN to detect landmines with pipeline B-scan images [25].

Although the previous studies have focused on the use of GPR B-scan images, it is often difficult to classify a specific target underground object by using only GPR B-scan images. In particular, the GPR B-scan images often tend to be similar among various underground objects such as cavities, manholes, pipes, electrical lines, gravels, concrete blocks and so on in complex urban areas. To enhance the classification performance, additional GPR C-scan images were simultaneously considered during the classification procedure using CNN [26,27]. Although the developed CNN using the combination of B- and C-scan images increased the classification accuracy compared to the conventional CNN using only B-scan images, it turned out that it still has difficulty differentiating underground cavities from chunks of gravel especially in complex urban areas due to their similar GPR reflection features. To address the misclassification issue caused by the underground chunks of gravel, Park et al. recently proposed a phase analysis technique of GPR data [16]. However, the temporal and spatial resolutions of the 3D GPR data are often insufficient for the precise phase analysis. Since the phase analysis results highly depend on a few pixel differences of the GPR images, the lack of GPR image resolution may cause false alarms during the phase analysis.

In this paper, a 3D GPR image-based underground cavity detection network (UcNet), which consists of CNN and the phase analysis of super-resolution (SR) GPR images, is newly proposed to enhance underground cavity detectability. By retaining the advantages of both CNN and phase analysis, underground cavities can be automatically classified with minimized false alarms. In particular, a deep learning-based SR network used for GPR image resolution enhancement significantly reduces the misclassification between the underground cavity and chunk of gravel. To examine the performance of the proposed UcNet, comparative study results on cavity detectability between the conventional CNN and the newly proposed UcNet are presented using in-situ 3D GPR data obtained from complex urban roads in Seoul, South Korea.

This paper is organized as follows. Section 2 explains the proposed UcNet, which is comprised of CNN, SR image generation, and phase analysis. Then, the 3D GPR data collection procedure from urban roads and experimental validation are described in Section 3. In particular, the comparative study results between the conventional CNN and newly proposed UcNet are addressed. Finally, Section 4 concludes the paper with a brief discussion.

2. Development of UcNet

Once 3D GPR data are obtained from a target area with various underground objects, the corresponding 2D GPR grid images comprised of the B- and C-scan images can be reconstructed for network training and testing as displayed in Figure 1 [26]. Figure 1a shows the representative 2D grid image of the cavity case, which have parabola and circle features that can be observed on the B- and C-scan images, respectively. The similar features are also revealed in the manhole case of

Figure 1b, although they have a different pattern and amplitude. On the other hand, no significant features appear in the subsoil background case as shown in Figure 1c, because there is no abrupt permittivity change. Thanks to these distinguishable features, underground objects can be classified well with the conventional CNNs. However, the underground gravel case of Figure 1d shows the similar morphological GPR B- and C-scan features as the cavity ones, which may cause false alarms. Thus, UcNet is newly proposed to minimize the false alarms.

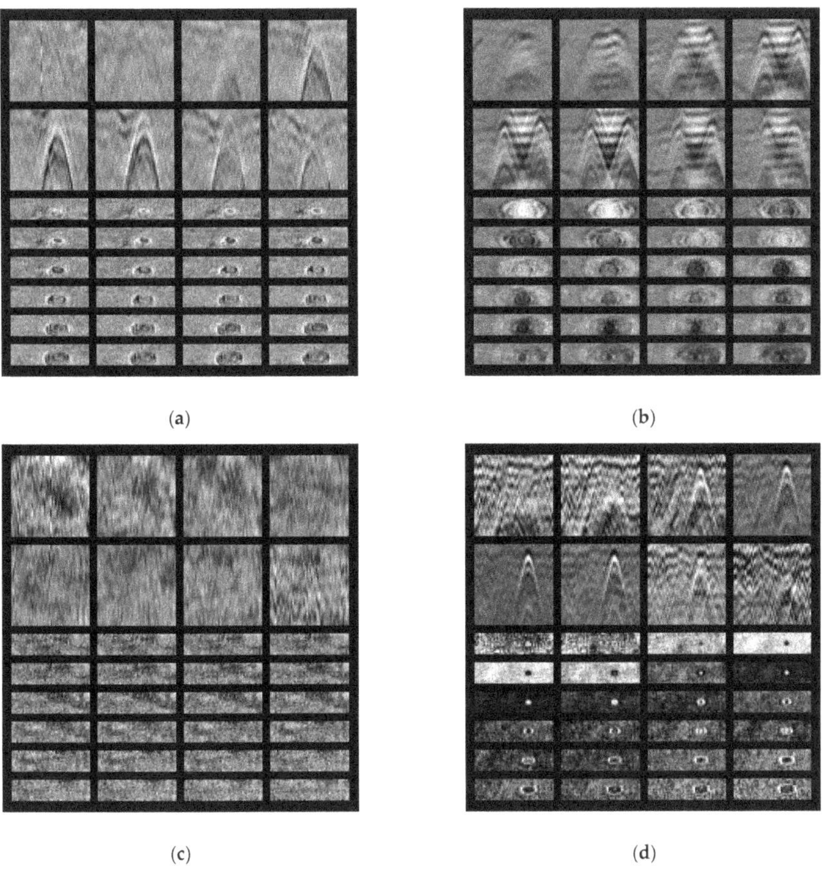

Figure 1. Representative 2D ground penetrating radar (GPR) grid images reconstructed from 3D GPR data: (**a**) cavity, (**b**) manhole, (**c**) subsoil background, and (**d**) gravel.

Figure 2 shows the overview of proposed UcNet consisted of the three phases, i.e., Phase I: CNN, Phase II: SR image generation, and Phase III: Phase analysis. First, the reconstructed 2D GPR grid images are fed into CNN for data classification. Subsequently, the cavity and gravel data classified in Phase I are transmitted into Phase II to generate their SR images from original low-resolution (LR) images. Finally, the phase analysis of the SR images is conducted in Phase III to update the classification results obtained in Phase I. Through these sequential processes, misclassification between the cavity and gravel cases can be minimized. The details of each phase are as follows.

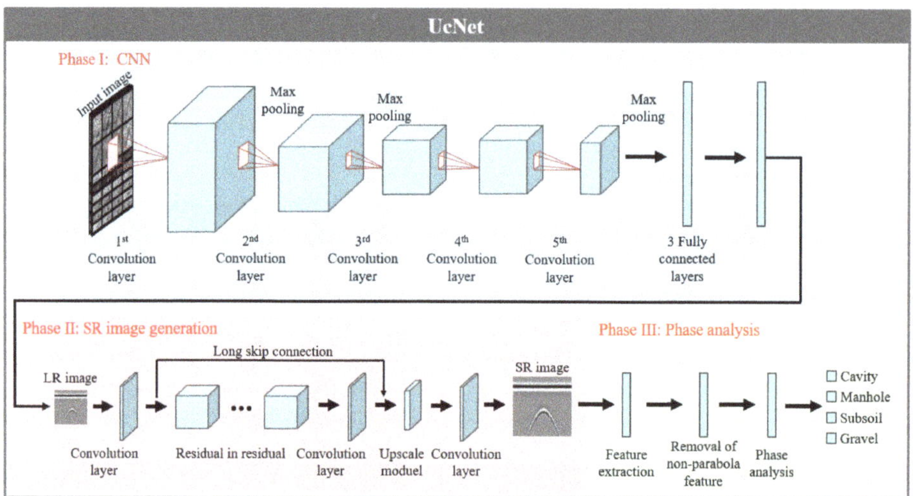

Figure 2. Overview of underground cavity detection network (UcNet): LR and SR denote the low-resolution and super-resolution, respectively.

Phase I: CNN is established by transfer learning from AlexNet [28], which is one of the widely used pre-trained CNN models for image classification, in this study. The modified CNN consists of five convolutional layers, three fully connected layers, three max pooling layers and 1000 softmax neurons, containing 650,000 neurons and 60 million parameters. To train CNN, the 2D GPR grid images are fed to the input layer consisted of the image size of 227 × 227 × 3 pixels, and the pixel features are then extracted through the convolutional layers. Figure 3 shows the representative training GPR 2D grid images. The 1st convolutional layer uses the kernel of 11 × 11 × 3 pixels with a stride of four. Consequently, the layer creates 96 feature maps and has the output of 55 × 55 × 96 pixels. The max pooling layer, which is one of sub sampling techniques, is then arranged after the 1st convolutional layer to reduce the size of feature maps. Next, the 2nd convolution layer is operated with the kernel of 5 × 5 pixels and creates 27 × 27 × 256 pixels. The max pooling layer is again arranged after the 2nd convolutional layer. The following convolutional layers are operated with the kernel of 3 × 3 pixels and create 13 × 13 × 256 pixels. The max pooling layer is once again arranged after the 5th convolutional layer. Once the features are extracted and shrunken on the convolutional layers, the feature maps are fed to fully connected layers. To avoid an overfitting issue, dropout layers are arranged after 1st and 2nd fully connected layers. In addition, the rectified linear unit is selected as an activation function. Finally, the output of the last fully connected layer is fed into a softmax layer having four probabilities, i.e., cavity, manhole, subsoil background, and gravel in this study.

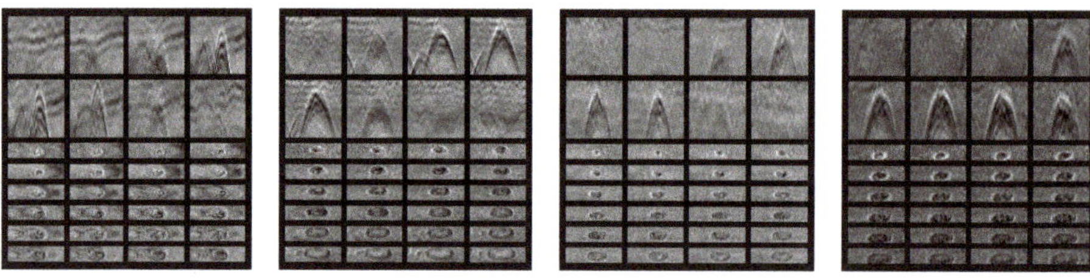

(a)

Figure 3. *Cont.*

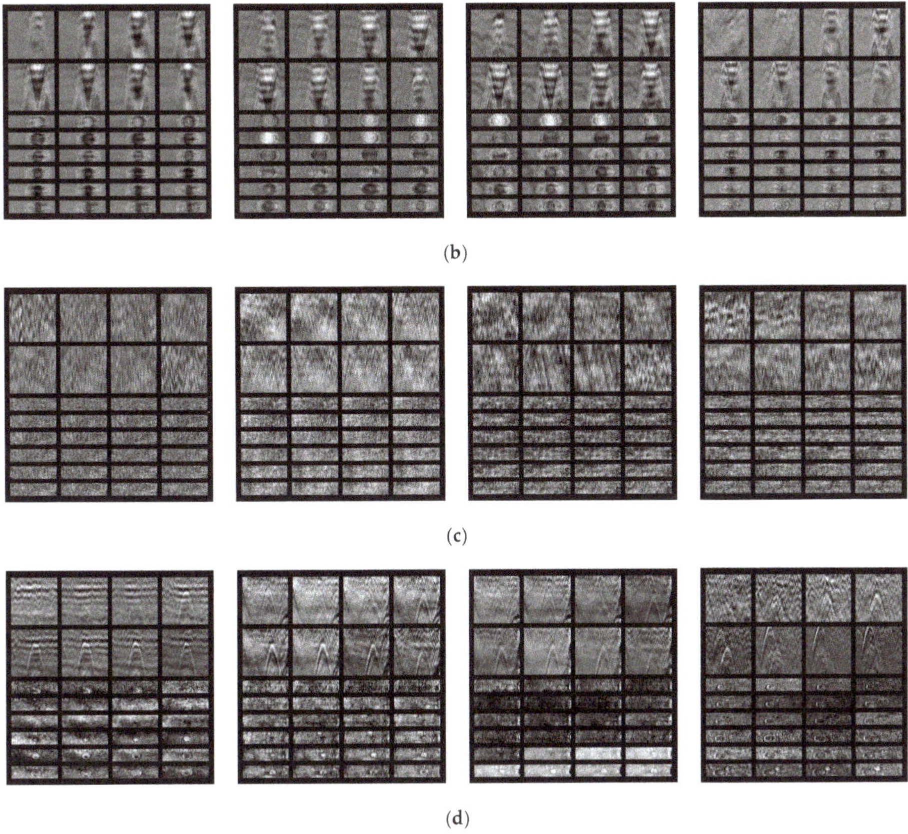

(b)

(c)

(d)

Figure 3. Representative training 2D GPR grid images of a (**a**) cavity, (**b**) manhole, (**c**) subsoil background, and (**d**) gravel.

Phase II: Once the underground objects are classified by Phase I, the cavity and gravel images are automatically transferred to Phase II. The reconstructed GPR grid image is comprised of eight B-scan and 24 C-scan images. The original sizes of B- and C-scan images are 50 × 50 and 50 × 13, respectively. To proceed Phase II, the 5th representative B-scan image is selected from each reconstructed GPR 2D grid image. Since the resolution of the selected B-scan image is not enough for Phase II as shown in Figure 4, the subsequent SR image generation process is necessary.

The SR image generation network was constructed by using the residual channel attention network [29], which is one of the deep learning networks comprised of 500 layers and 1.6 M parameters for image resolution enhancement. This network utilizes the residual in residual structure and the channel attention mechanism to enhance the feature learning of high frequency channels, which is useful for reconstructing high-resolution images among the various channels that make up image data. The residual channel attention network consists of the four main parts, i.e., the convolution layer, residual in residual structure, upscale module, and last convolution layer. First, the convolution layer is shallow feature extraction for the input image. Then, the residual in residual structure extracts deep features through the high frequency information learning. The residual in residual structure is the very deep structure comprised of 10 residual groups, and each residual group consists of 20 residual blocks. Each residual group is connected by a long skip connection as shown in Phase II of Figure 2. This residual in residual structure allows the residual channel attention network to learn more effective high frequency information by skipping the low frequency through the skip connection. The shallow and deep feature data, which have passed through each residual block and group, are extended to the SR size through the upscale module. The upscale module is composed of a deconvolution layer consisting of 256 kernels with 3 × 3 size and a single stride, which increases the size of each pixel by four times in this study. Finally, it is restored as the SR image through the last convolution layer. The 800 training images from the DIV2K dataset, which is well known high-quality images with 2K resolution, were used as the training dataset [30]. Then, 100 images collected from Urban100 dataset were used as the validation dataset. The representative resultant GPR images between LR (50 × 50 pixels) and SR (200 × 200 pixels) are compared in Figure 5. The SR image is successfully generated without any information loss.

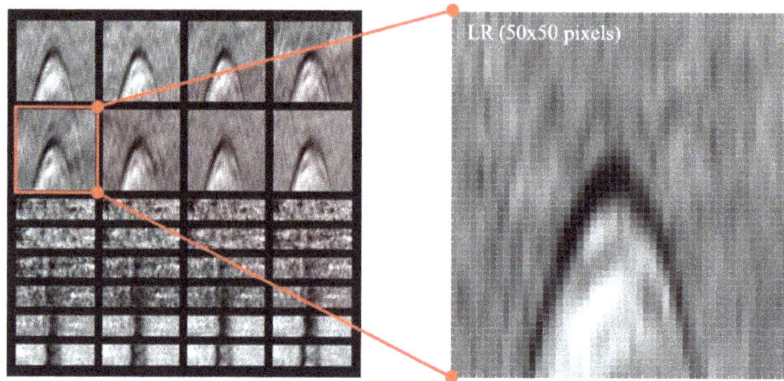

Figure 4. Representative LR GPR image for SR image generation input.

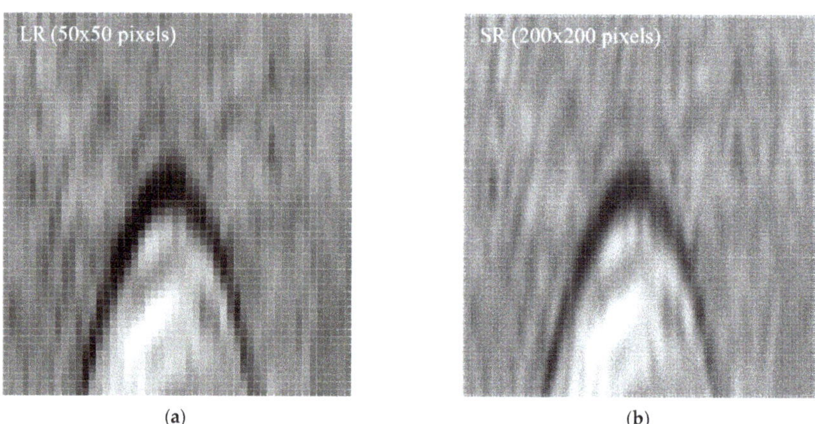

Figure 5. GPR image enhancement results: (**a**) LR and (**b**) SR images.

Phase III: Once the SR GPR images are generated in Phase II, the phase analysis is subsequently carried out in Phase III. To automate the phase analysis of the SR images, the feature extraction, removal of non-parabola features, and parabola boundary extraction are continuously proceeded as shown in Phase III of Figure 2. Figure 6a shows the representative SR B-scan image obtained from Phase II. Since the B-scan image includes a number of noise components, the feature extraction with noise removal procedure is sequentially performed. First, a median filter is applied to the B-scan image for removing pepper noise components, and the filtered image is then normalized with respect to the maximum amplitude. Subsequently, the extreme value distribution with 95% confidence interval is applied to obtain the dominant features. However, since the remaining features still contain non-parabola boundaries, as shown in Figure 6b, it should be removed. The unwanted spotted and non-continuous features can be removed by eliminating non-continuous pixels less than 500, as displayed in Figure 6c. Note that the target detectable cavity size of 30 cm is equivalent to 500 pixels in the SR image. Next, the parabola feature can be extracted by using the gradient between the remaining objects' centroid and the extremum of left- and right-side pixels. Since the empirical gradient value of typical parabolas is larger than 20°, only the parabola feature remains, as displayed in Figure 6d.

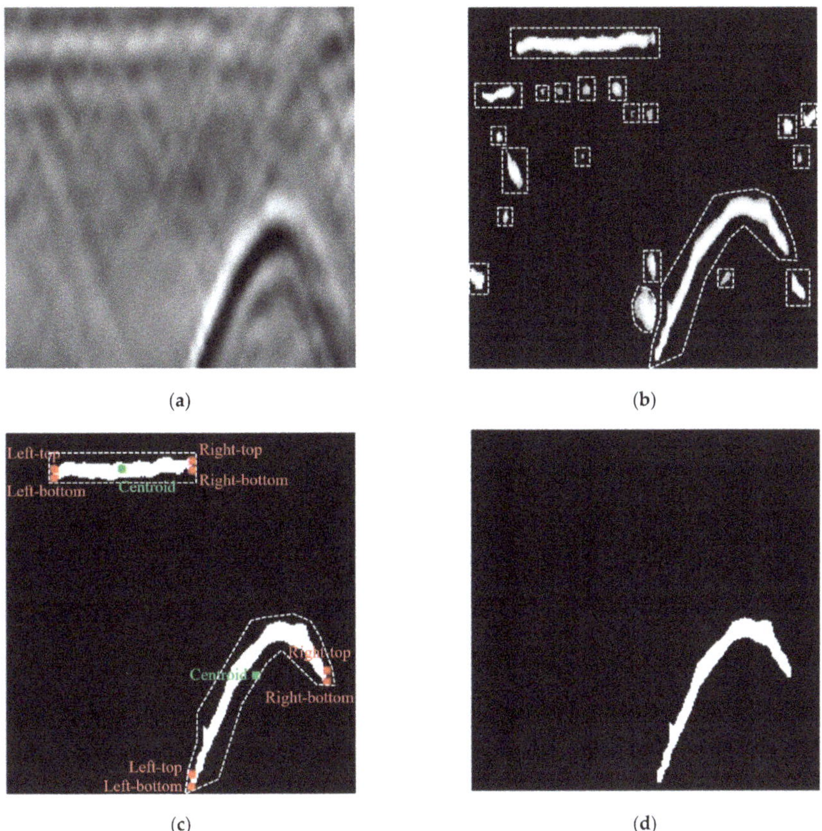

Figure 6. Representative (**a**) SR B-scan image, (**b**) feature extracted image, (**c**) noise removal image, and (**d**) parabola boundary extraction image.

Once the parabola boundary is automatically extracted, its phase can be analyzed as the final procedure. The extracted boundary value is converted to phase information using Equations (1) and (2).

$$H(x,z) = \frac{1}{\pi} P \int_{-\infty}^{\infty} \frac{I(x,z)}{z-\tau} d\tau, \quad (1)$$

where P denotes the Cauchy principal value. $I(x,z)$ is the GPR A-scan data. x and z are the spatial coordinates along the scanning and depth directions, respectively. The instantaneous phase value at each spatial point can be calculated by:

$$\theta(x,z) = \tan^{-1}\left(\frac{Im[H(x,z)]}{Re[H(x,z)]}\right), \quad (2)$$

where *Re* and *Im* represent the real and imaginary components of a complex value, respectively.

If the relative permittivity of the underground object is lower than that of the surroundings, the reflected electromagnetic waves are in-phase with the radiated waves. Otherwise, the reflected electromagnetic waves will be indicated out-of-phase [16]. The phase change ratio of the extracted parabola boundary can be expressed by:

$$\Delta\theta = \frac{\theta(x, lp) - \theta(x, fp)}{lp - fp}, \tag{3}$$

where fp and lp denote the first and last pixels along the A-scan of the extracted phase boundary.

To numerically validate the phase analysis of Phase III, the cavity and gravel cases were modeled using gprMax [31] as shown in Figure 7a. The target model was comprised of 1×0.8 m^2 soil layer and 1×0.2 m^2 air layer. Then, the underground cavity and gravel were respectively inserted inside the soil layer with a depth of 0.5 m as depicted in Figure 7a. Here, the relative permittivity values of the soil, cavity and gravel were designed as 5, 1, and 6, respectively. The transmitter was 40 mm apart from the receiver, and the finite difference time domain method was used to simulate electromagnetic wave propagation. The used electromagnetic wave was the normalized first derivative of a Gaussian curve with a center frequency of 1.6 GHz. Once the simulation models, i.e., the cavity and gravel cases, are prepared, the transmitter and receiver scan along the soil layer surface with spatial intervals of 20 mm for each model. Figure 7b,c shows the resultant images of the cavity and gravel cases, respectively. As shown in Figure 7b, the $\Delta\theta$ value of the radiated waves expressed by the dashed blue box has the positive value. Similarly, the cavity with the line red box has the positive value, which means in-phase with respect to the $\Delta\theta$ value of the radiated waves. On the other hand, Figure 7c shows the $\Delta\theta$ value of the gravel with the dotted green box is out-of-phase compared with the $\Delta\theta$ value of the radiated waves. Through this precise phase analysis in Phase III, the classification results obtained by considering only shape and amplitude recognition in Phase I can be updated.

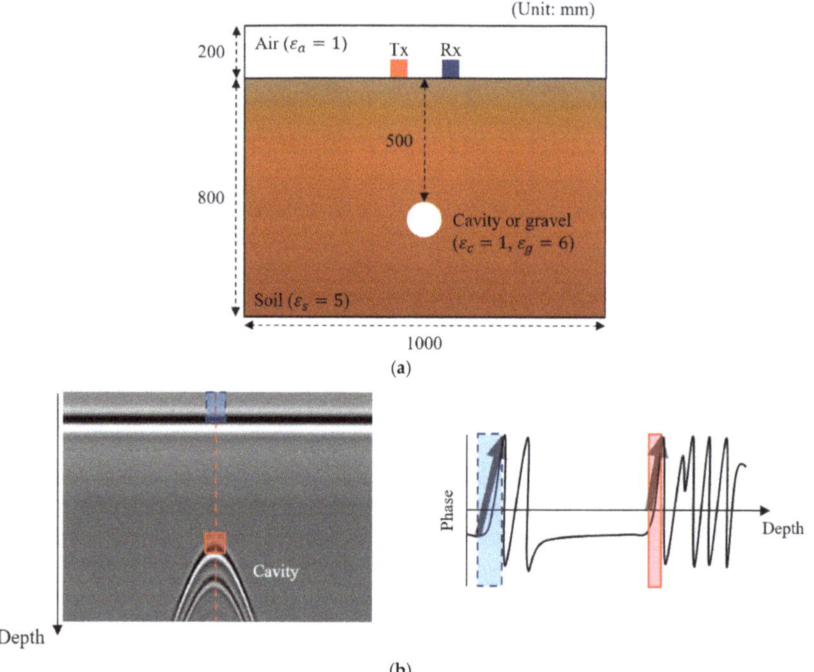

(b)

Figure 7. *Cont.*

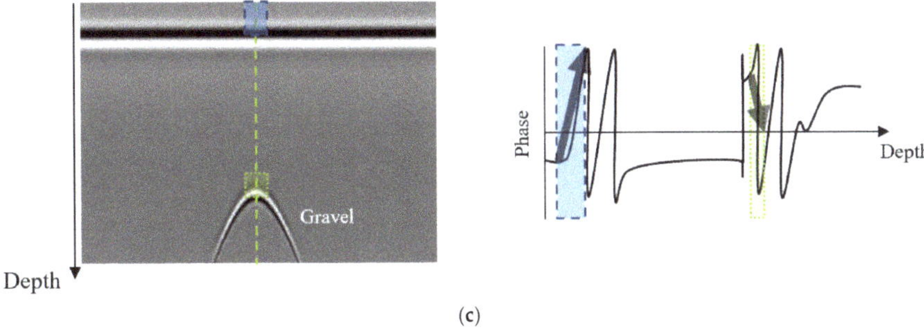

(c)

Figure 7. Numerical simulation of phase analysis: (**a**) 2D simulation model, (**b**) cavity case, and (**c**) gravel case. The dashed blue box, line red box, and dotted green box in the B-scan images correspond to the locations of the radiated wave, cavity, and gravel boundaries, respectively. (Tx: Transmitter antenna, Rx: Receiver antenna).

3. Experimental Validation

The newly proposed UcNet was experimentally validated using 3D GPR data obtained by a multi-channel GPR-mounted van at urban roads in Seoul, South Korea. Figure 8a shows the multi-channel GPR (DXG 1820, 3d-Radar company) [32] has 20 channels transmitting and receiving antennas, which is devised with 0.075 m interval of each channel that is able to cover 1.5 m scanning width at once. The multi-channel GPR designed consists of bow-tie monopole antennas. The multi-channel GPR has a frequency range of 200–3000 MHz with a step-frequency input wave, and the data acquisition system of GeoScopeTM Mk IV (Figure 8b) acquires the GPR data in real time with a time resolution of 0.34 ns and a maximum scanning rate of 13,000 Hz [32].

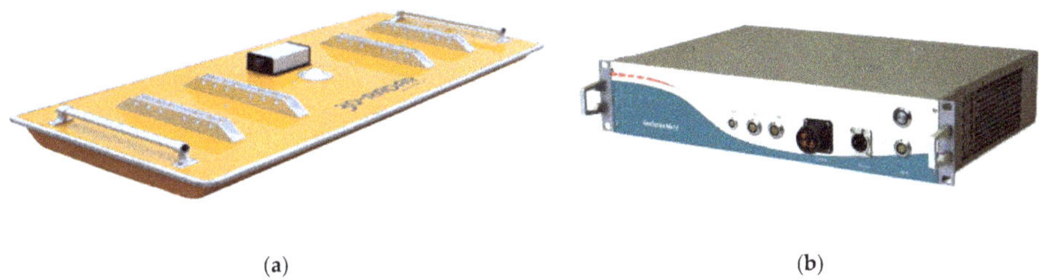

(**a**) (**b**)

Figure 8. (**a**) Multi-channel GPR and (**b**) data acquisition system.

As for the training dataset, several tens of kilometers of GPR data, including cavities, manholes, gravels and subsoil background were collected from 17 different regions in Seoul. To clearly confirm the underground objects, the pavement core drilling machine with a portable endoscope was used. Figure 9a shows the multi-channel GPR-mounted van used for field data collection and Figure 9b shows the confirmation process of verifying the underground objects found by the multi-channel GPR-mounted van. The representative confirmed underground cavities and gravels by portable endoscope are shown in Figure 10. A total of 1056 GPR grid images of 256 cavities, 256 manholes, 256 subsoil backgrounds, and 256 gravels cases were used for network training. Here, 20 training epochs and 0.001 initial learning rate were used in this study.

A total of 1056 GPR grid images of 256 cavities, 256 manholes, 256 subsoil backgrounds, and 256 gravels cases, which were not used for UcNet training, were used for blind testing in this study. Figure 11 shows the representative testing 2D GPR grid images of cavities, manholes, subsoil backgrounds, and gravels. As shown in Figure 11b, manholes generally have a distinguishable feature, which has double parabola shape with high intensity compared to the surrounding soil and upper pavement layer. On the other hand, there is no remarkable feature in B- and C-scan images of the subsoil background case as shown in Figure 11c. However, it can be observed that the B- and C-scan images of the cavity and gravel cases respectively show similar parabola and circular features by comparing between Figure 11a,d.

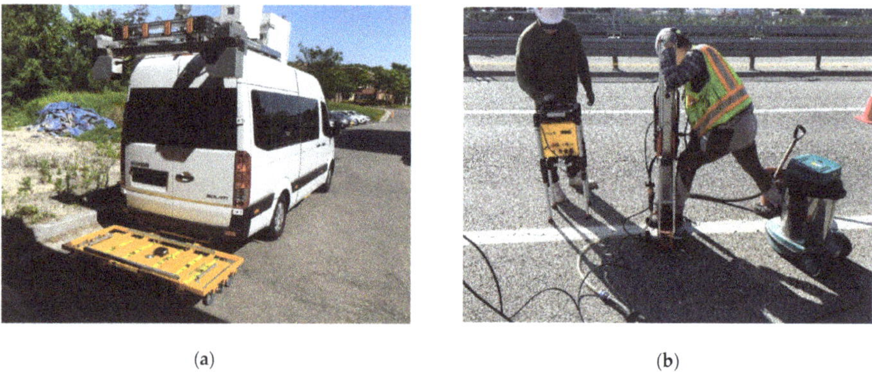

(a) (b)

Figure 9. In-situ validation tests with (**a**) multi-channel GPR-mounted van and (**b**) core drilling.

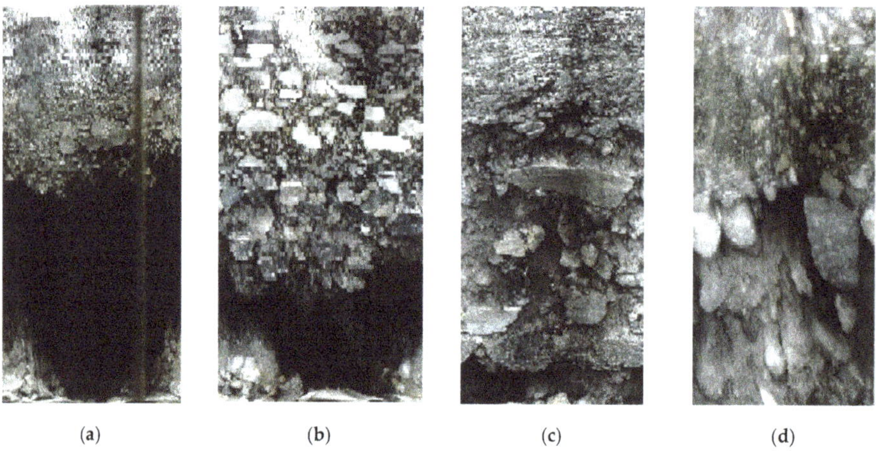

(a) (b) (c) (d)

Figure 10. Representative core drilling results of (**a**,**b**) cavities and (**c**,**d**) gravels.

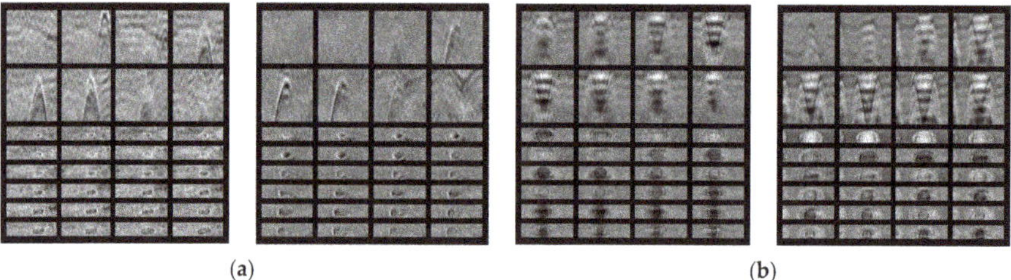

(a) (b)

Figure 11. *Cont.*

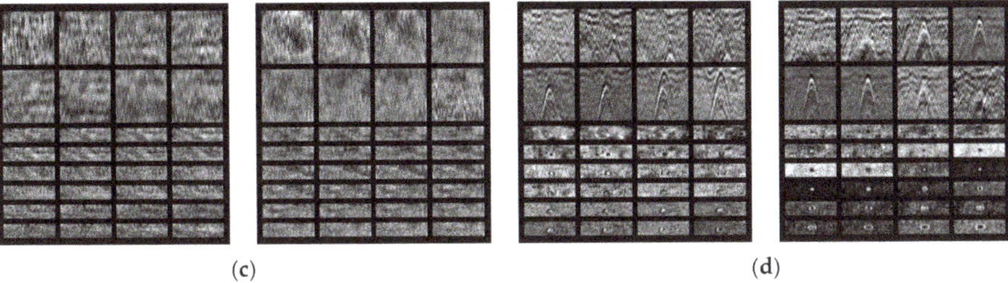

Figure 11. The representative testing images of (**a**) cavities, (**b**) manholes, (**c**) subsoil backgrounds, and (**d**) gravels.

3.1. Conventional CNN-based Underground Object Classification

To validate the effectiveness of UcNet, the two experimental validation results of the conventional CNN and newly developed UcNet were compared. Figure 12 shows the conventional CNN-based underground object classification results. Since Phase I of UcNet is equivalent to the conventional CNN, the processing results up to Phase I were considered as the conventional CNN one. As expected, manhole and subsoil background, which have significant features of 2D GPR grid images, are correctly classified compared with the ground truth confirmed by the portable endoscope. However, 11.74% of cavities and 33.73% of gravels are misclassified as each other due to their similar morphological features. The classification performance of the conventional CNN was evaluated by calculating statistical indices called precision and recall using the following equation:

$$\text{Precision} = \frac{\textit{true positive}}{\textit{true positive} + \textit{false positive}} \text{Recall} = \frac{\textit{true positive}}{\textit{true positive} + \textit{false negative}}. \quad (4)$$

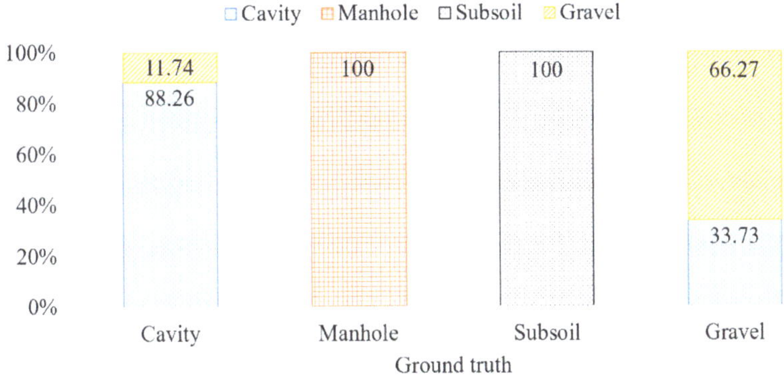

Figure 12. The results of conventional convolutional neural network (CNN)-based underground object classification.

Table 1 summarizes the precision and recall values obtained from the conventional CNN results. As for the manhole and subsoil background cases, the precision and recall values are 100%, indicating that they are properly classified by the conventional CNN. On the other hand, 88.26% and 66.27% of the precision values in the cavity and gravel cases, physically meaning that false positive occurs. Similarly, the relatively low recall values of the cavity and gravel cases means the false negative alarm due to the misclassification between the cavity and gravel cases.

Table 1. Statistical results obtained from conventional CNN.

Underground Object	Precision (%)	Recall (%)
Cavity	88.26	72.36
Manhole	100	100
Subsoil background	100	100
Gravel	66.27	84.95

3.2. Newly Developed UcNet

From the Phase I results described in Figure 12, Phases II and III were subsequently carried out. Figure 13a–d shows the representative SR B-scan images which are especially misclassified in Phase I. Figure 13a,b indicates the representative cavities cases misclassified as gravels, and Figure 13c,d shows vice versa. The misclassification results show very similar geometric features to each other, but the phase information at the parabola boundaries are distinctive between the cavity and gravel cases. In particular, although the LR GPR B-scan images has ambiguous pixel-level boundary information, the SR images show that much clearer parabola boundary information, making it possible to conduct the precise phase analysis in the subsequent Phase II.

Figure 14a,b indicates the procedure of parabola boundary extraction results with SR and noise removal image corresponding to Figure 13a,b. All parabola boundaries are clearly extracted from the SR images even though unwanted noise and non-parabola features coexist. Similarly, Figure 14c,d shows that the distinctive parabola boundaries are successfully extracted, which correspond to Figure 13c,d.

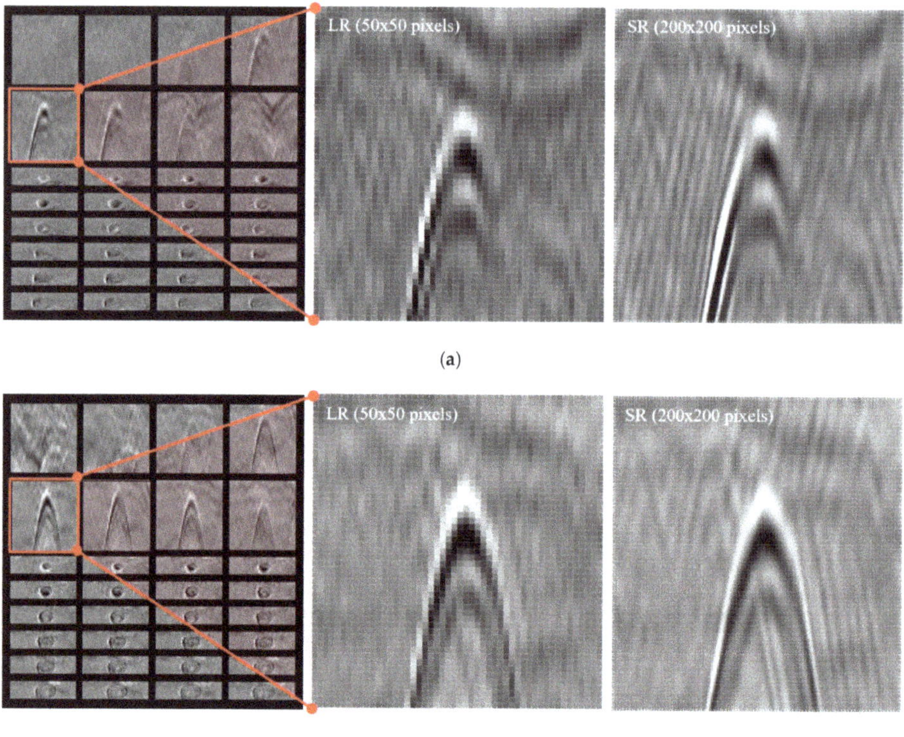

(a)

(b)

Figure 13. *Cont.*

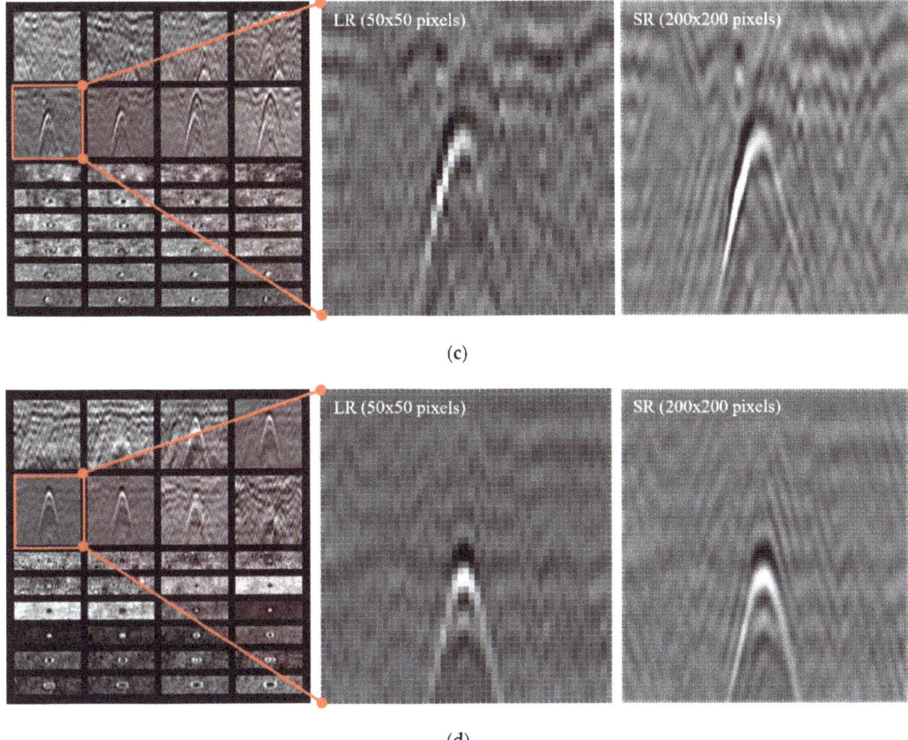

Figure 13. Representative LR and SR B-scan images of (**a**,**b**) cavity cases misclassified as gravels and (**c**,**d**) gravel cases misclassified as cavities.

Figure 15 shows the phase analysis results corresponding to the extracted parabola boundary information in Figure 14. As shown in Figure 15a, $\Delta\theta$ of the radiated waves has 0.062 which is positive value. Then, the $\Delta\theta$ values of 0.0481 and 0.0336 shown in Figure 15b,c indicate that they can be considered as underground cavities, not gravel. Conversely, the $\Delta\theta$ values of the misclassified cases from gravels to cavities have −0.0803 and −0.1073 as shown in Figure 15d,e, respectively. These out-of-phase information physically imply the high permittivity of the object in comparison with the surrounding soil, meaning that they are most likely gravel in the designed category of UcNet.

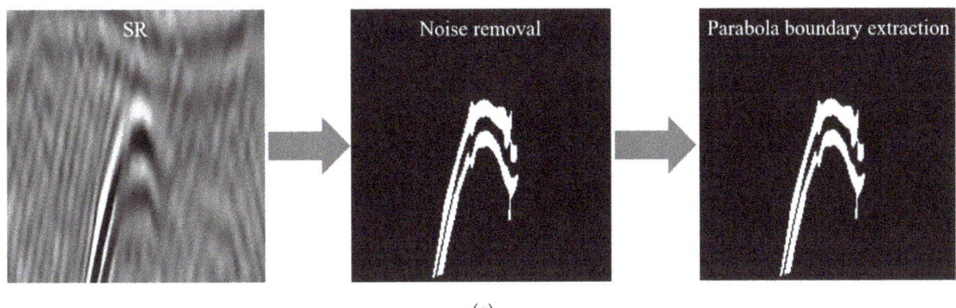

Figure 14. *Cont.*

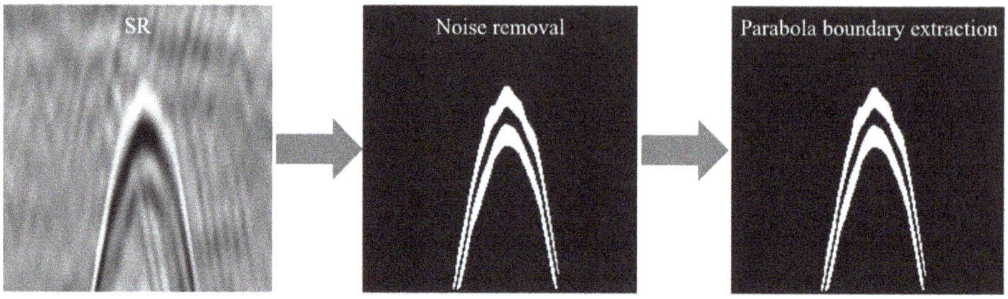

(b)

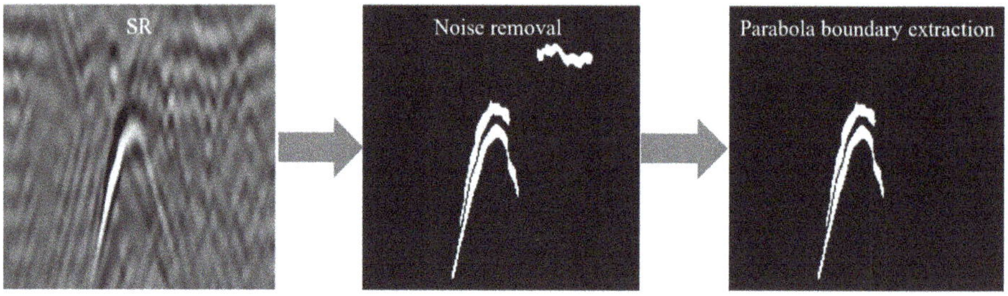

(c)

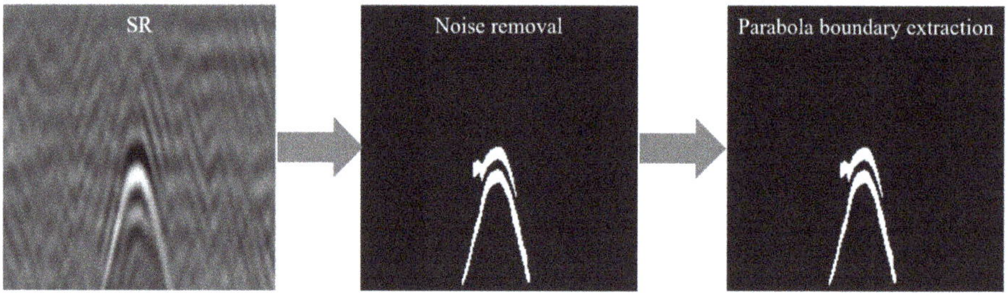

(d)

Figure 14. The parabola boundary extraction results of (**a**,**b**) cavity cases misclassified as gravels and (**c**,**d**) gravel cases misclassified as cavities.

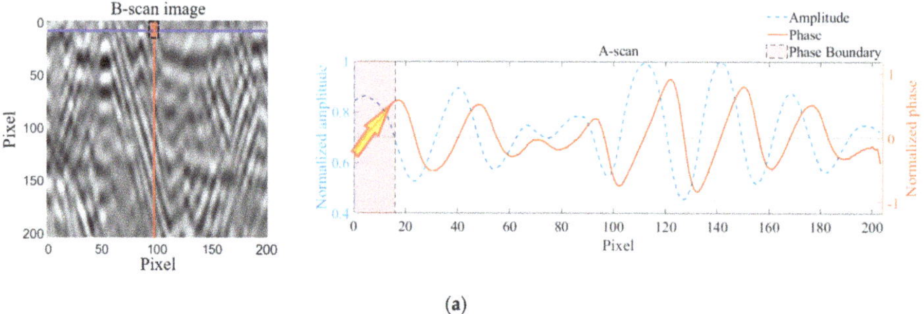

(**a**)

Figure 15. *Cont.*

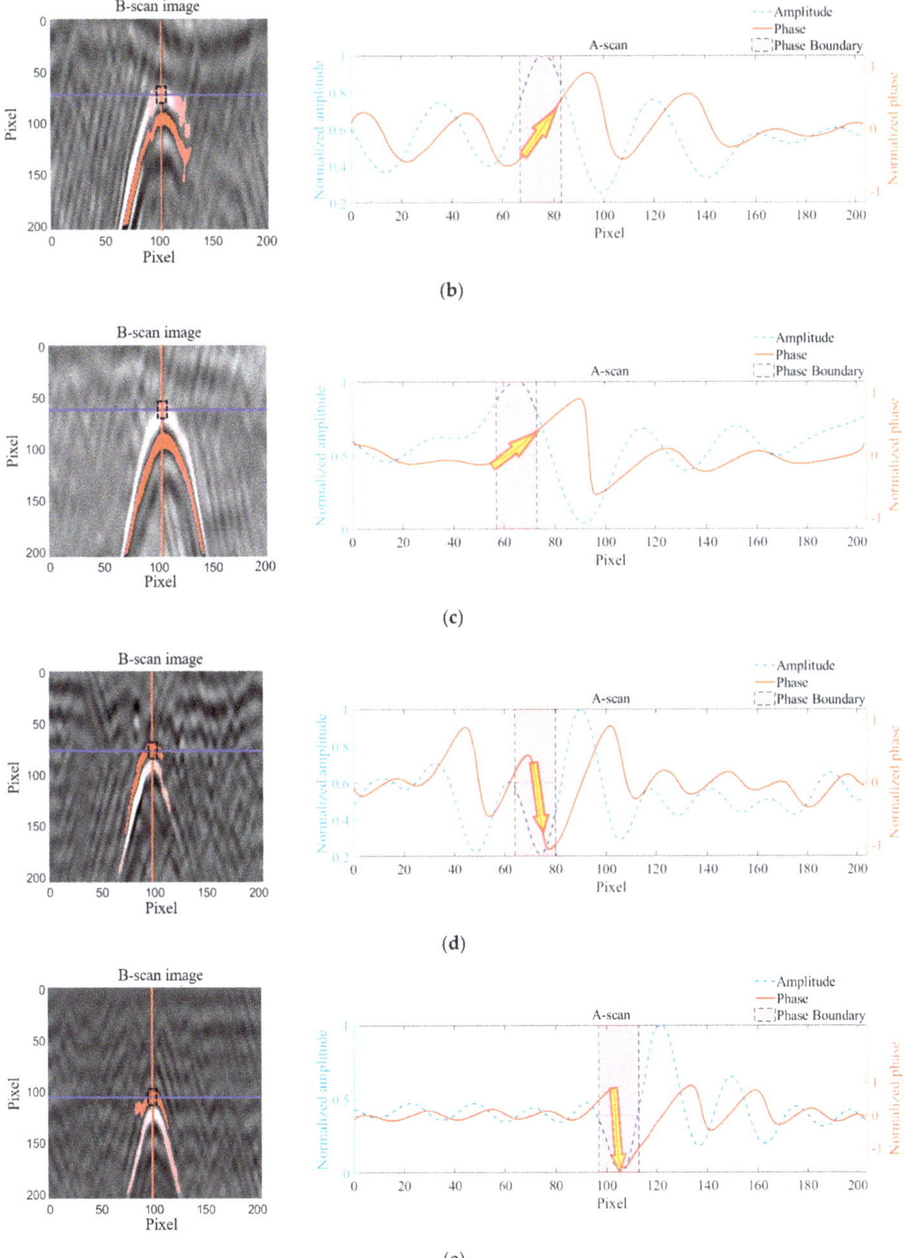

Figure 15. Phase analysis results at the extracted boundaries of (**a**) radiated wave, (**b**,**c**) cavities, and (**d**,**e**) gravels, respectively.

Based on the phase analysis results of Figure 15, the object classification results of Figure 11 were updated as shown in Figure 16. It can be easily observed that the misclassified cavities and gravels are properly updated without false alarms. Since all of misclassification cavity and gravel cases are correctly classified, the statistical precision and recall are increased to 100%.

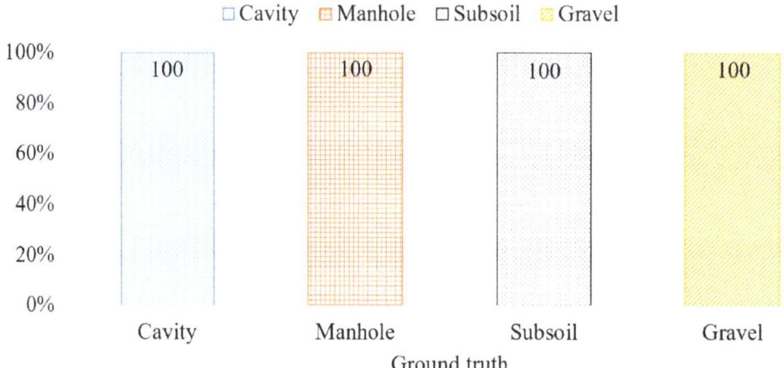

Figure 16. Updated underground object classification results using UcNet.

4. Conclusions and Discussion

This study newly proposes an underground cavity detection network (UcNet) for enhancing the cavity classification capability. Although convolutional neural networks (CNNs) utilized ground penetrating radar (GPR) triplanar images to classify underground objects, misclassification often occurs due to similar morphological features in B- and C-scan GPR images. This misclassification may lead to a substantial increase of maintenance cost and time. The proposed UcNet overcomes the existing technical hurdle through precise and reliable interpretation of GPR data without expert intervention. In particular, UcNet minimizes the misclassification between cavities and gravel chunks using the conventional CNNs. The effectiveness of the proposed UcNet was experimentally validated using in-situ GPR data obtained on real complex urban areas in Seoul, South Korea. Although the proposed UcNet works well with the validation datasets considered in this study, further investigations on other types of underground objects such as concrete dummies and underground pipes under various in-situ road and underground conditions are warranted. In particular, the authors are now creating our own deep classification network to directly handle 3D GPR data as well as constructing a GPR data library.

Author Contributions: M.-S.K. and Y.-K.A. conceived and designed this study. M.-S.K. and N.K. acquired experimental data and performed conventional CNN classification processing. S.B.I., J.-J.L. performed newly proposed UcNet classification processing and analysis. M.-S.K. and Y.-K.A. wrote the entire manuscript and designed the processing code. Y.-K.A. also helped to design the processing code and provided comments.

Funding: This work was supported by the National Research Foundation of Korea (NRF) grant funded by the Korea government (MSIT) (2018R1A1A1A05078493).

Acknowledgments: The authors would like to thank "Development of Evaluation and Analysis Technologies for Road Sink Research Team" for providing the GPR devices.

Conflicts of Interest: The authors declare no conflict of interest.

References

1. Brinkmann, R.; Parise, M.; Dye, D. Sinkhole distribution in a rapidly developing urban environment: Hillsborough Country, Tampa Bay area, Florida. *Eng. Geol.* **2008**, *99*, 169–184. [CrossRef]
2. Massive Sinkhole Nearly Swallows Home in Pennsylvania. Available online: https://philadelphia.cbslocal.com/2017/01/25/massive-sinkhole-opens-up-in-cheltenham-twp/ (accessed on 5 April 2019).
3. Huge Sinkholes Are Now Appearing in the Wrong Places. Available online: https://phys.org/news/2017-05-huge-sinkholes-wrong.html (accessed on 18 April 2019).

4. Watch: Four Killed after Massive Sinkhole Opens in Chinese City. Available online: https://www.newschannel5.com/news/world/china-sinkhole-video (accessed on 12 March 2019).
5. Huge Sinkhole Opens up on a Busy Chinese Road and Engulfs a Tree and a Minivan. Available online: https://www.dailymail.co.uk/news/article-4593858/Huge-sinkhole-opens-busy-Chinese-road.html (accessed on 4 April 2019).
6. Huge Sinkhole Swallows Street in Fukuoka, Japan. Available online: https://www.bbc.com/news/world-asia-37906065# (accessed on 5 April 2019).
7. Sinkhole in Seoul Forces 200 Residents to Evacuate. Available online: http://koreabizwire.com/sinkhole-in-seoul-forces-200-residents-to-evacuate/123512 (accessed on 5 April 2019).
8. Intrieri, E.; Gigli, G.; Nocentini, M.; Lombardi, L.; Mugnai, F.; Fidolini, F.; Casagli, N. Sinkhole monitoring and early warning: An experimental and successful GB-InSAR application. *Geomorphology* **2015**, *241*, 304–314. [CrossRef]
9. Sevil, J.; Gutierrez, F.; Zarroca, M.; Desir, G.; Carbonel, D.; Guerrero, J.; Linares, R.; Roque, C.; Fabregat, I. Sinkhole investigation in an urban area by trenching in combination with GPR, ERT and high-precision leveling. Mantled evaporate karst of Zaragoza city, NE Spain. *Eng. Geol.* **2017**, *231*, 9–20. [CrossRef]
10. Sun, H.; Pashoutani, S.; Zhu, J. Nondestructive evaluation of concrete bridge decks with automated acoustic scanning system and ground penetrating radar. *Sensors* **2018**, *18*, 1955. [CrossRef] [PubMed]
11. Ho, K.C.; Collins, L.M.; Huettel, L.G.; Gader, P.D. Discrimination mode processing for EMI and GPR sensors for hand-held land mine detection. *IEEE Trans. Geosci. Remote Sens.* **2004**, *42*, 249–263. [CrossRef]
12. Basile, V.; Carrozzo, M.T.; Negri, S.; Nuzzo, L.; Quarta, T.; Villani, A.V. A ground-penetrating radar survey for archaeological investigations in an urban area (Lecce, Italy). *J. Appl. Geophys.* **2000**, *44*, 15–32. [CrossRef]
13. Daniel, D.J. *Ground Penetrating Radar*, 2nd ed.; the Institution of Electrical Engineers: London, UK, 2004.
14. Groenenboom, J.; Yarovoy, A. Data processing and imaging in GPR system dedicated for landmine detection. *SSTA* **2002**, *3*, 387–402.
15. Sharma, P.; Kumar, B.; Singh, D.; Gaba, S.P. Critical analysis of background subtraction techniques on real GPR data. *Def. Sci. J.* **2017**, *67*, 559–571. [CrossRef]
16. Park, B.J.; Kim, J.G.; Lee, J.S.; Kang, M.S.; An, Y.K. Underground object classification for urban roads using instantaneous phase analysis of GPR data. *Remote Sens.* **2018**, *10*, 1417. [CrossRef]
17. Simi, A.; Bracciali, S.; Manacorda, G. Hough transform based automatic pipe detection for array GPR: Algorithm development and on-site tests. In Proceedings of the IEEE Radar Conference, Rome, Italy, 26–30 May 2008.
18. Li, W.; Cui, X.; Guo, L.; Chen, J.; Chen, X.; Cao, X. Tree root automatic recognition in ground penetrating radar profiles based on randomized Hough transform. *Remote Sens.* **2016**, *8*, 430. [CrossRef]
19. Noreen, T.; Khan, U.S. Using pattern recognition with HOG to automatically detect reflection hyperbolas in ground penetrating radar data. In Proceedings of the International Conference on Electrical and Computing Technologies and Applications, Ras Al Khaimah, UAE, 21–23 November 2017.
20. Torrione, P.A.; Morton, K.D.; Sakaguchi, R.; Collins, L.M. Histogram of oriented gradients for landmine detection in ground penetrating radar data. *IEEE Trans. Geosci. Remote Sens.* **2013**, *52*, 1539–1550. [CrossRef]
21. Mazurkiewicz, E.; Tadeusiewicz, R.; Tomecka-Suchon, S. Application of neural network enhanced ground penetrating radar to localization of burial sites. *Appl. Artif. Intell.* **2016**, *30*, 844–860. [CrossRef]
22. Travassos, X.L.; Avila, S.L.; Ida, N. Artificial neural networks and machine learning techniques applied to ground penetrating radar: a review. *Appl. Comput. Inform.* **2018**, in press. [CrossRef]
23. Zhang, Y.; Huston, D.; Xia, T. Underground object characterization based on neural networks for ground penetrating radar data. In Proceedings of the SPIE Smart Structures and Materials + Nondestructive Evaluation and Health Monitoring, Las Vegas, NV, USA, 20–24 March 2016.
24. Al-Nuaimy, W.; Huang, Y.; Nakhkash, M.; Fang, M.T.C.; Nguyen, V.T.; Eriksen, A. Automatic detection of buried utilities and solid objects with GPR using neural networks and pattern recognition. *J. Appl. Geophys.* **2000**, *43*, 157–165. [CrossRef]
25. Lameri, S.; Lombardi, F.; Bestagini, P.; Lualdi, M.; Tubaro, S. Landmine detection from GPR data using convolutional neural networks. In Proceedings of the European Signal Processing Conference, Kos, Greece, 28 August–2 September 2017.
26. Kim, N.G.; Kim, S.H.; An, Y.K.; Lee, J.J. A novel 3D GPR image arrangement for deep learning-based underground object classification. *Int. J. Pavement Eng.* **2019**. [CrossRef]

27. Kang, M.S.; Kim, N.G.; Lee, J.J.; An, Y.K. Deep learning-based automated underground cavity detection using three-dimensional ground penetrating radar. *Struct. Health Monit.* **2019**, 1475921719838081. [CrossRef]
28. Kizhevsky, A.; Sutskever, I.; Hilton, G. ImageNet classification with deep convolutional neural networks. In Proceedings of the NIPS'12, Lake Tahoe, NV, USA, 3–6 December 2012.
29. Zhang, Y.; Li, K.; Li, K.; Wang, L.; Zhong, B.; Fu, Y. Image super-resolution using very deep residual channel attention networks. In Proceedings of the European Conference on Computer Vision, Munich, Germany, 8–14 September 2018.
30. Timofte, R.; Gu, S.; Gool, L.V.; Zhang, L.; Yang, M.H. NTIRE 2018 challenge on single image super-resolution: Methods and results. In Proceedings of the Conference on Computer Vision and Pattern Recognition, Salt Lake City, UT, USA, 18–22 June 2018.
31. Warren, C.; Giannopoulos, A.; Giannakis, I. gprMax: Open source software to simulate electromagnetic wave propagation for ground penetrating radar. *Comput. Phys. Commun.* **2016**, *209*, 163–170. [CrossRef]
32. 3D-RADAR. Available online: http://3d-radar.com/system/ (accessed on 1 October 2019).

Article

Methodology for Processing of 3D Multibeam Sonar Big Data for Comparative Navigation

Andrzej Stateczny [1,*], Wioleta Błaszczak-Bąk [2], Anna Sobieraj-Żłobińska [1], Weronika Motyl [3] and Marta Wisniewska [3]

[1] Department of Geodesy, Faculty of Civil and Environmental Engineering, Gdansk University of Technology, Narutowicza 11-12, 80-233 Gdansk, Poland; anna.sobieraj@pg.edu.pl
[2] Institute of Geodesy, Faculty of Geodesy, Geospatial and Civil Engineering, University of Warmia and Mazury in Olsztyn, Oczapowskiego 2, 10-719 Olsztyn, Poland; wioleta.blaszczak@uwm.edu.pl
[3] Marine Technology Ltd., Roszczynialskiego 4/6, 81-521 Gdynia, Poland; w.motyl@marinetechnology.pl (W.M.); m.wisniewska@marinetechnology.pl (M.W.)
* Correspondence: andrzej.stateczny@pg.edu.pl

Received: 30 July 2019; Accepted: 24 September 2019; Published: 26 September 2019

Abstract: Autonomous navigation is an important task for unmanned vehicles operating both on the surface and underwater. A sophisticated solution for autonomous non-global navigational satellite system navigation is comparative (terrain reference) navigation. We present a method for fast processing of 3D multibeam sonar data to make depth area comparable with depth areas from bathymetric electronic navigational charts as source maps during comparative navigation. Recording the bottom of a channel, river, or lake with a 3D multibeam sonar data produces a large number of measuring points. A big dataset from 3D multibeam sonar is reduced in steps in almost real time. Usually, the whole data set from the results of a multibeam echo sounder results are processed. In this work, new methodology for processing of 3D multibeam sonar big data is proposed. This new method is based on the stepwise processing of the dataset with 3D models and isoline maps generation. For faster products generation we used the optimum dataset method which has been modified for the purposes of bathymetric data processing. The approach enables detailed examination of the bottom of bodies of water and makes it possible to capture major changes. In addition, the method can detect objects on the bottom, which should be eliminated during the construction of the 3D model. We create and combine partial 3D models based on reduced sets to inspect the bottom of water reservoirs in detail. Analyses were conducted for original and reduced datasets. For both cases, 3D models were generated in variants with and without overlays between them. Tests show, that models generated from reduced dataset are more useful, due to the fact, that there are significant elements of the measured area that become much more visible, and they can be used in comparative navigation. In fragmentary processing of the data, the aspect of present or lack of the overlay between generated models did not relevantly influence the accuracy of its height, however, the time of models generation was shorter for variants without overlay.

Keywords: 3D sonar; bathymetry; data reduction; autonomous navigation

1. Introduction

Unmanned underwater vehicles, also known as underwater robots, have developed rapidly over the past few years. These systems supersede previously used methods of the underwater exploration of Earth, such as, e.g., hydrographic measuring units with human crew. The trend in unmanned systems development is toward the execution of underwater tasks, including hydrographical surveys, near the bottom by underwater robots, such as remotely operated vehicles remotely controlled by an operator and autonomous underwater vehicles (AUVs) operated without operator input. Underwater positioning methods are not keeping pace with the fast development of AUVs and measurement tools. The main global navigational satellite system (GNSS) positioning method for submersible vehicles is limited to situations where the submersible vehicle can raise an antenna above the surface of the water. However, some AUVs need the more independent method of comparative (terrain) navigation via digital terrain models (DTMs) [1–7].

New methods of spatial data measurement using interferometric multibeam echosounders (MBES), high-frequency side scan sonar, and integrated MBES with sonars require new data processing methods. These new methods may also be suitable for creating autonomous navigation systems for unmanned underwater platforms based on the development of comparative navigation, which uses redundant positioning sources based on navigational radar and electronic navigational charts.

Comparative (terrain reference) navigation is an alternative method for position determination where the GNNS signal is unsuitable or unavailable. This type of navigation is based on searching for matches between a reference image prepared for a specific area (reference map) and a recorded image of a specific, small area, recorded in real time and used to generate a fragment of an area to compare with the reference map.

In comparative navigation, the ship's or vehicle's position is plotted by comparing a dynamically registered image with a pattern image. The pattern images can be bathymetric electronic navigational charts (bENCs), digital radar charts, sonar images, aerial images, or images from other sensors, such as magnetometers or gravimeters, suitably prepared for comparison with radar, sonar, aerial, or other images, respectively. The most frequently registered images at the sea are radar images, whereas the pattern is a numeric radar chart generated from topographic and hydrometeorological data or previous radar observations.

Many scientists globally are working on comparative (terrain reference) navigation [8–10]. Most studies have analyzed the shape of the bottom of bodies of water obtained from the depth of the basin. For example, in [11,12] the authors presented an autonomous underwater vehicle optimal path planning method for seabed terrain matching navigation to avoid these areas. In [13] authors present an application for the practical use of priors and predictions for large-scale ocean sampling. The proposed method takes advantage of the reliable, short-term predictions of an ocean model, and the utility of priors used in terrain-based navigation over areas of significant bathymetric relief to bound uncertainty error in dead-reckoning navigation. Another author [14,15] proposes a comprehensive evaluation method for terrain navigation information and constructs an underwater navigation information analysis model, which is associated with topographic features. Similar problems are presented in [16–18]. In [17], a tightly-coupled navigation is presented to successfully estimate critical sensor errors by incorporating raw sensor data directly into an augmented navigation system. Furthermore, three-dimensional distance errors are calculated, providing measurement updates through the particle filter for absolute and bounded position error. All these solutions are time consuming because they use a big data sets. MBES big data processing [19–30] has also been investigated. In [19], authors propose algorithm CUBE (combined uncertainty and bathymetry estimator). A model monitoring scheme CUBE ensures that inconsistent data are maintained as separate but internally consistent with the depth hypotheses. The other method is presented in [29]. The main purpose of the presented reduction algorithm is that, the position of point and depth value at this point will not be an interpolated. In the article, the author focused on the importance of neighborhood parameters during clustering of bathymetric data using neural networks (self-organizing maps).

Big data problems are closely related to the idea of single-beam echosounders measurements [31] and Light Detection and Ranging (LiDAR) [32–38].

The method of comparative underwater navigation presented in the work compares depth area images registered in semi-real time with depth areas in bENCs. The construction of bENCs for comparative navigation has been described previously [39].

A ship's position can be plotted by comparative methods using one of three basic methods [40].

- Determining the point of best match of the image with the pattern. The logical conjunction algorithm is used and it finds the point of best match of images recorded as a digital matrix. The comparison between the registered real image and the source image (in this case bENCs) as a whole is done using a method that can determine global difference or global similarity between the images.
- Using previously registered real images associated with the position of their registration. This method uses an artificial neural network (ANN) trained by a sequence created from vectors representing the compressed images and the corresponding position of the vehicle.
- Using the generated map of patterns. An ANN is trained with a representation of selected images corresponding to the potential positions of the vehicle. The patterns are generated based on a numerical terrain model, knowledge of the effect of the hydrometeorological conditions and the observation specificity of the selected device.

In addition to ANN, the literature also provides other solutions that can be used in comparative navigation. One possible solution is the application of a system based on an algorithm of multi-sensor navigational data fusion using a Kalman filter [40]. The said solution is intended to be implemented in a navigational decision support system on board a sea-going vessel. The other possible solution is comprehensive testing and analysis of a particle filter framework for real-time terrain navigation on an autonomous underwater vehicle [41].

Deterministic methods include comparative navigation, which is mainly performed using distance and proximity functions, as well as correlation and logical conjunction methods [42].

The idea of using ANN for position plotting is particularly intriguing. The teaching sequence of the ANN consists of registered images correlated with their positions. Teaching is performed in advance and can take as long as necessary. During the use of the trained network, the dynamic registered images are passed to the network input, and the network interpolates the position based on recognized images closest to the analyzed image. A merit of this method is that the network is trained with real images, including their disturbances and distortions, which are similar to those that are used in practice. The main problem with this method is that it requires previous registration of numerous real images in various hydrometeorological conditions, and the processing and compressing of images. After compressing the analyzed image, a teaching sequence for the neural network designed to plot the vehicle's position is constructed. The task of the network is to construct a mapping function associating the analyzed picture with the position.

Regardless of the method of comparative navigation, the basic problem is registration, filtering, and reduction of measurement data.

The standard methodology of the development of MBES big data, in general, consists of following stages: (1) Obtaining a whole 3D multibeam sonar data set, (2) pre-processing (including, among others the filtration process, noise removal and data reduction), (3) main processing (including, among others, DTM construction and development of bathymetric maps), (4) visualization, (5) analysis.

In this work, we present a new approach of acquiring and simultaneously processing a set of bathymetric observations. This is a different approach than presented in the literature on the subject [19,27,29]. The approach includes fragmentary data acquisition, and fast reduction (the optimum dataset method—OptD [35]) within acquired measuring strips in almost real time, and generation of DTMs. The OptD method was modified for this purpose. This modification relies on introducing in the OptD method a loop (FOR instruction) for fragmentary data processing. All these processes in our approach were performed in a first stage under acquisition of data, during measurement, whole data set was not obtained, but a fragment of the data set. The approach was considered where measuring strips were obtained without overlay and where measuring strips had overlay between each other. The proposed approach was compared with the method that uses full sets of bathymetric data. The results showed that our approach quickly obtained, reduced, and generated DTMs in almost real time for comparative navigation.

The originality of this paper was a new approach for 3D multibeam data processing. Reduction and 3D model generation in almost real time is an important research subject in the context of comparative navigation. The navigators need to have, in short time, the results of measurement for opportunity to compare generated isolines map (or DTMs) with existing maps (or DTMs). In this way, the navigators can detect differences in depth, and recognize obstacles at the bottom of the water reservoir.

2. Materials and Methods

While preparing a master image, such as the depth area taken from a bENC, the problem of data processing time is not critical. For recording and processing images intended for dynamic comparison with the master image, data reduction is the key problem. A dynamically registered image should be processed for a DTM generated in close to real time. DTM construction is an important step in generating the depth area of the acquired image to be compared with the master image of the bottom. A DTM of the bottom can be generated based on data obtained from GNSS measurements connected to a single-beam echosounder or a MBES, which is currently the most popular technology. The acquired data is used to create a DTM of the bottom of the river, lake, channel, or harbor area. The DTM not only models bottom area processes but also detects objects under the water surface and eventually helps with their visual inspection. Handling such a large volume of data is time-consuming and labor-intensive. Therefore, we propose the use of the OptD method [35–37] to reduce the data set. Reducing the number of observations allows the 3D model and depth area to be generated much faster. Moreover, the OptD method allows the data set to be divided into points representing the bottom of the river, lake, or channel and points representing objects that are not the bottom (items under the water surface).

2.1. Instrument Description

The 3D Sidescan 3DSS-DX-450 sonar system (Ping DSP) (Figure 1) uses a state-of-the-art acoustic transducer array, SoftSonar electronics, and advanced signal processing to produce superior swath bathymetry and 3D side-scan imagery. This system resolves multiple concurrent acoustic arrivals, separating backscatter from the seabed, sea surface, water column, and multipath arrivals to produce 3D side-scan image spanning the entire water column. High-resolution swath bathymetry coverage of up to 14 times altitude is achieved. The system operates on frequencies of 450 kHz and the maximum power consumption is 18 W. The dimensions of the sonar head are 57 × 9.8 cm, and its weight in air is 8 kg. The device generates a beam with a width of 0.4° and the maximum number of soundings per ping is 1440 across the swath [43].

Figure 1. Photograph of the 3D Sidescan 3DSS-DX-450 sonar system (photograph: A. Stateczny).

2.2. Test Area Characteristics

The measurements were carried out by HydroDron-1 [44] on Klodno Lake, which is a gutter-type lake and located in the Kashubian Lake District in Chmielno Commune, in the Kartuzy administrative unit (Pomorskie Province), Poland. Klodno Lake is one of the three Chmielenskie Lakes and lies on the Kolko Radunskie waterway. The Kashubian Route runs along the southern and western coastline of the lake, and the lake itself is connected to Male Brodne Lake and Radunski Dolny Lake through narrow waterways. Radunia river flows through the lake. The total area of the lake is 134.9 ha, the length is 2.0 km, and the maximum depth is 38.5 m. Data presented in the article were taken during a 9 day measurement campaign in April and May 2019.

2.3. Methodology

2.3.1. Methodology v1

In methodology v1, the test area was processed and divided into strips, and tests without and with overlay between the strips were performed. For the tests, the following assumptions were made.

(a) The test area was divided into strips p_i without overlay between them, where p_i was a strip with observations, $i= 1, 2, 3 \ldots , m$, and m was the number of strips.
(b) The test area was divided into strips po_i with 25–30% overlay between them, where po_i was a strip with observations, $i = 1, 2, 3 \ldots , m$, and m is the number of strips.

For strips without overlay (methodology v1.1) and strips with overlay (methodology v1.2.), DTMmv1.1 and DTMmv1.2, respectively, were generated using the kriging method, and the results of processing all strips combined together were whole DTMv1.1 and whole DTMv1.2, respectively.

2.3.2. Methodology v2

Methodology v2 differed from methodology v1 in that the DTMs were generated based on a reduced set. The reduction was performed by the OptD method. The tests were performed using similar assumptions to methodology v1.

(a) The test area was divided into strips p_i without overlay between them and the set was reduced by the OptD method.
(b) The test area was divided into strips po_i with 25–30% overlay between them and the set was reduced by the OptD method.

DTMmv2.1 and DTMmv2.2 were obtained for methodologies v2.1 and v2.2, and DTMv2.1 and DTMv2.2 were obtained as a sum of partial DTMs, respectively. The scheme for the proposed methodologies is presented in Figure 2.

Additionally, for comparison, DTMs were generated from all the data (100%) and from all the data after reduction (2%):

(a) DTM100% = whole DTMv1.1 = whole DTMv1.2.
(b) DTM2% = whole DTMv2.1 = whole DTMv2.2.

Our approach used the OptD reduction method and the kriging interpolation method.

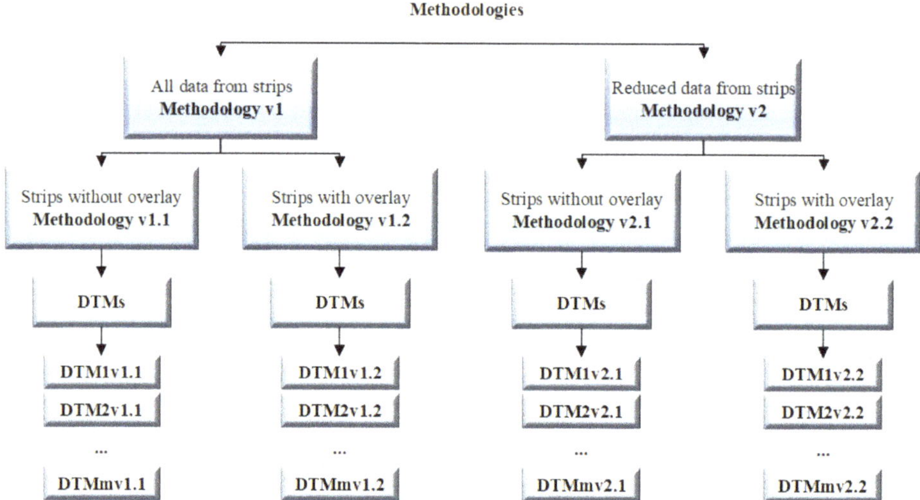

Figure 2. Scheme of the methodologies.

2.3.3. OptD method

The main aim of OptD method was the reduction of the set of measurement observations. The degree of reduction was determined by setting prior, the reduction optimization criterion (*c*) (e.g., the number of observations in the dataset that the user required after reduction). Reduction itself was based on the cartographic generalization method. The area of interest was divided on measuring strips *L*. Within each *L*, the relative position of the points to each other was considered. The way how the points were tested in the context of being removed or preserved in the dataset depended on the tolerance range *t* related to the chosen cartographic generalization method. The width of *L* and *t* are iteratively changed until the optimization criterion was achieved. In result, there were different levels of reduction in the individual parts of the processing area: There were more points in the detailed part of the scanned object and much less within uncomplicated structures or areas. Only those points that were significant remained. This method has been described in detail in [35–37].

Previous applications of the OptD method consisted of processing the entire data set (airborne laser scanning—ALS, terrestrial laser scanning—TLS, mobile laser scanning—MLS). In the case of MBES, the strips with observations were reduced in almost real time and happened in stages. The OptD method was modified for this purpose. This modification relied on introducing in the OptD method a loop (FOR instruction) for fragmentary data processing. The methodology of processing MBES based on the modified OptD method is presented in Figure 3.

The strip's width can be determined or set in relation to the measuring speed. The first measurement strip is reduced while the next strip is acquired. The second strip is attached to the previous reduced strip, and then the second is reduced while the third is obtained and so on, until the measurement is finished. Reduction conducted within each of the separated strips is based on the Douglas–Peucker cartographic generalization method [45,46]. The process can be performed for strips without overlay (methodology v2.1) or strips with overlay (methodology v2.2). Finally, we obtained a whole data set consisting of reduced strips.

(a) For methodology v2.1, the whole dataset after reduction = p_1 after OptD + p_2 after OptD + … + p_m after OptD
(b) For methodology v2.2, the whole dataset after reduction = po_1 after OptD + po_2 after OptD + … + po_m after OptD

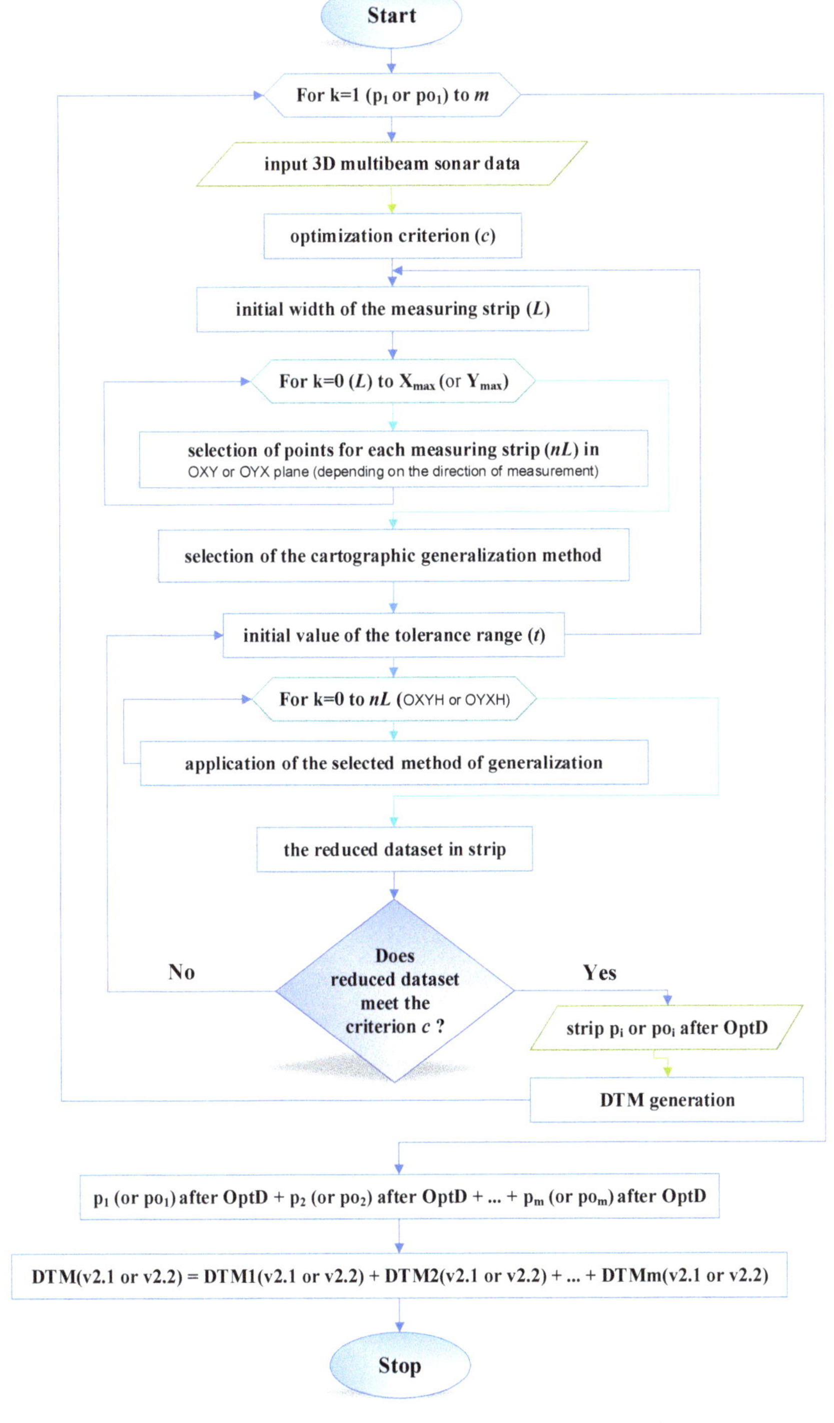

Figure 3. Processing of multibeam echosounders (MBES) data based on the Optimum Dataset OptD method.

In both versions, the optimization criterion, c, was adopted, which controls the reduction rate. For simplicity, this criterion was given as a percentage of points in the set after reduction. In this work, $c = 2\%$ was used, which is a high reduction rate. For almost real-time processing, processing time was important. Reduction decreased the number of observations, which substantially shortened the next process, DTM, and depth area generation. To generate DTM in almost real time, the kriging method was used.

In methodology v2.1, DTM1v2.1 was generated based on a reduced set with observations from the first measurement strip, p_1. DTM2v2.1 was generated from p_2 after the reduction, and the last interpolated node points were in the place where the DTM2v2.1 and DTM1v2.1 nodes coincided. The DTM3v2.1 nodes coincided with the nodes of the next DTM, DTM4v2.1, and the previous DTM, DTM2 v2.1, and so on. Methodology v2.2 used a similar process for strips with overlay.

Finally, the method gave the following models.

(a) For v2.1, the whole DTMv2.1 = DTM1v2.1 + DTM2v2.1 + ... + DTMmv2.1
(b) For v2.2, the whole DTMv2.2 = DTM1v2.2 + DTM2v2.2 + ... + DTMmv2.2

In addition to the DTMs generated by methodologies v2.1 and v2.2, DTMv2.1 and DTMv2.2 were obtained from the whole dataset after reduction. For comparison, using methodology v1, DTMmv1.1 and DTMmv1.2 and DTMv1.1 and DTMv1.2 were also generated.

2.3.4. Reduction

The methodology v1 test area was divided into strips with no overlay between them. The scheme for this division (methodology v1.1) is shown in Figure 4a and that for methodology v1.2, in which strips are selected with overlay, is shown in Figure 4b.

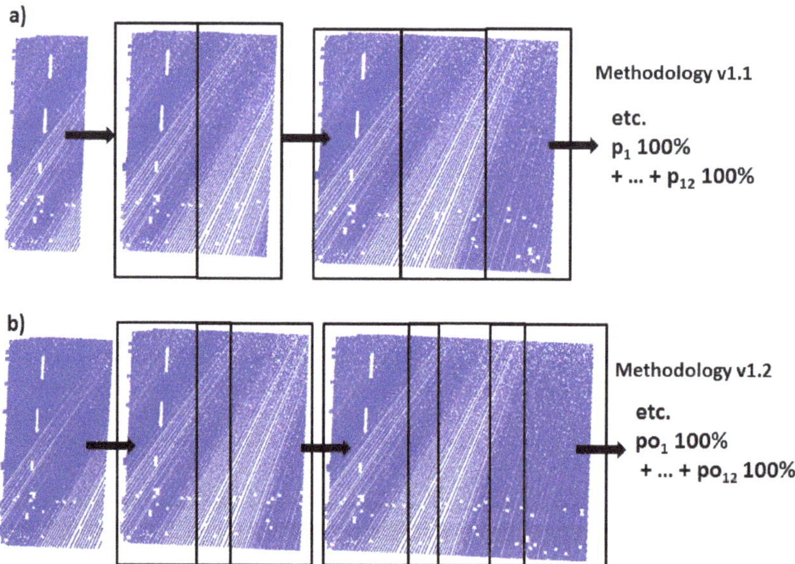

Figure 4. Scheme of test area division into strips (**a**) with and (**b**) without overlay.

After the measurement was completed, we obtained a set of observations (Figure 5).

Figure 5. Whole test area (694,185 points).

The statistical characteristics of the datasets obtained by methodology v1 are shown in Table 1. Each dataset represented an individual strip, p_i characterized by number of points in the dataset, and the minimum and maximum height of points. Additionally, information about range and standard deviation of height was included. They allowed us to initially assess the fragments of measured areas.

Table 1. Statistical characteristics for datasets in methodology v1. (Where: H—height, R—range, STD—standard deviation).

		Number of Points	H min. [m]	H max. [m]	R [m]	STD [m]
	Whole dataset	694185	15.08	23.65	8.57	2.44
Strips without overlay	p_1	54176	22.86	23.54	0.68	0.08
	p_2	35106	22.86	23.60	0.74	0.11
	p_3	44333	22.82	23.58	0.76	0.12
	p_4	53579	22.75	23.65	0.90	0.10
	p_5	52967	22.64	23.52	0.88	0.11
	p_6	55497	22.04	23.39	1.35	0.20
	p_7	70704	20.84	23.14	2.30	0.40
	p_8	84962	19.48	22.16	2.68	0.55
	p_9	79890	17.86	20.42	2.56	0.49
	p_{10}	53216	17.35	18.78	1.43	0.27
	p_{11}	55373	16.36	18.04	1.68	0.33
	p_{12}	54382	15.08	17.03	1.95	0.31
Strips with overlay	po_1	72718	22.86	23.54	0.68	0.09
	po_2	56075	22.82	23.60	0.78	0.12
	po_3	72478	22.75	23.65	0.90	0.11
	po_4	80936	22.73	23.65	0.92	0.10
	po_5	80642	22.40	23.52	1.12	0.14
	po_6	90022	21.50	23.39	1.89	0.29
	po_7	115584	20.18	23.16	2.98	0.62
	po_8	122894	18.75	22.21	3.46	0.75
	po_9	101810	17.71	20.42	2.71	0.56
	po_{10}	84810	16.92	18.78	1.86	0.41
	po_{11}	84551	15.82	18.04	2.22	0.44
	po_{12}	54592	15.08	17.04	1.96	0.31

In methodology v2, the original dataset was optimized using the OptD method. As in the previous case, the test area was divided into strips with and without overlay, and the variants are shown in Figures 6 and 7.

Figures 6 and 7 show the difference between methodology v2.1 and v2.2. In the case of v. 2.1, there are more points where the strips contact each other than in the middle of the strips. However, in the case of v2.2 more points are in the entire overlay area.

The appearance of the whole dataset after reduction conducted using the same optimization criterion (c = 2%) is shown in Figure 8. The time required for the reduction of the whole set to 2% of the original set was about 20 s, which is acceptable for comparative navigation.

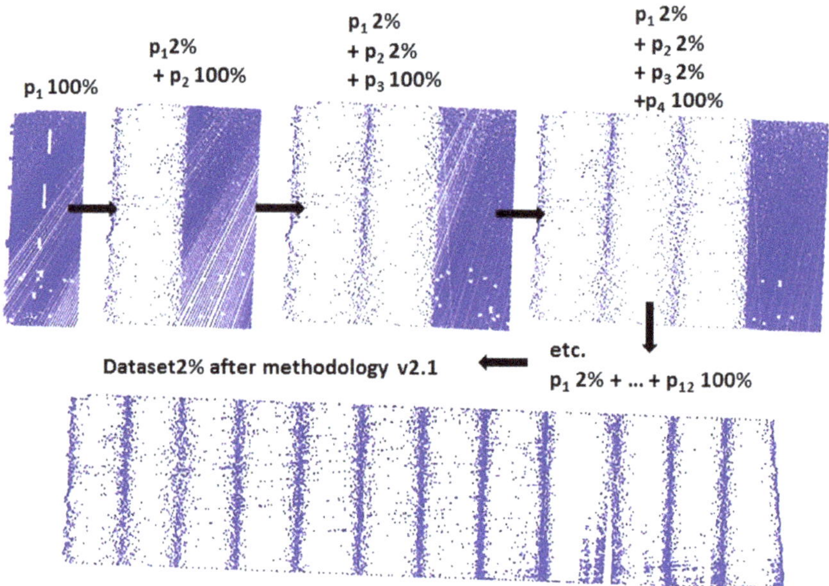

Figure 6. Scheme of the test area division into strips without overlay (methodology v2.1).

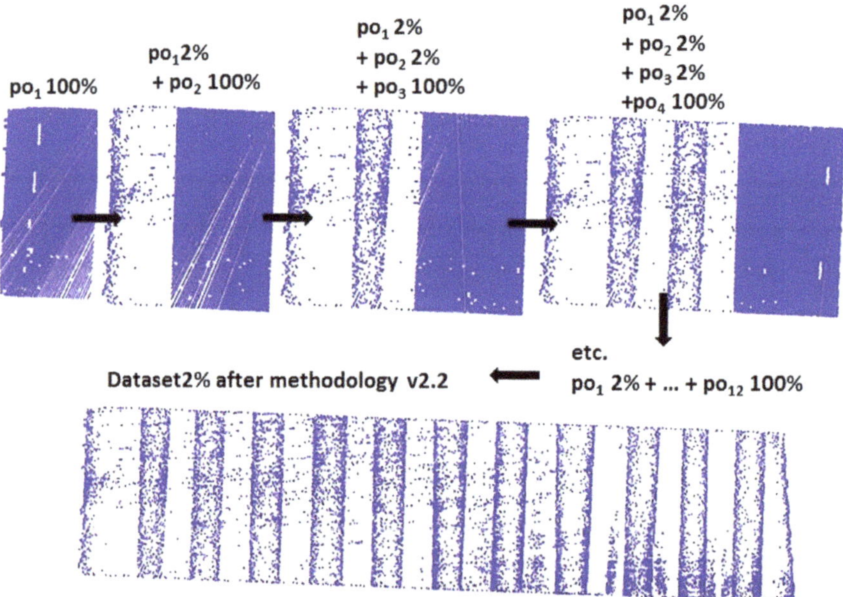

Figure 7. Scheme of test area division into strips with overlay (methodology v2.2).

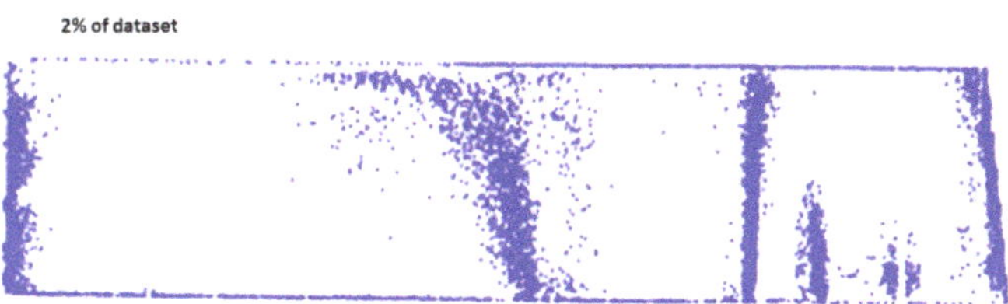

Figure 8. Optimized test area (13,976 points).

In the case of reducing the whole data set presented in Figure 8, we observed that the points remained in characteristic places of the studied area.

The statistical characteristics of the datasets obtained by methodology v2 are shown in Table 2.

Table 2. Statistical characteristics for datasets in methodology v2. (Where: H—height, R—range, STD—standard deviation).

		Number of Points	H min. [m]	H max. [m]	R [m]	STD [m]
	Optimized dataset	13976	15.10	23.57	8.47	2.86
Strips without overlay	p_1	1091	22.86	23.53	0.67	0.11
	p_2	702	22.86	23.57	0.71	0.13
	p_3	888	22.85	23.57	0.72	0.13
	p_4	1071	22.75	23.65	0.90	0.13
	p_5	1055	22.72	23.52	0.80	0.14
	p_6	1111	22.05	23.35	1.30	0.25
	p_7	1403	20.84	23.10	2.26	0.55
	p_8	1710	19.50	22.16	2.66	0.78
	p_9	1605	17.86	20.38	2.52	0.71
	p_{10}	1066	17.40	18.76	1.36	0.36
	p_{11}	1116	16.36	18.00	1.64	0.50
	p_{12}	1079	15.10	17.01	1.91	0.46
Strips with overlay	po_1	1446	22.86	23.54	0.68	0.13
	po_2	1132	22.86	23.60	0.74	0.13
	po_3	1444	22.81	23.65	0.84	0.14
	po_4	1633	22.77	23.65	0.88	0.14
	po_5	1619	22.42	23.52	1.10	0.21
	po_6	1794	21.56	23.38	1.82	0.42
	po_7	2321	20.26	23.11	2.85	0.86
	po_8	2456	18.76	22.19	3.43	1.08
	po_9	2018	17.71	20.42	2.71	0.80
	po_{10}	1691	16.92	18.77	1.85	0.62
	po_{11}	1689	15.91	18.01	2.10	0.65
	po_{12}	1094	15.10	16.96	1.86	0.46

The average data acquisition time in strips of 20 m at a measuring unit speed of 4 knots was about 20 s. The reduction within strips without overlay took 4–7 s, whereas for strips with overlay it took 6–9 s. The data processing time was much faster than for that of obtaining one strip.

3. Results

Each dataset representing strips p_i and po_i was used for DTM generation. The DTMs generated for strips p_1, p_2, and p_3 are shown in Figures 9–11, respectively. Next to DTMs, corresponding to them isoline maps are attached. They show, how the fragment of measured bottom of the lake look alike, when methodologies v1.1 and v2.1 were applied.

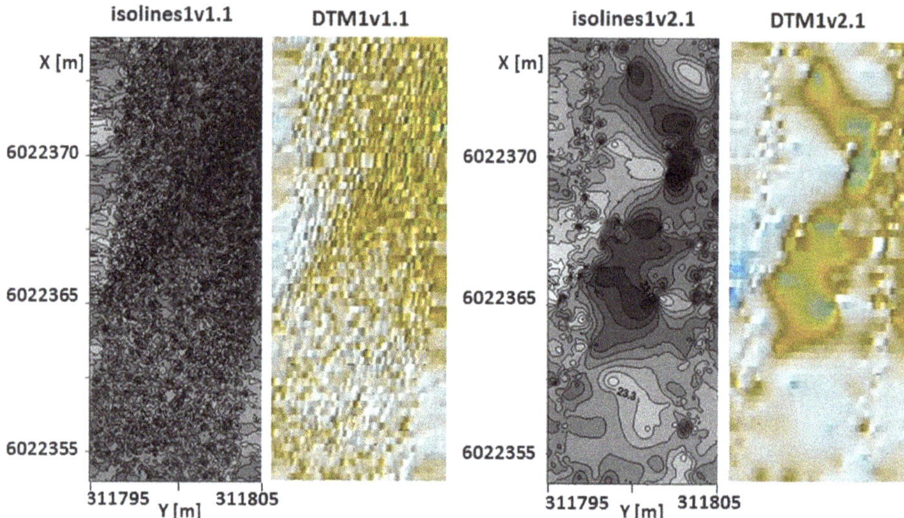

Figure 9. Isolines1v1.1 and digital terrain model (DTM)1v1.1, and isolines1v2.1 and DTM1v2.1 generated for strip 1 (p_1) by methodologies v1.1 and v2.1, respectively.

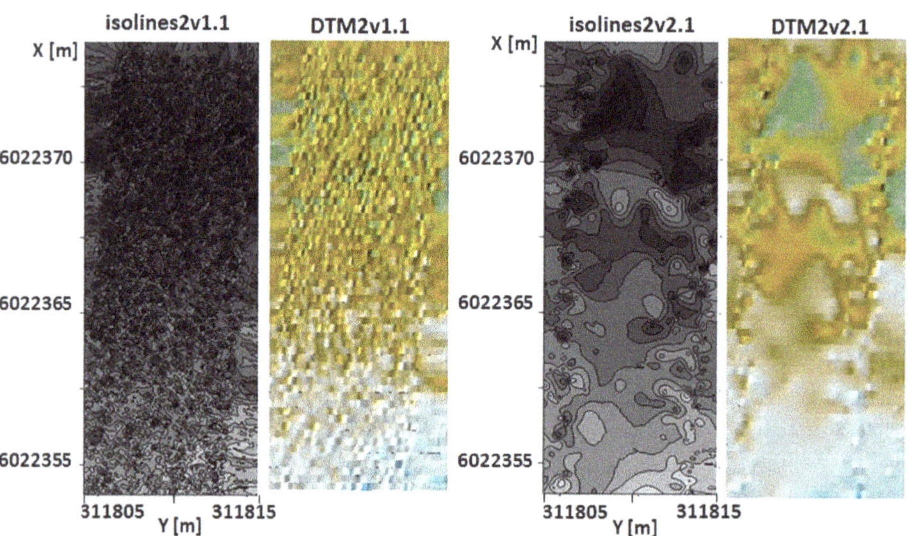

Figure 10. Isolines2v1.1 and DTM2v1.1, and isolines2v2.1 and DTM2v2.1 generated for strip 2 (p_2) by methodologies v1.1 and v2.1, respectively.

The generated DTMs and isolines maps were more readable in the case of v2.1. Fewer data has made the isoline image easier to read. Visibility of places with great depths was definitely better. Therefore, it was easier to assess the nature of the bottom from the DTMs generated by methodology v2.1. The statistical characteristics of the DTMs are presented in Table 3. As can be seen, DTMs from both methodologies v1.1 and v2.1 do not show significant statistical differences. In height, observed differences usually were about 2–3cm. For DTM4 and DTM6 they equaled −6 cm and 5 cm, respectively. This may indicated, on existence of some items on the bottom of the lake, that reduction allowed us to notice. The standard deviations calculated for DTMs generated from original dataset was usually smaller, about 1 cm in comparison to standard deviations corresponding to DTMs obtained from reduced datasets.

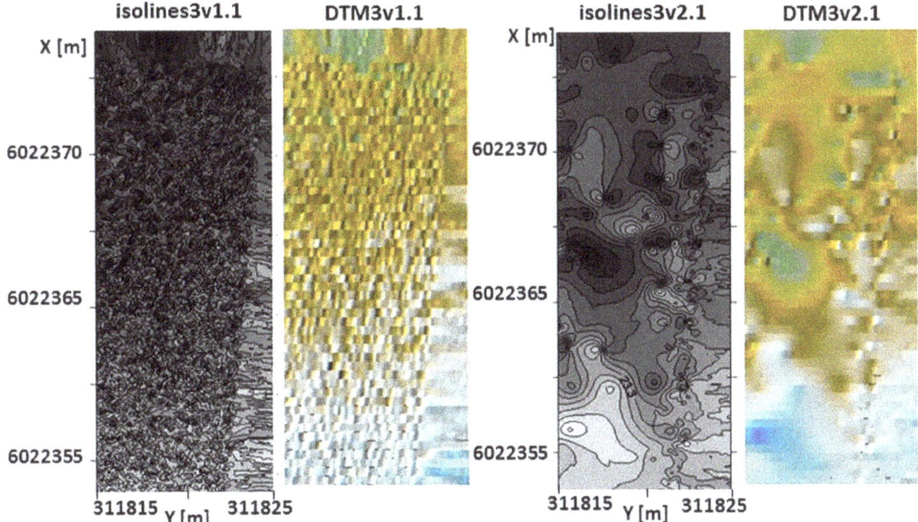

Figure 11. Isolines3v1.1 and DTM3v1.1, and isolines3v2.1 and DTM3v2.1 generated for strip 3 (p_3) by methodologies v1.1 and v2.1, respectively.

Table 3. Statistical characteristics for DTMmv1.1 and DTMmv2.1. (Where: H—height, STD—standard deviation).

DTM	H min. [m]	H max. [m]	STD [m]	Time [s] for Generated DTMs
DTM1v1.1	22.88	23.48	0.08	14
DTM1v2.1	22.85	23.47	0.09	14
DTM2v1.1	22.87	23.57	0.12	12
DTM2v2.1	22.87	23.53	0.12	10
DTM3v1.1	22.83	23.54	0.13	11
DTM3v2.1	22.83	23.54	0.14	9
DTM4v1.1	22.80	23.54	0.13	15
DTM4v2.1	22.79	23.60	0.12	12
DTM5v1.1	22.67	23.51	0.12	13
DTM5v2.1	22.65	23.51	0.13	11
DTM6v1.1	22.70	23.38	0.22	14
DTM6v2.1	22.67	23.33	0.24	10
DTM7v1.1	20.79	23.12	0.44	13
DTM7v2.1	20.77	23.12	0.47	9
DTM8v1.1	19.43	22.09	0.58	14
DTM8v2.1	19.45	22.09	0.59	10
DTM9v1.1	17.99	20.40	0.55	16
DTM9v2.1	17.96	20.37	0.55	12
DTM10v1.1	17.35	18.76	0.31	12
DTM10v2.1	17.37	18.74	0.32	10
DTM11v1.1	16.37	18.02	0.39	12
DTM11v2.1	16.36	17.99	0.40	9
DTM12v1.1	15.16	17.01	0.29	13
DTM12v2.1	15.15	17.01	0.29	10

The total generation time for DTMv1.1 was 159 s, whereas that for DTMv2.1 was 126 s.

The DTMs generated for strips p_1, p_2, and p_3 are presented Figures 11–14.

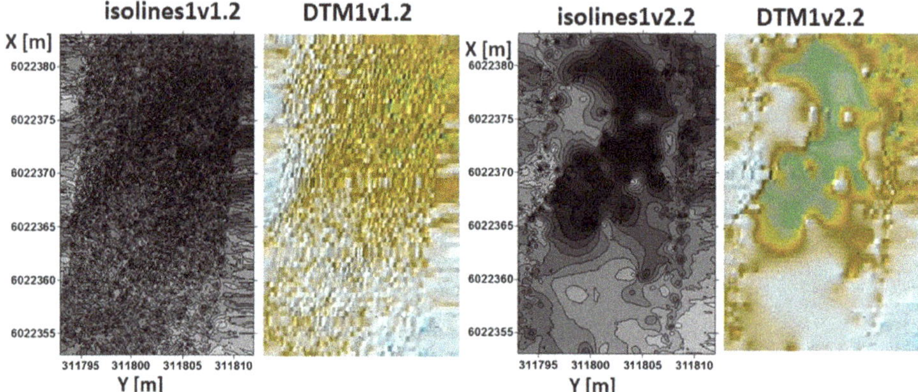

Figure 12. Isolines1v1.2 and DTM1v1.2, and isolines1v2.2 and DTM1v2.2 generated for strip 1 (po_1) by methodologies v1.2 and v2.2, respectively.

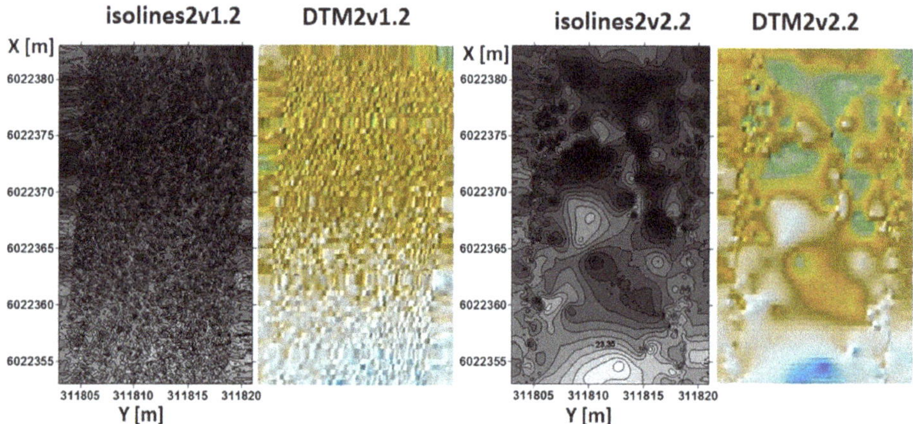

Figure 13. Isolines2v1.2 and DTM2v1.2, and isolines2v2.2 and DTM2v2.2 generated for strip 2 (po_2) by methodologies v1.2 and v2.2, respectively.

Analyzing Figures 12–14, it can be stated, as in the case of v2.1, that methodology v2.2 gave better results in terms of visibility and effectiveness of generated isolines maps and DTMs. In the figures showing the results of processing with the new v2.2 methodology, it was easier to read shallow and deep places. Methodology v1.2 figures were hard to read, and the isolines map were difficult to analyze.

The statistical characteristics of the DTMs are presented in Table 4.

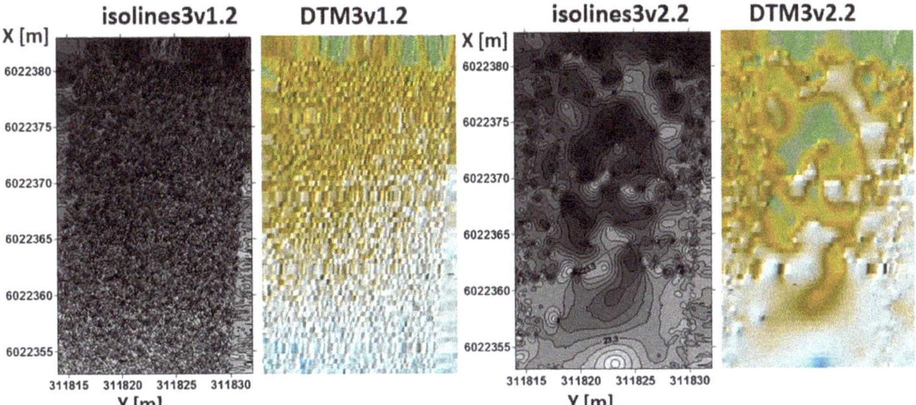

Figure 14. Isolines3v1.2 and DTM3v1.2, and isolines3v2.2 and DTM3v2.2 generated for strip 3 (po_3) by methodologies v1.2 and v2.2, respectively.

Table 4. Statistical characteristics for DTMmv1.2 and DTMmv2.2. (Where: H—height, STD—standard deviation).

DTM	H min. [m]	H max. [m]	STD [m]	Time [s] for Generated DTMS
DTM1v1.2	22.87	23.47	0.09	20
DTM1v2.2	22.86	23.50	0.12	19
DTM2v1.2	22.86	23.52	0.12	21
DTM2v2.2	22.87	23.53	0.13	18
DTM3v1.2	22.80	23.54	0.12	23
DTM3v2.2	22.82	23.55	0.12	19
DTM4v1.2	22.82	23.54	0.11	22
DTM4v2.2	22.79	23.55	0.13	18
DTM5v1.2	22.43	23.50	0.15	21
DTM5v2.2	22.46	23.51	0.16	17
DTM6v1.2	21.52	23.41	0.31	21
DTM6v2.2	21.53	23.39	0.32	16
DTM7v1.2	20.23	23.12	0.67	22
DTM7v2.2	20.25	23.09	0.68	16
DTM8v1.2	18.76	22.15	0.81	22
DTM8v2.2	18.76	22.16	0.82	15
DTM9v1.2	17.75	20.39	0.63	24
DTM9v2.2	17.71	20.42	0.65	16
DTM10v1.2	16.95	18.76	0.43	19
DTM10v2.2	16.96	18.75	0.45	14
DTM11v1.2	15.81	18.02	0.49	23
DTM11v2.2	15.83	18.00	0.50	16
DTM12v1.2	15.16	17.02	0.29	22
DTM12v2.2	15.17	16.99	0.28	16

The generation time for DTMv1.2 was 260 s, whereas that for DTMv2.2 was 201 s. Analyzing the statistical characteristics of DTMs obtained in methodologies v1.2 and v2.2, the trend can be observed. DTMs generated on the basis of the reduced dataset were about 1 cm higher than corresponding DTMs obtained from original measurement data.

DTM 100% and DTM 2% were also generated (Figures 15 and 16).

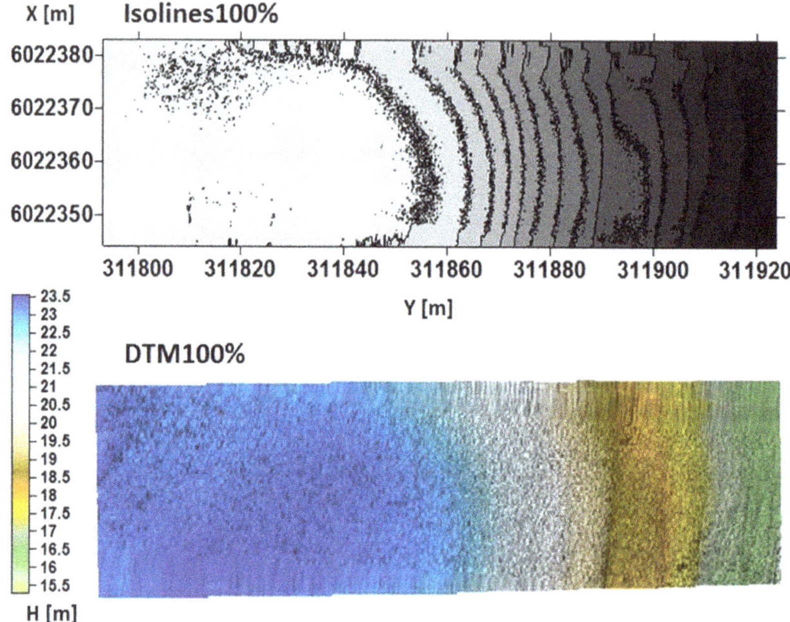

Figure 15. Isolines100% and DTM100%.

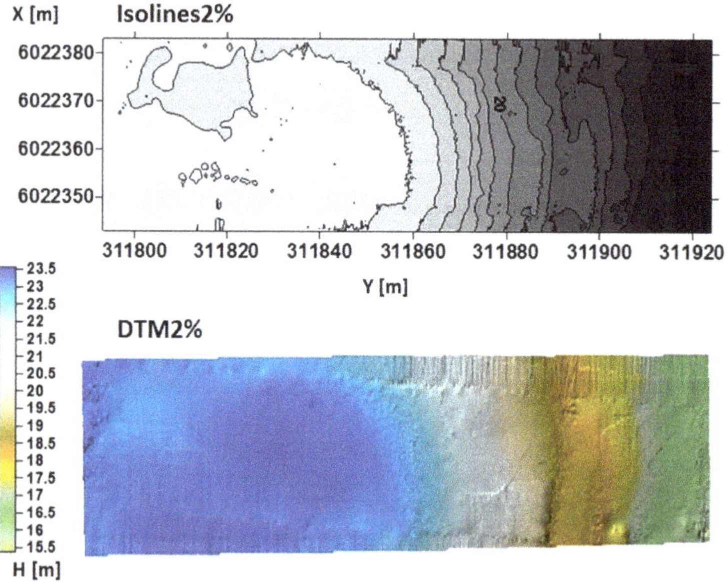

Figure 16. Isolines2% and DTM2%.

In Figures 15 and 16, the conclusion about a more readable isoline map was repeated. Figure Isolines2% (Figure 16) shows areas of depth in the test area better than Isolines100% in Figure 15.

The statistical characteristics of the isolines2%, DTM2%, isolines100% and DTM100% are presented in Table 5.

Table 5. Statistical characteristics for DTM100% and DTM2%. (Where: H—height, STD—standard deviation).

DTM	H min. [m]	H max. [m]	STD [m]	Time [s] for Generated DTMS
DTM100%	15.29	23.56	2.50	268
DTM2%	15.32	23.56	2.54	150

The total development time of the whole set was 508 s, consisting of 240 s acquisition time of the whole set and 268 s DTM100% generation time. The total development time of the reduced set was 410 s, consisting of 240 s acquisition time of the whole set, 20 s reduction of the set to 2% of the original set, and 150 s DTM2% generation time.

To assess how the DTM strips fit together, the height differences at the corresponding nodes between adjacent strips were calculated. The results for methodologies v1 and v2 are shown in Tables 6 and 7, respectively.

Table 6. Height differences between strips in methodology v1. (Where: H—height, STD—standard deviation).

		ΔH min. [m]	ΔH max. [m]	ΔH_{mean} [m]	STD [m]
Strips without overlay	p_1–p_2	−0.32	0.32	0.01	0.08
	p_2–p_3	−0.22	0.27	0.00	0.02
	p_3–p_4	−0.37	0.30	−0.01	0.08
	p_4–p_5	−0.49	0.28	−0.02	0.08
	p_5–p_6	−0.36	0.30	−0.02	0.07
	p_6–p_7	−0.40	0.32	−0.03	0.08
	p_7–p_8	−0.60	0.29	−0.09	0.13
	p_8–p_9	−0.58	0.27	−0.06	0.11
	p_9–p_{10}	−0.80	0.23	−0.06	0.11
	p_{10}–p_{11}	−0.52	0.19	−0.08	0.09
	p_{11}–p_{12}	−0.43	0.22	−0.03	0.07
Strips with overlay	po_1–po_2	−0.26	0.22	0.00	0.05
	po_2–po_3	−0.33	0.33	0.00	0.05
	po_3–po_4	−0.31	0.25	−0.01	0.04
	po_4–po_5	−0.26	0.24	0.00	0.04
	po_5–po_6	−0.28	0.24	0.00	0.04
	po_6–po_7	−0.37	0.28	−0.03	0.07
	po_7–po_8	−0.56	0.21	−0.04	0.10
	po_8–po_9	−0.43	0.20	−0.03	0.07
	po_9–po_{10}	−0.52	0.24	−0.03	0.07
	po_{10}–po_{11}	−0.38	0.25	−0.02	0.05
	po_{11}–po_{12}	−0.43	0.42	−0.01	0.06

Both methodologies gave similar results; the differences between almost all the statistical characteristics were close to zero. However, the difference for ΔH min. was larger (from −0.25 to 0.14 m) because some points representing an object with various values of H may be near the area where adjacent strips are coincident. Therefore, the content of the set processed by the OptD method was different. Nonetheless, data reduction by the OptD method made the main features in the modeled areas clearer (Figures 9–14).

Table 7. Height differences between strips in methodology v2. (Where: H—height, STD—standard deviation).

		ΔH min. [m]	ΔH max. [m]	ΔH_{mean} [m]	STD [m]
Strips without overlay	p_1–p_2	−0.21	0.34	0.03	0.07
	p_2–p_3	−0.29	0.28	0.00	0.11
	p_3–p_4	−0.22	0.25	0.00	0.07
	p_4–p_5	−0.34	0.21	−0.02	0.07
	p_5–p_6	−0.31	0.23	−0.02	0.09
	p_6–p_7	−0.24	0.24	−0.03	0.07
	p_7–p_8	−0.44	0.26	−0.08	0.11
	p_8–p_9	−0.55	0.28	−0.10	0.13
	p_9–p_{10}	−0.49	0.27	−0.04	0.08
	p_{10}–p_{11}	−0.40	0.18	−0.06	0.08
	p_{11}–p_{12}	−0.44	0.17	−0.04	0.07
Strips with overlay	po_1–po_2	−0.21	0.27	−0.01	0.08
	po_2–po_3	−0.20	0.05	−0.06	0.07
	po_3–po_4	−0.26	0.26	0.07	0.07
	po_4–po_5	−0.29	0.20	−0.02	0.08
	po_5–po_6	−0.20	0.20	0.03	0.06
	po_6–po_7	−0.31	0.33	−0.03	0.12
	po_7–po_8	−0.52	0.24	−0.11	0.14
	po_8–po_9	−0.70	0.28	−0.06	0.19
	po_9–po_{10}	−0.35	0.26	0.02	0.08
	po_{10}–po_{11}	−0.39	0.47	0.03	0.16
	po_{11}–po_{12}	−0.37	0.53	0.05	0.15

The statistical characteristics of the height differences for methodologies v1 and v2 are shown in Tables 8 and 9, respectively.

Table 8. Statistical characteristics for height differences between strips (methodology v1). (Where: H—height, STD—standard deviation).

		Methodology v1			
		ΔH min. [m]	ΔH max. [m]	ΔH_{mean} [m]	STD [m]
Strips without overlay	Min.	−0.80	0.19	−0.09	0.02
	Max.	−0.22	0.32	0.01	0.13
	Mean	−0.46	0.29	−0.03	0.08
	Standard deviation	0.16	0.04	0.03	0.03
Strips with overlay	Min.	−0.56	0.20	−0.04	0.04
	Max.	−0.26	0.42	0.00	0.10
	Mean	−0.38	0.26	−0.01	0.06
	Standard deviation	0.10	0.06	0.01	0.02

Table 10 shows the differences between statistical characteristics for height differences between methodologies v1 and v2.

The differences in statistical characteristics for height differences between using strips with and without overlay for methodology v2 are shown in Table 11. The majority of values were from −0.01 to 0.08 m, indicating that there were no significant differences between the approaches. However, the processing time for strips with overlay was longer than for strips without overlay. Therefore, the methodology based on data reduction and the variant that uses strips without overlays are suitable for depth area calculation.

Table 9. Statistical characteristics for height differences between strips (methodology v2). (H—height, STD—standard deviation).

		Methodology v2			
		ΔH min. [m]	ΔH max. [m]	ΔH_{mean} [m]	STD [m]
Strips without overlay	Min.	−0.55	0.17	−0.10	0.07
	Max.	−0.21	0.34	0.03	0.13
	Mean	−0.34	0.26	−0.03	0.09
	Standard deviation	0.12	0.05	0.04	0.02
Strips with overlay	Min.	−0.70	0.05	−0.11	0.06
	Max.	−0.20	0.53	0.07	0.19
	Mean	−0.35	0.28	−0.01	0.11
	Standard deviation	0.15	0.13	0.05	0.04

Table 10. Differences between statistical characteristics for height differences between methodologies. (Where: H—height, STD—standard deviation).

		Methodology v1—Methodology v2			
		ΔH min. [m]	ΔH max. [m]	ΔH_{mean} [m]	STD [m]
Strips without overlay	Min.	−0.25	0.02	0.01	−0.05
	Max.	−0.01	−0.02	−0.02	0.01
	Mean	−0.12	0.02	0.00	0.00
	Standard deviation	0.04	−0.01	0.00	0.01
Strips with overlay	Min.	0.14	0.15	0.07	−0.03
	Max.	−0.05	−0.11	−0.07	−0.09
	Mean	−0.03	−0.02	−0.01	−0.05
	Standard deviation	−0.05	−0.07	−0.04	−0.03

Table 11. Differences in statistical characteristics for height differences between strips with and without overlay. (Methodology v2). (where: H—height, STD—standard deviation).

	Strips with Overlay—Strips without Overlay			
	ΔH min. [m]	ΔH max. [m]	ΔH_{mean} [m]	STD [m]
Min.	−0.15	−0.12	−0.01	0.00
Max.	0.01	0.19	0.05	0.06
Mean	0.00	0.02	0.02	0.02
Standard deviation	0.04	0.08	0.02	0.02

4. Discussion

The new approach of methodology for processing MBES big data proposed by the authors was based on fragmentary 3D multibeam sonar data processing conducted in almost real time. All stages of standard methodology were performed not after acquisition of the whole dataset but in time, the fragments of data were acquired. While the one fragment of data was processed (execution of all stages: Reduction, DTM generation, isolines generation, analysis) the next fragment was obtained.

The most important step during the processing was reduction, because a reduced number of data allowed faster 3D bottom model generation, which can be compared with other types of data within terrain reference navigation.

Various tests can be found in the literature to speed up the calculation time of big data, e.g., parallel programming can be used with compute unified device architecture (CUDA). Using CUDA in processing of big datasets was tested, among others, by [47–49]. Within these works, tests on the possibility of using CUDA to generate a digital elevation model were performed. To speed up calculations there was also possibility to use artificial neural networks for modeling sea bottom shape, as these also continually implemented a surface approximation process [50,51]. These methods processed the entire dataset upon completion of the measurement. The use of these methods in almost real time was difficult, so in the new development methodology, we proposed using the OptD method.

The time needed to reduce the 3D multibeam sonar dataset based on OptD method with optimization criterion of 2% in strips was 4–7 s, whereas for strips with overlay it took 6–9 s. Such time can be considered as insignificant compared to the entire time needed for processing the whole data set. Moreover, the benefit of the reduction was a shorter time needed to generate the model. The times were as follows:

1. The total generation time for DTMv1.1 was 159 s, whereas that for DTMv2.1 was 126 s.
2. The generation time for DTMv1.2 was 260 s, whereas that for DTMv2.2 was 201 s.
3. The time needed for DTM generation was 268 s for DTM100% and 150 s for DTM2%.

Thus, the longest time was needed to generate a DTM100%. In all other cases, the time was shorter. The shortest time was needed for DTMv2.1 generation (the version with strips without overlay and with processing based on OptD method). The processing time depended on performed computer equipment and software. It is important, however, that the reduction algorithms, whose task is to speed up the development time, were uncomplicated and easy to implement.

The proposed solution also enables ongoing control during measurement. Acquired data were observed and initially analyzed in almost real time, therefore, if there was need, measurement can be repeated, completed or omitted in the selected area. Therefore, the presented approach can save time, labor, space on disks, etc.

5. Conclusions

For comparative navigation, data from MBES was processed by a new methodology which consisted of the OptD method to reduce the number of observations and generate DTMs representing measured fragments of the bottom of the area in almost real time. The data was then used to perform depth area calculations. The methodologies were based on fragmentary processing of observations organized in strips with or without overlay. Our analysis showed that using strips without overlay and with reduction by the OptD method (methodology 2.1) was an efficient, fast way to obtain data appropriate for 3D model generation that can be compared with a reference chart, such as bENCs. A major advantage of our method is that only points containing relevant information about depth differences are used for DTM construction and unimportant points belonging to flat areas are omitted. The resulting depth model of the bottom forms the first layer of a multi-layered model of the reference image bottom, which in many methods there is the only one layer. For comparative navigation based on the depth model above the flat bottom, the system cannot determine the position and additional information is required. For example, subsequent layers could be a bottom object layer and a layer containing information about the type of bottom. The layer containing characteristic points and bottom objects will use the same reduced points as the depth layer, allowing the analysis of data in semi-real time.

The general conclusions can be formulated as follows:

1. The new methodology is dedicated for 3D multibeam sonar data.
2. The new approach consists of the following steps: Acquisition the fragment of data, reducing data, and 3D model generation.
3. At the same time, the one fragment of data was processed with a new methodology, the next fragment of data was measured. This approach allows fast processing.

4. The generated DTMs or isolines maps can be simultaneously compared with existing maps (for example bENCs).
5. The time needed for fragmentary processing of 3D multibeam sonar data is shorter than the time needed for processing the whole data set.
6. The navigator has full control over the number of observations and the obtained DTMs are of good quality. In the case of isolines, mapping the obtained results shows that isolines generated by way of the OptD method are more readable and these isolines present more visible depths.

Author Contributions: Conceptualization, A.S. and A.S.-Ż.; methodology, W.B.-B.; bibliography review, A.S. and W.B.-B.; acquisition, analysis, and interpretation of data, A.S., W.M. and M.W.; writing—original draft preparation, A.S., A.S.-Ż. and W.B.-B.; writing—review and editing, A.S., W.M. and M.W.

Funding: This study was funded by the European Regional Development Fund under the 2014–2020 Operational Programme Smart Growth as part of the project "Developing of autonomous/remote operated surface platform dedicated hydrographic measurements on restricted reservoirs" implemented as part of the National Centre for Research and Development competition, INNOSBZ.

Acknowledgments: National Centre for Research and Development No. POIR.01.02.00-00-0074/16.

Conflicts of Interest: The author(s) declare(s) that they have no conflicts of interest regarding the publication of this paper.

References

1. Chen, P.; Li, Y.; Su, Y.; Chen, X.; Jiang, Y. Review of AUV Underwater Terrain Matching Navigation. *J. Navig.* **2015**, *68*, 1155–1172. [CrossRef]
2. Chen, P.-Y.; Li, Y.; Su, Y.-M.; Chen, X.-L.; Jiang, Y.-Q. Underwater terrain positioning method based on least squares estimation for AUV. *China Ocean Eng.* **2015**, *29*, 859–874. [CrossRef]
3. Claus, B.; Bachmayer, R. Terrain-aided Navigation for an Underwater Glider. *J. Field Robot.* **2015**, *32*, 935–951. [CrossRef]
4. Hagen, O.; Anonsen, K.; Saebo, T. Toward Autonomous Mapping with AUVs—Line-to-Line Terrain Navigation. In Proceedings of the Oceans 2015-MTS/IEEE Washington, Washington, DC, USA, 19–22 October 2015.
5. Jung, J.; Li, J.; Choi, H.; Myung, H. Localization of AUVs using visual information of underwater structures and artificial landmarks. *Intell. Serv. Robot.* **2017**, *10*, 67–76. [CrossRef]
6. Salavasidis, G.; Harris, C.; McPhail, S.; Phillips, A.B.; Rogers, E. Terrain Aided Navigation for Long Range AUV Operations at Arctic Latitudes. In Proceedings of the 2016 IEEE/OES Autonomous Underwater Vehicles (AUV), Tokyo, Japan, 6–9 November 2016; pp. 115–123.
7. Li, Y.; Ma, T.; Wang, R.; Chen, P.; Zhang, Q. Terrain Correlation Correction Method for AUV Seabed Terrain Mapping. *J. Navig.* **2017**, *70*, 1062–1078. [CrossRef]
8. Dong, M.; Chou, W.; Fang, B. Underwater Matching Correction Navigation Based on Geometric Features Using Sonar Point Cloud Data. *Sci. Program.* **2017**, *2017*, 7136702. [CrossRef]
9. Song, Z.; Bian, H.; Zielinski, A. Application of acoustic image processing in underwater terrain aided navigation. *Ocean Eng.* **2016**, *121*, 279–290. [CrossRef]
10. Ramesh, R.; Jyothi, V.B.N.; Vedachalam, N.; Ramadass, G.; Atmanand, M. Development and Performance Validation of a Navigation System for an Underwater Vehicle. *J. Navig.* **2016**, *69*, 1097–1113. [CrossRef]
11. Li, Y.; Ma, T.; Chen, P.; Jiang, Y.; Wang, R.; Zhang, Q. Autonomous underwater vehicle optimal path planning method for seabed terrain matching navigation. *Ocean Eng.* **2017**, *133*, 107–115. [CrossRef]
12. Li, Y.; Ma, T.; Wang, R.; Chen, P.; Shen, P.; Jiang, Y. Terrain Matching Positioning Method Based on Node Multi-information Fusion. *J. Navig.* **2017**, *70*, 82–100. [CrossRef]
13. Stuntz, A.; Kelly, J.S.; Smith, R.N. Enabling Persistent Autonomy for Underwater Gliders with Ocean Model Predictions and Terrain-Based Navigation. *Front. Robot. AI* **2016**, *3*, 23. [CrossRef]
14. Wang, L.; Yu, L.; Zhu, Y. Construction Method of the Topographical Features Model for Underwater Terrain Navigation. *Pol. Marit. Res.* **2015**, *22*, 121–125. [CrossRef]

15. Wei, F.; Yuan, Z.; Zhe, R. UKF-Based Underwater Terrain Matching Algorithms Combination. In Proceedings of the 2015 International Industrial Informatics and Computer Engineering Conference, Xi'an, China, 10–11 January 2015; pp. 1027–1030.
16. Zhou, L.; Cheng, X.; Zhu, Y. Terrain aided navigation for autonomous underwater vehicles with coarse maps. *Meas. Sci. Technol.* **2016**, *27*, 095002. [CrossRef]
17. Zhou, L.; Cheng, X.; Zhu, Y.; Dai, C.; Fu, J. An Effective Terrain Aided Navigation for Low-Cost Autonomous Underwater Vehicles. *Sensors* **2017**, *17*, 680. [CrossRef] [PubMed]
18. Zhou, L.; Cheng, X.; Zhu, Y.; Lu, Y. Terrain Aided Navigation for Long-Range AUVs Using a New Bathymetric Contour Matching Method. In Proceedings of the 2015 IEEE International Conference on Advanced Intelligent Mechatronics (AIM), Busan, Korea, 7–11 July 2015.
19. Calder, B.R.; Mayer, L.A. Automatic processing of high-rate, high-density multibeam echosounder data. *Geochem. Geophys. Geosyst.* **2003**, *4*, 1048. [CrossRef]
20. Kulawiak, M.; Lubniewski, Z. Processing of LiDAR and multibeam sonar point cloud data for 3D surface and object shape reconstruction. In Proceedings of the 2016 Baltic Geodetic Congress (BGC Geomatics), Gdańsk, Poland, 2–4 June 2016. [CrossRef]
21. Maleika, W. Moving Average Optimization in Digital Terrain Model Generation Based on Test Multibeam Echosounder Data. *Geo Mar. Lett.* **2015**, *35*, 61–68. [CrossRef]
22. Maleika, W. The Influence of the Grid Resolution on the Accuracy of the Digital Terrain Model Used in Seabed Modelling. *Mar. Geophys. Res.* **2015**, *36*, 35–44. [CrossRef]
23. Wlodarczyk-Sielicka, M. Interpolating Bathymetric Big Data for an Inland Mobile Navigation System. *Inf. Technol. Control.* **2018**, *47*, 338–348. [CrossRef]
24. Wlodarczyk-Sielicka, M.; Wawrzyniak, N. Problem of Bathymetric Big Data Interpolation for Inland Mobile Navigation System. In *Communications in Computer and Information Science, Proceedings of the 23rd International Conference on Information and Software Technologies (ICIST 2017), Druskininkai, Lithuania, 12–14 October 2017*; Springer: Cham, Switzerland, 2017; Volume 756, pp. 611–621. [CrossRef]
25. Wlodarczyk-Sielicka, M.; Lubczonek, J. The Use of an Artificial Neural Network to Process Hydrographic Big Data during Surface Modeling. *Computer* **2019**, *8*, 26. [CrossRef]
26. Rezvani, M.-H.; Sabbagh, A.; Ardalan, A.A. Robust Automatic Reduction of Multibeam Bathymetric Data Based on M-estimators. *Mar. Geod.* **2015**, *38*, 327–344. [CrossRef]
27. Yang, F.; Li, J.; Han, L.; Liu, Z. The filtering and compressing of outer beams to multibeam bathymetric data. *Mar. Geophys. Res.* **2013**, *34*, 17–24. [CrossRef]
28. Zhang, T.; Xu, X.; Xu, S. Method of establishing an underwater digital elevation terrain based on kriging interpolation. *Measurement* **2015**, *63*, 287–298. [CrossRef]
29. Wlodarczyk-Sielicka, M. Importance of Neighborhood Parameters During Clustering of Bathymetric Data Using Neural Network. In *Communications in Computer and Information Science, Proceedings of the 22nd International Conference on Information and Software Technologies (ICIST 2016), Druskininkai, Lithuania, 13–15 October 2016*; Springer: Cham, Switzerland, 2016; Volume 639, pp. 441–452. [CrossRef]
30. Lubczonek, J.; Borawski, M. A New Approach to Geodata Storage and Processing Based on Neural Model of the Bathymetric Surface. In Proceedings of the 2016 Baltic Geodetic Congress (BGC Geomatics), Gdansk, Poland, 2–4 June 2016; pp. 1–7. [CrossRef]
31. Specht, C.; Świtalski, E.; Specht, M. Application of an Autonomous/Unmanned Survey Vessel (ASV/USV) in Bathymetric Measurements. *Pol. Marit. Res.* **2017**, *24*, 36–44. [CrossRef]
32. Moszynski, M.; Chybicki, A.; Kulawiak, M.; Lubniewski, Z. A novel method for archiving multibeam sonar data with emphasis on efficient record size reduction and storage. *Pol. Marit. Res.* **2013**, *20*, 77–86. [CrossRef]
33. Kogut, T.; Niemeyer, J.; Bujakiewicz, A. Neural networks for the generation of sea bed models using airborne lidar bathymetry data. *Geod. Cartogr.* **2016**, *65*, 41–54. [CrossRef]
34. Aykut, N.O.; Akpınar, B.; Aydın, Ö. Hydrographic data modeling methods for determining precise seafloor topography. *Comput. Geosci.* **2013**, *17*, 661–669. [CrossRef]
35. Blaszczak-Bak, W. New Optimum Dataset method in LiDAR processing. *Acta Geodyn. Geomater.* **2016**, *13*, 379–386. [CrossRef]
36. Błaszczak-Bąk, W.; Koppanyi, Z.; Toth, C. Reduction Method for Mobile Laser Scanning Data. *ISPRS Int. J. Geo Inf.* **2018**, *7*, 285. [CrossRef]

37. Błaszczak-Bąk, W.; Sobieraj-Żłobińska, A.; Kowalik, M. The OptD-multi method in LiDAR processing. *Meas. Sci. Technol.* **2017**, *28*, 75009. [CrossRef]
38. Kazimierski, W.; Wlodarczyk-Sielicka, M. Technology of Spatial Data Geometrical Simplification in Maritime Mobile Information System for Coastal Waters. *Pol. Marit. Res.* **2016**, *23*, 3–12. [CrossRef]
39. Stateczny, A.; Gronska-Sledz, D.; Motyl, W. Precise Bathymetry as a Step Towards Producing Bathymetric Electronic Navigational Charts for Comparative (Terrain Reference) Navigation. *J. Navig.* **2019**. [CrossRef]
40. Borkowski, P.; Pietrzykowski, Z.; Magaj, J.; Mąka, M. Fusion of data from GPS receivers based on a multi-sensor Kalman filter. *Transp. Probl.* **2008**, *3*, 5–11.
41. Donovan, G.T. Position Error Correction for an Autonomous Underwater Vehicle Inertial Navigation System (INS) Using a Particle Filter. *IEEE J. Ocean. Eng.* **2012**, *37*, 431–445. [CrossRef]
42. Wawrzyniak, N.; Stateczny, A. MSIS Image Positioning in Port Areas with the Aid of Comparative Navigation Methods. *Pol. Marit. Res.* **2017**, *24*, 32–41. [CrossRef]
43. Ping DSP, Products Description. Available online: http://www.pingdsp.com/3DSS-DX-450 (accessed on 9 May 2019).
44. Stateczny, A.; Wlodarczyk-Sielicka, M.; Gronska, D.; Motyl, W. Multibeam Echosounder and Lidar in Process of 360° Numerical Map Production for Restricted Waters with Hydrodron. In Proceedings of the 2018 Baltic Geodetic Congress (BGC Geomatics) Gdansk, Olsztyn, Poland, 21–23 June 2018. [CrossRef]
45. Douglas, D.; Peucker, T. Algorithms for the reduction of the number of points required to represent a digitized line or its caricature. *Cartographica* **1973**, *10*, 112–122. [CrossRef]
46. Fei, L.; Jin, H. A three-dimensional Douglas–Peucker algorithm and its application to automated generalization of DEMs. *Int. J. Geogr. Inf. Sci.* **2009**, *23*, 703–718. [CrossRef]
47. Zeng, X.; He, W. GPGPU Based Parallel processing of Massive LiDAR Point Cloud. In Proceedings of the MIPPR 2009: Medical Imaging, Parallel Processing of Images, and Optimization Techniques. International Society for Optics and Photonics, Yichang, China, 30 October–1 November 2009; Volume 7497.
48. Chen, Y. High Performance Computing for Massive LiDAR Data Processing with Optimized GPU Parallel Programming. Master's Thesis, The University of Texas at Dallas, Richardson, TX, USA, 2012.
49. Cao, J.; Cui, H.; Shi, H.; Jiao, L. Big Data: A Parallel Particle Swarm Optimization-Back-Propagation Neural Network Algorithm Based on MapReduce. *PLoS ONE* **2016**, *11*, e0157551. [CrossRef] [PubMed]
50. Liu, S.; Wang, L.; Liu, H.; Su, H.; Li, X.; Zheng, W. Deriving Bathymetry from Optical Images with a Localized Neural Network Algorithm. *IEEE Trans. Geosci. Remote Sens.* **2018**, *56*, 5334–5342. [CrossRef]
51. Lubczonek, J. Hybrid neural model of the sea bottom surface. In *Lecture Notes in Computer Science, Proceedings of the International Conference on Artificial Intelligence and Soft Computing (ICAISC), Zakopane, Poland, 7–11 June 2004*; Springer: Berlin/Heidelberg, Germany, 2004; Volume 3070, pp. 1154–1160.

Article

Efficient Algorithm for SAR Refocusing of Ground Fast-Maneuvering Targets

Jun Wan [1], Yu Zhou [1,*], Linrang Zhang [1], Zhanye Chen [2] and Hengli Yu [1]

1 National Laboratory of Radar Signal Processing, Xidian University, Xi'an 710071, China; xidianwanjun@163.com (J.W.); lrzhang@xidian.edu.cn (L.Z.); xdhenry@163.com (H.Y.)
2 School of Microelectronics and Communication Engineering, Chongqing University, Chongqing 400044, China; czy@cqu.edu.cn
* Correspondence: zhouyu@mail.xidian.edu.cn; Tel.: +86-029-88203414

Received: 16 August 2019; Accepted: 19 September 2019; Published: 22 September 2019

Abstract: The synthetic aperture radar (SAR) image of moving targets will defocus due to the unknown motion parameters. For fast-maneuvering targets, the range cell migration (RCM), Doppler frequency migration and Doppler ambiguity are complex problems. As a result, focusing of fast-maneuvering targets is difficult. In this work, an efficient SAR refocusing algorithm is proposed for fast-maneuvering targets. The proposed algorithm mainly contains three steps. Firstly, the RCM is corrected using sequence reversing, matrix complex multiplication and an improved second-order RCM correction function. Secondly, a 1D scaled Fourier transform is introduced to estimate the remaining chirp rate. Thirdly, a matched filter based on the estimated chirp rate is proposed to focus the maneuvering target in the range–azimuth time domain. The proposed method is computationally efficient because it can be implemented by the fast Fourier transform (FFT), inverse FFT and non-uniform FFT. A new deramp function is proposed to further address the serious problem of Doppler ambiguity. A spurious peak recognition procedure is proposed on the basis of the cross-term analysis. Simulated and real data processing results demonstrate the validity of the proposed target focusing algorithm and spurious peak recognition procedure.

Keywords: complex Doppler ambiguity; fast-maneuvering target refocusing; non-uniform FFT (NUFFT); 1D scaled Fourier transform (1D SCFT); synthetic aperture radar (SAR)

---

1. Introduction

Synthetic aperture radar (SAR) can image the scenes of interest during the day and night regardless of weather conditions, which attracts considerable attention worldwide [1–10]. SAR has been widely used in numerous remote sensing applications, such as marine observation, traffic monitoring and antiterrorism. The growing demand for surveillance of moving targets has made imaging these targets a major task for modern SAR systems [11–15]. Nevertheless, the unknown motion parameters between the SAR platform and the moving target result in range cell migration (RCM) and Doppler frequency migration (DFM) [16,17]. These factors lead to the defocused image of moving targets. Thus, the defocusing effects induced by the RCM and DFM should be effectively removed.

Several methods have been presented to remove the RCM with the SAR system. The Hough/Radon transforms [18,19] were utilised to search the trajectory and correct the RCM. However, they suffer from high computational complexity due to the searching of the trajectory. On this basis, the first-order keystone transform (FOKT) [20,21], second-order keystone transform (SOKT) [22,23] and Doppler keystone transform [24] were proposed to remove the corresponding RCM without knowing a priori knowledge of the moving target. Although these transforms avoid searching of the target trajectory, their performance is limited by the effects of DFM and Doppler ambiguity. Methods, such as FOKT-based methods [25,26], stationary phase-based methods [27,28], joint time–frequency analysis-based methods [17,29] and modified SOKT-based methods [30], were presented to deal with these issues. However, these methods ignore the acceleration motion and only consider the moving target with a low-order (i.e., second-order) phase model. The acceleration motion and third-order phase should be further considered for the fast-maneuvering target [16,31,32]. Thus, the aforementioned methods may be inappropriate.

The components of RCM and DFM become increasingly complex due to the acceleration motion and third-order phase. Different from the RCM and DFM for moving targets with a low-phase model, which only includes first-order RCM (FRCM), second-order RCM (SRCM) and linear DFM (LDFM), the third-order RCM (TRCM) and quadratic DFM (QDFM) should be included for fast-maneuvering targets. The first-order phase of the target signal induces the Doppler centre shift; then, the Doppler centre ambiguity emerges due to the limitation of pulse repetition frequency (PRF) for the SAR system [25,27]. The complex azimuth Doppler spectrum induced by Doppler centre shift and DFM may distribute into one, two or multiple PRF bands. When the azimuth Doppler spectrum occupies two or multiple PRF bands, the target spectrum split occurs, which induces the Doppler spectrum ambiguity. The TRCM, QDFM, Doppler centre blur and spectrum ambiguity lead to the difficulty in the focusing of fast-maneuvering targets.

The axis mapping-based coherently integrated cubic phase function (CICPF) method [31] was introduced in consideration of the acceleration motion and third-order phase. However, this method directly applies SOKT to remove the SRCM and ignores the complex Doppler spectrum ambiguity (i.e., Doppler spectrum spanning over two or multiple PRF bands). If the Doppler spectrum is not located entirely on one PRF band, then the target trajectory will split into multiple parts after performing this method. A SOKT-based generalised Hough-high-order ambiguity function (SOKT-GHHAF) method [32] was proposed to focus the maneuvering targets for dealing with the aforementioned issue. This method uses the operation of Doppler centre shifting by PRF/2 to eliminate the effect of Doppler spectrum split. However, if the target spectrum bandwidth is larger than PRF/2, then the operation of Doppler centre shifting by PRF/2 will be invalid, and the effect of Doppler spectrum split will still persist [25]. The parameter searching-based methods [16,33] were introduced without the effect of Doppler spectrum ambiguity. Although these algorithms are effective, they suffer from large computational complexity induced by a brute-force parameter searching procedure.

We present a new computationally efficient algorithm for refocusing of ground fast-maneuvering targets on the basis of the previous works. In this algorithm, the RCM is corrected using sequence reversing, matrix complex multiplication and an improved SRCM correction function in the range frequency and azimuth slow time domain. The Doppler centre shift is removed simultaneously. Then, a 1D scaled Fourier transform (SCFT) with the constant factor ε is used to estimate the chirp rate of the target signal. Thereafter, a matched filter based on the estimated chirp rate is presented to focus the moving target in the range–azimuth time domain. In addition, a new deramp function with the constant factor φ is proposed to further deal with the Doppler spectrum ambiguity. Then, the operation of combining the new deramp function and SOKT is introduced to address the mismatch of the improved SRCM correction function. The cross-term interference for multiple targets is analysed, and a spurious peak recognition procedure is proposed. The simulated and real data processing results verify the proposed target focusing algorithm and spurious peak recognition procedure.

The main contributions of this work are listed as follows: (1) the proposed algorithm can achieve a well-focused result in the range–azimuth time domain because the acceleration motion and third-order phase of the fast-maneuvering target are considered; (2) the presented algorithm has low computational complexity given that it can be implemented by the fast Fourier transform (FFT), inverse FFT (IFFT) and non-uniform FFT (NUFFT); (3) two constant factors, namely, ε and φ, are introduced to expand the applicability of the proposed algorithm; (4) a new deramp function is introduced to further deal with the complex Doppler ambiguity; and (5) a spurious peak recognition procedure is presented to address the cross-term interference.

The rest of the paper is organised as follows: Section 2 provides the signal model and characteristics. Section 3 describes the proposed algorithm. Section 4 gives specific analysis related to the proposed algorithm. Section 5 presents the simulated and real data processing results. Section 6 gives the discussion of the proposed algorithm. Section 7 provides the final conclusions.

2. Signal Model and Characteristics

2.1. Signal Model

The motion geometry between the SAR platform under the side-looking strip-map mode and the ground maneuvering target on a slant-rang plane is shown in Figure 1. During the synthetic time T_a, the SAR platform flies with constant velocity v. The maneuvering target with cross-track velocity v_c, cross-track acceleration a_c, along-track velocity v_a and along-track acceleration a_a moves from point A to point B. R_0 and $R_s(t_n)$ denotes the nearest and instantaneous slant ranges between the SAR platform and the maneuvering target. t_n represents the azimuth slow time.

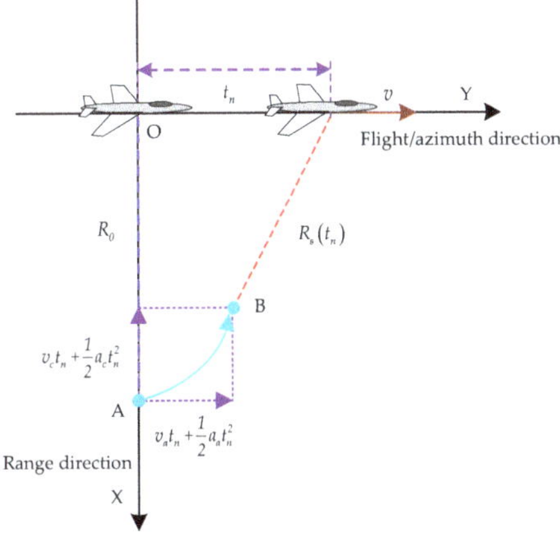

Figure 1. Motion geometry between the synthetic aperture radar platform and the ground maneuvering target on a slant-rang plane.

In accordance with the motion geometry described in Figure 1, $R_s(t_n)$ is expressed as

$$R_s(t_n) = \sqrt{\left(vt_n - v_a t_n - \frac{1}{2}a_a t_n^2\right)^2 + \left(R_0 - v_c t_n - \frac{1}{2}a_c t_n^2\right)^2}. \tag{1}$$

The instantaneous slant range $R_s(t_n)$ can be expanded on the basis of the Taylor series expansion. Considering the accuracy of the range model, a third-order range model is used as follows [16,31–33]:

$$R_s(t_n) \approx R_0 - v_c t_n + \frac{(v - v_a)^2 - R_0 a_c}{2R_0} t_n^2 + \frac{v_c (v - v_a)^2 + R_0 a_a (v_a - v)}{2R_0^2} t_n^3. \tag{2}$$

We suppose that the radar transmits the linear frequency modulation (LFM) signal with the form as follows [20]:

$$s_t(t) = \mathrm{rect}(\frac{t}{T_p}) \exp(j2\pi f_c t + j\pi\gamma t^2), \tag{3}$$

where t indicates the range time, rect$(\cdot)$ is the rectangle window function, T_p denotes the pulse length, γ is the chirp rate of the transmitted signal and f_c represents the carrier frequency. The received baseband signal is written as

$$s_{base}(t, t_n) = \sigma \text{rect}\left[\frac{t - 2R_s(t_n)/c}{T_p}\right] w_a(t_n) \exp\left\{j\pi\gamma\left[t - \frac{2R_s(t_n)}{c}\right]^2\right\} \exp\left[-j\frac{4\pi}{\lambda}R_s(t_n)\right], \tag{4}$$

where σ is the backscattering coefficient of the moving target, c denotes the speed of electromagnetic wave, $w_a(\cdot)$ is the azimuth window function and λ is the wavelength of the transmitted signal.

By substituting Equations (2) into (4) and performing the range compression [20,22], the received signal omitting the envelope in the range frequency and azimuth time domain yields

$$\begin{aligned} s_1(f, t_n) &= \text{rect}\left(\frac{f}{B}\right) w_a(t_n) \exp\left[-\frac{j4\pi}{c}(f + f_c)\right. \\ &\left.\cdot\left(R_0 - v_c t_n + \frac{(v-v_a)^2 - R_0 a_c}{2R_0} t_n^2 + \frac{v_c(v-v_a)^2 + R_0 a_a(v_a - v)}{2R_0^2} t_n^3\right)\right], \end{aligned} \tag{5}$$

where f denotes the range frequency variable and $B = \gamma T_p$ is the bandwidth of the transmitted signal.

After the range IFFT is applied to Equation (5), the received signal in the range and azimuth slow time domain is expressed as

$$\begin{aligned} s_1(t, t_n) &= \text{sinc}\left\{B\left[t - 2\left(R_0 - v_c t_n + \frac{(v-v_a)^2 - R_0 a_c}{2R_0} t_n^2 + \frac{v_c(v-v_a)^2 + R_0 a_a(v_a - v)}{2R_0^2} t_n^3\right)/c\right]\right\} \\ &\cdot w_a(t_n) \exp\left[-\frac{j4\pi}{\lambda}\left(R_0 - v_c t_n + \frac{(v-v_a)^2 - R_0 a_c}{2R_0} t_n^2 + \frac{v_c(v-v_a)^2 + R_0 a_a(v_a - v)}{2R_0^2} t_n^3\right)\right], \end{aligned} \tag{6}$$

where $\text{sinc}(x) = \sin(\pi x)/(\pi x)$ denotes the sinc function.

2.2. Signal Characteristics

In accordance with Equation (5), the range frequency variable f is coupled with the azimuth slow time variable t_n. Not only the coupling effects caused by the low-order terms, namely, the t_n- and t_n^2-term, but also those induced by the high-order one, namely, the t_n^3-term, exist. Therefore, the range position and Doppler frequency of the moving target change with the azimuth slow time.

As described in the sinc function term of Equation (6), the t_n-term induces FRCM, the t_n^2-term causes SRCM and the t_n^3-term leads to TRCM in the range dimension. According to the last exponential term of Equation (6), the t_n-term, t_n^2-term and t_n^3-term result in Doppler centre shift, LDFM and QDFM, respectively, in the Doppler frequency dimension. The RCM and DFM are severe for fast-maneuvering targets. This condition makes the trajectory span over multiple ranges and Doppler frequency cells. Thus, the complex RCM and DFM should be effectively removed to focus the moving target.

The target azimuth Doppler spectrum distribution must be further studied to obtain a well-focused result [34]. The Doppler centre shift of the fast-maneuvering target is larger than PRF/2, and the target shows Doppler centre blur. In the 2D spectrum dimension, the potential azimuth spectrum distributions caused by Doppler centre shift and DFM consist of the following cases. When $f_B < PRF/2$, where f_B denotes the azimuth spectrum bandwidth of the target signal, two azimuth spectrum distributions are introduced: case I: the azimuth spectrum is located entirely on one PRF band, as shown in Figure 2a, where f_{dc} represents the Doppler centre shift in the figure; case II: the azimuth spectrum spans over two PRF bands, as shown in Figure 2b. When $PRF/2 < f_B < PRF$, two other azimuth spectrum distributions are obtained: case III: the azimuth spectrum still occupies one PRF band, as displayed in Figure 2c; case IV: the azimuth spectrum also distributes into two PRF bands, as shown in Figure 2d. When $f_B > PRF$, as shown in Figure 2e, the azimuth spectrum distributes into several PRF bands. When the azimuth spectrum does not entirely occupy one PRF band, namely, cases II, IV and V, the target spectrum split will occur; this phenomenon induces the azimuth Doppler spectrum ambiguity [34,35].

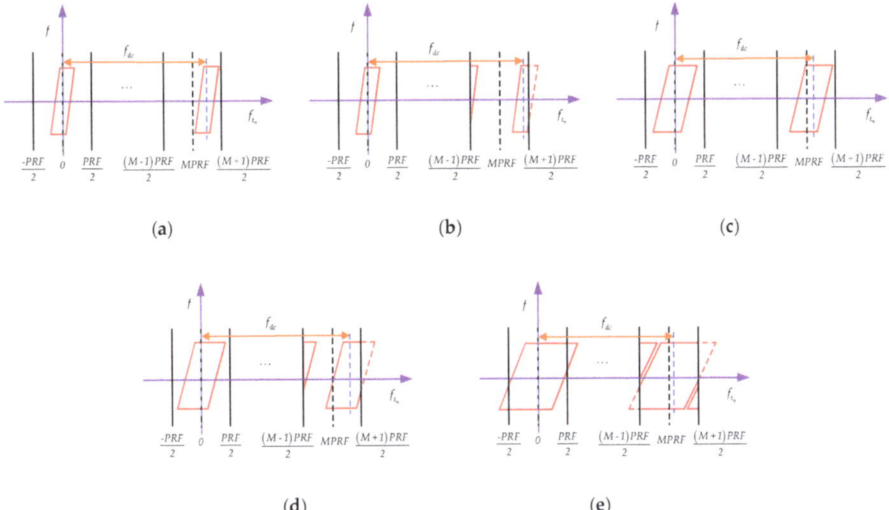

Figure 2. Potential azimuth Doppler spectrum distributions: (**a**) case I: $f_B < PRF/2$, spectrum entirely in one pulse repetition frequency (PRF) band; (**b**) case II: $f_B < PRF/2$, spectrum spanning two PRF bands; (**c**) case III: $PRF/2 < f_B < PRF$, spectrum entirely in one PRF band; (**d**) case IV: $PRF/2 < f_B < PRF$, spectrum spanning two PRF bands; (**e**) case V: $f_B > PRF$, spectrum spanning several PRF bands.

In summary, the complex RCM and DFM lead to severe integration loss and defocusing of the target image. In the existing methods, the Doppler ambiguity number searching [25,34] and Doppler centre shifting by PRF/2 operations [26,27,32] were usually used to deal with the Doppler centre blur and spectrum ambiguity, respectively. However, the Doppler ambiguity number searching operation increases the computational complexity. If the spectrum distribution belongs to case IV or V, then the Doppler centre shifting by PRF/2 operation will be invalid. Accordingly, the moving targets become difficult to be focused in the presence of complex Doppler ambiguity by applying transitional FOKT or SOKT-based [20–26,31,32] and stationary phase-based [27,28] methods. We present a new algorithm for refocusing of fast-maneuvering targets to deal with aforementioned problems.

3. Proposed Algorithm Description

The RCM, DFM, Doppler centre shift and Doppler ambiguity are the key issues for refocusing of fast-maneuvering targets according to the target signal properties described in the previous section. Therefore, a new fast algorithm is presented in this section.

3.1. RCM, Doppler Centre Shift and QDFM Compensation

In accordance with Equation (5), the discrete form of $s_1(f, t_n)$ omitting the azimuth window function is expressed as

$$\begin{aligned} s_1(m,n) &= \mathrm{rect}\left(\frac{m\Delta f}{B}\right)\exp\Big[-\frac{j4\pi}{c}(m\Delta f + f_c) \\ &\cdot\left(R_0 - v_c n\Delta t_n + \frac{(v-v_a)^2 - R_0 a_c}{2R_0} n^2 \Delta t_n^2 + \frac{v_c(v-v_a)^2 + R_0 a_a(v_a - v)}{2R_0^2} n^3 \Delta t_n^3\right)\Big], \end{aligned} \tag{7}$$

where $m (m = -M/2, -M/2+1, \cdots, -M/2-1, M/2)$ denotes the discrete range frequency number index related to continuous range frequency f, M is assumed to be an even integer, Δf denotes the range frequency sampling interval, $n (n = -N/2, -N/2+1, \cdots, N/2-1, N/2)$ represents the discrete slow time number index related to continuous slow time t_n, N is assumed to be an even integer and Δt_n indicates the pulse repetition interval.

Given the symmetrical property of the discrete slow time sequence, a new signal after reversing the discrete slow time sequence for each range frequency yields [36,37]

$$\begin{aligned} s_1(m,\overleftarrow{n}) &= s_1(m,-n) = \mathrm{rect}\left(\tfrac{m\Delta f}{B}\right)\exp\left[-\tfrac{j4\pi}{c}(m\Delta f + f_c)\right. \\ &\cdot\left.\left(R_0 + v_c n\Delta t_n + \tfrac{(v-v_a)^2 - R_0 a_c}{2R_0}n^2\Delta t_n^2 - \tfrac{v_c(v-v_a)^2 + R_0 a_a(v_a - v)}{2R_0^2}n^3\Delta t_n^3\right)\right], \end{aligned} \tag{8}$$

where "←" denotes the sequence reversing operation. As shown in Equations (7) and (8), the effects of t_n-term and t_n^3-term can be removed by multiplying Equation (7) by Equation (8). Thus, the corresponding result yields

$$\begin{aligned} s_2(m,n) &= s_1(m,n)\cdot s_1(m,\overleftarrow{n}) \\ &= \mathrm{rect}\left(\tfrac{m\Delta f}{B}\right)\exp\left[-\tfrac{j8\pi}{c}(m\Delta f + f_c)\left(R_0 + \tfrac{(v-v_a)^2 - R_0 a_c}{2R_0}n^2\Delta t_n^2\right)\right]. \end{aligned} \tag{9}$$

As described in Equation (9), the FRCM, TRCM and QDFM are effectively compensated. In the meantime, the Doppler centre blur and spectrum distributions belonging to cases II and IV are avoided given that the Doppler centre shift is effectively removed. Nevertheless, the effect induced by t_n^2-term still exists. Thus, the SRCM and FDFM should be effectively eliminated.

As verified in [16,33,38,39], the SRCM of the common metre-level range resolution SAR system depends on the SAR platform velocity given that the target along-track velocity and cross-track acceleration are considerably smaller than the velocity of SAR platform. Therefore, the SRCM can be removed by constructing the correction function on the basis of the SAR platform velocity v as long as the residual SRCM correction error is smaller than one range resolution bin. In accordance with Equation (9), an improved SRCM correction function is constructed as follows:

$$H_{RCM}(m,n) = \exp\left(\frac{j4\pi m\Delta f v^2 n^2 \Delta t_n^2}{cR_0}\right). \tag{10}$$

The SRCM correction error by applying the correction function in Equation (10) is obtained as

$$\Delta R_{RCM} = \left|\left[\frac{\left(v_a^2 - 2vv_a\right) - R_0 a_c}{4R_0}\right](T_a)^2\right|. \tag{11}$$

We suppose that a maneuvering target, which is denoted by Target A, is considered. The target parameters are $v_a = 30$ m/s, $a_a = -1$ m/s^2, $v_c = 32$ m/s and $a_c = -1$ m/s^2. The main radar parameters are $f_c = 10$ GHZ, $B = 80$ MHZ, $PRF = 1400$ HZ, $v = 250\ m/s$, $R_0 = 6000$ m and $T_a = 1.2$ s. The SRCM correction error ΔR_{RCM} is calculated as 0.486 m, which is smaller than one range resolution bin. Example A without noise is also presented. Figure 3a shows the target trajectory after range compression. The target suffers from severe RCM effect. As illustrated in Figure 3b, only the SRCM remains after FRCM and TRCM correction. Figure 3c exhibits the result of SRCM compensation by using the correction function in Equation (10). The SRCM is effectively removed, and the trajectory of the moving target is located on the same range bin. The residual SRCM correction error by using the SAR platform velocity can be ignored and the correction function in Equation (10) is considered effective for the common metre-level range resolution SAR system.

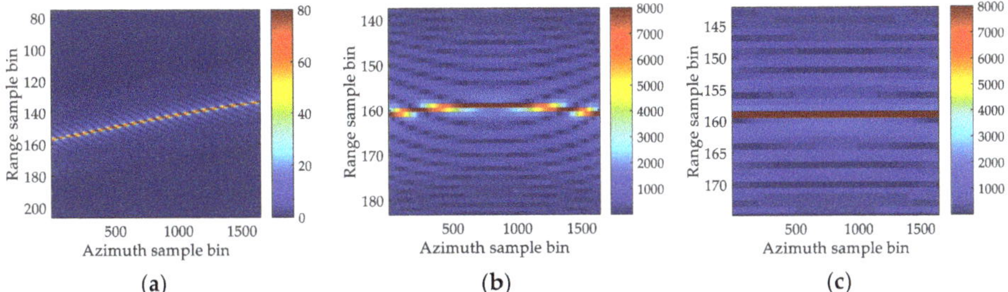

Figure 3. Results of Example A: (**a**) trajectory after range compression; (**b**) trajectory after first-order range cell migration and third-order range cell migration correction; (**c**) result of second-order range cell migration correction by using the correction function in Equation (10).

After multiplying Equation (9) by Equation (10), we have

$$\begin{aligned} s_3(m,n) &= s_2(m,n)\cdot H_{RCM}(m,n) \\ &\approx \mathrm{rect}\left(\frac{m\Delta f}{B}\right)\exp\left(-\frac{j8\pi}{c}(m\Delta f + f_c)R_0\right)\exp\left[-\frac{j8\pi}{\lambda}\frac{(v-v_a)^2 - R_0 a_c}{2R_0}n^2\Delta t_n^2\right]. \end{aligned} \tag{12}$$

After the range IFFT is applied to Equation (12), the continuous time expression is obtained as

$$s_3(t,t_n) = \mathrm{sinc}\left[B\left(t - \frac{4R_0}{c}\right)\right]\exp\left[-\frac{j8\pi}{\lambda}\frac{(v-v_a)^2 - R_0 a_c}{2R_0}t_n^2\right]. \tag{13}$$

According to Equation (13), the moving target is focused in the same range bin, and the target signal along azimuth slow time dimension can be modelled as a second-order phase signal (i.e., an LFM signal). The chirp rate (i.e., second-order phase coefficient) of the target signal in Equation (13) should be effectively estimated and compensated to focus the moving target.

3.2. Estimation of Chirp Rate by 1D SCFT

Several typical methods, such as, discrete chirp-Fourier transform (DCFT) [34,40], fractional Fourier transform (FRFT) [29,41], Lv's distribution (LVD) [26,42], and CICPF [17,31], have been proposed to estimate the chirp rate of the target signal in Equation (13). As for the parameter searching-based methods, DCFT and FRFT are computationally prohibitive due to the brute-force grid searching operation. Subsequently, the LVD and CICPF have been proposed without the searching process. However, these methods must transform the 1D LFM signal into a 2D parameter space to obtain the final estimation result. These methods still have a large computational burden due to the 2D data processing operation. Considering that the chirp rate of signal in Equation (13) has been doubled, the constant factor ε is introduced to expand the application scope. The corresponding 1D SCFT with constant factor ε yields

$$\begin{aligned} s_3\left(\tfrac{4R_0}{c}, f_{t_n^2}\right) &= \int s_3\left(\tfrac{4R_0}{c}, t_n\right)\exp\left(-j2\pi f_{t_n^2}\varepsilon t_n^2\right)dt_n^2 \\ &= \delta\left(f_{t_n^2} + \frac{2(v-v_a)^2 - 2R_0 a_c}{\varepsilon\lambda R_0}\right), \end{aligned} \tag{14}$$

where $s_3\left(\frac{4R_0}{c}, t_n\right) = \exp\left[-\frac{j8\pi}{\lambda}\frac{(v-v_a)^2 - R_0 a_c}{2R_0}t_n^2\right]$, $f_{t_n^2}$ is the azimuth scaled frequency variable corresponding to t_n^2 and $\delta(\cdot)$ denotes the Dirac delta function. The selection criterion of constant factor ε is discussed in Appendix A.

In accordance with Equation (14), the chirp rate is estimated as follows:

$$\frac{2(v-v_a)^2-2R_0a_c}{\lambda R_0}=-\varepsilon\hat{f}_{t_n^2}, \tag{15}$$

where $\hat{f}_{t_n^2}$ denotes the peak position in the azimuth scaled frequency domain.

Example B without noise is provided to validate the constant factor ε. We consider a target that is denoted by Target B. The parameters of Target B are as follows: $v_{a2}=-29.5\ \mathrm{m/s}$ and $a_{c2}=0.8\ \mathrm{m/s^2}$. The radar parameters are the same as those of Example A. Figure 4a,b show the result using the 1D SCFT with the constant factor $\varepsilon=1$. The defocusing result is obtained in Figure 4a,b because the chirp rate of the signal for Target B exceeds the scope of parameter estimation. Figure 4c,d depict the result using the 1D SCFT with the constant factor $\varepsilon=2$. Given that the constant factor $\varepsilon=2$ expands the parameter estimation scope, a clear peak appears in Figure 4c,d. Therefore, the results of Example B demonstrate the validity of the constant factor ε.

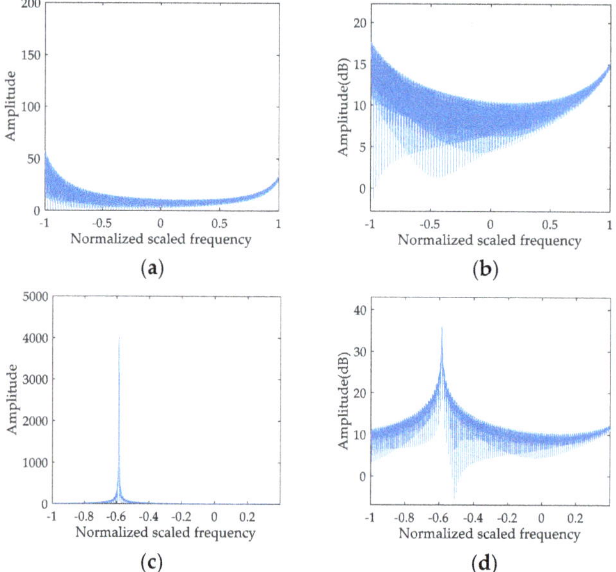

Figure 4. Results of Example B: (**a**) result using constant factor $\varepsilon=1$; (**b**) dB version of amplitude for Figure 4a; (**c**) result using constant factor $\varepsilon=2$; (**d**) dB version of amplitude for Figure 4c.

3.3. Azimuth Focusing by Matched Filter Based on Estimated Chirp Rate

The signal in Equation (9) is transformed into the range and azimuth frequency domain yields

$$s_2(f,f_{t_n})=\mathrm{rect}\left(\frac{f}{B}\right)\exp\left[-\frac{j8\pi}{c}(f+f_c)R_0\right]\exp\left\{j\pi\frac{cR_0}{4(f+f_c)\left[(v-v_a)^2-R_0a_c\right]}f_{t_n}^2\right\}. \tag{16}$$

By substituting Equation (15) into Equation (16), we have

$$s_2(f,f_{t_n})=\mathrm{rect}\left(\frac{f}{B}\right)\exp\left[-\frac{j8\pi}{c}(f+f_c)R_0\right]\exp\left\{-j\pi\frac{c}{2(f+f_c)\varepsilon\lambda\hat{f}_{t_n^2}}f_{t_n}^2\right\}. \tag{17}$$

In accordance with Equation (17), the matched filter based on the estimated chirp rate is constructed as follows:

$$H_1\left(f, f_{t_n}, \hat{f}_{t_n^2}\right) = \exp\left\{j\pi \frac{c}{2(f + f_c)\varepsilon\lambda \hat{f}_{t_n^2}} f_{t_n}^2\right\}. \tag{18}$$

After Equation (17) is multiplied by Equation (18) and 2D IFFT is performed, we have

$$s_4(t, t_n) = \text{sinc}\left[B\left(t - \frac{4R_0}{c}\right)\right]\text{sinc}(f_D t_n), \tag{19}$$

where f_D denotes the bandwidth of the signal in Equation (9).

As shown in Equation (19), the RCM, Doppler centre shift and DFM of the maneuvering target are effectively compensated. After 2D IFFT is performed, the moving target is refocused in the range–azimuth time domain; accordingly, the subsequent moving target processing operations are facilitated [22].

Figure 5 provides the flowchart of the presented algorithm. As shown in Figure 5, the proposed algorithm after range compression mainly includes three steps. The steps are summarised as follows:

(1) RCM correction

- (1.1) Apply the range FFT to the range compressed signal $s_1(t, t_n)$, and obtain $s_1(f, t_n)$.
- (1.2) Perform the sequence reversing operation to $s_1(f, t_n)$, and calculate Equation (9) to obtain $s_2(f, t_n)$.
- (1.3) Construct the SRCM correction function $H_{RCM}(f, t_n)$, and calculate Equation (12) to obtain $s_3(t, t_n)$.

(2) Estimate the chirp rate by using the 1D SCFT, and obtain $\hat{f}_{t_n^2}$.

(3) Azimuth focusing

- (3.1) Construct the azimuth match filter $H_1(f, f_{t_n}, \hat{f}_{t_n^2})$, and multiply Equation (16) by $H_1(f, f_{t_n}, \hat{f}_{t_n^2})$ to obtain $s_4(f, f_{t_n})$.
- (3.2) Perform 2D IFFT to $s_4(f, f_{t_n})$, and obtain the final focused result $s_4(t, t_n)$,

where step 1.1 is used to transform the range compressed signal into the range frequency domain, step 1.2 is applied to correct FRCM and TRCM, step 1.3 is performed to remove SRCM, step 3.1 is utilised to compensate LDFM and step 3.2 is used to obtain the final refocused result.

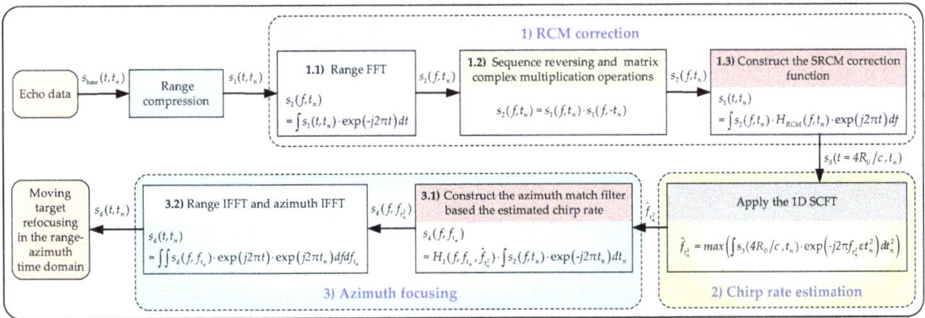

Figure 5. Flowchart of the presented algorithm.

4. Analysis Related to the Proposed Algorithm

4.1. Analysis of the SRCM Correction Function Mismatch

Considering that the SRCM correction error ΔR_{RCM} may be larger than one range resolution cell in some high-range resolution SAR systems, the mismatch of SRCM correction function in Equation (10) will appear. As for this case, we can utilise the SOKT to correct the SRCM. However, the Doppler bandwidth of Equation (9) is easily larger than one PRF band because the chirp rate has been doubled after the previous processing steps. Then, the target trajectory will split into multiple parts after directly using SOKT due to the Doppler spectrum ambiguity (i.e., the azimuth spectrum distribution belonging to case V). Therefore, the Doppler bandwidth of Equation (9) should be effectively compressed. In [25,34], the deramp functions are constructed to compress the azimuth spectrum. However, these deramp functions ignore the effects of the target unknown motion parameters; this condition leads to the performance degradation, especially for fast-maneuvering targets. In accordance with Equation (9), the Doppler bandwidth is expressed as follows:

$$f_D = \left|\frac{4}{\lambda}\left[\frac{(v-v_a)^2 - R_0 a_c}{R_0}\right]T_a\right|. \tag{20}$$

A new deramp function with constant factor φ is created to compress the Doppler bandwidth as follows:

$$H_{deramp}(f, t_n, \varphi) = \exp\left[\frac{j\pi}{c}(f+f_c)\frac{\varphi PRF\lambda}{T_a}t_n^2\right]. \tag{21}$$

By multiplying Equation (9) by Equation (21), we have

$$\begin{aligned} s_5(f, t_n) &= s_2(f, t_n)\cdot H_{deramp}(f, t_n, \varphi) \\ &= \mathrm{rect}\left(\frac{f}{B}\right)\exp\left[-\frac{j8\pi}{c}(f+f_c)R_0\right]\exp\left[-\frac{j8\pi}{c}(f+f_c)\left(\frac{(v-v_a)^2 - R_0 a_c}{2R_0} - \frac{\varphi PRF\lambda}{8T_a}\right)t_n^2\right], \end{aligned} \tag{22}$$

In accordance with Equation (22), the residual Doppler bandwidth is written as

$$f_{D\text{-}pre} = \left|\frac{4}{\lambda}\left[\frac{(v-v_a)^2 - R_0 a_c}{R_0} - \frac{\varphi PRF\lambda}{4T_a}\right]T_a\right| = \left|f_D - \varphi PRF\right|. \tag{23}$$

As shown in Equation (23), the main part of Doppler bandwidth is removed. The Doppler bandwidth f_D is extremely compressed. An appropriate constant factor φ is chosen to make the remaining Doppler bandwidth f_{D-pre} less than one PRF. The value of constant factor φ depends on the Doppler bandwidth of signal in Equation (9). If $\omega PRF < f_D < (\omega+1)PRF, \omega = 1, 2, 3, \cdots$, then the constant factor φ will be chosen as ω. After the pre-processing step in Equation (23), the SOKT (i.e., $(f+f_c)t_n^2 = f_c\xi^2$ [22,23]) can be applied to remove the SRCM. As a result, we have

$$s_5(f, \xi) = \mathrm{rect}\left(\frac{f}{B}\right)\exp\left[-\frac{j8\pi}{c}(f+f_c)R_0\right]\exp\left[-\frac{j8\pi}{\lambda}\left(\frac{(v-v_a)^2 - R_0 a_c}{2R_0} - \frac{\varphi PRF\lambda}{8T_a}\right)\xi^2\right]. \tag{24}$$

As exhibited in Equation (24), the coupling influence between the t_n^2-term and range frequency variable f is eliminated. The SRCM is accurately removed after the SOKT is used, which is beneficial to the subsequent refocusing processing of a moving target. The target trajectory split is avoided because the Doppler bandwidth is effectively compressed. The Doppler spectrum ambiguity is further eliminated by performing a new deramp function. In the meantime, the constant factor φ is introduced to change the Doppler bandwidth of the deramp function. Thus, the new deramp function has a wide applicability.

Some noise-free simulation results are presented to validate the above-mentioned processing step. The range bandwidth of the simulated radar is set to 300 MHz. Other main radar parameters are the same as those of Example A. In Example C, a target, which is denoted by Target C, is considered. The parameters of the target are as follows: $v_a = -6.6$ m/s, $a_a = 1$ m/s^2, $v_c = 24$ m/s and $a_c = -2.5$ m/s^2.

Figure 6 shows the result of Example C. An evident SRCM appears in Figure 6a. Figure 6b displays the result of SRCM compensation by using the correction function in Equation (10). The residual SRCM correction error cannot be ignored because it is larger than one range bin. Therefore, the mismatch of the SRCM correction function in Equation (10) appears. The Doppler spectrum distribution of Target C after FRCM and TRCM correction is presented in Figure 6c. The Doppler spectrum bandwidth is larger than one PRF and the Doppler spectrum occupies several PRF bands because the chirp rate has been doubled after previous processing steps. The trajectory splits into several parts after directly performing the SOKT, as shown in Figure 6d. Figure 6e illustrates the result of applying the new deramp function with constant factor $\varphi = 1$. The target Doppler spectrum bandwidth is extremely compressed, and the target Doppler spectrum is smaller than one PRF band. Figure 6f exhibits the result of performing the deramp and SOKT operations. Given that the target spectrum is located on one PRF band, the trajectory is focused in the same range bin after applying SOKT. Thus, the simulation results verify the new deramp function and SOKT operations.

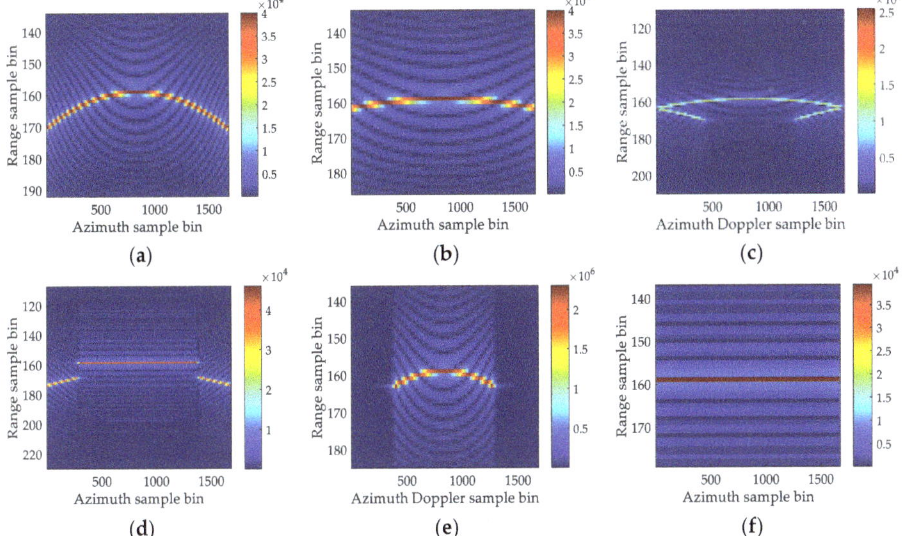

Figure 6. Results of Example C: (**a**) trajectory of Target C after FRCM and TRCM compensation; (**b**) result using the correction function in (10); (**c**) trajectory of Target C in the range and Doppler domain; (**d**) result by directly using the second-order keystone transform (SOKT); (**e**) result after applying the new deramp function in Equation (21); (**f**) result after performing new deramp function and SOKT.

4.2. Analysis of Multiple Target Focusing

In the previous section, the case of one moving target is considered. However, multiple targets may be present in the observation scene. In the case of multiple targets, the effect of cross term induced by nonlinear operation in Equation (9) should be discussed. We assume that the number of targets is D. Then, the range compressed signal in Equation (5) is expressed as

$$s_{mul,1}(f,t_n) = \sum_{i=1}^{D} \mathrm{rect}\left(\frac{f}{B}\right) \exp\left[-j\frac{4\pi}{c}(f+f_c)\left(R_{i,0}+\beta_{i,1}t_n+\beta_{i,2}t_n^2+\beta_{i,3}t_n^3\right)\right], \tag{25}$$

where $R_{i,0}$ is the nearest slant range of the ith target. $\beta_{i,1} = -v_{i,c}$, $\beta_{i,2} = \left((v - v_{i,a})^2 - R_{i,0}a_{i,c}\right)/(2R_{i,0})$ and $\beta_{i,3} = \left(v_{i,c}(v - v_{i,a})^2 + R_{i,0}a_{i,a}(v_{i,a} - v)\right)/\left(2R_{i,0}^2\right)$ represent the first-, second- and third-order phases of the ith target, respectively.

The target signal after RCM correction (RCMC) is written as follows:

$$s_{mul,2}(f, t_n) = \sum_{i=1}^{D} \mathrm{rect}\left(\frac{f}{B}\right) \exp\left(-j\frac{8\pi}{\lambda}\beta_{i,2}t_n^2\right) \exp\left[-j\frac{8\pi}{c}(f + f_c)R_{i,0}\right] + s_{cross,2}(f, t_n), \tag{26}$$

where

$$\begin{aligned} s_{cross,2}(f, t_n) &= \sum_{i=1}^{D} \sum_{j=1, i\neq j}^{D} \mathrm{rect}\left(\frac{f}{B}\right) \exp\left[-j\frac{4\pi}{c}(f + f_c)\left(R_{i,0} + R_{j,0}\right)\right] \\ &\cdot \exp\left[-j\frac{4\pi}{c}(f + f_c)\left(\beta_{i,1} - \beta_{j,1}\right)t_n\right] \exp\left[-j\frac{4\pi f_c}{c}\left(\beta_{i,2} + \beta_{j,2}\right)t_n^2\right] \\ &\cdot \exp\left[-j\frac{4\pi}{c}(f + f_c)\left(\beta_{i,3} - \beta_{j,3}\right)t_n^3\right], \end{aligned} \tag{27}$$

and $s_{cross,2}(f, t_n)$ denotes the cross terms of the signal in Equation (26). Given that the proposed method contains the nonlinear operation, the signal in Equation (26) includes auto and cross terms. According to Equations (26) and (27), the RCMs of auto terms are effectively eliminated, and only the second-order phase remains. Considering that the 1D SCFT is a linear transform, the second-order phases of the auto terms can be effectively estimated by the 1D SCFT. The matched filtering operation in Equation (18) is also a linear processing step. Thus, the auto terms can be refocused using the corresponding matched filters. With regard to the cross terms, $\beta_{i,1} \neq \beta_{j,1}$ and $\beta_{i,3} \neq \beta_{j,3}$ generally hold because $v_{i,c}$, $a_{i,c}$, $v_{i,a}$ and $a_{i,a}$ usually differ. Not only the second-order phases, but also the first- and third-order phases of the cross terms remain. Therefore, the RCMs of the cross terms still exist, and the energy of cross terms spreads along the range dimension. The cross terms are typically defocused in the 1D SCFT domain. In summary, the cross terms may not affect the refocusing of auto terms in this case.

However, moving targets may have the same first- and third-order phases, namely, $\beta_{i,1} = \beta_{j,1}$ and $\beta_{i,3} = \beta_{j,3}$, respectively, in a particular case. Therefore, the signal in Equation (26) is simplified and transformed into range and azimuth slow time domain as follows:

$$\begin{aligned} s_{mul,2}(t, t_n) &= \sum_{i=1}^{D} \mathrm{sinc}\left[B\left(t - \frac{4R_{i,0}}{c}\right)\right] \exp\left(-j\frac{8\pi}{\lambda}\beta_{i,2}t_n^2\right) \\ &+ \sum_{i=1}^{D} \sum_{j=1, i\neq j}^{D} \mathrm{sinc}\left\{B\left[t - \frac{2\left(R_{i,0} + R_{j,0}\right)}{c}\right]\right\} \exp\left[-j\frac{4\pi}{\lambda}\left(\beta_{i,2} + \beta_{j,2}\right)t_n^2\right]. \end{aligned} \tag{28}$$

In accordance with Equation (28), the RCMs of auto and cross terms are all removed. The auto terms are focused at the positions of $t = 4R_{i,0}/c$, and the cross terms are focused at the positions of $t = 2\left(R_{i,0} + R_{j,0}\right)/c$. Then, the 1D SCFT is performed to the auto and cross terms. Accordingly, we have

$$s_{auto,mul,2}\left(\frac{4R_{i,0}}{c}, f_{t_n^2}\right) = \sum_{i=1}^{D} \delta\left(f_{t_n^2} + \frac{4\beta_{i,2}}{\varepsilon\lambda}\right), \tag{29}$$

$$s_{cross,mul,2}\left[\frac{2\left(R_{i,0} + R_{j,0}\right)}{c}, f_{t_n^2}\right] = \sum_{i=1}^{D} \sum_{j=1, i\neq j}^{D} \delta\left[f_{t_n^2} + \frac{2\left(\beta_{i,2} + \beta_{j,2}\right)}{\varepsilon\lambda}\right]. \tag{30}$$

According to Equations (29) and (30), the peak positions of the auto and cross terms in the 1D SCFT domain are $f_{t_n^2} = -4\beta_{i,2}/\varepsilon\lambda$ and $f_{t_n^2} = -2\left(\beta_{i,2} + \beta_{j,2}\right)/\varepsilon\lambda$, respectively.

The peaks of the cross terms may not affect the determination of the auto term peaks according to the previous analysis, but they may lead to spurious peaks. As shown in Equations (29) and (30), the peak positions of the auto terms, namely, $t = 4R_{i,0}/c, f_{t_n^2} = -4\beta_{i,2}/\varepsilon\lambda$ and

$t = 4R_{j,0}/c$, $f_{t_n^2} = -4\beta_{j,2}/\varepsilon\lambda$, are symmetric with respect to the peak positions of the cross terms, that is, $t = 2(R_{i,0} + R_{j,0})/c$, $f_{t_n^2} = -2(\beta_{i,2} + \beta_{j,2})/\varepsilon\lambda$, in the range time and 1D SCFT domain.

In summary, Figure 7 shows the symmetric characteristics between the auto and the cross terms. As shown in Figure 7a, trajectories of auto terms A and B are represented by a straight purple line, and that of the corresponding cross term C is denoted by a straight red line. After RCMC, auto terms A and B and cross term C are focused in the range time domain. The positions of the auto terms in the range time domain, namely, $t = 4R_{i,0}/c$ and $t = 4R_{j,0}/c$, are symmetric with respect to the position of corresponding cross term $t = 2(R_{i,0} + R_{j,0})/c$. Figure 7b–d depict the 1D SCFT result of auto term A, cross term C and auto term B, respectively. The positions of the auto terms in the 1D SCFT frequency domain, namely, $f_{t_n^2} = -4\beta_{i,2}/\varepsilon\lambda$ and $f_{t_n^2} = -4\beta_{j,2}/\varepsilon\lambda$, are symmetric with respect to the position of the cross term, that is, $f_{t_n^2} = -2(\beta_{i,2} + \beta_{j,2})/\varepsilon\lambda$. Thus, the symmetric properties of the auto and cross terms in the range time and 1D SCFT domain can be used to preliminarily identify the potential spurious peak induced by the cross term.

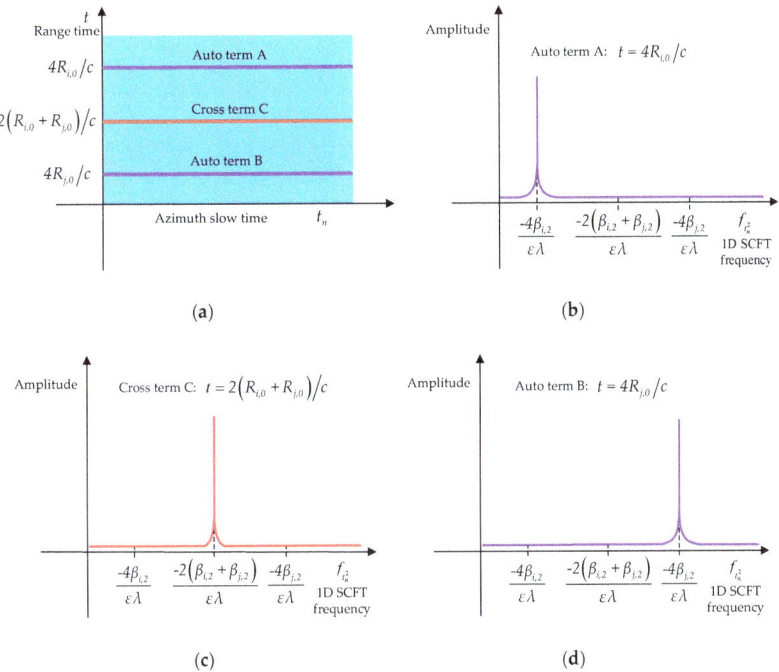

Figure 7. Schematic of the symmetric characteristics between the auto and cross terms: (**a**) result of the auto and cross terms after range cell migration correction (RCMC); (**b**) 1D scaled Fourier transform (SCFT) result of auto term A; (**c**) 1D SCFT result of cross term C; (**d**) 1D SCFT result of auto term B.

A spurious peak recognition procedure based on the 1D SCFT is presented to confirm the potential spurious peak caused by the cross term. Firstly, a recognition function is proposed in the range frequency and azimuth slow time domain as follows:

$$\begin{aligned} s_{re-mul}(f,t_n) &= s_{1-mul}(f,t_n)\cdot s^*_{1-mul}(f,t_n) \\ &= \sum_{i=1}^{D} \mathrm{rect}\left(\tfrac{f}{B}\right) + \sum_{i,j=1,i\neq j}^{D} \mathrm{rect}\left(\tfrac{f}{B}\right)\exp\left[-j\tfrac{4\pi}{c}(f+f_c)\left(R_{i,0}-R_{j,0}\right)\right] \\ &\cdot \exp\left[-j\tfrac{4\pi}{c}(f+f_c)\left(\beta_{i,1}-\beta_{j,1}\right)t_n\right]\exp\left[-j\tfrac{4\pi}{c}(f+f_c)\left(\beta_{i,2}-\beta_{j,2}\right)t_n^2\right] \\ &\cdot \exp\left[-j\tfrac{4\pi}{c}(f+f_c)\left(\beta_{i,3}-\beta_{j,3}\right)t_n^3\right]. \end{aligned} \quad (31)$$

After the SOKT is applied to the signal in Equation (31), we have

$$s_{re-mul}(f,\xi) = \sum_{i=1}^{D} \mathrm{rect}\left(\frac{f}{B}\right) + s_{re-cross}(f,\xi), \tag{32}$$

where

$$\begin{aligned} s_{re-cross}(f,\xi) = & \sum_{i,j=1,i\neq j}^{D} \mathrm{rect}\left(\tfrac{f}{B}\right)\exp\left[-j\tfrac{4\pi}{c}(f+f_c)\left(R_{i,0}-R_{j,0}\right)\right] \\ & \cdot\exp\left[-j\tfrac{4\pi}{c}\left(\beta_{i,1}-\beta_{j,1}\right)(f_c)^{\frac{1}{2}}(f+f_c)^{\frac{1}{2}}\xi\right]\exp\left[-j\tfrac{4\pi f_c}{c}\left(\beta_{i,2}-\beta_{j,2}\right)\xi^2\right] \\ & \cdot\exp\left[-j\tfrac{4\pi}{c}\left(\beta_{i,3}-\beta_{j,3}\right)(f+f_c)^{-\frac{1}{2}}(f_c)^{\frac{3}{2}}\xi^3\right]. \end{aligned} \tag{33}$$

The signal in Equation (32) contains D auto terms and $D(D-1)$ cross terms. After the range IFFT is performed, we have

$$s_{re-mul}(t,\xi) = \sum_{i=1}^{D} \mathrm{sinc}(Bt) + s_{re-cross}(t,\xi), \tag{34}$$

where $s_{re-cross}(t,\xi)$ represents the result of transforming the cross terms in Equation (33) into the range and azimuth slow time domain. If the first- and third-order phases do not satisfy $\beta_{i,1}=\beta_{j,1}$ and $\beta_{i,3}=\beta_{j,3}$, as shown in Equation (33), then not only the second-order phases, but also the first and third-order phases of cross terms remain for the recognition function. Therefore, the RCMs and DFMs of the cross terms exist, and the energy of cross terms is still defocusing. The first-, second- and third-order phases of the auto terms are removed. The auto terms are focused at the position of $t=0$ in the range time domain.

As for the special case, namely, $\beta_{i,1}=\beta_{j,1}$ and $\beta_{i,3}=\beta_{j,3}$, Equation (32) is rewritten as follows after the range IFFT is applied:

$$s_{re-mul}(t,\xi) = \sum_{i=1}^{D} \mathrm{sinc}(Bt) + \sum_{i,j=1,i\neq j}^{D} \mathrm{sinc}\left[B\left(t-\frac{2\left(R_{i,0}-R_{j,0}\right)}{c}\right)\right]\exp\left[-j\frac{4\pi f_c}{c}\left(\beta_{i,2}-\beta_{j,2}\right)\xi^2\right]. \tag{35}$$

According to Equation (35), the auto terms of the recognition function in Equation (35) are focused at $t=0$ and the cross terms of the recognition function in Equation (35) are focused at $t=2\left(R_{i,0}-R_{j,0}\right)/c$ and $t=-2\left(R_{i,0}-R_{j,0}\right)/c$ in the range domain. Then, the 1D SCFT is applied to the cross terms of the recognition function in Equation (35). As a result, we have

$$s_{cross,re-mul}\left[\frac{2\left(R_{i,0}-R_{j,0}\right)}{c}, f_{\xi^2}\right] = \sum_{i=1}^{D}\sum_{j=1,i\neq j}^{D} \delta\left[f_{\xi^2}+\frac{2\left(\beta_{i,2}-\beta_{j,2}\right)}{\varepsilon\lambda}\right]. \tag{36}$$

As shown in Equation (36), the peak positions of cross terms are $f_{\xi^2}=-2\left(\beta_{i,2}-\beta_{j,2}\right)/(\varepsilon\lambda)$ and $f_{\xi^2}=2\left(\beta_{i,2}-\beta_{j,2}\right)/(\varepsilon\lambda)$, respectively, in the 1D SCFT domain.

Therefore, the peak positions of the cross terms of the recognition function in Equation (35), namely, $t=2\left(R_{i,0}-R_{j,0}\right)/c, f_{\xi^2}=-2\left(\beta_{i,2}-\beta_{j,2}\right)/(\varepsilon\lambda)$ and $t=-2\left(R_{i,0}-R_{j,0}\right)/c, f_{\xi^2}=2\left(\beta_{i,2}-\beta_{j,2}\right)/(\varepsilon\lambda)$, are symmetric with respect to the position of $t=0, f_{\xi^2}=0$.

In summary, Figure 8 shows that, if the peak is the spurious peak, then the recognition function will have evident symmetrical peaks with the same amplitudes at positions of $t=2\left(R_{i,0}-R_{j,0}\right)/c, f_{\xi^2}=-2\left(\beta_{i,2}-\beta_{j,2}\right)/(\varepsilon\lambda)$ and $t=-2\left(R_{i,0}-R_{j,0}\right)/c, f_{\xi^2}=2\left(\beta_{i,2}-\beta_{j,2}\right)/(\varepsilon\lambda)$, which are symmetric with respect to the position of $t=0, f_{\xi^2}=0$. Otherwise, Figure 9 depicts that, if the peak is the target peak, then the evident symmetrical peaks at positions of $t=2\left(R_{i,0}-R_{j,0}\right)/c, f_{\xi^2}=-2\left(\beta_{i,2}-\beta_{j,2}\right)/(\varepsilon\lambda)$ and $t=-2\left(R_{i,0}-R_{j,0}\right)/c, f_{\xi^2}=2\left(\beta_{i,2}-\beta_{j,2}\right)/(\varepsilon\lambda)$ are absent in the output result of the recognition function.

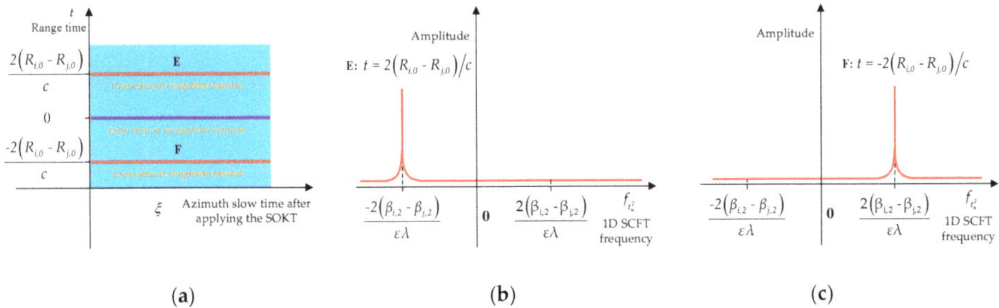

Figure 8. Result of spurious peak recognition function for spurious peak: (**a**) RCMC result; (**b**) 1D SCFT result for signal in $t = 2(R_{i,0} - R_{j,0})/c$; (**c**) 1D SCFT result for signal in $t = -2(R_{i,0} - R_{j,0})/c$.

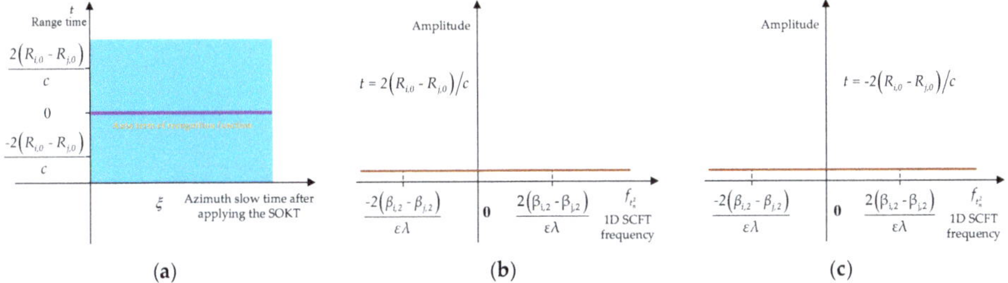

Figure 9. Result of spurious peak recognition function for target peak: (**a**) RCMC result; (**b**) 1D SCFT result for signal in $t = 2(R_{i,0} - R_{j,0})/c$; (**c**) 1D SCFT result for signal in $t = -2(R_{i,0} - R_{j,0})/c$.

The identification procedure of the potential spurious peak is summarised as follows:

Step 1) Extract all peak positions, which are denoted by $\left(t_z, f_{z,t_n^2}\right), z = 1 \cdots Z$, where Z is the number of peaks.

Step 2) Determine whether $t_w = \frac{t_u + t_x}{2}$ and $f_{w,t_n^2} = \frac{f_{u,t_n^2} + f_{x,t_n^2}}{2}$, $u = 1 \cdots Z$, $x = 1 \cdots Z$, $w = 1 \cdots Z$, $w \neq u \neq x$ is satisfied; if so, then the peak at position of $\left(t_w, f_{w,t_n^2}\right)$ may be the spurious peak of the cross term. Apply the recognition function (i.e., Steps 3–5) to identify the potential spurious peak; if not, then all peaks are the target peaks.

Step 3) Calculate Equation (31), and obtain $s_{re-mul}(f, t_n)$.

Step 4) Perform the SOKT and range IFFT to $s_{re-mul}(f, t_n)$, and obtain $s_{re-mul}(t, \xi)$.

Step 5) Apply azimuth 1D SCTF to $s_{re-mul}(t_w - t_u, \xi)$ and $s_{re-mul}(t_w - t_x, \xi)$. If evident symmetrical peaks with the same amplitudes described in Figure 8 are present, then the peak at $\left(t_w, f_{w,t_n^2}\right)$ mentioned in Step 2 will be a spurious peak. Otherwise, the peak at $\left(t_w, f_{w,t_n^2}\right)$ is the target peak.

The flowchart of potential spurious peak recognition procedure is given in Figure 10.

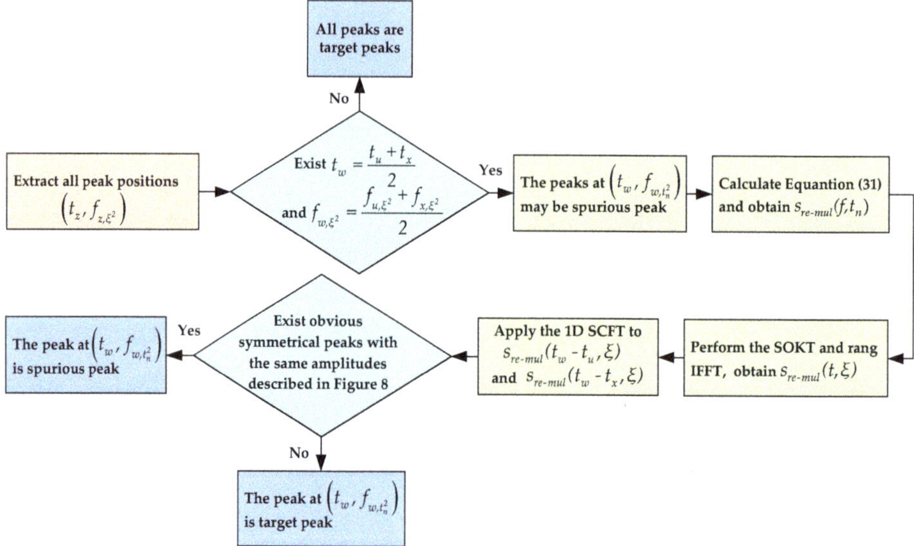

Figure 10. Flowchart of the potential spurious peak identification procedure.

Two examples are provided to validate the analysis of multiple target focusing and spurious peak recognition procedure. The radar parameters are the same as those of Example A. The signal-to-noise (SNR) is 7 dB. In Example D, we consider three targets with different motion parameters, and that are represented by Targets D, E and F. The phase parameters of these target signals are

$$\begin{gathered}\beta_{D,1} = -19.8 \text{ m/s } \beta_{D,2} = 1.2 \text{ m/s}^2, \text{and } \beta_{D,3} = 0.5 \text{ m/s}^3, \\ \beta_{E,1} = 15.6 \text{ m/s}, \beta_{E,2} = 2.4 \text{ m/s}^2, \text{and } \beta_{E,3} = -0.6 \text{ m/s}^3, \\ \beta_{F,1} = 30.5 \text{ m/s}, \beta_{F,2} = 3.6 \text{ m/s}^2, \text{and } \beta_{F,3} = 1.2 \text{ m/s}^3.\end{gathered}$$

Figure 11 displays the processing results of Example D. Figure 11a depicts the RCMC result. The background noise is removed from the obtained result to show the target trajectories clearly. Three straight trajectories related to Targets D, E and F appear because the RCM is effectively corrected. However, the cross terms are still defocused due to the serious effect of the RCM, thereby helping in suppressing the interference of the cross term. Figure 11b–d exhibit the 1D SCFT results for Targets D, E and F in the 147th, 177th and 207th range sample bins, respectively. Evident peaks with respect to Targets D, E and F appear in the corresponding figure. Figure 11e,f illustrate the 1D SCFT results of data in the 157th and 187th range sample bins, respectively, to indicate the defocusing of cross terms without loss of generality. The evident peaks are absent in the processing results. The cross terms still suffer from the effect of defocusing induced by the residual first- and third-order phase errors. The positions of peaks in Figure 11b–d satisfy the symmetric properties described in Figure 7. The potential spurious peak recognition procedure is utilised to identify whether the peak in Figure 11c is a spurious peak. Figure 12a,b depict the corresponding recognition result. Considering that clear peaks, which satisfy the symmetric feature shown in Figure 8, do not emerge in the 30th and −30th range time sample bins of the recognition function, the peak in Figure 11c is determined as the target peak.

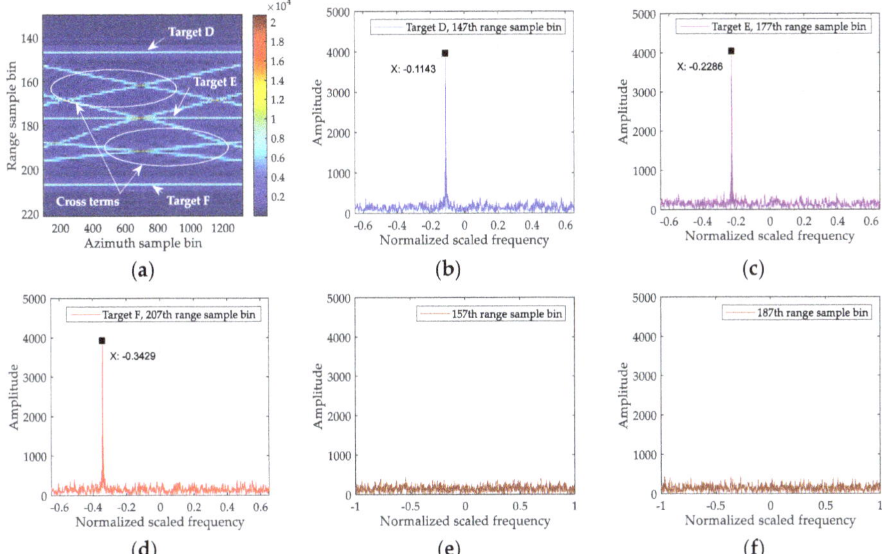

Figure 11. Results of Example D: (**a**) result after RCMC; (**b**) 1D SCFT result for Target D in the 147th range sample bin of Figure 11a; (**c**) 1D SCFT result for Target E in the 177th range sample bin of Figure 11a; (**d**) 1D SCFT result for Target F in the 207th range sample bin of Figure 11a; (**e**) 1D SCFT result of data in the 157th range sample bin of Figure 11a; (**f**) 1D SCFT result of data in the 187th range sample bin of Figure 11a.

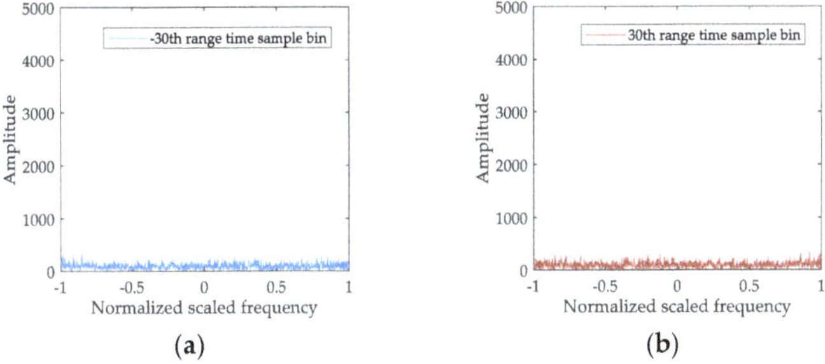

Figure 12. Potential spurious peak recognition results of Example D: (**a**) 1D SCFT result of the −30th range time sample bin of the recognition function; (**b**) 1D SCFT result of the 30th range time sample bin of the recognition function.

In Example E, we consider two targets that are denoted by Targets G and H. The phase parameters of these target signals are

$$\beta_{G,1} = 32.6\ \mathrm{m/s}, \beta_{G,2} = 1.2\ \mathrm{m/s^2}, \text{and} \beta_{G,3} = 0.8\ \mathrm{m/s^3},$$
$$\beta_{H,1} = 32.6\ \mathrm{m/s}, \beta_{H,2} = 3.6\ \mathrm{m/s^2}, \text{and} \beta_{H,3} = 0.8\ \mathrm{m/s^3}.$$

Figure 13 shows the obtained results of Example E. The RCMC result without noise is exhibited in Figure 13a to display the target trajectories clearly. Three straight trajectories are observed in Figure 13a because the first- and third-order phases of Target G equal those of Target H. Figure 13b–d illustrate the 1D SCFT result of the data in the 147th, 177th, and 207th range sample bins, respectively. Considering that peak positions in Figure 13b–d satisfy the symmetric characteristics described in Figure 7, the peak in Figure 13c may be a spurious peak. Figure 14 depicts the potential spurious peak recognition results. Given that evident symmetric peaks with the same amplitudes appear in the 30th and −30th range time sample bins of the recognition function, as shown in Figure 14a,b, the peak in Figure 13c is confirmed as the spurious peak.

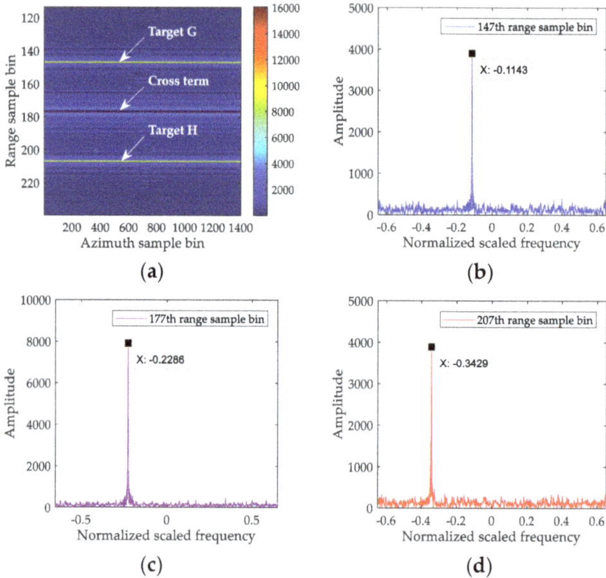

Figure 13. Results of Example E: (**a**) result after RCMC; (**b**) 1D SCFT result of Target G in the 147th range sample bin of Figure 13a; (**c**) 1D SCFT result of cross term in the 177th range sample bin of Figure 13a; (**d**) 1D-SCFT result of Target H in the 207th range sample bin of Figure 13a.

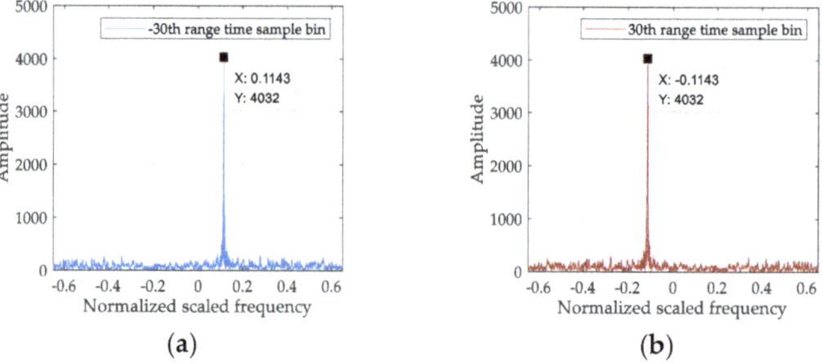

Figure 14. Potential spurious peak recognition results of Example E: (**a**) 1D SCFT result of the −30th range time sample bin of the recognition function; (**b**) 1D SCFT result of the 30th range time sample bin of the recognition function.

5. Experimental Results

In this section, several simulation experimental results in the presence of Gaussian background and real data processing results are analysed to verify the proposed method.

5.1. Simulation Experimental Result Analysis

The simulation radar parameters are listed in Table 1. Two maneuvering targets, which are denoted by T1 and T2, are considered. The motion parameters of the two targets are summarised in Table 2. The SNR after range compression is 8 dB. T1 is a Doppler spectrum ambiguity target, and its azimuth Doppler spectrum bandwidth is larger than PRF/2. The azimuth Doppler spectrum of T1 distributes into two PRF bands. The Doppler ambiguity number of T2 is −1, and its azimuth Doppler spectrum is still located on one PRF band. The constant factor is set to $\varepsilon = 2$ following the constant factor selection strategy in Appendix A. The FOKT-based [25], stationary phase-based [28] and SOKT-GHHAF methods [32] are used for comparison.

Table 1. Basic simulation radar parameters.

Parameters	Value
Carrier frequency	10 GHz
Range bandwidth	80 MHz
Pulse duration time	1 μs
Pulse repetition frequency	1400 Hz
Radar platform velocity	250 m/s
Nearest slant range	6000 m
Dwell time	1 s

Table 2. Simulation parameters for two targets.

	Cross-Track Velocity (m/s)	Cross-Track Acceleration (m/s^2)	Along-Track Velocity (m/s)	Along-Track Acceleration (m/s^2)
T1	36.8	−3.6	25.2	−4.5
T2	−26.5	1.6	5.9	0.6

Figure 15a depicts the result of range compression. Two curved trajectories are observed in the figure. Target energy also distributes into several range sample bins, thereby leading to severe RCM. The result by directly applying azimuth FFT is illustrated in Figure 15b. Notably, the energy of targets also spreads along the azimuth Doppler dimension, which induces serious DFM. The RCM and DFM result in severe defocusing effects. In addition, the azimuth Doppler spectrum of T1 spans over two PRF bands. The azimuth Doppler spectrum of T2 occupies one PRF. Figure 15c shows the result after RCMC, and the background is removed to indicate the trajectories of two targets clearly. The RCM is effectively eliminated, and the trajectory split is avoided. The energy of the target is focused in the corresponding range bin. In the meantime, the RCMs of cross terms remain, and cross terms still suffer from the effect of defocusing, thereby helping in suppressing the cross terms. Figure 15d exhibits the 1D SCFT result of T1. An evident peak with respect to T1 appears in Figure 15d. From the peak position, a well-focused result is obtained in the range–azimuth time domain by using the corresponding matched filter, as shown in Figure 15e–h. For the same reason, T2 is also accumulated as a peak in Figure 15i. With the peak position, T2 is well focused in the range–azimuth time domain by the matched filtering, as exhibited in Figure 15j–m.

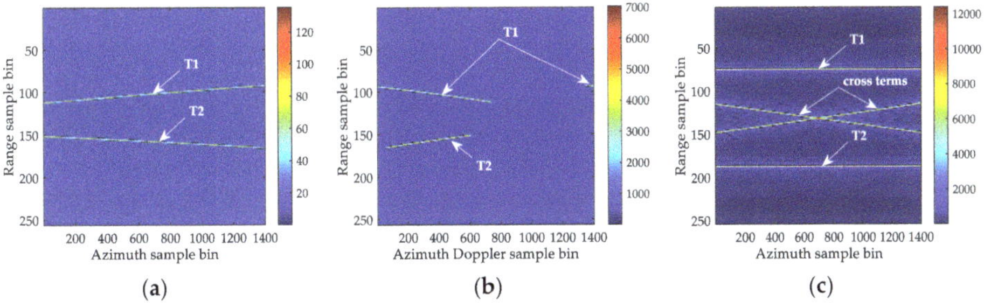

Figure 15. *Cont.*

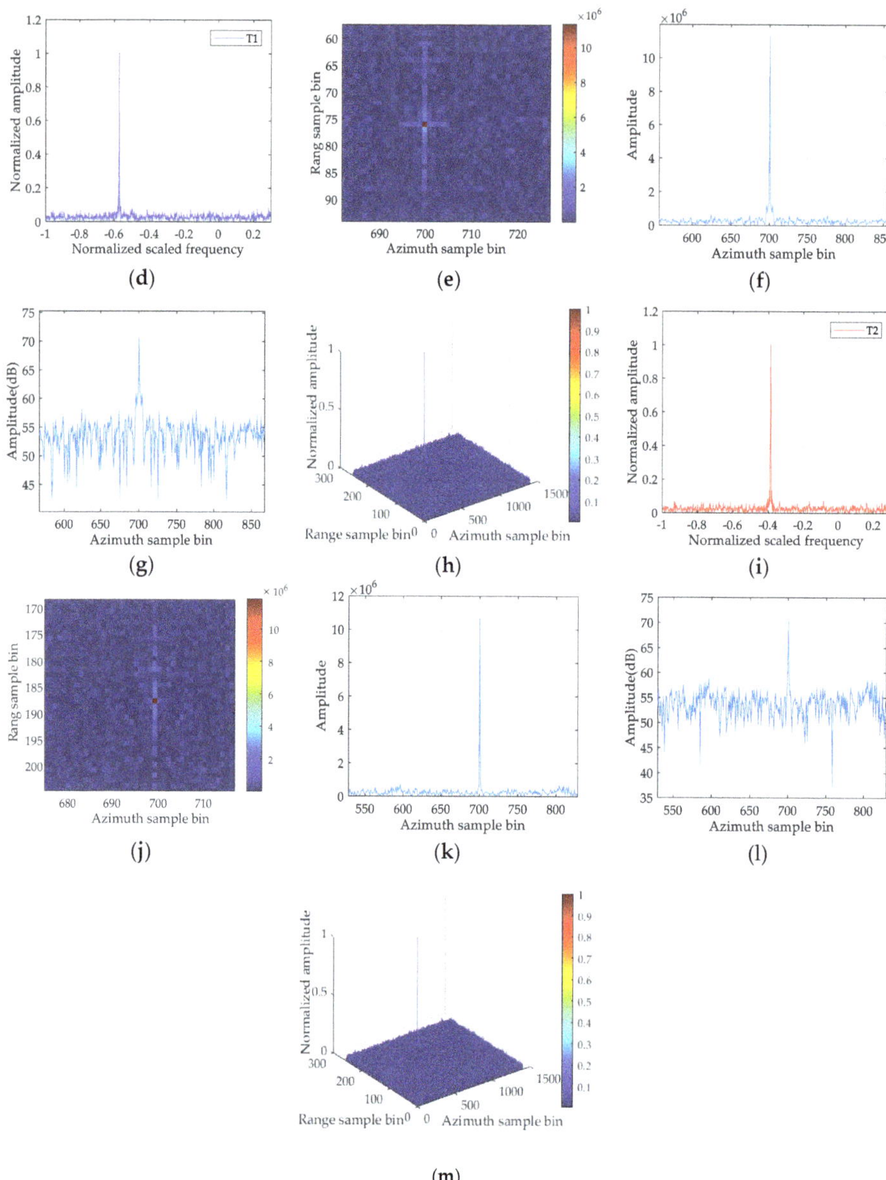

Figure 15. Results of simulation experiment: (**a**) range compression result; (**b**) azimuth Doppler spectrum distributions of two targets; (**c**) RCMC result; (**d**) 1D SCFT result of T1; (**e**) focusing result of T1 by using the proposed method; (**f**) profile for focusing result of T1; (**g**) dB version of amplitude for Figure 15f; (**h**) stereogram of Figure 15e; (**i**) 1D SCFT result of T2; (**j**) focusing result of T2 by performing the proposed method; (**k**) profile for focusing result of T2; (**l**) dB version of amplitude for Figure 15k; (**m**) stereogram of Figure 15j.

Figure 16a,b plot the SRCM correction result using the SOKT-GHHAF method before and after Doppler centre shifting by PRF/2 operation. The background is removed from the result to illustrate the target trajectory clearly. As presented in Figure 16a, the trajectory of T1 splits into two parts due to the azimuth Doppler spectrum of T1 spanning over two PRF bands. Given that the Doppler spectrum bandwidth is larger than PRF/2, the operation of Doppler centre shifting by PRF/2 in SOKT-GHHAF method is invalid. Figure 16b exhibits that the trajectory still splits into two parts, and this condition affects the performance of RCMC and induces the coherent integration loss. Figure 16c–e show the focusing result of T1 by applying the FOKT-based method. A defocused result appears in Figure 16c–e because the along-track velocity and acceleration motions are ignored. As presented in Figure 16f–h, the focusing performance of the stationary phase-based method deteriorates significantly given that the third-order phase is neglected.

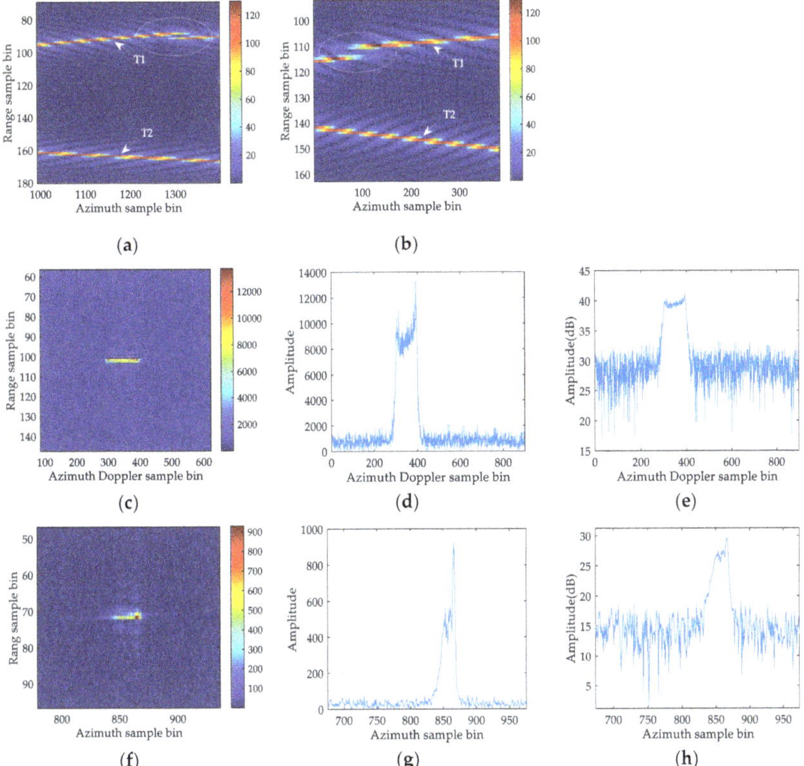

Figure 16. Processing results of compared methods for T1: (**a**) SRCM compensation result using second-order keystone transform-based generalised Hough-high-order ambiguity function (SOKT-GHHAF) method before Doppler centre shifting by PRF/2; (**b**) SRCM correction result using SOKT-GHHAF method after Doppler centre shifting by PRF/2; (**c**) focusing result by performing the first-order keystone transform-based method; (**d**) profile for focusing result of FOKT-based method; (**e**) dB version of amplitude for Figure 16d; (**f**) focusing result by applying the stationary phase-based method; (**g**) profile for focusing result of the stationary phase-based method; (**h**) dB version of amplitude for Figure 16f.

In summary, the result of simulation experiment validates that the proposed method can effectively compensate the RCM and DFM of the maneuvering target, and a well-focused result can be achieved regardless of Doppler ambiguity including Doppler centre blur and spectrum ambiguity. The performance of presented method is better than that of the FOKT-based and stationary phase-based methods. This result is attributed to the fact that the proposed method considers the along-track velocity and acceleration motions and can deal with the high-order RCM and DFM induced by the third-order phase. The proposed method is also more robust to Doppler ambiguity than the SOKT-GHHAF method because it can effectively solve the problems of Doppler centre shift and Doppler spectrum broadening.

5.2. Real Data Processing Result Analysis

In this section, two parts of RADARSAT-1 Vancouver scene data [1,26] are utilised to validate the presented algorithm. These real radar data are recorded by a C-band space-borne SAR system. The main radar system parameters are summarised in Table 3. The detailed parameters of these real data are given by [1,26]. The azimuth pulse number of selected data is 1500.

Table 3. Main radar parameters for RADARSAR-1.

Parameters	Value
Carrier frequency	5.3 GHz
Range bandwidth	30.116 MHz
Pulse duration time	41.74 μs
Pulse repetition frequency	1256.98 Hz

5.2.1. Processing Result of a Single Target

Figure 17a shows the image of the selected scene, where the interested target is highlighted in the figure. Figure 17b depicts the trajectory for the target of interest after range compression. The trajectory distributes into multiple range sample bins due to the serious RCM, which indicates the typical defocusing. Figure 17c illustrates the result of RCMC in the range–Doppler domain. The target energy is focused in the same range bin after RCMC, but the effect of DFM still remains. As shown in Figure 17d, an evident peak appears in the 1D SCFT domain. According to the peak position in Figure 17d, the target can be focused by using the proposed method, as exhibited in Figure 17e. Therefore, the above-mentioned real data processing results demonstrate the effectiveness of the presented method.

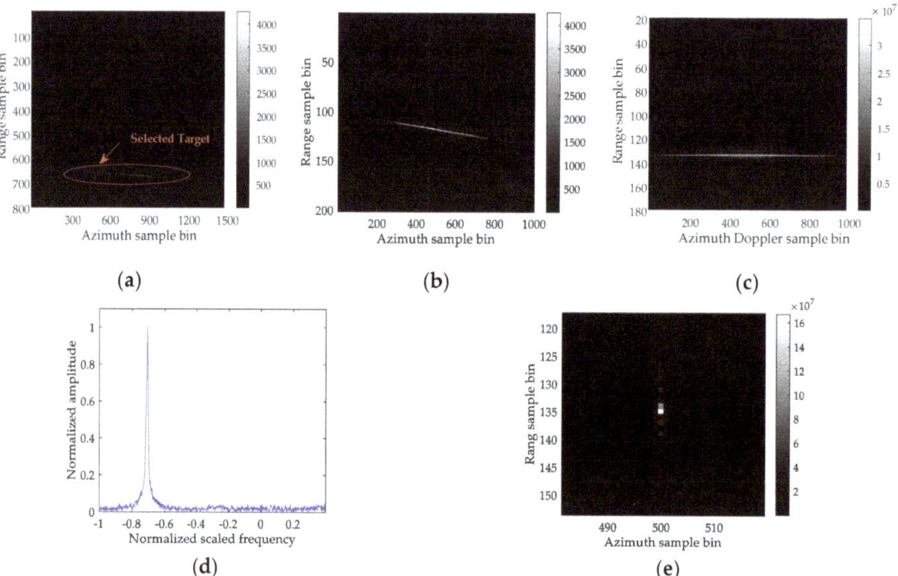

Figure 17. Real data processing results of a single target: (**a**) image of the selected scene; (**b**) trajectory of interested target after range compression; (**c**) result after RCMC; (**d**) 1D SCFT result; (**e**) focusing result for a single target of interest.

5.2.2. Processing Result of Two Targets

Figure 18a displays the image of the selected scene, where the two targets of interest, which are denoted by Target 1 and Target 2, are marked in the figure. Figure 18b shows the trajectories of Target 1 and Target 2 after range compression. The profile along the 420th azimuth sample bin of Figure 18b is depicted in Figure 18c,d. Two trajectories span over several range sample bins due to the serious RCM. Figure 18e depicts the result of RCMC in the range–Doppler domain. The profile along the 420th azimuth Doppler sample bin of Figure 18e is shown in Figure 18f,g. The RCM is effectively removed, but the DFM still exists. The trajectory in the middle of trajectories for Target 1 and Target 2 is a potential cross term according to the analysis in Section 4.2. Figure 18h–j illustrate 1D SCFT results of data in the 69th range sample bin (Target1), 79th range sample bin (cross term) and 89th range sample bin (Target2), respectively. Given that the peak positions in Figure 18h–j satisfy the symmetric features described in Figure 7, the peak shown in Figure 18i is preliminarily identified as a potential spurious peak. The spurious peak recognition results are shown in Figure 19a,b. The peak illustrated in Figure 18i is confirmed as the spurious peak because the symmetric peaks with the same amplitudes appear in 10th and −10th range time sample bins of the recognition function. The cross term is removed, and the focused results of Target 1 and Target 2 are depicted in Figure 19c,d, respectively. These real data processing results verify that the proposed method can be used to focus multiple targets and validate the proposed spurious peak recognition procedure.

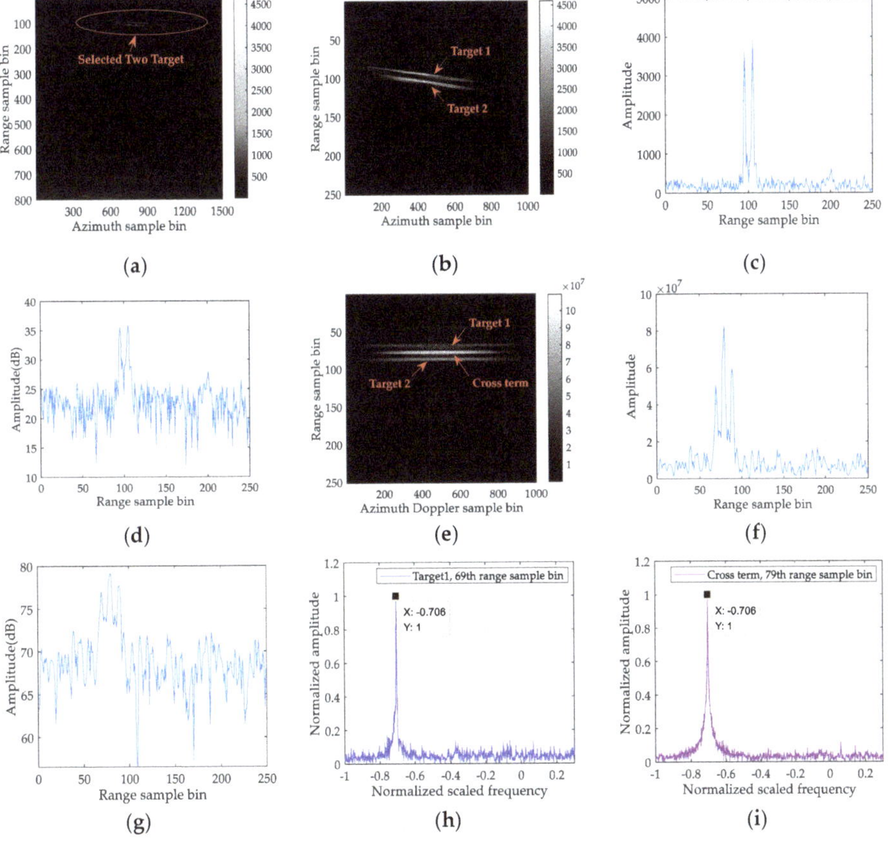

Figure 18. *Cont.*

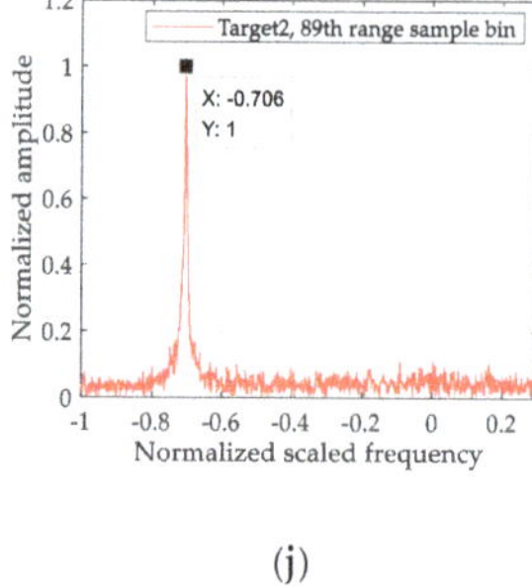

(j)

Figure 18. RCMC and 1D SCFT results: (**a**) image of the selected scene; (**b**) range compression result for two targets of interest; (**c**) profile along the 420th azimuth sample bin of Figure 18b; (**d**) dB version of amplitude for Figure 18c; (**e**) result after RCMC; (**f**) profile along the 420th azimuth Doppler sample bin of Figure 18e; (**g**) dB version of amplitude for Figure 18f; (**h**) 1D SCFT result of the 69th range sample bin (Target 1); (**i**) 1D SCFT result of the 79th range sample bin (cross term); (**j**) 1D SCFT result of the 89th range sample bin (Target 2).

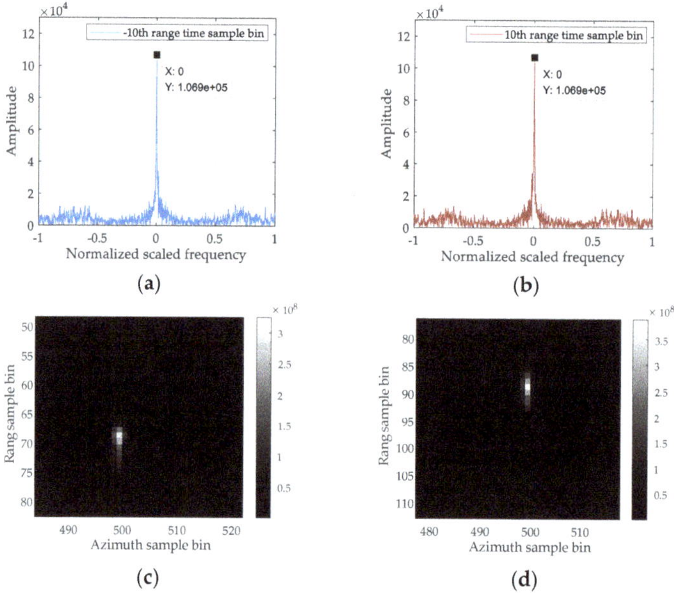

Figure 19. Potential spurious peak recognition and final focusing results: (**a**) 1D SCFT result for the −10th range time sample bin of the recognition function; (**b**) 1D SCFT result for the 10th range time sample bin of the recognition function; (**c**) focusing result of Target 1; (**d**) focusing result of Target 2.

6. Discussion

6.1. Computational Complexity

In this section, we discuss the computational complexity of the proposed method, FOKT-based method [25], stationary phase-based method [28] and SOKT-GHHAF method [32]. Similar to [32], the number of complex multiplications is utilised to indicate the computational complexity. We suppose that G represents the number of range bins and P denotes the number of pulses. For convenience, we assume that the SRCM correction function in Equation (10) is effective. The computational burden of the proposed algorithm includes a range FFT operation, a $G \times P$

point matrix complex multiplication, an SRCM correction operation, a 1D SCFT processing, an azimuth FFT operation, a matched filtering processing and a 2D IFFT operation. Notably, the 1D SCFT in Equation (14) can be easily implemented using the NUFFT of low computational burden. The detailed analysis of the NUFFT has been provided in [43,44]. The computational complexity of the NUFFT-based 1D SCFT is obtained using $O(Plog_2P)$ [43,44]. Thus, the total computational cost of the proposed method is denoted as $O(2PGlog_2G) + O(2GPlog_2P) + O(Plog_2P) + 3GP$. We assume that the searching times of β_2 and Doppler ambiguity number are represented by I_2 and I_d, respectively. The computational cost of the SOKT-GHHAF method is denoted as $O\left[GP^2 + P(P-1)G\right] + O\left(GP^2\right) + O\left(P^3\right) + O\left(I_2Plog_2^P\right)$ [32]. The computational burden for stationary phase-based method is obtained using $O[I_dI_2(PGlog_2G + GPlog_2P)]$ [28]. The computational complexity of the FOKT-based method is represented as $O[(I_d+1)(GPlog_2P + PGlog_2G)] + I_dGP$ [25].

In the case of the SRCM correction function mismatch, the SOKT operation should be added in the proposed method. The chirp-z-based SOKT is applied to compensate the SRCM. The computational cost of chirp-z-based SOKT is represented as $O(GPlog_2P)$ [45]. The total computational burden is denoted as $O(2PGlog_2G) + O(3GPlog_2P) + O(Plog_2P) + 3GP$. In this case, the computational complexity of the proposed algorithm is slightly increased. However, the proposed method still has low computational complexity.

Table 4 exhibits the detailed computational costs of the above-mentioned algorithms. The table shows that proposed method[1] denotes the computational complexity of the proposed method in the case of SRCM correction function matching, and proposed method[2] represents the computational complexity of the proposed method in the case of SRCM correction function mismatch. In summary, the computational complexity of the proposed algorithm is lower than that of the SOKT-GHHAF, stationary phase-based and FOKT-based methods because it can be implemented using FFT, IFFT and NUFFT. The parameter searching procedure is avoided.

Table 4. Comparison of computational complexities.

Methods	Computational Complexity
FOKT method [25]	$O[(I_d+1)(GPlog_2P + PGlog_2G)] + I_dGP$
Stationary phase-based method [28]	$O[I_dI_2(PGlog_2G + GPlog_2P)]$
SOKT-GHHAF method [32]	$O\left[GP^2 + P(P-1)G\right] + O\left(GP^2\right) + O\left(P^3\right) + O\left(I_2Plog_2^P\right)$
Proposed method [1]	$O(2PGlog_2G) + O(2GPlog_2P) + O(Plog_2P) + 3GP$
Proposed method [2]	$O(2PGlog_2G) + O(3GPlog_2P) + O(Plog_2P) + 3GP$

6.2. Some Remarks

Remark 1. *The different moving targets may have various scattering intensities for multiple target focusing. If the intensities of these targets are significantly different, then the target with higher intensity may submerge the target with a lower value; this condition affects the performance of the presented method. In this case, the CLEAN technique in [46,47] can be provided to remove the strong target effect. The strong and weak moving targets can be focused iteratively.*

Remark 2. *The proposed method has a relatively high demand on the target input SNR/signal-to-clutter and noise ratio (SCNR) because its processing procedure contains a nonlinear operation. Therefore, the presented algorithm is suitable for the fast realisation for refocusing of fast-maneuvering targets in the case of relatively high SNR/SCNR. However, a slow and weak moving target may be drowned by the strong clutter background. In this case, the performance will degrade. At this time, many excellent clutter rejection methods, such as displaced phase centre antenna [48] and space-time adaptive processing [49,50], can be performed to reject the clutter. After clutter suppression, the proposed method can be used to refocus the moving target given that the target input SCNR is significantly increased. The effectiveness of clutter suppression in the moving target refocusing application has been validated in previous studies [16,17,25,27,29,32]. Interested readers may refer to [27,29]* *for a detailed analysis about clutter rejection in the moving target refocusing applications. The fast realisation for refocusing of moving targets with a low SNR/SCNR in strong or extremely heterogeneous background is still a challenging work and will be investigated in the future.*

7. Conclusions

The unknown motion parameters of ground fast-maneuvering targets induce high-order RCM and DFM (i.e., CRCM and QDFM), which make the target energy seriously defocused. Fast-maneuvering targets easily exhibit complex Doppler ambiguity due to the limitation of PRF for SAR systems. These factors result in the difficulty in focusing of moving targets. In this work, a new computationally efficient algorithm is proposed to focus fast-maneuvering targets. The characteristics of the presented algorithm are summarised as follows: (1) the presented algorithm can effectively focus fast-maneuvering targets in the range–azimuth time domain because the acceleration and third-order phase are considered; (2) the proposed method is computationally efficient; (3) the proposed algorithm has a wide applicability because two constant factors ε and φ are introduced; (4) a new deramp function is proposed to further address the complex Doppler ambiguity including Doppler centre blur and spectrum ambiguity; (5) the cross-term interference for multiple target focusing is analysed, and a corresponding recognition procedure is proposed to identify the spurious peak. The effectiveness of the moving target refocusing algorithm and spurious peak recognition procedure has been confirmed by simulated and real data processing results. However, the proposed method introduces the nonlinear operation due to fast implementation for refocusing of fast-maneuvering targets; this condition weakens the performance in the case of low SNR/SCNR. This problem will be investigated in the future.

Author Contributions: Conceptualization, J.W. and L.Z.; Data curation, J.W. and Y.Z.; Formal analysis, J.W. and Z.C.; Funding acquisition, Y.Z. and L.Z.; Investigation, J.W. and H.Y.; Methodology, J.W. and Z.C.; Project administration, Y.Z. and L.Z.; Supervision, L.Z.; Validation, J.W., Y.Z. and Z.C.; writing—original draft, J.W.; writing—review and editing, Y.Z., L.Z., Z.C. and H.Y.

Funding: This research was funded by the National Natural Science Foundation of China, Grant Nos. 61871305, 61671361, and 61731023; and the APC was funded by the National Natural Science Foundation of China, Grant No. 61871305.

Acknowledgments: The authors would like to thank the anonymous reviewers for their valuable and useful comments and suggestions that helped improve the paper.

Conflicts of Interest: The authors declare no conflict of interest.

Appendix A

In this appendix, the selection criterion of the constant factor ε for 1D SCFT in our proposed method is discussed. In accordance with the peak in Equation (14), the equation is obtained as follows:

$$\hat{f}_{t_n^2} = -\frac{2(v-v_a)^2 - 2R_0 a_c}{\varepsilon \lambda R_0}. \tag{A1}$$

To ensure the constant factor ε matching, the following inequality should be satisfied:

$$f_{t_n^2 max} \geq \left| \frac{2(v-v_a)^2 - 2R_0 a_c}{\varepsilon \lambda R_0} \right|, \tag{A2}$$

where $f_{t_n^2 max}$ denotes the maximum value of $\left| f_{t_n^2} \right|$.

We assume that the value scopes of target along-track velocity v_a and cross-track acceleration a_c are $[-v_{amax}, v_{amax}]$ and $[-a_{cmax}, a_{cmax}]$. Accordingly, the following equation is obtained:

$$\left| \frac{2(v-v_a)^2 - 2R_0 a_c}{\varepsilon \lambda R_0} \right|_{max} = \frac{2(v+v_{amin})^2 + 2R_0 a_{cmax}}{\varepsilon \lambda R_0}. \tag{A3}$$

By substituting Equation (A3) into Equation (A2), we obtain:

$$f_{t_n^2 max} \geq \frac{2(v+v_{amin})^2 + 2R_0 a_{cmax}}{\varepsilon \lambda R_0}. \tag{A4}$$

In accordance with Equation (A4), the selection scope of constant factor ε is obtained as follows:

$$\varepsilon \geq \frac{2(v+v_{amin})^2 + 2R_0 a_{cmax}}{f_{t_n^2 max} \lambda R_0}. \tag{A5}$$

The constant factor should satisfy the inequality in Equation (A5). According to Equation (15), if a large constant factor is chosen, then the estimation error will be increased. Therefore, a smaller constant factor can be selected to improve estimation accuracy under the condition described in Equation (A5).

References

1. Cumming, I.G.; Wong, F.H. *Digital Processing of Synthetic Aperture Radar Data: Algorithm and Implementation*; Artech House: Norwood, MA, USA, 2005.
2. Moreira, A.; Iraola, P.P.; Younis, M.; Krieger, G.; Hajnsek, I.; Papathanassiou, K.P. A tutorial on synthetic aperture radar. *IEEE Geosci. Remote Sens. Mag.* **2013**, *1*, 6–43. [CrossRef]
3. Gomba, G.; Parizzi, A.; Zan, F.D.; Eineder, M.; Bamler, R. Toward operational compensation of ionospheric effects in SAR interferograms: The split-spectrum method. *IEEE Trans. Geosci. Remote Sens.* **2016**, *54*, 1446–1461. [CrossRef]
4. Bovenga, F.; Giacovazzo, V.M.; Refice, A.; Veneziani, N. Multichromatic analysis of InSAR data. *IEEE Trans. Geosci. Remote Sens.* **2013**, *51*, 4790–4799. [CrossRef]
5. Bovenga, F.; Derauw, D.; Rana, F.M.; Barbier, C.; Refice, A.; Veneziani, N.; Vitulli, R. Multi-chromatic analysis of SAR images for coherent target detection. *Remote Sens.* **2014**, *6*, 8822–8843. [CrossRef]
6. Filippo, B. COSMO-SkyMed staring spotlight SAR data for micro-motion and inclination angle estimation of ships by pixel tracking and convex optimization. *Remote Sens.* **2019**, *11*, 766. [CrossRef]
7. Filippo, B.; Pia, A.; Danilo, O.; Carmine, C. Micro-motion estimation of maritime targets using pixel tracking in Cosmo-Skymed synthetic aperture radar data-an operative assessment. *Remote Sens.* **2019**, *11*, 1637.
8. Iglesias, R.; Fabregas, X.; Aguasca, A.; Mallorqui, J.J.; López-Martínez, C.; Gili, J.A.; Corominas, J. Atmospheric phase screen compensation in ground-based SAR with a multiple-regression model over mountainous regions. *IEEE Trans. Geosci. Remote Sens.* **2014**, *52*, 2436–2449. [CrossRef]
9. Qin, M.; Li, D.; Tang, X.; Cao, Z.; Li, W.; Xu, L. A fast high-resolution imaging algorithm for helicopter-borne rotating array SAR based on 2-D chirp-z transform. *Remote Sens.* **2019**, *11*, 1669. [CrossRef]
10. Tang, S.; Zhang, L.; So, H.C. Focusing high-resolution highly-squinted airborne SAR data with maneuvers. *Remote Sens.* **2018**, *10*, 862. [CrossRef]
11. Baumgartner, S.V.; Krieger, G. Simultaneous high-resolution wide-swath SAR imaging and ground moving target indication: Processing approaches and system concepts. *IEEE J. Sel. Top. Appl. Earth Observ. Remote Sens.* **2015**, *8*, 5015–5029. [CrossRef]
12. Chen, Z.; Zhou, Y.; Zhang, L.; Wei, H.; Lin, C.; Liu, N.; Wan, J. General range model for multi-channel SAR/GMTI with curvilinear flight trajectory. *Electron. Lett.* **2019**, *55*, 111–112. [CrossRef]
13. Huang, Y.; Liao, G.; Xu, J.; Li, J.; Yang, D. GMTI and parameter estimation for MIMO SAR system via fast interferometry RPCA method. *IEEE Trans. Geosci. Remote Sens.* **2018**, *56*, 1774–1787. [CrossRef]
14. Rahmanizadeh, A.; Amini, J. An integrated method for simulation of synthetic aperture radar (SAR) raw data in moving target detection. *Remote Sens.* **2017**, *9*, 1009. [CrossRef]
15. Chen, Z.; Zhang, L.; Zhou, Y.; Lin, C.; Tang, S.; Wan, J. A non-adaptive space-time clutter canceller for multi-channel synthetic aperture radar. *IET Signal Process.* **2019**, *13*, 472–479. [CrossRef]
16. Huang, P.; Liao, G.; Yang, Z.; Ma, J. An approach for refocusing of ground fast-moving target and high-order motion parameter estimation using Radon-high-order time-chirp rate transform. *Digit. Signal Process.* **2016**, *48*, 333–348. [CrossRef]

17. Li, D.; Zhan, C.; Ma, H.; Liu, H.; Su, J.; Liu, Q. An efficient SAR ground moving target refocusing method based on PPFFT and coherently integrated CPF. *IEEE Access.* **2019**, *7*, 114102–114115. [CrossRef]
18. Oveis, A.H.; Sebt, M.A. Coherent method for ground-moving target indication and velocity estimation using Hough transform. *IET Radar Sonar Navig.* **2017**, *11*, 646–655. [CrossRef]
19. Zeng, H.; Chen, J.; Wang, P.; Yang, W.; Liu, W. 2-D coherent integration processing and detecting of aircrafts using GNSS-based passive radar. *Remote Sens.* **2018**, *10*, 1164. [CrossRef]
20. Perry, R.P.; DiPietro, R.C.; Fante, R.L. SAR imaging of moving targets. *IEEE Trans. Aerosp. Electron. Syst.* **1999**, *35*, 188–200. [CrossRef]
21. Zhu, D.; Li, Y.; Zhu, Z. A Keystone transform without interpolation for SAR ground moving-target imaging. *IEEE Geosci. Remote Sens. Lett.* **2007**, *4*, 18–22. [CrossRef]
22. Zhou, F.; Wu, R.; Xing, M. Approach for single channel SAR ground moving target imaging and motion parameter estimation. *IET Radar Sonar Navig.* **2007**, *1*, 59–66. [CrossRef]
23. Kirkland, D. Imaging moving targets using the second-order keystone transform. *IET Radar Sonar Navig.* **2011**, *5*, 902–910. [CrossRef]
24. Li, G.; Xia, X.G.; Peng, Y. Doppler keystone transform: An approach suitable for parallel implementation of SAR moving target imaging. *IEEE Geosci. Remote Sens. Lett.* **2008**, *5*, 573–577. [CrossRef]
25. Sun, G.; Xing, M.; Xia, X.G.; Wu, Y.; Bao, Z. Robust ground moving-target imaging using deramp-keystone processing. *IEEE Trans. Geosci. Remote Sens.* **2013**, *51*, 966–982. [CrossRef]
26. Tian, J.; Cui, W.; Xia, X.G.; Wu, S. Parameter estimation of ground moving targets based on SKT-DLVT processing. *IEEE Trans. Comput. Imaging* **2016**, *2*, 13–26. [CrossRef]
27. Zhu, S.; Liao, G.; Qu, Y.; Zhou, Z.; Liu, X. Ground moving targets imaging algorithm for synthetic aperture radar. *IEEE Trans. Geosci. Remote Sens.* **2011**, *49*, 462–477. [CrossRef]
28. Zhu, S.; Liao, G.; Yang, D.; Tao, H. A new method for radar high-speed maneuvering weak target detection and imaging. *IEEE Geosci. Remote Sens. Lett.* **2014**, *11*, 1175–1179.
29. Huang, P.; Xia, X.G.; Gao, Y.; Liu, X.; Liao, G.; Jiang, X. Ground moving target refocusing in SAR imagery based on RFRT-FrFT. *IEEE Trans. Geosci. Remote Sens.* **2019**, *57*, 5476–5492. [CrossRef]
30. Wan, J.; Zhou, Y.; Zhang, L.; Chen, Z. Ground moving target focusing and motion parameter estimation method via MSOKT for synthetic aperture radar. *IET Signal Process.* **2019**, *13*, 528–537. [CrossRef]
31. Su, J.; Tao, H.; Xie, J.; Rao, X.; Guo, X. Imaging and Doppler parameter estimation for maneuvering target using axis mapping based coherently integrated cubic phase function. *Digit. Signal Process.* **2017**, *62*, 112–124. [CrossRef]
32. Huang, P.; Liao, G.; Yang, Z.; Xia, X.G.; Ma, J.; Zheng, J. Ground maneuvering target imaging and high-order motion parameter estimation based on second-order keystone and generalized Hough-HAF transform. *IEEE Trans. Geosci. Remote Sens.* **2017**, *55*, 320–335. [CrossRef]
33. Yu, W.; Su, W.; Gu, H. Ground moving target motion parameter estimation using radon modified lv's distribution. *Digit. Signal Process.* **2017**, *69*, 212–223. [CrossRef]
34. Chen, Z.; Zhou, Y.; Zhang, L.; Lin, C.; Huang, Y.; Tang, S. Ground moving target imaging and analysis for near-space hypersonic vehicle-borne synthetic aperture radar system with squint angle. *Remote Sens.* **2018**, *10*, 1966. [CrossRef]
35. Wan, J.; Zhou, Y.; Zhang, L.; Chen, Z. A Doppler ambiguity tolerated method for radar sensor maneuvering target focusing and detection. *IEEE Sens. J.* **2019**, *19*, 6691–6704. [CrossRef]
36. Wu, Y.; So, H.C.; Liu, H. Subspace-based algorithm for parameter estimation of polynomial phase signals. *IEEE Trans. Signal Process.* **2008**, *56*, 4977–4983.
37. Huang, P.; Liao, G.; Yang, Z.; Shu, Y.; Du, W. Approach for space-based radar maneuvering target detection and high-order motion parameter estimation. *IET Radar Sonar Navig.* **2015**, *9*, 732–741. [CrossRef]
38. Yang, J.; Liu, C.; Wang, Y. Imaging and parameter estimation of fast-moving targets with single-antenna SAR. *IEEE Geosci. Remote Sens. Lett.* **2013**, *11*, 529–533. [CrossRef]
39. Zhang, X.; Liao, G.; Zhu, S.; Gao, Y.; Xu, J. Geometry-information-aided efficient motion parameter estimation for moving-target imaging and location. *IEEE Geosci. Remote Sens. Lett.* **2015**, *12*, 155–159. [CrossRef]
40. Xia, X.G. Discrete Chirp-Fourier transform and its application to chirp rate estimation. *IEEE Trans. Signal Process.* **2000**, *48*, 3122–3133.
41. Almeida, L.B. The fractional Fourier transform and time–frequency representations. *IEEE Trans. Signal Process.* **1994**, *42*, 3084–3091. [CrossRef]

42. Lv, X.; Bi, G.A.; Wan, C.; Xing, M. Lv's distribution: Principle, implementation, properties, and performance. *IEEE Trans on Signal Process.* **2011**, *59*, 3576–3591. [CrossRef]
43. Liu, Q.H.; Nguyen, N. An accurate algorithm for nonuniform fast Fourier transforms (NUFFT's). *IEEE Microw. Guided Wave Lett.* **1998**, *8*, 18–20. [CrossRef]
44. Song, J.Y.; Liu, Q.H.; Torrione, P.; Collins, L. Two-dimensional and three-dimensional NUFFT migration method for landmine detection using ground-penetrating radar. *IEEE Trans. Geosci. Remote Sens.* **2006**, *44*, 1462–1469. [CrossRef]
45. Zhang, J.; Su, T.; Li, Y.; Zheng, J. Radar high-speed maneuvering target detection based on joint second-order keystone transform and modified integrated cubic phase function. *J. Appl. Remote Sens.* **2016**, *10*, 035009. [CrossRef]
46. Misiurewicz, J.; Kulpa, K.S.; Czekala, Z.; Filipek, T.A. Radar detection of helicopters with application of CNEAN method. *IEEE Trans. Aerosp. Electron. Syst.* **2012**, *48*, 3525–3537. [CrossRef]
47. Li, X.; Kong, L.; Cui, G.; Yi, W. CLEAN-based coherent integration method for high-speed multi-targets detection. *IET Radar Sonar Navig.* **2016**, *10*, 1671–1682. [CrossRef]
48. Maori, D.C.; Sikaneta, I. A generalization of DPCA processing for multichannel SAR/GMTI radars. *IEEE Trans. Geosci. Remote Sens.* **2013**, *51*, 560–572. [CrossRef]
49. Klemn, R. Introduction to space-time adaptive processing. *Electron. Commun. Eng. J.* **1999**, *11*, 5–12. [CrossRef]
50. DiPietro, R.C. Extended factored space-time processing for airborne radar systems. In Proceedings of the Twenty-Sixth Asilomar Conference on Signals, Systems & Computers, Pacific Grove, CA, USA, 26–28 October 1992; pp. 425–430.

Article

An Improved Generalized Chirp Scaling Algorithm Based on Lagrange Inversion Theorem for High-Resolution Low Frequency Synthetic Aperture Radar Imaging

Xing Chen, Tianzhu Yi, Feng He, Zhihua He and Zhen Dong *

College of Electronic Science, National University of Defense Technology, No. 109 De Ya Road, Changsha 410073, China

* Correspondence: dongzhen@nudt.edu.cn; Tel.: +86-156-1622-6428

Received: 2 June 2019; Accepted: 7 August 2019; Published: 10 August 2019

Abstract: The high-resolution low frequency synthetic aperture radar (SAR) has serious range-azimuth phase coupling due to the large bandwidth and long integration time. High-resolution SAR processing methods are necessary for focusing the raw data of such radar. The generalized chirp scaling algorithm (GCSA) is generally accepted as an attractive solution to focus SAR systems with low frequency, large bandwidth and wide beam bandwidth. However, as the bandwidth and/or beamwidth increase, the serious phase coupling limits the performance of the current GCSA and degrades the imaging quality. The degradation is mainly caused by two reasons: the residual high-order coupling phase and the non-negligible error introduced by the linear approximation of stationary phase point using the principle of stationary phase (POSP). According to the characteristics of a high-resolution low frequency SAR signal, this paper firstly presents a principle to determine the required order of range frequency. After compensating for the range-independent coupling phase above 3rd order, an improved GCSA based on Lagrange inversion theorem is analytically derived. The Lagrange inversion enables the high-order range-dependent coupling phase to be accurately compensated. Imaging results of P- and L-band SAR data demonstrate the excellent performance of the proposed algorithm compared to the existing GCSA. The image quality and focusing depth in range dimension are greatly improved. The improved method provides the possibility to efficiently process high-resolution low frequency SAR data with wide swath.

Keywords: synthetic aperture radar (SAR); low frequency; high-resolution; large bandwidth; improved generalized chirp scaling (GCS); Lagrange inversion theorem; range-dependent coupling

---

1. Introduction

Higher spatial resolution is an important development direction of synthetic aperture radar (SAR). Recent SAR systems are capable of resolutions in the decimeter regime. This requires the usage of large range bandwidth and wide azimuth beamwidth. The high-resolution, together with the all-weather day-and-night imaging capabilities, is turning SAR into an ideal tool for regular mapping and monitoring applications [1,2]. Moreover, microwaves can penetrate into vegetation and even the ground up to a certain depth [3]. The penetration capabilities depend on the carrier frequencies as well as on the complex dielectric constants, densities and conductivities of the observed targets. The high frequencies, like the X-band (8–12 GHz), show typically a high attenuation and are mainly backscattered on the top of the vegetation. Low frequencies, like P- and L-band (0.23$\sim$1 GHz and 1$\sim$2 GHz, respectively) [4], usually penetrate deep into vegetation, snow and ice. A high-resolution low frequency SAR system refers to a SAR system which operates with a low frequency (P- or

L-band) signal with a large fractional bandwidth (>0.2, i.e., the ultra-wideband SAR [2,5–8]) and a wide antenna beamwidth (corresponding to high azimuth resolution in the decimeter regime). The fractional bandwidth is defined by the ratio of the signal bandwidth to the center frequency. The combination of low frequency with large bandwidth and wide beam allows SAR to obtain high-resolution images of concealed targets, with the capability of penetrating the ground or foliage surface, thus it has broad applications for both military and civil purposes in recent years [9–12]. However, the large fractional bandwidth and long azimuth integration time used in high-resolution low frequency SAR bring new challenges to get high-quality images by the conventional image formation.

A crucial problem is the serious coupling between the range and azimuth frequencies in the phase of high-resolution low frequency SAR transfer function [13]. In two-dimension (2D) frequency domain, the phase is range-dependent and can be decomposed into two parts: the range-independent terms and the range-dependent terms. Different algorithms make different approximations of these two parts. At low frequencies, many of the simplifying assumptions made by traditional algorithms are not valid, such as range-Doppler algorithm [14] and chirp scaling algorithm (CSA) [15], resulting in serious image degradations as blurring and resolution loss. The problem stems from high-order range-azimuth phase coupling. Several approaches have been used in the processing of this type of SAR system. The time-domain algorithms and the wavenumber-domain Omega-K ($\omega - k$) algorithm [16–18] are two common options. The time-domain algorithms, often referred to as backprojection (BP) class algorithms [19], are most accurate and can easily adapt to all SAR configurations. Due to their computational complexity and the poor ability to integrate accurate autofocus algorithm into its imaging process, their use is restricted. The $\omega - k$ algorithm is an ideal solution without approximation in range cell migration (RCM) correction, which can focus data up to very high-resolution values regardless of their azimuth and range bandwidth. However, it is only applicable to spaceborne SAR data with a straight sensor trajectory and can only perform the range-independent motion compensation (MoCo). In addition, the Stolt interpolation makes it to be time-consuming . The extended Omega-K algorithm (EOKA) [20,21] is proposed to integrate the high precise range-dependent MoCo but it is still inefficient due to the Stolt interpolation.

For efficiency reasons, the chirp scaling class algorithms are still attractive. Efforts have been made to modify the CSA to process the low frequency SAR data. Without the interpolation, the chirp scaling class algorithms are effective and phase-preserving. The nonlinear chirp scaling algorithm (NCSA) [13] is proposed to take into account the cubic range-independent coupling phase and the range dependence of secondary range compression (SRC) term, which has better performance than CSA on processing the raw data of highly squint or low frequency SAR cases. Whereas the range dependence of cubic- and higher order terms are neglected. Some modified NCSAs are proposed in References [22–24], which resolve the high-order range-independent coupling terms. However, the cubic and higher order range-dependent coupling terms are still neglected. Besides, the first order approximation of range frequency modulation (FM) rate will introduce a quadratic phase error in the spectrum and degrades the focusing quality of image. Thus, the range focusing depth is restricted, which shows that the NCSA may not be suitable for the high-resolution low frequency SAR processing. A helpful comparison of the BP class algorithm, $\omega - k$ algorithm, EOKA and NCSA can be found in References [22,25]. In References [26,27], a generalized chirp scaling algorithm (GCSA) is developed for the SAR systems operating on wide bandwidths at low frequencies. The GCSA is an efficient arbitrary-order CSA that processes the data using the appropriate number of the approximation terms. Both the higher range-independent terms and range-dependent terms are considered. The GCSA efficiently extends the utility of frequency domain processing for high-resolution low frequency SAR systems. However, the imaging quality of GCSA also decreases as the fractional bandwidth get larger and beamwidth gets wider. In addition, the improvement of focus quality is not significant when the order is greater than 3rd and the edge targets of the range swath still have obvious degradation. Two main reasons lead to this phenomenon. One is that the residual coupling terms are still significant and the other is the error caused by the linear approximation of stationary phase point when solving the fast Fourier transform (FFT) expression. The linear approximation makes the range-dependent high-order phase terms not effectively compensated, even if a higher-order model is used.

The aim of this study is to overcome these two limits of GCSA and propose a more accurate approach than the GCSA for processing wide-swath, high-resolution low frequency SAR data. To our knowledge, the Lagrange inversion [28–31] gives the power series representation of the inverse of an analytic function, which is quite suitable for calculating the expression of stationary phase point. This paper utilizes Lagrange inversion to calculate a more precise expression of stationary phase point, while compensating for all range-independent coupling terms above 3rd order. In our approach, the high-order chirp scaling technique is extended to achieve the effect of the range-variant filtering required in high order phase terms. The experimental results show that the improved algorithm has a better focusing performance than the original GCSA. The resolution, sidelobe level and range focusing depth are significantly improved.

This paper is organized as follows. In Section 2, the signal model of high-resolution low frequency SAR is analyzed and the limitations of existing GCSA are briefly described. Then, a principle to determine the required order of range-dependent coupling phase is presented. In Section 3, an improved GCSA based on the Lagrange inversion theorem is introduced to focus the high-resolution low frequency SAR data. Focused results obtained by the conventional GCSA and the improved GCSA are presented and analyzed to verify our analysis in Section 4. A discussion is givne in Section 5. Finally, conclusions are drawn in Section 6.

2. Background and Problem Statement

2.1. Signal Model

The 2D spectrum is a key for the frequency domain algorithm development. In our analysis, we only consider the phase terms of the SAR signal and ignore the initial phase. The range-dependent SAR transfer function in the 2D frequency domain can be expressed as [14]

$$\Phi\left(f_\tau, f_\eta; R_0\right) = -\frac{4\pi R_0 f_0}{c}\sqrt{D^2(f_\eta) + \frac{2f_\tau}{f_0} + \frac{f_\tau^2}{f_0^2}} - \frac{\pi f_\tau^2}{K_r} \tag{1}$$

where f_τ and f_η represent the range frequency and the azimuth frequency, respectively. R_0 is the closest slant range from a point target to the radar trajectory, f_0 is the carrier frequency, c is the speed of light, K_r is the range chirp rate, $D(f_\eta) = \sqrt{1 - \frac{c^2 f_\eta^2}{4V_r^2 f_0^2}}$ is the migration factor, V_r is the moving velocity of the radar platform.

The first term in Equation (1) represents the coupling relationship between the range frequency and the azimuth frequency, which is called the range-azimuth coupling term. It varies with the slant range. The second term is the range modulation term. The square root term can be expanded into a Taylor series with respect to f_τ and kept up to the nth order,

$$p\left(f_\tau, f_\eta; n\right) = D(f_\eta) + \frac{f_\tau}{f_0 D(f_\eta)} + \frac{D^2(f_\eta) - 1}{2f_0^2 D^3(f_\eta)} f_\tau^2 + \sum_{i=3}^{n} \gamma_i f_\tau^i \tag{2}$$

where γ_i denotes the coefficient of the ith term and is given in Appendix A. The first term in Equation (2) corresponds to the azimuth modulation. The second term corresponds to the RCM (first-order coupling term). The third term denotes the SRC (second-order coupling term). The remainder higher order terms donates the high-order range-azimuth coupling. Different frequency algorithms are based on specific order approximations of this equation. For example, the CSA uses a 2nd order model and the NCSA uses a 3rd moder model. For the low frequency SAR with a small bandwidth and narrow beamwidth, quadratic approximation is enough, whereas for large bandwidth the high-order terms become significant.

Reconsider the phase in Equation (1). Actually, the coupling phase term can be decomposed into two parts: the range-independent terms and the range-dependent terms, namely,

$$\Phi\left(f_\tau, f_\eta; R_0, n\right) = -\frac{4\pi R_c f_0}{c} p\left(f_\tau, f_\eta; n\right) - \frac{4\pi \Delta R f_0}{c} p\left(f_\tau, f_\eta; n\right) - \frac{\pi f_\tau^2}{K_r} \tag{3}$$

where R_c represents the slant range of scene center, $\Delta R = R_0 - R_c$ represents the difference in slant range between the point target and the scene center point. The first term donates the range-independent coupling phase and the second term donates the range-dependent coupling phase, which varies with the slant range R_0.

2.2. The Limitations of the Conventional GCSA

The phase error due to the Taylor approximation can be expressed as

$$\Phi_{error}\left(f_\tau, f_\eta; R_0, n\right) = -\frac{4\pi R_0 f_0}{c}\left(\sqrt{D^2(f_\eta) + \frac{2f_\tau}{f_0} + \frac{f_\tau^2}{f_0^2}} - p(f_\tau, f_\eta; n)\right) \tag{4}$$

Note that this error gets larger when the slant range R_0 increases, the maximum range frequency $f_{\tau,\max}$ increases, the maximum azimuth frequency $f_{\eta,\max}$ increases, while the center frequency f_0 decreases. In high-resolution, low frequency and far range situations, the approximation error is large and a high order model is required.

To illustrate the phase error of Taylor expansion, a numerical analysis is carried out. Assume that the center frequency f_0 equals to 600 MHz, bandwidth B_r is 300 MHz, beamwidth is 29° and the target slant range R_0 is 12 km. Figure 1 shows the relationship between the phase errors and the range frequency for different order approximations. The maximal phase error of 4th order model is 1025° and the maximal phase error of 6th order approximation is about 81.48°, which indicates that the coupling phase terms above 6th order still have an important influence on the image quality. High-resolution imaging methods should take these terms into account.

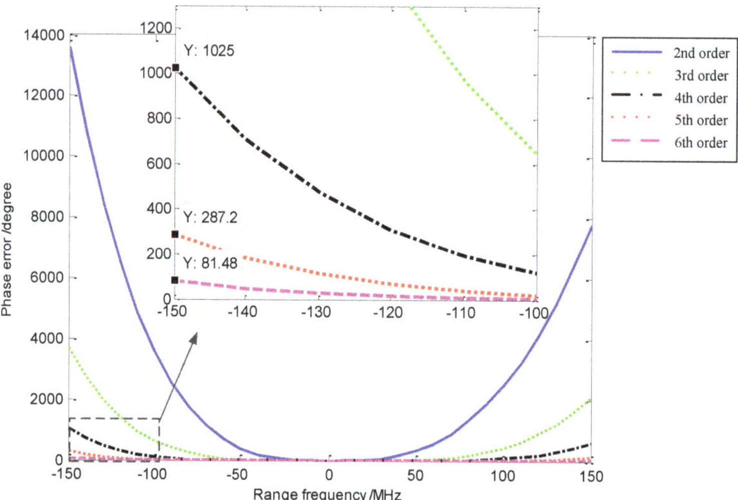

Figure 1. Phase errors of 2nd to 6th order Taylor series approximation. The point target at a range of 12 km with a center frequency of 600 MHz, a bandwidth of 300 MHz and a beamwidth of 29°.

In Reference [26], Zaugg et al. gives a guideline to determine required order nth from Equation (4) for proper focusing: If less than 30% of the support band has a phase error greater than $\pi/10$, then one can predict less than a 20% loss in the azimuth resolution defocusing. The conventional GCSA first compensates the 3rd to nth order range-independent coupling terms and then uses the high-order chirp scaling filter to compensate the range-dependent coupling phase. The coupling phase terms above nth order are neglected.

In the case of low-frequency high-resolution SAR with large bandwidth and wide beam, the coupling phase above 6th order still has a large proportion. Therefore, a higher order model is needed. However, when we use a model higher than 3rd order, the linear approximation of the stationary phase point when using the principle of stationary phase (POSP) will bring serious errors, which severely degrades the actual performance of high-order models in conventional GCSA. This error makes the high-order range-dependent coupling phase not to be accurately compensated, even though a higher order model is used. This is because the nonlinear FM component of signal cannot be neglected. Besides, the complexity of algorithm design will increase when using a higher order model.

Using the same parameters presented in the previous analysis, Figure 2 shows the residual range migrations of a point target at a range of 12 km after processing by the conventional GCSA. It is assumed here that the reference slant range is 10 km. The range migration crosses several range gates even using a 7th order model. This is because the conventional algorithm does not accurately compensate the range-dependent coupling terms.

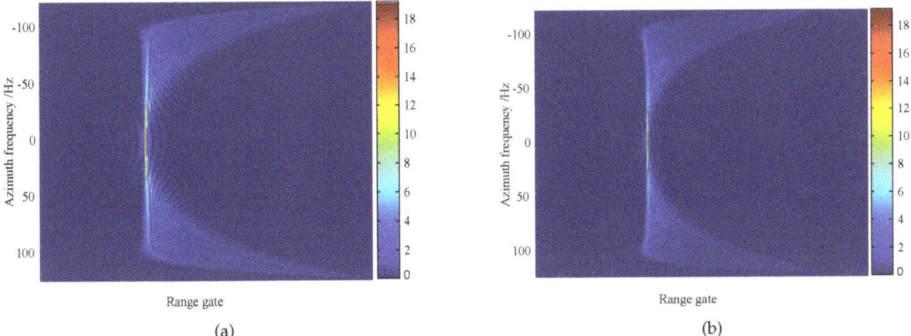

Figure 2. Range migrations of point target at a range of 12 km after RCMC. (**a**) Algorithm in Reference [26] (6th order model). (**b**) Algorithm in Reference [26] (7th order model).

2.3. New Principle to Determine the Required Order of Range Frequency

According to the previous analysis, the phase error of 6th order model is still significant. In fact, the range-independent couplings can be compensated by the reference range phase. Figure 3 shows the range-dependent coupling phase errors of different order models. If the range-independent coupling terms are firstly compensated, the maximum range-dependent phase error of 6th order model is 13.58°, which has a small effect on the imaging results.

Therefore, if the range-independent coupling terms are firstly compensated, we should only consider the range-dependent coupling terms to determine the required order. Thus, a new principle of the proposed method to determine the required order is given here. Assume that the Mth order range-dependent coupling term should be taken into consideration, the phase error should meet

$$-\frac{4\pi\Delta R f_0}{c}\left(\sqrt{D^2(f_\eta)+\frac{2f_\tau}{f_0}+\frac{f_\tau^2}{f_0^2}}-p\left(f_\tau,f_\eta;M\right)\right)\Bigg|_{f_\tau=\frac{B_r}{2},f_\eta=\frac{2f_0V_r}{c}\sin\theta}\leq\delta\pi \quad (5)$$

where θ is the beamwidth, δ equals to $1/10$ in this paper and it can be set to a larger value when the imaging accuracy is not high.

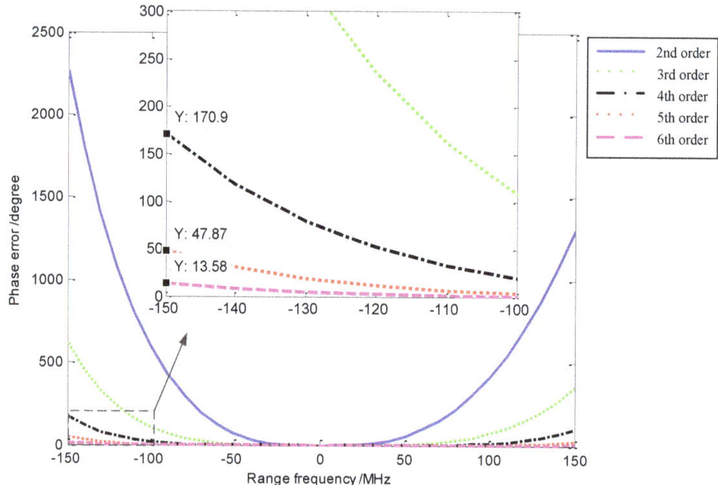

Figure 3. Range-dependent coupling phase errors of 2nd to 6th order Taylor series approximation. The point target at the slant range 12 km with a center frequency of 600 MHz, a bandwidth of 300 MHz, a beamwidth of 29° and the reference slant range of 10 km.

Inequality (5) is an important basis for the proposed algorithm. It can be seen that the phase error is positively correlated with the range bandwidth B_r, the beamwidth θ and the scene size $2\Delta R$ and negatively correlated with the center frequency f_0. If the center frequency and scene size are given, the required order is determined by the range bandwidth and the azimuth beamwidth.

Figure 4 shows an example of a P-band SAR system. The center frequency is 800 MHz and the scene size equals to 600 m. The required order varies with different beamwidth and fractional bandwidth. As can be seen from Figure 4, under the given parameters, the 2nd order model can only process data with a fractional bandwidth of less than 0.34. As the beamwidth increases, the fractional bandwidth value that can be processed is smaller. For low frequency SAR systems with large bandwidths and wide beams, a higher order range-dependent coupling phase should be considered.

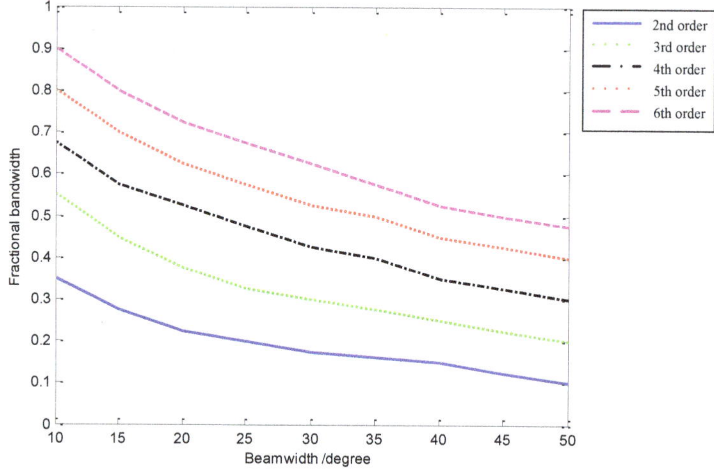

Figure 4. An example of the required order of different beamwidth and fractional bandwidth in a P-band synthetic aperture radar (SAR) system.

3. The Improved GCSA Based on Lagrange Inversion Theorem

3.1. Procedure of Algorithm

Based on the analysis in Section 2, we proposed an improved GCSA in this section. The flow chart of the proposed algorithm is shown in the Figure 5. There are two main improvements to the algorithm. The first is to perform all the high-order (≥3rd) reference phase compensation in the 2D frequency domain. The second is to use the Lagrange inversion theorem to solve the expression of the range FFT and inverse FFT (IFFT). The derivation of the proposed algorithm is given in Section 3.2. The steps of the proposed algorithm are as follows:

Step 1: Select the parameter M according to the principle of Equation (5). Calculate $q_2 \sim q_M$, $X_3 \sim X_M$ and $C_2 \sim C_M$ according to Equations (25) and (26). Then, 2D FFT is implemented to transfer the data into 2D frequency domain.

Step 2: Multiply the range-independent high-order phase correction (HOPC) and perturbation equations in the 2D frequency domain. And a range IFFT is carried out to transfer the data into the range-Doppler (RD) domain.

Step 3: Multiply the high-order chirp scaling phase function. Then, the data are transformed into 2D frequency domain by range FFT.

Step 4: Multiply the range compression function to perform the RC, SRC and bulk RCMC. Then, the data are transferred into RD domain by IFFT along the range.

Step 5: Multiply the azimuth compression and residual phase correction function. And the images can be obtained by azimuth IFFT.

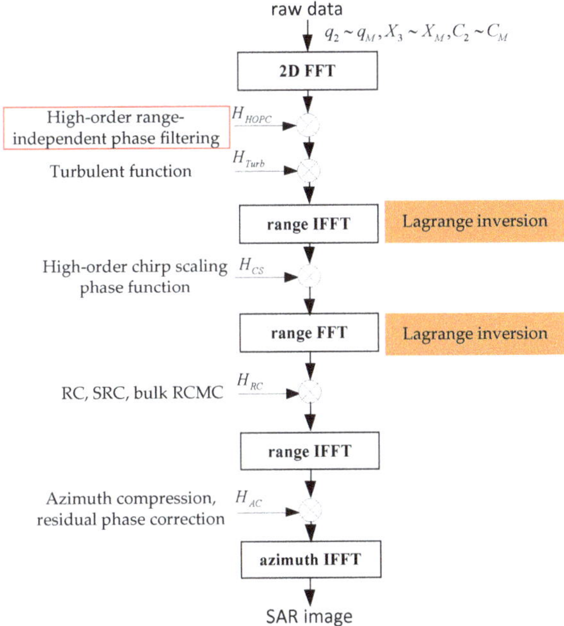

Figure 5. The flow chart of the proposed algorithm.

3.2. Theoretical Formulation

As mentioned in Section 2, the high-order terms of high-resolution low frequency SAR account for a large proportion. If it can not be eliminated, the image will be deteriorated dramatically. In order to reduce the phase error and improve imaging accuracy, the range-independent high-order (≥3rd) coupling phases of Equation (3) is firstly compensated in the 2D frequency domain. The 3rd order is chosen because of the need to preserve the chirp information of the range signals. The HOPC function at the reference range is given by

$$H_{HOPC} = \exp\left[\frac{4\pi R_c f_0}{c}\left(\sqrt{D^2(f_\eta) + \frac{2f_\tau}{f_0} + \frac{f_\tau^2}{f_0^2}} - D(f_\eta) - \frac{f_\tau}{f_0 D(f_\eta)} - \frac{D^2(f_\eta) - 1}{2f_0^2 D^3(f_\eta)} f_\tau^2\right)\right] \tag{6}$$

Suppose that the Mth range-dependent coupling should be taken into account. The residual signal after HOPC can be expressed as

$$\Phi(f_\tau, f_\eta; R_0) = -\frac{4\pi R_0 f_0}{c} D(f_\eta) - \frac{4\pi R_0}{cD(f_\eta)} f_\tau - \frac{\pi}{K_m} f_\tau^2 - \frac{4\pi \Delta R f_0}{c} \sum_{i=3}^{M} \gamma_i f_\tau^i \tag{7}$$

where K_m donates the range FM rate and can be expressed as

$$K_m = \frac{K_r}{1 - K_r \frac{cR_0 f_\eta^2}{2V_r^2 f_0^3 D(f_\eta)^3}} \tag{8}$$

It can be found that the residual high-order coupling becomes zero at the reference range but the residual high-order phase at other ranges increases as the target away from the reference range. Then we filter the data with a turbulent function

$$H_{Turb} = \exp\left(j\pi \sum_{i=3}^{M} X_i f_\tau^i\right) \tag{9}$$

This filtering step provides an accurate accommodation of the range dependence of the high-order terms. After the turbulent compensation, the phase can be expressed as

$$\begin{aligned} \Phi_1(f_\tau, f_\eta; R_0) &= -\tfrac{4\pi R_0 f_0}{c} D(f_\eta) - \tfrac{4\pi R_0}{cD(f_\eta)} f_\tau - \tfrac{\pi}{K_m} f_\tau^2 + \pi \sum_{i=3}^{M} \left(X_i - \tfrac{4\pi \Delta R f_0}{c} \gamma_i\right) f_\tau^i \\ &= \phi_0 + \phi_1 f_\tau + \phi_2 f_\tau^2 + \sum_{i=3}^{M} \phi_i f_\tau^i \end{aligned} \tag{10}$$

with

$$\begin{aligned} \phi_0 &= -\tfrac{4\pi R_0 f_0}{c} D(f_\eta) \\ \phi_1 &= -\tfrac{4\pi R_0}{cD(f_\eta)} = -2\pi\tau_d \\ \phi_2 &= -\tfrac{\pi}{K_m} \\ \phi_i &= \pi\left[X_i - 2f_0 D(f_\eta)\gamma_i \Delta\tau\right] (3 \le i \le M) \end{aligned} \tag{11}$$

Then the range IFFT is performed along the range direction. Based on the POSP, the relationship between τ and f_τ can be expressed as

$$\tau = -\frac{1}{2\pi}(\phi_1 + 2\phi_2 f_\tau + \ldots + M\phi_M f_\tau^{M-1}) \tag{12}$$

In Reference [26], only the first order term of f_τ is retained when solving the stationary phase point, that is, $f_\tau = -\frac{\pi}{\phi_2}\left(\tau + \frac{\phi_1}{2\pi}\right)$ .This approximation is effective in most cases. However, as the center frequency decreases, the beamwidth and bandwidth increase, this approximation introduces severe phase errors, resulting in significant degradation throughout the image. To solve this problem, we use the Lagrange inversion to find a more accurate solution for f_τ.

In mathematical analysis, the Lagrange inversion theorem gives the Taylor series expansion of the inverse function of an analytic function and it can be expressed as Theorem 1 [28–31].

Theorem 1. *Suppose w is defined as a function of z by an equation of the form $w = h(z)$, where h is analytic at a point z_0 and $h'(z_0) \neq 0$. Then it is possible to invert or solve the equation for z, expressing it in the form $z = g(w)$ given by a power series*

$$g(w) = z_0 + \sum_{n=1}^{\infty} g_n (w - h(z_0))^n \tag{13}$$

where

$$g_n = \frac{1}{n!} \lim_{z \to z_0} \frac{d^{n-1}}{dz^{n-1}} \left[\left(\frac{z - z_0}{h(z) - h(z_0)} \right)^n \right] \tag{14}$$

$g(w)$ represents an analytic function of w in a neighbourhood of $w = h(z_0)$.

As is shown in Equation (12), τ is a power series of f_τ, that is, $\tau = h(f_\tau)$. Let $z_0 = 0$ and $h(z_0) = -\frac{\phi_1}{2\pi}$. According to the Lagrange inversion theorem, the stationary phase point can be given by

$$\begin{aligned} f_\tau = & -\tfrac{\pi}{\phi_2}\left(\tau + \tfrac{\phi_1}{2\pi}\right) - \tfrac{3\pi^2\phi_3}{2\phi_2^3}\left(\tau + \tfrac{\phi_1}{2\pi}\right)^2 + \tfrac{\pi^3\left(-9\phi_3^2+4\phi_2\phi_4\right)}{2\phi_2^5}\left(\tau + \tfrac{\phi_1}{2\pi}\right)^3 \\ & -\tfrac{5\pi^4\left(27\phi_3^3-24\phi_2\phi_3\phi_4+4\phi_2^2\phi_5\right)}{8\phi_2^7}\left(\tau + \tfrac{\phi_1}{2\pi}\right)^4 + \tfrac{3\pi^5\left(-189\phi_3^4+252\phi_2\phi_3^2\phi_4-60\phi_2^2\phi_3\phi_5+8\phi_2^2\left(-4\phi_4^2+\phi_2\phi_6\right)\right)}{8\phi_2^9}\left(\tau + \tfrac{\phi_1}{2\pi}\right)^5 + \ldots \end{aligned} \tag{15}$$

The detailed derivation of Equation (15) is given in Appendix C. Therefore, the IFFT expression of Equation (10) can be expressed as

$$\Phi_2(\tau, f_\eta; R_0) = \phi_0 + A_2(\tau - \tau_d)^2 + A_3(\tau - \tau_d)^3 + \ldots + A_M(\tau - \tau_d)^M \tag{16}$$

with (Here, only $A_2 \sim A_6$ are given due to space restraints.)

$$\begin{cases} A_2 = -\frac{\pi^2}{\phi_2} \\ A_3 = -\frac{\pi^3\phi_3}{\phi_2^3} \\ A_4 = \frac{\pi^4\left(-9\phi_3^2+4\phi_2\phi_4\right)}{4\phi_2^5} \\ A_5 = -\frac{\pi^5\left(27\phi_3^3-24\phi_2\phi_3\phi_4+4\phi_2^2\phi_5\right)}{4\phi_2^7} \\ A_6 = \frac{\pi^6\left(-189\phi_3^4+252\phi_2\phi_3^2\phi_4-60\phi_2^2\phi_3\phi_5+8\phi_2^2\left(-4\phi_4^2+\phi_2\phi_6\right)\right)}{8\phi_2^9} \\ \ldots. \end{cases} \tag{17}$$

where $\tau_d = h(z_0) = 2R_0/(cD(f_\eta))$ is the time delay in RD domain. The variation of τ_d with f_η is called range migration, which must be removed before azimuth compression. The shape of the range migration trajectory depends on the target slant R_0.

The high-order CS function can be given by

$$H_{CS}(\tau, f_\eta) = \exp\left[j\pi q_2\left(\tau - \tau_{ref}\right)^2 + j\pi \sum_{i=3}^{M} q_i\left(\tau - \tau_{ref}\right)^i\right] \tag{18}$$

where $\tau_{ref} = 2R_c/(cD(f_\eta))$ is the reference trajectory. This step aims to compensate the range-dependent coupling caused by the large fractional bandwidth and wide beamwidth. The phase after high-order CS filter can be expressed as

$$\Phi_3(\tau, f_\eta; R_0) = \phi_0 + A_2(\tau - \tau_d)^2 + \sum_{i=3}^{M} A_i(\tau - \tau_d)^i + \pi q_2\left(\tau - \tau_{ref}\right)^2 + \pi \sum_{i=3}^{M} q_i\left(\tau - \tau_{ref}\right)^i \tag{19}$$

According to the chirp scaling principle, the desired trajectory τ_s of the target located at range R_0 has the same shape as the reference trajectory and the relationship between τ_s, τ_{ref} and τ_d can be expressed as

$$\begin{aligned} \tau_{ref} &= \tau_s - \alpha \Delta\tau \\ \tau_d &= \tau_s - (\alpha - 1)\Delta\tau \end{aligned} \tag{20}$$

where $\Delta\tau = 2\Delta R / (cD(f_\eta))$ and $\alpha = D(f_\eta)/D(f_{\eta c})$. Using the relationship in Equation (20) and expand Equation (19) at τ_s, we obtain

$$\Phi_4(\tau, f_\eta; R_0) = \phi_0 + \pi C_0(\Delta\tau) + \pi \sum_{i=1}^{M} C_i (\tau - \tau_s)^i \tag{21}$$

The expressions for coefficients C_i are given in Appendix B. As can be seen form Equation (11), parameters A_i and C_i imply the range FM rate K_m, which is dependent on azimuth frequency and slant range. Imaging performance is affected by approximations to K_m. To model the range range-dependence of K_m, we expand it at the reference slant range with Taylor series and keep up to the second order,

$$K_{m,app} \approx K_f + K_s K_f^2 \Delta\tau + K_s^2 K_f^3 \Delta\tau^2 \tag{22}$$

where K_f represents the FM rate at the reference slant range and can be expressed as

$$K_f = \frac{K_r}{1 - K_r \tau_{ref} K_s} \tag{23}$$

with

$$K_s = \frac{c^2 f_\eta^2}{4v^2 f_0^3 D(f_\eta)^2} \tag{24}$$

According to Appendix B, the expressions of $q_i (i \geq 2)$ and $X_i (i \geq 3)$ can be expressed as

$$\begin{cases} q_2 = K_f \frac{1-\alpha}{\alpha} \\ q_3 = \frac{K_f^2 K_s (1-\alpha)}{3\alpha} \\ X_3 = \frac{K_s(\alpha-2)}{3K_f(\alpha-1)} \\ q_4 = -\frac{K_f^3 \left(9K_f K_s(\alpha-1) X_3 + 2Ks^2 - 6D(f_\eta) f_0 (\alpha-1)\gamma_3\right)}{12\alpha} \\ X_4 = \frac{9K_f K_s(\alpha-2)X_3 - 27K_f^2(\alpha-1)X_3^2 + 2Ks^2 - 6D(f_\eta) f_0(\alpha-1)\gamma_3}{12K_f(\alpha-1)} \\ q_5 = -\frac{45K_f^6 K_s(\alpha-1)X_3^2 + 12K_f^5 X_3\left(K_s^2 - 3D(f_\eta) f_0(\alpha-1)\gamma_3\right)}{20\alpha} \\ -\frac{16K_f^5 K_s(\alpha-1)X_4 - 4K_f^4 D(f_\eta) f_0(3K_s\gamma_3 + 2(\alpha-1)\gamma_4)}{20\alpha} \\ X_5 = -\frac{-45K_f^2 K_s(\alpha-2)X_3^2 + 12K_f X_3\left(-K_s^2 + 10K_f(\alpha-1)X_4 + 3D(f_\eta) f_0(\alpha-2)\gamma_3\right)}{20K_f(\alpha-1)} \\ -\frac{4D(f_\eta) f_0(3K_s\gamma_3 + 2(\alpha-2)\gamma_4) - 16K_f K_s(\alpha-2)X_4 + 135K_f^3(\alpha-1)X_3^3}{20K_f(\alpha-1)} \\ \ldots \end{cases} \tag{25}$$

And coefficients C_i becomes

$$\begin{cases} C_1 = 0 \\ C_2 = q_2 + K_f \\ C_3 = q_3 + K_f^3 X_3 \\ C_4 = q_4 + \frac{9}{4} K_f^5 X_3^2 + K_f^4 X_4 \\ C_5 = q_5 + \frac{27}{4} K_f^7 X_3^3 + 6K_f^6 X_3 X_4 + K_f^5 X_5 \\ \ldots \end{cases} \tag{26}$$

After the chirp scaling operation, a range FFT is carried to transform Equation (21) into the 2D frequency domain. Similarly, f_τ is a power series of τ,

$$f_\tau = \frac{1}{2}\left(2C_2\left(\tau - \tau_s\right) + 3C_3(\tau - \tau_s)^2 + ...MC_M(\tau - \tau_s)^{M-1}\right) \tag{27}$$

Use the Lagrange inversion theorem again. Let $z_0 = \tau_s$ and $h(z_0) = 0$, the stationary phase point is given by

$$\begin{aligned}\tau = \tau_s + \tfrac{1}{C_2}f_\tau - \tfrac{3C_3}{2C_2^3}f_\tau^2 + \frac{\left(\frac{9C_3^2}{2} - 2C_2C_4\right)}{C_2^5}f_\tau^3 - \frac{\left(\frac{135C_3^3}{8} - 15C_2C_3C_4 + \frac{5}{2}C_2^2C_5\right)}{C_2^7}f_\tau^4 \\ + \frac{\left(\frac{567C_3^4}{8} - \frac{189}{2}C_2C_3^2C_4 + 12C_2^2C_4^2 + \frac{45}{2}C_2^2C_3C_5 - 3C_2^3C_6\right)}{C_2^9}f_\tau^5 + ...\end{aligned} \tag{28}$$

Thus, the phase after range FFT can be expressed as

$$\Phi_5(f_\tau, f_\eta) = \phi_0 + \pi C_0(\Delta\tau) - 2\alpha\tau_d f_\tau + \pi\sum_{i=1}^{M} E_i f_\tau^i \tag{29}$$

with

$$\begin{cases} E_1 = -2\tau_{ref}(1 - \alpha) \\ E_2 = -\frac{1}{C_2} \\ E_3 = \frac{C_3}{C_2^3} \\ E_4 = \frac{\left(-9C_3^2 + 4C_2C_4\right)}{4C_2^5} \\ E_5 = \frac{\left(27C_3^3 - 24C_2C_3C_4 + 4C_2^2C_5\right)}{4C_2^7} \\ ... \end{cases} \tag{30}$$

The first term of Equation (29) represents the azimuth compression phase, the second term is the residual phase, the third term is the linear phase corresponding to the target position, the remainder are related to the range compression (RC), SRC and bulk RCM, which are range invariant. Thus, the bulk RCM, SRC and RC can be compensated by a conjugate multiply in the 2D frequency domain. The filtering function can be expressed as

$$H_{RC}(f_\tau, f_\eta) = \exp(-j\pi\sum_{i=1}^{M} E_i f_\tau^i) \tag{31}$$

After the compensation, the phase of signal can be expressed as

$$\Phi_6(f_\tau, f_\eta) = \phi_0 + \pi C_0(\Delta\tau) - 2\pi\alpha\tau_d f_\tau \tag{32}$$

An IFFT is carried out along the range direction and the signal phase becomes

$$\Phi_7(\tau, f_\eta) = \phi_0 + \pi C_0(\Delta\tau) \tag{33}$$

Therefore, the azimuth compression function is given by

$$H_{AC}(\tau, f_\eta) = \exp\left[j\pi\left(\frac{4\pi R_0 f_0}{c}\left(D(f_\eta) - 1\right) - C_0(\Delta\tau)\right)\right] \tag{34}$$

Finally, the echo data are transformed into 2D time domain by a azimuth IFFT and the focused image is obtained.

4. Experiment Results and Analysis

In this section, we provide some imaging results to demonstrate the performance of proposed algorithm and the analysis of principle. The system parameters are listed in Table 1. The parameters of platform velocity, pulse duration and center slant range of each SAR system are set to the same value. The reference slant range is selected as the scene center slant range. The theoretical azimuth and range resolutions are evaluated by $\rho_a = 0.886c/(4f_0\sin(\theta/2))$ and $\rho_r = 0.886c/(2B_r)$. The maximum Doppler frequency can be expressed as

$$f_D = \frac{2V_r}{\lambda_{\min}}\sin\left(\frac{\theta_{\max}}{2}\right) \tag{35}$$

where $\lambda_{\min}$ and $\theta_{\max}$ are the minimum signal wavelength and the maximum azimuth angle. The PRF must not be chosen less than two times of f_D. The range oversampling rate is set to 1.2.

Table 1. SAR system simulation parameters.

Parameters	P-band	L-band
Center frequency f_0 (MHz)	600	1360
Bandwidth B_r (MHz)	300	272/544/816/1088
Fractional bandwidth (%)	50	20/40/60/80
Beamwidth θ (°)	29	11
Azimuth resolution ρ_a (m)	0.44	0.5
PRF (Hz)	240	240
Velocity of platform V_r (m/s)	100	100
Pulse duration T_r (us)	10	10
Center slant range R_c (km)	10	10

Firstly, to investigate the effects of two improvements in the proposed algorithm, a P-band SAR with a center frequency of 600 MHz was simulated. The fractional bandwidth is set to 50% (corresponding to a range resolution of 0.44 m) and the beamwidth is set to 29° (corresponding to an azimuth resolution of 0.44 m). Nine targets are arranged in the illuminated scene along the azimuth center at different distances form the reference range with an interval of 200 m. The conventional GCSA in Reference [26] is used in comparison with the proposed algorithm. According to the principle in Section 2.3, the data is processed with a 6th order model. To highlight the effect of Lagrange inversion, the conventional GCSA with Lagrange inversion (ignoring the range-independent coupling terms above 6th order) is also used in comparison. The 2D focused images of targets at the ranges R_c, $R_c + 800$ m and $R_c + 1600$ m are shown in Figure 6. In these figures, contour maps donate the 2D focusing quality. To evaluate the quality of these images quantitatively, the resolution (Res), peak sidelobe ratio (PSLR) and integrated sidelobe ratio (ISLR) along the range and azimuth directions are presented in Table 2. No wighting function or sidelobe control approach is used to obtain a fair comparison. Note that the ideal response is not completely symmetric in the range direction due to the significant range-azimuth coupling.

Figure 6a–c and Table 2 illustrate that the conventional GCSA causes deterioration of the images. The images of three targets are defocused in two directions, even at the scene center targets. The Res, PSLR and ISLR degrade considerably. This issue becomes increasingly serious as the distance increases. This problem is caused by the residual high-order range-independent coupling terms and the phase error introduced by approximate of stationary phase point. As the fractional bandwidth increases, even the scene center point has a severe degradation. Especially, the asymmetric sidelobe in range dimension is obvious. Figure 6d–f shows the imaging results obtained by the GCSA with Lagrange inversion. It can be seen that all three targets are effectively focused, which indicates that the Lagrange inversion can eliminate the range-dependent coupling terms. However, since the range-independent coupling terms above 6th order are neglected, the sidelobes in range direction are severe and the images have a certain quality degradation. The imaging results obtained by the proposed algorithm are shown in Figure 6g–i, with better quality than Figure 6d–f. From the quality indices presented in Table 2, the imaging coherency is good over the entire swath. These benefits stem from the compensation of coupling terms above 6th order and Lagrange inversion.

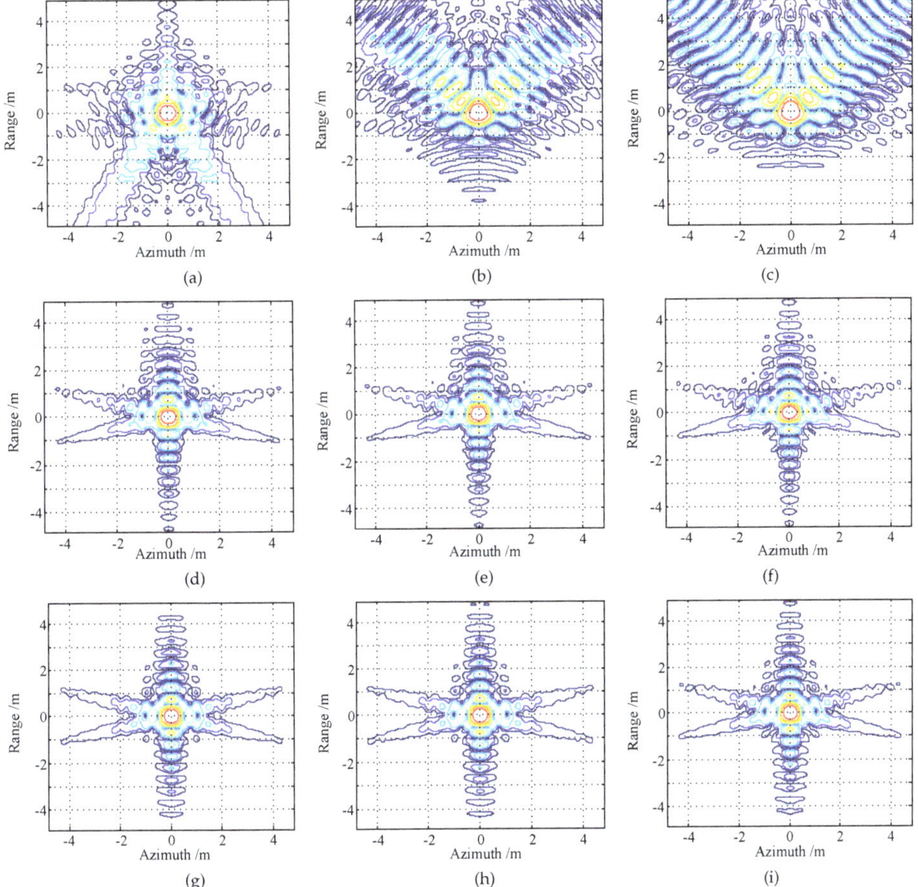

Figure 6. Focused results by different algorithm for P-band SAR data with a center frequency of 600 MHz. (**a–c**) Conventional generalized chirp scaling algorithm (GCSA). (**d–f**) Conventional GCSA with Lagrange inversion. (**g–i**) Proposed algorithm. The three subgraphs of each row correspond to contours of the targets located at R_c, $R_c + 800$ m and $R_c + 1600$ m, respectively. The dynamic range of contour is -35 dB$\sim$0 dB.

To evaluate the accuracy of proposed algorithm better, we measure the Res and differential resolutions (DRES) [32] at different slant range. Figure 7 shows the Res and DRES of nine point targets processed in different algorithms. The references for the DRES measurements are the range and azimuth resolutions obtained by $\omega - k$ algorithm. It can seen that the range and azimuth resolutions loss of GCSA is greater than 13% compared with the $\omega - k$ algorithm but the resolutions loss of the proposed algorithm is less than 1%. We can conclude that the focusing performances of the proposed algorithm are much better than the ones of GCSA and closed to the ones of $\omega - k$ algorithm. The image quality and focusing depth in range dimension are greatly improved.

Table 2. Measured parameters of the imaging results for Figure 6.

Method	$R_0 - R_c$	Azimuth			Range		
		Res /m	PSLR /dB	ISLR /dB	Res /m	PSLR /dB	ISLR /dB
Conventional GCSA in Reference [26]	0	0.4927	−16.9620	−14.2797	0.5221	−14.8340	−9.9040
	800 m	0.5344	−21.0146	−15.0018	0.5690	−17.1714	−11.5202
	1600 m	0.5615	−20.7017	−14.6937	0.6341	−16.6871	−10.2194
Conventional GCSA+Lagrange	0	0.4385	−15.0555	−13.7173	0.4505	−12.4212	−9.7310
	800 m	0.4385	−15.1025	−13.7213	0.4505	−12.4613	−9.7596
	1600 m	0.4427	−14.8451	−13.2582	0.4518	−12.8161	−10.1929
Proposed algorithm	0	0.4365	−15.1755	−13.9167	0.4479	−12.9714	−10.2222
	800 m	0.4365	−15.1673	−13.9233	0.4479	−13.0231	−10.2356
	1600 m	0.4406	−15.0641	−13.5990	0.4492	−13.2844	−10.5722

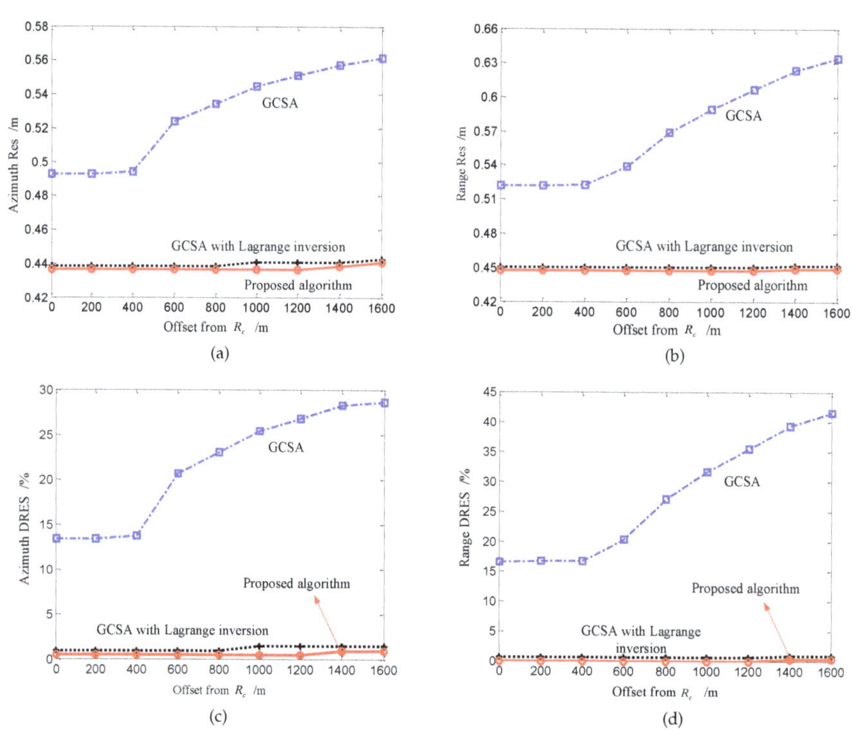

Figure 7. Resolutions (Res) and differential resolutions (DRES) in azimuth and range given by different algorithm, where the resolutions obtained by $\omega - k$ algorithm are references. The DRES presents the loss in spatial resolutions. Nine targets are arranged in the illuminated scene along the azimuth center at different distances form the reference range with an interval of 200 m. (**a**) Azimuth Res. (**b**) Range Res. (**c**) Azimuth DRES. (**d**) Range DRES.

Secondly, in order to better validate the performance of the proposed algorithm, the point scatterer was placed at the edge of the swath, where $R_0 - R_c$ = 2 km. A typical L-band SAR system with a center frequency of 1.36 GHz was simulated. The beamwidth is set to 11° (corresponding to an azimuth resolution of 0.5 m). The fractional bandwidth is set to 20%, 40%, 60% and 80%, respectively. According to Equation (5), four sets of data are focused by the 3rd order, 4th order, 6th order and 7th order models, respectively.

Figure 8 shows the Res and DRES versus fractional bandwidth for the GCSA and proposed algorithm. The resolutions obtained by $\omega - k$ algorithm are references. When the fractional bandwidth is 20%, both the GCSA and proposed algorithm can achieve good focusing performance. As the fractional bandwidth increases, the performance of GCSA drops dramatically. If the resolution loss threshold is 10%, the GCSA can only process the data where the fractional bandwidth is less than 30%. However, the proposed algorithm achieves nearly the theoretically resolutions for fractional bandwidths up to 80% for L-band. The resolution broadening is less than 1%, which shows the performance is greatly improved over that of the original GCSA.

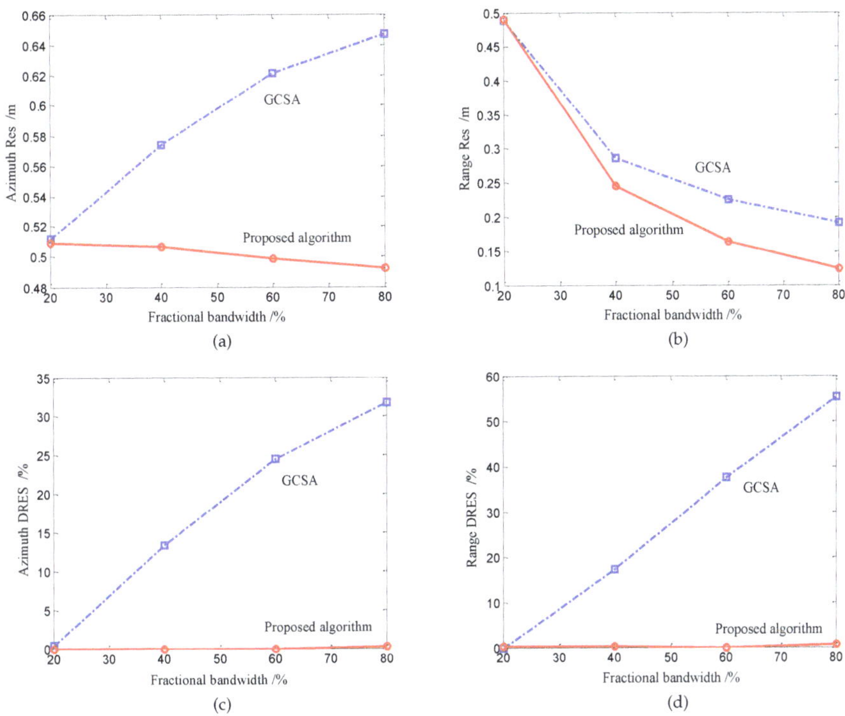

Figure 8. Resolutions (Res) and differential resolutions (DRES) in azimuth and range given by the generalized chirp scaling algorithm (GCSA) and proposed algorithm, where the resolutions obtained by $\omega - k$ algorithm are references. (**a**) Azimuth Res. (**b**) Range Res. (**c**) Azimuth DRES. (**d**) Range DRES.

Next, to illustrate the worst case shape of the proposed image of the point target, Figure 9 illustrates the contour plots, range profiles and azimuth profiles of the processed images at 80% fractional bandwidth. In both figures, the contour maps denote the 2D focusing quality and the profiles represent the focusing quality along the azimuth and range directions. The measured parameters are shown in Table 3. As can be seen, the results of GCSA in this case suffers form severe distortion and broadening. The signal in both range and azimuth are almost defocused. However, the proposed algorithm preserves the focusing performance. It is easy to recognize that the images obtained by the proposed algorithm are well focused, as are shown in Figure 9d–f. The nearly theoretical values of spatial resolution, PSLR and ISLR are obtained, which demonstrates the validity of the proposed algorithm. The good performance is given by the high-order range-independent phase filtering and the Lagrange inversion, which greatly reduces the range-dependent phase error. By comparing the range and azimuth profiles, it is evident that the Lagrange inversion makes the range-dependent coupling terms effectively compensated. The focusing depth in range dimension is greatly improved. The improved algorithm is consistent with the conventional GCSA in terms of computational complexity but the focusing performance is significantly improved. The improved method provides an attractive solution for processing the low frequency large bandwidth and wide beamwidth SAR data.

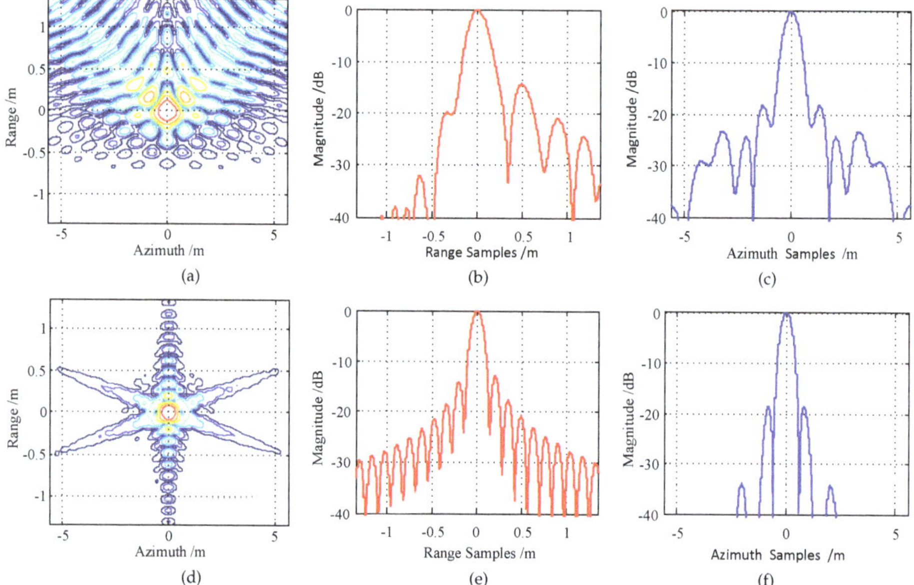

Figure 9. Focused results of point scatterers for L-band SAR data at 80% fractional bandwidth, using the generalized chirp scaling algorithm (GCSA) and proposed algorithm. (**a**–**c**) Conventional GCSA. (**d**–**f**) Proposed algorithm. The three subgraphs of each row correspond to the contour maps, range profiles and azimuth profiles, respectively. The dynamic range of contour is −35 dB∼0 dB.

Table 3. Measured parameters of imaging results for Figure 9.

Method	Azimuth			Range		
	Res /m	**PSLR /dB**	**ISLR /dB**	**Res /m**	**PSLR /dB**	**ISLR /dB**
Conventional GCSA in Reference [26]	0.6471	−18.3522	−12.7106	0.1915	−14.2602	−9.0208
Proposed algorithm	0.4922	−18.5128	−16.9421	0.1239	−12.9655	−9.5501

Finally, to further test the analysis presented in this paper, a SAR real image (Longmen, Henan, China) is used as the input radar cross section to generate SAR echo. The center frequency of the transmitted signal is 400 MHz, the bandwidth is 250 MHz, the beamwidth is 25°, the velocity is 120 m/s, the PRF is 200 Hz, the pulse duration is 10 us and the center slant range is 10 km. The scene size is 3.0 km in range and 1.5 km in azimuth. The resolutions are 0.53 m in range and 0.76 m in azimuth. In the simulation, the input image is a real complex image. Each cell of the complex image is treated as a point scatterer (this actually contains the target signal and noise). No additional noise was added during the simulation. Figure 10 shows the imaging results processed by the conventional GCSA and the improved algorithm.

As is shown in Figure 10, at the center range region, the focusing quality of the two images is nearly the same. However, at the near and far range region, the texture character of the image obtained by improved GCSA is clearer than that obtained by conventional GCSA. It is obvious that better focusing results are obtained by the proposed method. The image has very high quality. The roads, rivers and farmland can be clearly distinguished. The edge scene along range dimension is well focused. It is evident that the range focusing depth is greatly improved.

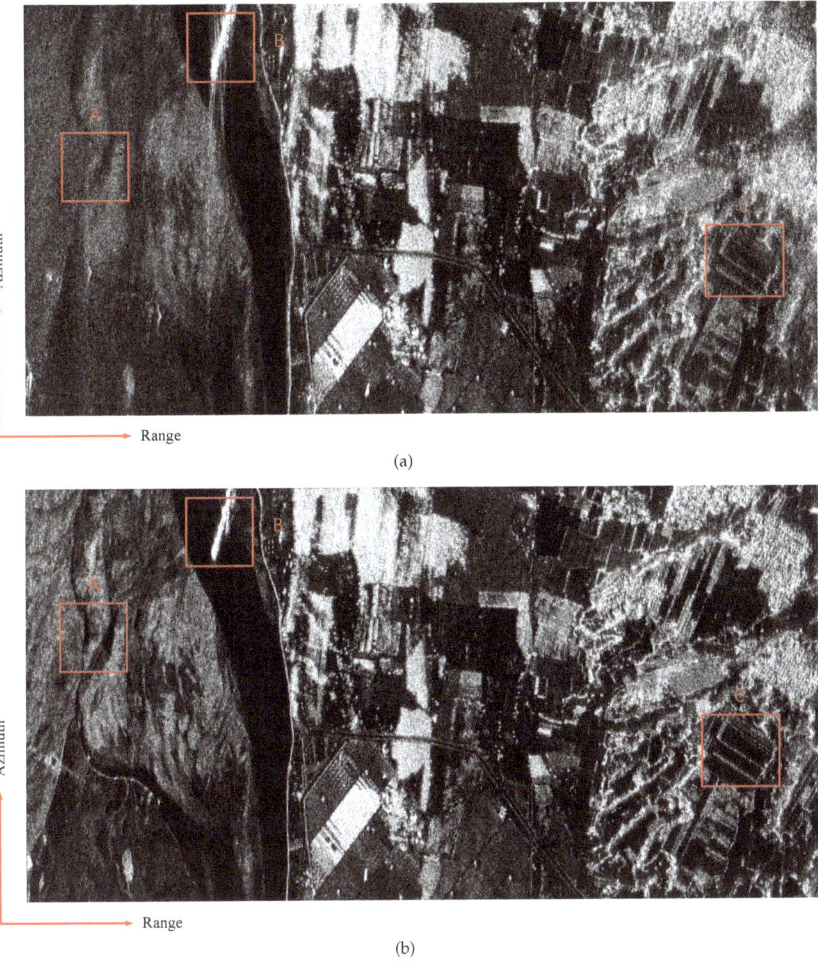

Figure 10. Comparison of imaging results of real data processed by different algorithms. (**a**) Conventional generalized chirp scaling algorithm (GCSA). (**b**) Proposed algorithm.

In order to have a distinct contrast, three subregions marked by red solid rectangle are extracted and analyzed in detail. The zooms regions are shown in Figure 11a–f. The entropy [33,34] of images is calculated to compare the focus quality and shown in Table 4. It is generally acknowledged that SAR images with better quality focus have smaller entropy. It is easy to find that images obtained by proposed method have smaller entropy. Therefore, the effectiveness of proposed algorithm is again validated.

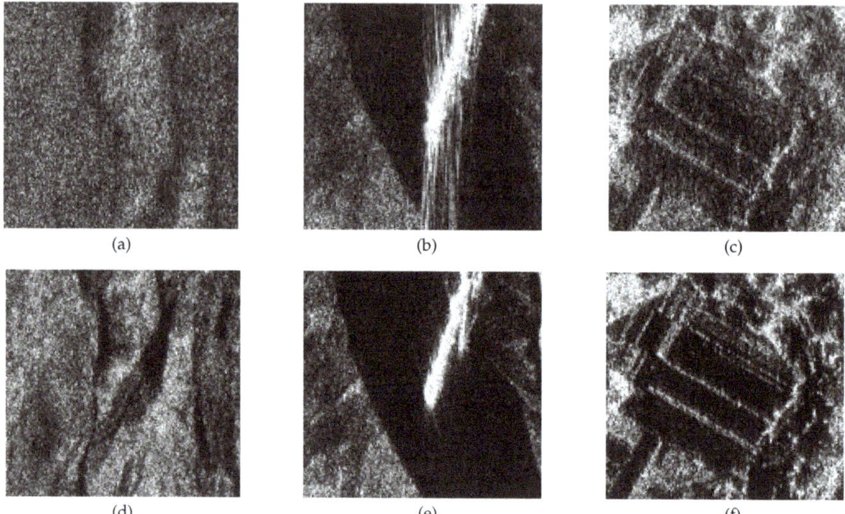

Figure 11. Zoom images. (**a–c**) Zoom images of subregions extracted from Figure 10a. (**d–f**) Zoom images of subregions extracted from Figure 10b.

Table 4. Image entropy of the zoom images.

Method	Region A	Region B	Region C
Conventional GCSA in Reference [26]	7.6037	7.1572	7.6275
Proposed algorithm	7.5957	6.9543	7.2874

5. Discussion

The performance of the proposed algorithm is mainly limited by the approximation error of range FM rate K_m (Equation (22)). This approximation will introduce a quadratic phase error (QPE) in the spectrum and degrades the quality of the image. In the CSA, the range dependence of SRC is neglected and the FM rate is calculated at the reference range $K_{m,app} = K_f$. The NCSA uses a linear approximation of the FM rate and the GCSA uses a 2nd order approximation model.

The QPE can be expressed as

$$\Phi_{QPE}(f_\tau, f_\eta; R_0) = \pi f_\tau^2 \left(\frac{1}{K_m} - \frac{1}{K_{m,app}} \right) \tag{36}$$

From Equation (8), we can see that the Taylor expansion is feasible only under the following condition

$$\left| \frac{K_r c R_0 f_\eta^2}{2V_r^2 f_0^3 D(f_\eta)^3} \right| \neq 1 \tag{37}$$

Let $G(K_r, R_0, V_r, f_0, f_\eta) = \frac{K_r c R_0 f_\eta^2}{2V_r^2 f_0^3 D(f_\eta)^3}$, this function is positively related to K_r, R_0, f_η and negatively related to V_r, f_0. And $K_r = B_r/T_r$, where T_r is the pulse duration. The azimuth frequency varies within the following range $-\frac{PRF}{2} + f_{\eta c} \leq f_\eta \leq f_{\eta c} + \frac{PRF}{2}$, where PRF is the pulse repetition frequency. Thus, G is an even function about f_η and inequality (37) actually implied condition $G < 1$. This inequality is satisfied in most L- and P-band SAR systems. However, as the center frequency decreases, the maximum azimuth frequency increases and the slant range increases, this inequality may not be satisfied. At this time, the Taylor approximation of range FM rate will introduce a large error and worsen the performance of algorithm.

It should be noted that $G < 1$ is required for all frequency domain approximation algorithms (such as the chirp scaling class and the range-Doppler class algorithms) [14]. This is satisfied for a linear FM signal with a large time bandwidth product (TBP). The TBP is defined as the product of the pulse duration T_r and bandwidth B_r. For example, assuming the SAR system with a following parameters: center frequency f_0 = 400 MHz, bandwidth B_r = 200 MHz, beamwidth θ = 29°, velocity V_r = 100 m/s, pulse duration T_r = 2 us (TBP = 400), target slant range R_0 = 12 km and center slant range R_c = 10 km. Figure 12a shows the variation of G with azimuth frequency f_η. Figure 12c–e show the QPEs in zero-order, 1st-order and 2nd-order approximations, respectively. It can be seen that G is not less than 1 at this time. Due to the existence of breakpoints, the QPEs of 1st-order and 2nd-order approximations are larger than zero-order approximation. The maximum QPE of 2nd-order model is about 5000°, which will seriously deteriorate the image quality.

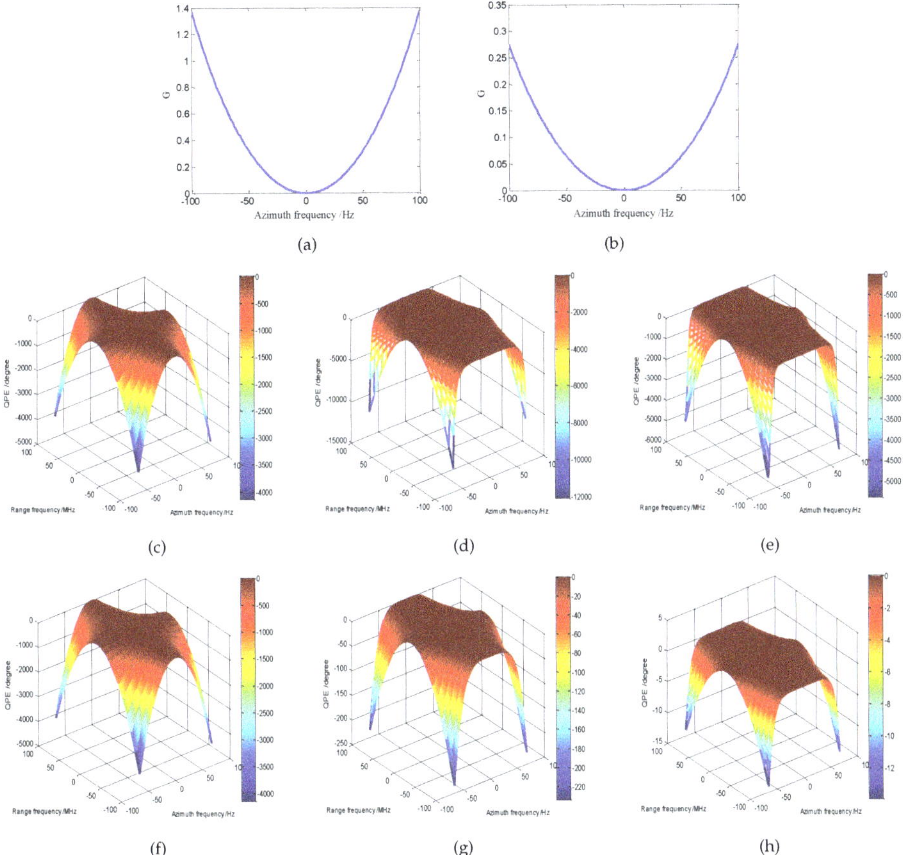

Figure 12. Parameters G and quadratic phase error (QPE). (**a**) Parameter G when T_r = 2 us (TBP = 400). (**b**) Parameter G when T_r = 10 us (TBP = 2000). (**c**–**e**) QPEs of zero-order, 1st-order and 2nd-order approximations corresponding to (**a**). (**f**–**h**) QPEs of zero-order, 1st-order and 2nd-order approximations corresponding to (**b**).

Change the pulse duration T_r = 10 us (TBP = 2000), the variation of G with azimuth frequency is show in Figure 12b. The QPEs of different approximations are also shown in Figure 12f–h. G is far less than 1, meeting the inequality (37). At this time, the maximum QPE of 1st-order model is 230° and the maximum QPE of 2nd-order model is 14°, which has a limit effect on the imaging results.

By comparison, it can be concluded that the proposed algorithm has the best performance in frequency domain approximation algorithms when Equation (37) is satisfied. If Equation (37) is not satisfied, the performance of the approximation algorithm is degraded and the $\omega - k$ algorithm and time-domain algorithms are a better choice. At the same time, it should be pointed out that most of the existing airborne SAR systems have a large TBP [35,36], which shows that the proposed algorithm still has a broad application prospects. In particular, the proposed algorithm can be applied to high-resolution highly squint SAR imaging, where the phase coupling is also severe.

6. Conclusions

The high-resolution low frequency SAR has the characteristics of large bandwidth and long integration time. This trait will cause serious range-azimuth phase coupling, which limits the performance of conventional GCSA and results in image defocusing. The longer the integration time or larger bandwidth is, the more serious the deterioration will be. This paper proposes an improved GCSA based on Lagrangian inversion theorem for high-resolution low frequency SAR data processing.

Through the theoretical analysis, we find two main reasons about the defocusing. Firstly, the influence of the residual high-order phase is still significant when the fractional bandwidth is large and/or integration time is long. Secondly, the linear approximation of stationary phase point will make the high-order range-dependent phase coupling not effectively compensated. Aim to solve these two problems, this paper firstly proposes a new criterion for determining the order of Taylor expansion. The range-independent coupling phase terms above 3rd order are first compensated. Moreover, the Lagrange inversion theorem is introduced to obtain a more accurate stationary phase point. The performance and accuracy of the improved GCSA has been demonstrated using the simulated data in P- and L-band. The experimental results show that for P-band SAR systems with resolutions of 0.44 m, the proposed algorithm can focused full-swath targets with a resolution loss of less than 1%. For L-band SAR system with an azimuth resolution of 0.5 m, the edge point scatterer can be focused well even at 80% fractional bandwidth. The proposed algorithm has similar performance to the $\omega - k$ algorithm and is significantly better than the GCSA.The improved method provides the possibility to efficiently process full-swath high-resolution low frequency SAR data.

Author Contributions: All authors have made great contributions to the work. X.C. and T.Y. carried out the theoretical framework. X.C., T.Y., Z.H. and F.H. conceived and designed the experiments; X.C. performed the experiments and wrote the manuscript; X.C. and T.Y. analyzed the data; Z.D. gave insightful suggestions for the work and the manuscript.

Acknowledgments: This research was supported by the National Natural Science Found of China under Grant 61771478. The authors would like to thank all the anonymous reviewers for their valuable comments and helpful suggestions which lead to substantial improvements of this paper.

Conflicts of Interest: The authors declare no conflict of interest.

Appendix A

Here, some expressions are given for some characteristics of the SAR signal. Using the Taylor expansion principle, the coefficients γ_i in Equation (2) can be expressed as

$$\begin{aligned}
\gamma_3 &= -\frac{D^2(f_\eta)-1}{2D^5(f_\eta)f_0^3} \\
\gamma_4 &= -\frac{(D^2(f_\eta)-1)(D^2(f_\eta)-5)}{8D^7(f_\eta)f_0^4} \\
\gamma_5 &= \frac{(D^2(f_\eta)-1)(3D^2(f_\eta)-7)}{8D^9(f_\eta)f_0^5} \\
\gamma_6 &= \frac{(D^2(f_\eta)-1)(D^4(f_\eta)-14D^2(f_\eta)+21)}{16D^{11}(f_\eta)f_0^6}
\end{aligned} \tag{A1}$$

Appendix B

In this Appendix, we will derive the calculation of $q_i(i \geq 2)$, $X_i(i \geq 3)$ and $C_i(i \geq 0)$.

Based on Equations (11), (17) and (20), we expand Equation (19) into a Taylor series with respect to τ_s, the Equation (21) is obtained. The coefficient C_0 is a Mth order polynomial of $\Delta\tau$ and it can be given by

$$\begin{aligned} C_0(\Delta\tau) = &\left[\alpha^2 q_2 + (\alpha-1)^2 K_m\right]\Delta\tau^2 + \left[\alpha^3 q_3 + (\alpha-1)^3 K_m^3 X_3\right]\Delta\tau^3 \\ &+\tfrac{1}{8}\left[8\alpha^4 q_4 + 2(\alpha-1)^4 K_m^4\left(9K_m X_3^2 + 4X_4\right) - 16D(f_\eta) f_0 (\alpha-1)^3 K_m^3 \gamma_3\right]\Delta\tau^4 + \ldots \end{aligned} \tag{A2}$$

In order to model the rang dependence of coefficients $C_i(i > 0)$, they are also approximated as a second order series in $\Delta\tau$,

$$\begin{aligned} &C_1 = 2\pi\left[(\alpha-1)K_m + \alpha q_2\right]\Delta\tau + 3\pi\left[\alpha^2 q_3 + (\alpha-1)^2 K_m^3 X_3\right]\Delta\tau^2 \\ &C_2 = \pi\left(K_m + q_2\right) + 3\pi\left[\alpha q_3 + (\alpha-1) K_m^3 X_3\right]\Delta\tau \\ &+\tfrac{3}{2}\pi\left[4\alpha^2 q_4 + (\alpha-1)K_m^3\left(9(\alpha-1)K_m^2 X_3^2 + 4(\alpha-1)K_m X_4 - 4D(f_\eta) f_0 \gamma_3\right)\right]\Delta\tau^2 \\ &C_3 = \pi\left(q_3 + K_m^3 X_3\right) + \pi\left[4\alpha q_4 + (\alpha-1)K_m^4\left(9K_m X_3^2 + 4X_4\right) - 2D(f_\eta) f_0 K_m^3 \gamma_3\right]\Delta\tau \\ &+\tfrac{1}{8}\pi\left[80\alpha^2 q_5 + 20(\alpha-1)^2 K_m^5\left(27K_m^2 X_3^3 + 24K_m X_3 X_4 + 4X_5\right)\right. \\ &\left.-32D(f_\eta) f_0 (\alpha-1) K_m^4\left(9K_m X_3 \gamma_3 + 2\gamma_4\right)\right]\Delta\tau^2 \\ &\ldots \end{aligned} \tag{A3}$$

For each C_i, the higher order terms of $\Delta\tau$ are very small and can be neglected. Combining Equation (22) and (A3), the coefficients $C_i(i > 0)$ can be rewritten as

$$\begin{aligned} &C_1 = C_{11}\Delta\tau + C_{12}\Delta\tau^2 \\ &C_2 = C_{20} + C_{21}\Delta\tau + C_{22}\Delta\tau^2 \\ &C_3 = C_{30} + C_{31}\Delta\tau + C_{32}\Delta\tau^2 \\ &\ldots \\ &C_M = C_{M0} + C_{M1}\Delta\tau + C_{M2}\Delta\tau^2 \end{aligned} \tag{A4}$$

with

$$\begin{aligned} &C_{11} = 2\left[\alpha q_2 + K_f(\alpha-1)\right] \\ &C_{12} = 3\alpha^2 q_3 + K_f^2(\alpha-1)\left[2K_s + 3K_f(\alpha-1)X_3\right] \\ &C_{20} = K_f + q_2 \\ &C_{21} = 3\alpha q_3 + K_f^2\left[K_s + 3K_f(\alpha-1)X_3\right] \\ &C_{22} = \tfrac{1}{2}\left[12\alpha^2 q_4 + K_f^3\left(18K_f K_s(\alpha-1)X_3 + 27K_f^2(\alpha-1)^2 X_3^2 + 2\left(K_s^2 + 6K_f(\alpha-1)^2 X_4\right)\right.\right. \\ &\left.\left.-12D(f_\eta) f_0 (\alpha-1)\gamma_3\right)\right] \\ &C_{30} = q_3 + K_f^3 X_3 \\ &C_{31} = 4\alpha q_4 + 3K_f^4 K_s X_3 + K_f^4(\alpha-1)\left(9K_f X_3^2 + 4X_4\right) - 2D(f_\eta) f_0 K_f^3 \gamma_3 \\ &C_{32} = \tfrac{1}{8}\left[48K_f^5 K_s^2 X_3 + 20K_f^5(\alpha-1)^2\left(27K_f^2 X_3^3 + 24K_f X_3 X_4 + 4X_5\right) - 48K_f^4 K_s D(f_\eta) f_0 \gamma_3\right. \\ &\left.+80\alpha^2 q_5 + 8K_f^5 K_s(\alpha-1)\left(45K_f X_3^2 + 16X_4\right) - 32D(f_\eta) f_0 K_f^4(\alpha-1)\left(9K_f X_3 \gamma_3 + 2\gamma_4\right)\right] \\ &\ldots \end{aligned} \tag{A5}$$

According to (A4), the coefficients C_i are range-dependent. The expression for these coefficients contains linear and quadratic terms in $\Delta\tau$. To remove the range dependence of these terms, it is required to set the coefficients of $\Delta\tau$ to zero. Let C_{11} in (A5) be zero, we can obtain

$$q_2 = K_f \frac{1-\alpha}{\alpha} \tag{A6}$$

Let C_{12} and C_{21} be zero, we can obtain

$$\begin{aligned} q_3 &= \frac{\frac{K_f^2 K_s(1-\alpha)}{3\alpha}}{} \\ X_3 &= \frac{K_s(\alpha-2)}{3K_f(\alpha-1)} \end{aligned} \tag{A7}$$

Similarly, $C_{M-1,2}$ and C_{M1} be zero, the expressions of q_M and X_M can be obtained. And the coefficients C_i becomes

$$\begin{aligned} C_1 &= 0 \\ C_2 &= C_{20} \\ C_3 &= C_{30} \\ &\dots \\ C_M &= C_{M0} \end{aligned} \tag{A8}$$

Thus, Equations (25) and (26) are obtained.

Appendix C

This Appendix is used to explain how to use the Lagrange inversion theorem to solve the stationary phase point. According to Equations (12) and (13), τ can be expressed as a series form of f_τ. A Lagrange inversion expression for a general power series is given here. For a function expressed in a series

$$\omega = h(z) = a_0 + a_1(z - z_0) + \dots + a_n(z - z_0)^n \tag{A9}$$

$$f(z) = \frac{z - z_0}{h(z) - a_0} = \frac{1}{a_1 + a_2(z - z_0) + \dots + a_n(z - z_0)^{n-1}} \tag{A10}$$

Thus, we can get

$$g_1 = f(z)|_{z=z_0} = \frac{1}{a_1} \tag{A11}$$

$$g_2 = \frac{1}{2}\frac{df^2(z)}{dz}\bigg|_{z=z_0} = -\frac{a_2}{a_1^3} \tag{A12}$$

$$g_3 = \frac{1}{6}\frac{d^2 f^3(z)}{dz^2}\bigg|_{z=z_0} = \frac{2a_2^3}{a_1^5} - \frac{a_3}{a_1^4} \tag{A13}$$

$$g_4 = \frac{1}{24}\frac{d^3 f^4(z)}{dz^3}\bigg|_{z=z_0} = -\frac{5{a_2}^3 - 5a_1a_2a_3 + {a_1}^2a_4}{{a_1}^7} \tag{A14}$$

$$g_5 = \frac{1}{120}\frac{d^4 f^5(z)}{dz^4}\bigg|_{z=z_0} = \frac{14a_2^4 - 21a_1a_2^2a_3 + 3a_1^2a_3^2 + 6a_1^2a_2a_4 - a_1^3a_5}{a_1^9} \tag{A15}$$

$$g_6 = \frac{1}{720}\frac{d^5 f^6(z)}{dz^5}\bigg|_{z=z_0} = \frac{1}{a_1^{11}}\left(-42a_2^5 + 84a_1a_2^3a_3 - 28a_1^2a_2^2a_4 + 7a_1^2a_2\left(-4a_3^2 + a_1a_5\right) + a_1^3\left(7a_3a_4 - a_1a_6\right)\right) \tag{A16}$$

Equations (12) and (27) are special forms of Equation (A9) and the inversion expressions can be easily obtained according to Equation (14).

References

1. Reigber, A.; Scheiber, R.; Jager, M. Very-High-Resolution Airborne Synthetic Aperture Radar Imaging: Signal Processing and Applications. *Proc. IEEE* **2013**, *101*, 759–783. [CrossRef]
2. Xie, H.; An, D.; Huang, X. Spatial resolution analysis of low frequency ultrawidebeam-ultrawideband synthetic aperture radar based on wavenumber domain support of echo data. *J. Appl. Remote Sens.* **2015**, *9*, 095033. [CrossRef]
3. Moore, R.K. *Microwave Remote Sensing*; Advanced Book Program; Addison-Wesley Pub. Co.: Upper Saddle River, NJ, USA, 1999.
4. Schlund, M.; Davidson, M.W.J. Aboveground Forest Biomass Estimation Combining L- and P-Band SAR Acquisitions. *Remote Sens.* **2018**, *10*, 1151. [CrossRef]
5. Taylor, J.D. *Ultrawideband Radar: Applications and Design*; CRC Press: Boca Raton, FL, USA, 2012.
6. Mokole, E.L.; Sabath, F. Examples of Ultrawideband Definitions and Waveforms. In *Principles of Waveform Diversity and Design*; Wicks, M.C., Ed.; Scitech Publishing: Raleigh, NC, USA, 2010.
7. Rau, R.; Mcclellan, J.H. Analytic models and postprocessing techniques for UWB SAR. *IEEE Trans. Aerosp. Electron. Syst.* **2002**, *36*, 1058–1074.
8. Immoreev, I.Y. Ultrawideband radars: Features and capabilities. *J. Commun. Technol. Electron.* **2009**, *54*, 1–26. [CrossRef]
9. Potsis, A.; Reigber, A.; Mittermayer, J. Improving the Focusing Properties of SAR Processors for Wide-band and Wide-beam Low Frequency Imaging. In Proceedings of the IEEE 2001 International Geoscience and Remote Sensing Symposium, Sydney, NSW, Australia, 9–13 July 2001; pp. 3047–3049.
10. Vu, V.T.; Sjogren, T.K.; Pettersson, M.I. Detection of Moving Targets by Focusing in UWB SAR—Theory and Experimental Results. *IEEE Trans. Geosci. Remote Sens.* **2010**, *48*, 3799–3815. [CrossRef]
11. Ressler, M.; Happ, L.; Nguyen, L. The Army Research Laboratory ultra-wide band testbed radars. In Proceedings of the IEEE International Radar Conference, Alexandria, VA, USA, 8–11 May 1995.
12. Hellsten, H. CARABAS-an UWB low frequency SAR. In Proceedings of the IEE MTT-S International Microwave Symposium Digest, Albuquerque, NM, USA, 1–5 June 1992; pp. 1495–1498.
13. Davidson, G.W.; Cumming, I.G.; Ito, M.R. A chirp scaling approach for processing squint mode SAR data. *IEEE Trans. Aerosp. Electron. Syst.* **1996**, *32*, 121–133. [CrossRef]
14. Cumming, I.G.; Wong, F.H. *Digital Signal Processing of Synthetic Aperture Radar Data: Algorithms and Implementation*; Artech House: Norwood, MA, USA, 2005.
15. Raney, K.; Runge, H.; Bamler, R. Precision SAR Processing Using Chirp Scaling. *IEEE Trans. Geosci. Remote Sens.* **1994**, *32*, 786–799. [CrossRef]
16. Cafforio, C.; Prati, C.; Rocca, F. SAR data focusing using seismic migration techniques. *IEEE Trans. Aerosp. Electron. Syst.* **1991**, *27*, 194–207. [CrossRef]
17. Bamler, R. A Comparison of Range-Doppler and Wavenumber Domain SAR Focussing Algorithms. *IEEE Trans. Geosci. Remote Sens.* **1992**, *30*, 706–713. [CrossRef]
18. Potsis, A.; Reigber, A. Comparison of chirp scaling and wavenumber domain algorithms for airborne low-frequency SAR. In Proceedings of the SPIE—SAR Image Analysis, Modeling, and Techniques V, Silsoe, UK, 19–23 September 2003; pp. 11–19.
19. Ulander, L.M.H.; Hellsten, H.; Stenstrom, G. Synthetic-aperture radar processing using fast factorized back-projection. *IEEE Trans. Aerosp. Electron. Syst.* **2003**, *39*, 760–776. [CrossRef]
20. Reigber, A.; Alivizatos, E.; Potsis, A. Extended wavenumber-domain synthetic aperture radar focusing with integrated motion compensation. *IEE Proc. Radar Sonar Navig.* **2006**, *153*, 301–310. [CrossRef]
21. Madsen, S.N. Motion compensation for ultra wide band SAR. In Proceedings of the IEEE 2001 International Geoscience and Remote Sensing Symposium, Sydney, NSW, Australia, 9–13 July 2001; pp. 1436–1438.
22. An, D.; Li, Y.; Huang, X. Performance Evaluation of Frequency-Domain Algorithms for Chirped Low Frequency UWB SAR Data Processing. *IEEE J. Sel. Top. Appl. Earth Obs. Remote Sens.* **2014**, *7*, 678–690.
23. Chen, L.; An, D.; Huang, X. A NLCS focusing approach for Low Frequency UWB One- Stationary Bistatic SAR. In Proceedings of the 2015 16th International Radar Symposium (IRS), Dresden, Germany, 24–26 June 2015.
24. Sun, G.; Xing, M.; Liu, Y. Extended NCS Based on Method of Series Reversion for Imaging of Highly Squinted SAR. *IEEE Geosci. Remote Sens. Lett.* **2011**, *8*, 446–450. [CrossRef]

25. Vu, V.T.; Sjogren, T.K.; Pettersson, M.I. A Comparison between Fast Factorized Backprojection and Frequency-Domain Algorithm in UWB Low frequency SAR. In Proceedings of the 2008 IEEE International Geoscience and Remote Sensing Symposium (IGARSS 2008), Boston, MA, USA, 7–11 July 2008; pp. 1284–1287.
26. Zaugg, E.C.; Long, D.G. Generalized Frequency-Domain SAR Processing. *IEEE Trans. Geosci. Remote Sens.* **2009**, *47*, 3761–3773. [CrossRef]
27. Yi, T.; He, Z.; He, F.; Dong, Z.; Wu, M. Generalized Chirp Scaling Combined with Baseband Azimuth Scaling Algorithm for Large Bandwidth Sliding Spotlight SAR Imaging. *Sensors* **2017**, *17*, 1237–1256. [CrossRef]
28. Vu, V.T.; Sjogren, T.K.; Pettersson, M.I. Two-Dimensional Spectrum for BiSAR Derivation Based on Lagrange Inversion Theorem. *IEEE Geosci. Remote Sens. Lett.* **2014**, *11*, 1210–1214. [CrossRef]
29. Gessel, I.M. Lagrange inversion. *J. Comb. Theory Ser. A* **2016**, *144*, 212–249. [CrossRef]
30. Gessel, I.M. A combinatorial proof of the multivariable lagrange inversion formula. *J. Comb. Theory Ser. A* **1987**, *45*, 178–195. [CrossRef]
31. Novelli, J.C.; Thibon, J.Y. Noncommutative Symmetric Functions and Lagrange Inversion. *Adv. Appl. Math.* **2008**, *40*, 8–35. [CrossRef]
32. Vu, V.T.; Sjogren, T.K. Ultrawideband Chirp Scaling Algorithm. *IEEE Geosci. Remote Sens. Lett.* **2010**, *7*, 281–285.
33. Zeng, T.; Wang, R.; Li, F. SAR Image Autofocus Utilizing Minimum-Entropy Criterion. *IEEE Geosci. Remote Sens. Lett.* **2013**, *10*, 1552–1556. [CrossRef]
34. Wang, J.; Liu, X. SAR Minimum-Entropy Autofocus Using an Adaptive-Order Polynomial Model. *IEEE Geosci. Remote Sens. Lett.* **2006**, *3*, 512–513. [CrossRef]
35. Damini, A.; McDonald, M.; Haslam, G.E. X-band wideband experimental airborne radar for SAR, GMTI and maritime surveillance. *IEE Proc. Radar Sonar Navig.* **2003**, *150*, 305–312. [CrossRef]
36. Ender, J.H.G.; Brenner, A.R. PAMIR—A wideband phased array SAR/MTI system. *IEE Proc. Radar Sonar Navig.* **2003**, *150*, 165–172. [CrossRef]

Article

Analytical Approximation Model for Quadratic Phase Error Introduced by Orbit Determination Errors in Real-Time Spaceborne SAR Imaging

Xiaoyu Yan [1,*], Jie Chen [1], Holger Nies [2] and Otmar Loffeld [2]

1 School of Electronic and Information Engineering, Beihang University, 37 Xueyuan Rd., Haidian Dist., Beijing 100191, China
2 Center for Sensor Systems (ZESS), University of Siegen, Paul-Bonatz-Strasse 9-11, 57076 Siegen, Germany
* Correspondence: yanxy@buaa.edu.cn

Received: 9 June 2019; Accepted: 10 July 2019; Published: 12 July 2019

Abstract: Research on real-time spaceborne synthetic aperture radar (SAR) imaging has emerged as satellite computation capability has increased and applications of SAR imaging products have expanded. The orbit determination data of a spaceborne SAR platform are essential for the SAR imaging procedure. In real-time SAR imaging, onboard orbit determination data cannot achieve a level of accuracy that is equivalent to the orbit ephemeris in ground-based SAR processing, which requires a long processing time using common ground-based SAR imaging procedures. It is important to study the influence of errors in onboard real-time orbit determination data on SAR image quality. Instead of the widely used numerical simulation method, an analytical approximation model of the quadratic phase error (QPE) introduced by orbit determination errors is proposed. The proposed model can provide approximation results at two granularities: approximations with a satellite's true anomaly as the independent variable and approximations for all positions in the satellite's entire orbit. The proposed analytical approximation model reduces simulation complexity, extent of calculations, and the processing time. In addition, the model reveals the core of the process by which errors are transferred to QPE calculations. A detailed comparison between the proposed method and a numerical simulation method proves the correctness and reliability of the analytical approximation model. With the help of this analytical approximation model, the technical parameter iteration procedure during the early-stage development of an onboard real-time SAR imaging mission will likely be accelerated.

Keywords: quadratic phase error; SAR; approximation; spaceborne real-time SAR imaging; orbit determination error

---

1. Introduction

Real-time synthetic aperture radar (SAR) imaging has always been a research focus in certain applications. Real-time imaging enhances the effectiveness and widens the application range of the SAR technique. In airborne SAR, real-time sensing plays an important role, and there are many well-established systems [1–5] and techniques [6–11]. Real-time processing techniques are also available for unmanned aerial vehicle (UAV) SAR applications [12–14]. For spaceborne SAR, because of its unique configuration and onboard computation capabilities, real-time SAR imaging systems and techniques are still being developed for both digital [15–17] and optical [18] methods.

For the purpose of implementing onboard real-time spaceborne SAR imaging, a set of Doppler parameters with high precision and accuracy must be calculated, and accurate position and velocity vectors are needed for the Doppler parameter calculation. In typical methods used to process SAR echo data, a high-accuracy orbit ephemeris is applied in the ground-based processing system, and state vectors, i.e., position and velocity vectors of the SAR satellite, are determined with high accuracy. However, this kind of high-accuracy orbit ephemeris is only generated hours, days, or even weeks after a single SAR observation [19]. Thus, this kind of high-accuracy orbit ephemeris cannot be obtained during real-time SAR imaging. Nevertheless, we are still able to achieve onboard orbit determination using GNSS-based receivers and appropriate algorithms [20,21]. It is acknowledged that this approach to orbit determination has relatively low accuracy in the state vectors of SAR satellites compared with the high-accuracy orbit ephemeris. However, it remains the best, and possibly the only, onboard orbit determination method that is available for real-time SAR imaging.

SAR mission designers must account for the fact that errors in such onboard orbit determination data will adversely affect the quality of the SAR imaging product through inaccurate Doppler parameter calculation. Although estimation algorithms have been developed to generate the Doppler centroid and Doppler rate with high accuracy for SAR imaging, they are time-consuming in practice. Real-time spaceborne SAR imaging certainly does not belong on a mission that is not time-sensitive. The product of real-time SAR imaging may be the source data for other mission types, such as onboard deformation monitoring or target recognition. Given these requirements, the estimation algorithm should not be built into the processing workflow because of its long processing time. Thus, the concept of quadrature phase error (QPE) is introduced to evaluate the influence of onboard orbit determination errors on spaceborne SAR imaging products. The azimuth impulse response width of a SAR image has a certain relationship with QPE, and it provides a more intuitive evaluation standard of the errors.

Errors of an onboard orbit determination system are often given in the form of probabilities, along with the type of probability distributions and the corresponding parameters. The most common case for these errors is a normal distribution with an expected value of zero and a standard deviation σ. Therefore, the probability distribution of QPE needs to be examined. In practice, there exist two major methods that have been applied to QPE analysis, i.e., the extreme value method and the numerical simulation method. The extreme value method is the simplest way to investigate QPE; it includes the maximum possible error in the calculation and ignores the probability distribution of the errors. This method faces challenges in revealing the QPE probability distribution, which means it only reflects the extreme situation in practice. For the numerical simulation methods, Monte Carlo simulations are often applied [22]. However, this approach requires a huge number of generated error samples and repeated experiments.

In this paper, an analytical approximation model of the QPE introduced by orbit determination errors is proposed. With the a priori probability of the onboard orbit determination system and SAR satellite orbit elements, together with the observing geometry of the spaceborne SAR and appropriate approximations, a series of equations describing the parameters of the QPE probability distribution are presented. This analytical approximation model can provide approximation results at two granularities, i.e., the approximations with the satellite's true anomaly as the independent variable and the approximations for all positions of the satellite during its entire orbit.

Following this section, Section 2 describes the QPE analytical approximation model in general. Section 3 derives the key variables in the analytical approximation model and reveals the appropriate approximations in the model. This is followed by the evaluation of the analytical approximation model by comparing it with numerical simulation results in Section 4, and a related discussion is presented in Section 5. Finally, Section 6 provides a brief conclusion regarding the analytical approximation model.

2. Proposed Model: QPE Analytical Approximation Model

The QPE analytical approximation model requires four groups of parameters as input, i.e., the Earth's physical parameters, the satellite platform orbit parameters, the SAR payload key parameters, and the a priori probability of the onboard orbit determination system. In this section, the coordinate systems, vector in the line-of-sight method, and QPE approximation method in the proposed model are described.

2.1. Coordinate Systems for Modeling

Two main coordinate systems are used in the derivation of the analytical approximation model. These two coordinate systems can be helpful in generating a clear description of the relative movement of the spaceborne SAR and the point of interest on the surface of the Earth.

2.1.1. Earth-Centered Inertial Coordinate System

Generally, the QPE calculation requires the position, velocity, and acceleration coordinates of the spaceborne SAR platform and the SAR antenna aiming point on the Earth's surface, together with other variables. A widely used coordinate system in modeling the relative motion between the spaceborne SAR and the point of interest is the Earth-Centered Inertial (ECI) coordinate system. This system has an x-axis and z-axis aligned with the mean equinox and the celestial North Pole, respectively. The y-axis forms a right-handed coordinate system that joins the x-axis and z-axis.

In common practice, the onboard GNSS-based orbit determination system of a spaceborne satellite calculates its positions in a given coordinate system using the GNSS broadcast message [21]. Each GNSS in service now has its own unique coordinate system, e.g., WGS 84 in GPS, CGCS2000 in BDS, and GTRF in Galileo. For modeling the orbit determination errors in the proposed analytical approximation model, all real-time-measured position and velocity coordinates of the spaceborne SAR platform are transformed into the ECI system in this paper.

2.1.2. Orbital Plane Coordinate System

The movement parameters of a spaceborne SAR satellite can be described in the orbital plane coordinate system. In this system, the Earth is located at one of the foci of the elliptical orbit, and the satellite can be regarded as a point. Polar coordinates are used to describe the movement parameters.

There exist six orbital elements that uniquely identify a specific orbit of the satellite: the semi-major axis a, eccentricity e, and true anomaly ν are used in the orbital plane coordinate system to determine the satellite's movement at a certain time. The inclination i, the longitude of the ascending node Ω, and the argument of periapsis ω are necessary to construct the transformation matrix between the orbital plane coordinate system and the ECI system.

The orbital plane coordinate system and the ECI system are summarized in Figure 1.

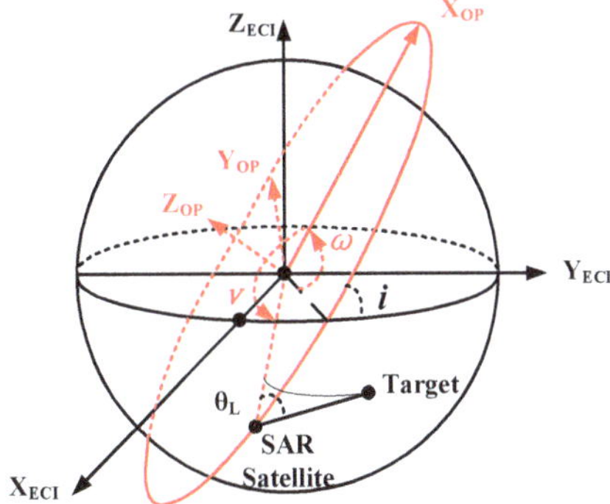

Figure 1. The orbital plane (OP) coordinate system and the ECI system. Elements related to the OP coordinate system are in red. ω is the argument of periapsis, i is the orbit eccentricity, ν is true anomaly, and θ_L is the center off-nadir angle in SAR.

2.2. Vector in the Line-of-Sight Method

In the analytical approximation model presented in this paper, the attitude errors of the spaceborne SAR satellite are not taken into consideration. The true value of the attitude of the platform is used in both the analytical approach and the following numerical verification.

With this assumption, the vector in the line-of-sight (VLS) method can be used to calculate the state vectors of the point of interest. The relative range from the SAR antenna phase center to the point of interest can be derived from the time delay of the echo signal. For spaceborne SAR geometry, the echo signal enters the receiver after several pulse repetition intervals (PRIs) until the moment of transmission, and then it is demodulated and sampled by the receiver. In each range profile of the sampled data, each sampled time delay can be determined by the index number of its range gate. The signal transmission delay from the antenna phase center to the sampling ADCs in the receiver can be accurately measured in the state-of-the-art spaceborne SAR system. t_{delay} represents the delayed time of the echo of the point-of-interest signal from its transmission time, and c represents the velocity of light, whose relative range is described by

$$\rho = \frac{ct_{delay}}{2} \tag{1}$$

A unit line-of-sight vector $\hat{\mathbf{u}}_{a,ECI}$ in the ECI system can be created; it is perpendicular to the aperture of the SAR antenna and points to the surface of the Earth. The position vector $\mathbf{P}_{tar,ECI}$ and velocity vector $\mathbf{V}_{tar,ECI}$ of the point on the surface of the Earth at which the antenna points can be shown in stripmap SAR geometry as

$$\mathbf{P}_{tar,ECI} = \mathbf{P}_{sat,ECI} + \rho \cdot \hat{\mathbf{u}}_{a,ECI} = \begin{bmatrix} P_{tar,ECI,x} \\ P_{tar,ECI,y} \\ P_{tar,ECI,z} \end{bmatrix} \tag{2}$$

$$\mathbf{V}_{tar,ECI} = \begin{bmatrix} -\omega_e \cdot P_{tar,ECI,y} \\ \omega_e \cdot P_{tar,ECI,x} \\ 0 \end{bmatrix} \tag{3}$$

in which ω_e is the angular speed of the Earth's rotation.

Because there are errors in the satellite's position and velocity measurements from the onboard GNSS-based orbit determination system, the calculated $\mathbf{P}_{tar,ECI}$ and $\mathbf{V}_{tar,ECI}$ will also contain errors at the same time. The relative range ρ remains unaffected because errors exist in measurements while the real geometry remains unchanged, which means that the time delay remains equal to the accuracy value.

2.3. QPE Approximation

QPE can be expressed in radians as [23]

$$\mathrm{QPE} = \pi \Delta f_r \left(\frac{T_{int}}{2}\right)^2 \tag{4}$$

in which f_r is the Doppler rate of the echo signal, and T_{int} represents the integration time, also known as the synthetic-aperture time. In stripmap SAR geometry, the integration time is the time it takes the point of interest to enter and exit the entire 3 dB edge of the SAR antenna's beam illumination.

The variables $\mathbf{R}_{st}$, $\mathbf{V}_{st}$, and $\mathbf{A}_{st}$ represent the relative position, velocity, and acceleration of the spaceborne SAR satellite, respectively. In the ECI system, for the antenna aiming point on the ground, the commonly used f_r expression in stripmap mode when the squint angle equals zero is

$$f_r = \frac{2}{\lambda}\left[\frac{\mathbf{V}_{st}\cdot\mathbf{V}_{st}}{|\mathbf{R}_{st}|} + \frac{\mathbf{A}_{st}\cdot\mathbf{R}_{st}}{|\mathbf{R}_{st}|} - \frac{(\mathbf{V}_{st}\cdot\mathbf{R}_{st})^2}{|\mathbf{R}_{st}|^3}\right] \tag{5}$$

The f_r expression in Equation (5) is an approximation expression, and it can achieve high precision if more terms remain. In the analytical approximation model, only the first two terms in Equation (5) are analyzed. This simplifies the model while retaining a relatively high precision.

The orbit determination system measurements contain errors, which will lead to errors in $\mathbf{R}_{st}$, $\mathbf{V}_{st}$, and $\mathbf{A}_{st}$. The terms that define the differences between the state vectors derived from measurements and the true state vectors are

$$\Delta\mathbf{R}_{st} = \mathbf{R}_{st,e} - \mathbf{R}_{st} \tag{6}$$

$$\Delta\mathbf{V}_{st} = \mathbf{V}_{st,e} - \mathbf{V}_{st} \tag{7}$$

$$\Delta\mathbf{A}_{st} = \mathbf{A}_{st,e} - \mathbf{A}_{st} \tag{8}$$

where the subscript e indicates that the vectors contain measurement errors. Thus, the difference between the measured f_r and true f_r value is

$$\Delta f_r = f_{r,e} - f_r \tag{9}$$

The parameter f_r in Equation (5) is often used to determine the precise value for SAR raw data simulation or imaging algorithms. For the approximate expression of QPE, considering the magnitude of each term in f_r, some terms can be omitted while still achieving an acceptable approximation of QPE. The VLS method in Section 2.2 maintains the real value of the relative range $\rho = |\mathbf{R}_{st}|$. Thus, Equation (4) combined with Equation (9) gives the QPE approximation expression

$$\begin{aligned}\mathrm{QPE} &= \pi\left(f_{r,e} - f_r\right)\left(\frac{T_{int}}{2}\right)^2 \\ &\approx \frac{2\pi}{\lambda\rho}\left(2\Delta\mathbf{V}_{st}\cdot\mathbf{V}_{st} + \Delta\mathbf{A}_{st}\cdot\mathbf{R}_{st}\right)\left(\frac{T_{int}}{2}\right)^2\end{aligned} \tag{10}$$

For convenience, the terms $\Delta\mathbf{V}_{st}\cdot\mathbf{V}_{st}$ and $\Delta\mathbf{A}_{st}\cdot\mathbf{R}_{st}$ are named *velocity* and *acceleration vector term* in the following sections.

The approximation of QPE in Equation (10) is the basis of the QPE analytical approximation function and maximum QPE derivation.

3. Key Variables in the Analytical Approximation Model

For a given true anomaly ν, the distance between the satellite and the Earth's center is

$$r = \frac{a\left(1-e^2\right)}{1+e\cos\nu} = \frac{p}{1+e\cos\nu} \tag{11}$$

where $p = a\left(1-e^2\right)$ in Equation (11) is the semi-latus rectum of the elliptic orbit. The position, velocity, and acceleration vectors of the satellite in the orbital plane (OP) are

$$\mathbf{P}_{sat,OP} = r \begin{bmatrix} \cos\nu \\ \sin\nu \\ 0 \end{bmatrix} \tag{12}$$

$$\mathbf{V}_{sat,OP} = \sqrt{\frac{\mu}{p}} \begin{bmatrix} -\sin\nu \\ e + \cos\nu \\ 0 \end{bmatrix} \tag{13}$$

$$\mathbf{A}_{sat,OP} = -\frac{\mu\left(1 + e\cos\nu\right)^2}{p^2} \begin{bmatrix} \cos\nu \\ \sin\nu \\ 0 \end{bmatrix} \tag{14}$$

where μ is the standard gravitational parameter of the Earth.

The transformation matrix A_{O2E} is used to convert the vector in the orbital plane to the ECI system:

$$\mathbf{A}_{O2E} = \begin{bmatrix} \cos\omega & -\sin\omega & 0 \\ \sin\omega & \cos\omega & 0 \\ 0 & 0 & 1 \end{bmatrix} \begin{bmatrix} 1 & 0 & 0 \\ 0 & \cos i & -\sin i \\ 0 & \sin i & \cos i \end{bmatrix} \begin{bmatrix} \cos\Omega & -\sin\Omega & 0 \\ \sin\Omega & \cos\Omega & 0 \\ 0 & 0 & 1 \end{bmatrix} \tag{15}$$

The parameter Ω in $\mathbf{A}_{O2E}$ results in rotation with the z-axis in the ECI system. Considering the geometry and the Earth's ellipsoid reference, in the following discussion, we set $\Omega = 0$ for convenience, and this will not result in an error value in the QPE approximation.

The onboard orbit determination system measurements are given in ECI terms, which can be regarded as the true state vector plus an error vector. With the assumption that the measurement errors of the satellite's position and velocity follow a normal distribution with an expected value of 0, the error vectors can be described as

$$\Delta\mathbf{P} = \begin{bmatrix} \Delta p_x \\ \Delta p_y \\ \Delta p_z \end{bmatrix} \sim \begin{bmatrix} N\left(0, \sigma_{p,x}^2\right) \\ N\left(0, \sigma_{p,y}^2\right) \\ N\left(0, \sigma_{p,z}^2\right) \end{bmatrix} \tag{16}$$

$$\Delta\mathbf{V} = \begin{bmatrix} \Delta v_x \\ \Delta v_y \\ \Delta v_z \end{bmatrix} \sim \begin{bmatrix} N\left(0, \sigma_{v,x}^2\right) \\ N\left(0, \sigma_{v,y}^2\right) \\ N\left(0, \sigma_{v,z}^2\right) \end{bmatrix} \tag{17}$$

Then, the state vector measurements of the satellite are

$$\mathbf{P}_{sat,ECI,e} = \mathbf{A}_{O2E}\mathbf{P}_{sat,OP} + \Delta\mathbf{P} \tag{18}$$

$$\mathbf{V}_{sat,ECI,e} = \mathbf{A}_{O2E}\mathbf{V}_{sat,OP} + \Delta\mathbf{V} \tag{19}$$

From Equations (2), (3), (18) and (19), the position vectors of the antenna aiming point are

$$\mathbf{P}_{tar,ECI,e} = \mathbf{P}_{sat,ECI,e} + \rho \cdot \hat{u}_{a,ECI} \tag{20}$$

With the help of Equations (3), (19) and (20), the difference in the relative velocity from the satellite to the antenna aiming point can be given as

$$\Delta \mathbf{V}_{st} = \left(\mathbf{V}_{sat,ECI,e} - \mathbf{V}_{tar,ECI,e}\right) - \left(\mathbf{V}_{sat,ECI} - \mathbf{V}_{tar,ECI}\right) = \begin{bmatrix} \Delta v_x + \omega_e \Delta p_y \\ \Delta v_y - \omega_e \Delta p_x \\ \Delta v_z \end{bmatrix} \tag{21}$$

3.1. The Velocity Vector Term in the Approximate Doppler Rate

With Equation (21), we have $\Delta \mathbf{V}_{st}$. However, we still need the expression of $\mathbf{V}_{st}$ to calculate the velocity vector term $\Delta \mathbf{V}_{st} \cdot \mathbf{V}_{st}$. It is worth recalling that we need the approximate Doppler rate expression. To calculate the velocity vector term in Equation (10), we propose using the velocity vector of a satellite traveling in a circular orbit $\mathbf{V}_s$ in our analytical approximation model rather than using $\mathbf{V}_{st}$. Thus, we have

$$\begin{aligned} \mathbf{V}_{st} &= \mathbf{A}_{O2E} \mathbf{V}_{sat,OP}\big|_{e=0} \\ &= \sqrt{\frac{\mu}{p}} \begin{bmatrix} -\sin\left(\nu + \omega\right) \\ \cos\left(\nu + \omega\right)\cos i \\ \cos\left(\nu + \omega\right)\sin i \end{bmatrix} \end{aligned} \tag{22}$$

The velocity vector term results in a Doppler rate difference by

$$\begin{aligned} \Delta f_{r,v} &= \frac{2}{\lambda \rho} \cdot 2 \Delta \mathbf{V}_{st} \cdot \mathbf{V}_{st} \\ &= \frac{4}{\lambda \rho} \sqrt{\frac{\mu}{p}} \left[-\left(\Delta v_x + \omega_e \Delta p_y\right) \sin\left(\nu + \omega\right) \right. \\ &\left. + \left(\Delta v_y - \omega_e \Delta p_x\right) \cos\left(\nu + \omega\right) \cos i + \Delta v_z \cos\left(\nu + \omega\right) \sin i \right] \end{aligned} \tag{23}$$

Assuming that the variances of the position and velocity errors are the same standard deviation in the x-, y-, and z-direction in the ECI system, i.e., $\sigma_{p,x} = \sigma_{p,y} = \sigma_{p,z} = \sigma_p$ and $\sigma_{v,x} = \sigma_{v,y} = \sigma_{v,z} = \sigma_v$, we can find the expected value and standard deviation of $\Delta f_{r,v}$ by

$$\mathbb{E}\left[\Delta f_{r,v}\right] = 0 \tag{24}$$

$$\sigma\left[\Delta f_{r,v}\right] = \frac{4}{\lambda \rho} \sqrt{\frac{\mu}{p}} \sqrt{\sigma_v^2 + \omega_e^2 \left[\sin^2\left(\nu + \omega\right) + \cos^2\left(\nu + \omega\right)\cos^2 i\right] \sigma_p^2} \tag{25}$$

We also need an approximate maximum value of the standard deviation of $\Delta f_{r,v}$ to calculate the approximate maximum value of QPE. We suppose that $\cos i$ has a maximum value of $\cos i = 1$, and

$$\sigma\left[\Delta f_{r,v}\right]_{max} = \frac{4}{\lambda \rho} \sqrt{\frac{\mu}{p}} \cdot \sqrt{\omega_e^2 \sigma_p^2 + \sigma_v^2} \tag{26}$$

3.2. True Anomaly

Equation (21) gives the expression of difference vectors in the velocity vector term in Equation (10). However, the difference vector in the acceleration vector term, i.e., $\Delta \mathbf{A}_{st}$, is given its expression with the help of true anomaly calculation.

The accurate acceleration vector term in the orbital plane is given in Equation (14). True anomaly ν should be calculated using the onboard-measured satellite's position and velocity vectors. A common way to solve ν is given in [24]

$$\nu = \text{atan2}\left(\sqrt{\frac{p}{\mu}}\left(\mathbf{V}_{sat,ECI} \cdot \mathbf{R}_{sat,ECI}\right), p - \left|\mathbf{R}_{sat,ECI}\right|\right) \tag{27}$$

where the atan2 (Y, X) function returns an unambiguous result of arctan (Y/X) with a range of $(-\pi, \pi]$. Obviously, the measured state vectors of the satellite $\mathbf{P}_{sat,ECI,e}$ and $\mathbf{V}_{sat,ECI,e}$ will result in a value of ν_e that is not accurate. The probably distribution parameters of ν_e should be determined in order to analyze the term $\Delta\mathbf{A}_{st}$ and its influence on QPE.

The detailed derivation is in Appendix A. Here, the standard deviation and its maximum value of the difference between the calculated ν_e and its accurate value $\Delta\nu = \nu_e - \nu$ are given.

$$\sigma\left[\Delta\nu\right] = \left|\cos\nu\right| \cdot \frac{1 + e\cos\nu}{\sqrt{\mu a e^2\left(1 - e^2\right)}} \cdot \sqrt{\frac{\mu}{p}\left(1 + e^2 + 2e\cos\nu\right)\sigma_p^2 + \frac{p^2}{\left(1 + e\cos\nu\right)^2}\sigma_v^2} \tag{28}$$

$$\sigma\left[\Delta\nu\right]_{\max} = \frac{1}{e}\sqrt{\frac{1}{a^2}\sigma_p^2 + \frac{a}{\mu}\sigma_v^2} \tag{29}$$

We assume that the variances of the position and velocity errors are the same standard deviation in the x-, y-, and z-direction in the ECI system, i.e., $\sigma_{p,x} = \sigma_{p,y} = \sigma_{p,z} = \sigma_p$ and $\sigma_{v,x} = \sigma_{v,y} = \sigma_{v,z} = \sigma_v$. Equation (29) can provide a good approximation when orbit eccentricity e is rather small.

It is important to point out that $\sigma\left[\nu\right]_{\max}$ is not equal to the exact maximum value of $\sigma\left[\nu\right]$ for $\nu \in [0, 2\pi]$. This approximate maximum value is subsequently used to calculate the maximum value of QPE. With this assumption, it would be clearer to identify the core error terms of each variable. A more general expression without the assumption can be found in Appendix A.

3.3. *The Acceleration Vector Term in the Approximate Doppler Rate*

With the expression $\sigma\left[\Delta\nu\right]$, the standard deviation of the difference between ν calculated from state vector measurements and its true value is addressed, and the acceleration vector term can now can be dealt with. As $\mathbf{R}_{st}$ needs to be analyzed, the traditional yaw steering method [25] is applied here.

$$\theta_{yaw} = \arctan\left[\frac{\sin i}{N - \cos i}\cos\left(\nu + \omega\right)\right] \tag{30}$$

where N is the number of revolutions per day.

In this approximation model, we replace $\Delta\mathbf{A}_{st}$ with $\Delta\mathbf{A}_s$ only and ignore the errors in the target acceleration vector while keeping the satellite's part. In the meantime, ν is replaced by $\nu + \Delta\nu$ in the row vector of Equation (14). We can combine Equations (14), (18) and (20) with

$$\hat{\mathbf{u}}_{a,ECI} = \begin{bmatrix} -\sin\nu & -\cos\nu & 0 \\ \cos\nu & -\sin\nu & 0 \\ 0 & 0 & 1 \end{bmatrix} \begin{bmatrix} \cos\theta_{yaw} & 0 & -\sin\theta_{yaw} \\ 0 & 1 & 0 \\ \sin\theta_{yaw} & 0 & \cos\theta_{yaw} \end{bmatrix} \begin{bmatrix} 1 & 0 & 0 \\ 0 & \cos\theta_L & \sin\theta_L \\ 0 & -\sin\theta_L & \cos\theta_L \end{bmatrix} \begin{bmatrix} 0 \\ 1 \\ 0 \end{bmatrix} \tag{31}$$

where θ_L is the center off-nadir angle in stripmap SAR. The acceleration vector term results in a Doppler rate difference of

$$\Delta f_{r,a} = \frac{2}{\lambda\rho}\Delta\mathbf{A}_{st}\cdot\mathbf{R}_{st}$$
$$= k_a\left[\cos\theta_L + \sqrt{\sin^2\theta_L\sin^2\theta_{yaw}+\cos^2\theta_L}\,\sin\left(\Delta\nu - \arctan\frac{\cos\theta_L}{\sin\theta_L\sin\theta_{yaw}}\right)\right] \tag{32}$$

in which

$$k_a = \frac{2\mu\left(1+e\cos\nu\right)^2}{\lambda p^2} \tag{33}$$

In common SAR operations, θ_{yaw} is usually small. Then, we have the following expansion without cubic and higher-order terms.

$$\arctan\frac{\cos\theta_L}{\sin\theta_L\sin\theta_{yaw}} = \frac{\pi}{2} - \arctan\frac{\sin\theta_L\sin\theta_{yaw}}{\cos\theta_L}$$
$$\approx \frac{\pi}{2} - \tan\theta_L\sin\theta_{yaw} \tag{34}$$

With the result from Appendix B and Equations (32)–(34), the expected value and standard deviation of $\Delta f_{r,a}$ are

$$\mathbb{E}\left[\Delta f_{r,a}\right] = k_a\cos\theta_L + k_a\sqrt{\sin^2\theta_L\sin^2\theta_{yaw}+\cos^2\theta_L}\left(1-\frac{\mathrm{Var}\left[\Delta\nu\right]}{2}\right)\left(\frac{1}{2}\tan^2\theta_L\sin^2\theta_{yaw}-1\right) \tag{35}$$

$$\sigma\left[\Delta f_{r,a}\right] = k_a\sqrt{\sin^2\theta_L\sin^2\theta_{yaw}+\cos^2\theta_L}\sqrt{\mathrm{Var}\left[\Delta\nu\right]\left(\tan^2\theta_L\sin^2\theta_{yaw}+\frac{\mathrm{Var}\left[\Delta\nu\right]}{2}\right)} \tag{36}$$

Equation (35) shows an interesting result. Instead of the expected zero value, we have $\mathbb{E}\left[\Delta f_{r,a}\right]$, which means there will be bias in the expected value of the Doppler rate f_r due to the calculated true anomaly. This bias can be and should be removed when calculating the Doppler parameters for the imaging algorithm.

However, we need the approximate maximum value of $\sigma\left[\Delta f_{r,a}\right]$. If we put aside the quadratic term of $\mathrm{Var}\left[\Delta\nu\right]$ in Equation (36), the leftover part is modulated by $\mathrm{Var}\left[\Delta\nu\right]$ and $\sin^2\theta_{yaw}$ at each ν point. When x is relatively small, $\arctan x \propto x$. Combined with Equation (30), we can ascertain that $\sin^2\theta_{yaw} \propto \sin^2\left[\cos\left(\nu+\omega\right)\right] \propto \cos^2\left(\nu+\omega\right)$. It can also be found that, from Equation (28), if we take the appropriate approximation, then $\sigma\left[\Delta\nu\right] \propto \cos\nu$. From these approximations, ν in $f\left(\nu\right) = \cos\left(\nu+\omega\right)\cos\nu$ when it reaches its maximum value $f\left(\nu\right)_{max}$, and the very same ν will also make $\sigma\left[\Delta f_{r,a}\right]$ reach its maximum. Obviously, at this moment,

$$\nu = k\pi - \frac{\omega}{2},\ \text{for } k\in\mathbb{Z}_{>0} \tag{37}$$

If $k=1$, then

$$\mathrm{Var}\left[\Delta\nu\right]_{\max}\big|_{\nu=\pi-\omega/2} \approx \left(\left|\cos\frac{\omega}{2}\right|\sigma\left[\Delta\nu\right]_{\max}\right)^2$$
$$= \frac{\cos^2\frac{\omega}{2}}{e^2}\left(\frac{1}{a^2}\sigma_p^2 + \frac{a}{\mu}\sigma_v^2\right) \tag{38}$$

together with

$$k_{a,max} = \frac{2\mu\left(1+e\right)^2}{\lambda p^2} \tag{39}$$

and $\theta_{yaw} = \theta_{yaw,max}$ in Equation (36). Thus, we have

$$\sigma\left[\Delta f_{r,a}\right]_{max} = k_{a,max}\sqrt{\sin^2\theta_L \sin^2\theta_{yaw,max} + \cos^2\theta_L} \cdot \sqrt{\mathrm{Var}\left[\Delta v\right]_{\max}\left(\tan^2\theta_L \sin^2\theta_{yaw,max} + \frac{\mathrm{Var}\left[\Delta v\right]_{\max}}{2}\right)} \quad (40)$$

3.4. Integration Time, Slant Range, and Approximate QPE

For a given SAR configuration and true anomaly, the integration time is used to calculate QPE, while the slant range contributes both to the integration time and QPE. In our proposed analytical approximation model for QPE, both the accurate and approximate integration time and slant range are considered. The former ones are applied to generate the approximate QPE with true anomaly as the independent variable, and the latter ones are merged into the maximum of the QPE calculation.

In the accurate calculation, the slant ranges are obtained by solving a quadratic equation:

$$\frac{\left(\rho\hat{u}_{a,ECI,x} + R_{sat,ECI,x}\right)^2 + \left(\rho\hat{u}_{a,ECI,y} + R_{sat,ECI,y}\right)^2}{E_a^2} + \frac{\left(\rho\hat{u}_{a,ECI,z} + R_{sat,ECI,z}\right)^2}{E_b^2} = 1 \quad (41)$$

where $E_a = 6{,}378{,}137.000$ m and $E_b = 6{,}356{,}752.314$ m are the semi-major axis and semi-minor axis of the ellipsoid reference in WGS 84, respectively. Then, the integration time is

$$T_{int} = \frac{\lambda}{L_a}\frac{\left|\mathbf{R}_{st}\right|}{\left|\mathbf{V}_{st}\right|}\frac{\left|\mathbf{R}_s\right|}{\left|\mathbf{R}_t\right|} \quad (42)$$

In the approximate maximum QPE calculation, we use a slant range that is based on the circular Earth with a radius of $E_{mean} = \left(2E_a + E_b\right)/3$ and

$$\rho_{mean} = E_{mean}\frac{\sin\left[\theta_L + \pi - \arcsin\left(a\sin\theta_L / E_{mean}\right)\right]}{\sin\theta_L} \quad (43)$$

With the satellite's velocity at apogee only instead of $\left|\mathbf{V}_{st}\right|$, together with a instead of $\left|\mathbf{R}_s\right|$, we get the mean integration time.

$$T_{int,mean} = \frac{a\lambda\rho_{mean}}{E_{mean}L_a}\sqrt{\frac{a\left(1+e\right)}{\mu\left(1-e\right)}} \quad (44)$$

Thus, we have the approximate expected value and standard deviation of QPE and its maximum expression:

$$\mathbb{E}\left[\mathrm{QPE}\right] = \pi\mathbb{E}\left[\Delta f_{r,a}\right]\left(\frac{T_{int}}{2}\right)^2 \quad (45)$$

$$\begin{aligned}\sigma\left[\mathrm{QPE}\right] &= \pi\sqrt{\sigma^2\left[\Delta f_{r,v}\right] + \sigma^2\left[\Delta f_{r,a}\right]}\left(\frac{T_{int}}{2}\right)^2 \\ &= \sqrt{\sigma^2\left[\mathrm{QPE}_v\right] + \sigma^2\left[\mathrm{QPE}_a\right]}\end{aligned} \quad (46)$$

$$\sigma\left[\mathrm{QPE}\right]_{max} = \pi\sqrt{\sigma^2\left[\Delta f_{r,v}\right]_{max} + \sigma^2\left[\Delta f_{r,a}\right]_{max}}\left(\frac{T_{int,mean}}{2}\right)^2 \quad (47)$$

where $\sigma\left[\mathrm{QPE}_v\right]$ and $\sigma\left[\mathrm{QPE}_a\right]$ are the standard deviations of QPE introduced by the velocity vector term and acceleration vector term, respectively.

4. Evaluation and Results

In this section, the analytical approximation model is evaluated by comparing it with the Monte Carlo simulation approach.

4.1. Evaluation Model

For the purpose of validating the proposed analytical approximation model for QPE, a Monte Carlo-based numerical simulation is introduced as a reference for validating the proposed method. In the simulation approach, large samples of error data are generated first according to the specified probability distribution parameters. Then, these error samples are input to a simulation program. For each sample of error data in the orbit determination data, the simulation program calculates the Doppler parameters of the SAR imaging procedure with and without the given error sample set and records the difference between these two situations. The probability distribution of the difference from the Doppler parameter simulation is analyzed to provide a reference for the validation of the analytical approximation model.

In this simulation program, there are no approximations such as those in the analytical approximation model. For beam steering, the simulation uses the total zero Doppler steering (TZDS) method [26] of TerraSAR-X, which has a joint yaw and pitch steering to minimize the Doppler centroid residence. The ellipsoid reference in WGS 84 is chosen to be the Earth's surface mode in the simulation.

The QPE approximation has the same slant range and integration time data as the Monte Carlo simulation in order to get a more precise approximation result for $\nu \in [0, 2\pi]$. The QPE maximum value calculation works with the mean slant range and integration time for a much simpler computation.

The Monte Carlo method generates 30,000 error samples of the position and velocity vectors per true anomaly, and there are 1000 true anomaly calculation points in $\nu \in [0, 2\pi]$.

The inputs of the simulation are provided in Table 1. These parameters can be divided into four groups. The first group includes the Earth's physical parameters E_a and E_b, which represent the semi-major and semi-minor axes of the Earth ellipsoid, respectively. The second group includes the satellite platform orbit parameters, namely, the semi-major axes of the orbit a, the argument of periapsis ω, the orbital inclination i, and the eccentricity e. The third group contains the SAR payload parameters, i.e., the center frequency of the transmitting signal, the center off-nadir angle, and the azimuth length of the antenna. The last group represents the probability distribution of the errors in the real-time orbit determination data. The errors have a normal distribution with an expected value of zero, thus only the standard deviations are listed in the table. σ_p and σ_v represent the standard deviations of the errors in the position and velocity data from the real-time orbit determination system. A comparison of the 3D positioning residuals (RMS) using broadcast ephemerides results in around 1.5 m in the reference [21]. We selected a 3 m standard deviation in the real-time positioning errors. The standard deviation in the real-time velocity error is approximated by the results in [20].

Table 1. Simulation Parameters.

Parameter	Value	Parameter	Value
E_a	6,378,137.000 m	E_b	6,356,752.314 m
a	6,778,140.000 m	ω	90.00°
i	97.42°	e	0.0011
SAR Center Frequency	9.60 GHz	σ_p	3 m
Center off-nadir angle	33.8°	σ_v	0.1 m/s
Antenna Azimuth Length	1.92 m		

4.2. Approximation of the Velocity Vector Term

Figure 2 shows the result of comparing the statistical parameters of $\Delta f_{r,v}$ and QPE_v between the Monte Carlo simulation and the analytical approximation model. $\mathbb{E}\left[\text{QPE}_v\right]$ shows oscillations around zero, and these are due to the Monte Carlo simulation method. With more error samples processed in the numerical simulation, the oscillation would be much smaller, which confirms the effectiveness of the analytical approximation model. The mean differences in the results between the numerical simulation and analytical approximation model in Figure 2c,d are 1.64×10^{-3} Hz/s and 0.093°. The difference of $\sigma\left[\Delta f_{r,v}\right]$ may be high for the algorithm method, but it is a satisfactory result for the approximate QPE calculation. The integration time is not a consistent value, which leads to the differently shaped curves in Figure 2c,d.

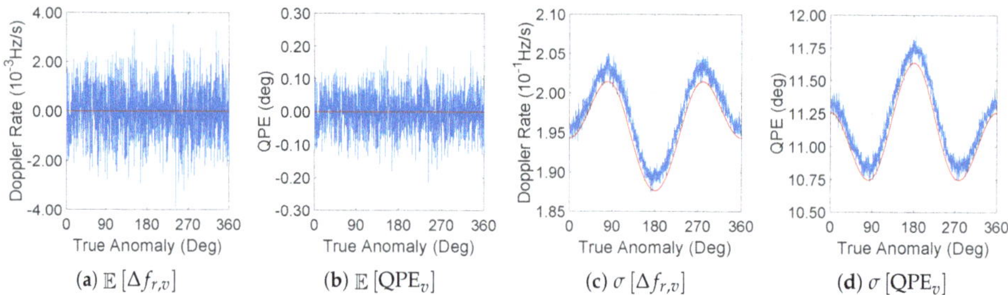

(**a**) $\mathbb{E}\left[\Delta f_{r,v}\right]$ (**b**) $\mathbb{E}\left[\text{QPE}_v\right]$ (**c**) $\sigma\left[\Delta f_{r,v}\right]$ (**d**) $\sigma\left[\text{QPE}_v\right]$

Figure 2. Velocity vector term: numerical simulation results (blue) and analytical approximation results (red). The figures show that the results from the analytical approximation model are consistent with the numerical simulation results. (**a**,**b**) The expected values have no significant relationship with the variation in the true anomaly and are positioned at a relatively low level, i.e., around zero. The differences between the two sets of results in (**c**,**d**) are due to the ignored terms in the approximation derivations.

4.3. True Anomaly Calculation

$\sigma\left[\Delta\nu\right]$ plays an important role in the analytical approximation model of QPE, and Figure 3 presents the results from both the numerical simulation and the analytical approximation model. Figure 3a indicates that the results of the analytical approximation model are almost identical to those of the numerical simulation. However, in the analytical approximation model, at 90° and 270°, $\sigma\left[\Delta\nu\right]$ reaches zero according to Equation (28), while the numerical simulation result is not zero at that position. In the proposed model, we assume that there are no errors in the term $p - \left|\mathbf{R}_{sat,ECI}\right|$ in Equation (27), which leads to a value of zero for $\sigma\left[\Delta\nu\right]$ when the satellite is positioned at true anomalies 90° and 270°. In fact, at those positions, the errors in the measurements start playing a major role in the $\sigma\left[\Delta\nu\right]$ calculation. However, at those two specific positions, $\sigma\left[\Delta\nu\right]$ reaches its minimum, thus it does not introduce much error to the analytical approximation model.

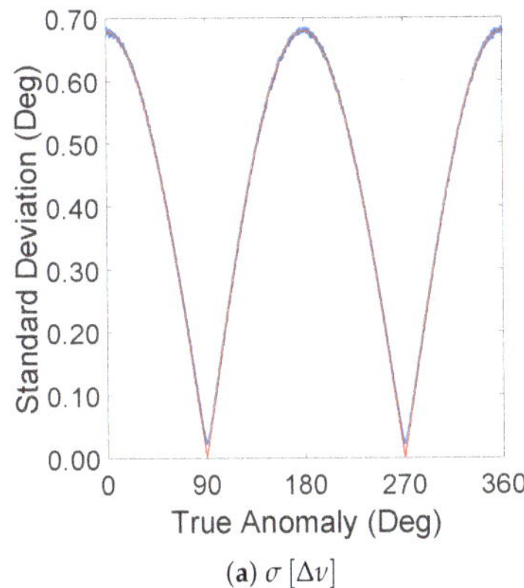

(**a**) $\sigma[\Delta\nu]$

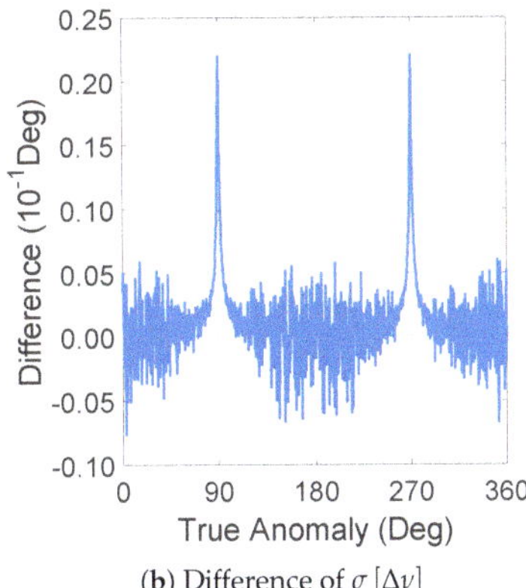

(**b**) Difference of $\sigma[\Delta\nu]$

Figure 3. The true anomaly result from measurements between numerical simulation and the analytical approximation model. (**a**) The analytical approximation model (red) fits the numerical simulation (blue) in most areas of true anomaly. However, at certain true anomalies, which are the unique positions of the satellite in its orbit, the difference between the two sets of results increases. (**b**) At around 90° and 270°, significant peaks arise for approximately 0.023° in the standard deviation of $\Delta\nu$.

4.4. Approximation of the Acceleration Vector Term

The expected value of the Doppler rate by the acceleration vector term $\mathbb{E}[\Delta f_{r,a}]$ shows different characteristics from those of the velocity vector terms. Figure 4a no longer has an expected value of zero compared with Figure 2a. This non-zero $\mathbb{E}[\Delta f_{r,a}]$ value is due to the calculation of true anomaly from onboard state vector measurements, which also result in a non-zero $\mathbb{E}[\mathrm{QPE}_a]$. In addition, the analytical approximation model fits the numerical simulation result for all $\nu \in [0, 2\pi]$.

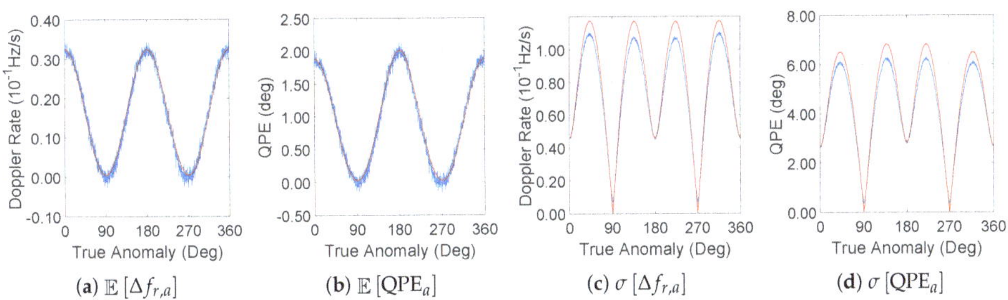

(**a**) $\mathbb{E}[\Delta f_{r,a}]$ (**b**) $\mathbb{E}[\mathrm{QPE}_a]$ (**c**) $\sigma[\Delta f_{r,a}]$ (**d**) $\sigma[\mathrm{QPE}_a]$

Figure 4. Acceleration vector term: numerical simulation results (blue) and analytical approximation results (red). The analytical approximation results are consistent with the numerical simulation results. The expected values in (**a**,**b**) no longer have an expected value of zero, indicating different characteristics from those of the velocity vector terms. The differences around 90° and 270° of true anomaly between the two sets of results in (**c**,**d**) are due to the ignored terms in the calculated true anomaly approximation derivations.

For the comparison between $\sigma[\Delta f_{r,a}]$ and $\sigma[\mathrm{QPE}_a]$, at most positions, the analytical approximation model fits the numerical simulation point by point. Apart from some simulation points, it maintains accuracy. From the comparison of Figure 4c,d, the influence of the integration time may not seem significant in contrast to the comparison of Figure 2c,d. In fact, both $\sigma[\mathrm{QPE}_a]$ and $\sigma[\mathrm{QPE}_v]$ are calculated with the same integration time series. This major difference in the curve appearance is caused by the relatively narrow fluctuation interval of $\sigma[\mathrm{QPE}_v]$.

4.5. QPE Approximation

The Monte Carlo simulation and analytical approximation model are illustrated in Figure 5. Figure 5a shows exactly the same result as that in Figure 4b, which confirms Equation (45) in the analytical approximation model. The results of $\sigma\,[\mathrm{QPE}]$ in Figure 5b, however, are presented in a slightly complex form. This is because of the combination of the properties of $\sigma\,[\mathrm{QPE}_v]$ and $\sigma\,[\mathrm{QPE}_a]$.

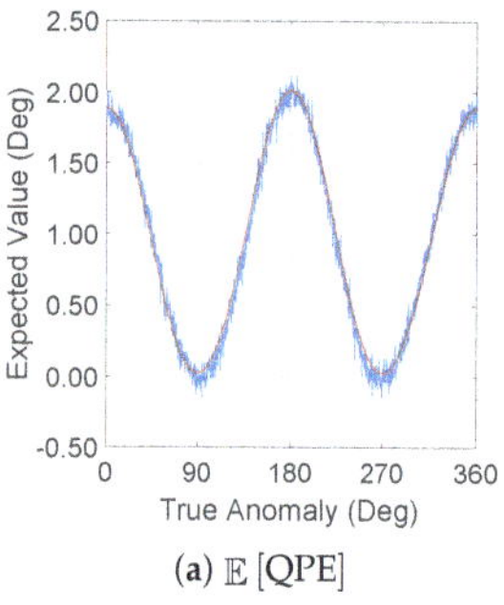

(a) $\mathbb{E}\,[\mathrm{QPE}]$

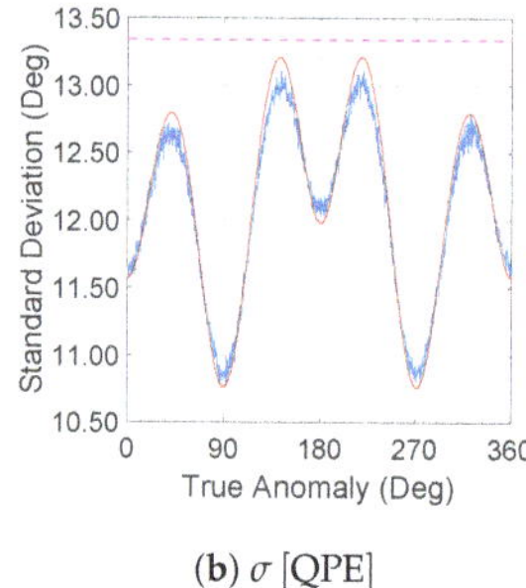

(b) $\sigma\,[\mathrm{QPE}]$

Figure 5. QPE result from numerical simulation (blue) compared with that from the analytical approximation model (red). The expected value and standard deviation of QPE are shown in (**a**) and (**b**), respectively. The dashed magenta line in (**b**) is the value of $\sigma\,[\mathrm{QPE}]_{max}$ from Equation (47).

The dashed magenta line in Figure 5b is $\sigma\,[\mathrm{QPE}]_{max}$. This maximum value does not equal the exact maximum of $\sigma\,[\mathrm{QPE}]$; instead, it has a 1.80% relative error with the given simulation parameters. The most important advantage of $\sigma\,[\mathrm{QPE}]_{max}$ is that researchers can calculate this value easily with the aid of only a calculator and the appropriate parameters, combined with Equation (47) and related equations, and an acceptable approximation maximum value can still be determined.

5. Discussion

We designed an analytical approximation model of the QPE introduced by orbit determination errors. Our aim is to reduce the calculations and the processing time while revealing the core of the process by which error is transferred to QPE calculations. The technical parameter iteration period during early-stage development for an onboard real-time SAR imaging mission will benefit from the proposed methods.

SAR imaging requires the relative position, velocity, and acceleration vectors between the radar antenna and the targets to estimate the Doppler centroid and Doppler rate. Currently, in many ground-based processing systems for existing SAR missions, high-accuracy orbit determination data are applied to estimate the Doppler parameters. Such data have a time delay from the SAR raw data acquisition period, which is usually in the range of several hours or more. For real-time onboard SAR imaging missions, it is not realistic to wait for such high-accuracy orbit determination data to meet the "real-time" requirement. Real-time onboard orbit determination data have to be applied in the real-time onboard SAR imaging procedure. Current spaceborne SAR missions are usually based on a professional remote sensing satellite platform, and they are not as sensitive to perturbation. For a spaceborne SAR payload mounted on a small satellite platform to perform a constellation mission, such as MirrorSAR proposed by the DLR [27], there will be some differences (i.e., it is not as steady as a large remote sensing satellite platform) in the movement of the actual target satellite. In these situations, the errors in the real-time onboard orbit determination data should be evaluated in order to examine whether they will jeopardize the new real-time SAR imaging mission concept.

There exist other terms of phase error in the context of SAR data processing [28]. Besides QPE, there are linear, higher-order (more than cubic), sinusoidal, and random phase errors, among others. We selected QPE as the error of interest because it has defocus and loss-of-resolution effects on the image. An acceptable QPE should be seen as a tool that is used out of necessity but is not adequate for new SAR missions.

Research on the prediction of SAR imaging quality from the precision of the orbit ephemeris has been carried out [29]: the upper limit expressions of multiple phase errors and their sensitivity to the orbit ephemeris covariance have been previously established. However, this method has two major limitations when applied to spaceborne real-time SAR imaging missions. Firstly, those upper limits only reflect the extreme situation, which will not always occur during a mission. We need to examine the whole situation of the phase error, as well as its probability form. Currently, the numerical simulation method can produce the probability of the phase error, but it is time-consuming. Secondly, in spaceborne real-time SAR imaging missions, it is impossible to get the orbit ephemeris directly, and it is calculated from onboard orbit determination data. The direct link between the precision of the onboard orbit determination data and the phase error in SAR imaging is still unclear. The proposed analytical approximation model is a potential solution to the two limitations above. With the precision of the onboard orbit determination system, which can be measured on the ground, the proposed analytical approximation model gives the direct QPE probability distribution result, which can simplify the analysis.

Compared with existing methods, the proposed analytical approximation model reduces the total calculations and processing time while revealing the core of the process by which errors are transferred to QPE calculations. This proposed analytical approximation model can reveal the QPE distribution, compared with the extreme value method mentioned in Section 1. For the numerical simulation method, with M error samples for each true anomaly calculation point and a total of N true anomaly calculation points, the calculation complexity reaches $O(MN)$. It is important to point out that M is usually a large number for better statistical results. No error sample set is generated in the analytical approximation model; thus, the calculation complexity is reduced. The QPE distribution result for N true anomaly calculation points is $O(N)$, and for the distribution parameters, the validated true anomaly $\nu \in [0, 2\pi]$ is $O(1)$, which can significantly reduce the total calculation and processing time. Spaceborne real-time SAR imaging mission designers can quickly determine the maximum QPE with the help of only a calculator and the appropriate parameters. Therefore, this model will accelerate the technical parameter iteration procedure during the early-stage development of an onboard real-time SAR imaging mission and provide strong proof for the feasibility of such spaceborne real-time SAR missions.

There will always be some processing blocks in the design of a spaceborne SAR system. For each processing block, the errors introduced by processing must be limited for optimal performance of the whole system. From the design practice of the existing system, an azimuth impulse response width of 1.02 can be an acceptable limitation for a single processing block, which equals a QPE of no more than 0.27π [23]. The results in Section 4 are used here as an example. With the analytical approximation model, the result, and the three-sigma rule, it is clear that, with the simulation parameters given, if $\mathbb{E}[\Delta f_{r,a}]$ is removed during the Doppler rate calculation, then the QPE introduced by the onboard orbit determination system will be less than 40.02°, with a probability of 99.73%. This result equals an azimuth impulse response width of less than 2%, with a probability of 99.73%.

In its current form, the proposed analytical approximation model is given when the attitude error is ignored; i.e., the orbit determination error and the attitude error of the platform are decoupled. The next generation of the analytical approximation model will take the attitude error into consideration in order to examine the coupled relationship. In addition, for spaceborne bistatic real-time SAR imaging missions, a bistatic form of the model will be needed. These two directions will be our future research topics.

6. Conclusions

As the errors in the state vectors from an onboard orbit determination system for a spaceborne SAR affect real-time SAR imaging quality, this paper proposes an analytical approximation model to identify the introduction of QPE by state vector errors. This analytical approximation model can provide approximation results at two granularities, i.e., the approximation with the satellite's true anomaly as the independent variable and the approximation for $\nu \in [0, 2\pi]$. Compared with numerical simulation methods such as Monte Carlo, the proposed analytical approximation model reduces the total calculations and processing time. Another benefit of the proposed analytical approximation model is that it reveals the core of the error transfer, so each parameter's role in the introduced QPE is clear. Spaceborne real-time SAR imaging mission designers can quickly determine QPE with the help of only a calculator and the appropriate parameters. In some ways, this will accelerate the technical parameter iteration period during early-stage development of an onboard real-time SAR imaging mission and provide convincing evidence of the feasibility of spaceborne real-time SAR missions.

Author Contributions: Conceptualization, X.Y. and J.C.; Funding acquisition, J.C. and O.L.; Investigation, X.Y. and H.N.; Methodology, X.Y.; Project administration, J.C. and O.L.; Resources, X.Y., J.C., H.N., and O.L.; Supervision, J.C. and O.L.; Validation, X.Y. and H.N.; Visualization, X.Y. and H.N.; Writing—original draft, X.Y.; and Writing—review & editing, X.Y. and H.N.

Funding: This research was funded by the National Natural Science Foundation of China under Grant 61861136008 and Grant 61671043. The author Xiaoyu Yan was funded by China Scholarship Council (CSC) under Grant 201706020033.

Acknowledgments: The authors thank Yue Fang from Beihang University for useful discussions. The authors are grateful to Xi (Cathy) Chen from Northwestern University for her understanding and useful suggestions.

Conflicts of Interest: The authors declare no conflict of interest.

Appendix A. Errors in Calculating True Anomaly

Let us have

$$Y = \sqrt{\frac{p}{\mu}} \mathbf{V}_{sat,ECI} \cdot \mathbf{R}_{sat,ECI} \tag{A1}$$

$$X = p - |\mathbf{R}_{sat,ECI}| \tag{A2}$$

with the assumption that the position error has no influence on term X in Equation (A2). Thus, we have

$$\Delta \frac{Y}{X} = \sqrt{\frac{p}{\mu}} \cdot \frac{1}{p-r} \Delta \mathbf{V} \cdot \mathbf{R} \tag{A3}$$

The term $\Delta \mathbf{V} \cdot \mathbf{R}$ is the difference in the term $\mathbf{V}_{sat,ECI} \cdot \mathbf{R}_{sat,ECI}$ in Equation (A1) between the measured data and the accurate value.

$$\begin{aligned} \Delta \mathbf{V} \cdot \mathbf{R} =& \mathbf{V}_{sat,ECI,e} \cdot \mathbf{R}_{sat,ECI,e} - \mathbf{V}_{sat,ECI} \cdot \mathbf{R}_{sat,ECI} \\ =& \left(\mathbf{A}_{O2E}\mathbf{V}_{sat,OP} + \Delta \mathbf{V}\right) \cdot \left(\mathbf{A}_{O2E}\mathbf{P}_{sat,OP} + \Delta \mathbf{P}\right) - \mathbf{A}_{O2E}\mathbf{V}_{sat,OP} \cdot \mathbf{A}_{O2E}\mathbf{P}_{sat,OP} \\ \approx& c_p \sqrt{\frac{\mu}{p}} + c_v \frac{a\left(1-e^2\right)}{1+e\cos\nu} \end{aligned} \tag{A4}$$

where the position error times the velocity error are ignored and

$$c_p = -\Delta p_x \left[\sin(\nu+\omega) + e\sin\omega\right] + \Delta p_y \left[\cos(\nu+\omega) + e\cos\omega\right]\cos i + \Delta p_z \left[\cos(\nu+\omega) + e\cos\omega\right]\sin i \tag{A5}$$

$$c_v = \Delta v_x \cos(\nu+\omega) + \Delta v_y \sin(\nu+\omega)\cos i + \Delta v_z \sin(\nu+\omega)\sin i \tag{A6}$$

The c_p and c_v parameters in Equations (A5) and (A6) can be regarded as the core function of the error introduced by position and velocity errors when calculating true anomaly. The calculated true anomaly's variance depends directly on the variance of the terms c_p and c_v, which link directly to the variance of the measurement errors of position and velocity in each axis of the ECI system.

Considering the derivative of Equation (27), with $Y/X = \tan\nu$, we have the difference in calculated ν and its real value.

$$\begin{aligned}\Delta\nu &= \left(\operatorname{atan2}\frac{Y}{X}\right)' \cdot \Delta\frac{Y}{X} \\ &= \frac{1}{1+(Y/X)^2} \cdot \sqrt{\frac{p}{\mu}} \cdot \frac{1}{p-r} \cdot \Delta\mathbf{V} \cdot \mathbf{R} \\ &= \frac{\cos\nu\,(1+e\cos\nu)}{e\sqrt{\mu a\,(1-e^2)}} \left[c_p\sqrt{\frac{\mu}{p}} + c_v\frac{a\,(1-e^2)}{1+e\cos\nu}\right]\end{aligned} \tag{A7}$$

The variance of $\Delta\nu$ is

$$\operatorname{Var}[\Delta\nu] = \cos^2\nu \cdot \frac{(1+e\cos\nu)^2}{\mu a e^2\,(1-e^2)} \cdot \left[\frac{\mu}{p}\operatorname{Var}[c_p] + \left(\frac{a\,(1-e^2)}{1+e\cos\nu}\right)^2 \operatorname{Var}[c_v]\right] \tag{A8}$$

A general expression of the variance of c_p and c_v is

$$\begin{aligned}\operatorname{Var}[c_p] = \operatorname{Var}[\Delta p_x]\,[\sin(\nu+\omega)+e\sin\omega]^2 &+ \operatorname{Var}[\Delta p_y]\,[\cos(\nu+\omega)+e\cos\omega]^2\cos^2 i \\ &+ \operatorname{Var}[\Delta p_z]\,[\cos(\nu+\omega)+e\cos\omega]^2\sin^2 i\end{aligned} \tag{A9}$$

$$\operatorname{Var}[c_v] = \operatorname{Var}[\Delta v_x]\cos^2(\nu+\omega) + \operatorname{Var}[\Delta v_y]\sin^2(\nu+\omega)\cos^2 i + \operatorname{Var}[\Delta v_z]\sin^2(\nu+\omega)\sin^2 i \tag{A10}$$

If all the variances of position and velocity errors are the same standard deviation in the x-, y-, and z-directions in the ECI system, i.e., $\sigma_{p,x} = \sigma_{p,y} = \sigma_{p,z} = \sigma_p$ and $\sigma_{v,x} = \sigma_{v,y} = \sigma_{v,z} = \sigma_v$, simpler expressions for c_p and c_v can be given by

$$\operatorname{Var}[c_p] = \sigma_p^2\left(1+e^2+2e\cos\nu\right) \tag{A11}$$

$$\operatorname{Var}[c_v] = \sigma_v^2 \tag{A12}$$

With Equations (A8), (A11) and (A12), we can derive Equation (28) in Section 3.2. This expression can be used to determine the dependent variable $\sigma[\Delta\nu]$ with the independent variable ν. In addition, we need a simpler expression of the maximum $\sigma[\Delta\nu]$ in order to support the quick calculation method for QPE. Beginning with Equation (28), we can ignore some terms in the equation related to orbit eccentricity e when its value is rather small (less than 0.01). Thus, we have

$$\sigma[\Delta\nu]_{\max} = \frac{1}{e}\sqrt{\frac{\sigma_p^2}{a^2} + \frac{a\sigma_v^2}{\mu}} \tag{A13}$$

which is Equation (29) in Section 3.2.

Appendix B. Expected Value and Variance of the Trigonometric Function of a Variable with a Normal Distribution

Suppose $\theta \sim N(0,\sigma^2)$. Then, consider another variable,

$$e^{j\theta} = \cos\theta + j\sin\theta \tag{A14}$$

where j is the imaginary unit. In the meantime, the moment-generating function of θ with a normal distribution is

$$\mathbb{E}\left[e^{j\theta}\right] = e^{0+(j\sigma)^2/2} = e^{-\sigma^2/2} \tag{A15}$$

which means

$$\mathbb{E}\left[\cos\theta\right] = e^{-\sigma^2/2} \tag{A16}$$

$$\mathbb{E}\left[\sin\theta\right] = 0 \tag{A17}$$

The variance of $\cos\theta$ and $\sin\theta$ is

$$\begin{aligned}
\mathrm{Var}\left[\cos\theta\right] &= \mathbb{E}\left[\cos^2\theta\right] - \mathbb{E}\left[\cos\theta\right]^2 \\
&= \frac{1}{2}\left(1+\mathbb{E}\left[\cos 2\theta\right]\right) - \left(e^{-\sigma^2/2}\right)^2 \\
&= \frac{1}{2}\left(1-e^{-\sigma^2}\right)^2
\end{aligned} \tag{A18}$$

$$\begin{aligned}
\mathrm{Var}\left[\sin\theta\right] &= \mathbb{E}\left[\sin^2\theta\right] - \mathbb{E}\left[\sin\theta\right]^2 \\
&= \frac{1}{2}\mathbb{E}\left[1-\cos 2\theta\right] \\
&= \frac{1}{2}\left(1-e^{-2\sigma^2}\right)
\end{aligned} \tag{A19}$$

For a new variable $z \sim N\left(\mu,\sigma^2\right)$, $\theta = z-\mu$,

$$\begin{aligned}
\mathbb{E}\left[\cos z\right] &= \mathbb{E}\left[\cos\left(\theta+\mu\right)\right] \\
&= \mathbb{E}\left[\cos\theta\cos\mu - \sin\theta\sin\mu\right] \\
&= \cos\mu\,\mathbb{E}\left[\cos\theta\right] - \sin\mu\,\mathbb{E}\left[\sin\theta\right] \\
&= e^{-\sigma^2/2}\cos\mu
\end{aligned} \tag{A20}$$

$$\begin{aligned}
\mathbb{E}\left[\sin z\right] &= \mathbb{E}\left[\sin\left(\theta+\mu\right)\right] \\
&= \mathbb{E}\left[\sin\theta\cos\mu + \cos\theta\sin\mu\right] \\
&= \cos\mu\,\mathbb{E}\left[\sin\theta\right] + \sin\mu\,\mathbb{E}\left[\cos\theta\right] \\
&= e^{-\sigma^2/2}\sin\mu
\end{aligned} \tag{A21}$$

The variance of $\cos z$ and $\sin z$ is

$$\begin{aligned}
\mathrm{Var}\left[\cos z\right] &= \mathbb{E}\left[\cos^2 z\right] - \mathbb{E}\left[\cos z\right]^2 \\
&= \frac{1}{2} + \frac{1}{2}\mathbb{E}\left[\cos\left(2\theta+2\mu\right)\right] - \mathbb{E}\left[\cos z\right]^2 \\
&= \frac{1}{2} + \frac{1}{2}e^{-2\sigma^2}\cos 2\mu - e^{-\sigma^2}\cos^2\mu
\end{aligned} \tag{A22}$$

$$\begin{aligned}
\mathrm{Var}\left[\sin z\right] &= \mathbb{E}\left[\sin^2 z\right] - \mathbb{E}\left[\sin z\right]^2 \\
&= \frac{1}{2} - \frac{1}{2}\mathbb{E}\left[\cos\left(2\theta+2\mu\right)\right] - \mathbb{E}\left[\sin z\right]^2 \\
&= \frac{1}{2} - \frac{1}{2}e^{-2\sigma^2}\cos 2\mu - e^{-\sigma^2}\sin^2\mu
\end{aligned} \tag{A23}$$

References

1. Moreira, A.; Brodacholl, A.; Dom, J.; Kochsiek, F.; Potzsch, W. Airborne real-time SAR processing activities at DLR. In Proceedings of the IGARSS '92 International Geoscience and Remote Sensing Symposium, Houston, TX, USA, USA, 26–29 May 1992; Volume 1, pp. 412–414.

2. Tian, H.; Chang, W.; Li, X. Integrated system of mini-SAR real-time signal processing and data storage. In Proceedings of the 2015 IEEE Radar Conference, Johannesburg, South Africa, 27–30 October 2015; pp. 354–358.
3. Jia, G.; Buchroithner, M.; Chang, W.; Li, X. Simplified Real-Time Imaging Flow for High-Resolution FMCW SAR. *IEEE Geosci. Remote Sens. Lett.* **2015**, *12*, 973–977.
4. Gu, C.; Chang, W.; Li, X.; Jia, G.; Liu, Z. A compact FMCW SAR real-time imaging system and its performance analysis. In Proceedings of the IET International Radar Conference, Hangzhou, China, 14–16 October 2015.
5. Li, C.; Jie, L.; Cui, L.; Yao, D. The Squinted-looking SAR real-time imaging Based on Multi-core DSP. In Proceedings of the IET International Radar Conference, Hangzhou, China, 14–16 October 2015; p. 4. [CrossRef]
6. Moreira, A. Real-time synthetic aperture radar (SAR) processing with a new subaperture approach. *IEEE Trans. Geosci. Remote Sens.* **1992**, *30*, 714–722. [CrossRef]
7. Palm, S.; Wahlen, A.; Stanko, S.; Pohl, N.; Wellig, P.; Stilla, U. Real-time onboard processing and ground based monitoring of FMCW-SAR videos. In Proceedings of the EUSAR 2014 10th European Conference on Synthetic Aperture Radar, Berlin, Germany, 3–5 June 2014; pp. 1–4.
8. Que, R.; Ponce, O.; Scheiber, R.; Reigber, A. Real-time processing of SAR images for linear and non-linear tracks. In Proceedings of the 2016 17th International Radar Symposium (IRS), Krakow, Poland, 10–12 May 2016; pp. 1–4.
9. Que, R.; Ponce, O.; Baumgartner, S.V.; Scheiber, R. Multi-mode real-time SAR on-board processing. In Proceedings of the EUSAR 2016—11th European Conference on Synthetic Aperture Radar, Hamburg, Germany, 6–9 June 2016; pp. 1–6.
10. Gu, C.; Chang, W.; Li, X.; Jia, G.; Luan, X. A new distortion correction method for FMCW SAR real-time imaging. *IEEE Geosci. Remote Sens. Lett.* **2017**, *14*, 429–433. [CrossRef]
11. Sun, G.C.; Liu, Y.; Xing, M.; Wang, S.; Guo, L.; Yang, J. A Real-Time Imaging Algorithm Based on Sub-Aperture CS-Dechirp for GF3-SAR Data. *Sensors* **2018**, *18*, 2562. [CrossRef] [PubMed]
12. Walker, B.; Sander, G.; Thompson, M.; Burns, B.; Fellerhoff, R.; Dubbert, D. A High Resolution, Four-Band SAR Testbed with Real-Time Image Formation. In Proceedings of the International Geoscience and Remote Sensing Symposium, Lincoln, NE, USA, 31–31 May 1996; Volume 3, pp. 1881–1885.
13. Wang, J.; Xue, G.; Zhou, Z.; Song, Q. A new subaperture nonlinear chirp scaling algorithm for real-time UWB-SAR imaging. In Proceedings of the 2006 CIE International Conference on Radar, Shanghai, China, 16–19 October 2006; pp. 1–4.
14. Malanowski, M.; Krawczyk, G.; Samczyński, P.; Kulpa, K.; Borowiec, K.; Gromek, D. Real-time high-resolution SAR processor using CUDA technology. In Proceedings of the 2013 14th International Radar Symposium (IRS), Dresden, Germany, 19–21 June 2013; Volume 2, pp. 673–678.
15. Tang, H.; Li, G.; Zhang, F.; Hu, W.; Li, W. A spaceborne SAR on-board processing simulator using mobile GPU. In Proceedings of the 2016 IEEE International Geoscience and Remote Sensing Symposium (IGARSS), Beijing, China, 10–15 July 2016; pp. 1198–1201.
16. Ding, Z.; Xiao, F.; Xie, Y.; Yu, W.; Yang, Z.; Chen, L.; Long, T. A Modified Fixed-Point Chirp Scaling Algorithm Based on Updating Phase Factors Regionally for Spaceborne SAR Real-Time Imaging. *IEEE Trans. Geosci. Remote Sens.* **2018**, 56, 7436–7451. [CrossRef]
17. Sugimoto, Y.; Ozawa, S.; Inaba, N. Near Real-Time SAR Image Focusing Onboard Spacecraft. In Proceedings of the IGARSS 2018—2018 IEEE International Geoscience and Remote Sensing Symposium, Valencia, Spain, 22–27 July 2018; pp. 8038–8041.
18. Bergeron, A.; Doucet, M.; Harnisch, B.; Suess, M.; Marchese, L.; Bourqui, P.; Desnoyers, N.; Legros, M.; Guillot, L.; Mercier, L.; et al. Satellite on-board real-time SAR processor prototype. In Proceedings of the International Conference on Space Optics—ICSO 2010, International Society for Optics and Photonics, Rhodes Island, Greece, 4–8 October 2010; Volume 10565, p. 105652W.
19. Montenbruck, O.; Yoon, Y.; Gill, E.; Garcia-Fernandez, M. Precise orbit determination for the TerraSAR-X mission. In Proceedings of the International Symposium on Space Technology and Science, Kanazawa, Japan, 4–11 June 2006; Volume 25, p. 613.
20. Yu, Z.; You, Z. Real-time onboard orbit determination using GPS navigation solutions. In Proceedings of the 2011 First International Conference on Instrumentation, Measurement, Computer, Communication and Control, Beijing, China, 21–23 October 2011; pp. 949–952.

21. Yang, Y.; Yue, X.; Dempster, A.G. GPS-based onboard real-time orbit determination for LEO satellites using consider Kalman filter. *IEEE Trans. Aerosp. Electron. Syst.* **2016**, *52*, 769–777. [CrossRef]
22. Yan, X.; Chen, J.; Yang, W. Monte Carlo Analysis of Orbital Station Motion Parameter Errors Influence on SAR Azimuth Resolution Degradation. In Proceedings of the IGARSS 2018—2018 IEEE International Geoscience and Remote Sensing Symposium, Valencia, Spain, 22–27 July 2018; pp. 7805–7808.
23. Cumming, I.; Wong, F. *Digital Processing of Synthetic Aperture Radar Data: Algorithms and Implementation*; Artech House Remote Sensing Library; Artech House: Norwood, MA, USA, 2005.
24. Crassidis, J.; Junkins, J. *Optimal Estimation of Dynamic Systems*; Chapman & Hall/CRC Applied Mathematics & Nonlinear Science; CRC Press: Boca Raton, FL, USA, 2004.
25. Raney, R. Doppler properties of radars in circular orbits. *Int. J. Remote Sens.* **1986**, *7*, 1153–1162. [CrossRef]
26. Fiedler, H.; Boerner, E.; Mittermayer, J.; Krieger, G. Total zero Doppler steering-a new method for minimizing the Doppler centroid. *IEEE Geosci. Remote Sens. Lett.* **2005**, *2*, 141–145. [CrossRef]
27. Krieger, G.; Zonno, M.; Mittermayer, J.; Moreira, A.; Huber, S.; Rodriguez-Cassola, M. MirrorSAR: A Fractionated Space Transponder Concept for the Implementation of Low-Cost Multistatic SAR Missions. In Proceedings of the EUSAR 2018 12th European Conference on Synthetic Aperture Radar, Aachen, Germany, 4–7 June 2018; pp. 1–6.
28. Carrara, W.; Goodman, R.; Majewski, R. *Spotlight Synthetic Aperture Radar: Signal Processing Algorithms*; Artech House Remote Sensing Library; Artech House: Norwood, MA, USA, 1995.
29. Jin, M.Y. Prediction of SAR focus performance using ephemeris covariance. *IEEE Trans. Geosci. Remote Sens.* **1993**, *31*, 914–920. [CrossRef]

Article

Obtaining High-Resolution Seabed Topography and Surface Details by Co-Registration of Side-Scan Sonar and Multibeam Echo Sounder Images

Xiaodong Shang [1], Jianhu Zhao [1,2,*] and Hongmei Zhang [3]

[1] School of Geodesy and Geomatics, Wuhan University, 129 Luoyu Road, Wuhan 430079, China
[2] Institute of Marine Science and Technology, Wuhan University, 129 Luoyu Road, Wuhan 430079, China
[3] School of Power and Mechanical Engineering, Wuhan University, 8 South Donghu Road, Wuhan 430072, China
* Correspondence: jhzhao@sgg.whu.edu.cn; Tel.: +86-1379-708-8531

Received: 20 May 2019; Accepted: 21 June 2019; Published: 26 June 2019

Abstract: Side-scan sonar (SSS) is used for obtaining high-resolution seabed images, but with low position accuracy without using the Ultra Short Base Line (USBL) or Short Base Line (SBL). Multibeam echo sounder (MBES), which can simultaneously obtain high-accuracy seabed topography as well as seabed images with low resolution in deep water. Based on the complementarity of SSS and MBES data, this paper proposes a new method for acquiring high-resolution seabed topography and surface details that are difficult to obtain using MBES or SSS alone. Firstly, according to the common seabed features presented in both images, the Speeded-Up Robust Features (SURF) algorithm, with the constraint of image geographic coordinates, is adopted for initial image matching. Secondly, to further improve the matching performance, a template matching strategy using the dense local self-similarity (DLSS) descriptor is adopted according to the self-similarities within these two images. Next, the random sample consensus (RANSAC) algorithm is used for removing the mismatches and the SSS backscatter image geographic coordinates are rectified by the transformation model established based on the correct matched points. Finally, the superimposition of this rectified SSS backscatter image on MBES seabed topography is performed and the high-resolution and high-accuracy seabed topography and surface details can be obtained.

Keywords: side-scan sonar; multibeam echo sounder; initial image matching with constraint; dense local self-similarity; superimposition

---

1. Introduction

Seabed topography and morphology provide fundamental information for marine scientific research and ocean engineering [1–4]. Currently, the most widely used sensors in conducting ocean surveying and mapping are side-scan sonar (SSS) and multibeam echo sounder (MBES) [5,6].

As for the SSS, there are two types of operations: towing operation and hull-mounted [5]. The first type is more popular given that the second type of operation may degrade the SSS backscatter image quality because of the platform movements caused in the surveying process [5,7]. When towing, the SSS is usually installed in a towfish and towed by a cable behind a surveying vessel (Figure 1a). It emits a wide-angle beam and receives thousands of seabed echoes at fixed time intervals to form a high-resolution seabed image [5,8]. This kind of operation can minimize the effects of platform movements on the SSS backscatter images. More importantly, as the towfish is near the seabed, it can be used for deep seafloor observation [5]. Thus, the SSS backscatter image geographic coordinates are not accurate because of towing, potential currents dragging and flat bottom assumption [9–12]. By using the Ultra Short Base Line (USBL) or Short Base Line (SBL) in the surveying process, the accuracy of towfish positions can be improved. However, these two devices are not used in most cases because of high cost. Moreover, three-dimensional (3D) topography is not available by adopting SSS [5,13]. Although the emergence of interferometric SSS (ISSS) can acquire bathymetry, the measurement quality is very sensitive to underwater noise and seriously degrades for echoes close to nadir [14]. In addition to the ISSS, there also exist some other kinds of sonar systems that can acquire bathymetry data, such as Ping DSP 3DSS, Edgetech 6205 and Klein HydroChart 3500 [15], but they can only be used for hydrographic surveys in restricted areas, such as roadsteads or lakes [15]. The output from 3DSS is in the form of an intensity point cloud, which can be post-processed either as bathymetry, backscatter seabed image or a combination of both. However, this kind of data has been found less capable in the detection of seabed targets [16]. As for the Edgetech 6205 or Klein HydroChart 3500, they are mainly used in shallow waters, less than 35 m or 50 m respectively [17,18]. These new sonar systems have restrictions and their performances in the field of bathymetry measurement are usually inferior to those obtained by MBES [6].

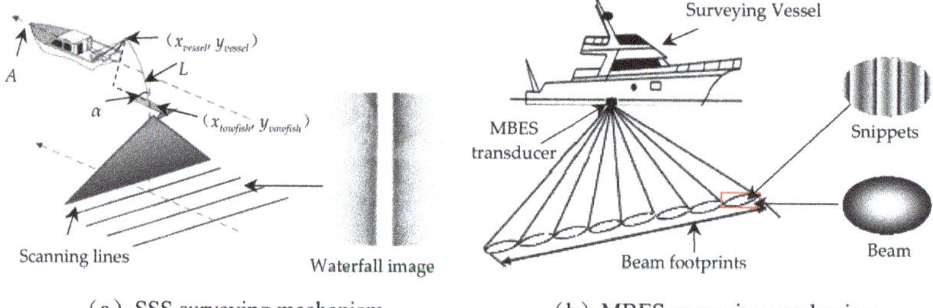

Figure 1. SSS (**a**) and MBES (**b**) surveying mechanism. $(x_{towfish}, y_{towfish})$ and (x_{vessel}, y_{vessel}) are respectively the coordinates of the towfish and vessel tow point, L is the cable length, α is the angle between the cable and horizontal direction, A is the vessel heading.

MBES is designed for high-accuracy bathymetric work and modern MBES can receive hundreds of echoes from the seabed for one ping and provide a highly detailed backscatter image simultaneously. The MBES is usually installed on a surveying vessel (Figure 1b). Although it can be used for large-scale bathymetric measurement, the interval of footprints in a ping will be enlarged and the bathymetric resolution will be decreased as the water depth and beam incident angle grow. When producing a MBES image, backscatter data are extracted from the time series traces contained within a beam are recorded [19,20]. The central point of a beam is optimally positioned with minimal geometric distortions and the image pixels are distributed around it. With seabed topography, vessel attitudes and the refraction of acoustic waves considered, the positions of MBES image pixels can be accurately computed [19]. By superimposing the MBES image on the seabed topography, the comprehensive seabed topography and surface features can be presented together [21–23]. However, the resolution, signal to noise ratio (SNR) and intensity contrast of MBES images are lower than those of SSS backscatter images [6]. The complementarity of SSS and MBES data provides a way to obtain detailed seabed features by superposing a two-dimensional (2D) SSS backscatter image onto 3D MBES bathymetric terrain.

However, because SSS backscatter image geographic coordinates are inaccurate, it is a challenge to conduct the superimposition of these two categories of data directly. Much research has been carried out in this field. LeBas et al. [24] created synthetic image from the seabed topography and adopted the hierarchical chamfer matching method to match the SSS backscatter and synthetic images. According to the matched result, the SSS backscatter image geographic coordinates were rectified and the obtained SSS backscatter image was superposed on the seabed topography. Yang et al. [25] used the similarity between MBES topography isobaths and SSS backscatter image contours to carry out image matching and the following superimposition operation. The two methods need sufficient topological variations on the seafloor for creating distinct edges or feature points in sonar images and may fail when dealing with a flat seabed with various sediments. According to similar seabed topography characteristics, textures, targets and sediment distributions being reflected on both SSS and MBES images, Zhao et al. [26] adopted the improved Speeded-Up Robust Features (SURF) algorithm to extract common feature points and conducted image registration, then rectified the SSS backscatter image geographic coordinates, referring to the MBES image based on the matched results. To obtain more correct matches, they conducted seabed classification for SSS and MBES images and used the classification images for image matching followed by a superimposition operation [27]. This method can improve matching performance but its results depend heavily on the seabed classification accuracy. The above methods ignore the imaging mechanism differences between SSS and MBES images, which often lead to inaccurate descriptions of the feature points and result in inaccurate image matching. Considering the limitations of the existing methods, this paper proposes a new SSS and MBES image matching method for acquiring high-resolution and high-accuracy seabed topography and surface details, which can overcome the limitations of adopting a single MBES or SSS for seabed mapping. The remainder of this paper is organized as follows: Section 2 describes the superimposition method in detail; Section 3 designs the experiments to verify the proposed method; Section 4 analyzes the results; Section 5 discusses the proposed method; and Section 6 presents the conclusions according to the experiments and discussions.

2. Method

To get high-resolution seabed topography and surface backscatter data, the proposed method is shown in Figure 2.

Firstly, both SSS and MBES data are processed to form the geocoded images.

Secondly, the SSS and MBES image matching based on common features is conducted, which includes the initial matching with the SURF algorithm with image geographic coordinate constraint and finer matching by using dense local self-similarity (DLSS) descriptors.

Thirdly, the random sample consensus (RANSAC) algorithm is used for removing mismatches and then the SSS backscatter image geographic coordinates are rectified according to the relationship model established by the correct matched points.

Finally, according to the consistent geographic coordinates, the rectified SSS backscatter image is superposed on the MBES bathymetric terrain.

The proposed method is described in detail in the following paragraphs and the image matching program was written using Matlab in a computer with the i7, 3.40GHz Intel Core and 8.00 GB RAM.

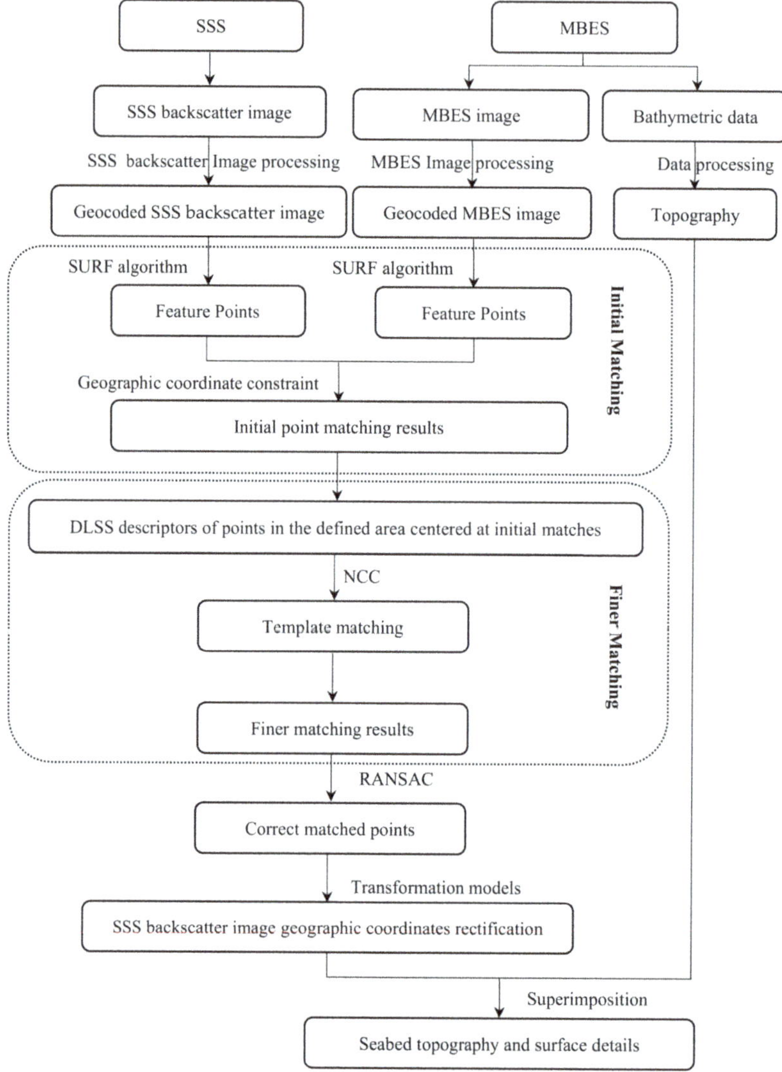

Figure 2. The acquisition of high-resolution seabed topography and surface details.

2.1. SSS and MBES Data Processing

The original SSS data were first decoded to form waterfall images ping by ping. In a complete process of SSS data, bottom tracking, radiometric correction, geometric correction and geocoding,and so forth, are included [5,28–32]. This SSS data processing was conducted by self-developed software and the processing procedure is shown in Figure 3a. Because thousands of echoes exist in a ping scanning line, a high-resolution SSS backscatter image could be obtained after the above data processing. By referring to Figure 1a, the positions of SSS towfish were reckoned by combining vessel-mounted GPS positioning results, cable length and the towfish heading, and thus became inaccurate due to the towing operation, the flat bottom assumption, towing distance and bearing controlled by the towing speed and the currents. Therefore, the geographic coordinates of high-resolution SSS backscatter image needed to be rectified.

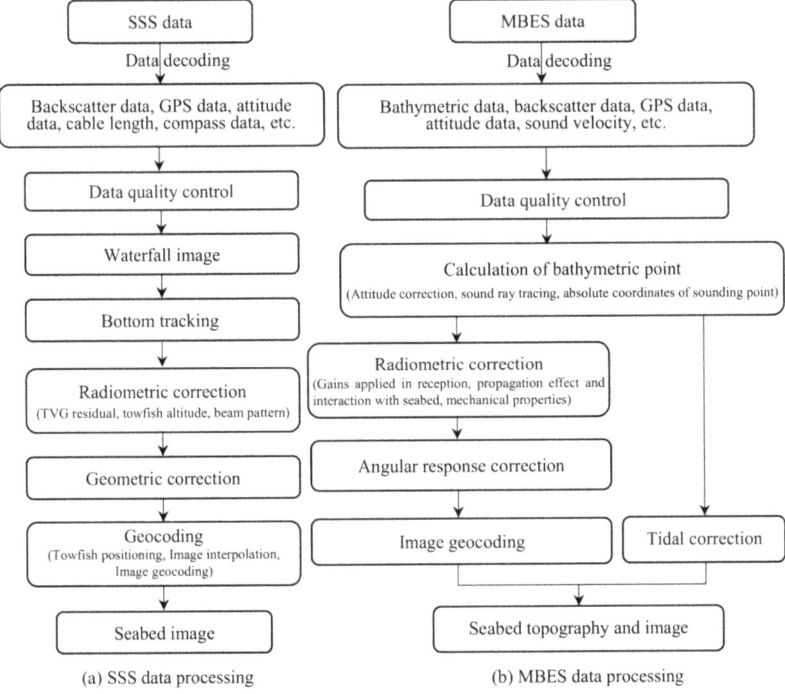

Figure 3. The flow charts of SSS (**a**) and MBES (**b**) data processing.

The original MBES data were first decoded to get the bathymetric data, backscatter data, GPS data, attitude data, sound velocity, and so forth. Quality control for these raw data was done first. Then, calculation of bathymetric point was conducted which includes attitude correction, sound ray tracing, absolute coordinates obtaining of bathymetric point, and so forth. After getting the locations of beam footprints, the authors conducted the radiometric correction, angular response correction and geocoding for the MBES backscatter data, and could then obtain the seabed image. Finally, after tidal correction for the sound ray tracing results, the absolute 3D coordinates of each bathymetric point were obtained and thus the seabed topography could be achieved. The procedure of MBES backscatter and bathymetric data processing was displayed in Figure 3b. To avoid gaps on the MBES image, the survey vessel velocity should be determined appropriately. According to the MBES data processing procedure and MBES measurement principle (Figure 1b), the geographic coordinates of bathymetric and backscatter data were accurate. However, the seabed topography and image obtained by using MBES were of low resolution, as only hundreds of echoes exist in one ping. As for the MBES backscatter, there were three types of acquired data: single beam intensity, individual beam time series and integrated time series. Although there exist a few to dozens of snippets in each beam footprint, the resolution of MBES image was still far lower than that of the SSS backscatter image. Thus, the MBES image is usually used for reflecting the large-scale seabed surface features instead of high-resolution visualization.

2.2. SSS and MBES Image Matching

Though with different characteristics (e. g. resolution, position accuracy and intensity contrast), both SSS and MBES images can reflect seabed features of the same area. Thus, the SSS and MBES image matching can be conducted by searching for common features in both images. To get accurate matched points, the matching method is proposed, which includes initial matching using the SURF algorithm with image geographic coordinate constraint, and the finer matching by using dense local self-similarity (DLSS) descriptors. The initial image matching step is to detect image feature points in the first place and use the SURF descriptors for feature-based image matching with the constraint of image coordinates. Since the SURF algorithm is not robust to nonlinear intensity differences between SSS and MBES images, some mismatches may arise in the initial matching step. To obtain more accurate matched points, the DLSS descriptor was used in the finer matching step and a template matching strategy was adopted. This two-step image matching process is described in detail in the following paragraphs.

2.2.1. Initial Image Matching Using SURF Algorithm with Image Geographic Coordinate Constraint

The SURF algorithm, which is invariant to rotation, scale, brightness and contrast [33], was adopted for detecting candidate feature points in SSS and MBES images to produce the SURF descriptors. By concatenating a 4D descriptor vector V of each sub-region, which underlies intensity structure for the square region of 4×4, the SURF descriptor of length 64 can be obtained [33].

$$V = \left(\sum d_x, \sum d_y, \sum |d_x|, \sum |d_y|\right) \tag{1}$$

where, d_x and d_y are respectively the Haar wavelet responses in horizontal and vertical directions. Since using the SURF descriptors for multi-model image matching directly may lead to many mismatches, some valid constraints can be added for specific applications [34,35].

For geocoded SSS and MBES images, the accuracies of their geographic coordinates are different and the geographic coordinates of the MBES image are more accurate than those of SSS because the SSS surveying is easily affected by the towing operation and many other factors. This inconsistency of image positional accuracy leads to the fact that the geographic coordinates of the same features presented in both images will be different. This kind of difference is within a range and can be determined by SSS location error, which is discussed in detail in Section 5.1. Thus, when the distances of the detected feature points in both images are too large, they will certainly not be the correct matches. Accordingly, the SURF algorithm for SSS and MBES image matching can be conducted when the distances of detected feature points in both images are within a threshold *Dis*, which can reduce the mismatches compared with using the SURF algorithm directly. With this constraint, the initial feature-based image matching was conducted as shown in Equation (2).

$$if \begin{array}{c} D_i \leq Dis \\ otherwise \end{array} then \begin{array}{c} SURF\ matching \\ Refused \end{array} \tag{2}$$

where, D_i is the distance between the candidate matched points. The determination of threshold *Dis* is also described in detail in Section 5.1.

2.2.2. Finer Image Matching Using DLSS Descriptors

Because the SSS backscatter reflects relative backscatter levels and the MBES images are usually normalized to dB scale, there is no linear relationship between their backscatter intensity levels. Meanwhile, as the resolution, SNR and intensity contrast of SSS backscatter image are higher than those of the MBES image, the same feature points of the seabed may be presented more clearly in SSS backscatter image than in the MBES image. As a result, some feature points can be detected in the SSS backscatter image but not in the MBES image. The SURF algorithm may detect incorrect matches when dealing with the non-linear intensity differences because it mainly focuses on the intensity variations of feature points and spatial distributions of their gradient information [33–36]. Moreover, as the number of detected feature points in the two images were different, the finally obtained number of matched points was limited. To get better matched points, the shape properties of distinct targets, topography changes or sediment distributions in these images can be used.

The DLSS descriptor can represent these shape properties and is robust against significant nonlinear intensity differences by integrating the local self-similarity (LSS) descriptors of multiple local image regions in a dense sampling grid [37]. LSS was calculated by correlating the image patch centered at a point q with a larger surrounding image region, resulting in a correction surface $S_q(x,y)$ as below [38].

$$S_q(x,y) = \exp\left(-\frac{SSD_q(x,y)}{\max(\text{var}_{noise}, \text{var}_{auto}(q))}\right) \tag{3}$$

where, $SSD_q(x,y)$ is the sum of square differences between image patch intensities, var_{noise} is a constant corresponding to intensity variations and var_{auto} is the maximal SSD of all patches within a very small neighborhood of q. Then, the $S_q(x,y)$ was transformed into a log-polar representation and partitioned into bins. The maximal correction value of each bin was selected to generate the LSS descriptor associated with the pixel q. In order to be invariant to the intensity variations of the image, the LSS was linearly stretched into the range [0, 1] [38]. An example of the LSS descriptor is shown in Figure 4, which indicates that the LSS descriptors of feature points are similar when the points share a common layout.

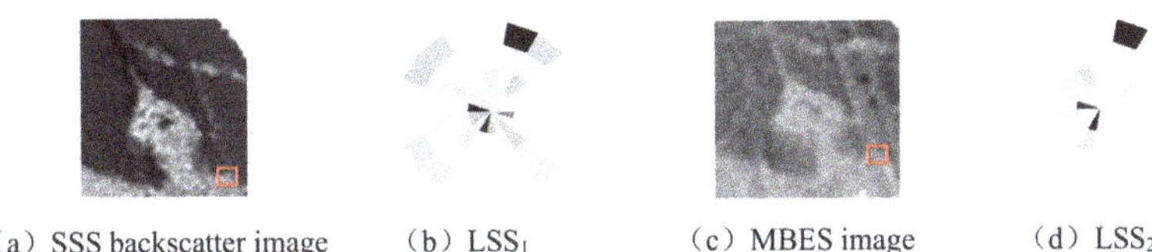

(a) SSS backscatter image (b) LSS_1 (c) MBES image (d) LSS_2

Figure 4. An example of local self-similarity (LSS) descriptor. The (**a**) and (**c**) are SSS and MBES image; red rectangles in (**a**) and (**c**) are the locations of the feature points; (**b**) and (**d**) are LSS descriptors of the two points.

By collecting the LSS descriptors from spatial regions within a dense overlapping grid covering a defined square area, the DLSS was produced as a combined feature descriptor [37]. Using the DLSS descriptors, a template matching strategy was conducted based on the initial image matching results.

First, the DLSS descriptors of initial matched points in the SSS backscatter image were produced and a searching area centered at the initial matched points in the MBES image with the radius r was defined. Meanwhile, the DLSS descriptors of every pixel in this defined area were calculated.

Secondly, calculating the similarity between the DLSS descriptors of the SSS backscatter image feature points and those of the points still unprocessed in the above defined searching area in the MBES image. The normalized cross correlation (NCC) of DLSS descriptors as shown in Equation (4) was adopted as the similarity metric.

$$NCC = \frac{\sum_{i=1}^{n} (DLSS1_i - mean(DLSS1))(DLSS2_i - mean(DLSS2))}{\sqrt{\sum_{i=1}^{n} (DLSS1_i - mean(DLSS1))^2 \sum_{i=1}^{n} (DLSS2_i - mean(DLSS2))^2}} \tag{4}$$

where, $DLSS1$ and $DLSS2$ are the DLSS descriptors of points in SSS and MBES images, N is the dimension of DLSS descriptor.

Finally, by choosing maximum NCCs and selecting the corresponding points in the defined area of the MBES image, the finer matches could be obtained. After conducting the above process for all the initial matched points in the SSS backscatter and MBES images, more correct matched points were obtained.

2.3. SSS Backscatter Image Geographic Coordinates Rectification Based on the Matched Points

Even after the finer matching step, mismatches still existed. The reason is that, affected by the measurement environment (e.g., the currents, the surveying vessel velocity and attitude changes), the seabed features presented in the SSS and MEBS images may not be identical. To eliminate these mismatches, the random sample consensus (RANSAC) algorithm was adopted. Since the RANSAC algorithm was conducted by selecting some existing matched points to build a geometric model [39] and the local geometric distortions in different parts of the SSS backscatter image were different, the SSS backscatter image could be segmented into different blocks according to feature distributions and towfish attitudes and the RANSAC algorithm could be conducted separately in each block [40–42]. When segmenting the SSS backscatter image, enough matched points should be contained in each block and the split lines are supposed to lie on the edge of the targets [41,42]. Meanwhile, to ensure the continuity between adjacent blocks, an overlapping ratio (e.g., one seventh of the block) can be set between them. To further ensure the remaining matches were correct, a geometric transformation model was established by using the matched points after application of the RANSAC algorithm and the coordinates of feature points in the SSS backscatter image were rectified. By referring to the matched feature points in the MBES image, the corresponding feature points in the rectified SSS backscatter image were considered as correct matches when their coordinate errors were less than double standard deviations. After this process, the correct matched points were obtained.

Based on the correct matched points between SSS and MBES images in each block, a geometric transformation model could be built to rectify the geographic coordinates of the SSS backscatter image by referring to those of the MBES image. Since the geographic distortions in local regions of the SSS backscatter image can be modeled by the affine function [43], the affine transformation model as shown in Equation (5) can be built within each block.

$$\begin{aligned} x_m &= a_0 + a_1 x_s + a_2 y_s \\ y_m &= b_0 + b_1 x_s + b_2 y_s \end{aligned} \tag{5}$$

where, (x_s, y_s) and (x_m, y_m) are geographic coordinates of matched feature points in SSS and MBES images, a_0, a_1, a_2 and b_0,b_1,b_2 are the model parameters to be calculated. After obtaining the geometric transformation models within each segmented block, the SSS backscatter image geographic coordinates could be rectified.

2.4. Superimposition of Rectified SSS Backscatter Image on MBES Terrain

After the rectification, the geographic coordinates of the SSS backscatter image were inconsistent with those of MBES image. Thus, the rectified SSS backscatter image could be superposed on the MBES terrain. This process was conducted by following steps:

(1) Based on the matched points, the geographic coordinates of the SSS backscatter image were rectified by referring to those of the MBES image.

(2) The 3D seabed topography was obtained by using the regular square grid method, which has the same resolution as the SSS backscatter image.

(3) According to the geographic coordinates, each grid point of the 3D topography was attributed with the corresponding grey value of SSS backscatter image.

After this process, the high-resolution 3D seabed topography and surface details can be presented together.

3. Experiments and Results

3.1. Experimental Data

To validate the proposed method, an experiment was carried out in Meizhou Bay, China with the area of 1.4 × 0.3 km. In this water area, the depth ranges from 10 to 14 m; seabed sampling showed sediments mainly contain silt and sand. In this experiment, EdgeTech 4100P with the slant range of 112 m and 7502 sampling points for each ping was adopted for the SSS measurement. For the MBES measurement, EM 3002 with a maximum angular sector of 130° and 131 beams was adopted. The MBES data were collected using the Seabed Information System (SIS) and restored in "*.all" files. The SSS data were collected using the EdgeTech Discover Software and restored in "*.xtf" files. After collecting these data, both SSS and MBES data were processed by self-developed software. The processing of SSS and MBES data is shown in Figure 3. After data processing, the geocoded SSS backscatter image, MBES image and seabed topography were created, as shown in Figure 5a–c. The coordinates are in WGS84. Considering the sensitivity of geographical data, the coordinates have been subtracted by a constant. Meanwhile, the MBES and SSS backscatter images and the MBES topography were resampled in the same ground sample distance (GSD) of 1m. Even with the same GSD, the SSS backscatter image could reflect clearer contours of seabed surface features because there exist many more sampling points for one ping of the SSS measurement than that of the MBES measurement. For example, the pipeline is presented more clearly in the SSS backscatter image than in the MBES image. Meanwhile, the topography shown in Figure 5c mainly reflects the 3D seabed undulations but with less detail.

Thus, the superimposition of the high-resolution SSS backscatter image on high-accuracy seabed terrain can take advantage of both datasets, which can capture both the 3D seabed topography and the detailed surface features together.

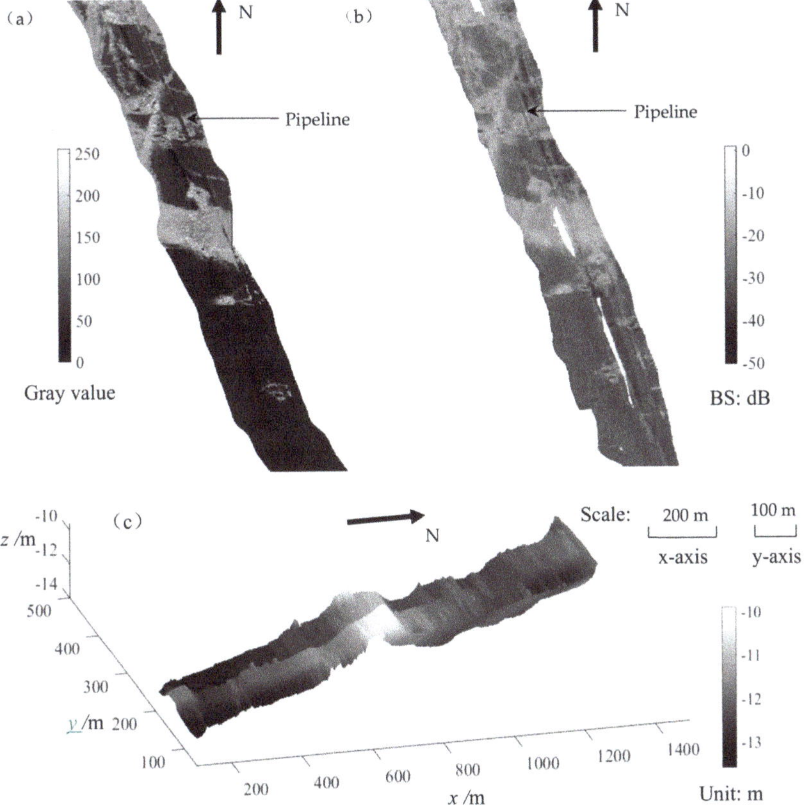

Figure 5. The SSS backscatter image (**a**), MBES image (**b**) and seabed topography (**c**) of the experimental area.

3.2. Experimental Procedure and Results

According to the proposed method depicted in Section 2, the SURF algorithm was first used for detecting feature points in SSS and MBES images. In a unified geographical framework, the image geographic coordinates can serve as the constraint in the initial image matching step as described in Section 2.2.1. This matching result is shown in Figure 6b. Compared with the matching results shown in Figure 6a, which were obtained by the classical SURF algorithm without constraint, the initial matching result with constraint seems better. The reason is that the SURF algorithm is not robust when dealing with the nonlinear intensity difference between SSS and MBES images. Meanwhile, the features existed on the seabed are relatively simple and the detected edge and point features can be represented by similar SURF descriptors, which may decrease the distinctiveness of SURF descriptors. Because of the two factors, even when two feature points in the SSS and MBES images are distant, they may be regarded as matches when they are represented by similar SURF descriptors. When using the image geographic coordinates as constraint in the initial matching step, the initial image matching is only conducted within a valid distance, which avoids mismatches when the geographic coordinates of two feature points are too far away.

Moreover, since there are many more sampling points for SSS measurement than for MBES measurement, the seabed features in the MBES image may not be as clear as those in the SSS backscatter image. As a result, some feature points may be detected in the SSS backscatter image but cannot be presented in the MBES image. Consequently, it was found that several detected feature points in the SSS backscatter image are matched to one feature point in the MBES image, particularly in zoomed area (b1) of Figure 6b. This several-to-one matching problem is obviously incorrect, which will be settled in the finer image matching step.

As image geographic coordinates are used as a constraint in the initial image matching step, the correct matched points in the MBES image are supposed to be located within an area centered at the initial matched ones. Thus, the optimization search operation described in Section 2.2.2 was conducted and the better matched points are shown in Figure 6c. In this process, a template matching strategy using DLSS descriptors was conducted. Compared with the SURF descriptor, the DLSS descriptor is robust to the nonlinear intensity difference between SSS and MBES images because it reflects the shape properties of feature points and their surrounding areas. Meanwhile, this finer image matching using a template matching strategy can also help find the same number of feature points in the MBES image as these in the SSS backscatter image. As a result, more correct matched points can be obtained by adopting the finer image matching step. Compared with Figure 6b, obtained from the initial image matching step, the several-to-one matching phenomena was removed after the finer matching step, which is clearly shown in the zoomed area (c1) of Figure 6c. This improvement of matching performance is also shown in Table 1. Table 1 lists the numbers of all the matches obtained by using SURF, initial matching and finer matching and that of the correct matches after the application of the RANSAC algorithm.

Even if the finer image matching step is followed, there still exist mismatches in Figure 6c. To eliminate theses mismatches, the SSS backscatter image was segmented into blocks and the RANSAC algorithm was adopted in each block. In the segment process, it can be seen that there exist three main independent areas with abundant seabed features. Thus, the SSS backscatter image was segmented into three blocks according to the feature distributions and towfish attitudes. Meanwhile, the overlapping ratio of one seventh between adjacent blocks was set and the split lines lay on the edge of isolated targets, which can be seen in the red rectangles in Figures 6 and 7. The image matching results after this process are shown in Figure 7 and Table 1. It can be seen that even after using the RANSAC algorithm, some mismatches still exist in Figure 7a, because too many mismatches occurred due to the use of the SURF algorithm directly and the RANSAC algorithm is sensitive to the high outlier rate. As a result, these mismatches cannot be effectively eliminated by using the RANSAC algorithm. This phenomenon also proves that adding the constraint of image geographic coordinates in the initial matching step was necessary, which improved the matching ratio from 8% to 55% as shown in Table 1. Even so, mismatches still exist after the initial step, which can be seen in the zoomed area (b1) of Figure 7b. After the finer image matching step, the matching ratio rose to 86%. More importantly, these matches are correct, as shown in the zoomed area (c1) of Figure 7c. This is because the combination of both the initial and finer image matching not only makes full use of the intensity variations and gradient information of feature points, but also considers their geographic coordinates and the shape properties of the area around them.

Using the correct matched points that were finally obtained in each block, the transformation model shown in Equation (5) could be established. According to the obtained transformation model, the SSS backscatter image geographic coordinates could be rectified following the method in Section 2.3. The accuracy of the rectified SSS backscatter image geographic coordinates is discussed in Section 4.

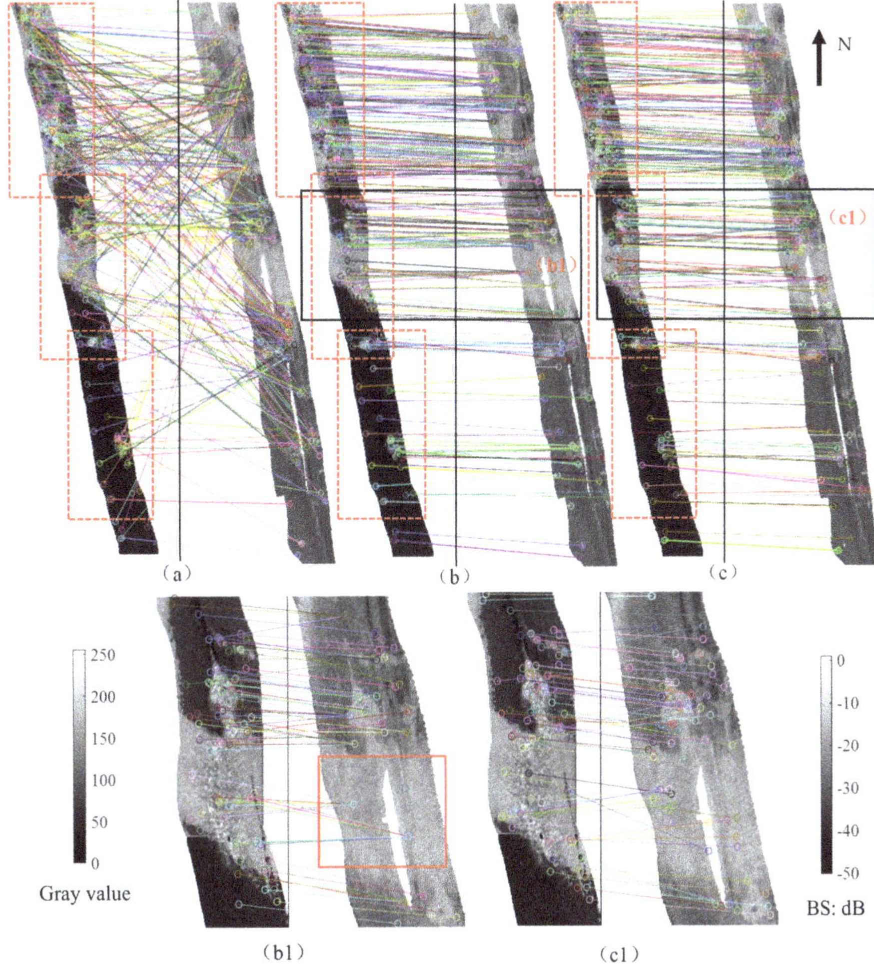

Figure 6. Image matching results by the Speeded-Up Robust Features (SURF) algorithm (**a**), initial matching with constraint (**b**) and finer matching (**c**). The colored full lines in (**a**) (**b**) (**c**) (**b1**) and (**c1**) are used to connect matched points. The red dotted rectangles in (**a**) (**b**) (**c**) are the segmented blocks. The black rectangles in (**b**) and (**c**) are the zoomed areas (**b1**) and (**c1**). The red rectangle in (**b1**) shows the several-to-one matching phenomena.

Table 1. The matched results of SURF algorithm, initial matching with constraint and finer matching before and after using the RANSAC.

	Matches	Correct Matches	Percentage
SURF	341	28	8%
Initial matching	231	128	55%
Fine matching	231	198	86%

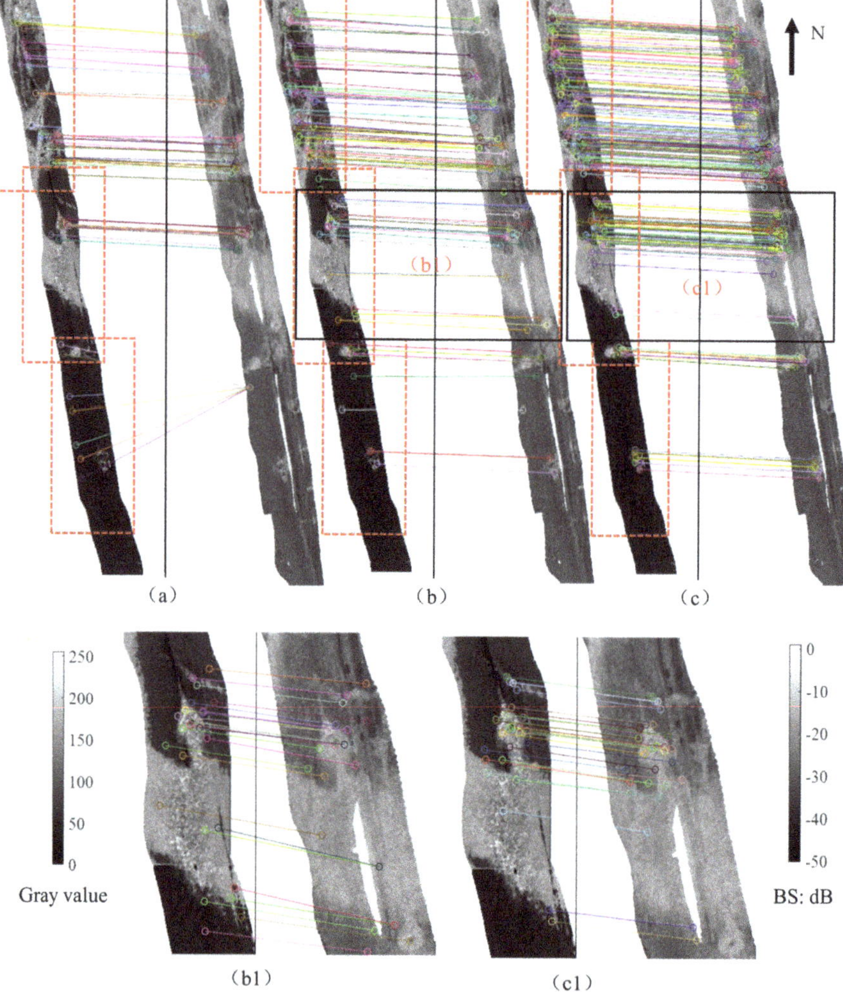

Figure 7. Image matching results by SURF algorithm (**a**), initial matching with constraint (**b**) and finer matching (**c**) after the application of random sample consensus (RANSAC) algorithm. The colored full lines in (**a**) (**b**) (**c**) (**b1**) and (**c1**) are used to connect matched points. The red rectangles in (**a**) (**b**) (**c**) are the segmented blocks. The black rectangles in (**b**) and (**c**) are the zoomed areas (**b1**) and (**c1**).

Superimposing the rectified SSS backscatter image on seabed topography, the comprehensive presentation of 3D topography and surface details is shown in Figure 8. Compared with

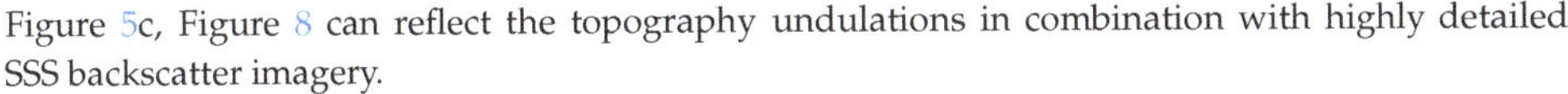
Figure 5c, Figure 8 can reflect the topography undulations in combination with highly detailed SSS backscatter imagery.

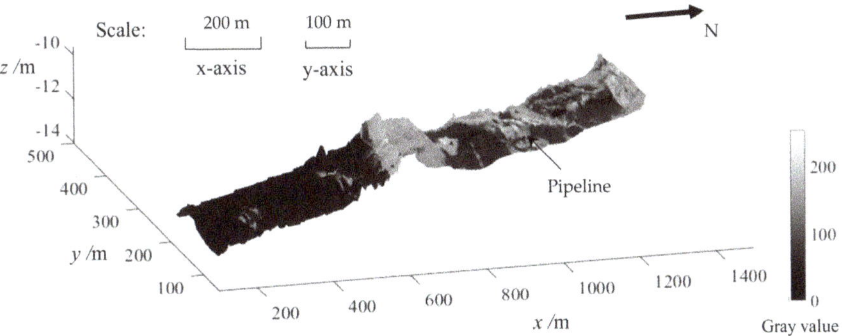

Figure 8. Representation of seabed topography and surface details.

4. Evaluation and Analysis

To assess the accuracy of the rectified SSS backscatter image geographic coordinates, the interior checking and exterior checking were conducted. The interior checking involves calculating the geographic coordinate biases of matched feature points used to build the transformation model before and after the rectification, while the exterior checking involves calculating the matches not used to build the transformation model. Meanwhile, the statistical results of these biases were also calculated. In this experiment, two-thirds of matched points were used for establishing the transformation model and the interior checking; the remaining one-third of matched points were consequently used for the exterior checking. The statistical results and probability distribution function (PDF) curves of each checking are shown in Table 2, Table 3, Figures 9 and 10. In Tables 2 and 3, "Raw" means the coordinate biases of matched feature points in the raw SSS and MBES images; "Rectified" means those in the rectified SSS and MBES images; "dy" means those of the east direction, and "dx" means those of the north direction.

Table 2. The statistical results of interior checking (Unit: m).

	Bias	Max	Min	Mean	Std.
Raw	dy	7.79	−19.88	−4.32	3.67
	dx	8.57	−17.87	−5.98	4.45
Rectified	dy	9.61	−7.69	0.01	2.95
	dx	9.43	−7.35	0.00	3.17

Table 3. The statistical results of exterior checking (Unit: m).

	Bias	Max	Min	Mean	Std.
Raw	dy	1.68	−19.93	−5.06	3.41
	dx	8.47	−15.79	−6.17	4.83
Rectified	dy	7.97	−7.12	−0.48	2.58
	dx	9.88	−7.28	−0.27	3.53

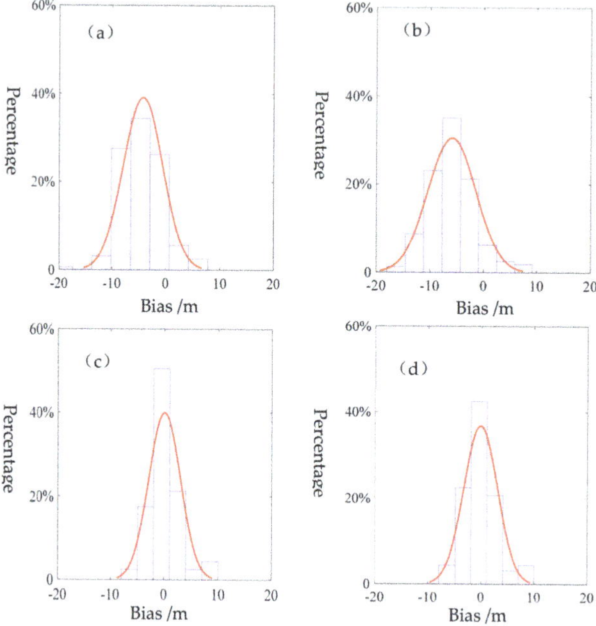

Figure 9. The probability distribution function (PDF) curves of interior checking. (**a**) and (**c**) are PDF curves of coordinate biases in y direction before and after rectification; (**b**) and (**d**) are those of coordinate biases in x direction.

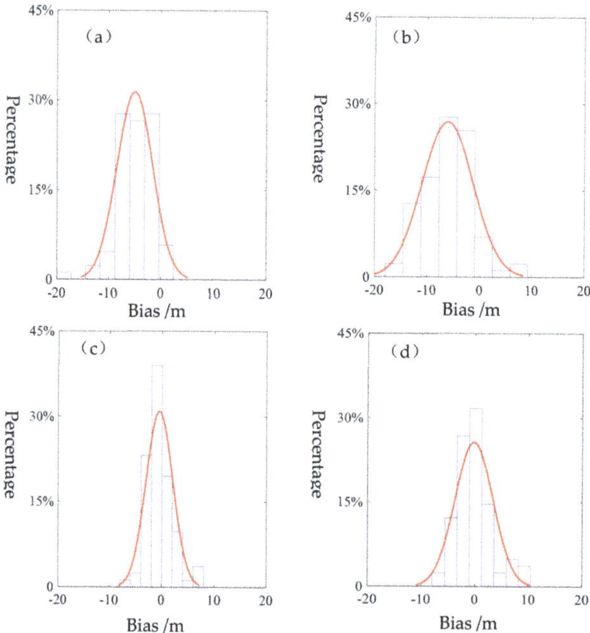

Figure 10. The PDF curves of exterior checking. (**a**) and (**c**) are PDF curves of coordinate biases in y direction before and after rectification; (**b**) and (**d**) are those of coordinate biases in x direction.

From Table 2, Table 3, Figures 9 and 10, it can be seen that obvious systemic errors and large standard deviations exist between the raw geographic coordinates of matched points. Tables 2 and 3 show the systemic errors in the two directions are 4.32 m and 5.98 m for interior checking and 5.06 m and 6.17 m for exterior checking. After using the transformation model, these systemic errors were removed. Meanwhile, the standard deviations became smaller, which can be seen in Table 2, Table 3, Figures 9 and 10. These results prove that the accuracy of the SSS backscatter image geographic coordinates improved after the rectification. Moreover, the accuracy of both checkings were consistent with each other, which indicates that the proposed method is robust in building the transformation model.

5. Discussions

5.1. Determination of the Threshold Dis and Searching Radius

In the initial image matching process, the image geographic coordinates served as the constraint and the threshold *Dis* were determined by SSS location error. Usually, SSS is installed in a towfish and towed by a cable behind a surveying vessel for seabed surveying, which is shown in Figure 1a. During this process, the towfish positions can be estimated by Equation (6).

$$\begin{pmatrix} x_{towfish} \\ y_{towfish} \end{pmatrix} = \begin{pmatrix} x_{vessel} \\ y_{vessel} \end{pmatrix} + \begin{pmatrix} L\cdot \cos\alpha\cdot \cos(A+\pi) \\ L\cdot \cos\alpha\cdot \sin(A+\pi) \end{pmatrix} \tag{6}$$

where, $(x_{towfish}, y_{towfish})$ and (x_{vessel}, y_{vessel}) are respectively the coordinates of the towfish and vessel tow point, L is the cable length, α is the angle between the cable and horizontal direction, A is the vessel heading. Affected by the vessel velocity variation, the accuracies of vessel positions and heading data, the towfish position error $d_{towfish}$ can be determined by Equation (7).

$$d_{towfish} = \sqrt{dx_{towfish}{}^2 + dy_{towfish}{}^2} = \sqrt{dx_{vessel}{}^2 + dy_{vessel}{}^2 + L^2\cdot \sin^2\alpha\cdot d\alpha^2 + L^2\cdot \cos^2\alpha\cdot dA^2} \tag{7}$$

where, $(dx_{towfish}, dy_{towfish})$ and $(dx_{vessel}, dy_{vessel})$ are respectively the towfish and tow point position errors in x and y directions, $d\alpha$ is the variation range of the angle α, dA is the heading difference between the vessel and towfish. For most SSS measurements in offshore waters, the vessel position error is about 1~3m; the length of the cable is about 10~20m; the angle α is about 30° and its variation range is about 10°; the heading difference between the vessel and towfish is about 1°–5° [44]. After substituting these values into Equation (7), the $d_{towfish}$ ranges roughly from 10 to 100 m. The above parameters will become bigger with the increase of surveying vessel size in open sea. In this experiment, the sea was calm and the vessel's course was steady. Thus, the $d_{towfish}$ was about 18 m, which served as a referenced threshold in the initial image matching. To determine the optimal threshold, different thresholds in Equation (2) were used by referring to $d_{towfish}$ and the matched results are shown in Table 4, which shows that the detected matched points increased as the threshold became larger but the number of correct ones peaked when the threshold was 20. Thus, the initial matched result was obtained with this threshold and followed by the finer matching.

Table 4. The matching results of different thresholds.

Threshold (m)	Detected Matches	Correct matches	Percentage
10	109	92	84%
20	231	128	55%
30	285	120	42%
40	317	102	32%
50	327	100	30%

In the finer matching process, the template matching strategy was used and the searching radius for better matches was determined. A small searching radius is not enough for finding better matches while a large one will cost more time and the matched results may also not be the optimal. During the finer matching process, different searching radiuses were used and the matched results are shown in Table 5. This indicated that 20 m could be adopted as a suitable searching radius in the finer matching step as most correct matches were obtained when using this value.

Table 5. The matching results of different searching radiuses.

Searching Radius (m)	Detected Matches	Correct Matches	Percentage
10	231	167	72%
20	231	198	86%
30	231	174	75%
40	231	162	70%
50	231	146	63%

5.2. Method Repeatability: Applicability of the Proposed Method in another Situation

To further verify the performance of the proposed method in other waters when adopting different kinds of sonars, another experiment was conducted in BeiBu Gulf. In this area, the depth ranges from 40 to 50 m; seabed sampling shows that sediments mainly consist of sand, which has two forms of existence: sand waves and ripples. Klein 3000 and EM710 were separately adopted for SSS and MBES measurement. After processing the obtained measurement data—the seabed topography is shown in Figure 11—the SSS and MBES images were presented in Figure 12. It can be seen that the seabed topography and the MBES image can only reflect the sand waves but the SSS backscatter image can also capture the small scale sand ripples clearly. This difference can be seen in areas (a1) and (b1) of Figure 11. This difference can also be seen in zoomed areas (a1) and (b1) of Figure 12.

Based on the common sand wave features reflected by both images, the proposed image matching method described in Section 2.2 was performed and the matched result shown in Figure 12 proved that the method can also be effective when using other kinds of measurement instruments in a different water area. An area with sharper edges is zoomed in for better visualization, as shown in Figure 12c. Using the correct matches that were finally obtained to rectify the SSS backscatter image geographic coordinates, the obtained SSS backscatter image was superposed on the seabed topography. The comprehensive presentation of 3D topography and surface details is shown in Figure 13. Compared with Figure 11, it can reflect both the obvious sand waves and detailed sand ripples.

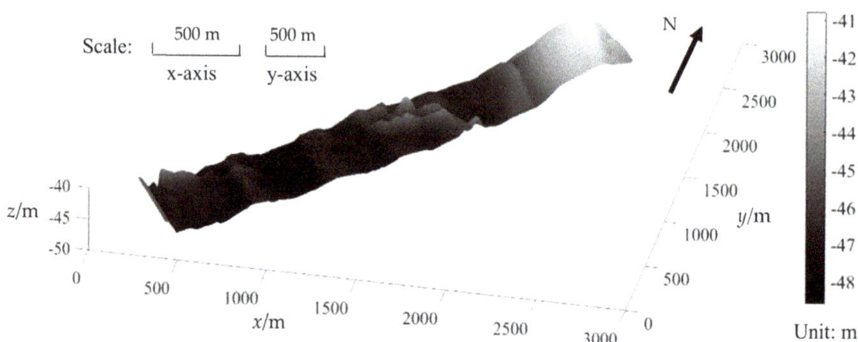

Figure 11. Seabed topography of the new water area.

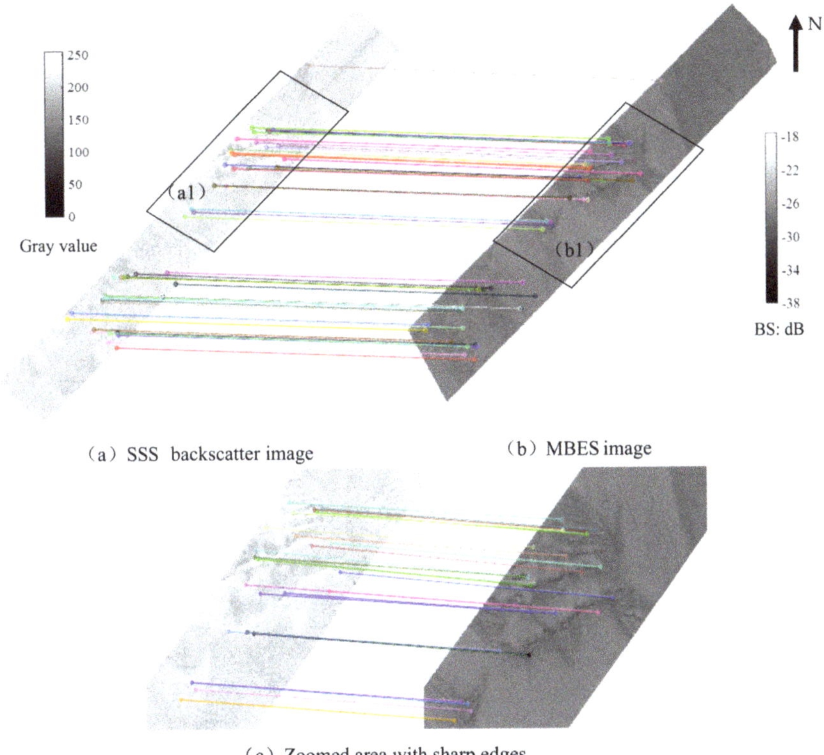

Figure 12. The matched result of SSS backscatter image (**a**) and MBES image (**b**) of the new water area. The (**a1**) and (**b1**) are the zoomed areas; (**c**) is the example with sharper edges; the colored full lines in (**a**) (**b**) and (**c**) are used to connect matched points.

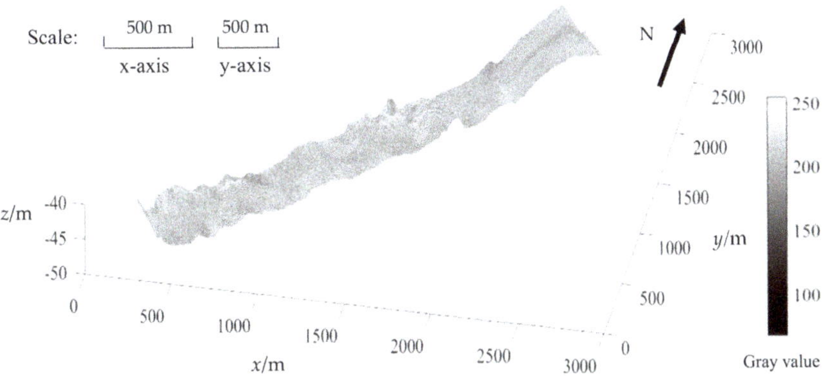

Figure 13. Comprehensive seabed topography and surface details of the new water area.

5.3. Sonar Frequencies

The acoustic waves emitted by a sonar propagate through sea water to seabed and the backscattered echoes are received by the transducer to form seabed backscatter images. When using different frequencies of acoustic wave, the impacts of surface and volume scatter are different [45–47]. Taking the sandy sediments as an example, penetration depth is limited to 12mm for 500 kHz, while 100 kHz may penetrate 90mm into the subsurface [48]. In addition, the acoustic frequency affects sonar backscatter image texture presentations due to changing surveying parameters, such as footprint sizes and different sensitivity to seafloor features [23,45]. As a result, the presentation of seabed features may be different when using different SSS and MBES frequencies [23,45,47]. As a result, finding matches might become difficult.

At specific frequencies, the sonar backscatter images can always reflect seabed features and the relative positions of these features are stable. Therefore, based on the geometrical structure formed by the association of three detected features, image matching can still be conducted [43]. Meanwhile, deep learning has also been applied in the sonar image processing field [49,50] and some related image matching methods [51] can be used for sonar image matching.

Recently, multi-frequency SSS and MBES have been applied, such as EM2040 and Edgetech 4200SP. As for multi-frequency sonar backscatter data, they can be rendered as acoustically colorful images [23,46]. The formed colorful image can reflect more seabed features compared with those obtained from monochromatic sonar with a single center frequency. Consequently, more features can be detected from the colorful image and can be used for image matching.

6. Conclusions

Despite some new sonar systems appearing that can provide co-registered seabed images and topography, they can only be used in restricted waters. The SSS and MBES are still the most effective sonars for obtaining seabed images and topography, especially for deep seafloor observation. Based on the complementarity of SSS and MBES data, this paper proposes a new method for acquiring high-resolution seabed topography and surface details by combining SSS and MBES data. Through taking the image geographic coordinates as the constraint when using the SURF algorithm for initial image matching, the authors have obtained more correct initial matched points compared to that without constraint. Then, the finer matching step is conducted by adopting a template matching strategy, which uses the DLSS descriptor to reflect the shape properties of the area centered feature points. The combination of the initial and finer matching steps can help improve the matching performance and the percentage of correct matched points rose to 86%. Based on the obtained matched points and considering the difference of geographic distortions in different parts of the SSS backscatter image, the SSS backscatter image was segmented into several blocks and the geometric transformation model within each block can be established to rectify its geographic coordinates. Subsequently, the rectified SSS backscatter image can be superposed on the topography, which can reflect both the 3D seabed undulations in combination with a high detailed SSS backscatter imagery.

Experiments have verified this method. By this method, the high-resolution and high-accuracy seabed topography and surface details can be represented together, which is meaningful for understanding and interpreting seabed topography. Meanwhile, this paper discusses the accuracy of the reckoned SSS positions and uses it as a reference threshold in the image matching process. In addition, this paper discusses the impact of sonar frequency on the sonar backscatter image and provides some useful suggestions when dealing with multi-frequency sonar image matching.

Author Contributions: Conceptualization, X.S. and J.Z.; Data curation, X.S. and J.Z.; Formal analysis, X.S., J.Z. and H.Z.; Funding acquisition, J.Z. and H.Z.; Methodology, X.S. and J.Z.; Supervision, J.Z. and H.Z.; Visualization, X.S., J.Z. and H.Z.; Writing—original draft, X.S.; Writing—review & editing, X.S., J.Z. and H.Z.

Funding: This research was funded by the National Natural Science Foundation of China, grant numbers 41576107, 41376109 and 41606114.

Acknowledgments: The authors would like to thank the editors and anonymous reviewers for their comments and suggestions. The data used in this study were provided by the Guangzhou Marine Geological Survey Bureau. The authors are grateful for their support.

Conflicts of Interest: The authors declare no conflict of interest.

References

1. Mayer, L.; Jakobsson, M.; Allen, G.; Dorschel, B.; Falconer, R.; Ferrini, V.; Lamarche, G.; Snaith, H.; Weatherall, P. The Nippon Foundation—GEBCO Seabed 2030 Project: The Quest to See the World's Oceans Completely Mapped by 2030. *Geosciences* **2018**, *8*, 63. [CrossRef]
2. DeSanto, J.B.; Sandwell, D.T.; Chadwell, C.D. Seafloor geodesy from repeated sidescan sonar surveys. *J. Geophys. Res. Solid Earth* **2016**, *121*, 4800–4813. [CrossRef]
3. Powers, J.; Brewer, S.K.; Long, J.M.; Campbell, T. Evaluating the use of side-scan sonar for detecting freshwater mussel beds in turbid river environments. *Hydrobiologia* **2015**, *743*, 127–137. [CrossRef]
4. Lurton, X. Forty years of progress in multibeam echosounder technology for ocean investigation. *J. Acoust. Soc. Am.* **2017**, *141*, 3948. [CrossRef]
5. Blondel, P. *The Handbook of Sidescan Sonar*; Springer: Berlin/Heidelberg, Germany; New York, NY, USA, 2009; ISBN 978-3-540-42641-7.
6. Lurton, X.; Jackson, D. *An Introduction to Underwater Acoustics*, 2nd ed.; Springer-Praxis: New York, NY, USA, 2008; ISBN 3540429670.
7. Tamsett, D.; Hogarth, P. Sidescan sonar beam function and seabed backscatter functions from trace amplitude and vehicle roll data. *IEEE J. Ocean. Eng.* **2015**, *411*, 155–163. [CrossRef]
8. Bell, J.M.; Linnett, L.M. Simulation and analysis of synthetic sidescan sonar images. *IEE Proc. Radar Sonar Navig.* **1997**, *144*, 219–226. [CrossRef]
9. Cobra, D.T.; Oppenheim, A.V.; Jaffe, J.S. Geometric distortions in side-scan sonar images: A procedure for their estimation and correction. *IEEE J. Ocean. Eng.* **1992**, *17*, 252–268. [CrossRef]
10. Clarke, J. Dynamic Motion Residuals in Swath Sonar Data: Ironing out the Creases. *Int. Hydrogr. Rev.* **2003**, *4*, 6–23.
11. Cervenka, P.; Moustier, C.D.; Lonsdale, P.F. Geometric corrections on sidescan sonar images based on bathymetry: Application with SeaMARC II and Sea Beam data. *Mar. Geophys. Res.* **1995**, *17*, 217–219. [CrossRef]
12. Cervenka, P.; de Moustier, C. Postprocessing and corrections of bathymetry derived from sidescan sonar systems: Application with SeaMARC II. *IEEE J. Ocean. Eng.* **1994**, *19*, 619–629. [CrossRef]
13. Coiras, E.; Petillot, Y.; Lane, D.M. Multiresolution 3-D reconstruction from side-scan sonar images. *IEEE Trans. Image Process.* **2007**, *16*, 382–390. [CrossRef] [PubMed]
14. Fezzani, R.; Zerr, B.; Mansour, A.; Legris, M.; Vrignaud, C. Fusion of Swath Bathymetric Data: Application to AUV Rapid Environment Assessment. *IEEE J. Ocean. Eng.* **2019**, *44*, 111–120. [CrossRef]
15. Stateczny, A.; Gronska, D.; Motyl, W. HydroDron—New step for professional hydrography for restricted waters. *BGC Geomat.* **2018**, 226–230. [CrossRef]
16. Crawford, A.; Connors, W. Performance Evaluation of a 3-D Sidescan Sonar for Mine Countermeasures. In Proceedings of the OCEANS 2018 MTS, Charleston, SC, USA, 22–25 October 2018.
17. Ai, Y.; Armstrong, S.; Fleury, D. Evaluation of the Klein Hydrochart 3500 Interferometric Bathymetry Sonar for Noaa Sea Floor Mapping. In Proceedings of the OCEANS 2015 MTS, Washington, DC, USA, 18–21 May 2015.
18. Brisson, L.; Wolfe, D.; Staley, M. Interferometric swath bathymetry for large scale shallow water hydrographic surveys. In Proceedings of the Canadian Hydrographic Conference, St. John's, NL, Canada, 14–17 April 2014.
19. Schimel, A.C.; Beaudoin, J.; Parnum, I.; LeBas, T.P.; Schmidt, V.; Keith, G.; Ierodiaconou, D. Multibeam sonar backscatter data processing. *Mar. Geophys. Res.* **2018**, *39*, 121–137. [CrossRef]
20. Lucieer, V.; Roche, M.; Degrendele, K.; Malik, M.; Dolan, M.; Lamarche, G. User expectations for multibeam echo sounders backscatter strength data-looking back into the future. *Mar. Geophys. Res.* **2018**, *39*, 23–40. [CrossRef]
21. LeBas, T.P.; Huvenne, V. Acquisition and processing of backscatter data for habitat mapping–comparison of multibeam and sidescan systems. *Appl. Acoust.* **2009**, *70*, 1248–1257. [CrossRef]
22. Mitchell, G.A.; Orange, D.L.; Gharib, J.J.; Kennedy, P. Improved detection and mapping of deepwater hydrocarbon seeps: Optimizing multibeam echosounder seafloor backscatter acquisition and processing techniques. *Mar. Geophys. Res.* **2018**, *39*, 323–347. [CrossRef]

23. Fakiris, E.; Blondel, P.; Papatheodorou, G.; Christodoulou, D.; Dimas, X.; Georgiou, N.; Kordella, S.; Dimitriadis, C.; Rzhanov, Y.; Geraga, M.; et al. Multi-Frequency, Multi-Sonar Mapping of Shallow Habitats—Efficacy and Management Implications in the National Marine Park of Zakynthos, Greece. *Remote Sens.* **2019**, *11*, 461. [CrossRef]
24. LeBas, T.P.; Mason, D.C. Automatic registration of TOBI side-scan sonar and multi-beam bathymetry images for improved data fusion. *Mar. Geophys. Res.* **1997**, *19*, 163–176. [CrossRef]
25. Yang, F.; Wu, Z.; Du, Z.; Jin, X. Co-registering and fusion of digital information of multibeam sonar and side-scan sonar. *Geomat. Inf. Sci. Wuhan Univ.* **2006**, *31*, 740–743. [CrossRef]
26. Zhao, J.; Wang, A.; Guo, J. Study on fusion method of the block image of MBS and SSS. *Geomat. Inf. Sci. Wuhan Univ.* **2013**, *38*, 287–290. [CrossRef]
27. Yan, J. Acquisition and superposition of the high-quality measurement information of multibeam echo sonar. *Acta Geod. Et Cartogr. Sin.* **2019**, *48*, 400. [CrossRef]
28. Fakiris, E.; Papatheodorou, G. Quantification of regions of interest in swath sonar backscatter images using grey-level and shape geometry descriptors: The TargAn software. *Mar. Geophys. Res.* **2012**, *33*, 169–183. [CrossRef]
29. Wang, A.; Zhang, H.; Wang, X.; Shang, X. Processing Principles of Side-scan Sonar Data for Seamless Mosaic Image. *J. Geomat.* **2017**, *42*, 26–29. [CrossRef]
30. Cervenka, P.; de Moustier, C. Sidescan sonar image processing techniques. *IEEE J. Ocean. Eng.* **1993**, *18*, 108–122. [CrossRef]
31. Ye, X.; Yang, H.; Li, C.; Jia, Y.; Li, P. A Gray Scale Correction Method for Side-Scan Sonar Images Based on Retinex. *Remote Sens.* **2019**, *11*, 1281. [CrossRef]
32. Capus, C.G.; Banks, A.C.; Coiras, E.; Ruiz, I.T.; Smith, C.J.; Petillot, Y.R. Data correction for visualisation and classification of sidescan SONAR imagery. *IET Radar Sonar Navig.* **2008**, *2*, 155–169. [CrossRef]
33. Bay, H.; Ess, A.; Tuytelaars, T.; Gool, L. Speeded-Up Robust Features (SURF). *Comput. Vis. Image Underst.* **2008**, *110*, 346–359. [CrossRef]
34. Sedaghat, A.; Mohammadi, N. High-resolution image registration based on improved SURF detector and localized GTM. *Int. J. Remote Sens.* **2019**, *40*, 2576–2601. [CrossRef]
35. Tola, E.; Lepetit, V.; Fua, P. Daisy: An efficient dense descriptor applied to wide-baseline stereo. *IEEE Trans. Pattern Anal. Mach. Intell.* **2009**, *32*, 815–830. [CrossRef]
36. Li, J.; Hu, Q.; Ai, M.; Zhong, R. Robust feature matching via support-line voting and affine-invariant ratios. *ISPRS J. Photogramm. Remote Sens.* **2017**, *132*, 61–76. [CrossRef]
37. Ye, Y.; Shen, L.; Hao, M.; Wang, J.; Xu, Z. Robust optical-to-SAR image matching based on shape properties. *IEEE Geosci. Remote Sens. Lett.* **2017**, *14*, 564–568. [CrossRef]
38. Shechtman, E.; Irani, M. Matching local self-similarities across images and videos. In Proceedings of the IEEE Conference on Computer Vision and Pattern Recognition, Minneapolis, MN, USA, 17–22 June 2007; pp. 1–8.
39. Fischler, M.A.; Bolles, R.C. Random sample consensus: A paradigm for model fitting with applications to image analysis and automated cartography. *Commun. ACM* **1981**, *24*, 381–395. [CrossRef]
40. Chailloux, C.; Caillec, J.; Gueriot, D.; Zerr, B. Intensity-Based Block Matching Algorithm for Mosaicing Sonar Images. *IEEE J. Ocean. Eng.* **2011**, *36*, 627–645. [CrossRef]
41. Zhao, J.; Wang, A.; Zhang, H.; Wang, X. Mosaic method of side-scan sonar strip images using corresponding features. *IET Image Process.* **2013**, *7*, 616–623. [CrossRef]
42. Wang, A.; Zhao, J.; Guo, J.; Wang, X. Elastic Mosaic Method in Block for Side Scan Sonar Image Based on Speeded-Up Robust Features. *Geomat. Inf. Sci. Wuhan Univ.* **2018**, *43*, 697–703. [CrossRef]
43. Daniel, S.; Leannec, F.L.; Roux, C.; Solaiman, B.; Maillard, E.P. Side-Scan Sonar Image Matching. *IEEE J. Ocean. Eng.* **1998**, *23*, 245–259. [CrossRef]
44. Yang, L.; Jiao, Y.; Xu, J. Underwater target positioning technology of side scan sonar based on ultra short baseline. *China Harb. Eng.* **2017**, *37*, 6–9. [CrossRef]
45. Feldens, P.; Schulze, I.; Papenmeier, S.; Schönke, M.; Schneider von Deimling, J. Improved Interpretation of Marine Sedimentary Environments Using Multi-Frequency Multibeam Backscatter Data. *Geosciences* **2018**, *8*, 214. [CrossRef]
46. Tamsett, D.; McIlvenny, J.; Watts, A. Colour Sonar: Multi-Frequency Sidescan Sonar Images of the Seabed in the Inner Sound of the Pentland Firth, Scotland. *J. Mar. Sci. Eng.* **2016**, *4*, 26. [CrossRef]

47. Hughes Clarke, J.E. Multispectral Acoustic Backscatter from Multibeam, Improved Classification Potential. In Proceedings of the US Hydrographic Conference, National Harbor, MD, USA, 16–19 March 2015; pp. 1–18.
48. Huff, L.C. Acoustic Remote Sensing as a Tool for Habitat Mapping in Alaska Waters. In *Marine Habitat Mapping Technology for Alaska*; Reynolds, J.R., Greene, H.G., Eds.; Alaska Sea Grant, University of Alaska Fairbanks: Fairbanks, AK, USA, 2008; pp. 29–46.
49. Chen, J.L.; Summers, J.E. Deep neural networks for learning classification features and generative models from synthetic aperture sonar big data. *Acoust. Soc. Am. J.* **2016**, *140*, 3423. [CrossRef]
50. Song, Y.; He, B.; Liu, P.; Yan, T. Side scan sonar image segmentation and synthesis based on extreme learning machine. *Appl. Acoust.* **2019**, *146*, 56–65. [CrossRef]
51. Chen, Z.; Liu, L.; Sa, I.; Ge, Z.; Chli, M. Learning context flexible attention model for long-term visual place recognition. *IEEE Robot. Autom. Lett.* **2018**, *3*, 4015–4022. [CrossRef]

Article

A Gray Scale Correction Method for Side-Scan Sonar Images Based on Retinex

Xiufen Ye [1,*], Haibo Yang [1], Chuanlong Li [1], Yunpeng Jia [1] and Peng Li [2]

1 College of Automation, Harbin Engineering University, Harbin 150001, China; 1902599290@hrbeu.edu.cn (H.Y.); lcl@hrbeu.edu.cn (C.L.); jiayunpeng@hrbeu.edu.cn (Y.J.)
2 School of Management, Harbin University of Commerce, Harbin 150028, China; lipeng@hrbcu.edu.cn
* Correspondence: yexiufen@hrbeu.edu.cn; Tel.: +86-451-82518891

Received: 24 April 2019; Accepted: 25 May 2019; Published: 29 May 2019

Abstract: When side-scan sonars collect data, sonar energy attenuation, the residual of time varying gain, beam patterns, angular responses, and sonar altitude variations occur, which lead to an uneven gray level in side-scan sonar images. Therefore, gray scale correction is needed before further processing of side-scan sonar images. In this paper, we introduce the causes of gray distortion in side-scan sonar images and the commonly used optical and side-scan sonar gray scale correction methods. As existing methods cannot effectively correct distortion, we propose a simple, yet effective gray scale correction method for side-scan sonar images based on Retinex given the characteristics of side-scan sonar images. Firstly, we smooth the original image and add a constant as an illumination map. Then, we divide the original image by the illumination map to produce the reflection map. Finally, we perform element-wise multiplication between the reflection map and a constant coefficient to produce the final enhanced image. Two different schemes are used to implement our algorithm. For gray scale correction of side-scan sonar images, the proposed method is more effective than the latest similar methods based on the Retinex theory, and the proposed method is faster. Experiments prove the validity of the proposed method.

Keywords: side-scan sonar image; gray scale correction; Retinex; image enhancement

1. Introduction

With the continuous development of side-scan sonar technology, aspects of side-scan sonars, such as data acquisition stability, sonar image resolution, and image clarity have been improved, providing better technical support for hydrographic surveying and charting. Given the development of marine resource, it is necessary to scan the seabed with side-scan sonars to grasp the general information of the seabed scene and topography for many applications, such as offshore oil drilling, channel dredging, submarine pipeline detection, seabed structure detection, marine environment detection, marine archaeology, and detection and location of large-scale seabed targets [1–4].

In Figure 1, a side-scan sonar is scanning the seabed scene. The side-scan sonar transducer installed on both sides of the Autonomous Underwater Vehicle (AUV) emits spherical acoustic signals. After reflection from the seabed, the reflected signals are collected and received by the receiver according to the transmission time of the sonar signal. A side-scan sonar image is formed by converting the reflected signal intensity into the gray level. However, the energy of the sonar acoustic wave can attenuate in water. There are three main types of attenuation. The first category is physical attenuation, the second is the absorption of seawater and the third is echo attenuation [5–7]. In addition to the causes of sonar energy attenuation, side-scan sonar image is also affected by beam patterns, angular responses of different sediments, and changes in seabed topography [7–9]. Based on the above reasons, the gray-scale of side-scan sonar images is uneven.

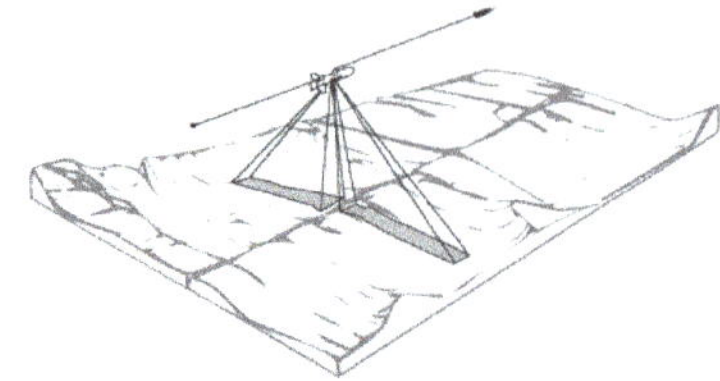

Figure 1. Working schematic diagram of a side-scan sonar.

As shown in Figure 2, the original side-scan sonar image has uneven gray distribution, which affects the interpretation of the side-scan sonar image and the subsequent image processing. Therefore, gray scale correction should be conducted before processing the side-scan sonar image, such as image matching, stitching, and target recognition [10–12].

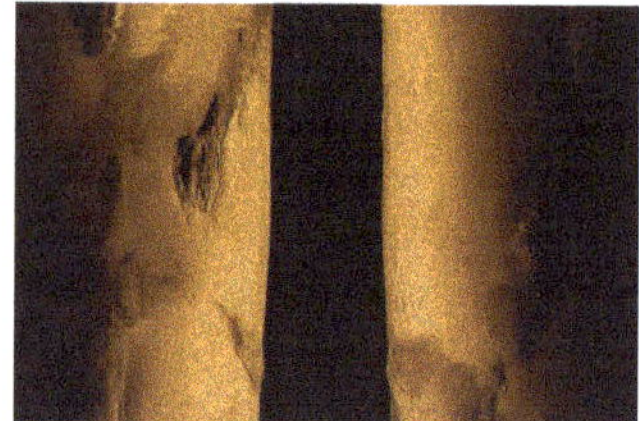

Figure 2. Original side-scan sonar image.

2. Current Gray Scale Correction Methods for Side-Scan Sonar Images

At present, many kinds of gray scale correction methods are available for side-scan sonar images, which can be categorized into six kinds of methods.

2.1. Time Variant Gain (TVG)

The TVG method is a commonly used method for gray scale correction of side-scan sonar images. The time variant gain method, adopted by Johnson et al., is based on the distance between each point of the seabed and the sonar array in the side-scan sonar image [9,13]. The side-scan sonar images are compensated using Equation (1):

$$EL = 2TL - TS = 30\lg R + 2\alpha R/10^3 - S_f \tag{1}$$

where EL is the compensation amount, TL is the energy loss caused by the propagation process, TS is the target strength, R is the propagation distance, α is the absorption coefficient, S_f is the seabed backscattering intensity. Since α and R cannot be easily obtained, their empirical values are required in the compensation. The TVG method is usually implemented using hardware. While the intensity can be compensated for, to a certain extent, it is impossible to mimic the same sonar energy attenuation process. Sometimes unrealistic gain parameters cause secondary gray distortion. Two problems arise if we use the algorithm in Equation (1) to gray correct side-scan sonar images. (1) It is difficult to determine the specific values of the parameters in Equation (1) only using side-scan sonar images, so imbalanced correction may occur. (2) Different side-scan sonar images require different parameters in the TVG equation to achieve better image enhancement, so the algorithm is not universal.

2.2. Histogram Equalization (HE)

Histogram equalization is used to improve the uniformity of the gray distribution of the side-scan sonar image by adjusting the gray distribution of the entire image. The essence of histogram equalization involves enlarging the gray level difference of the image, and the overall contrast of the image is improved after equalization. HE has been widely used because of its simplicity and directness [14]. However, HE often increases image noise, and because some gray scale merging results in blurring the weak edges of the image, it leads to over-enhancement in the regions with large histogram peaks, which is extremely disadvantageous to side-scan sonar image processing. Therefore, histogram equalization is not the best method to address gray distortion in side-scan sonar images.

2.3. Nonlinear Compensation

Nonlinear compensation involves dividing the range of 0 to 255 gray levels into many parts, and then to compensate the gray value of different parts with a piecewise function. However, this leads to excessive gray correction in side-scan sonar images, which may distort the original information of the side-scan sonar image [15].

2.4. Function Fitting

Function fitting method involves using N Ping side-scan sonar data as the selected image. Then, according to the average value of pixels in the column of the selected sonar image, the gray change curve is obtained in the row direction and it is fit with a function. Next, the image is compensated and corrected according to the function obtained by fitting. The representative methods are mixed exponential regression analysis (MIRA) [16] and the method was proposed by Al-Rawi et al. [17]. The MIRA method uses an exponential function, such as Equation (2), to fit the gray level change of the side-scan sonar image, and then compensates the image by normalization.

$$f(z) = ae^{bz} + ce^{dz} \tag{2}$$

where, a, b, c, and d are the four weights representing the echo decay for each ping signal, and z is the spatial location (or index) of each sample within the ping. Shippey et al. [14] stated that the gray level distribution of side-scan sonar image per ping is closer to a Rayleigh distribution than exponential distribution. Therefore, the cubic spline model was used to compensate and correct the side-scan sonar image by the fitting polynomial function.

2.5. Sonar Propagation Attenuation Model

Burguera et al. [18] analyzed the side-scan sonar propagation model, which can be expressed by

$$I(\mathrm{p}) = K \cdot \Phi(p) \cdot R(p) \cdot \cos(\beta(p)) \tag{3}$$

where $I(\mathrm{p})$ is the echo intensity, which is the received side-scan sonar data intensity; K is the normalized coefficient; $\Phi(p)$ is the acoustic penetration intensity; $R(p)$ is the reflection intensity of the acoustic wave on the seabed; and $\beta(p)$ is the incidence angle of the sonar. As the seabed reflection intensity is needed, $I(\mathrm{p})$ can be obtained from the propagation model formula. The acoustic penetration intensity model can be derived from the sensitivity model proposed by Kleeman and Kuc [19], but the influence of the incident angle of the sonar on the intensity of side-scan sonar is excluded in Burguera et al. [18]. Thus, the correction effect is not good, and the parameter information in the calculation equation needs the parameter information and propagation information of side-scan sonar, so this method is not suitable for gray scale correction of a single side-scan sonar image.

2.6. Beam Pattern

The beam pattern is determined by the working characteristics and physical design of the sonar sensor array [7], but it is also one of the reasons for the uneven gray level of side-scan sonar images. Chang and colleagues [6,7,9] determined the energy distribution function relative to the angle by summing up the energy levels for each angle over the whole data series. Then, according to the statistical results, the average energy of each angle can be obtained. Finally, the inverse of this average can be applied as a correcting factor to individual datum in the time series. However, this method needs to consider the change in seabed topography and seabed sediments; otherwise, the correction image will be poor.

3. Gray Scale Correction Method Based on Retinex

The image processing algorithm based on Retinex theory is a commonly used optical image defogging and low illumination image enhancement algorithm, which was proposed by Land in 1963 [20]. It decomposes an image into an illumination map and a reflectance map, expressed as

$$S(x,y) = R(x,y) * L(x,y) \tag{4}$$

where, *S(x,y)* is the original image, *R(x,y)* is the reflectance map, *L(x,y)* is the illumination map, and the operator * means element-wise multiplication. The reflectance map reflects the essential information of the scene in the image, and the illuminated map reflects the brightness information of the environment in the image. The change in brightness information results in the change in the gray value of the image. Therefore, to ensure that the image can normally reflect the scene information, we need to reduce the influence of illumination change on the original image.

The commonly used image enhancement methods based on Retinex include Single Scale Retinex (SSR), Multi-Scale Retinex (MSR), Multi-Scale Retinex with Color Restoration (MSRCR), and Multiscale Retinex with Chromaticity Preservation (MSRCP) [21–24].

The SSR method uses Retinex to deform Equation (4) by calculating the logarithm to produce the reflection map:

$$r(x,y) = \log R(x,y) = \log(\frac{S(x,y)}{L(x,y)}) \tag{5}$$

Firstly, we get the low-pass function that is calculated with Equation (6), then use the low-pass function to estimate the illumination map that corresponds to the low frequency part of the original image. Thus, the reflection map represented by the high frequency component of the original image can be obtained with Equation (7). Finally, the obtained logarithmic reflectance map is restored to the corrected image.

$$F(x,y) = \lambda e^{-\frac{(x^2+y^2)}{c^2}} \tag{6}$$

$$r(x,y) = \log S(x,y) - \log[F(x,y) \otimes S(x,y)] \tag{7}$$

where c is the Gaussian circumference scale, λ is a scale, and $\otimes$ represents convolution operation.

The image corrected by the SSR algorithm may cause blurring and excessive correction, so the original image is processed by multi-scale low-pass function in the MSR algorithm. The multi-scale low-pass function is the weighted sum of multi-scale low-pass function in SSR algorithm. Thus, we can implement the algorithm with Equation (8).

$$r(x,y) = \sum_k^K W_k\{\log S(x,y) - \log[F_k(x,y) \otimes S(x,y)]\} \tag{8}$$

where, K is the number of *F(x,y)*, when $K = 1$, MSR is SSR. W_k is the weight. The value of K is usually 3 and $W_1 = W_2 = W_3 = 1/3$.

As images processed with the MSR algorithm have a color imbalance, MSRCR and MSRCCP algorithms were developed on the basis of MSR. The MSRCR algorithm uses a color restoration factor to avoid the color imbalances caused by image local contrast enhancement. MSRCP uses the MSR algorithm and intensity information of each channel of image to enhance image.

At present, some new algorithms are based on Retinex. Guo et al. [25] proposed a simplified enhancement model called low-light image enhancement (LIME). They used max-RGB technology to estimate the illumination map, which takes the maximum value of the three channels of color image R,

G, and B, then uses structure to refine the illumination map, uses gamma correction to re-estimate the non-linearity of the fine illumination map as the illumination map, and finally uses Retinex to obtain the enhanced image. The naturalness preserved enhancement (NPE) [26] algorithm is a non-linear uniform illumination map enhancement method. The image is decomposed into an illumination map and a reflection map by a filter, then the illumination map is transformed, and finally the illumination map and reflection map are merged again as the final enhancement image. Simultaneous reflection and illumination estimation (SRIE) [27] is a weighted variable model for simultaneous estimation of reflected and illuminated images. The estimated illuminated map is processed to enhance the image. Fu et al. [28] proposed a multi-deviation fusion method (MF) to adjust the illumination by fusing multiple derivations of the initially estimated illumination map.

4. Our Method

The key to the image enhancement algorithm based on Retinex lies in the acquisition of illumination map and the restoration of the reflection map and image color. Firstly, we obtain the smoothed image by smoothing the original image, and adding a constant value to the smoothed image as the illumination map $L(x,y)$. The constant value is added to avoid noise in the enhanced image. Then, the reflection map $R(x,y)$ is acquired based on an element-wise division according to Retinex with Equation (4).

As the gray value of the reflected map pixels obtained is low, we need to restore the illumination and color of the reflected map. We multiply the reflected map $R(x,y)$ directly by a constant coefficient. The bigger the constant coefficient, the bigger the gray value of the corrected image, and the brighter the enhanced image. Experiments were conducted afterward.

Side-scan sonar images are different from natural images, so we considered the following when designing the image enhancement algorithm for side-scan sonar images: (1) Since the side-scan sonar image is originally a gray-scale image and the color side-scan sonar image is the result of pseudo-color processing, we did not use max-RGB technology to obtain illumination map, but directly converted the pseudo-color sonar image into gray-scale. (2) As there is no large change in the gray gradient in side-scan sonar images, we used the mean filter or bilateral filter to directly smooth the gray image of the side-scan sonar image, which not only meets the requirements of gray scale correction of side-scan sonar images, but also improves the speed of the algorithm. (3) In side-scan sonar images, there may be a large area of black area nearby because of a hilltop or raised terrain. Therefore, we added a constant to the smoothed image as the illumination image L that avoids much of the noise or "cartoon" phenomena in the black area of the enhanced image. In summary, considering the characteristics of the side-scan sonar image, we propose a side-scan sonar image enhancement algorithm based on Retinex.

According to our proposed algorithm, we use Equation (9) to correct the gray scale of side-scan sonar images:

$$S'(x,y) = A\frac{S(x,y)}{(S(x,y) \otimes F(x,y) + a)} \tag{9}$$

where $S(x,y)$ represents the original image, $S(x,y) \otimes F(x,y) + a$ is the illumination map L, $F(x,y)$ is a smoothing filter function, and a is a constant. The constant is mainly used to suppress noise. A is a constant coefficient through which the gray value of the corrected image can be adjusted to change the brightness of the corrected image.

The filtering function $F(x,y)$ can be used in many filtering methods. Due to the characteristics of side-scan sonar images and considering the time complexity of the algorithm, we choose two methods: Mean filter and bilateral filter. Mean filter is the simplest linear filtering operation. Each pixel of the output image is the average value of the corresponding pixel of the input image in the core window. Mean filtering is the simplest linear filtering operation. Each pixel of the output image is the average value of the corresponding pixel of the input image in the core window. The mean filtering is fast, but does not protect image details well. Bilateral filtering is a non-linear filtering method that combines the spatial proximity and the pixel value similarity of the image. It also considers the spatial information and gray level similarity. Its advantage is that the algorithm is simple and can protect the image edge, but its disadvantage is that it is slower than mean filtering.

In our algorithm, the parameters A and α are constant. We will provide experiments to analyze the influence of the changes of A and α on the enhancement of side-scan sonar image, and how to select and adjust their respective values.

Our method and SSR method have similarities and differences. The similarities are that both methods are based on Retinex and use smoothing to obtain illumination map. The differences are as follows: (1) Firstly, the SSR method calculates the logarithm of the original image, and then smooths the image with a Gaussian low-pass function to obtain the illumination map. However, our method is to directly smooth the original image with mean filtering or bilateral filtering to obtain the illumination map. (2) The SSR method takes the anti-logarithm operations to restore the illumination of the image. However, the enhanced image often looks unnatural and frequently appears to be over-enhanced. Our method is to multiply the reflectance map directly by a constant to restore the illumination of the image. (3) Our method adds a constant to the smoothed image as the illumination map, which avoids a lot of noise in the enhanced image, while the SSR method fails to avoid the problem of noise generation. Therefore, our method is more suitable for side-scan sonar image enhancement. The following experiments support our conclusion.

Considering the time complexity of the algorithm and the characteristics of side-scan sonar images, we implemented our method by means of mean filtering and bilateral filtering, then analyzed and evaluated the two smoothing schemes through experiments.

5. Experiments and Analysis

In this section, we provide the overall framework of the proposed algorithm, as shown in Figure 3. First, we obtain the gray image from the original side-scan sonar data, or gray the pseudo-color side-scan sonar image to obtain the gray image of side-scan sonar image. Then, the gray image of the side-scan-sonar image is filtered to obtain the smooth image. The constant coefficient a is added to the smoothed image as the illumination map L in the Retinex model, and then the reflected map R is obtained based on an element-wise division using Equation (4). The reflected map R is multiplied by a constant coefficient A as the enhanced image. Finally, the enhanced gray image is pseudo-color processed to obtain the final gray scale corrected side-scan sonar image.

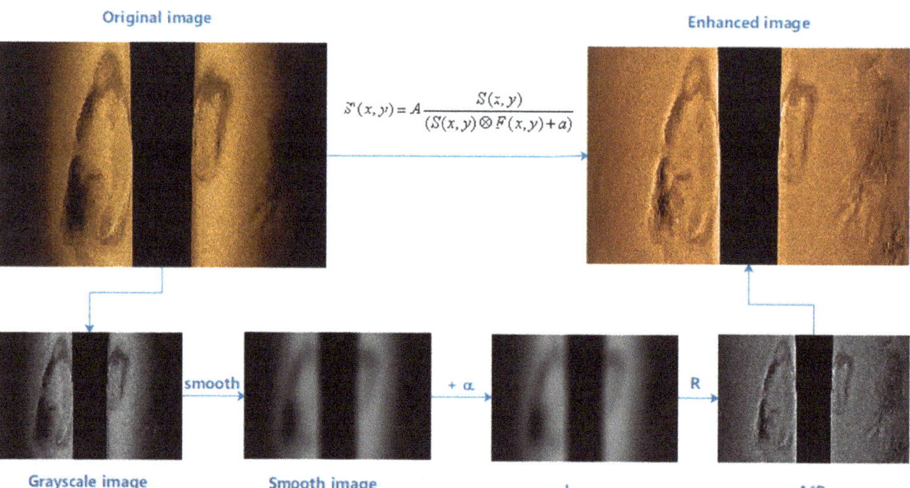

Figure 3. The overall framework of the algorithm in this paper.

5.1. Experiments and Analysis of Parameter A

The constant coefficient A can be adjusted using this method. If a brighter gray scale correction image is needed, the value of A can be set higher, and vice versa. We analyzed it through experiments. In the experiment, the size of the original sonar image was 800×525 pixels. We set the constant a to 15. Then, we smoothed the image using the mean filter. The size of the filter template was set to 1/17 of the size of original image, and A was set to 80, 140, and 200.

The experimental results are shown in Figure 4. The results show that the side-scan sonar image corrected using this method can correct the original gray distortion image, so that the image scene information scanned by sonar can be displayed normally. The size of A only affects the overall brightness of the enhanced image and does not cause secondary distortion of the enhanced image. The enlarged detail image shown in Figure 5 shows that the gray distortion of the original image is serious. After gray scale correction, the gray distortion of the image disappears, and the texture information of the image is displayed normally. In the experiment, the bigger the value of A, the higher the gray value of the corrected image, and the clearer the image. The adjustment of A does not affect the overall gray distribution of the enhanced image, but only the overall brightness of the enhanced image. However, when the value of A is set too large, the image is too bright, and the enhanced brightness is not suitable for human perception. We did not fix A to a size. If different brightness enhancement images are required, adjust the value of A. If A must have a fixed value, we recommend A = 140, because we do a lot of experiments in Appendices A and B, A = 140 meets the experimental requirements.

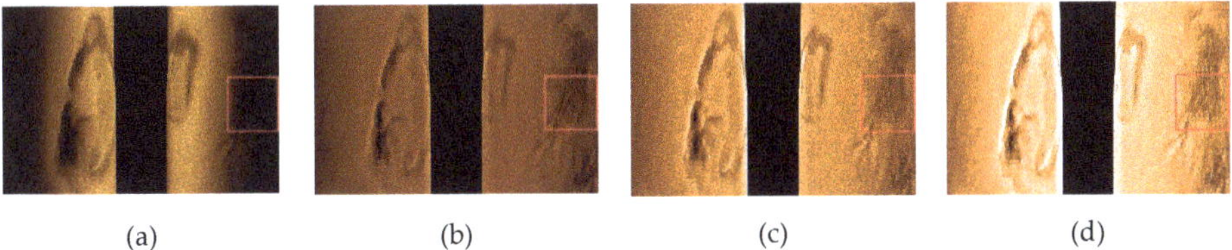

Figure 4. Experimental comparison using mean filter. (**a**) Original image and corrected image with (**b**) $A = 80$, (**c**) $A = 140$, and (**d**) $A = 200$.

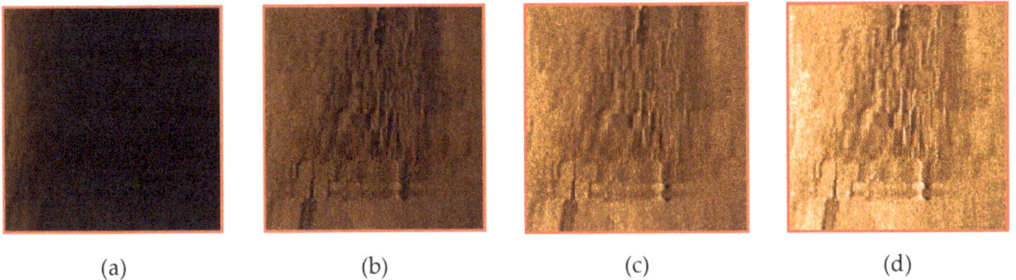

Figure 5. Local enlargement of Figure 4. (**a**) Original image and corrected image with (**b**) $A = 80$, (**c**) $A = 140$, and (**d**) $A = 200$.

5.2. Experiments and Analysis of Parameter a

In our algorithm, constant a is used to suppress the noise. We experimentally analyzed the selection of the a value. In the experiment, we set the constant A to 140. Then, we smoothed the image using the mean filter. The size of the filter template was set to 1/17 of the size of original image, and a was set to 0, 5, 10, 15, 30, 40, and 60.

As shown in Figure 6, when we changed the value of a while the other parameters remained unchanged, the noise in the dark area of the enhanced image was amplified with the decrease in the value of a, and vice versa. However, since we add a to the smoothed original image as the illumination estimation of the original image, when a increases excessively, the estimated illumination cannot accurately represent the illumination map of the original image. If the a value is too large, the effect of image correction worsens, as supported by our experimental results. The parameters setting of the side-scan sonar may be different in different batches of sonar data, which leads to differences in the image characteristics of side-scan sonars. Thus, the value of a is an empirical value. According to our experiment and as shown in the experimental results in Appendix B, the value of a is about 15. The value of a does not need to be adjusted for the same batch of side-scan sonar images, and different batches may require fine-tuning.

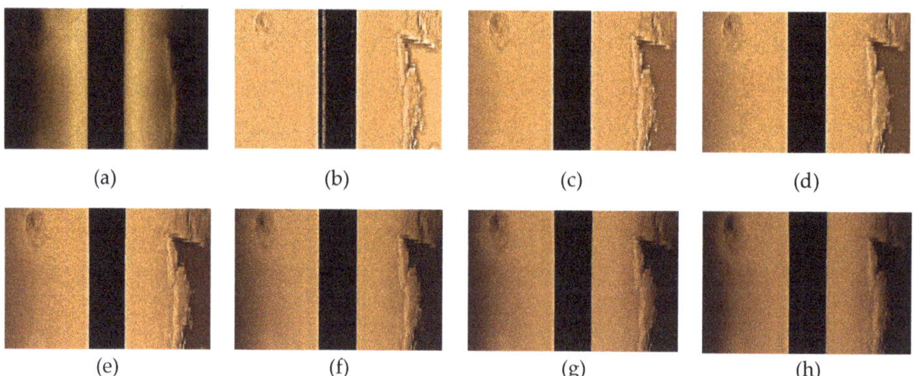

Figure 6. Experimental comparison using mean filter. (**a**) Original image and corrected image with and (**b**) $a = 0$, (**c**) $a = 5$, (**d**) $a = 10$, (**e**) $a = 15$, (**f**) $a = 30$, (**g**) $a = 40$, and (**h**) $a = 60$.

5.3. Experiments and Analysis of Smoothing Function

In this algorithm, the gray image from a side-scan sonar can be smoothed using many methods. Bilateral filtering was used to smooth side-scan sonar images, which was compared with the experimental results of mean filtering. In the contrast experiment, A was set to 140 and a to 15. Figure 7 compares the experimental results produced when using the mean filter and bilateral filter.

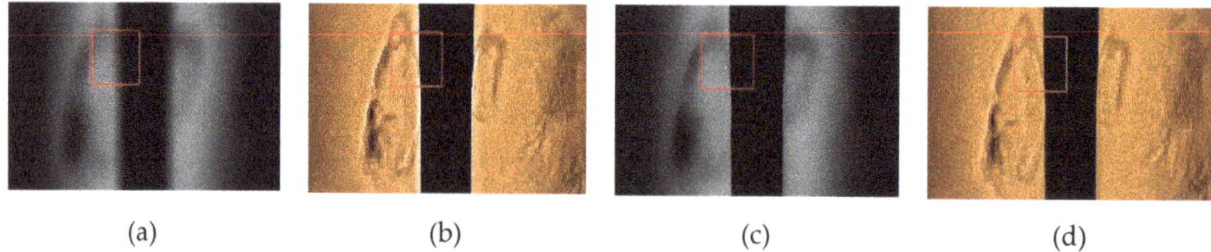

Figure 7. Experimental comparison of images. (**a**) Smoothed image using mean filter, (**b**) mean method results, (**c**) smoothed image using bilateral filter, and (**d**) bilateral method image.

As shown in Figure 7, the enhanced images produced by these two methods are similar overall, but in terms of image details, the experimental results show that the illumination map L produced by bilateral filtering is clearer than the illumination map L obtained by mean filtering in the water column area, and the mean filtering is blurred. The smoothed image obtained by bilateral filtering reflects the illumination distribution of the original image better, and the corrected image obtained in the experiment is more stable and clear. To see more clearly, we enlarged part of Figure 7. As shown in Figure 8, because the gray image is smoothed with only the mean filter, the gray image after smoothing has an unclear gray boundary in the region with a large gray gradient. The illumination map smoothed by mean filter cannot accurately express the illumination distribution of the original image, especially around the region with a large gray gradient. Therefore, the corrected image obtained using the mean filter will show some over-enhancement and the halo phenomena in the areas where the gray level of the original image changes too much. The corrected image produced using the bilateral filter is more normal, and there is no halo phenomenon because the map produced with bilateral filtering is more in line with the actual gray distribution of the original image.

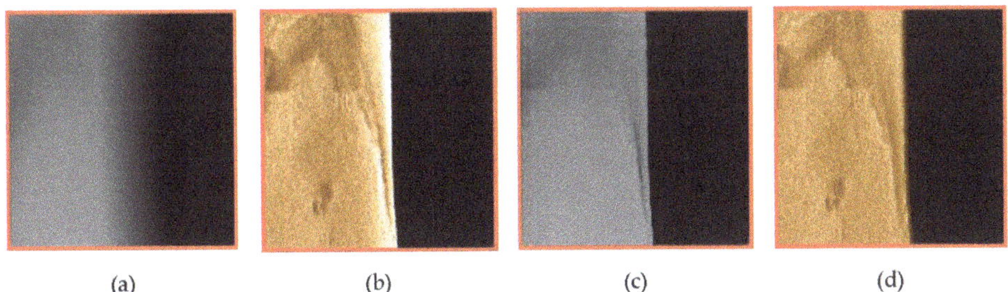

Figure 8. Comparison of experimental details: (**a**) smoothed image using mean filter, (**b**) mean method results, (**c**) smoothed image using bilateral filter, and (**d**) bilateral method results.

5.4. Comparative Experiments and Analysis of Other Methods Based on Retinex

Considering several commonly used image enhancement methods based on Retinex, we used different original side-scan sonar images to verify the stability and advantages of our proposed image enhancement algorithm, and used SSR, MSR, MSRCR, and MSRCP of four commonly used image enhancement methods based on Retinex for comparison with our methods. The parameters of bilateral filter were set as follows: diameter range of each pixel neighborhood was set to 35, sigma-color is set to one-seventh of the height of the original image, and sigma-color represented the sigma value of the color space filter. The larger the value of this parameter, the wider the colors in the neighborhood of the pixel are mixed together. Sigma-space was set to one-seventh of the width of the original image and sigma-spaces represent the sigma values of filters in coordinate space. The larger the sigma values, the more distant the pixels interact with each other. Our other experimental parameters were as follows: A was set to 140 and a was set to 15. The results are shown in Figure 9. For better illustration and display, we enlarged a local area of Figure 9, as shown in Figure 10.

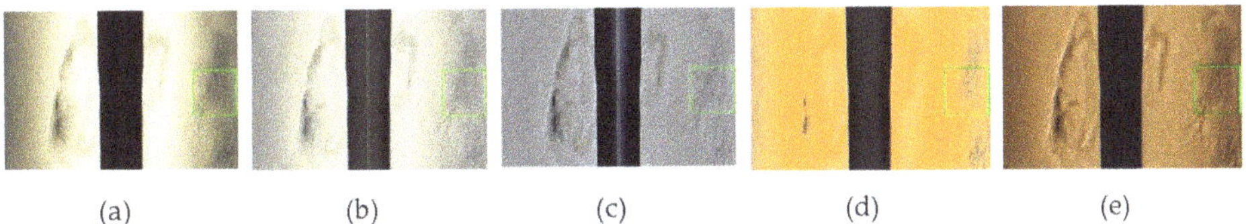

Figure 9. Comparative experiment of common methods based on Retinex. Images enhanced with (**a**) SSR, (**b**) MSR, (**c**) MSRCR, (**d**) MSRCP, and (**e**) using our bilateral filtering method.

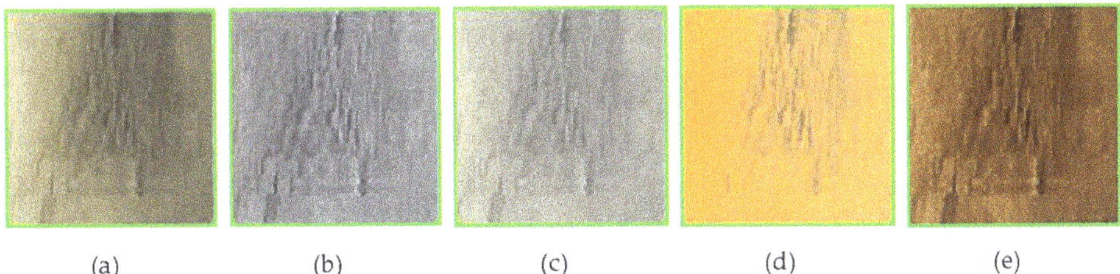

Figure 10. Comparison of experimental details. Images enhanced with (**a**) SSR, (**b**) MSR, (**c**) MSRCR, (**d**) MSRCP, and (**e**) using our bilateral filtering method.

The experimental results show that SSR, MSR, MSRCR, and MSRCP produce considerable noise and the image sharpness of the enhancement is very poor and the image corrected by the SSR method cause blurred and excessive corrections. Therefore, these four methods are not suitable for gray correction of side-scan sonar images. The enhanced image color produced using the MSRCR algorithm is more consistent with the original side-scan sonar image color than using the MSR algorithm. The enhancement effect of our method is better than that of the other four methods. The enlarged detail map shows that the detail map of our method is more clear and stable.

To fairly evaluate the image enhancement algorithm, we used the peak signal-to-noise ratio (PSNR), information entropy, standard deviation, and average gradient as the evaluation indexes. PSNR represents the ability of an image enhancement algorithm to suppress noise. The larger the value, the better the ability to noise suppression and the smaller the image distortion. PNSR can be calculated with Equations (10) and (11).

$$MSE = \frac{1}{H \times W} \sum_{i=1}^{H} \sum_{j=1}^{W} (X(i,j) - Y(i,j))^2 \tag{10}$$

$$PSNR = 10 \log_{10}\left(\frac{(2^n - 1)^2}{MSE}\right) \tag{11}$$

where MSE represents the mean square error of the current image X and the reference image Y; H and W are the height and width of the image, respectively; n is the number of bits per pixel, which is generally 8, meaning the gray scale of the pixel is 256; and the unit of PSNR is dB.

Information entropy is a measure of image information richness and is calculated as

$$H(X) = -\sum_{k=0}^{L=1} P_k \lg P_k \tag{12}$$

where L is the maximum gray level of image X, and P_k is the number of pixels whose gray value of image X is K.

The standard deviation reflects the dispersion of all image pixel values to the mean value; the smaller its value, the more balanced the gray distribution of the image. We can calculate standard deviation using Equation (13), where $v(x_i)$ is the gray value of the pixels in the image, $\overline{v(x)}$ is the average gray level of the image, and n is the number of pixel points in the image:

$$I_{std} = \sqrt{\frac{1}{n} \sum_{1}^{n} \left(v(x_i) - \overline{v(x)}\right)^2} \tag{13}$$

The average gradient represents the ability to express the details of the image, the image sharpness, and texture changes. The bigger the average gradient, the sharper the edge of the image, and the clearer the image. Average gradient can be calculated with

$$\overline{G} = \frac{1}{M \times N} \sum_{i=1}^{M} \sum_{j=1}^{N} \sqrt{\frac{\Delta I_x^2 + \Delta I_y^2}{2}} \tag{14}$$

where ΔI_x^2 is the gradient in the horizontal direction, ΔI_y^2 is the gradient in the vertical direction, and M and N are the height and width of the image, respectively.

According to the experimental results provided in Table 1, the proposed algorithm is superior to the other four algorithms in PNSR, information entropy, image standard deviation, and average gradient. So, we proved that the enhanced image produced using our algorithm is clearer, the ability to suppress noise is stronger, the gray distribution is more balanced, the image information is richer, and the details of the image are enhanced. Therefore, the image enhancement algorithm in this paper is better than the other three algorithms.

Table 1. Objective Evaluation Index of Image Enhancement Algorithms.

Method	PSNR	Information Entropy	Standard Deviation	Average Gradient
Original image		6.81746	42.1187	5.89745
SSR	7.50039	6.80343	74.8485	6.4373
MSR	8.42826	5.8048	54.2663	9.2899
MSRCR	6.48304	6.54661	54.3578	6.29672
MSRCP	7.22485	5.66164	43.4622	5.42281
Ours	14.5954	7.58250	43.8442	9.48556

We compared the latest image enhancement algorithms based on Retinex theory. The experimental parameters of the algorithms were obtained from the parameters set by the respective authors. We did not change the parameters of the algorithm. We used the smoothing filter and bilateral filter to implement our method, and the experimental parameters were the same as those used above. The experimental code was realized using MATLAB (MathWorks, Natick, US). All the experiments were conducted on a PC running Windows 10 (Microsoft, Redmond, US) OS with 4 G RAM and a 2.4 GHz CPU. In this paper, three different side-scan sonar images were used for experiments, representing three types of side-scan sonar images: (a) An image with a large area of black due to the occlusion of seabed hills (Figure 11a), (b) an image with no black area and more texture (Figure 11b,c) an image with black areas, textures, and hills (Figure 11c).

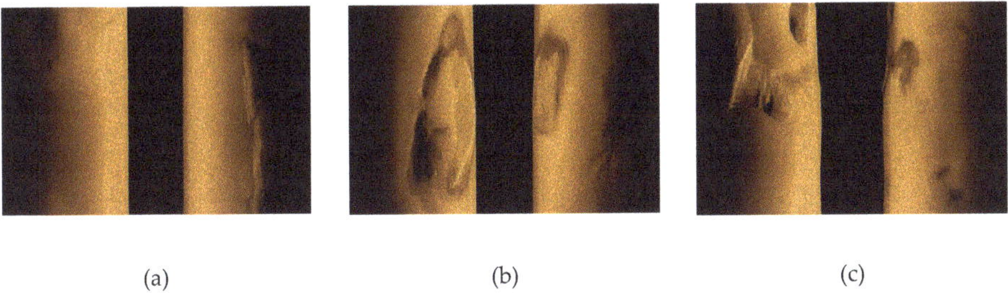

(a) (b) (c)

Figure 11. Original side-scan sonar images: (**a**) a large area of black area, (**b**) an image with no black area and more texture, and (**c**) image with black areas, textures, and hills.

Three different types of side-scan sonar images were compared using low-light image enhancement (LIME), naturalness preserved enhancement (NPE), simultaneous reflection and illumination estimation (SRIE), multi-deviation fusion method (MF), and our two methods with mean filtering and bilateral filtering methods. The experimental results are shown in Figure 12. In terms of enhancement effect, LIME, NPE, SRIE, MF, and the two methods in this paper obviously enhance side-scan sonar images, but the image enhanced by NPS method produces a lot of noise in the dark area of the image. As shown in Figure 13, we enlarged the images of different methods for the second side-scan sonar image. The local enlarged image shows that there are some inadequate corrections at the left and right ends of the image enhanced with the LIME, NPE, SRIE, and MF methods, but no such situation was observed using the two methods in this paper.

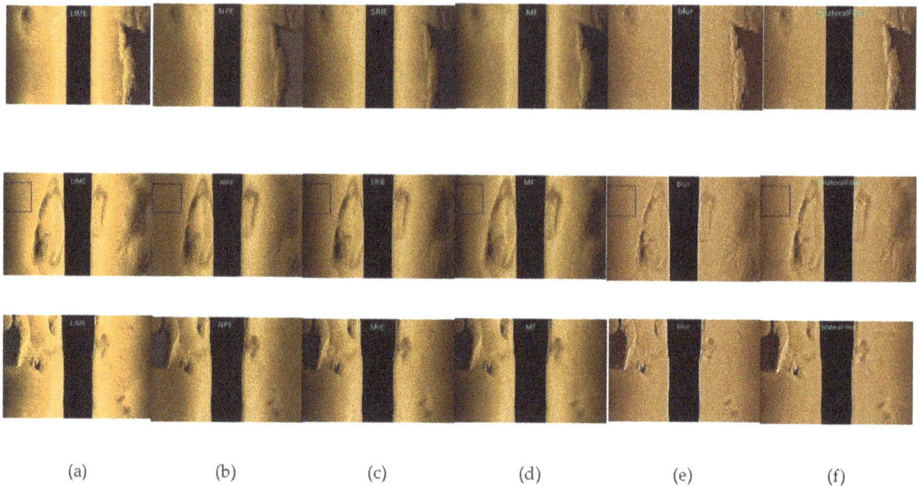

(a) (b) (c) (d) (e) (f)

Figure 12. Side-scan sonar images enhanced using (**a**) LIME, (**b**) NPE, (**c**) SRIE, (**d**) MF, and our method with (**e**) mean filtering, and (**f**) bilateral filtering.

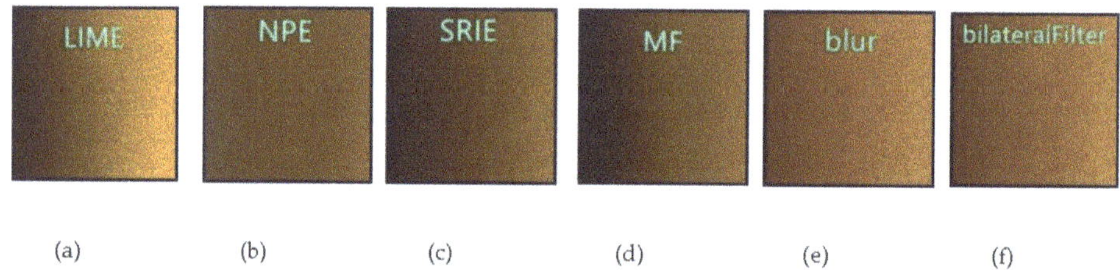

(a) (b) (c) (d) (e) (f)

Figure 13. Local enlargement of the second kind of enhanced image (**a**) LIME, (**b**) NPE, (**c**) SRIE, (**d**) MF, and our method with (**e**) mean filtering, and (**f**) bilateral filtering.

As shown in Figure 14, the histogram gray value of the enlarged image of the four methods, LIME, NPE, SRIE, and MF, has obvious peaks in the low gray part. However, the two methods in this paper have no obvious peaks in the low gray part of the histogram. LIME, NPE, SRIE, and MF do not effectively enhance the image at both ends and the enhancement effect is not as good as the two methods in this paper.

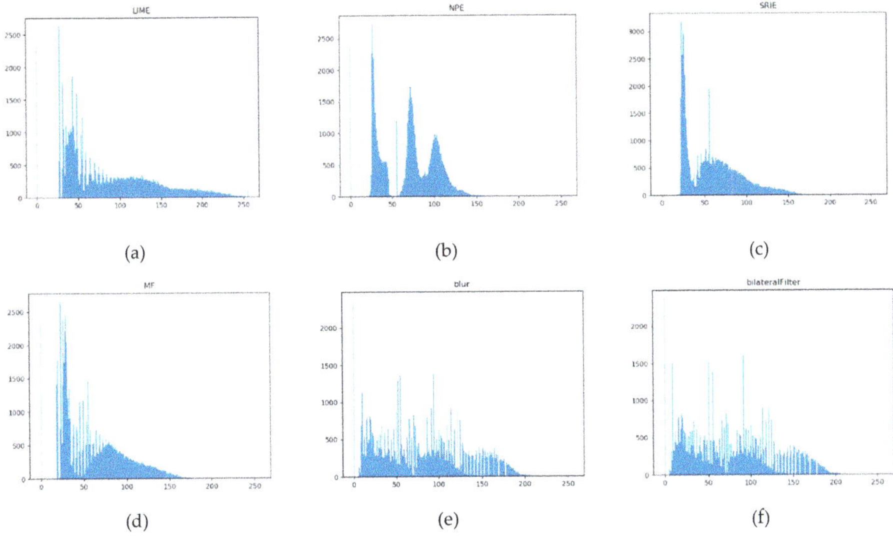

Figure 14. The histograms corresponding to Figure 13: (**a**) LIME, (**b**) NPE, (**c**) SRIE, (**d**) MF, and our method with (**e**) mean filtering, and (**f**) is bilateral filtering.

To better evaluate the performance of the algorithm, we used PSNR, information entropy, standard deviation, average gradient, and algorithm running time to evaluate the image enhancement algorithm as a whole. We calculated the index using the whole enhanced image instead of the local enlarged image. The evaluation index table is shown in Table 2.

Table 2. Objective evaluation index of image enhancement algorithms.

Original Image	Algorithm	PNSR	Information Entropy	Standard Deviation	Average Gradient	Algorithm Time-Consuming(s)
Image1	LIME	12.8283	7.59424	54.9971	11.3484	2.15712
	NPE	16.9553	7.21194	45.2014	8.33524	20.9530
	SRIE	19.1414	7.28538	46.8345	7.37814	35.7950
	MF	19.3875	7.23853	46.1300	8.39727	3.0470
	our method of mean filter	14.8291	7.50084	45.5599	9.76665	0.5430
	our method of bilateral filter	15.0356	7.50243	44.6718	9.06058	1.7860
Image2	LIME	12.5039	7.55372	53.4062	11.6171	2.0779
	NPE	17.4068	7.17702	14.9912	8.01561	20.6560
	SRIE	18.9174	7.26646	43.9448	7.51642	33.6090
	MF	18.6418	7.25066	42.7528	8.66020	1.5150
	our method of mean filter	14.3681	7.56064	44.7653	10.3218	0.5440
	our method of bilateral filter	14.5954	7.58250	43.8442	9.48556	1.6880
Image3	LIME	12.9156	7.52301	56.4356	11.98260	2.40157
	NPE	18.0214	7.36287	45.1654	8.42337	20.5220
	SRIE	19.3221	7.37251	47.1540	7.91938	35.2410
	MF	19.0227	7.33752	46.9760	8.49770	1.4690
	our method of mean filter	14.9149	7.57515	46.7814	10.30420	0.6860
	our method of bilateral filter	15.1388	7.61042	45.8554	9.41506	1.5670

Among the PSNR, information entropy, standard deviation and average gradient of the evaluation indexes, our algorithm in this paper is similar to the other four latest algorithms. The information entropy of image enhancement based on bilateral smoothing is the highest, and the information entropy of image enhancement based on mean filtering method is the second. Our algorithm is faster. Our mean filter is the fastest, followed by our bilateral filter, then the MF algorithm, in which our mean filter method is at least twice as fast as the other methods. The speed of our bilateral filtering method is equal to MF, which is faster than LIME, NPE, and SRIE. We can use the mean filter and bilateral filter to smooth the gray image of side-scan sonar image separately and use the two schemes to enhance side-scan sonar images. The evaluation index shows that the enhancement effect of the bilateral filtering method in this paper is slightly better than that of the mean filtering method, but the mean filtering method takes less time than that of the bilateral filtering method. Therefore, the mean filtering enhancement algorithm is more suitable for online processing situations when the image processing speed is more important, and the bilateral filtering method is more suitable for side-scan sonar image enhancement that requires better image enhancement during offline processing. The side-scan sonar image enhancement algorithm in this paper is simple but not inferior to other high-quality algorithms. The reasons for this situation are as follows: (1) The original side-scan sonar image is a gray-scale image, and there are no R, G, and B channels. The color side-scan sonar image is pretreated by pseudo-color application, so we do not need to use max-RGB technology to obtain the illumination map of the original side-scan sonar image. (2) The key to using the Retinex image enhancement algorithm is to obtain an accurate illumination map. The illumination map needs to smooth the details of the original image as much as possible while maintaining the boundaries of the gray distribution of the original image. So, the other four methods use more complex algorithms to obtain an accurate illumination map, which results in an increase in the running time. However, side-scan sonar images are different from natural illumination images. The gray gradient of the side-scan sonar image changes little, and there is no case of too large a gray gradient. Therefore, it is not necessary to use a complex algorithm to obtain a fine illumination image to ensure that the enhanced image does not display a halo phenomenon. To suppress halo generation, we can achieve the desired effect by increasing the value of A. (3) The illumination map produced by the other four methods is not smooth enough, which leads to poor gray level correction and enhancement effect on the left and right ends of the side-scan sonar waterfall image.

5.5. Comparative Experiments and Analysis of Gray Scale Correction Methods for Side-Scan Sonar Images

The existing gray level correction methods for side-scan sonar images are realized by experiments. As the TVG method and sonar propagation attenuation model needs to know some side-scan sonar parameters, we only compared the histogram method, the non-linear gray level compensation method, and the function fitting method with our mean filtering method. Figure 15 shows that various methods improve the gray distortion of the original side-scan sonar image after gray correction. The corrected images obtained by the histogram and non-linear compensation methods show that some areas of the image are too strong, then the left and right ends of the image are too weak, which leads to an unsatisfactory enhancement effect of the whole corrected image. The function fitting method and the method proposed in this paper are better than the other two methods.

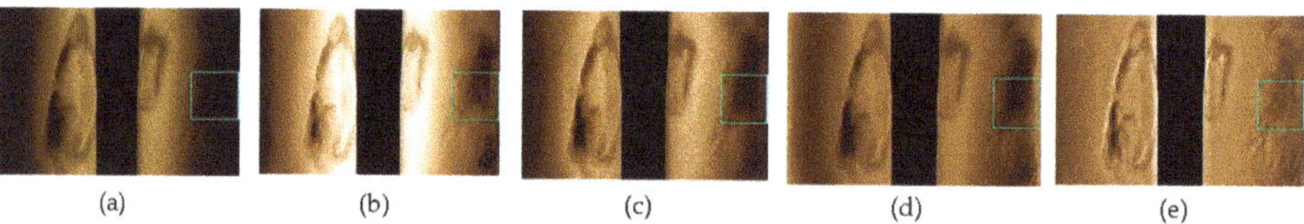

Figure 15. Comparison experiments of common methods used to correct side-scan sonar images: (**a**) The original side-scan sonar image, (**b**) histogram, (**c**) non-linear compensation, (**d**) function fitting, and (**e**) our method.

Figure 16 depicts a histogram comparison of the enhanced images. The histogram correction method results in overweight correction, which produces a particularly high gray value in some areas of the image. The non-linear compensation and function fitting methods have too many parts with low gray values, which proves that the image correction is inadequate. The histogram of the side-scan sonar image enhancement method proposed in this paper shows that the gray value distribution of the side-scan sonar image after correction is uniform, and the correction result of the algorithm is satisfactory.

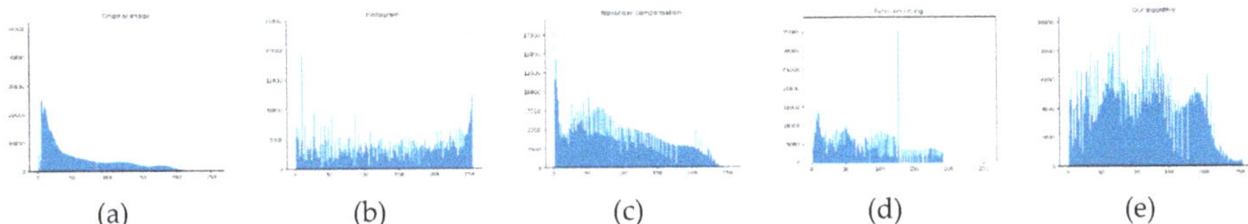

Figure 16. Histograms corresponding to the images in Figure 15: (**a**) the original side-scan sonar image (**b**) histogram, (**c**) non-linear compensation, (**d**) function fitting, and (**e**) our method.

To better analyze the contrast effect of the different enhancement methods, we enlarged the enhanced image locally. The enlarged image is shown in Figure 17. We found that the texture of the side-scan sonar image is destroyed by the function fitting method. The enhanced side-scan sonar images obtained by function fitting, histogram, and non-linear compensation methods have low gray values at the left and right ends of the image, and the effect of gray correction is not obvious. However, our method still enhances the local enlargement image very clearly.

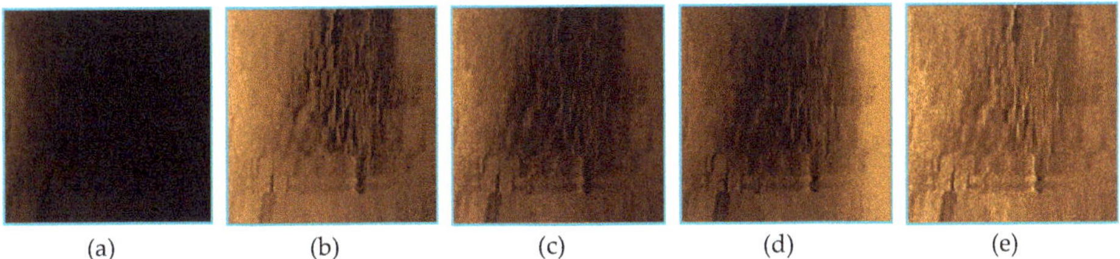

Figure 17. Local enlargement of Figure 15: (**a**) the original side-scan sonar image, and the images produced using the (**b**) histogram, (**c**) non-linear compensation, (**d**) function fitting, and (**e**) our methods.

Only visually observing the corrected image may not be sufficient to differentiate the methods. We also contrasted the image enhancement indexes for the local enlarged image, as shown in Table 3. The enhanced image in this paper is superior to other methods in information entropy and average gradient.

Table 3. Indicators comparison of local magnification maps.

Method	PSNR	Information Entropy	Standard Deviation	Average Gradient
Original image		6.1333	11.0744	2.55545
HE	9.82301	7.26025	45.695	9.87564
Non-linear compensation	11.4153	7.17749	45.9537	8.71157
Function fitting	9.91988	6.67453	48.7124	8.34709
Our method	8.41441	7.41057	38.7549	13.6594

Our method was compared with the commonly used gray level correction method for side-scan sonar images. According to the experimental results, our method is superior to the other methods. Compared with other common methods and the latest optical image methods based on Retinex, the gray scale correction effect of side-scan sonar images using our method is better than those of these methods, as shown by the experimental results and data indicators.

6. Expansion of Our Method

The proposed method is not only suitable for gray level correction of side-scan sonar images, but can also be used to enhance low illumination optical color images. The steps for low illumination color image enhancement are shown in Figure 18. Firstly, we separate the three channels of the color image into R, G, and B channels. Then, we use the above-mentioned method (gray image smoothing filtering), and add a constant a as the illumination image L. Using Equation (15), the three channels are separately removed from the illumination image L and multiplied by constant coefficient A to obtain the enhanced images of the three channels. Finally, the three channels are merged into the final color enhanced image.

$$S'_{rgb} = A * \frac{S_{rgb}}{L} \tag{15}$$

To verify the speed of our proposed algorithm for low illumination color image enhancement, we conducted an experiment. The smoothing method used in the experiment was mean filtering. The size of the filter template was one-seventh of the original image. A was set to 160 and a was set to 17. All the experiments were conducted on a PC running Windows 10 (Microsoft, Redmond, US) OS with 8 G RAM and a 3.7 GHz CPU. Our code was implemented using C++ and OpenCV, which is an image processing library.

As shown in Figure 19, we selected four low illumination color images of different sizes for enhancement. Table 4 shows the time required for low illumination color image enhancement with different-sized images. Because of its speed, this algorithm can meet the real-time video processing requirements. If GPU is used to accelerate the processing, the algorithm will be faster.

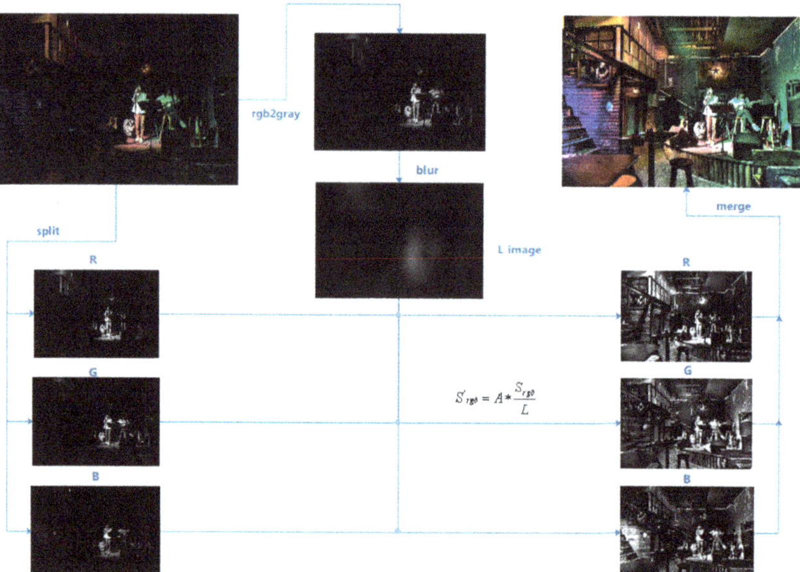

Figure 18. Low illumination color image enhancement process.

(a) (b) (c) (d)

Figure 19. Enhancement effect of four low illumination color scene images: (**a**) scene 1, (**b**) scene 2, (**c**) scene 3, and (**d**) scene 4.

Table 4. Time required to enhance different-sized low illumination color images.

Scene	Image Size	Image Occupied Storage	Time-Consuming (s)
1	370×415	450 KB	0.035
2	560×420	689 KB	0.061
3	720×610	1.4 MB	0.104
4	1024×683	963 KB	0.206

As shown in Figure 19, the first three low-illumination color images are better enhanced by our image enhancement method, but the fourth image is enhanced with an obvious halo phenomenon. The halo phenomenon occurs after the image is enhanced in areas where the gray gradient changes greatly. This mainly occurs because the smoothing method adopted in this paper is too simple, which results in the obtained illumination map being too rough and the illumination distribution of the original image is not well expressed. Thus, our enhancement method is more suitable for side-scan sonar images that where the gray gradients changes are not so abrupt.

7. Conclusions

Side-scan sonars are widely used in ocean exploration. As the original side-scan sonar images are affected by gray distortion, gray scale correction is needed before any further image processing. Considering the difference between side-scan sonar images and visible images, we proposed a gray correction method for side-scan sonar images based on Retinex, which is simple and easy to implement. Compared with the commonly used gray scale correction methods for side-scan sonar images, this method avoids the limitations of the current gray scale correction methods, such as the need to know the side-scan sonar parameters, the need to recalculate or reset the parameters for different side-scan sonar image processing, and the poor image enhancement effect. Compared with the latest image enhancement algorithms based on Retinex, our proposed methods have similar image enhancement indexes, and our method is the fastest. When it is necessary to adjust the brightness of the corrected image, only the magnitude of constant coefficient A in the algorithm needs to be adjusted. Our method can also be used to enhance low illumination color images, and the experimental results show that the algorithm is fast.

Author Contributions: Conceptualization, X.Y. and H.Y.; Data curation, H.Y.; Formal analysis, H.Y.; Funding acquisition, X.Y.; Investigation, X.Y. and C.L.; Methodology, X.Y.; Project administration, P.L.; Resources, C.L. and

Y.J.; Software, H.Y., Y.J. and P.L.; Supervision, C.L.; Validation, H.Y.; Visualization, Y.J.; Writing – original draft, X.Y.; Writing – review & editing, X.Y.

Funding: This work was supported by the National Natural Science Foundation of China (Grant No. 41876100), the National key research and development program of China (Grant No. 2018YFC0310102 and 2017YFC0306002), the State Key Program of National Natural Science Foundation of China (Grant No.61633004), the Development Project of Applied Technology in Harbin (Grant No.2016RAXXJ071) and the Fundamental Research Funds for the Central Universities (Grant No. HEUCFP201707).

Acknowledgments: To verify the adaptability of the parameters of our algorithm, we used side-scan sonar images from previous studies. Thank you to the authors of these papers.

Conflicts of Interest: The authors declare no conflict of interest.

Appendix A

To verify the generality of our method, we used our method (bilateral filtering) to enhance several original side-scan sonar images and fixed A to 140 and a to 15. The results are shown in the following.

Appendix B

To verify that our method is also suitable for correction of side-scan sonar images in other batches, we used our algorithm (bilateral filtering) to enhance them. Because we lacked other batches of side-scan sonar images, we were only able to obtain images by clipping them from previous studies [7–9] and the quality of the side-scan sonar images was poor. However, the effects of the enhanced images verified the adaptivity of our algorithm.

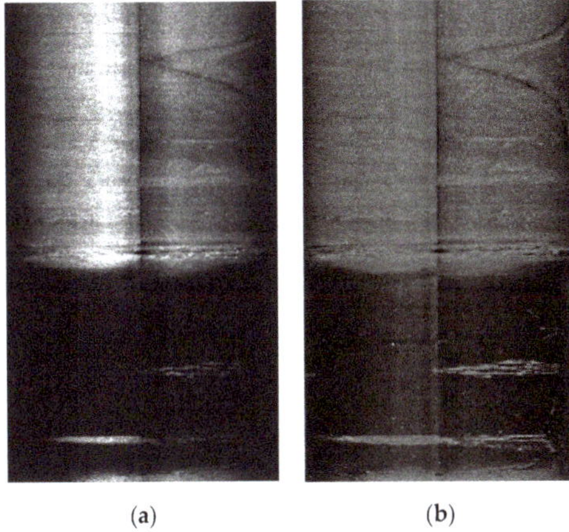

(a) (b)

Figure A1. Side-scan sonar image enhancement from Zhao, J et al. [9]: (**a**) original image with changes in sediments and (**b**) enhanced image (bilateral filtering, $A = 140$, $a = 17$).

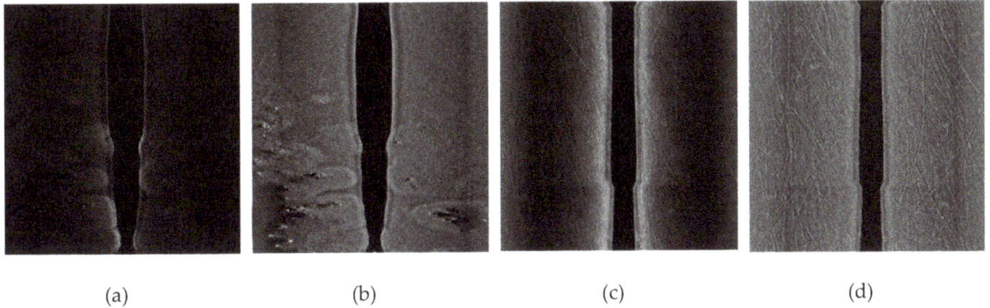

(a) (b) (c) (d)

Figure A2. Side-scan sonar image enhancement from previous studies [7,8]: (**a**) original and (**b**) enhanced image from Capus, C et al. [7]; (**c**) original image and (**d**) enhanced image from Capus, C et al. [8] (bilateral filtering, $A = 140$, $a = 15$).

References

1. Dura, E.; Zhang, Y.; Liao, X.; Dobeck, G.J.; Carin, L. Active learning for detection of mine-like objects in side-scan sonar imagery. *IEEE J. Ocean. Eng.* **2015**, *30*, 360–371. [CrossRef]
2. Reggiannini, M.; Salvetti, O. Seafloor analysis and understanding for underwater archeology. *J. Cult. Herit.* **2017**, *24*, 147–156. [CrossRef]
3. Hovland, M.; Gardner, J.V.; Judd, A.G. The significance of pockmarks to understanding fluid flow processes and geohazards. *Geofluids* **2002**, *2*, 127–136. [CrossRef]
4. Kaeser, A.J.; Litts, T.L.; Tracy, T.W. Using low-cost side-scan sonar for benthic mapping throughout the lower flint river. *River. Res. Appl.* **2013**, *29*, 634–644. [CrossRef]
5. Blondel, P. *The Handbook of Sidescan Sonar*; Springer: Berlin/Heidelberg, Germany, 2009; pp. 62–66. ISBN 978-3-540-42641-7.
6. Chang, Y.C.; Hsu, S.K.; Tsai, C.H. Sidescan sonar image processing: Correcting Brightness Variation and Patching Gaps. *J. Mar. Sci Tech.-Taiw.* **2010**, *18*, 785–789.
7. Capus, C.G.; Banks, A.C.; Coiras, E.; Ruiz, I.T.; Smith, C.J.; Petillot, Y.R. Data correction for visualisation and classification of sidescan SONAR imagery. *IET Radar Sonar Navig.* **2008**, *2*, 155–169. [CrossRef]
8. Capus, C.; Ruiz, I.T.; Petillot, Y. Compensation for Changing Beam Pattern and Residual TVG Effects with Sonar Altitude Variation for Sidescan Mosaicing and Classification. In Proceedings of the 7th European Conference Underwater Acoustics, Delft, The Netherlands, 5–8 July 2004.

9. Zhao, J.; Yan, J.; Zhang, H.M.; Meng, J.X. A new radiometric correction method for side-scan sonar images in consideration of seabed sediment variation. *Remote Sens.* **2017**, *9*, 575. [CrossRef]
10. Ye, X.F.; Li, P.; Zhang, J.G.; Shi, J.; Gou, S.X. A feature-matching method for side-scan sonar images based on nonlinear scale space. *J. Mar. Sci. Technol.* **2016**, *21*, 38–47. [CrossRef]
11. Wang, A.; Zhang, H.; Wang, X. Processing principles of side-scan sonar data for seamless mosaic image. *J. Geomat.* **2017**, *42*, 26–33.
12. Zhao, J.; Shang, X.; Zhang, H. Reconstructing seabed topography from side-scan sonar images with self-constraint. *Remote Sens.* **2018**, *10*, 201. [CrossRef]
13. Johnson, H.P.; Helferty, M. The geological interpretation of side-scan sonar. *Rev. Geophys.* **1990**, *28*. [CrossRef]
14. Shippey, G.; Bolinder, A.; Finndin, R. Shade correction of side-scan sonar imagery by histogram transformation. In Proceedings of the Oceans, Brest, France, 13–16 September 1994; IEEE: New York, NY, USA; pp. 439–443.
15. Li, P. Research on Image Matching Method of the Side-Scan Sonar Image. Doctoral Dissertation, Harbin Engineering University, Harbin, China, 2016.
16. Al-Rawi, M.S.; Galdran, A.; Yuan, X. Intensity Normalization of Sidescan Sonar Imagery. In Proceedings of the Sixth International Conference on Image Processing Theory, Tools and Applications (IPTA), Oulu, Finland, 12–15 December 2016; pp. 12–15.
17. Al-Rawi, M.S.; Galdran, A.; Isasi, A. Cubic Spline Regression Based Enhancement of Side-Scan Sonar Imagery. In Proceedings of the Oceans IEEE, Aberdeen, UK, 19–22 June 2017; pp. 19–22.
18. Burguera, A.; Oliver, G. Intensity Correction of Side-Scan Sonar Images. In Proceedings of the IEEE Emerging Technology & Factory Automation, Barcelona, Spain, 16–19 September 2014; pp. 16–19.
19. Kleeman, L.; Kuc, R. *Springer Handbook of Robotics*; Springer: Berlin/Heidelberg, Germany, 2008; pp. 491–519. ISBN 978-3-319-32552-1.
20. Land, E.H. The retinex theory of color vision. *Sci. Am.* **1997**, *237*, 108–128. [CrossRef]
21. Jobson, D.J.; Rahman, Z.; Woodell, G.A. A multiscale retinex for bridging the gap between color images and the human observation of scenes. *IEEE Trans. Image Process.* **1997**, *6*, 965–976. [CrossRef] [PubMed]
22. Barnard, K.; Funt, B. Analysis and Improvement of Multi-Scale Retinex. In Proceedings of the 5th Color and Imaging Conference Final Program and Proceedings, Scottsdale, AZ, USA, 1–5 January 1997; pp. 221–226.
23. Jobson, D.J.; Woodell G, G.A. Retinex processing for automatic image enhancement. *Electron. Imaging.* **2004**, *13*, 100–110. [CrossRef]
24. Jin, L.; Miao, Z. Research on the Illumination Robust of Target Recognition. In Proceedings of the IEEE International Conference on Signal Processing, Chengdu, China, 6–10 December 2016; pp. 811–814.
25. Guo, X.; Li, Y.; Ling, H. LIME: Low-light image enhancement via illumination map estimation. *IEEE Trans. Image Process.* **2017**, *26*, 982–993. [CrossRef] [PubMed]
26. Wang, S.; Zheng, J.; Hu, H.M. Naturalness preserved enhancement algorithm for non-uniform illumination images. *IEEE Trans. Image Process.* **2013**, *22*, 3538–3548. [CrossRef] [PubMed]
27. Fu, X.; Zeng, D.; Huang, Y.; Zhang, X.; Ding, X. A Weighted Variational Model for Simultaneous Reflectance and Illumination Estimation. In Proceedings of the IEEE Conference on Computer Vision and Pattern Recognition (CVPR), Las Vegas, NV, USA, 27–30 June 2016; pp. 2782–2790.
28. Fu, X.; Zeng, D.; Huang, Y.; Liao, Y.; Ding, X.; Paisley, J. A fusion-based enhancing method for weakly illuminated images. *Signal Process.* **2016**, *129*, 82–96. [CrossRef]

Article

Estimation of Translational Motion Parameters in Terahertz Interferometric Inverse Synthetic Aperture Radar (InISAR) Imaging Based on a Strong Scattering Centers Fusion Technique

Ye Zhang, Qi Yang *, Bin Deng, Yuliang Qin and Hongqiang Wang

College of Electronic Science and Technology, National University of Defense Technology, Changsha 410073, China; zhangye10@nudt.edu.cn (Y.Z.); dengbin@nudt.edu.cn (B.D.); qinyuliang@nudt.edu.cn (Y.Q.); wanghongqiang@nudt.edu.cn (H.W.)

* Correspondence: yangqi08@nudt.edu.cn; Tel.: +86-731-8457-5714

Received: 25 April 2019; Accepted: 20 May 2019; Published: 23 May 2019

Abstract: Translational motion of a target will lead to image misregistration in interferometric inverse synthetic aperture radar (InISAR) imaging. In this paper, a strong scattering centers fusion (SSCF) technique is proposed to estimate translational motion parameters of a maneuvering target. Compared to past InISAR image registration methods, the SSCF technique is advantageous in its high computing efficiency, excellent antinoise performance, high registration precision, and simple system structure. With a one-input three-output terahertz InISAR system, translational motion parameters in both the azimuth and height direction are precisely estimated. Firstly, the motion measurement curves are extracted from the spatial spectrums of mutually independent strong scattering centers, which avoids the unfavorable influences of noise and the "angle scintillation" phenomenon. Then, the translational motion parameters are obtained by fitting the motion measurement curves with phase unwrapping and intensity-weighted fusion processing. Finally, ISAR images are registered precisely by compensating the estimated translational motion parameters, and high-quality InISAR imaging results are achieved. Both simulation and experimental results are used to verify the validity of the proposed method.

Keywords: interferometric inverse synthetic aperture radar (InISAR); image registration; translational motion parameters estimation; strong scattering centers fusion; terahertz radar imaging

1. Introduction

Historically, the application of inverse synthetic aperture radar (ISAR) imaging in target recognition is limited, owing to the fact that the technique only captures the projected two-dimensional (2-D) characteristics of the target. To overcome this drawback, interferometric ISAR (InISAR) that provides three-dimensional (3-D) information of the target has been developed [1–11]. InISAR systems are generally composed of several fixed channels, with one as both a transmitter and receiver and the rest as full-time receivers. The 3-D geometry of the target can be reconstructed based on the phase difference of different ISAR images. Until now, nearly all researches regarding InISAR imaging have been carried out in the microwave band, while limited research about InISAR imaging with terahertz (THz) radars is available in the current literature. Compared to InISAR imaging with microwave radars, THz radars can more easily achieve a higher carrier frequency and wider absolute bandwidth, which provide higher spatial resolution and more detailed information of the target. Furthermore, THz InISAR imaging holds large potential in maneuvering target surveillance and recognition in space and near space, and it is of great significance to study and advance this technique.

With the noticeable progress of THz sources and detectors over the past few decades, it has become possible to achieve imaging and recognition with a THz radar system. For now, we can summarize the imaging radar system in the THz band into three main categories: raster-scanning radar system, mixed-scanning radar system, and an SAR/ISAR system. The raster-scanning radar system focuses the beam to a fixed area using several lenses, and the 3-D imaging result is obtained by recording the data of each scanning area. Representatives of such systems include Jet Propulsion Laboratory [12–17] and Pacific Northwest National Laboratory [18,19]. The imaging frame rate is determined by the number of scanning pixels and oscillation frequency of the scanning mirrors, which makes it time-consuming to obtain a target image. Besides, the system structure is complex and can be easily damaged. Compared to the raster-scanning radar system, the mixed-scanning radar system substitutes one-dimensional (1-D) raster scanning with 1-D electrical scanning or 1-D mechanical scanning, and the imaging speed can be greatly improved. Representatives of mixed-scanning radar systems include Chinese Academy of Sciences [20–22] and China Academy of Engineering Physics [23,24]. The mixed-scanning radar system can achieve 3-D imaging near real time, but some defects still exist (e.g., the imaging field of view is limited, and the target needs to be stationary). The SAR/ISAR system acquires the target image through relative motion between radar and target. The resolution depends on the bandwidth of the sweep signal and the length or relative rotation angle of the synthetic aperture. Representatives include FGAN-FHR [25–27], the US Defense Advanced Research Projects Agency (DARPA) [28], and so on [29–33]. Compared to the former two THz radar systems, the SAR/ISAR system has a relatively simple system structure and low hardware cost, and it has no limitation of target distance. However, the SAR/ISAR system can only achieve 2-D imaging. The THz InSAR/InISAR system has the advantages of the SAR/ISAR system and also has the ability to achieve 3-D imaging. Thus, it is meaningful to establish an InSAR/InISAR system in the THz band.

As we know, the translational motion of a target will lead to image misregistration in InISAR imaging, and an image registration process is necessary to be carried out first. In recent years, some research regarding image registration in InISAR systems have been presented. These image registration methods can be summarized into three categories. The first is based on the correlation coefficient of ISAR images such as the time domain correlation method and the frequency domain searching method [4]. However, in order to guarantee registration precision, the searching step of motion parameters should be controlled within a very limited range in the THz band, which significantly increases the computational complexity. The second is based on the selection of a reference distance, such as the respective reference distance compensation method and the reference distance deviation compensation method [5,6]. In these methods, the reference distance is chosen as the total distance from the transmitter to reference center and the reference center to corresponding receiver. There is no need to estimate the motion parameters, and the method is suitable for both three-antenna and multiantenna configurations. However, in a real InISAR imaging scene, it is highly challenging to acquire the accurate distances from the reference center to the other receivers (except the one as both a transmitter and receiver). The third is based on estimation of the motion parameters of target. Reference [7] presents a procedure based on a multiple antenna-pair InISAR imaging system with nine antennas to estimate angular motion parameters. The angular motion parameters measurement is based on the spatial spectrums of the whole target, and the phase wrapping of motion measurement curves is avoided by using a pair of antennas with a short baseline. However, there are some shortcomings in this method. Firstly, the multiple antenna-pair configuration increases the system complexity and hardware cost, especially in high-frequency applications. Secondly, the spatial spectrums in a fixed range cell usually contain information of several scattering centers, which would introduce the "angle glint" phenomenon to the motion measurement curves. Thirdly, the method does not consider the influence of noise, but the motion parameter measurement precision of the weak scattering centers is sensitive to noise. In addition to the limitations of the above methods, most research just gives the simulation results and are short on experimental validation.

Unlike SAR images, high-frequency ISAR images of moving targets usually consist of several dominating reflectors such as corner reflectors formed by the tail, fuselage, and wings of the aircraft. Taking the computing efficiency, practicability, robustness, and precision into consideration, a strong scattering centers fusion (SSCF) technique is proposed in this paper to estimate translational motion parameters using these dominating reflectors, which overcomes the defects in the aforementioned image registration methods. With a one-input three-output THz InISAR system, translational motion parameters both in the azimuth and height directions are accurately estimated. Strong scattering centers (SSCs) are extracted in the image domain with a rectangular filter operation, which eliminates most of the noise. Motion measurement curves are derived from the spatial spectrums of mutually independent SSCs so that the "angle scintillation" phenomenon can be effectively suppressed. Then, the translational motion parameters are obtained by fitting the motion measurement curves with phase unwrapping and intensity-weighted fusion processing. Finally, image registration is achieved by compensating the estimated motion parameters to the radar echoes, and the InISAR imaging results are obtained with a simple interference operation.

This paper is organized as follows: in Section 2, the signal model is established. The SSCF technique is described in detail in Section 3. In Section 4, the simulations of the point targets model under different signal-to-noise ratio (SNR) and equivalent verification experiments with a multichannel THz radar system are carried out. A discussion is given in Section 5. Conclusions are presented in in Section 6.

2. Signal Model of Interferometric Inverse Synthetic Aperture Radar (InISAR) Imaging

The configuration of the InISAR system is demonstrated in Figure 1. Antenna O acts as both the transmitter and receiver, while antennas A and B operate in the receiving mode only. L_1 and L_2 denote the lengths of the baselines OA and OB, respectively. A target coordinate system xoz is built referencing the right-angle layout of the three antennas. $P(x_p, y_p, z_p)$ is an arbitrary scattering center located on the target whose projections on planes xoy and yoz are P_1 and P_2, respectively. y_0 denotes the initial distance from antenna O to the reference center o, and R_{AP}, R_{BP}, and R_{OP} denote the initial distances from P to three antennas, respectively. Suppose the transmitting linear frequency modulated (LFM) signal from antenna O is

$$s(\hat{t}, t_m) = \mathrm{rect}(\frac{\hat{t}}{T_p}) \exp[j2\pi(f_c t + \gamma \hat{t}^2/2)], \tag{1}$$

where

$$\mathrm{rect}\,(u) = \begin{cases} 1 & |u| \leq 0.5 \\ 0 & |u| > 0.5 \end{cases}, \tag{2}$$

then, T_p is the pulse width, f_c is the carrier frequency, γ is the chirp rate, t_m is the slow time, $\hat{t}$ is the fast time, and $t = t_m + \hat{t}$ is the full time.

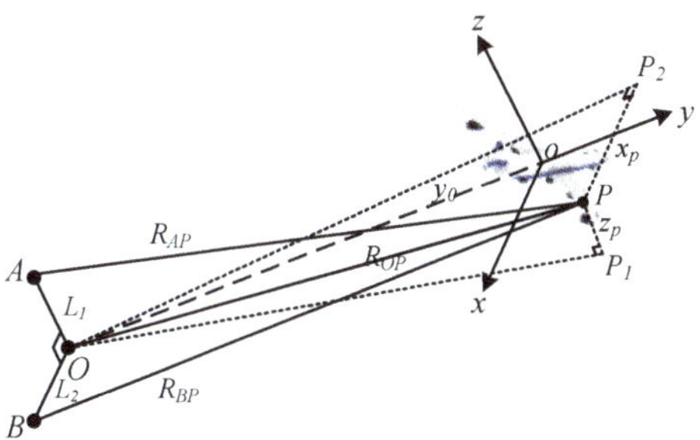

Figure 1. Schematic of the three-receiver interferometric inverse synthetic aperture radar (InISAR) imaging configuration.

Ignoring the signal envelope, the received signal at receiver i (i = O, A, B) from P is

$$s_i(\hat{t}, t_m) = \sigma_P \exp\{j2\pi[f_c(t - \frac{R_{OP}(\hat{t}, t_m) + R_{iP}(\hat{t}, t_m)}{c}) + \frac{\gamma}{2}(\hat{t} - \frac{R_{OP}(\hat{t}, t_m) + R_{iP}(\hat{t}, t_m)}{c})^2]\}, \tag{3}$$

where c is the wave propagation velocity, σ_P is the reflection coefficient of P, and $R_{iP}(\hat{t}, t_m)$ represents the distance from receiver i to P at time t. These distances are:

$$R_{OP}(t) = \sqrt{(x_P + \Delta R_x(t))^2 + (y_P + y_0 + \Delta R_y(t))^2 + (z_P + \Delta R_z(t))^2}, \tag{4}$$

$$R_{AP}(t) \approx R_{OP}(t) + \frac{L_1^2 - 2L_1(x_P + \Delta R_x(t))}{2y_0}, \tag{5}$$

$$R_{BP}(t) \approx R_{OP}(t) + \frac{L_2^2 - 2L_2(z_P + \Delta R_z(t))}{2y_0}, \tag{6}$$

where $\Delta R_x(t)$, $\Delta R_y(t)$, and $\Delta R_z(t)$ represent the displacement of P from time instant 0 to time t in the x, y, and z directions, respectively. The approximations of $R_{AP}(t)$ and $R_{BP}(t)$ are based on the assumption that the imaging scene satisfies the far-field condition.

Based on the 'stop–go' model approximation and de-chirping processing, the received signal after range compression is

$$\begin{aligned} s_i(y, t_m) = \quad & \sigma_P \mathrm{sin\,c}\{\tfrac{2\gamma T_p}{c}[y - \tfrac{(R_{OP}(t_m) + R_{iP}(t_m) - 2R_{Oo}(t_m))}{2}]\} \\ & \times \exp\{-j\tfrac{2\pi}{\lambda}[R_{OP}(t_m) + R_{iP}(t_m) - 2R_{Oo}(t_m)]\} \end{aligned}, \tag{7}$$

where $R_{Oo}(t_m)$ is the reference distance

$$R_{Oo}(t_m) = \sqrt{\Delta R_x{}^2(t_m) + (y_0 + \Delta R_y(t_m))^2 + \Delta R_z{}^2(t_m)}. \tag{8}$$

Within the short time of data acquisition, the translational velocities of the aircraft are assumed as constant (i.e., $\Delta R_x(t_m) = V_x t_m$, $\Delta R_y(t_m) = V_y t_m$, $\Delta R_z(t_m) = V_z t_m$). After migration through range cell (MTRC) correction and azimuth compression, the ISAR images of three channels can be obtained as

$$s_O(y, f_a) = A_P \mathrm{sin\,c}\{\frac{2\gamma T_p}{c}(y - y_P)\}\mathrm{sin\,c}\{T_a[f_a + \frac{2\mathbf{PV}}{\lambda}]\}\exp(j\varphi_0), \tag{9}$$

$$s_A(y, f_a) = A_P \mathrm{sin\,c}\{\frac{2\gamma T_p}{c}(y - y_P)\}\mathrm{sin\,c}\{T_a[f_a + \frac{2\mathbf{PV}}{\lambda y_0} - \frac{L_1 V_x}{\lambda y_0}]\}\exp[j(\varphi_0 + \varphi_1)], \tag{10}$$

$$s_B(y, f_a) = A_P \mathrm{sin\,c}\{\frac{2\gamma T_p}{c}(y - y_P)\}\mathrm{sin\,c}\{T_a[f_a + \frac{2\mathbf{PV}}{\lambda y_0} - \frac{L_2 V_z}{\lambda y_0}]\}\exp[j(\varphi_0 + \varphi_2)], \tag{11}$$

where A_P is the scattering intensity of P in the ISAR images, T_a is the total data acquisition time, $\mathbf{P} = (x_P, y_P, z_P)$ is the coordinate vector, and $\mathbf{V} = (V_x, V_y, V_z)$ is the translational motion vector. φ_0 and φ_i $(i = 1, 2)$ denote the constant phase and interferometric phases, respectively, and they are expressed as

$$\varphi_0 = -\frac{2\pi}{\lambda}(y_P + \frac{x_P^2 + y_P^2 + z_P^2}{y_0}), \tag{12}$$

$$\varphi_1 = \frac{2\pi}{\lambda}\frac{2L_1 x_P - L_1^2}{2y_0}, \tag{13}$$

$$\varphi_2 = \frac{2\pi}{\lambda}\frac{2L_2 z_P - L_2^2}{2y_0}. \tag{14}$$

From Equations (9) to (11), it can be seen that the Doppler shifts of ISAR images are $L_1 V_x T_a / \lambda y_0$ and $L_2 V_z T_a / \lambda y_0$, respectively. Obviously, the Doppler shifts will lead to pixels misregistration between ISAR images. Therefore, image registration must be accomplished before interferometric imaging. It is clear that image registration is essentially the compensation of the translational motion parameters V_x and V_z. Once these motion parameters are achieved, the Doppler shifts can be eliminated by compensating the radar echoes of antenna A and antenna B with them.

3. The Strong Scattering Centers Fusion (SSCF) Technique

In this subsection, the proposed SSCF technique is described in detail. This method can settle problems such as the "angle glint", sensitivity to noise, and phase wrapping in motion parameter estimation.

The details of the proposed SSCF technique are described as follows.

Step 1) After data preprocessing (i.e., range alignment, autofocus [34,35], and MTRC correction), the spatial spectrums of radar echoes are obtained as $s_O(m,k)$, $s_A(m,k)$, and $s_B(m,k)$, $1 \le m \le M$ and $1 \le k \le N$, where M and N denote the number of samples and the number of pulses, respectively. Then, the ISAR images can be obtained as $s_O(m,n)$, $s_A(m,n)$, and $s_B(m,n)$ after azimuth compression, where $1 \le n \le N$.

Step 2) Based on a fixed threshold, strong scattering areas on the object can be extracted from the three ISAR images. Firstly, the strongest scattering center in the first strong scattering area of antenna O can be easily found. Subsequently, the corresponding scattering centers in ISAR images of antenna A and antenna B can be confirmed, since they are distributed in the same range cell. This scattering center is then extracted in the image domain with a rectangular filter, whose length is determined by the main lobe width (3 dB) of the extracted scattering center. The noise is filtered in this step. This process is iterated to search for the localized strongest scattering center in all the strong scattering areas until the strongest scattering center in the last strong scattering area has been extracted.

Step 3) The extracted data of all SSCs in the image domain are rearranged to form new image matrices $s_O(m',n)$, $s_A(m',n)$, and $s_B(m',n)$, where $1 \le m' \le M'$, and M' denotes the number of SSCs. By performing an inverse Fourier transformation on variable n, the new spatial spectrums are obtained as $s_O(m',k)$, $s_A(m',k)$, and $s_B(m',k)$, respectively. Here, each row of the spatial spectrums contains information of only one scattering center. The phase difference curves $s_O^*(m',k)s_A(m',k)$ and $s_O^*(m',k)s_B(m',k)$ can be calculated as

$$\varphi_{xm'}(k) = -\frac{2\pi}{\lambda}\frac{L_1^2 - 2L_1(x_{m'} + V_{xm'}k\Delta t)}{2y_0}, \tag{15}$$

$$\varphi_{zm'}(k) = -\frac{2\pi}{\lambda}\frac{L_2^2 - 2L_2(z_{m'} + V_{zm'}k\Delta t)}{2y_0},$$

where Δt is the pulse repetition interval, and $V_{xm'}$ and $V_{zm'}$ are the estimated translation velocity of the m'th strong scattering center.

Step 4) In the THz band, the wavelength is very short. On the other hand, in order to guarantee the precision of altitude measurement, a relative long baseline is required [7]. Therefore, the values of $\varphi_{xm'}(k)$ and $\varphi_{zm'}(k)$ usually exceed the range $[-\pi,\pi]$, and the phase unwrapping operation is necessary. Here, the one-dimensional path integral method is adapted to achieve phase unwrapping of the motion measurement curves, and the theory is

$$\begin{gathered}\Delta\varphi(k) = \begin{cases} \varphi(k+1) - \varphi(k) & |\varphi(k+1) - \varphi(k)| \le \pi \\ \varphi(k+1) - \varphi(k) + 2\pi & \varphi(k+1) - \varphi(k) < -\pi \\ \varphi(k+1) - \varphi(k) - 2\pi & \varphi(k+1) - \varphi(k) > \pi \end{cases}, \\ \varphi_{\text{new}}(k+1) = \varphi_{\text{new}}(k) + \Delta\varphi(k) \end{gathered} \tag{17}$$

where $\varphi(k)$ and $\varphi_{\text{new}}(k)$ are phases before and after phase unwrapping, respectively.

Step 5) The motion measurement curves along the x axis and z axis are calculated as

$$R_{xm'}(k) = \varphi_{xm'\text{new}}(k) \cdot \frac{\lambda y_0}{2\pi L_1} + \frac{L_1}{2} = x_{m'} + V_{xm'} k \Delta t, \tag{18}$$

$$R_{zm'}(k) = \varphi_{zm'\text{new}}(k) \cdot \frac{\lambda y_0}{2\pi L_2} + \frac{L_2}{2} = z_{m'} + V_{zm'} k \Delta t. \tag{19}$$

The estimated velocities of each SSC can be obtained by fitting these time-dependent curves. Finally, the estimation values of translational motion parameters are acquired by intensity-weighted fusion of all SSCs:

$$V_x = \frac{\sum\limits_{m'=1}^{M} V_{xm'} A_{m'}}{\sum\limits_{m'=1}^{M} A_{m'}}, \; V_z = \frac{\sum\limits_{m'=1}^{M} V_{zm'} A_{m'}}{\sum\limits_{m'=1}^{M} A_{m'}}; \tag{20}$$

where $A_{m'}$ is the mean scattering intensity of the m'th SSC in the ISAR images.

By compensating the radar echoes of antenna A and antenna B with the estimated motion parameters, the coordinates of the target are finally obtained from $s_O^*(y, f_a)\ s_A(y, f_a)$ and $s_O^*(y, f_a) s_B(y, f_a)$,

$$x_P = \frac{\lambda y_0}{2\pi L_1}\varphi_1 + \frac{L_1}{2}, \; z_P = \frac{\lambda y_0}{2\pi L_2}\varphi_2 + \frac{L_2}{2}. \tag{21}$$

The derivation of the value of y_P is omitted here, as it can be directly obtained from the range scale in the ISAR images. To summarize the above analysis, a block scheme of the InISAR imaging procedure based on the SSCF technique is presented in Figure 2.

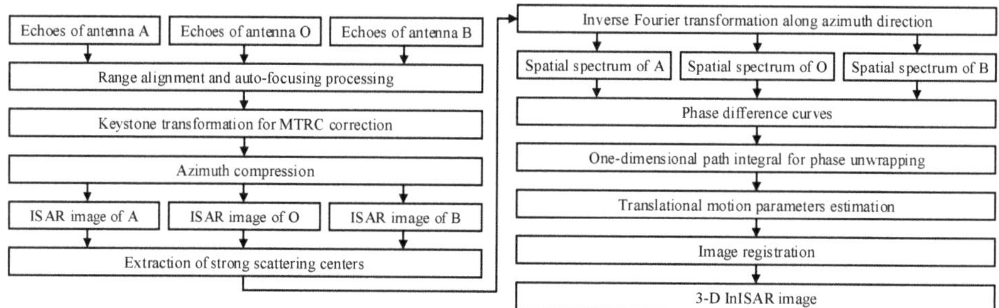

Figure 2. The flowchart of InISAR imaging based on the strong scattering centers fusion (SSCF) technique.

Regarding the InISAR imaging theory, the Doppler shift between ISAR images should be less than one-eighth of the Doppler cell after image registration to guarantee a relatively high InISAR imaging precision. Thus, the velocity estimation error ΔV should satisfy

$$|\Delta V| < \frac{y_0 \lambda}{8 L T_a}. \tag{22}$$

Equation (22) is applied as the criteria to evaluate the performance of the SSCF technique in this paper.

4. Results and Analysis

4.1. The Point Target Simulation Results

In order to verify the effectiveness of the proposed SSCF technique, an InISAR imaging simulation of a moving airplane model is presented. The parameters in the simulation are shown in Table 1. The airplane model contained 64 scattering centers, and the size of the airplane model was 21, 24, and 7.5 m in length, width, and height, respectively. For a vivid visualization, the model was depicted in four different views from four visual angles in Figure 3, with (a), (b), (c), and (d) corresponding to the 3-D view and projections on the *xoy*, *xoz*, and *yoz* planes, respectively. Three SSCs were assigned at the tail, wing, and fuselage, respectively, as highlighted with red circles in Figure 3. The ratio of scattering intensity between these SSCs and the others was 3:1. As a result, several ordinary scattering centers were distributed in the same range cell with the SSCs. The target was assigned to move at a constant velocity along the x direction, which led to image misregistration between ISAR images of antenna O and antenna A.

Table 1. Parameter settings of the radar system and target.

Parameter	Setting Value
Carrier frequency	220 GHz
Bandwidth	5 GHz
Pulse width	50 us
Pulse repetition frequency	1000 Hz
Pulse number	1000
Sampling frequency	20 MHz
Target speed	(300, 0, and 0 m/s)
Antenna O location	(0, 20,000, and 0 m)
Antenna A location	(0.5, 20,000, and 0 m)
Antenna B location	(0, 20,000, and 0.5 m)

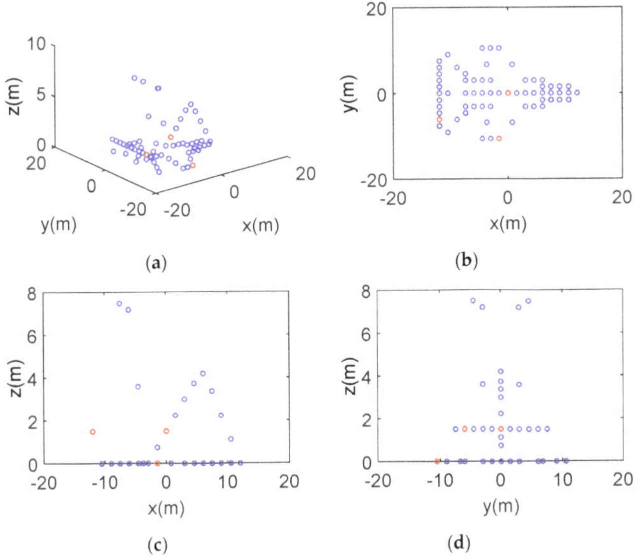

Figure 3. Four different views of the airplane model. (**a**) 3-D view; (**b**) *xoy* view projection; (**c**) *xoz* view projection; and (**d**) *yoz* view projection.

After range alignment, autofocus, and MTRC correction processing, three ISAR images were obtained. Based on the system parameters in Table 1, we calculated the Doppler shift between ISAR images of antenna A and antenna O at 5.5 cells. The range profiles of ISAR images at $y = 0$ of antenna A and antenna O are shown in Figure 4. It was obvious that there was a Doppler shift of six cells on the peak positions. To compare performances between the conventional method in [7] and our method, we added a pair of antennas with 0.04 m length of baseline to the conventional method because it used short baselines to achieve phase unwrapping of motion measurement curves.

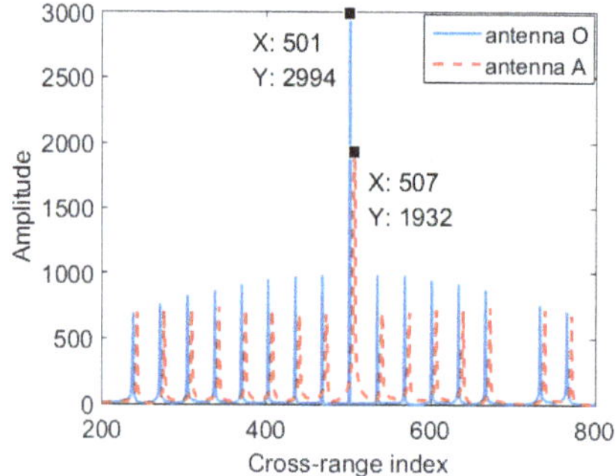

Figure 4. ISAR images at $y = 0$ of antenna O and antenna A before image registration.

Figure 5 shows the spatial spectrums of radar echoes, and Figure 6 shows the phase difference curves of the spatial spectrums, both with (a) corresponding to the conventional method and (b) corresponding to our method. The SNR was 0 dB in this simulation. It can be seen from Figure 5a that the multiple scattering centers in a fixed range cell introduced a serious "angle glint" phenomenon. The "angle glint" phenomenon was eliminated by extracting the SSCs, as shown in Figure 5b. The influence of the multiscattering centers in motion parameter estimations is visualized in Figure 6a. Firstly, the "angle glint" phenomenon and noise introduced serious nonlinearity to the phase difference curves, which destroyed the real phase difference relationship and deteriorated the estimation precision of motion parameters. Secondly, without the short baseline, different scattering centers had different phase wrapping positions, and the one-dimensional path integral method was not applicable in this condition. In contrast, the curves in Figure 6b were linear, and the phase wrapping positions of each curve were constant. Noise was effectively filtered with the filter operation in the image domain, which ensured an excellent linearity of the motion trajectory even under a low SNR.

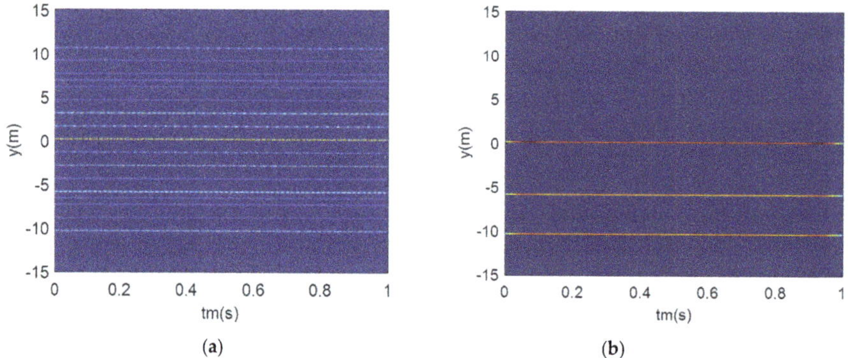

Figure 5. The spatial spectrums of radar echoes. (**a**) Conventional method. (**b**) Our method.

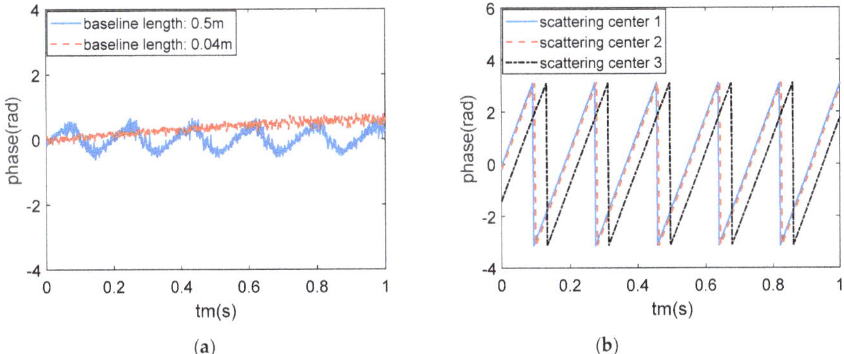

Figure 6. Phase difference curves of spatial spectrums. (**a**) Conventional method. (**b**) Our method.

Based on the one-dimensional path integral method, the phase difference curves of the three SSCs after phase unwrapping were acquired, as shown in Figure 7. All curves were absolutely linear with no fluctuation. Subsequently, translational motion parameters were easily estimated based on polynomial curve fitting and intensity-weighted fusion operation. In this simulation, the estimated velocities of the conventional method and our method were 68.678 and 300.057 m/s, respectively, which meant the conventional method was invalid under this low-SNR simulation environment. The estimated velocity of our method was nearly the same as the real velocity, which verified the effectiveness of the proposed SSCF technique. By compensating the estimated velocity of our method to the echo signals of antenna A, the final imaging results are shown in Figure 8, with (a) corresponding to the range profiles at $y = 0$ of antenna A and antenna O and (b) corresponding to the final InISAR imaging results. It can be seen from Figure 8a that the Doppler shift was eliminated, and image registration was achieved. In Figure 8b, the red circles are the real positions of the target, and the blue dots are the InISAR imaging results. It was clear that the InISAR image and the target model were precisely overlapped.

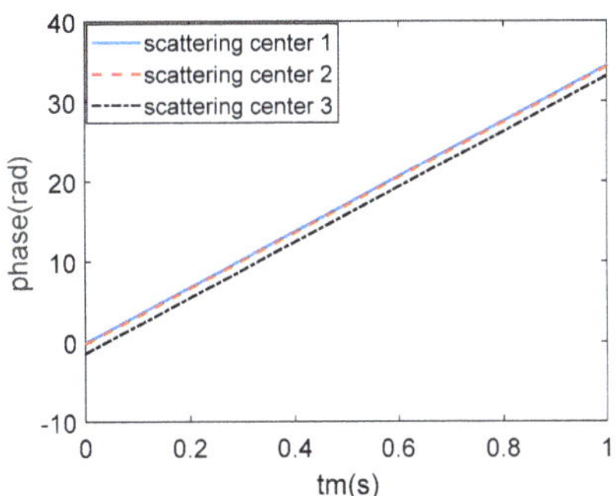

Figure 7. Phase difference curves after phase unwrapping based on the SSCF technique.

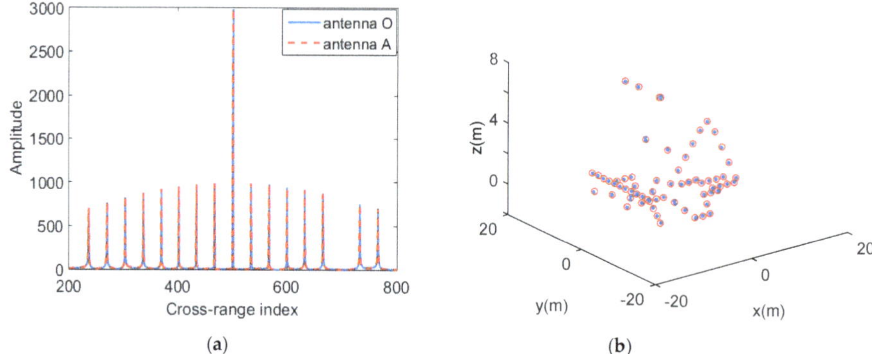

Figure 8. Imaging results after image registration. (**a**) ISAR images at $y = 0$ of antenna O and antenna A. (**b**) InISAR imaging results.

At the end of this simulation, the antinoise abilities of the two methods were evaluated. With our simulation parameters, the maximum allowed mean absolute error (MAE) of the estimated velocity in this simulation was derived as 6.8 m/s on the basis of Equation (22). The MAE of the estimated velocity is defined as

$$\text{MAE} = \sum_{i=1}^{N} |V_i - V|/N, \tag{23}$$

where V_i is the ith estimated velocity, V is the real velocity, and N is the number of simulation times. Here, the MAE of the estimated velocity was provided based on 100 Monte Carlo simulations under different SNR environments, as shown in Figure 9. The MAE of the conventional method was larger than 6.8 m/s when the SNR was under 30 dB, which meant the antinoise performance of the conventional method was very poor. This method was only suitable for extremely high SNR environments. The MAE of our method was under 6.8 m/s when the SNR was larger than −30 dB. The low MEA in estimated velocity proved the excellent antinoise ability of the proposed SSCF technique.

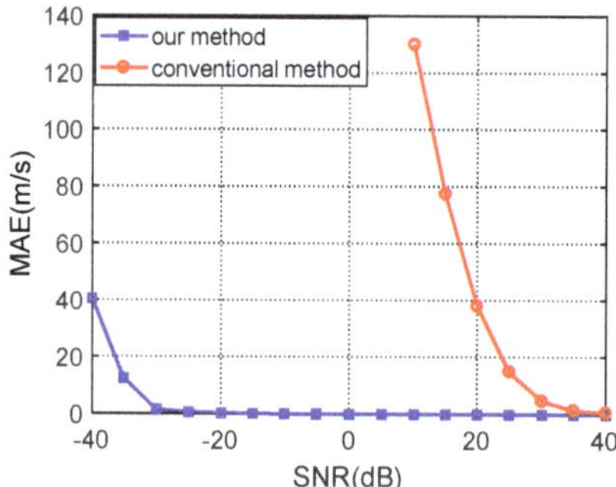

Figure 9. The mean absolute error (MAE) under different signal-to-noise ratio.

In order to show the superiority of the proposed SSCF technique over the past methods, performance comparisons of the frequency domain searching (FDS) method in [4], respective reference distance compensation (RRDC) method in [5] and [6], angular motion parameters estimation (AMPE) method in [7], and proposed SSCF technique in this paper are listed in Table 2. The algorithm reconstruction times (excluding the echo generation process) were obtained on a desktop computer with Intel core i7-7820X 3.60GHz CPU and 32GB RAM using Matlab codes. The reconstruction time of the FDS method depended on the initial value and searching step of motion parameters. In this experiment, the initial value and searching step were set as 200 and 3 m/s, respectively, and a typical single-parameter optimization algorithm was adapted. From Table 2, it was seen that the FDS method was too time-consuming to fulfill the requirement of real time imaging. The RRDC method had the fastest imaging speed, but the radar needed to have ranging function. Besides, it was highly challenging to acquire accurate distances from the reference center to the other receivers (except the one as both a transmitter and receiver). The AMPE method needed a complex system structure, and the antinoise ability was poor. The SSCF technique proposed in this paper was advantageous in its high computing efficiency, excellent antinoise performance, and simple system structure. Therefore, the SSCF technique was more suitable for image registration in InISAR imaging.

Table 2. Comparison of algorithm performances.

Method	Reconstruction Time	Number of Channels	Anti-Noise Ability	Need Ranging Function
Frequency domain Searching (FDS)	23.145 s	3	Strong	No
Respective Reference Distance Compensation (RRDC)	0.699 s	3	Strong	Yes
Angular Motion Parameters Estimation (AMPE)	0.744 s	9	Weak	No
Strong Scattering Centers Fusion (SSCF)	1.657 s	3	Strong	No

4.2. Experimental Results

In order to further verify the effectiveness of the proposed SSCF technique, equivalent verification experiments in the laboratory environment were carried out. The schematic diagram of the THz radar system and the photograph of the front-end setup is shown in Figure 10. The five antennas were arranged in two rows, with three receiving antennas in the upper row and one transmitting antenna and one receiving antenna in the other row. R1 and R2 formed the vertical interferometric baseline, and R2 and R3 formed the horizontal interferometric baseline. Both the vertical and horizontal baseline lengths were 2.1 cm. R4 was ignored in the InISAR experiment. Besides InISAR imaging, this system also had many other potential applications such as InSAR imaging, ViSAR imaging, and micromotion target 3-D imaging. The THz radar system was based on the linear frequency modulated pulse principle. A chirped signal ranging from 217.1 to 222.1 GHz was transmitted, and the echo signals were received by the four receiving antennas.

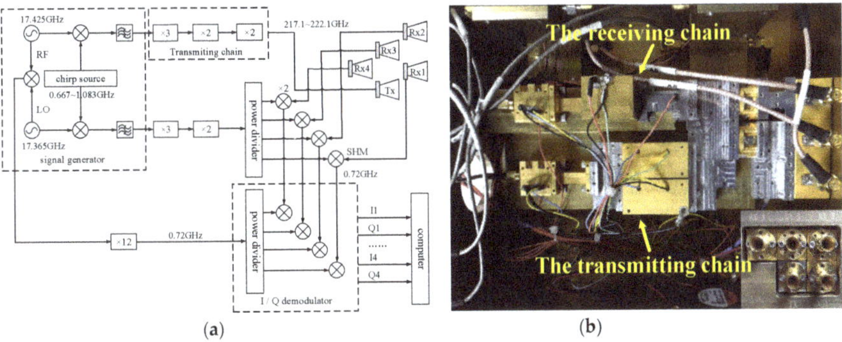

Figure 10. The terahertz radar system. (**a**) Schematic diagram. (**b**) Photograph of the front-end setup.

The experimental configuration is shown in Figure 11. In the experiment, the THz radar was put on a one-dimensional horizontal guide platform, and the velocity of radar was 1 m/s. The vertical distance between the radar and the target was 10 m, and the initial connection from the target to the radar platform was perpendicular to the motion direction. This was a typical SAR imaging scenario, but it was equivalent in that the radar was static, and the target moved along the horizontal direction, which matched the InISAR scenario described in this paper. To decrease the SNR of the radar echoes, the power transmitted from the radar was reduced to 10 mW. The pulse width was 163.8 μs, the pulse repetition frequency was 2500 Hz, the sampling frequency was 12.5 MHz, and the data acquisition time was 1 s. The target was an Airbus A380 model. The length, wingspan, and height of the model were 45, 52, and 17 cm, respectively.

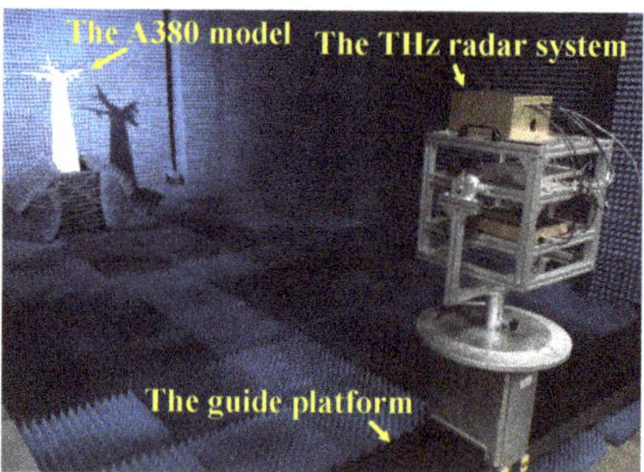

Figure 11. The experimental configuration.

The reflected signals received by R1, R2, and R3 were used to form the InISAR images, and the reference signal was the reflected signal of a corn reflector located at the same position of the airplane model received by R2. Thus, the imaging geometry is the same as the discussed L-shaped three-antenna configuration. During the imaging process, nonlinearity of the signal frequency and the inconsistencies of the amplitudes and phases among channels were compensated together with the reference signal, and a phase gradient autofocus algorithm [36] was adopted to compensate the influence of guide platform vibration. The ISAR images of channels R2 and R3 were interpolated three times and shown in Figure 12. Taking the strong scattering center at the right wing as an example, there are five Doppler cells that deviated along the azimuth direction.

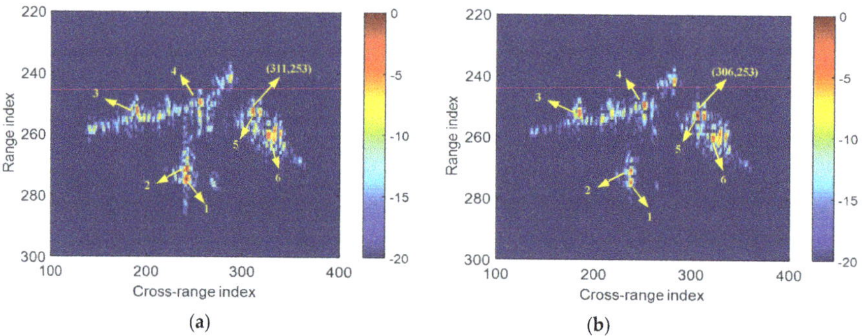

Figure 12. The ISAR images of A380 model. (**a**) Channel R2. (**b**) Channel R3.

Based on a threshold of 3 dB, the ISAR image of six SSCs were extracted, as indicated in Figure 12. The spatial spectrums of these SSCs were then acquired. The phase difference curves of spatial spectrums between channel R2 and R3 are shown in Figure 13, with (a) and (b) corresponding to the conditions before and after phase unwrapping processing, respectively. As illustrated in Figure 13a, all curves were linear, and the phase wrapping position of each curve was constant. With our experimental parameters, the maximum allowed error in velocity estimation for precise image registration was 0.0812 m/s on the basis of Equation (22). Based on the motion measurement curves in Figure 13b, velocity along the horizontal direction was estimated as 0.9737 m/s, which satisfied the estimated precision of parameters for image registration. After compensating the velocity to the echo signal of channel R3, image registration was achieved. Finally, the InISAR imaging results were obtained, as shown in Figure 14, with (a), (b), (c), and (d) corresponding to the 3-D view and projections on the *xoy*, *xoz*, and *yoz* planes, respectively. From the InISAR imaging results, we saw that the key parts in the A380 model such as the engine, wing, and vertical fin could be clearly identified, and the imaging results were clear and close to the real airplane model. These results further verified the effectiveness of the proposed SSCF method. In this chapter, we did not compare the performance between our method and the conventional method because the baseline length in this equivalent verification experiment could not be any shorter. Phase wrapping in the motion measurement curves was inevitable. In this condition, the conventional method was not applicable.

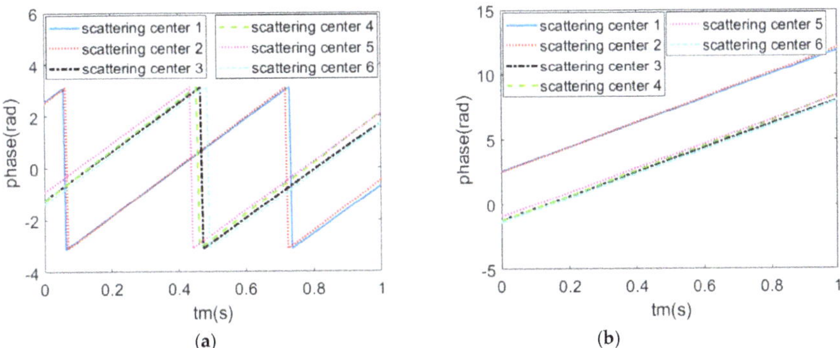

Figure 13. Phase difference curves of the spatial spectrum. (**a**) Before phase unwrapping. (**b**) After phase unwrapping.

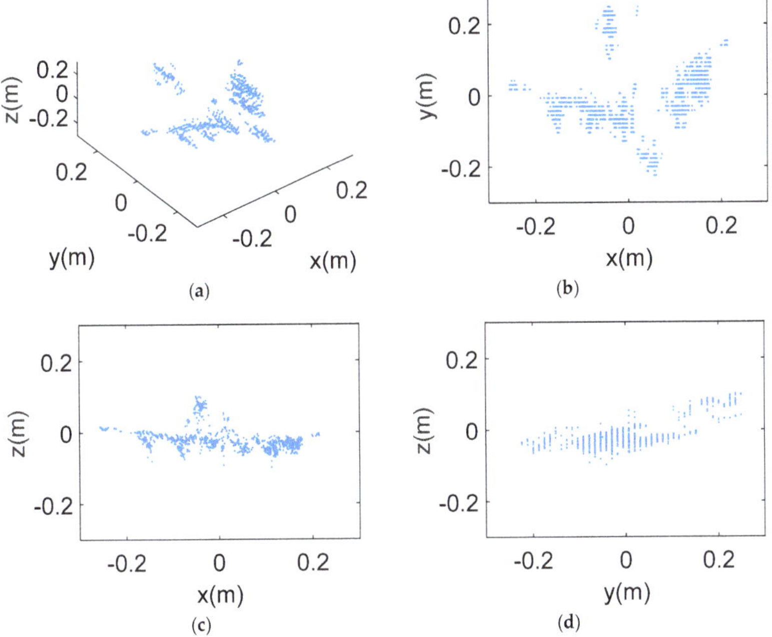

Figure 14. InISAR imaging results of the airplane model. (**a**) 3-D view; (**b**) *xoy* view projection; (**c**) *xoz* view projection; and (**d**) *yoz* view projection.

5. Discussion

Taking the algorithm's efficiency, practicability, robustness, and precision into consideration, the simulation and experimental results have verified that the SSCF technique proposed in this paper is the most suitable method for image registration in THz InISAR imaging. Until now, there has been limited research on InISAR imaging with terahertz radars. This paper takes the lead in putting forward a THz InISAR imaging system, and it has carried out InISAR experiments in the laboratory environment. In the next stage, long-distance experiments that connect a traveling-wave tube amplifier will be carried out to verify the SSCF technique proposed in this paper.

6. Conclusions

In this paper, a translational motion parameter estimation method based on SSCF technique was proposed to achieve image registration in InISAR imaging under a low-SNR environment. The "angle glint" phenomenon and phase wrapping in motion measurement curves were solved, and the interference of noise was also removed with the rectangular filter operation in image domain. Based on this method, the estimation accuracy of translational motion parameters can fulfill image registration requirements when the SNR was larger than −30 dB. Both simulation and experimental results are presented to verify the validity of the proposed method. The method also can be extended to target with a more complicated motion feature. Until now, the experiments were carried out in a laboratory environment, but the work in this paper can provide support to the remote application of THz InISAR imaging systems in the future.

Author Contributions: Y.Z. was responsible for the theoretical work and performed the experiments. Y.Q. and H.W. provided the experimental environment and equipment. B.D. and Q.Y. revised the paper. The authors also wish to extend their sincere thanks to editors and reviewers for their careful reading and fruitful suggestions.

Funding: This work was supported by the National Nature Science Foundation of China (Grant No. 61871386).

Conflicts of Interest: The authors declare no conflict of interest.

References

1. Xu, X.; Narayanan, R.M. Three-dimensional interferometric ISAR imaging for target scattering diagnosis and modeling. *IEEE Trans. Image Process.* **2001**, *10*, 1094–1102.
2. Wang, G.; Xia, X.; Chen, V.C. Three-dimensional ISAR imaging of maneuvering targets using three receivers. *IEEE Trans. Image Process.* **2001**, *10*, 436–447. [CrossRef] [PubMed]
3. Given, J.A.; Schmidt, W.R. Generalized ISAR—Part II: Interferometric techniques for three-dimensional location of scatterers. *IEEE Trans. Image Process.* **2005**, *14*, 1792–1797. [CrossRef]
4. Biao, T.; Na, L.; Yang, L. A Novel image registration method for InISAR imaging system. *SPIE Secur. Def.* **2014**, *9252*. [CrossRef]
5. Liu, C.; He, F.; Gao, X. Novel reference range selection method in InISAR imaging. *J. Syst. Eng. Electron.* **2012**, *23*, 512–521. [CrossRef]
6. Tian, B.; Shi, S.; Liu, Y. Image registration of interferometric inverse synthetic aperture radar imaging system based on joint respective window sampling and modified motion compensation. *J. Appl. Remote Sens.* **2015**, *9*, 095097. [CrossRef]
7. Zhang, Q.; Yeo, T.S.; Du, G. Estimation of three-dimensional motion parameters in interferometric ISAR imaging. *IEEE Trans. Geosci. Remote Sens.* **2004**, *42*, 292–300. [CrossRef]
8. Felguera-Martin, D.; Gonzalez-Partida, J.T.; Almorox-Gonzalez, P. Interferometric inverse synthetic aperture radar experiment using an interferometric linear frequency modulated continuous wave millimeter-wave radar. *IET Radar Sonar Navig.* **2011**, *5*, 39–47. [CrossRef]
9. Ng, W.H.; Tran, H.T.; Martorella, M. Estimation of the total rotational velocity of a non-cooperative target with a high cross-range resolution three-dimensional interferometric inverse synthetic aperture radar system. *IET Radar Sonar Navig.* **2017**, *11*, 1020–1029. [CrossRef]
10. Xu, G.; Xing, M.; Xia, X.G. 3D geometry and motion estimations of maneuvering targets for interferometric ISAR with sparse aperture. *IEEE Trans. Image Process.* **2016**, *25*, 2005–2020. [CrossRef]

11. Martorella, M.; Stagliano, D.; Salvetti, F. 3D interferometric ISAR imaging of non-cooperative targets. *IEEE Trans. Aerosp. Electron. Syst.* **2014**, *50*, 3102–3114. [CrossRef]
12. Dengler, R.J.; Cooper, K.B.; Chattopadhyay, G. 600 GHz Imaging Radar with 2 cm Range Resolution. In Proceedings of the IEEE/MTT-S International Microwave Symposium, Honolulu, HI, USA, 3–8 June 2007; pp. 1371–1374.
13. Cooper, K.B.; Dengler, R.J.; Chattopadhyay, G. A high-resolution imaging radar at 580 GHz. *IEEE Microw. Wirel. Compon. Lett.* **2008**, *18*, 64–66. [CrossRef]
14. Cooper, K.B.; Dengler, R.J.; Llombart, N. Penetrating 3-D imaging at 4 and 25 meter range using a submillimeter-wave radar. *IEEE Trans. Microw. Theory Tech.* **2008**, *56*, 2771–2778. [CrossRef]
15. Llombart, N.; Cooper, K.B.; Dengler, R.J. Time-delay multiplexing of two beams in a terahertz imaging radar. *IEEE Trans. Microw. Theory Tech.* **2010**, *58*, 1999–2007. [CrossRef]
16. Cooper, K.B.; Dengler, R.J.; Llombart, N. THz imaging radar for standoff personnel screening. *IEEE Trans. Terahertz Sci. Technol.* **2011**, *1*, 169–182. [CrossRef]
17. Blazquez, B.; Cooper, K.B.; Llombart, N. Time-delay multiplexing with linear arrays of THz radar transceivers. *IEEE Trans. Terahertz Sci. Technol.* **2014**, *4*, 232–239. [CrossRef]
18. Sheen, D.M.; Mcmakin, D.L.; Thomas, E.H. Active Millimeter-Wave Standoff and Portal Imaging Techniques for Personnel Screening. In Proceedings of the IEEE Conference on Technologies for Homeland Security, Waltham, MA, USA, 11–12 May 2009.
19. Sheen, D.M.; Hall, T.E.; Severtsen, R.H. Active wideband 350 GHz imaging system for concealed-weapon detection. *Proc. SPIE* **2009**, *7309*. [CrossRef]
20. Gao, X.; Li, C.; Gu, S. Design, Analysis and Measurement of a Millimeter Wave Antenna Suitable for Stand off Imaging at Checkpoints. *J. Infrared Millim. Terahertz Waves* **2011**, *32*, 1314–1327. [CrossRef]
21. Gu, S.; Li, C.; Gao, X. Terahertz aperture synthesized imaging with fan-beam scanning for personnel screening. *IEEE Trans. Microw. Theory Tech.* **2012**, *60*, 3877–3885. [CrossRef]
22. Sun, Z.; Li, C.; Gu, S. Fast Three-Dimensional Image Reconstruction of Targets Under the Illumination of Terahertz Gaussian Beams with Enhanced Phase-Shift Migration to Improve Computation Efficiency. *IEEE Trans. Terahertz Sci. Technol.* **2014**, *4*, 479–489. [CrossRef]
23. Cheng, B.B.; Cui, Z.M.; Lu, B. 340 GHz 3-D imaging radar with 4Tx-16Rx MIMO array. *IEEE Trans. Terahertz Sci. Technol.* **2018**, *5*, 509–519. [CrossRef]
24. Gao, J.; Cui, Z.M.; Cheng, B.B. Fast Three-Dimensional Image Reconstruction of a Standoff Screening System in the Terahertz Regime. *IEEE Trans. Terahertz Sci. Technol.* **2018**, *4*, 38–51. [CrossRef]
25. Essen, H.; Wahlen, A.; Sommer, R. High-Bandwidth 220 GHz Experimental Radar. *Electron. Lett.* **2007**, *43*, 1114–1116. [CrossRef]
26. Stanko, S.; Caris, M.; Wahlen, A. Millimeter Resolution with Radar at Lower Terahertz. In Proceedings of the IEEE International Radar Symposium, Dresden, Germany, 19–21 June 2013; pp. 235–238.
27. Caris, M.; Stanko, S.; Palm, S. 300 GHz radar for high resolution SAR and ISAR applications. In Proceedings of the 16th International Radar Symposium, Dresden, Germany, 24–26 June 2015; pp. 577–580.
28. Kim, S.; Fan, R.; Dominski, F. A 235 GHz Radar for Airborne Applications. In Proceedings of the 2018 IEEE Radar Conference, Oklahoma City, USA, 23–27 April 2018; pp. 1549–1554.
29. Cheng, B.B.; Jiang, G.; Wang, C. Real-time imaging with a 140 GHz inverse synthetic aperture radar. *IEEE Trans. Terahertz Sci. Technol.* **2013**, *3*, 594–605. [CrossRef]
30. Ding, J.; Kahl, M.; Loffeld, O. THz 3-D image formation using SAR techniques: Simulation, processing and experimental results. *IEEE Trans. Terahertz Sci. Technol.* **2013**, *3*, 606–616. [CrossRef]
31. Danylov, A.A.; Goyette, T.M.; Waldman, J. Terahertz Inverse Synthetic Aperture Radar (ISAR) Imaging with a Quantum Cascade Laser Transmitter. *Opt. Express* **2010**, *18*, 16264–16272. [CrossRef]
32. Zhang, B.; Pi, Y.; Li, J. Terahertz Imaging Radar with Inverse Aperture Synthesis Techniques: System Structure, Signal Processing, and Experiment Results. *IEEE Sens. J.* **2015**, *15*, 290–299. [CrossRef]
33. Yang, Q.; Qin, Y.L.; Zhang, K. Experimental research on vehicle-borne SAR imaging with THz radar. *Microw. Opt. Technol. Lett.* **2017**, *59*, 2048–2052. [CrossRef]
34. Ye, W.; Yeo, T.S.; Bao, Z. Weighted least square estimation of phase errors for SAR/ISAR autofocus. *IEEE Trans. Geosci. Remote Sens.* **1999**, *37*, 2487–2494. [CrossRef]

35. Cao, P.; Xing, M.; Sun, G. Minimum Entropy via Subspace for ISAR Autofocus. *IEEE Geosci. Remote Sens. Lett.* **2010**, *7*, 205–209. [CrossRef]
36. Wahl, D.E.; Eichel, P.H.; Ghiglia, D.C. Phase gradient autofocus—A robust tool for high resolution SAR phase correction. *IEEE Trans. Aerosp. Electron. Syst.* **1994**, *30*, 827–835. [CrossRef]

Article

The Empirical Application of Automotive 3D Radar Sensor for Target Detection for an Autonomous Surface Vehicle's Navigation

Andrzej Stateczny [1,*], Witold Kazimierski [2], Daria Gronska-Sledz [3] and Weronika Motyl [3]

1. Department of Geodesy, Gdansk University of Technology, 80-233 Gdansk, Poland
2. Faculty of Navigation, Institute of Geoinformatics, Maritime University of Szczecin, 70-500 Szczecin, Poland; w.kazimierski@am.szczecin.pl
3. Marine Technology Ltd., 71-248 Szczecin, Poland; d.gronska@marinetechnology.pl (D.G.-S.); w.motyl@marinetechnology.pl (W.M.)

* Correspondence: andrzej.stateczny@pg.edu.pl; Tel.: +48-609-568-961

Received: 17 April 2019; Accepted: 10 May 2019; Published: 14 May 2019

Abstract: Avoiding collisions with other objects is one of the most basic safety tasks undertaken in the operation of floating vehicles. Addressing this challenge is essential, especially during unmanned vehicle navigation processes in autonomous missions. This paper provides an empirical analysis of the surface target detection possibilities in a water environment, which can be used for the future development of tracking and anti-collision systems for autonomous surface vehicles (ASV). The research focuses on identifying the detection ranges and the field of view for various surface targets. Typical objects that could be met in the water environment were analyzed, including a boat and floating objects. This study describes the challenges of implementing automotive radar sensors for anti-collision tasks in a water environment from the perspective of target detection with the application for small ASV performing tasks on the lake.

Keywords: autonomous surface vehicles; anti-collision; automotive radar; target detection

---

1. Introduction

Unmanned vehicle technology and surface robots have been rapidly developed over the past few years. These systems supersede previously used methods for exploring the underwater parts of the Earth. Trends in the development of unmanned systems point clearly towards the future execution of underwater tasks, including hydrographical surveys, using the direct nearness to the bottom by autonomous surface vehicles (ASVs). The use of ASVs can supplement or replace many hours of measurements conducted by teams of hydrographers, especially in remote areas. Nowadays ASVs are used in many scientific and commercial implementations. They can be met for example in the army for reconnaissance and combat purposes, they serve as research units providing information on various aspects of the aquatic environment, and as carriers of measuring equipment for the inventory of watercourses and reservoirs.

The main tasks and challenges for ASVs depend on the kind of mission performed. However, some of them are common for all approaches and can be treated as a basis for specialized tasks. Among these for sure is the navigation itself and mission control with the use of telemetry, as the most basic priority for ship navigation is its safety. The navigation process can be however understood variably in different applications. In some approaches and applications, like simultaneous localization and mapping (SLAM), the term navigation also means getting information about the surrounding area. Various sensors for this purpose can be employed like 3D laser scanners [1]. These aspects, generally referred to as navigation, can also be found for example in [2,3], where lidar is also used for navigation purposes. Some authors also include the path planning process into navigation itself, while the others threat dynamic path planning as more collision avoidance tasks. A fine survey on this can be found for example in [4,5].

One of the most important safety tasks during the operation of USV is avoiding collisions. It might be treated more as a situation awareness task and not navigation itself. Nevertheless the process of automating collision avoidance is always a key issue for unmanned vehicles, as it directly influences its safety and ability to perform a mission. Maneuvering for anti-collision purposes consists of several steps, including target detection, movement vector estimation, identifying the correct collision avoidance maneuver based on navigation obstacles and other moving ships, and finally implementation. The first step is always to get information about the surrounding environment, which is done by on-board sensors, processing the information for the anti-collision module. The most obvious sensors for this case of robotic application are range finders or in more advanced applications laser scanners [6]. The use of other sensors require implementation of advanced fusion algorithms, like in [7], where it is proposed to enhance the laser system with cameras. The information gathered from sensors is then processed by the anti-collision system to find the most suitable track. A fine survey on this, together with the review of sensors used for anti-collision in USV can be found for example in [8]. There are also some examples in literature for using radar sensor as a core for anti-collision, mostly in maritime ASV. Such approach can be found for example in [9] or [10] in which typical pulse X-band radar is used or in [11] in which frequency-modulated continuous-wave radar (FMCW) X-band radar is proposed (it might also be an option in [9]). These approaches are suitable for maritime application, however typical marine sensors may be too big for the smaller USV in inland waters. The new contribution of this research is to indicate a new approach, which is the implementation of autonomous radar for water vehicles.

In this paper, we propose a new idea, to implement an automotive 3D radar sensor in the autonomous navigation system of an ASV. The proposed approach is a combination of the traditional approach met in the waters, namely the use of radar sensors, and the approach known from roads in which radars are used for car collision systems. This sensor works with a fixed antenna, unlike the traditional rotated radar antennas. The first step in target tracking by radar is target detection, especially small targets in close range observation, which is essential in restricted waters.

In this study, experiments were conducted in real inland waterway conditions with an automotive 3D radar sensor mounted on an ASV owned by Marine Technology Ltd., named HydroDron. The goal was to use the collected data to address the autonomous collision avoidance problem for future intelligent ASV systems. This will be the basis for the implementation of such an innovative system for real-time ASV missions.

It should be pointed out here, that the sensor and system proposed in this study is suitable for small ASVs performing their duties in inland water or in restricted harbor areas. The detection range of this sensor is too small to be useful for marine vessels and therefore marine applications are beyond the scope of this paper. The use cases covered by this research assume that the ASV is performing her autonomic mission (likely hydrographic, but can be any other), navigating in a lake, river or near-coast waters. Hydrographic surveys are often performed in the areas near recreational or fishing sites. The goal of the research was to present detection possibilities for typical objects that could also be met in these areas, which includes not only boats, but also other floating objects.

The paper is organized as follows—in Section 2 the idea and theory of the radar used is presented; Section 3 gives the details of the anti-collision system concept together with the review of related works and papers; Section 4 provides a description of the research; and Section 5 includes the conclusions.

2. Automotive Radar Sensors

Automotive radar is used to detect objects in the vicinity of a vehicle. The sensor consists of a transmitter and receiver. The transmitter emits radio waves that return to the receiver after bouncing from the target. By controlling the direction in which radio waves are sent and received, it is possible to determine the distance, speed, and direction of the objects.

There are two basic methods for measuring distances using radar. The first is known as the direct propagation method, which measures the delay associated with receiving the reflected signal. The delay is correlated with the distance of the reflecting object depending on the speed of light and period and the transmission and reception of waves. The second method is known as the indirect propagation method. In the case of indirect propagation, a modulated frequency is sent and received, and the frequency difference can be used to directly determine the distance and relative velocity of the object. This requires controllable antennas that can be automatically routed or receive signals simultaneously from several different directions.

2.1. FMCW Radars

There are two types of automotive radars, pulse radar or radar with continuous wave. The latter is termed the frequency-modulated continuous-wave radar (FMCW). The pulse radar sends short pulses and determines the distance by measuring the delay time between the transmitted and feedback signal [12]. The FMCW radar continuously sends a linearly modulated signal and determines the distance based on the difference in the transmitted and received frequencies, as shown in Figure 1.

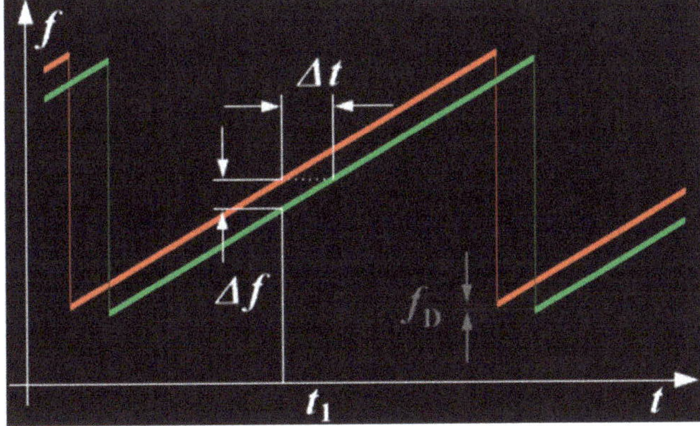

Figure 1. Ranging with an FMCW system [13].

Measuring very short time periods in electronics is difficult, which means that building a good resolution pulse radar is very expensive. However, the resulting resolution is relatively precise, e.g., the FMCW radar can easily have a resolution of 0.5 m [14].

Impulse radars are blind at short distances—for example, the 50–100 m in front of the radar is usually a blind spot. FMCW radars do not have this problem. However, for long-range targets, the pulse radar is better due to the narrower bandwidth and less noise.

In both methods, if the target is moving, the motion creates a Doppler shift in the frequency of the transmitted radar waves. This is an additional advantage for the impulse radar because it can also measure the relative target speed. In the FMCW radar, this is a problem because the distance is measured by measuring the frequency difference between the transmitted and received radar waves, and each additional frequency offset caused by the Doppler effect of the moving target "shifts" the measured distance of the object. To remedy this problem, FMCW radar systems use several different modulation schemes, including modulation with increasing frequency and frequency reduction. If these offsets are not alleviated by algorithms or have very fast frequency changes, this effect may cause the appearance of ghost targets on the FMCW radar [15].

Automotive radars are divided into three categories—long-range radar (LRR), medium range radar (MRR), and short-range radar (SRR). LRR is used to measure the distance and speed of other vehicles, MRR is used in the wider field of view, and SRR is used to detect objects near the vehicle. Two main frequency bands are used in the automotive radar systems—24 GHz and 77 GHz [16,17].

SRR for vehicles uses 24 GHz frequency because the band can detect objects at short and medium distances. A 24 GHz radar is also used to detect an object that can be obstructed or is located very close. Radar systems with the same repeatability can also be used to detect dead spots, which directly involves avoiding collisions. SRR sensors are not used to measure the angle of the detected objects and have a very wide side coverage. Usually, they are operated in pulse mode or in continuous wave mode. Small range radars require a controllable antenna with a large scan angle, creating a wide field of view [18].

While difficult to implement, LRR uses the higher permitted transmission power (77 GHz) to obtain better performance. It is easier to develop 24 GHz bands, but more difficult to integrate such radars systems into the vehicle due to their larger size. In addition, these sensors work with the same performance as the 77 GHz radars but with three times larger antennas. Therefore, the 77 GHz radar is smaller and, in contrast to the 24 GHz radar, is easier to integrate with a vehicle at a lesser cost. An additional and undeniable advantage of the wider 77 GHz band is that it provides drivers with a better resolution of objects by providing greater accuracy. The detection and reaction to the presence of both large and small objects are possible due to the clever signal processing. In the case of LRR, a higher resolution is also provided for a more limited scanning range, which requires a larger number of directional antennas [16,17].

In the 77–81 GHz range, bandwidths up to 4 GHz are available, while the bandwidth available in the 24 GHz band is 200 MHz. The difference between the frequency of the signal emitted by the transmitter and the frequency of the received reflected signal is linearly dependent on the distance from the transmitter to the object. The accuracy of measuring this distance and resolution are important. The resolution is understood as the minimum distance between the objects so that they can be distinguished as different. The transition from 24 GHz to 77 GHz results in a 20 times better performance in terms of resolution and accuracy. The resolution of the 77 GHz radar range can reach 4 cm. For comparison, the 24 GHz radar achieves a resolution of 75 cm. Therefore, the advantage of the 77 GHz system is that it can detect objects that are at a short distance away from each other. Finally, the wavelength of 77 GHz signals is a third the frequency of the 24 GHz system, which enables significantly smaller modules in the spatially limited areas of the vehicle. The relative antenna sizes are shown in Figure 2.

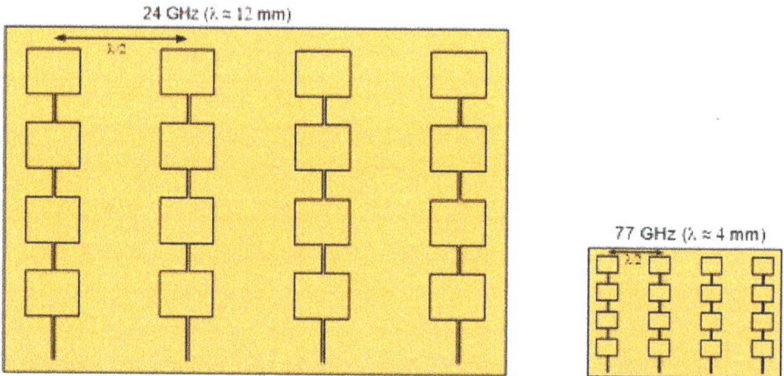

Figure 2. Antenna sizes for 24 GHz (left) and 77 GHz (right) radar systems [19].

Figure 3 shows the range and width of coverage of SRR, MRR, and LRR. LRR can detect objects in a wide area and can cover a range from 10 to 150 m at a beam width of 10°. MRR can cover a range up to 50 m with a beam width of 30°. In contrast, SRR can be used to track objects within a distance of 20 m from the vehicle with a beam width of 60° [20–22].

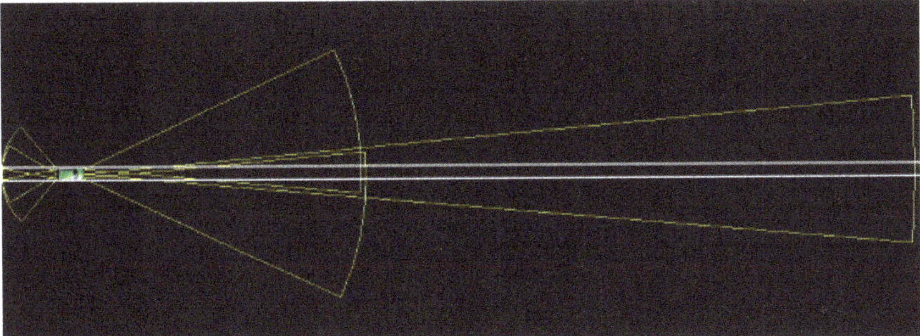

Figure 3. The range and width of coverage of short-range radar, medium-range radar and long-range radar [19].

2.2. Radar Used in this Study

The specifications of the radar used in this study are presented in Table 1. It is a 3D UMRR 42HD automotive radar with a 24 GHz microwave sensor. The type 42 antenna has a wide field of view. The sensor is a 24 GHz 3D/UHD radar for motion management and is able to operate under adverse conditions, measuring in parallel parameters such as angle, radial speed, range, and reflectivity. It is usually used as a standalone radar for detecting approaching and receding motion. More details on the sensor used can be found in [23].

Table 1. Specifications of the UMRR 0C Type 42 anti-collision radar.

Parameter	Characteristics
Sensor Performance	
Maximum range on passenger car	250 m (@20 dBm)/170 m (@12.7 dBm)
Maximum range on truck	340 m (@20 dBm)/280 m (@12.7 dBm)
Instrumented range	340 m
Minimum range	1.5 m
Range accuracy	Typically < ±2.5% or < ±0.25 m, whichever is greater
Radial speed interval	−88.8 to +88.8 m/s
Minimum absolute radial speed	0.1 m/s
Speed accuracy	Typically < ±0.28 m/s or ±1%, whichever is greater
Angle interval (total field of view)	−8° to +8° (elevation); −50 to +50 (azimuth)
Angle accuracy (horizontal)	<1°
Update time	57 ms
Environmental	
Ambient temperature	−40 to +74 °C
Shock	100 *G*
Vibration	14 *G*
IP	67
Pressure/transport altitude	0–10,000 m
Mechanical	
Weight	1290 g
Dimensions	21.3 cm × 15.5 cm × 4.0 cm
General	
Power supply	13–32 V DC
Frequency band	24.0–24.25 GHz
Bandwidth	<250 MHz
Maximum transmit power (EIRP)	<12.7 (<20 possible) dBm
Interfaces	CAN V2.0b (passive) RS485 full duplex 10/100 Ethernet

3. Sensors for ASV Autonomous Anti-collision

As it was said, the anti-collision systems in ASVs are part of the wider concept which aims to provide tools for the safe and reliable navigation of vessels. One of the key elements of such a systems are the sensors, which provide data about the environment for further processing.

3.1. Situation Awarness Systems for ASV

Previous research has addressed the various aspects of navigation of unmanned vehicles. Video data and LiDAR fusion are described in [24]. In [25], an algorithm using LiDAR and a camera for detecting and tracking surface obstacles using the Kalman filter is presented. The legal aspects of ASV navigation, including anti-collision, are described in [26]. An approach using artificial neural networks (ANN) to solve the ASV anti-collision problem is presented in [27–29], where ANN was used to control the autonomous robot. The 3D mobile (3D LiDAR) and GNSS applied to autonomous car navigation was presented in [30].

One of a few attempts to use both radar and LiDAR in the navigation of mobile robots was described in [31–33], which also highlighted new development directions for land mapping based on radar and LiDAR. An approach using radar and LiDAR fusion to detect obstacles was taken in [34], an attempt to replace the radar with LiDAR was shown in [35] and the aspects of obstacle sensing by synthetic aperture radar interferometry was presented in [36]. An interesting approach of the anti-collision system for ASV is described in [37] in which the gathering of situational awareness relies on GPS and AIS.

Target detection in close range observation is very important to provide the next step in ASV navigation, which can be achieved by developing an advanced fast filter to track targets at a close range using automotive 3D radar. Neural solutions for radar target tracking by maritime navigation radar have been previously described by the authors of this study [38,39].

3.2. Anticollision Based on Radar Systems

As it was mentioned in the introduction, there are also several examples in literature of using radar target detection and tracking for the anti-collision of ASV. These examples can be found for example in [9–11], but also in [40] in which the anti-collision system based on the sensors traditionally used on maritime ships is presented.

A radar sensor is a commonly used device for anti-collision at sea. Two kinds of solutions are used for the marine environment—X-band radars and S-band radars. Both of them have their advantages and disadvantages, however from the ASV point of view none of them are suitable. The reason for this is that they both require relatively large antennas to achieve reasonable resolution. Such antennas (of a few meters wide) cannot be mounted on the ASV, which are usually floating platforms of a few meters in length. On the other hand, the advantages of the radar technology and its usefulness for anti-collision purposes are hard to be overestimated. Radar waves are relatively resistant to environmental clutters and thus can be used in fog or even rainy conditions in which cameras and lidars are useless. Taking all this into account, as well as experience in radar data processing, we were looking for the possibilities of providing radar technology for anti-collision purposes in ASVs. Thus, an idea arose to use radars used in automotive applications for tracking and anti-collision in HydroDron. The main advantages of this proposed, novel solution are:

- Small antenna size (in comparison to marine and inland radars)
- Good detection ranges (up to 300 m for big targets) in comparison to rangefinders
- Better detection possibilities (in theory—to be checked empirically) in comparison to rangefinders
- Wider antenna angle in comparison to rangefinders.

The innovative approach is however as always burdened with some risk. The important questions are—how will the 77 GHz radar deal with a water environment, how will surface targets be detected in this type of radar, what will the detection ranges be in this implementation, and what particular processing techniques will be useful for this radar in this particular implementation. The answers for these questions have to be found and for this reason suitable research is needed. In this study, the goal was to empirically verify the detection possibilities using automotive 3D radar in the water environment, as most previous research using these types of radar systems have been conducted in onshore conditions. The radar observations of various targets on the water were collected using a radar mounted on an ASV.

4. Methodology and Research

The research presented in this paper aimed at providing empirical verification of the automotive anti-collision radar for target detection in a water environment. For fulfilling this aim a set of research was designed with the use of empirical measurements and statistical data evaluation. The radar device was mounted on an ASV and it was used in real time in pre-planned scenarios (stationary and including movement of the platform). The data was recorded and in a later processing stage, statistically analyzed. This section provides a description of the research concept, scenarios, research equipment, and statistical data evaluation, while the results are given in the next section.

4.1. Research Concept and Scenarios

To collect robust measurements, useful for evaluation, the study first identified the detection possibilities and range for the objects that were typically in inland waters and then determined the empirical field of detection, which was verified with a declarative beam pattern.

The data was collected in two scenarios, as stationary research and on a moving platform. In both scenarios, the radar was mounted on the HydroDron ASV, which is described in more detail in the next section. The research was conducted on Klodno Lake in northern Poland. In the first part, the HydroDron was moored at the end of the wooden jetty and the targets were moved to provide observational data for various targets and to characterize their detection parameters (see Figure 4).

Figure 4. Configuration for stationary research. Targets were moved along the "moving lane" in front of the stationary HydroDron.

The targets used in the first scenario are presented in Figure 5, including an airtoy (dragon), lifebuoy, small boat fender, swans, and a radar reflector. These targets were selected as typical objects that could act as obstacles on the water surface. The initial research undertaken for the moving ASV has shown that the detection ranges were small for most of these targets. Therefore, better characterizing their detection in the stationary scenario was deemed important. Additionally, detection of the floating radar reflector was tested to determine if the shape of the target had an influence on detection. It has to be pointed out that even such small objects can damage or impair valuable devices carried on board the ASV or the vehicle itself.

The second scenario was performed with a moving ASV and one target, a moving inflatable boat (Figure 6), as an example of a typical collision target on inland waters.

The goal of the second scenario was to analyze and characterize the detection while the target was moving in a typical way. The scenario was divided into three stages, corresponding to IMO Collision Regulations—head-on, crossing, and overtaking. During crossing and overtaking, different ranges were tested to obtain complex information regarding the angle of view. In general, more than 40,000 single radar measurements were collected and analyzed, representing 5 head-on, 11 crossing, and 10 overtaking situations.

Figure 5. Objects used in the first scenario as targets, (**a**) airtoy, (**b**) lifebuoy, (**c**) fender, (**d**) swans, (**e**) radar reflector.

Figure 6. Inflatable boat used in the second scenario as the target.

4.2. Research Equipment and Configuration

In this study, the autonomous anti-collision system for the HydroDron was an automotive radar sensor, a type 42 UMRR automotive 3DHD radar with a 24 GHz microwave sensor. The radar sensor was mounted on the unmanned surface vehicle HydroDron.

The HydroDron is an autonomous catamaran, 4 m long and 2 m wide, which is being developed as a prototype intelligent autonomous multipurpose surface vehicle dedicated to hydrographic measurements. The HydroDron can perform tasks in water areas that are inaccessible or difficult to access by larger vessels. The popularity of this type of vehicle is continually increasing because of its potential to install more sensors and perform surveys in the absence of an onboard operator. The weight and maneuverability of the vehicle have been reduced to meet the objectives of its deployment. The platform has a wide range of measuring equipment—an integrated bathymetric and sonar system, an external inertial navigation system, a sound velocity profiler and sound velocity sensor, a GPS receiver, a high-frequency single-beam echosounder, a single-beam dual-frequency echosounder, two cameras, LiDAR, and a UMRR 0C Type 42 radar.

One individual sensor (3D/UHD 24 GHz radar) is employed for traffic management, simultaneously measuring many parameters, such as angle, radial velocity, range, reflection coefficient, and the entire series of motionless and motion reflector parameters. It is possible to detect many reflectors that are simultaneously in the field of view; as many as 256 targets can be detected at once. This number can be halved depending on the selected communication interface. Sorting of these reflectors is based on the range; if more than 128 targets are detected, then short-range targets are reported first. The radar is suitable for determining the speed and lighting in red light, and the sensor can be used in approaching or in receding traffic mode [19]. Figure 7 shows the radar mounted on the HydroDron platform with sensors, including the rotating and stationary camera, mounted to increase visual data collection.

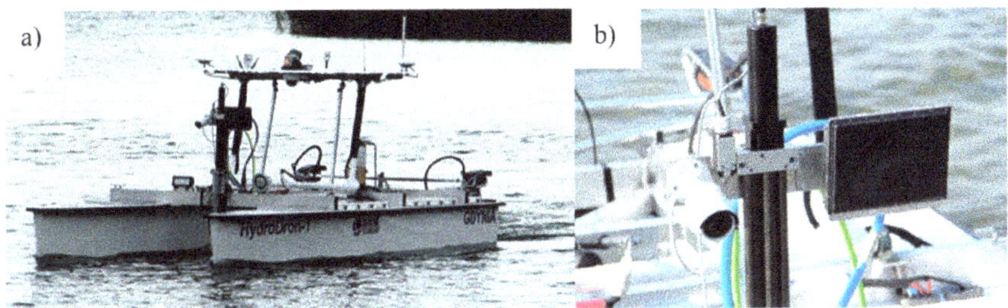

Figure 7. Experimental system (**a**) ASV HydroDron and (**b**) radar 3D/UHD 24 GHz sensor mounted on the HydroDron platform.

4.3. Data Evaluation

Several parameters were measured in both sets of scenario experiments. The most important were the ranges and relative bearings, i.e., relative azimuth angles. In general, the range of the first detection and last detection for each target was recorded. These were direct relative measurements of the sensor in the local coordinate system, related to sensor position. Measurements from several iterations (up to 11) were collected for each scenario to obtain good statistical evaluations. The Student's t-distribution was assumed because of the small sample size. The results showed that in this particular case it was a reasonable approach. Based on this statistical analysis, the average (Equation (1)), standard deviation (Equation (2)), and standard error (Equation (3)) were calculated for each scenario.

$$\overline{x} = \frac{\sum_{i=1}^{n} x_i}{n} \quad (1)$$

$$\sigma_x = t_{n-1,\alpha}\sqrt{\frac{\sum_{i=1}^{n}(x_i-\overline{x})}{n-1}} \tag{2}$$

$$S_{\overline{x}} = \frac{\sigma_x}{\sqrt{n}} \tag{3}$$

where:

x represents the measured value,
x_i represents measurement in *i-th* iteration,
n represents the number of iterations (number of measurements),
$\overline{x}$ is the mean value of x for n measurements,
σ_x is the standard deviation for n measurements of x,
$t_{n-1,\alpha}$ is the critical value in Student's t-distribution for n degrees of freedom and confidence, level α (68.3% in this study), and
$S_{\overline{x}}$ is the standard error of x (the standard deviation of the mean value).

Using this method, the minimum and maximum detection range can be obtained for various bearings, which provides an empirical field of view.

5. Results and Discussion

5.1. Stationary Scenario

In this scenario, measurements were performed individually for the five targets presented in Figure 5. The object travelled along the moving lane; this target was observed on the screen and the detection data were recorded. The measurement numbers varied from four to six, depending on the detection observation and results achieved; the results are compiled in Table 2. At least four in/out iterations were made for each artificial target. If the results were coherent, performing more measurements was pointless. The results achieved were very cohesive and as such, the number of observations performed was reduced. The data achieved were enough to determine if the particular object was well or poorly detected. The exemption to this rule were the swans, which are live animals and could not be steered. They approached the vehicle twice and moved away. Thus, two measurements for the minimum detection range and two for the maximum detection range were measured.

Table 2. Minimum and maximum detection distance results from the Stationary Scenario.

Target	**Swans**		**Airtoy**		**Lifebuoy**		**Radar Reflector**	
Number of Measurements	**2**		**4**		**6**		**4**	
detection range [m]	min	max	min	max	min	max	min	max
Mean	3.96	15.42	4.65	16.43	4.50	13.72	2.99	17.04
St. Dev.	0.11	0.93	0.88	1.02	0.59	1.41	0.34	0.18
St. Error	0.15	1.21	0.53	0.61	0.27	0.64	0.21	0.11

Table 2 presents the minimum and maximum distances of detection along the moving lane. The mean values as well as standard deviations and standard errors are provided for the minimum detection and the maximum detection range. The fender was not detected at all and thus it is not included in the table. The swans were observed for a few minutes and during this time two measurement lines were chosen. The best detection results were achieved for the radar reflector, wherein the maximum distance was determined at the end of measurement line although the real maximum distance was larger. The measured minimum distance was the best for all the analyzed targets. Despite the complicated direct target observations on the screen and the target being mixed with others, the post-processing of data allowed the airtoy detection to be extracted and good data were obtained. The lifebuoy target visibility was very good and the target was easily distinguishable. The swans were visibly distorted but observable.

The results indicate that most objects were visible at a distance of approximately 13–17 m. The minimum distance was generally 3–4 m based on a geometrical distribution. In summary, this stage of research shows that the radar can detect even small targets for anti-collision. However, small floating targets, such as fenders and small buoys, are not detected.

5.2. Moving Platform Scenario

In this scenario, three cases were analyzed when head-on, crossing, and overtaking motion. During the experiment, the ASV moved with a stable course and speed, while the inflatable boat moved according to the desired trajectory. Both boats were simulating typical collision situations that could reasonably occur while the ASV was deployed to collect measurements. In each of the three parts, the target was approaching its own ship from various relative bearings. In Figure 8, the situation is explained graphically, presenting the ranges of the relative bearing for each situation. It should be pointed out, that the terms head-on, corssing, and overtaking was slightly modified compared to the traditional understanding of IMO COLREG (Collision Regulations) requirements. For the purpose of this research, we proposed that head-on means the situation in which the target is approaching from the relative bearings in the range (−10°;10°); crossing means the situation in which the target is approaching from the relative bearings in the ranges (−90°; −8°) and (8°; 90°); and overtaking means the situation in which the target is approaching from the relative bearings in the ranges (80°; −80°). As it can be seen in Figure 8, the areas slightly overlap each other and the relative course of the object decides the type of movement. Such definition of the areas in the scenario ensures the analysis of the maximum detection range in the entire filed of view of the radar antenna.

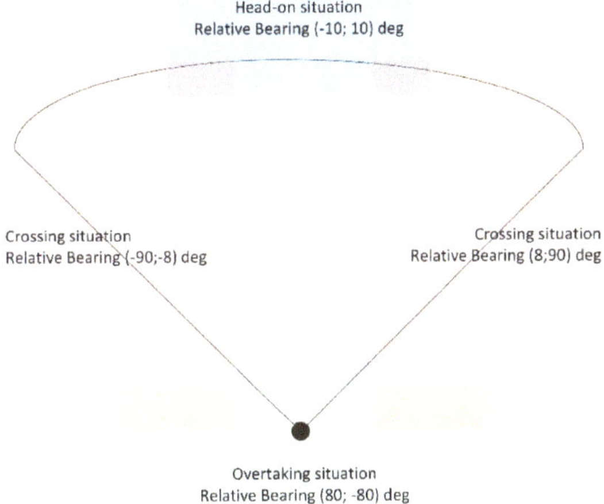

Figure 8. Situation areas in dynamic scenario.

For head-on motion, the boats were moving toward each other from a large distance. The goal was to determine the maximum detection distance for such a boat. According to the declarative field of view, the head-on course should provide the maximum detection distance. The tracks analyzed in this scenario (after entering the field of view) are presented in Figure 9. The own ship was located in the beginning of the coordinate system and the x-axis is oriented with a relative bearing of 0°. The relative tracks of the targets after entering the field of view are given.

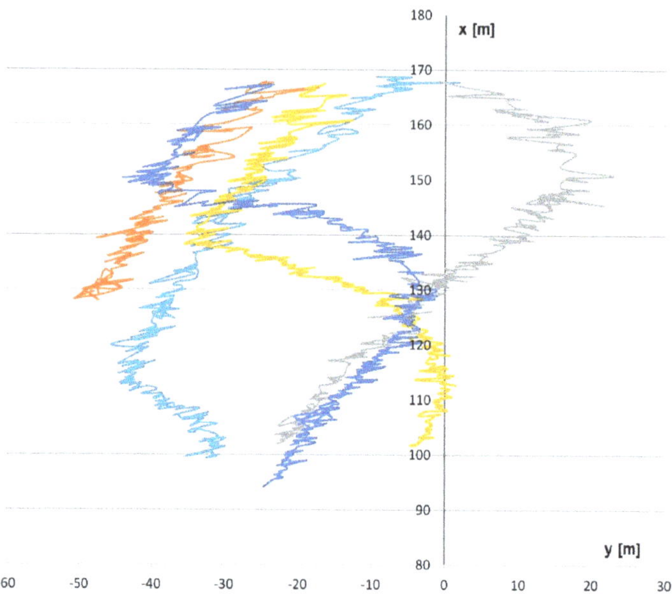

Figure 9. Moving Platform Scenario tracks for head-on motion experiments.

The results for the head-on situation are compiled in Table 3. Because of the good convergence of measurements, it was decided that five iterations were good enough in this situation.

Table 3. Moving Platform Scenario results for head-on motion.

Measurement Number	Range [m]	Relative Bearing [°]	Relative Radial Speed [m/s]
1	169.32	−8.00	−2.28
2	169.16	−1.44	−2.60
3	168.68	−1.44	−2.36
4	168.72	−6.40	−2.84
5	169.32	−7.84	−2.36
Mean	**169.04**	−5.02	−2.49
St. Dev.	0.38	4	0.28
St. Error	**0.17**	1.79	0.12

As shown in Table 3, the mean detection range for this type of boat is 169.04 m. This measurement was reproducible, such that according to the 3-sigma rule, the real detection range should vary from 168.5 to 169.5 m. This result generally confirms the declarative detection range for small cars onshore. The relative bearings confirmed the scenario assumptions (head-on situation) and the relative speed shows that the target approached at a nearly constant speed of 2.5 m/s.

In the crossing and overtaking motion, the main goal was to find the angles at which the target appeared in the field of view and then left it. Eleven tracks were recorded for both the crossing and overtaking motion experiments. The relative tracks are presented in Figure 8. The graphs show a plan view, in which the HydroDron is in the middle of the body frame coordinate system and the x-axis points toward the heading. The observed platform presented in Figure 6 (an inflatable boat) was moving according to the established patterns. The HydroDron was moving with a steady course and speed, while the target was maneuvering. As it can be seen in Figure 10a, the tracks were recorded in various distances, from a few meters up to 170 m (detection maximum for this type of target). In the case of overtaking movement (Figure 10b), only the moment of the first target detection is important and thus only the incoming target was taken into account. Notably, the tracks were selected to verify the angles over various distances, both smaller and bigger. The measurement results for crossing the tracks are provided in Table 4.

Table 4. Moving Platform Scenario results for the crossing motion experiments.

	Relative Bearing [°]		Relative Bearing [°]	
	Port Side		Starboard Side	
	Incoming Target	Outcoming Target	Incoming Target	Outcoming Target
Mean	−40.93	−45.38	44.13	43.26
St. Dev.	18.89	7.93	11.31	8.40
St. Error	8.45	3.55	5.06	3.76
Minimum	−12.00	−37.6	29.92	35.68
Maximum	−53.44	−50.56	52.16	53.76

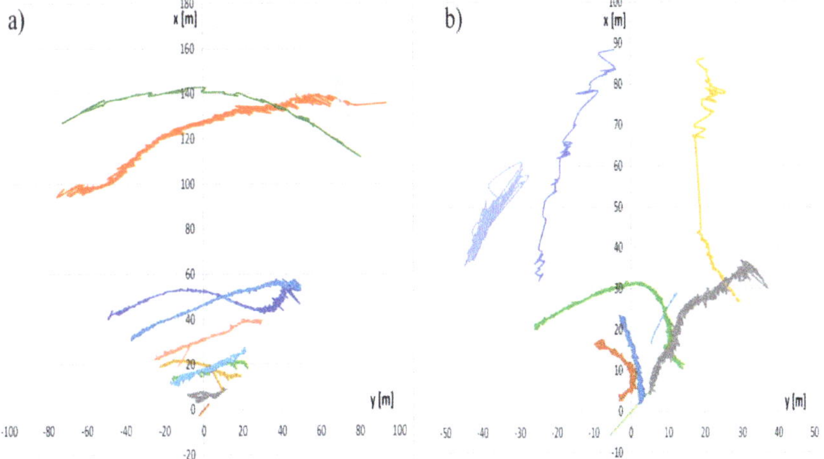

Figure 10. Moving Platform Scenario tracks for (**a**) crossing and (**b**) overtaking motion experiments.

As shown in Table 4, the statistics for the crossing motion experiments are divided into the port side and the starboard side of the ASV (and radar). Additionally, the results are presented separately for the target coming into the field of view and then leaving it. The sample size was small due to the complexity of the study, so the T-distribution was used. In each case presented in Table 3, the mean value is within (40°–45°), which can be treated as the typical angular restriction of the field of view. However, the maximum values are more than 50° and the distribution is more or less symmetrical. Furthermore, the standard deviation and standard error achieved in these experiments are relatively big because the crossing motion occurred at various ranges. Based on evaluation of the detailed data, we found that at larger ranges, the angular field of view was smaller and the crossing target entered the view later; for example, when the target appeared at 169.2 m, the angle was −12°.

The observations made in the crossing motion experiments were confirmed in the measurements for the overtaking situation, as shown in Table 5. Seven measurements are presented, together with statistics based on the T-distribution assumption. In these experiments, only incoming targets were analyzed, and the absolute value of the bearing was calculated without dividing it into portside and starboard side.

Table 5. Moving Platform Scenario results for the overtaking motion experiments.

Measurement Number	Range [m]	Absolute Value of Relative Bearing [°]
1	62.64	39.84
2	18.24	51.2
3	7.52	49.92
4	2.92	50.72
5	4.24	50.72
6	6.44	51.52
7	30.32	51.36
Mean	18.90	**49.33**
St. Dev.	23.98	4.68
St. Error	10.72	**2.09**

Overtaking usually occurs at relatively small distances, a situation that was reproduced in this research. This results in a better and more accurate mean value of nearly 50°. In the first measurement, where the distance was more than 60 m, the bearing was smaller. Summarizing these observations, the angular field of view should be determined as a function of range; for small ranges it is nearly linear (approximately 45–50°), but for bigger ranges the field of view falls exponentially. To verify this hypothesis, the relative bearing graphs are presented in Figures 11 and 12. Figure 11 presents the relationship between the range and relative bearings, wherein the minimum- and maximum-recorded bearings for each distance are plotted. The envelope for more than 40,000 measurements collected in the Moving Platform Scenario is presented.

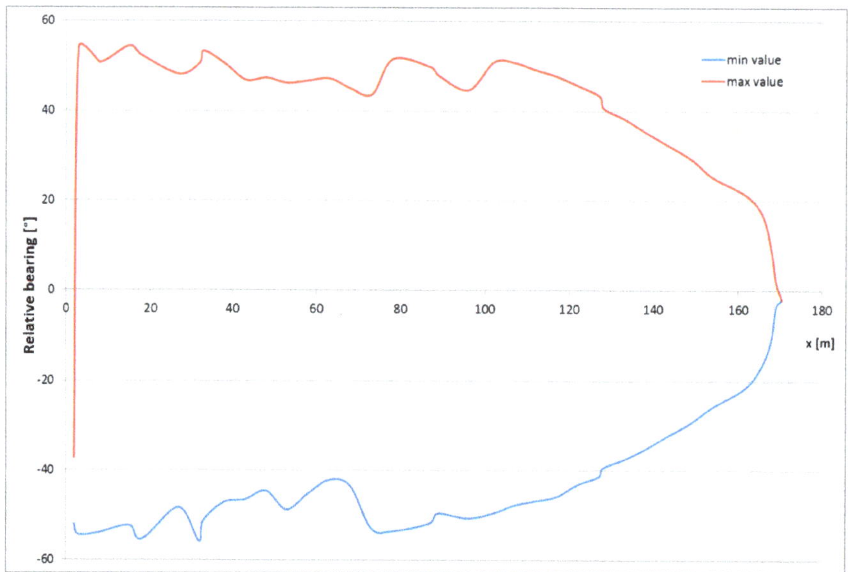

Figure 11. Relative bearing as a function of x-coordinate in the Moving Platform Scenario.

As shown in Figure 11, larger x-coordinates correspond to smaller bearings to detect the target. Although the maximum detection range is still approximately 170 m, the geometry of the sensor and experimental configuration suggests that targets at smaller distances will not be detected. For example, a target at a distance of 150 m at a relative bearing of 40° will not be detected by radar. This consideration leads directly to the detection area pattern presented in Figure 12. The graph shows the measurement points positions in the Cartesian coordinate body system. The vertical axis indicates the direction of

ASV movement with an envelope of detectable targets. The field of view appears as a quarter circle with a radius of almost 170 m. Up to approximately 120 m, the angular width of the field of view is almost identical (90–100°); at further distances, the effective width is smaller. Notably, no lobes are observed, which could be expected, based on the declarative beam pattern. The envelope presented in the graph is generally a smooth line that was created based on the minimum and maximum range values for each bearing with a resolution of 1°. A comparison with the measurement points indicates that this envelope is rather optimistic and the effective field of view is narrower. Some disturbances to the envelope can be observed at distances of about 100 m (x-axis), which are larger on the starboard side, where the graphed line is less smooth. One reason for this might be the mounting on the boat left of the echosounder pole. This hypothesis should be verified in the future.

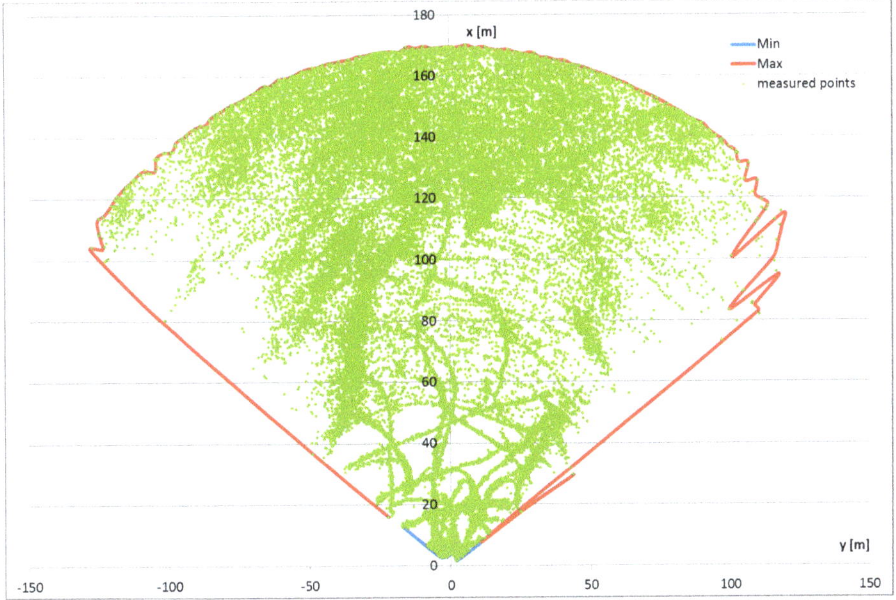

Figure 12. Empirical detection pattern based on measurements collected in the Moving Platform; measured points are in green and the envelope is shown with a red line.

6. Conclusions

The research in this study provides an empirical analysis of surface target detection in a water environment with automotive radar, which can be used for the future development of tracking and anti-collision systems for ASVs. The research focused on identifying the detection ranges and field of view for various targets. Typical objects that could be met in the water environment were analyzed, including a boat and floating objects.

The overarching goal of the research was to verify a novel approach for object detection in a water environment. The novelty was based on using radar sensor for this approach, which is usually implemented in cars for road situations. This approach may in the future overcome the disadvantages of other systems used for anti-collision in ASV, namely laser rangefinders, lidars, and cameras. The proposed system was verified in real conditions with the online recordings.

The research showed that the system was capable of detecting many small targets but some objects, such as a fender, were not detected. Therefore, detection depends both on the size of the target and the material. In general, objects that are air inflated, such as fenders or airtoys, show worse detectability than solid targets, such as lifebuoys. Detection can be significantly improved using a radar reflector, however these reflectors are not usually deployed in practice. In general it can be said that the maximum detection range of small targets is about 15 m, while in very short distance (less than 3 m in research configuration) they are in the shadow due to antenna mounting. It can be assumed that 15 m is a reasonable distance to perform hard anti-collision maneuvers (like full stop) for such a small target with good maneuverability, however this judgment has to be verified in future research.

The second part of the study was conducted using an inflatable boat as the model object in motion. A complex analysis of the field of view for this target was performed, including the radial distances at different angles for various movement parameters. In general, the empirical research confirmed the product limitations and performance declared by the producers, under the assumption that an inflatable boat could be treated as a small car based on their size similarities. The maximum detecting range, confirmed with statistical post-processing, was about 170 m, while the field of view was about 100° (50° for each side). These values seem to be reasonable for planning anti-collision maneuvers with moving targets.

In summary, for larger targets that represent the greatest risk, the radar system provides good detection for anti-collision purposes. For smaller targets, the detection ranges are smaller, although for most targets, it would be small enough for the ASV to maneuver around. Additionally, some small targets were not detected. Generally, the automotive radar system may be a good basis for an ASV anti-collision system; however it should be supplemented with the integration of additional sensors, such as laser rangefinders. In the future, the detection stability and additional small targets should be investigated. It would be also interesting to see how this kind of radar would react in a sea environment. ASVs used at sea might be also a possible target of implementation.

Author Contributions: Conceptualization, A.S. and W.K.; methodology, W.K.; bibliography review, A.S.; acquisition, analysis, and interpretation of data, W.K., D.G-S. and W.M.; writing—original draft preparation, W.K.; writing—review and editing, A.S.

Funding: This study was funded by the European Regional Development Fund under the 2014-2020 Operational Programme Smart Growth; as part of the project, "Developing of autonomous/remote operated surface platform dedicated hydrographic measurements on restricted reservoirs" implemented as part of the National Centre for Research and Development competition, INNOSBZ and under grant No 1/S/IG/16 financed from a subsidy of the Ministry of Science and Higher Education for statutory activities.

Conflicts of Interest: The author(s) declare(s) that they have no conflict of interest regarding the publication of this paper.

References

1. Droeschel, D.; Schwarz, M.; Behnke, S. Continuous mapping and localization for autonomous navigation in rough terrain using a 3D laser scanner. *Robot. Auton. Syst.* **2017**, *88*, 104–115. [CrossRef]
2. Williams, G.M. Optimalization of eyesafe avalanche photodiode lidar for automobile safety and autonomous navigation systems. *Opt. Eng.* **2017**, *56*, 031224. [CrossRef]
3. Huang, L.; Chen, S.; Zhang, J.; Cheng, B.; Liu, M. Real-Time Motion Tracking for Indoor Moving Sphere Objects with a LiDAR Sensor. *Sensors* **2017**, *17*, 1932. [CrossRef] [PubMed]
4. Song, R.; Liu, Y.; Bucknall, R. Smoothed A* algorithm for practical unmanned surface vehicle path planning. *Appl. Ocean Res.* **2019**, *83*, 9–20. [CrossRef]
5. Korayem, M.H.; Esfeden, R.A.; Nekoo, S.R. Path planning algorithm in wheeled mobile manipulators based on motion of Arms. *J. Mech. Sci. Technol.* **2015**, *29*, 1753–1763. [CrossRef]
6. Maclachlan, R.; Mertz, C. Tracking of Moving Objects from a Moving Vehicle Using a Scanning Laser Rangefinder. In Proceedings of the 2006 IEEE Intelligent Transportation Systems Conference Proceedeings, Toronto, ON, Canada, 17–20 September 2006; pp. 301–306. [CrossRef]
7. Wei, P.; Cagle, L.; Reza, T.; Ball, J.; Gafford, J. LiDAR and camera detection fusion in a real-time industrial multi-sensor collision avoidance system. *Electronics* **2018**, *7*, 6. [CrossRef]
8. Polvara, R.; Sharma, S.; Wan, J.; Manning, A.; Sutton, R. Obstacle avoidance approaches for autonomous navigation of unmanned surface vehicles. *J. Navig.* **2017**, *71*. [CrossRef]

9. Almeida, C.; Franco, T.; Ferreira, H.; Martins, A.; Santos, R.; Almeida, J.M.; Silva, E. Radar based collision detection developments on USV ROAZ II. In Proceedings of the Oceans 2009-Europe, Bremen, Germany, 11–14 May 2009; pp. 1–6.
10. Zhuang, J.Y.; Zhang, L.; Zhao, S.Q.; Cao, J.; Wang, B.; Sun, H.B. Radar-based collision avoidance for unmanned surface vehicles. *China Ocean Eng.* **2016**, *30*, 867–883. [CrossRef]
11. Eriksen, B.-O.H.; Wilthil, E.F.; Flåten, A.L.; Brekke, E.F.; Breivik, M. Radar-based Maritime Collision Avoidance using Dynamic Window. In Proceedings of the IEEE Aerospace Conference, Big Sky, MT, USA, 3–10 March 2018. [CrossRef]
12. Gresham, I.; Jain, N.; Budka, T. A 76–77GHz Pulsed-Doppler Radar Module for Autonomous Cruise Control Applications. In Proceedings of the IEEE MTT-S International Microwave Symposium (IMS2000), Boston, MA, USA, 11–16 June 2000; pp. 1551–1554.
13. Wolff, C. Frequency-Modulated Continuous-Wave Radar. Radar Tutorial. Available online: http://www.radartutorial.eu/02.basics/pubs/FMCW-Radar.en.pdf (accessed on 31 July 2018).
14. Ramasubramanian, K. *Using a Complex-Baseband Architecture in FMCW Radar Systems*; Texas Instrument: Dallas, TX, USA, 2017; Available online: http://www.ti.com/lit/wp/spyy007/spyy007.pdf (accessed on 3 August 2018).
15. Brückner, S. Maximum Length Sequences for Radar and Synchronization. Cuvillier. Available online: https://cuvillier.de/uploads/preview/public_file/9760/9783736991927_Leseprobe.pdf (accessed on 2 August 2018).
16. Automotive Radar—A Tale of Two Frequencies. life.augmented. Available online: https://blog.st.com/automotive-radar-tale-two-frequencies/ (accessed on 3 August 2018).
17. Sjöqvist, L.I. What Is an Automotive Radar? Gapwaves. Available online: http://blog.gapwaves.com/what-is-an-automotive-radar (accessed on 3 August 2018).
18. Schneider, M. Automotive Radar—Status and Trends. In Proceedings of the GeMiC 2005, Ulm, Germany, 5–7 April 2005; pp. 144–147.
19. Ramasubramanian, K.; Ramaiah, K.; Aginsky, A. *Moving from Legacy 24 GHz to State-Of-The-Art 77 GHz Radar*; Texas Instrument: Dallas, TX, USA, 2017; Available online: http://www.ti.com/lit/wp/spry312/spry312.pdf (accessed on 3 August 2018).
20. Matthews, A. What Is Driving the Automotive LiDAR and RADAR Market? Automotive Electronic Specifier, Kent, UK. 2017. Available online: https://automotive.electronicspecifier.com/sensors/what-is-driving-the-automotive-lidar-and-radar-market#downloads (accessed on 3 August 2018).
21. Bronzi, D.; Zou, Y.; Villa, F. Automotive Three-Dimensional Vision Through a Single-Photon Counting SPAD Camera. *IEEE Trans. Intell. Transp. Syst.* **2016**, *17*, 782–795. [CrossRef]
22. Mende, R.; Rohling, H. *New Automotive Applications for Smart Radar Systems*; Smartmicro Publications: Braunshweig, Germany; Available online: http://www.smartmicro.de/company/publications/ (accessed on 3 August 2018).
23. Smartmicro. Available online: http://www.smartmicro.de/fileadmin/user_upload/Documents/TrafficRadar/UMRR_Traffic_Sensor_Type_42_Data_Sheet.pdf (accessed on 22 August 2018).
24. Jo, J.; Tsunoda, Y.; Stantic, B. A Likelihood-Based Data Fusion Model for the Integration of Multiple Sensor Data: A Case Study with Vision and Lidar Sensors. In *Robot Intelligence Technology and Applications 4*; Springer International Publishing: Basel, Switzerland, 2017; Volume 447, pp. 489–500.
25. Jooho, L.; W, W.J.; Nakwan, K. Obstacle Avoidance and Target Search of an Autonomous Surface Vehicle for 2016 Maritime RobotX Challenge. In Proceedings of the IEEE OES International Symposium on Underwater Technology (UT), Busan, Korea, 21–24 February 2017.
26. Mei, J.H.; Arshad, M.R. COLREGs Based Navigation of Riverine Autonomous Surface Vehicle. In Proceedings of the IEEE 6TH International Conference on Underwater System Technology, Penang, Malaysia, 13–14 December 2016; pp. 145–149.
27. Barton, A.; Volna, E. Control of Autonomous Robot using Neural Networks. In Proceedings of the International Conference on Numerical Analysis and Applied Mathematics 2016 (ICNAAM-2016), Rhodes, Greece, 19–25 September 2016; Volume 1863.
28. Ko, B.; Choi, H.J.; Hong, C. Neural Network-based Autonomous Navigation for a Homecare Mobile Robot. In Proceedings of the 2017 IEEE International Conference on Big Data and Smart Computing (BIGCOMP), Jeju, Korea, 13–16 February 2017; pp. 403–406.

29. Praczyk, T. Neural anti-collision system for Autonomous Surface Vehicle. *Neurocomputing* **2015**, *149*, 559–572. [CrossRef]
30. Lil, J.; Bao, H.; Han, X. Real-time self-driving car navigation and obstacle avoidance using mobile 3D laser scanner and GNSS. *Multimed. Tools Appl.* **2016**, *76*, 23017–23039.
31. Guan, R.P.; Ristic, B.; Wang, L.P. Feature-based robot navigation using a Doppler-azimuth radar. *Int. J. Control* **2017**, *90*, 888–900. [CrossRef]
32. Guerrero, J.A.; Jaud, M.; Lenain, R. Towards LIDAR-RADAR based Terrain Mapping. In Proceedings of the 2015 IEEE International Workshop on Advanced Robotics and its Social Impacts (ARSO), Lyon, France, 30 June–2 July 2015.
33. Hollinger, J.; Kutscher, B.; Close, B. Fusion of Lidar and Radar for detection of partially obscured objects. In Proceedings of the Unmanned Systems Technology XVII, Baltimore, MD, USA, 22 May 2015; Volume 9468.
34. Mikhail, M.; Carmack, N. Navigation Software System Development for a Mobile Robot to Avoid Obstacles in a Dynamic Environment using Laser Sensor. In Proceedings of the SOUTHEASTCON 2017, Charlotte, NC, USA, 31 March–2 April 2017.
35. Jeon, H.C.; Park, Y.B.; Park, C.G. Robust Performance of Terrain Referenced Navigation Using Flash Lidar. In Proceedings of the 2016 IEEE/Ion Position, Location and Navigation Symposium (PLANS), Savannah, GA, USA, 11–16 April 2016; pp. 970–975.
36. Jiang, Z.; Wang, J.; Song, Q. Off-road obstacle sensing using synthetic aperture radar interferometry. *J. Appl. Remote Sens.* **2017**, *11*, 016010. [CrossRef]
37. Oh, H.N.H.; Tsourdos, A.; Savvaris, A. Development of Collision Avoidance Algorithms for the C-Enduro USV. *IFAC Proc. Vol.* **2014**, *47*, 12174–12181. [CrossRef]
38. Stateczny, A.; Kazimierski, W. Selection of GRNN network parameters for the needs of state vector estimation of manoeuvring target in ARPA devices. In *Photonics Applications in Astronomy, Communications, Industry, and High-Energy Physics Experiments IV, Proceedings of the Society of Photo-Optical Instrumentation Engineers (SPIE), Wilga, Poland, 30 May–2 June 2005*; Romaniuk, R.S., Ed.; SPIE: Bellingham, WA, USA, 2006; Volume 6159, p. F1591.
39. Kazimierski, W.; Zaniewicz, G.; Stateczny, A. Verification of multiple model neural tracking filter with ship's radar. In Proceedings of the 13th International Radar Symposium (IRS), Warsaw, Poland, 23–25 May 2012; pp. 549–553.
40. Xinchi, T.; Huajun, Z.; Wenwen, C.; Peimin, Z.; Zhiwen, L.; Kai, C. A Research on Intelligent Obstacle Avoidance for Unmanned Surface Vehicles. *Proc. Chin. Autom. Congr. (CAC)* **2018**. [CrossRef]

Article

On the Reliability of Surface Current Measurements by X-Band Marine Radar

Katrin G. Hessner [1,*], Saad El Naggar [2], Wilken-Jon von Appen [2] and Volker H. Strass [2]

1 OceanWaveS GmbH, 21339 Lüneburg, Germany
2 Alfred-Wegener-Institut Helmholtz-Zentrum für Polar- und Meeresforschung, 27515 Bremerhaven, Germany; Saad.El.Naggar@awi.de (S.E.N.); wilken-jon.von.appen@awi.de (W.-J.v.A.); Volker.Strass@awi.de (V.H.S.)
* Correspondence: hessner@oceanwaves.de; Tel.: +49-4131-699-5822

Received: 22 February 2019; Accepted: 18 April 2019; Published: 30 April 2019

Abstract: Real-time quality-controlled surface current data derived from X-Band marine radar (*MR*) measurements were evaluated to estimate their operational reliability. The presented data were acquired by the standard commercial off-the-shelf MR-based *sigma s6* WaMoS® II (WaMoS® II) deployed onboard the German Research vessel *Polarstern*. The measurement reliability is specified by an *IQ* value obtained by the WaMoS® II real-time quality control (*rtQC*). Data which pass the *rtQC* without objection are assumed to be reliable. For these data sets accuracy and correlation with corresponding vessel-mounted acoustic Doppler current profiler (ADCP) measurements are determined. To reduce potential misinterpretation due to short-term oceanic variability/turbulences, the evaluation of the WaMoS® II accuracy was carried out based on sliding means over 20 min of the reliable data only. The associated standard deviation $\sigma_{WaMoS} = 0.02$ m/s of the mean WaMoS® II measurements reflect a high precision of the measurement and the successful *rtQC* during different wave, current and weather conditions. The direct comparison of 7272 WaMoS® II/ADCP northward and eastward velocity data pairs yield a correlation of $r \geq 0.94$, with $|bias_{\Delta}| \leq 0.06$ m/s and $\sigma_S = 0.05$ m/s. This confirms that the MR-based surface current measurements are accurate and reliable.

Keywords: X-Band radar; marine radar current measurement; quality control; measurement reliability; accuracies; precision; WaMoS® II; vessel mounted acoustic Doppler current profiler

---

1. Introduction

Marine radars (MR) are designed for navigation and vessel traffic control. Depending on the physical environmental conditions given by precipitation, wind and waves, signatures of the sea surface commonly referred to as *sea clutter* become visible in the near range (<5 km) of the MR radar images. Regarded as a disruptive noise for navigational purposes, *sea clutter* is normally suppressed. Even though *sea clutter* signatures are well known, they are still not completely resolved, and are still under investigation both experimentally and theoretically. Nevertheless, it turns out that *sea clutter* includes valuable information on surface waves [1]. Following Bragg theory, *sea clutter* is caused by the backscatter of the transmitted electromagnetic waves from the short sea surface ripples in the range of half the electromagnetic wavelength (i.e., ~1.5 cm). Longer waves, such as wind sea (~10 m) and swell (~100 m), become visible as they modulate the *sea clutter* signal. Both surface currents and water depth affect the wave propagation [2,3]. As MRs image *sea clutter* simultaneously in time and space, this allows the derivation of multi-directional unambiguous wave information, surface currents, and (in shallow water) also water depth.

Driven by the growing need for precise information about waves and surface currents, commercially available MR-based wave and current monitoring devices, such as WaMoS® II, have been developed [4–6]. Their capability and performance in a wide range of different applications, ranging from coastal applications [7–9] to vessel-mounted applications [10–12], have been proven.

The *sea clutter* observations of MRs typically range up to 3–5 km, with spatial and temporal resolutions on the order of 7.5 m and 2 s, respectively. This allows MRs to monitor waves longer than 15 m and current conditions over an area of several km^2 in real time. As *sea clutter* is caused by the *Bragg* backscatter of the transmitted electromagnetic waves from the short sea surface ripple waves (~2 cm), a minimum wind speed of 2–3 m/s is required for its presence [13]. In calm periods in the absence of ripples, no *sea clutter* can be observed, thereby preventing MR sea state and current observations. Also, signatures of rain or snow (*weather clutter*), or other features in the radar image not related to *sea clutter*, can disturb MR wave and current observations. These environmental limitations reduce the confidence and acceptance of the MR-based measurements, and therefore need to be treated carefully.

The aim of this paper is to assess the present status of the MR-based WaMoS® II system, focusing on its data usability with respect to reliability and accuracy. For this purpose, current measurements obtained onboard the German research vessel *Polarstern* during the Atlantic transit cruise PS113 [14] between Punta Arenas, Chile, and Bremerhaven, Germany, in May 2018 are used. The outline of the paper is as follows: In Section 2, we give a brief introduction on the methods used to estimate the accuracy and precision of fluctuating measurements. Section 3 describes the sensors used, with a focus on WaMoS® II. In Section 4, we present the WaMoS® II real-lime quality control (*rtQC*) used to specify data reliability. Observations made during the Atlantic transit cruise PS113 are presented in Section 5. Results of the accuracy estimation and comparison with acoustic Doppler current profiler (ACDP) measurements are presented in Section 6. Finally, in Section 7, we give a summary and draw conclusions.

2. Methods: Accuracy and Precision

A common method to evaluate the accuracy and precision of measurements is to perform a direct comparison of data sets from different sensors. In the case of MR-based current measurements, corresponding reference measurements from in situ sensors like ADCPs are used [15]. The underlying assumption of this approach is that both sensors observe the same property (P), and it is assumed that spatial and temporal homogeneity and deviations between the data sets can be related directly to inaccuracies in the measurements. However, this method of comparison is limited in that observed deviations do not automatically relate to inaccuracies of the measurement technique [16]. The biggest contribution to independent sensor deviations can be attributed to the different measurement locations of the sensors. For example, an ADCP delivers subsurface current measurements in a limited local volume, while MR-based observation represents current measurements at the sea surface over a spatial domain of several hundreds of square meters. Due to different current structures (e.g., wind-forced surface Ekman flow and geostrophic current shear extending deeply over a large part of the water column), vertical homogeneity is not given at all times. Ref. [17] found that 80%, or more, of the observed deviations between ADCP and HF radar current measurements on the West Florida shelf were associated with horizontal and vertical separation between the measurements.

In addition, the informative value of the direct comparison of two independent data sets might be misleading, as it completely neglects the natural variability of the current as a vector, consisting of mean, oscillatory and chaotic contributions. This makes the results more difficult to compare and properly interpret [18]. Therefore, we use a combination of methods to evaluate the quality, reliability, precision, and accuracy of MR surface current measurements.

For practical handling of a fluctuating quantity, P, its temporal averages $\overline{P}$ over a suitable period (averaging time τ) are used. This allows to describe P as $P = \overline{P} + P'$, where P' is the fluctuation with $\overline{P'} = 0$. Based on this assumption, the resulting measurement is represented by the average $\overline{P}$, which depends, among other things, also on the used sampling and averaging intervals.

In this paper, we aim to evaluate the general performance of WaMoS® II data by directly comparing both the mean current, $\overline{U}$, and the corresponding standard deviation, σ_U, representing the short-term oceanic fluctuating component of the current. To evaluate the accuracy of the WaMoS® II measurements, a direct comparison of $\overline{U}$, with the corresponding ADCP measurements is carried out, where the accuracy is described by correlation coefficient (r), bias ($\overline{\Delta}$) and standard deviation (σ_Δ) of the difference:

$$r = \frac{\sum_{i=1}^{N}\left(X_i - \overline{X}\right)\left(Y_i - \overline{Y}\right)}{\sqrt{\sum_{i=1}^{N}\left(X_i - \overline{X}\right) \cdot \sum_{i=1}^{N}\left(Y_i - \overline{Y}\right)}} \quad (1)$$

$$\overline{\Delta} = \frac{1}{N}\sum_{i=1}^{N}\Delta_i, \text{ with } \Delta_i = X_i - Y_i \quad (2)$$

$$\sigma_\Delta = \sqrt{\frac{1}{N-1}\left(\sum_{i=1}^{N}(\Delta_i - \Delta)^2\right)} \quad (3)$$

where $\overline{X} = \frac{1}{N}\sum_{i=1}^{N} X_i$ and $\overline{Y} = \frac{1}{N}\sum_{i=1}^{N} Y_i$ represent the mean measurement of the data sets X and Y, respectively. The resulting combined standard deviation is defined as $\sigma_\Delta = \sqrt{\sigma_X^2 + \sigma_Y^2}$. Assuming that the measurement errors of the two sensors are uncorrelated and of equal magnitude, the individual (single) standard deviation $\sigma_s = \sigma_X = \sigma_Y$, and can hence be estimated by:

$$\sigma_S = \frac{1}{2}\sqrt{2}\,\sigma_\Delta. \quad (4)$$

Note that r, $\overline{\Delta}$, and σ_S include deviations related to horizontal ($\sigma_{\Delta h}$) and vertical ($\sigma_{\Delta v}$) gradients, as well as temporal variation (σ_t) of P, which are not related to inaccuracies of the measurement device. Using the mean instead of the instantaneous measurements leads to statistically more stable and reliable results as the effect of uncorrelated natural variability is minimized.

To evaluate the precision of an individual sensor itself, we use a more general approach. This approach is based on statistical analysis of a property P, represented by a statistical population $\{P_1, P_2, \ldots, P_N\}$. The precision of the measurement of P can be estimated by the standard deviation $\sigma\left(\overline{P}\right)$ of the mean $\overline{P}$, which is given by:

$$\sigma\left(\overline{P}\right) = \sqrt{\frac{1}{N-1}\sum_{i=1}^{N}\left(P_i - \overline{P}\right)^2} \quad (5)$$

where $i = 1, N$ denotes individual values over the averaging interval τ.

Following this strategy, the precision of a measurement is estimated by the standard deviation σ_P of the mean $\overline{P}$. Using an averaging interval of τ = 20–30 min in combination with typical update rates of WaMoS® II measurements ranging between 1–3 min allows us to obtain a sufficient number of independent measurements, and hence gives statistically significant results for our investigation.

3. Data

The data used for the WaMoS® II-ADCP comparison were acquired on board *Polarstern* during the Atlantic transit cruise PS 113 (May 2018) [14].

3.1. *Sigma S6 WaMoS® II*

The MR-based measurements were carried out by the *sigma* S6 WaMoS® II system. This standard commercial, off-the-shelf system consists of a high-speed video digitizing and storage device, which can be interfaced to most conventional analog and digital navigational X-Band radars. The *sigma* S6 system technology can be supplied with different software packages for various real-time applications like small target detection, oil spill detection and ice navigation and monitoring, as well as real-time sea state and current measurements (WaMoS® II).

The WaMoS® II system can be operated from fixed platforms and coastal sites, as well as from moving vessels. For the latter application, the horizontal vessel motion needs to be compensated. The large vertical beam width of MRs, the range of which is normally between 20° and 25°, depending on the used radar type, ensures the ability to scan the sea surface even when the ship is pitching and rolling [19]. Hence, we assume that vessel motions like pitch, roll and heave have no critical influence on the WaMoS measurements.

The horizontal vessel movement can be removed during data processing, either in the space-time or in the wave number-frequency domain. The compensation in the wave number-frequency domain requires that the vessel movement over ground is constant (no variation in speed or course) during radar data acquisition given by number of individual radar images times radar repetition rate. In this case, the vessel movement is related to a fixed Doppler shift ($\vec{k}\vec{V}_{ship}$), and can be separated from the Doppler shift related to the surface current ($\vec{k}\vec{U}_{current}$), where $\vec{k}$ is the wave number vector. The motion compensation in the space-time domain is performed by georeferencing [15,20]. Using GPS ship position and heading (gyro), orientation and position are estimated for every radar pulse. When transforming the sea clutter information from polar coordinates to Cartesian image sequences, each point of the resulting analysis area corresponds to a fixed position relative to the earth, independent of how the vessel is moving during the acquisition time. This method requires more computing time and limits the area available for analysis, but is independent of the ship movement. However, in cases when the vessel is moving very fast (>20 kn), this method might fail, and this occurs when the analysis area moves out of the radar view field. In both cases, very precise vessel heading is required as the error due to misalignment is amplified by the vessel speed [10]. For this application, WaMoS® II processing was set to *georeferencing* mode, as the vessel speed of *Polarstern* was <12 kn, in general. With the sampling strategy of 64 images per individual WaMoS® II measurement and a radar rotation time of 2.5 s, the maximum expected offset during data acquisition is about 1 km, and this is acceptably small compared to the WaMoS® II radar view range of 3 km.

The WaMoS® II system onboard *Polarstern* is connected to an analog SAM Radarpilot 1100 (9.4 GHz), with a rotation rate of RPT = 2.5 s. This radar is dedicated to the WaMoS® II application (in the following, this is referred to as the WaMoS® II radar). The mounted 5 ft antenna provides 1.5° angular resolution. Running in short pulse mode, with a pulse length of 80 nsec, the transceiver delivers data with 12 m range resolution. By oversampling of the radar raw data in direction and range, the *sigma* S6 digitizer delivers radar information with an angular resolution of 0.35° and 5.62 m in range (26.7 MHz). The WaMoS® II radar view field covers a range for 0.303–2.356 km from the antenna, with the second quadrant sector blanked due to the mast construction (Figure 1).

For one individual WaMoS® II measurement on board *Polarstern*, 64 consecutive radar images were analyzed, so that the WaMoS® II results represent temporal means of 2.67 min (64 × 2.5 s). To overcome the effects of the directional dependency of the wave imaging in radar images from radar look direction relative to wave and wind direction [10], the WaMoS® II analysis areas were placed all around the vessel. Figure 1 shows a vessel-oriented radar image, which was obtained by *sigma* S6 WaMoS® II on 12 May 2018, 12:00 UTC. At that time *Polarstern* was sailing northeastwards (42°) at 12 kn (6.2 m/s), while the wind was blowing from 271°, at about 14.4 m/s. The color refers to the measured radar return: black meaning no return level, and white indicating the maximum level. The radar return is digitized to 12 bits, which allows a signal strength ranging from 0–4095. To ensure no information is lost due to clipping at the lower limits, the digitizer is set below the noise level of the system. For the *Polarstern* system, a mean noise level of 500 was determined during system installation. To avoid reflections from the vessel superstructure in the near range, the system starts sampling after a dead range of 300 m. The analysis areas are the three grey rectangles indicating size (128×256 pixels) and alignment (35°, 255°, 325° relative to vessel heading) of the *sigma* S6 WaMoS® II (Figure 1). To overcome the directional dependency of the wave imaging in the radar images, the individual wave spectra of each analysis area are averaged. From the resulting spatially averaged spectrum, statistical wave parameters such as significant wave height (H_s), peak wave period (T_p), peak wave direction

(θ_p), etc., are derived. The WaMoS® II update rate onboard *Polarstern* is approximately 3 min, given by the time taken for 64 images to be acquired, multiplied by the radar repetition rate of 2.5 s.

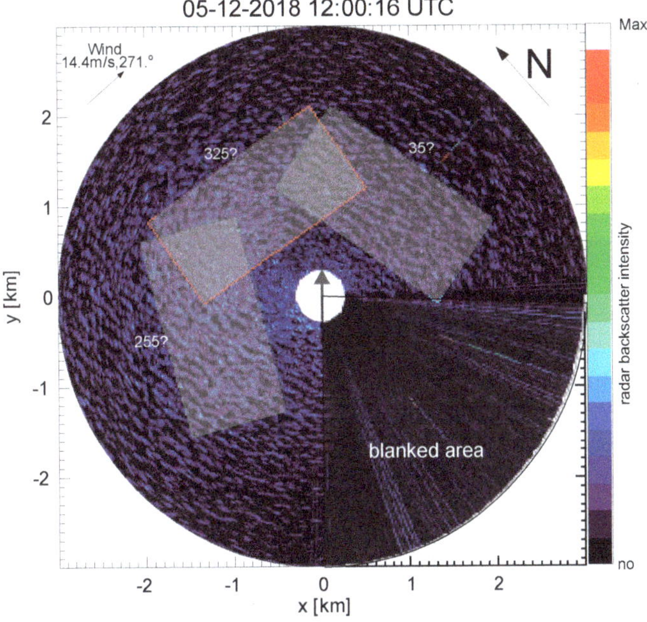

Figure 1. Vessel head-oriented WaMoS® II radar image acquired onboard the *Polarstern* on 12 May 2018, 12:00 UTC. The color scale refers to strength of the received radar return. The highlighted boxes indicate size and alignment of the three WaMoS® II wave analysis areas relative to the radar view field. The arrow in the center indicates the orientation of the *Polarstern*, which was moving at 12 kn and a course of 43° during data acquisition, and north is indicated by the arrow in the top right.

From the measurements of the phase speed (c) of the captured surface waves, the underlying ocean surface currents ($\vec{U_s}$) are derived by identifying deviations from the known dispersion relations of surface waves. Assuming that $\vec{U_s}$ is small compared to c, the depth-weighted effective surface current is given by:

$$\vec{U_s}(k) = 2k \int_0^H \vec{U}(z) \exp(-2kz) dz \tag{6}$$

where $\vec{U}(z)$ is the vertical current profile, with z being positive downwards and H being the water depth. As the influence decays exponentially with depth, the resulting current $\vec{U_s}$ represents a vertical average of the ocean currents within the wave-influenced surface layer D_W [21]. D_W varies depending on the wavelength ($\lambda = 2\pi/k$) and height of the captured waves. On average, D_W is assumed to range between 3 and 10 m depending on the predominant wave length, and this will be shallower for short wind sea waves than for long swell waves [15].

3.2. ADCP

As a reference, data from a vessel-mounted acoustic Doppler current profiler (ADCP) type Ocean Surveyor from Teledyne RD instruments [22] were used. Its transducers/receivers, operating at a nominal frequency of 150 kHz, are mounted in the hull of *Polarstern*, about 11 m below the water line. It was working in long-range mode with a vertical cell size of 4 m, a blanking distance of

4 m, and a maximum range of ~320 m. Heading, pitch, and roll from the ship's inertial navigation system (GPS and magnetically constrained "gyro") were used to convert the ADCP velocities to earth coordinates. The accuracy of the ADCP velocities mainly depends on the quality of the position fixes and the ship's heading data. Further errors stem from a misalignment of the transducer with the ship's centerline. The ADCP data were processed using the Ocean Surveyor Sputum Interpreter (OSSI) developed by GEOMAR, Helmholtz Centre for Ocean Research, Kiel ([22]), which corrects for a possible misalignment between the ADCP transducer orientation and the ship's forward direction.

To avoid interference with vessel-induced currents, the ADCP measurements are averaged over the 20–50 m depth range. For the data comparison, quality controlled ADCP current data with averages over 2 min were used. The quality filter is based on the statistical analysis, where data outliers exceeding the range of mean value and standard deviation of surface velocity are neglected. The ideal-theoretical precision $\sigma_{ADCP(ideal)}$ of the ADCP measurements can be estimated from the single ping/bin standard deviation of $\sigma_{ADCP(SP)} = 0.3$ m/s, given by the ADCP manufacturer. Neglecting natural variability and assuming vertical and temporal homogeneity and independence over 20–50 m (7 depth bins) and 2 min (100 pings), results in $\sigma_{ADCP(ideal)} = \sigma_{ADCP(SP)} / \sqrt{N} = 0.0113$ m/s.

4. WaMoS® II Quality Control

The detection of surface waves in radar images is sensitive to data acquisition and environmental conditions. To create *sea clutter*, a minimum wind speed is required [13]. Furthermore, a minimum wave height is required to significantly modulate the *sea clutter* information. In case of insufficient wave signatures in the radar images, the system cannot deliver reliable information. To indicate the reliability of a *sigma* S6 WaMoS® II measurement, real-time quality control (*rtQC*) is implemented. During processing, the *rtQC* is carried out in different steps, which are also summarized in Figure 2:

- Data acquisition check: This check verifies if all mandatory information is available.
- *Sea clutter* checks: In this step, the radar raw input is controlled. In cases of insufficient *sea clutter* information due to no sea state, rain, very low wind conditions (<3 m/s) or missing data sections, the data set is marked with a quality identifier IQ of $0 < IQ < 9$.
- SNR-checks: Here, the quality of the separation of signal and noise within the image spectrum is used to evaluate the quality of the current estimate and wave filtering. Data sets which do not pass these checks have potentially unreliable current estimates and are marked with $10 < IQ < 400$.
- Wave system checks: When deriving the standard sea state parameters from the frequency direction spectrum $E(f,\theta)$, the number of individual wave systems is determined. In cases of more than 5 wave systems, $E(f,\theta)$ can be regarded as too noisy to give reliable results. In these cases, the resulting data sets are marked with $400 < IQ < 1000$.
- Data basis check: For spatial and temporal averages, a minimum of individual measurements is required to obtain a statistically stable result and hence trustful data. All mean data sets with less measurements are marked with $0 < IQ < 10$.

The results of the different *rtQC* tests are accumulated and summarized in an individual IQi value, which is assigned to the measurement areas (i) (Figure 2). The IQ values are binary numbers where each digit (0 or 1) corresponds to whether the individual measurement has passed or failed a particular test. These binary numbers are then converted to decimals, which is what is given as output and discussed here. Based on the individual IQ value, it is possible to separate reliable data which pass the tests from unreliable data which do not pass all of the tests. Furthermore, the binary structure of the IQ makes it possible to backtrack the individual results of the different tests, and therefore enables validation of the significance of an individual test and the adjustment/optimization, if necessary, of thresholds for particular installations. For example, if the sea clutter test yields parts of the analysis area that contain rain signatures (see Figure 1), this leads to $IQ = 4$. If in the SNR check the data does not pass the significance test so that $IQ = 20$, this is added to give $IQ = 24$. Furthermore, if the wave system check failed as the resulting spectrum is too noisy to identify significant individual wave systems,

$IQ = 100$ is added. The accumulated result is then $IQ = 124$, indicating unreliable measurements due to insufficient radar data because of rain. For the presented data set, it turned out that the sea clutter checks do not indicate unreliable data on their own. This means that partly missing or disturbed sea clutter information alone does not automatically lead to identification as unreliable data. Only in combination with the subsequent quality checks can the results be regarded as unreliable. Even when the results of the sea clutter checks do not explicitly indicate insufficient data, it reveals the potential cause of measurement failures. This will be discussed in more detail with respect to rain signatures later on. For practical use a decimal IQ valid limit of 10 is set.

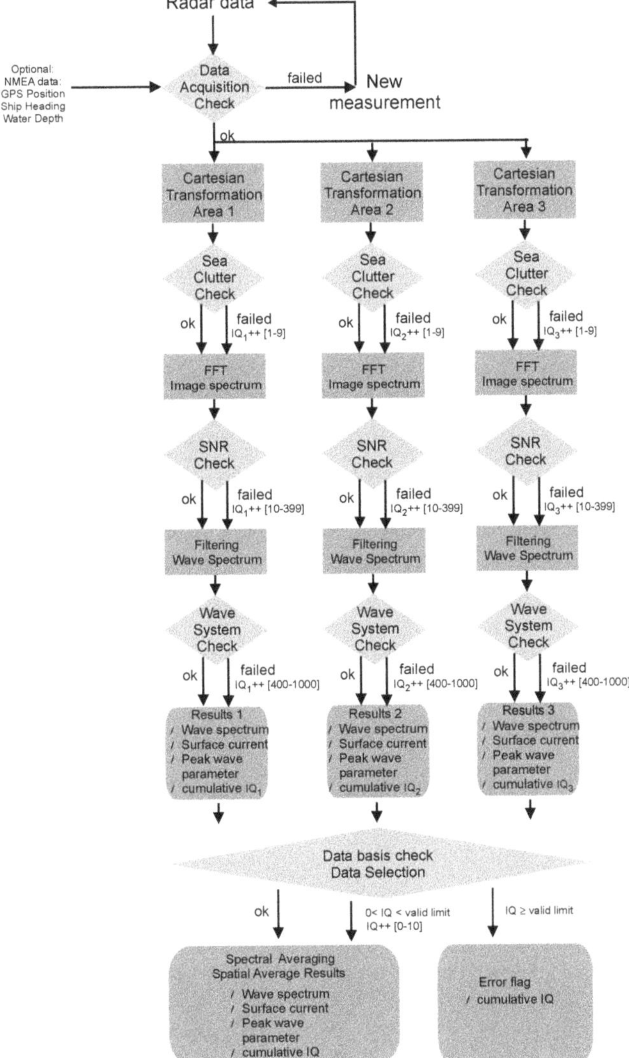

Figure 2. Flow chart of the WaMoS II real-time quality control (rtQC) The different quality checks (diamonds) are carried out with in the levels of the WaMoS processing chain. Boxes indicate key steps in the processing from the radar images in polar coordinates to the resulting wave spectra, peak wave and current parameter.

Figure 3a shows a radar image acquired during a calm period with wind speed <3 m/s. Due to the absence of ripple waves (~3 cm) no *sea clutter* is visible. Figure 3b shows a radar image acquired during heavy rain. The potential *sea clutter* is covered by the rain signatures visible as patches of high backscatter intensity generated by the rain drops in the air. In both cases, no sufficient *sea clutter* is visible, and the corresponding *rtQC* results in $IQ > 400$.

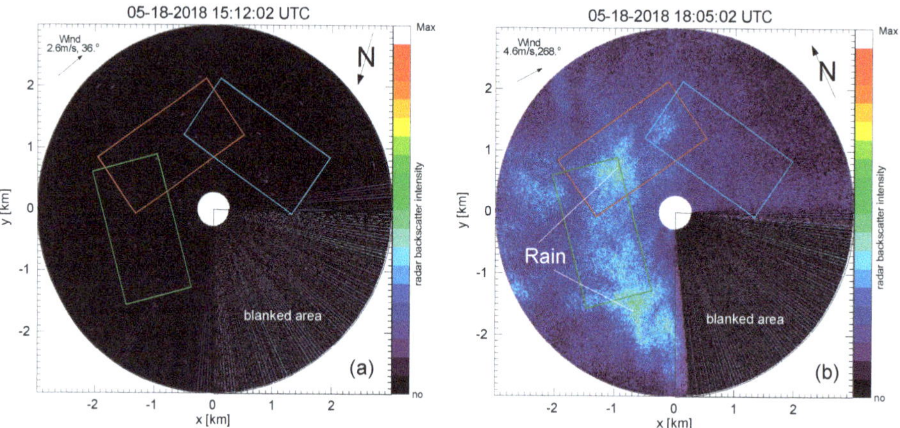

Figure 3. Radar images showing insufficient *sea clutter* information for wave and current retrieval. (**a**) Calm wind condition showing no *sea clutter*. (**b**) Heavy rain condition, where rain signatures covers *sea clutter* information. The color code refers to strength of the received radar return. The highlighted boxes indicate size and alignment of the three WaMoS® II analysis areas.

Figure 4 shows radar images under limited conditions due to moderate rain. The rain signatures are spatially localized (Figure 4a) or are weak (Figure 4b) and allow the detection of wave signatures and therefore also the wave and current analysis. The evaluation of the resulting data suggested that some of the wave parameters are more robust than others with respect to limited *sea clutter* information. The analysis shows that the directly measured parameters like wave period (T_p), wave direction (θ_p), and surface current ($\vec{U}$) are more robust than the indirect measurement of significant wave height (H_s). As long as no other *rtQC* check has been failed, these direct wave parameters can be assumed to be reliable when $IQ < 10$. This does not apply to the indirect estimates of H_s, which are based on the accurate determination of the signal to noise ratio (SNR). Missing *sea clutter* information leads immediately to missing signal intensity and hence to a lower SNR, which results in a decrease in H_s.

It needs to be stressed that the *rtQC* is performed in real time during data processing, and does not require post processing. Furthermore, the identification of insufficient radar information is carried out independently of external information on rain and wind. In cases where the internal WaMoS® II *rtQC* identifies insufficient *sea clutter* ($IQ > 10$), this information is aligned with wind information (if available) to give the combined information, to indicate that currently MR measurements are not possible due to the lack of sufficient wind. This keeps the WaMoS® II *rtQC* independent of additional external information and their reliability. The relation between *IQ* and wind information is an additional piece of information for the user and makes it possible to evaluate the actual threshold of the wind speed for the particular WaMoS® II set up (used radar, installation geometry, etc.).

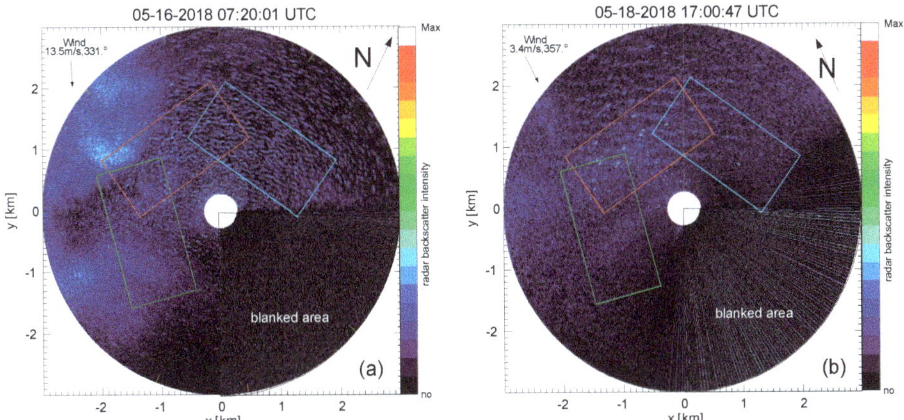

Figure 4. Radar images showing limited sea clutter information for wave and current retrieval. (**a**) Rain signature partly obscures *sea clutter*. (**b**) Weak rain signatures blur the *sea clutter*.

5. Observations During the Cruise

Figure 5 shows the cruise track of *Polarstern* during *PS113* for May, 2018 outside of exclusive economic zones (*EEZ*, 200 miles). The WaMoS® II recording started when *Polarstern* left the *EEZ* of Argentina, 11 May 2018. During the cruise different current regimes from wind-forced to density-driven currents up to more than 2 m/s were encountered. Besides different current regimes, various environmental conditions were experienced. These ranged from a storm event at the beginning of the cruise (May 12/13), with wind speeds up to 20–25 m/s and sea states up to 6–7 m significant wave height (H_s), to calm (wind speed <3 m/s) and rainy periods. The latter give the possibility of validating the proper *rtQC*. Even though no reference data for the WaMoS® II sea state measurements were available, unreliable data can be identified in the data set, as during insufficient *sea clutter* conditions, the peak wave direction and current shows an unrealistically high variance.

During the cruise, it turned out that the other X-Band radar onboard *Polarstern*, which is used for navigation, interfered strongly with the WaMoS® II radar. This led to partly corrupted radar image acquisition and gaps in the time series. These corrupted radar data were identified by WaMoS® II *rtQC*. To minimize and evaluate the impact of the radar interferences on the WaMoS® II measurements, corrupted images or sectors were replaced by blanked data, and such data sets are marked with $IQ = [1$ or $2]$, depending on the amount of missing data in the analysis sequence. This allowed us to evaluate the impact of missing sectors on the general performance of the system. It turned out that, when sufficient *sea clutter* was visible in the other parts of the image and no other *rtQC* test failed ($IQ > 10$), the direct measurement was undisturbed and reliable. The missing signal leads to an underestimation of the indirect measurement of H_s, which is indicated by $IQ > 0$.

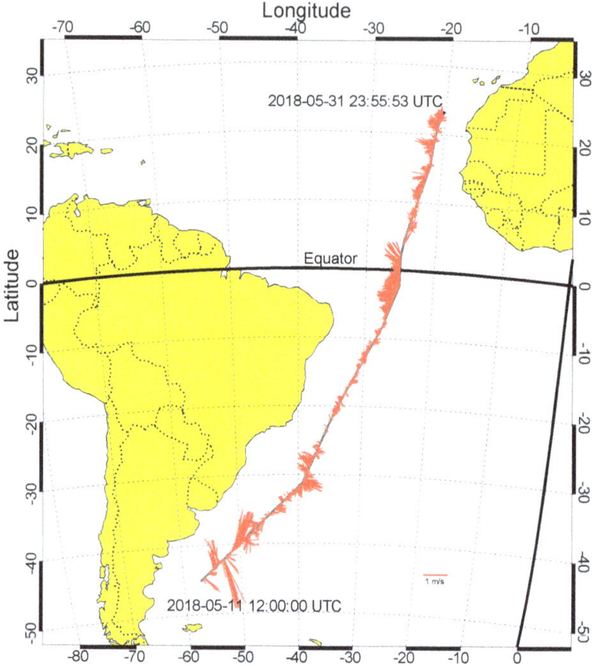

Figure 5. Cruise track (grey) of *Polarstern*. Stick plot (red) indicating the surface currents observed by WaMoS® II during PS113 (May data only). The length of the sticks is related to current speed, orientation represents current direction (going to).

The resulting data (Figure 6) analysis proved the performance of the *rtQC*. It successfully identified data sets with insufficient *sea clutter* during rain or no sufficient wind (>3 m/s) conditions, which are marked grey. Cases with insufficient wind speed were confirmed by independent direct wind speed measurements (German Weather Service, DWD) onboard. Rain cases were confirmed by visual observations and visual inspections of the corresponding radar image. Figure 6 shows the time series of wind speed (top panel in turquoise) and WaMoS® II current speed and direction (middle and lower panel). WaMoS® II data with $IQ < 10$ are marked in red, while data with $IQ > 10$ are marked in grey. The data sets, which were acquired during very low wind speeds (<3 m/s), show unrealistic scatter in the current speed and direction. These data sets were successfully identified ($IQ > 10$) by the WaMoS® II *rtQC*, and hence are displayed in grey. This evaluation confirms that a minimum wind speed of 3 m/s is required for reliable WaMoS® II measurements. In cases of too low wind speed, the radar images contain only random noise rather than wave information. Without the stringent quality control, the WaMoS algorithm outputs current solutions based on random noise which are completely unrelated to the real current conditions (e.g., May 18th). Please note that the WaMoS® II *rtQC* is independent of the wind measurements; hence, it is independent from their availability, accuracy and reliability.

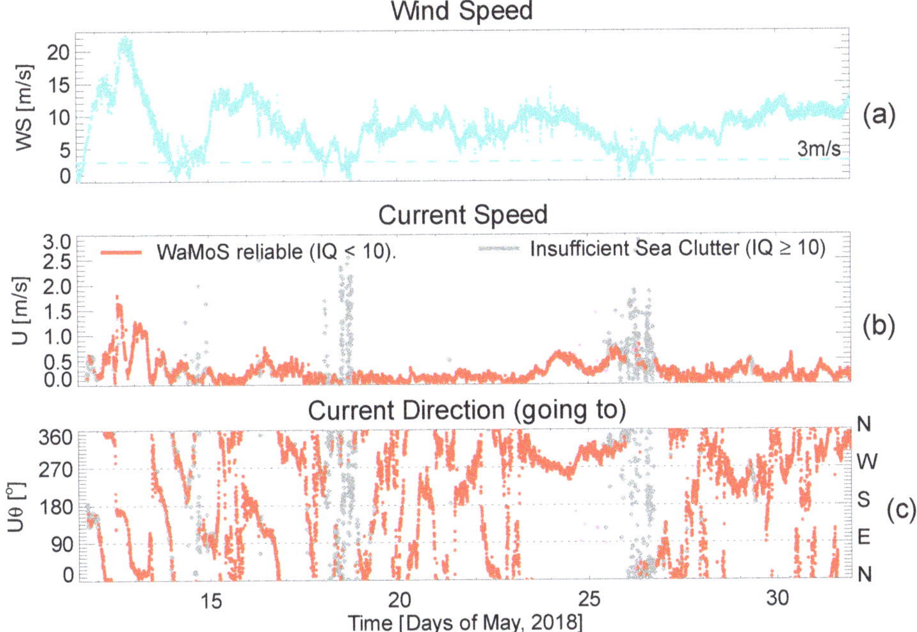

Figure 6. Time series of wind speed (**a**) and WaMoS® II surface current speed (**b**) and direction (**c**). Data which passed the WaMoS® II *rtQC* with $IQ < 10$ are in red, data with $IQ \geq 10$ are assumed to be unreliable and are shown in grey.

For the May 2018 period shown, WaMoS® II carried out 9727 individual instantaneous measurements. From this data set, about 80% pass the *rtQC* with $IQ < 10$ and can be accepted as reliable for direct wave and surface current measurements. The rest of the data is identified as unreliable because of insufficient *sea clutter* due to interferences with the navigation radar and/or environmental conditions (no sufficient wind, rain, very low sea state). About 10% of the WaMoS® II data had reduced quality, with $10 < IQ < 400$, characterized by noisy wave spectra. These data sets may include valuable information, but no statistically reliable estimates of integrated wave parameters can be derived. Especially rain induced noise interferes with *Hs* estimates, while more robust direct measurements like T_p or current $\vec{U}$ might contain valuable information. The final 10% of the data set with $IQ > 400$ does not contain any sufficient radar signals for WaMoS® II processing.

The results of the WaMoS® II surface current measurements were compared with the ADCP measurements (Figure 7). The visual comparisons of the measurements demonstrate the general agreement of the WaMoS® II surface and ADCP subsurface measurements, with exception of the equatorial region, where a significant vertical current shear does not allow a direct comparison.

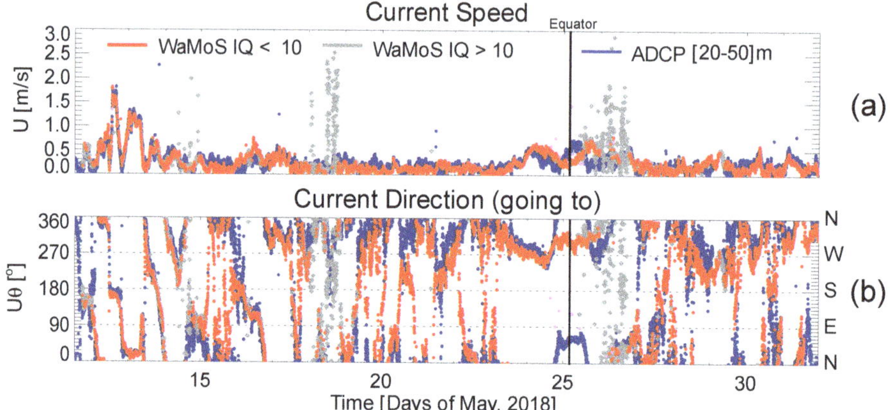

Figure 7. Time series of the current speed (**a**) and current direction (**b**) as derived by WaMoS® II (red/grey) and ADCP (blue) on board *Polarstern* during PS113 cruise. The grey values indicate WaMoS® II marked as unreliable ($IQ \geq 10$). The vertical line at May 25th marks the time when the equator was crossed.

The best agreements in current speeds and directions between WaMoS® II and ADCP were observed at the beginning of the cruise (May 12th–14th) in mid-latitudes (~40° S). Oceanographically, this region is characterized by the opposing northward oriented *Falkland-Malvinas Current* and the southward flowing *Brazil Current*. The area where both currents meet, the *Brazil Falkland-Malvinas Confluence*, is recognized as one of the most energetic in the world's ocean, with large-amplitude meanders and mesoscale eddies [23]. WaMoS® II as well as the ADCP, observed almost identical currents with maximum speeds up to 2 m/s, strongly varying in speed and direction. Leaving this zone, a region with current speed <0.5 m/s was passed. Here small deviations between surface WaMoS® II and sub-surface ADCP measurements can be observed. This is most likely due to vertically inhomogeneous current conditions. In the region of the equator (±2°, ~May 25th), the surface current measured by WaMoS® II and the subsurface current recorded by ADCP deviate. This deviation is primarily caused by the Equatorial Under Current (EUC) that occupies the depth range of 30–250 m with its strong eastward-directed velocities [14], opposing the north-westward-directed wind drift of the surface layer.

6. Results

In this section, we present the comparison of the quality-controlled WaMoS® II surface current data ($IQ < 10$) with the quality-controlled ADCP subsurface current data. The agreement of both data sets is estimated from the following statistical parameters: Bias, correlation coefficient (r) and standard deviation (σ_Δ, σ_s). For both data sets (WaMoS® II and ADCP), standard deviation of the mean (σ_{WaMoS}, σ_{ADCP}) for the individual measurements was determined over an averaging interval of 20 min. For the comparison, the data obtained near the equator was excluded, because vertical homogeneity was not given there [14]. Finally, 7272 individual data sets pass the *rtQC* and are used for this evaluation.

Figure 8 shows the direct comparison of the quality controlled WaMoS® II surface and ADCP subsurface current measurements for the eastward (*UE*) and northward (*UN*) components. For both components (*UE* and *UN*), the statistical results (Figure 8a,c) are in the same range (r: 0.94, 0.97; bias = −0.02 m/s, −0.06 m/s and σ_s: 0.05 m/s, 0.05 m/s, respectively). For the *UN* component, the correlation r as well as $bias_\Delta$ and σ_s are slightly higher. This is most likely related to the fact that higher absolute values of *UN* with speeds up to 1.5 m/s were observed, while *UE* speeds remained below 0.7 m/s during the entire cruise. The corresponding histograms of the signed differences (ΔUE

and ΔUN) between the WaMoS and ADCP current components are shown in Figures 8b and 8d. Both distributions can be fairly well approximated by a Gaussian distribution ($ae^{-(x-b)^2/c^2}$, with a the height, b the center, c the width (the standard deviation) of the Gaussian distribution). Given the fact that WaMoS and ADCP measure in different depths, and that ocean currents tend to be vertically sheared and to veer with depth geostrophically, following the Ekman spiral of the wind driven flow or the Stokes drift associated with surface waves, some differences between the two measurements are expected. Hence, a bias of the found magnitude and the slight deviations of the histograms from Gaussian shape do not necessarily signify a measurement error.

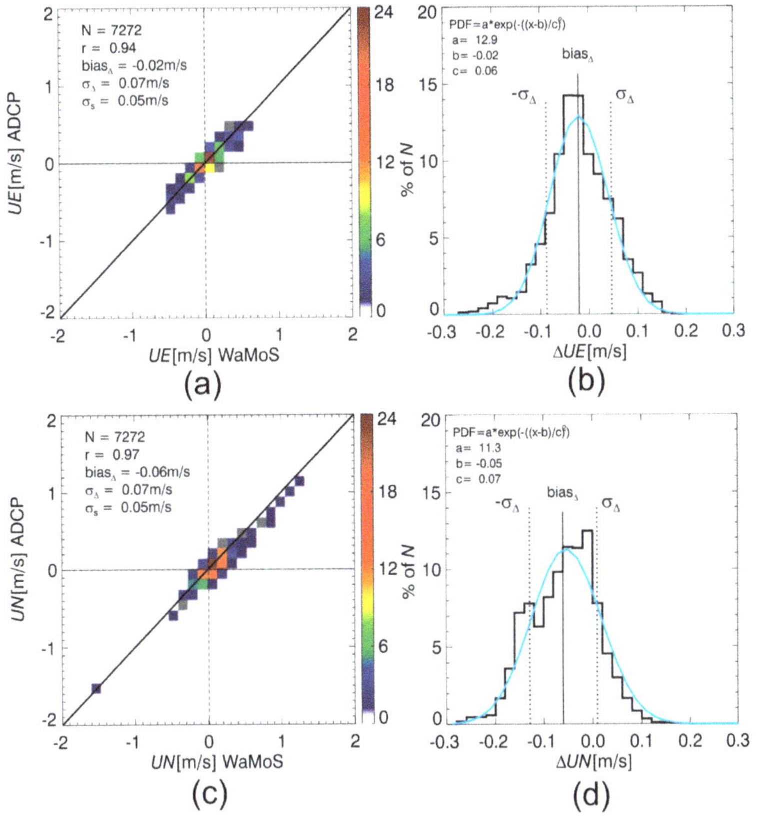

Figure 8. Comparison of the Eastward (UE) and Northward current components (UN) of WaMoS® II surface and ADCP subsurface (mean of the 20–50 m depth range) currents. (**a**,**c**) Scatter plot of WaMoS versus ADCP. (**b**,**d**) Histograms (black) and PDF (blue) of the differences between WaMoS and ADCP. N gives the number of data pairs, r the correlation coefficient, $\sigma_S = \frac{1}{2}\sqrt{2}\,\sigma_\Delta$ the standard deviation. The bin width is 0.1 m/s in the scatter plots and 0.02 m/s in the histograms.

The comparison of the absolute current speed ($US = \sqrt{UE^2 + UN^2}$) (Figure 9a,b), reveals the same agreement (r: 0.96 and σ_S: 0.05 m/s) with an almost perfect Gaussian distribution and no significant bias (−0.0004 m/s) of the differences. For the current direction ($U\theta$, Figure 9c,d) the correlation is slightly lower (r: 0.87). The bias of −6.88° suggests that the disagreement between the two systems is mostly due to differences in the orientation of the determined currents. However, we cannot completely rule out at present that one of the two measurement systems or both are subject to some small systematic errors. A bias in the same range as in our observations is reported from most studies comparing MR-derived surface current estimates with estimates from other devices (e.g., [10]). It is also possible that effects from a possible misalignment of the ADCP transducer are not completely removed during the ADCP data processing.

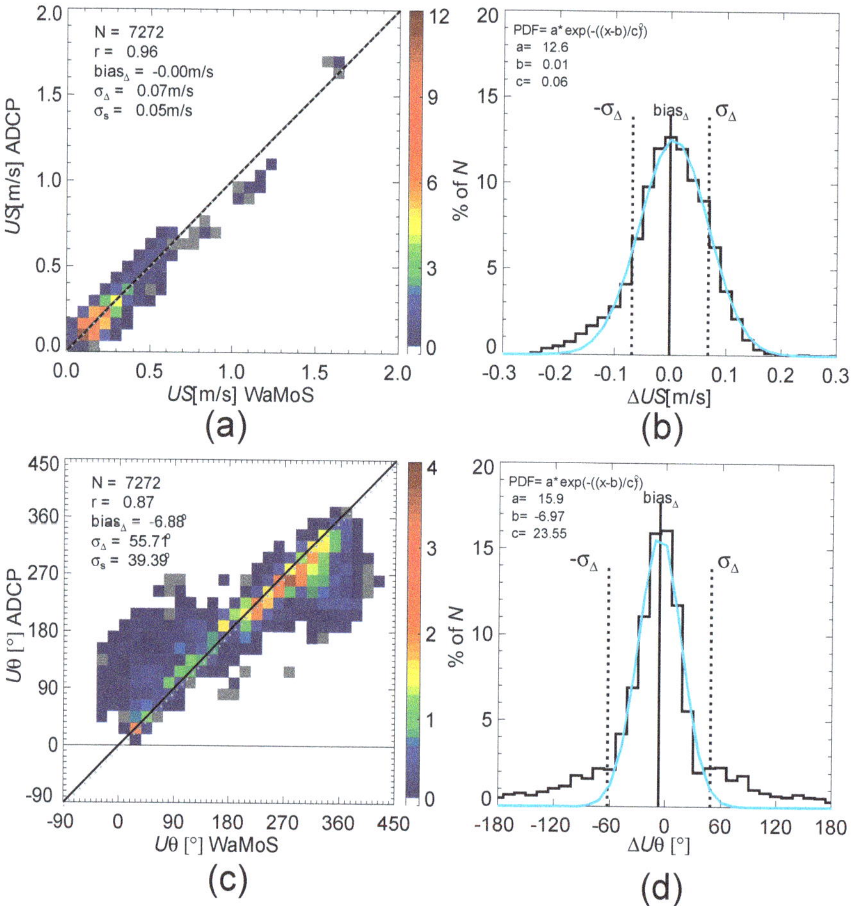

Figure 9. Comparison of the current speed (*US*) and current direction (*Uθ*) of WaMoS® II surface and ADCP subsurface (mean of the 20–50 m depth range) currents. $U\theta = 0$ refers to northward current. Please note that due to the angle discontinuity at North (360°/0°), $U\theta$ covers the range of [−90:450°]. (**a**,**c**): Scatter plot of WaMoS versus ADCP. (**b**,**d**): Histograms (black) and PDF (blue) of the differences between WaMoS and ADCP. *N* gives the number of data pairs, *r* the correlation coefficient, $\sigma_S = \frac{1}{2}\sqrt{2}\,\sigma_\Delta$ the standard deviation. The bin width for *US* is 0.1 m/s in the scatter plot (**a**) and 0.02 m/s in the histogram (**b**). The bin width for *Uθ* is 16° in the scatter plot (**c**) and 12° m/s in the histogram (**d**).

Histograms of the absolute deviations $|\Delta(UE)|$ and $|\Delta(UN)|$ (black) and of the individual standard deviations for WaMoS (red) and ADCP (blue) for both current components are shown in Figure 10. Again, the measurement differences for both current components reveal approximately the same distribution, covering the range up to 0.2 m/s and standard deviation of $\sigma_\Delta = 0.07$ m/s. The corresponding histograms for the standard deviations of the individual measurements, σ_{WaMoS} and σ_{ADCP} (WaMoS® II: red, ADCP: blue) show the same behavior: WaMoS exhibits most variation in the interval 0–0.02 m/s, decaying exponentially at higher values. Due to the natural variability of currents, $\sigma > 0$ must be expected, especially in areas with strong currents. Therefore, σ does not solely reflect precision of the measurements. Again, both components (*UE*, *UN*) show the same behavior. The distribution of the ADCP data has its peak at the 0.02–0.04 m/s bin and decays from there exponentially to higher values. For both sensors, the individual standard deviations σ_{WaMoS} and σ_{ADCP} are below the estimated common value, σ_Δ, obtained from the comparison: $\sigma_{WaMoS}, \sigma_{ADCP} < \sigma_S$. The square root of the sum of the squares of both standard deviations is also below the combined value: $\sqrt{\sigma^2_{WaMoS} + \sigma^2_{ADCP}} < \sigma_\Delta$. The slightly lower values of σ_{WaMoS} compared to σ_{ADCP} are assumed to result from the different spatial coverages. Since the ADCP delivers measurements that are locally more confined than those obtained by WaMoS® II, they can be assumed to be subject to more variability than the WaMoS® II data representing areal means over several square kilometers. In summary, most of the observed deviations between WaMoS and ADCP can be explained by the different observation volumes (vertical and horizontal extents) and the natural velocity variability contained therein. The results of the comparison between WaMoS and ADCP for the different current components are summarized in Table 1.

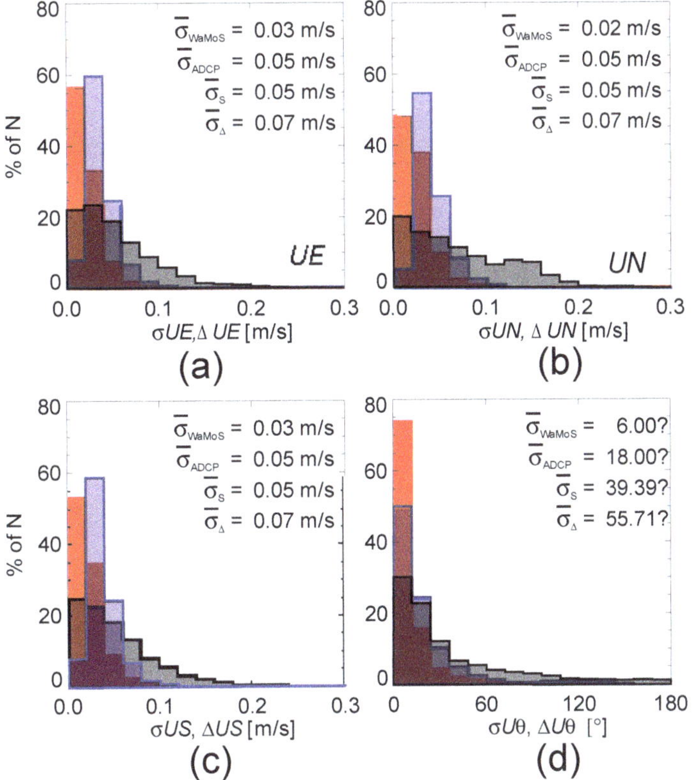

Figure 10. Histogram of the standard deviation of the mean WaMoS® II (red) and ADCP (blue) current component—(**a**) *UE*, (**b**) *UN*, (**c**) *US*, and (**d**) $U\theta$—and the absolute difference between the WaMoS and ADCP measurements (black). The bin width of the histograms is 0.02 m/s for the current components and 12° for the current direction.

Table 1. Results of the comparison between WaMoS® II surface and ADCP subsurface current measurements.

Parameter	*Symbol*	*UE*	*UN*	*US*	*Uθ*
Number of data sets	N			7272	
Correlation coefficient (Equation (1))	r	0.94	0.97	0.96	0.87
Bias (Equation (2))	$\overline{\Delta}$	−0.02 m/s	−0.06 m/s	−0.0004 m/s	−6.88°
Total standard deviation of the difference (Equation (3))	σ_Δ	0.07 m/s	0.07 m/s	0.07 m/s	55.71°
Individual standard deviation (Equation (4))	σ_S	0.05 m/s	0.05 m/s	0.05 m/s	39.39°
Standard deviation of the temporal mean ADCP measurements (Equation (5))	σ_{ADCP}	0.05 m/s	0.05 m/s	0.05 m/s	18.00°
Standard deviation of the temporal WaMoS®II measurements (Equation (5))	σ_{WaMoS}	0.03 m/s	0.02 m/s	0.03 m/s	6.00°

7. Summary and Conclusions

The key motivation of this analysis was to evaluate the usability of the MR-based *sigma* S6 WaMoS® II system with respect to the reliability, precision and eventually accuracy of the surface current measurements. As previous evaluations of WaMoS® II measurements were based on direct comparisons with ADCP measurements, the resulting accuracies often lead to misinterpretation. This is because the data sets used are results of different measurement principles, and also because of temporal and spatial misalignment of the data sets [24]. The fact that vertical and horizontal current shears, which lead to deviation of WaMoS® II surface and ADCP subsurface current measurements, do not automatically reflect an error in one of the measurements was occasionally mentioned in earlier work [15,25].

To reduce the effect of natural current variability on the data set comparison, temporal means over an averaging interval of 20 min were used. The averaging further allows one to estimate the standard deviation of the observed current. The mean current values of WaMoS® II and ADCP were then directly compared with standard statistical tools, such as correlation coefficient (r), bias ($|\overline{\Delta}|$), and standard deviation (σ_Δ, σ_s). The results of $r > 0.9$, $|\overline{\Delta}| < 0.06$ m/s and $\sigma_s = 0.05$ m/s reveal an excellent agreement between the two data sets and hence the validity of the measurements, especially when taking into account that these values may include deviations unrelated to errors or inaccuracies in the measurement devices but to vertical and horizontal inhomogeneities. Only data sets acquired in the equatorial region, where a large vertical current shear associated with the Equatorial Undercurrent exists, were excluded from this comparison. The standard deviation of the individual mean current value ($\sigma_{WaMoS}, \sigma_{ADCP}$) was assumed to reflect the precision of the individual measurements, even when this parameter is strongly linked to the natural variability of the currents. Here the results $\sigma_{WaMoS} = 0.02$ m/s for both current components reflect the high stability of the measurement during different current regimes and wave conditions. This value represents a combination of the natural variability of the flow and potential measurement errors. Due to the spatial character of the WaMoS® II measurement σ_{WaMoS} is lower than the equivalent value obtained for the ADCP, $\sigma_{ADCP} = 0.04$ m/s. The fact that both σ_{WaMoS} and σ_{ADCP} do not exceed the combined single standard deviation σ_s confirms the consistency of the validation. Assuming that the sensor-related error of the ADCP is equivalent to the theoretical error $\sigma_{ADCP(theoretical)} = 0.0113$ m/s the error related to the natural current variations $\sigma_{fluctuation}(ADCP) = \sigma_{ADCP} - \sigma_{ADCP(theoretical)} = 0.03$ m/s. The fact that $\sigma_{WaMoS} < \sigma_{fluctuation}(ADCP)$ is related to the larger integration domain of the WaMoS and that $\sigma_{ADCP(theoretical)}$ is an ideal value, which in reality is likely larger.

To ensure consistent data set precision, and that all WaMoS® II data sets satisfy this high validity, an internal quality control flag is set for each individual measurement. During *Polarstern* cruise PS113 different environmental conditions with high and low wind speeds and wave heights as well a different precipitation conditions were met, proving the proper performance of the WaMoS® II *rtQC*.

For operational use of X-Band radar wave and current observations, stringent quality control is advisable. Interferences in the radar images or in the absence of wind no reliable and accurate observation of waves and current are possible and can lead to inaccurate measurements. Therefore, the quality of the measurement needs to be indicated to the user.

For our comparison of WaMoS® II measurements with ADCP reference data, we unfortunately could not make use of ADCP currents above 17 m. This points to a general shortcoming of vessel-mounted ADCPs; namely, that no current measurements are taken from the water column above the keel depth of the ship plus some blanking distance. This shortcoming hampers scientific progress in the understanding of processes which govern the coupling of atmosphere and ocean. Having shown in our study that quality-controlled WaMoS® II measurements compare well with ADCP data in regions which are not obviously subject to strong near-surface vertical shear such as, for instance, the equatorial region with its undercurrent, suggests making use of WaMoS® everywhere when the quality control indicates valid measurements, and combine those with ADCP data in order to obtain a full vertical current profile reaching up to the surface.

Author Contributions: Conceptualization, K.G.H.; methodology, K.G.H. and W.-J.v.A.; software, K.G.H. and W.-J.v.A.; validation, K.G.H. and W.-J.v.A.; formal analysis, K.G.H. and W.-J.v.A.; investigation, K.G.H., S.E.N., W.-J.v.A., and V.H.S.; resources, K.G.H., S.E.N., W.-J.v.A., and V.H.S.; data curation, K.G.H. and W.-J.v.A.; writing—original draft preparation, K.G.H.; writing—review and editing, K.G.H., S.E.N., W.-J.v.A., and V.H.S.; visualization, K.G.H.; supervision, K.G.H.; project administration, S.E.N. and V.H.S.; funding acquisition, K.G.H., S.E.N., and V.H.S.

Funding: This research received no external funding.

Acknowledgments: The authors would like to thank the German Weather Service *DWD* for providing the wind information. Furthermore, we would like to thank Ralf Krocker (AWI) and Thomas Liebe and Johannes Rogenhagen (*F. Laeisz*) for supporting the WaMoS® II installation on board *Polarstern*. ADCP Software OSSI, version 1.9 provided by T. Fischer, GEOMAR, Kiel. We are moreover grateful to the comments provided by four reviewers, which helped to improve the manuscript.

Conflicts of Interest: The authors declare no conflict of interest.

References

1. Oudshoorn, H.M. The use of radar in hydrodynamic surveying. *Coast. Eng. Proc.* **1960**, 59–76. [CrossRef]
2. Young, I.; Rosenthal, W.; Ziemer, F. A Three-Dimensional Analysis of Marine Radar Images for the Determination of Ocean Wave Directionality and Surface Currents. *J. Geophys. Res.* **1985**, *90*, 1049–1059. [CrossRef]
3. Senet, C.; Seeman, J.; Ziemer, F. The near-surface current velocity determined from image sequences of the sea surface. *IEEE Trans. Geosci. Remote Sens.* **2001**, *39*, 492–505. [CrossRef]
4. Ziemer, F.; Dittmer, J. A system to monitor ocean wave fields. In Proceedings of the Oceans '94, Brest, France, 13–16 September 1994.
5. Borge, J.C.N.; Reichert, K.; Dittmer, J. Use of nautical radar as a wave monitoring instrument. *Coast. Eng.* **1999**, *37*, 331–342. [CrossRef]
6. Borge, J.C.N.; Rodrígzues, G.R.; Hessner, K.G.; González, P.I. Inverstion of marine radar images for surface wave analysis. *J. Atmos. Ocean. Technol.* **2004**, *21*, 1291–1300. [CrossRef]
7. Bell, P.S.; Lawrence, J.; Norris, J.V. Determining currents from marine radar data in an extreme current environment at a tidal energy test site. In Proceedings of the IEEE International Geoscience and Remote Sensing Symposium (IGARSS 2012), Munich, Germany, 22–27 July 2012.
8. Hessner, K.G.; Reichert, K.; Borge, J.C.N.; Stevens, C.; Smith, M. High-resolution X-Band radar measurements of currents, bathymetry and sea state in highly inhomogeneous coastal areas. *Ocean. Dyn.* **2014**, *64*, 989–998. [CrossRef]
9. Hessner, K.G.; Wallbridge, S.; Dolphin, T. Validation of Areal Wave and Current Measurements Based on X-Band Radar. In Proceedings of the 2015 IEEE/OES Eleveth Current, Waves and Turbulence Measurement (CWTM 2015), St. Petersburg, FL, USA, 2–6 March 2015.

10. Lund, B.; Graber, H.C.; Hessner, K.G.; Williams, N.J. On Shipbord Marin X-Band Radar Near-Surface Current "Calibration". *J. Athmos. Ocean. Technol.* **2015**, *32*, 1928–1944. [CrossRef]
11. Lund, B.; Collins, C.O., III; Tamura, T.; Graber, H.C. Multi-directional wave spectra from marine X-Band radar. *Ocean. Dyn.* **2016**, *66*, 973–988. [CrossRef]
12. Lund, B.; Zappa, C.J.; Graber, H.C.; Cifuentes-Lorenzen, A. Shipboard Wave Measurments in the Southern Ocean. *J. Athmos. Ocean. Technol.* **2017**, *34*, 2113–2126. [CrossRef]
13. Skolnik, M.I. *Introduction to Radar Systems*; McGraw-Hill: New York, NY, USA, 1981.
14. Strass, V. The Expedition PS113 of the Research Vessel POLARSTERN to the Atlantic Ocean in 2018. *Berichte zur Polar- und Meeresforschung* **2018**, *724*, 72.
15. Lund, B.; Graber, H.C.; Tamura, H.C.; Varlamow, S.M.; Collins, C.O., III. A new technique for the retrieval of near-surface vertical current shear from marine X-band radar images. *J. Geophys. Res.* **2015**, *120*, 8466–8486. [CrossRef]
16. Chapman, R.D.; Shay, L.K.; Graber, H.C.; Edson, J.B.; Karachintsev, A.; Trump, C.L.; Ross, D.B. On the accuracy of HR radar surface current measurments: Intercaomparison with ship-based sensors. *J. Geophys. Res.* **1997**, *102*, 18.737–18.748. [CrossRef]
17. Liu, Y.; Weisberg, R.H.; Merz, C.R. Assessment of codar seasonde and wera hf radars in mapping surface currents on the West Floria shelf. *J. Atmos. Ocean. Techol.* **2014**, *31*, 1363–1382. [CrossRef]
18. Williams, A.J. Standards for in Situ current measurement. In Proceedings of the 2015 IEEE/OES Eleveth Current, Waves and Turbulence Measurement (CWTM 2015), St. Petersburg, FL, USA, 2–6 March 2015.
19. Noris, A.; Wall, A.D.; Bole, A.G.; Dineley, W.O. *Radar and ARPA Manual: Radar and Target Tracking for Professional Mariners, Yachtsmen and Users of Marine Radar*, 2nd ed.; Elsevier Science: Amsterdam, The Netherlands, 2005; p. 544.
20. Bell, P.S.; Osler, J.C. Mapping bathymetry using X-band marine radar data recorded from a moving vessel. *Ocean. Dyn.* **2011**, *61*, 2141–2156. [CrossRef]
21. Dean, R.G.; Dalrymple, R.A. *Water Wave Meachanics for Engineers and Scientists*; World Scientific: Singapore, 1991.
22. Fischer, J.; Brandt, P.; Dengler, M.; Müller, M.; Symonds, D. Surveying the upper ocean with the ocean surveyor: A new phased array doppler current profiler. *J. Atmos. Ocean. Technol.* **2003**, *20*, 742–751. [CrossRef]
23. Gordon, A.L. Brazil—Malvinas Confluence. *Deep Sea Res.* **1989**, *36*, 359–384. [CrossRef]
24. Ardhuin, F.; Marié, L.; Rascle, N.; Forget, P.; Roland, A. Observation and estimation of Lagrangian, Stokes, and Eulerian currents induced by wind and waves at the sea surface. *J. Phys. Oceanogr.* **2009**, *19*, 2820–2838. [CrossRef]
25. Hessner, K.G.; El-Naggar, S.; Strass, V.; Krägefsky, S.; Witte, H. Evaluation of X-Band MR surfac and ADCP subsurface currents obtaine onboard the German Research Vessel Polarstern during ANT-XXXI expedition 2015/16. In Proceedings of the OCEANS 2017, Anchorage, AK, USA, 18–21 September 2017.

Article

An Imaging Algorithm for Multireceiver Synthetic Aperture Sonar

Xuebo Zhang [1,2], Cheng Tan [3,*] and Wenwei Ying [4]

1. College of Electrical and Information Engineering, Lanzhou University of Technology, Lanzhou 730050, China; xbzhang@scun.edu.cn
2. Laboratory of Underwater Acoustics, Zhanjiang 524022, China
3. Aerodynamics Research and Development Center, Computational Aerodynamics Institute, Mianyang 621000, China
4. Naval Research Academy, Beijing 102249, China; aayang@scun.edu.cn

* Correspondence: kevint@scun.edu.cn

Received: 20 February 2019; Accepted: 17 March 2019; Published: 20 March 2019

Abstract: For the multireceiver synthetic aperture sonar (SAS), the point target reference spectrum (PTRS) in the two-dimensional (2D) frequency domain and azimuth modulation in the range Doppler domain were first deduced based on a numerical evaluation method and accurate time delay. Then, the difference between the PTRS and azimuth modulation generated the coupling term in the 2D frequency domain. Compared with traditional methods, the PTRS, azimuth modulation and coupling term was better at avoiding approximations. Based on three functions, an imaging algorithm is presented in this paper. Considering the fact that the coupling term is characterized by range variance, the range-dependent sub-block processing method was exploited to perform the decoupling. Simulation results showed that the presented method improved the imaging performance across the whole swath in comparison with existing multireceiver SAS processor. Furthermore, real data was used to validate the presented method.

Keywords: synthetic aperture sonar (SAS); multireceiver; numerical evaluation; numerical transfer function; imaging algorithm

---

1. Introduction

Synthetic aperture sonar (SAS) [1] provides high resolution images via the coherent processing of successive echo data along the virtual aperture. This makes SAS a suitable technique for applications such as searching for small objects [2], imaging of wrecks [3], underwater archaeology [4] and pipeline inspection [5]. Additionally, it improves the classification and detection of objects based on SAS images [6]. Multireceiver SAS [7], as opposed to monostatic SAS constellation, offers a fast mapping rate at a given resolution. However, it does this at the cost of complicated signal processor.

For the multireceiver SAS, the point target reference spectrum (PTRS) [8] is a prerequisite of fast imaging algorithms. The two-way slant range of the multireceiver SAS consists of two hyperbolic range histories, which include the instantaneous range between the moving transmitter and target and that between the target and moving receiver. The two hyperbolic range histories make it difficult to deduce the point of stationary phase (PSP) and PTRS using the method of stationary phase [9]. In order to solve this problem, approximations are often exploited. In [10–12], the phase center approximation (PCA) was used to model a transducer located at the midpoint between the transmitter and receiver. With this, the echo data of multiple receivers is converted into the monostatic format. However, the preprocessing includes the compensation of phase errors [13]. Due to the space variance of approximation errors [13], it is difficult to compensate phase errors completely. Loffeld et al. [14] have presented an analytic PTRS. Their method was based on the approximation that the transmitter and receiver contribute equally to the Doppler frequency. Based on the method of stationary phase [9], two PSPs corresponding to the transmitter's phase and receiver's phase were deduced. The phase history of the transmitter and that of the receiver were expanded into a power series at their individual PSPs. Both phases were then combined to generate a quadratic function. It can be found that two approximations are exploited by this method. One is the equal Doppler contribution of the transmitter and receiver, and the other is the Taylor approximation of the transmitter's phase and receiver's phase. In general, this method only applies to the narrow beam case [14,15]. There are still some other methods deducing the analytic PTRS. The basic idea relies on series approximation. In [16], the quadratic approximation of the two-way range was exploited. This introduced a large residual error, which degraded the imaging performance at close range. Moreover, this method did not consider the compensation of the stop-and-hop error [12]. A single target suffers from the coordinate deviation in azimuth [17]. However, a distributed target suffers from the distortion. In [18,19], the two-way range was expanded into a power series with respect to the slow time. Additionally, the PSP was expanded into a power series based on the series reversion method. The accuracy of the two-way range and PSP was limited by the number of terms in the polynomial. With this method, the series approximation was used twice. The approximation error increased with the slow time. The accumulative error would be large when the SAS system works with the wide beam case. In [20,21], the instantaneous Doppler wavenumber was exploited to deduce the analytic PTRS. The two-way slant range was formulated as a function of equivalent bistatic squint angle and half bistatic angle [20,21]. Based on the method of stationary phase [9], the azimuth wavenumber can also be expressed as a function of equivalent bistatic squint angle and half bistatic angle. Then, the PTRS was expressed as a function of half bistatic angle, which should be analytically calculated. Considering that the triangle whose vertices were the transmitter, receiver and target, the fourth order equation with respect to half bistatic angle was obtained based on the theorem of sine and some basic algebra skills. In [21], this equation was solved using the series reversion method [18,19]. The accuracy of the PTRS is also limited by the number of terms in the power series.

Using an accurate time delay of the transmitted signal [12], the back projection (BP) algorithm [22] can provide high resolution results. In this paper, we present an imaging algorithm, which was also based on the accurate time delay of the transmitted signal. With the numerical evaluation method [23], we first calculated the PSP and azimuth PSP. The PTRS and azimuth modulation can be easily obtained based on their respective numerical PSPs. Then, we obtained the coupling term in the two-dimensional (2D) frequency domain using the phase difference between the numerical PTRS and azimuth modulation. The PTRS, coupling term and azimuth modulation avoiding approximations were further exploited to develop the imaging processor, which compensated the coupling phase based on the sub-block processing method.

This paper is organized as follows. In Section 2, the imaging geometry and signal model are introduced. Section 3 introduces the PTRS, azimuth modulation and coupling term based on the numerical evaluation method. Section 4 presents the imaging algorithm based on three functions. Section 5 compares the presented method with traditional methods, and highlights the advantages of the presented method. In Section 6, processing results of the simulated data and real data are used to validate the presented method. Finally, some conclusions are reported in the last section.

2. Imaging Geometry and Signal Model

The imaging geometry of the multireceiver SAS is shown in Figure 1. The linear array consists of a transmitter and receiver array including M uniformly spaced receivers. In Figure 1, the black rectangle denotes the transmitter. Each receiver has an integer index $i \in [1, M]$. For the i-th receiver, the distance between the receiver and transmitter is denoted by d_i. The linear array is aligned in the sonar moving direction, which is called the azimuth dimension. The horizontal direction represents the range dimension. Since the SAS configuration shown in Figure 1 is characterized by azimuth invariance, an ideal point target located at coordinates $(r, 0)$ is used. t denotes the slow time in the azimuth dimension. The fast time in the range dimension is represented by τ. c is the sound speed in water. v represents the velocity of the sonar platform.

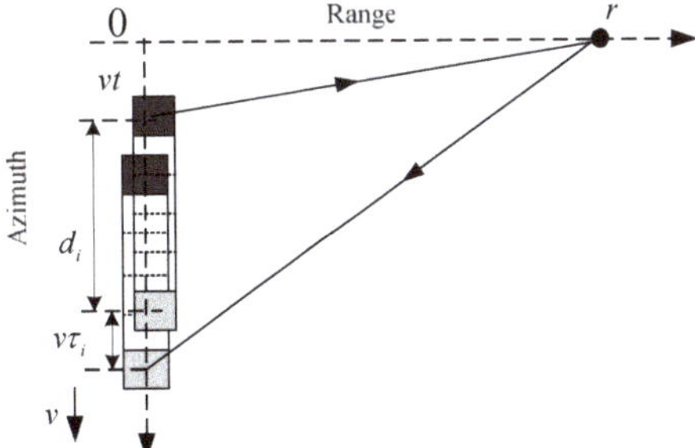

Figure 1. Imaging geometry of the multireceiver synthetic aperture sonar (SAS).

In Figure 1, the two-way slant range is from the transmitter to target and then back to the i-th receiver. When the transmitter moves to the position $v \cdot t$, a chirp signal $p(\tau)$ is transmitted. The accurate time delay [12] of the echo signal corresponding to the i-th receiver is given by:

$$\tau_i = \frac{v(v \cdot t + d_i) + c\sqrt{v^2t^2 + r^2}}{c^2 - v^2} + \frac{\sqrt{\left[v(v \cdot t + d_i) + c\sqrt{v^2t^2 + r^2}\right]^2 + (c^2 - v^2)(2v \cdot t \cdot d_i + {d_i}^2)}}{c^2 - v^2} \quad (1)$$

In (1), we consider the forward distance of the i-th receiver during the signal reception, because the sonar is continuously travelling along the azimuth dimension [17]. After demodulation, the echo signal corresponding to the i-th receiver is expressed as:

$$ss_i(\tau, t) = p(\tau - \tau_i)\omega_a(t)\exp\{-j2\pi f_c \tau_i\} \quad (2)$$

where f_c is the center frequency. The composite beam pattern corresponding to the i-th receiver and transmitter is represented by $\omega_a(\cdot)$. For simplicity, we neglect this beam pattern to concentrate on the phase processing.

3. PTRS, Azimuth Modulation and Coupling Term

The BP algorithm [22] should compute the instantaneous range from the moving transmitter to target and then back to moving receiver. Therefore, (1) is very suitable for the BP algorithm. Since (1) considers the influence of stop-and-hop assumption, BP algorithm can provide high resolution image. However, (1) cannot be used by traditional fast algorithms. For simplicity, the forward distance of the i-th receiver during the signal reception is approximated by $v\tau_i \approx 2vr/c$. This approximation degrades the imaging performance when the SAS system works with the wide beam case. In this paper, the accurate time delay shown as (1) is extended to fast imaging algorithms. The PTRS, azimuth modulation and coupling term play an important role in developing fast imaging algorithms. We start from the deduction of the PTRS, azimuth modulation and coupling term in this section.

3.1. PTRS

Based on the Fourier transformation (FT), the signal denoted by (2) is transformed into the 2D frequency domain. The expression is given by:

$$SS_i(f_\tau, f_t) = P(f_\tau)\int_{-T_s/2}^{T_s/2} \exp\{-j2\pi(f_c + f_\tau)\tau_i - j2\pi f_t t\}dt \quad (3)$$

where T_s denotes the integration time; $P(f_\tau)$ is the spectrum of the transmitted signal; f_τ and f_t are the instantaneous and Doppler frequencies.

Due to the complex expression of (1), it is difficult to calculate (3). To solve this problem, approximations are usually exploited by traditional methods. However, the residual errors would influence the imaging performance. Here, we present the numerical result, which avoids approximations.

The phase of the exponent in (3) is defined as:

$$\Psi_i(f_\tau, f_t) = -2\pi(f_c + f_\tau)\tau_i - 2\pi f_t t \tag{4}$$

Applying the method of stationary phase [9] to (4) yields:

$$\frac{\partial \tau_i(\widetilde{t}_i)}{\partial t} + \frac{f_t}{f_c + f_\tau} = 0 \tag{5}$$

where $\widetilde{t}_i \in [-T_s/2, T_s/2]$ represents the PSP; $\frac{\partial \tau_i}{\partial t}$ denotes the first derivative with respect to the slow time.

Using (1), the first derivative with respect to the slow time is given by:

$$\begin{aligned} \frac{\partial \tau_i}{\partial t} = \frac{v^2 + ctv^2\left[(vt)^2 + r^2\right]^{-0.5}}{c^2 - v^2} + \frac{\left\{\left\{v[(vt) + d_i] + c\sqrt{(vt)^2 + r^2}\right\}^2 + (c^2 - v^2)\left[2(vt)d_i + d_i^2\right]\right\}^{-0.5}}{c^2 - v^2} \times \\ \left\{\left[v^2 t + vd_i + c\sqrt{(vt)^2 + r^2}\right]\left\{v^2 + ctv^2\left[(vt)^2 + r^2\right]^{-0.5}\right\} + vd_i(c^2 - v^2)\right\} \end{aligned} \tag{6}$$

(5) cannot be solved analytically, as (6) is very complicated. Due to this, the numerical evaluation method is used to calculate the effective solution of (5). The effective solution called the PSP is denoted by $\widetilde{t}_i$. Substituting the numerical PSP into (4) yields:

$$\Psi_i(f_\tau, f_t; \widetilde{t}_i) = -2\pi(f_c + f_\tau)\tau_i(\widetilde{t}_i) - 2\pi f_t \widetilde{t}_i \tag{7}$$

Examining (7), we see that the numerical PTRS avoids approximations. Besides, the PTRS is a function of the instantaneous frequency f_τ, Doppler frequency f_t and range r. The space variance makes the development of fast imaging algorithms a challenge.

Based on the series expansion, the PTRS is decomposed into the azimuth modulation and coupling term. Using the coupling term, the decoupling operation between the range and azimuth dimensions is first carried out, and hence the imaging process is decomposed into two separate filtering processes in the range and azimuth dimensions. However, it is hard to obtain both terms using series expansion, because (7) is not an analytic expression.

3.2. Azimuth Modulation

After the decoupling operation, the azimuth compression is usually performed in the range Doppler domain. It is easily concluded that the filtering function related to the azimuth compression is only a function of the range r and Doppler frequency f_t [8,9]. In other words, the azimuth modulation is independent of the instantaneous frequency f_τ [8,9]. If the azimuth modulation were obtained, the phase difference between the PTRS and azimuth modulation denotes the coupling term. In other words, the azimuth modulation should be first derived. For conventional methods, the azimuth modulation and coupling term are simultaneously obtained based on the series approximation of the PTRS with respect to the instantaneous frequency. Inspecting (7), it is impossible to perform the series approximation, because the PTRS in this paper does not possess the explicit expression. To solve this problem, the numerical evaluation method is still used to calculate the azimuth modulation. Since the azimuth modulation is independent of the instantaneous frequency [8,9], we obtain the azimuth modulation by setting $f_\tau = 0$ in (7). This is given by:

$$\varphi_{ac\_i}(f_t; r) = \Psi_i(f_\tau, 0; \hat{t}_i) = -2\pi f_c \tau_i(\hat{t}_i) - 2\pi f_t \hat{t}_i \tag{8}$$

In (8), $\hat{t}_i \in [-T_s/2, T_s/2]$ represents the PSP used by the azimuth modulation. At this point, we call it the azimuth PSP. It is independent of the instantaneous frequency [8,9]. Although we present the azimuth modulation in (8), the azimuth PSP is not expressed explicitly. Considering the fact that the azimuth modulation is independent of the instantaneous frequency, we turn our attention to the deduction of the azimuth PSP. Setting $f_\tau = 0$ in (5) yields:

$$\frac{\partial \tau_i(\hat{t}_i)}{\partial t} + \frac{f_t}{f_c} = 0 \tag{9}$$

Based on the numerical method, the azimuth PSP $\hat{t}_i$ is calculated. Substituting the azimuth PSP into (8), we obtain the numerical expression of the azimuth modulation.

3.3. Coupling Term

Because of the relative motion between the sonar and target, the distance between them changes with time; hence, the time delay changes correspondingly. The effect is that the received echo from the same target at different azimuth sample times will distribute at different bins along the range direction. This phenomenon is called the range cell migration (RCM), which completely describes the coupling between the range and azimuth dimensions. Due to the coupling, the imaging process cannot be simply decomposed into two separate filtering processes in the range and azimuth dimensions. The direct processing scheme is to cancel the coupling before the azimuth matched filtering.

With traditional methods, the coupling term is obtained by using the series expansion of the PTRS with respect to the instantaneous frequency. Since the PTRS shown as (7) does not own an explicit expression, the series expansion method cannot be exploited. In practice, the PTRS consists of the coupling term and azimuth modulation. Therefore, the difference between the PTRS and azimuth modulation denotes the coupling term between the range and azimuth dimensions. It is expressed as:

$$\varphi_i(f_\tau, f_t; r) = \Psi_i(f_\tau, f_t; r) - \varphi_{ac\_i}(f_t; r) \tag{10}$$

Until now, the PTRS, coupling term and azimuth modulation are all deduced. The azimuth modulation is a wideband signal. After decoupling, the matched filtering is expected to perform the focusing in the azimuth dimension.

4. Imaging Algorithm

The key issue of the SAS imagery is the decoupling. Inspecting (10), the coupling phase is a function of the range, instantaneous frequency and Doppler frequency. Due to the range variance, we cannot perform the decoupling operation in the range Doppler domain. Based on the characteristic of the coupling phase, the sub-block processing method is used to perform the decoupling operation. In this section, important steps of the presented method are introduced in detail.

4.1. 2D FT

In this step, each receiver's data is transformed into 2D frequency domain based on the FT. Each receiver data is undersampling. The energy of the frequency $f_t' \in [-PRF/2, PRF/2]$ is a combination of all the energy related to the frequency points $(\cdots, f_t' - PRF, f_t', f_t' + PRF, \cdots) \in [-M \cdot PRF/2, M \cdot PRF/2]$. Here, $f_t' \in [-PRF/2, PRF/2]$ denotes the Doppler frequency related to the sampling rate of each receiver's data. In other words, the spectrum in the Doppler domain is aliased M times. Here the pulse repetition frequency is denoted by PRF, which is also the sampling frequency of each receiver's data in the azimuth dimension. The alias can be suppressed by the coherent processing of multireceiver data.

4.2. Decoupling

The coupling phase shown as (10) is characterized by range variance. Although the coupling phase is accurate, it is hard to compensate the coupling phase completely. Here, the sub-block processing method is exploited to perform the decoupling operation. Since it is hoped that the echo signal of a target is in the same range cell (usually at r), the range deviation from its desired position denotes the RCM. Based on (10), the RCM in the 2D frequency domain is expressed as:

$$\varphi_{\text{de}\_i}(f_\tau, f_t; r) = -\varphi_i(f_\tau, f_t; r) - 4\pi f_\tau \frac{r}{c} \tag{11}$$

From (11), we see that the RCM is a function of the range, instantaneous frequency and Doppler frequency. Based on this characteristic, the sub-block processing method rather than interpolation is exploited to carry out the range cell migration correction (RCMC). The decoupling is decomposed into two steps. One is the bulk decoupling, and the other is the differential decoupling. Based on (11), the filtering function of the bulk decoupling is given by:

$$H_{\text{bul}\_i} = \text{conj}\{P(f_\tau)\} \cdot \exp\{j\varphi_{\text{de}\_i}(f_\tau, f_t; r_c)\} \tag{12}$$

where r_c represents the center range of the mapping swath; conj$(\cdot)$ represents the complex conjugate.

The bulk decoupling simultaneously performs the range matched filtering. Considering targets at reference range, the coupling is completely removed after this step. However, other targets suffer from the residual error, which is $\varphi_{\text{de}\_i}(f_\tau, f_t; r) - \varphi_{\text{de}\_i}(f_\tau, f_t; r_c)$. Based on the sub-block processing method, the differential decoupling is used to solve this problem. Before performing the differential decoupling, the whole swath is virtually segmented into N sub-blocks in the range direction. Based on (11) and (12), the filtering function of the differential decoupling is written as

$$H_{i\_n} = \exp\{j\varphi_{\text{de}}(f_\tau, f_t; r_{\text{ref}\_n}) - j\varphi_{\text{de}\_i}(f_\tau, f_t; r_c)\} \tag{13}$$

where the center range of the n-th sub-block is used as the reference range, which is denoted by r_{ref_n}. The variable $n \in [1, N]$ denotes the sub-block index.

Based on (13), we remove the coupling between the range and azimuth dimensions for the data of the n-th sub-block. Then, the data is transformed into the time domain via the inverse Fourier transformation (IFT) in the range direction. For each sub-block, the differential decoupling is carried out using (13). The processed sub-blocks are extracted and stored in the range Doppler domain. The coupling between the range and azimuth dimensions is cancelled when N sub-blocks are reckoned into a new signal matrix.

4.3. Azimuth Compression

The sonar transmits and receives signals and then stores the received signal. This process is conducted at a set of locations along the moving trajectory. The signal collected at these locations forms different units of synthetic aperture called azimuth sample serials. By processing the data collected within the synthetic aperture time, an equivalent large sonar array can be obtained. By coherently processing the echo data, we obtain the high resolution in the azimuth dimension.

In practice, the azimuth echo of the SAS system can also be regarded as a wideband signal. The matched filter can still be used to compress the azimuth signal. According to the theory of the matched filter [9], it is necessary to set parameters of the matched filter to be the same as Doppler parameters of the azimuth signal. As a result, parameters should be adjusted at different ranges in order to get high resolution in azimuth across the whole swath. This is the main difference between the range compression and azimuth compression.

Inspecting (8), the azimuth modulation only depends on the range r and Doppler frequency f_t. It is not a function of the instantaneous frequency f_τ any more. This term is responsible for the azimuth matched filtering after performing the decoupling between the range and azimuth dimensions. Considering the new signal matrix in the range-Doppler domain, the azimuth compression is directly carried out based on the azimuth modulation. The filtering function is given by:

$$H_{ac\_i} = \exp\{-j\varphi_{ac\_i}(f_t; r)\} \tag{14}$$

After this step, the signal is compressed in azimuth. However, the azimuth spectrum of each receiver's data is aliased M times due to the $(1/M)$-th undersampling.

4.4. Coherent Superposition

Monostatic SAS system has a transducer, which is used as the transmitter and receiver at different times. The wide swath in range requires a low rate of PRF. However, the high resolution in azimuth requires high rate of PRF. In other words, the different requirements of the pulse repetition frequency in range and azimuth dimensions lead to a trade-off between swath width and azimuth resolution.

To solve this issue, the multireceiver SAS including a transmitter and receiver array is used. When the sampling of each individual receiver occurs at a rate of PRF, the effective sampling of the equivalent monostatic system is $M \cdot PRF$. After transmitting a pulse, M samples can be recorded. Based on the PCA method [12], the unambiguous spectrum satisfying the Nyquist rate is recovered before the SAS imagery. With the PCA method, the PTRS of multireceiver SAS is decomposed into two parts. One depends on d_i. The other is independent of d_i, and it is similar to the PTRS of monostatic SAS. When it comes to recover the unambiguous spectrum, the phase related to d_i must be simultaneously compensated. The subsequent processing can be done with the help of imaging algorithms based on monostatic SAS system. With the presented method, the PTRS cannot be decomposed into aforementioned two parts. According to the theory of the linear time invariant system [9], we can exchange the processing order between the SAS imagery and recovery of the unambiguous spectrum. In other words, each receiver's data can be focused before recovering the unambiguous spectrum. Considering each receiver's data, we obtain focusing results in the range Doppler domain by using the decoupling and azimuth compression. Since each receiver's data is sampled by the pulse repetition frequency PRF, the spectrum of a single receiver data must be aliased. Fortunately, the coherent processing of multireceiver data can suppress this alias. The spectrums of all receiver data are coherently superposed in the range Doppler domain. The resultant data would satisfy the Nyquist rate in azimuth, and the equivalent sampling rate is $M \cdot PRF$. The high resolution image is obtained after an azimuth IFT.

According to the presented steps, the block diagram of the proposed algorithm is shown in Figure 2.

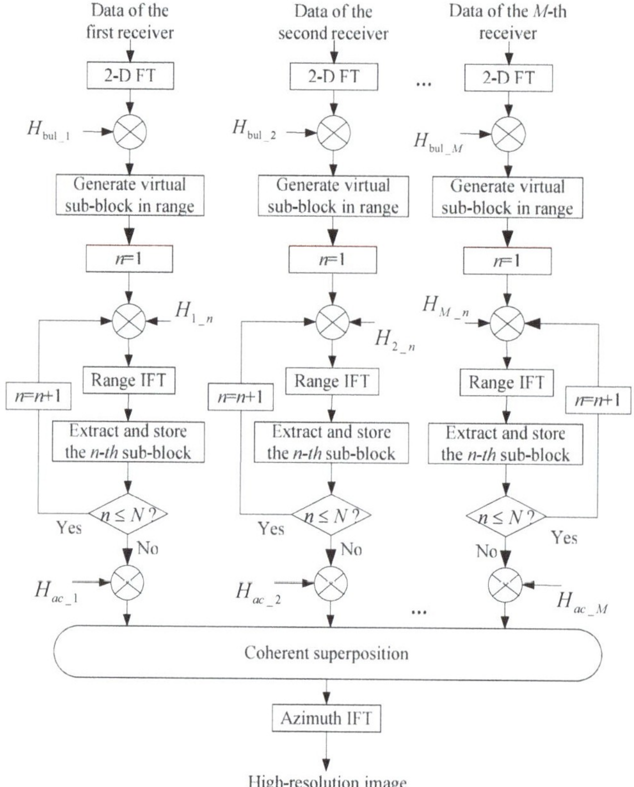

Figure 2. Block diagram of the presented processor. See the text (sections: PTRS, azimuth modulation and coupling term, and imaging algorithm) for all terms and full names of abbreviations used in the figure.

5. Comparison with Traditional Methods

In this section, we begin with the comparison of the PTRS, which is the basis of fast imaging algorithms. To directly use the method of stationary phase [9], the series expansion of the two-way slant range is often used by traditional methods [12,16,19]. From Figure 1, the two-way slant range can be formulated as:

$$R_i(t;r) = c\tau_i = \sqrt{r^2 + (vt)^2} + \sqrt{r^2 + (vt + v\tau_i + d_i)^2} \tag{15}$$

where $\sqrt{r^2 + (vt)^2}$ denotes the instantaneous range between the target and transmitter, and $\sqrt{r^2 + (vt + v\tau_i + d_i)^2}$ is the instantaneous range between the i-th receiver and target. Here, $v\tau_i$ represents the forward distance during the signal reception. Considering the complex expression of the accurate time delay, $v \cdot \tau_i$ is often approximated by $2vr/c$ for simplicity.

(15) is the sum of two square roots, which make it difficult to acquire the PTRS based on the method of stationary phase [9]. To solve this problem, (15) is expanded into a power series, which is given by:

$$R_i(t;r) \approx \sum_{q=0}^{Q} k_{i\_q} t^q \tag{16}$$

where k-coefficients in (16) can be calculated by the rule of series expansion [9]. To obtain the analytical PSP, the parameter Q often satisfies $2 \le Q \le 4$.

Inspecting (16), the series expansion would lead to the approximation error. Besides, the error of stop-and-hop approximation is not completely compensated. Since the presented method is based on the accurate time delay, both issues are successfully avoided.

We now discuss the link of the SAS imagery between the presented method and traditional methods [12,16,19]. Expanding the PTRS as a Taylor series with respect to the instantaneous frequency, various imaging algorithms are developed. The second order series approximation is exploited by the range-Doppler (R-D) algorithm [24] and chirp scaling (CS) algorithm [25]. The nonlinear CS algorithm [26,27] and quartic phase algorithm [28] are based on the third and fourth order approximation, respectively. The range migration algorithm (RMA) [29,30] requires that the PTRS is a linear function of the range. Generally, the series approximation of the PTRS is unavoidable if traditional fast algorithms were still exploited. Based on traditional methods, we conclude that the azimuth modulation is independent of the instantaneous frequency. With this characteristic, we first derive the azimuth modulation by setting $f_\tau = 0$ in (7). Then, the difference between the PTRS and azimuth modulation denotes the coupling term. Consequently, the Taylor expansion of the PTRS is avoided. Based on (7), (8) and (10), the block diagram related to the deduction of the PTRS, coupling term and azimuth modulation is shown in Figure 3a. The dashed rectangle in Figure 3a denotes important terms deduced by the presented method. Figure 3b shows the deduction of traditional methods. It can be found that traditional methods are very tedious. In this paper, we provide a novel aspect for the accurate deduction of the PTRS, azimuth modulation and coupling term.

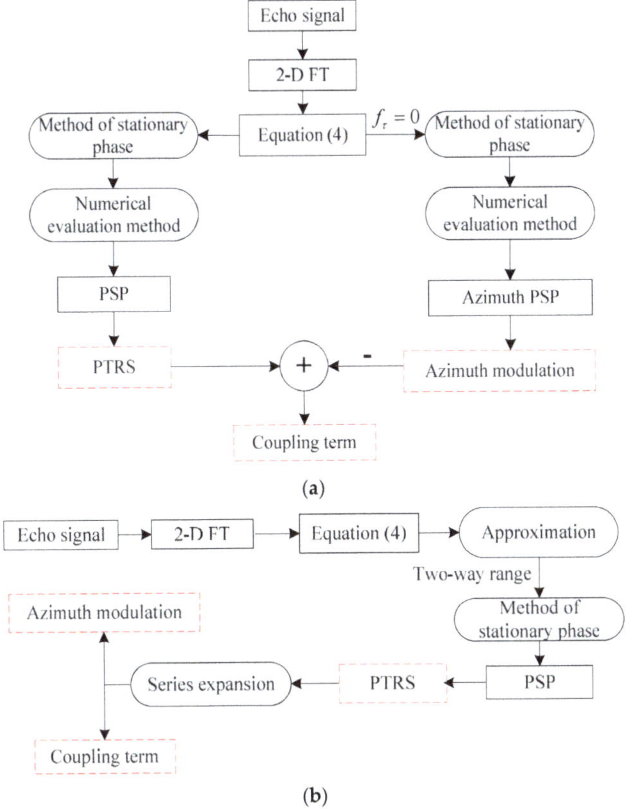

Figure 3. Deduction of three important functions. (**a**) Presented method; (**b**) traditional methods. See the text (section: Introduction) for all terms and full names of abbreviations used in the figure.

In general, our method owns two major advantages. Traditional fast imaging methods cannot directly use the accurate time delay, which is often exploited by the BP algorithm. Approximations are usually used to deduce the PTRS, azimuth modulation and coupling term. Unfortunately, the approximations degrade the imaging performance. In this paper, the PTRS, azimuth modulation and coupling term are deduced based on the accurate time delay. Consequently, these functions avoid approximations. It is the first advantage of the presented method.

The second advantage is that our imaging scheme can be simply extended to any other PTRS. The presented imaging scheme does not require the series expansion of the PTRS with respect to the instantaneous frequency. With the presented method, the key step is to deduce the azimuth modulation. Considering the analytic PTRS, the azimuth modulation can be directly obtained by setting $f_\tau = 0$ in the PTRS. When the PTRS is complicated, the azimuth modulation is deduced by using two steps. The first step should calculate the azimuth PSP by setting $f_\tau = 0$ in PSP. The subsequent step sets $f_\tau = 0$ in the PTRS and substitutes the azimuth PSP into the PTRS. Based on the PTRS and azimuth modulation, the coupling term can be calculated. After carrying out the decoupling operation based on the sub-block processing method, the imaging process is decomposed into two separate filtering processes in the range and azimuth dimensions.

6. Simulations and Real Data Processing

6.1. Simulation Results

In this section, simulations are exploited to validate the presented method. The SAS parameters are listed in Table 1.

Table 1. The SAS System Parameters.

Parameters	Value	Units
Center frequency	150	kHz
Bandwidth	20	kHz
Platform velocity	2	m/s
Receiver length in azimuth	0.02	m
Length of receiver array	0.6	m
Transmitter length in azimuth	0.04	m
Pulse repetition interval	0.3	s

6.1.1. Processing Results of Presented Method

To understand the presented method, the processing results of the main steps are discussed in detail. For clarity, we suppose that there is a point target located at coordinates (141 m, 17 m). Considering the first receiver's data shown in Figure 4a, we carry out the bulk decoupling in the 2D frequency domain. The resultant signal is shown in Figure 4b. Then, we perform the differential decoupling, and the result is shown in Figure 4c. After this step, the azimuth compression is conducted based on the azimuth modulation. Figure 4d depicts the result after the azimuth compression. Each receiver's data is undersampled using the pulse repetition frequency, which cannot satisfy the Nyquist rate in azimuth. Due to this reason, all results shown in Figure 4 are aliased in the azimuth dimension. Fortunately, the alias can be suppressed by processing the multireceiver data coherently.

Inspecting Figure 4c,d, we find that both results are visually indistinguishable. In fact, the result shown in Figure 4c is considered to be the input of the subsequent filter, i.e., azimuth compression. Since the azimuth compression performs the phase compensation in the frequency domain, the signal magnitudes are visually indistinguishable in the frequency domain. However, the major difference can be found in the space domain. Applying the azimuth IFT to Figure 4c,d, Figure 5 shows the results in the 2D space domain. From Figure 5b, the signal shown in Figure 5a is compressed in the azimuth dimension, and the circled part represents the recovered target. Each receiver's data sampled by the pulse repetition frequency is undersampling. Due to this reason, the ghost targets are introduced. The coherent processing of multireceiver data can solve this problem.

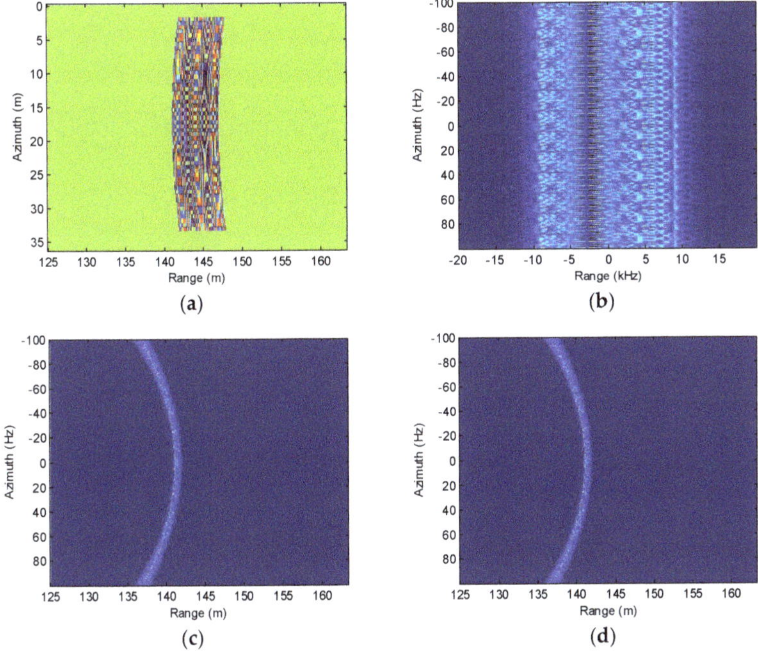

Figure 4. Processing results of a single receiver data. (**a**) Single receiver data; (**b**) bulk decoupling; (**c**) differential decoupling; (**d**) azimuth compression.

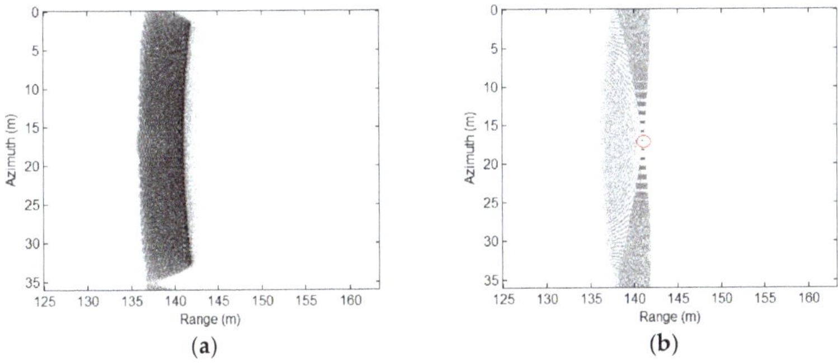

Figure 5. Results in the two-dimensional (2D) space domain. (**a**) Differential decoupling; (**b**) azimuth compression.

Based on the steps in Section 4, M results corresponding to M receiver data are obtained. Each result is similar to Figure 4d. We coherently superpose M results, and the signal is shown in Figure 6a. The coherent superposition is equivalent to the improvement of the sampling rate in azimuth. Therefore, the data shown in Figure 6a satisfies the Nyquist rate, which is increased to $M \cdot PRF$. Performing an azimuth IFT, we obtain the high resolution image, which is shown in Figure 6b. To visually examine the focusing performance, we depict the azimuth slice corresponding to Figure 6b. The azimuth slice is shown in Figure 6c. For comparison, the data shown in Figure 4a is directly processed by the presented method. The azimuth slice corresponding to Figure 5b is also depicted in Figure 6c. From Figure 6c, the azimuth slice related to a single receiver data is aliased, as each receiver data is undersampled by the pulse repetition frequency. By coherently processing multireceiver SAS data, we obtain the high resolution image. Therefore, we conclude that the presented method successfully focuses the point target.

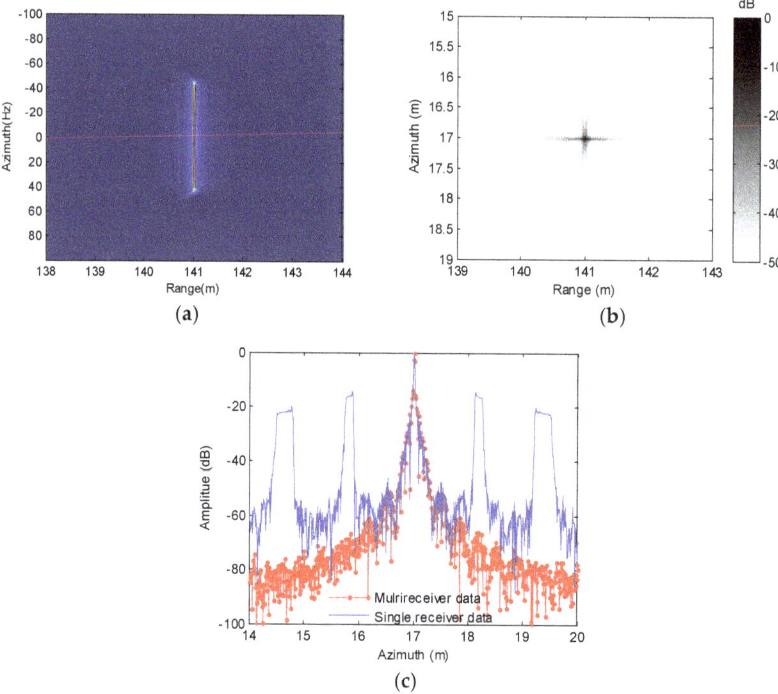

Figure 6. Processing results of all receiver data. (**a**) Coherent superposition; (**b**) focused target; (**c**) azimuth slice.

6.1.2. Influence of Sub-Block Width on Imagery

In general, the imaging algorithm can be decomposed into two steps. The first step is to derive the PTRS, coupling term and azimuth modulation. With the presented method, we accurately deduce three terms. The second step is to design the imaging algorithm based on the PTRS, coupling term and azimuth modulation. Since the coupling term in the 2D frequency domain is range variant, we perform the decoupling based on the sub-block processing method. However, it is hard to compensate the coupling term completely. In other words, the sub-block processing method results in the residual error. Here, we discuss the influence of the residual error on the SAS imagery. The imaging scenario consisting of 18-point targets is shown in Figure 7. The targets are marked by T1, T2, . . . , and T18, respectively. T1, T2, . . . , and T6 are located at close range. T7, T8, . . . , and T12 are located at medium range. The remaining targets are supposed to locate at far range. The coupling term is characterized by space variance, which may lead to the space variance of the optimal sub-block width. When the performance of the presented method is not inferior to that of traditional method, the sub-block width used by the presented method is defined as the optimal width of the sub-block. In Figure 7, we depict three sub-blocks, which are denoted by the red, blue and pink rectangles. Each sub-block consists of six targets.

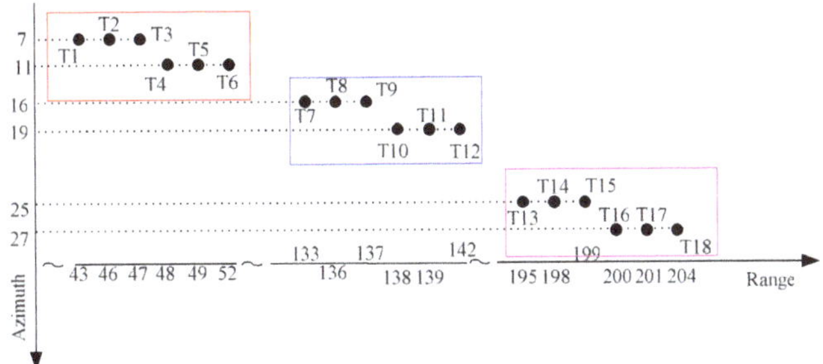

Figure 7. Simulated scenario with 18-point targets.

We first focus on the targets circled by the red rectangle in Figure 7. In this sub-block, the reference range used by the differential decoupling is 53 m. The difference between the target range and reference range denotes the half width of the sub-block. The BP algorithm [18] based on (1) is viewed as the precise method. Here, the results of the BP algorithm [22] are used as the criteria. For multireceiver SAS systems, the PCA method [12] is widely used. At this point, we mainly conduct the comparison between the presented method and PCA based R-D algorithm. Figure 8 shows the azimuth slices of T1, . . . , and T6.

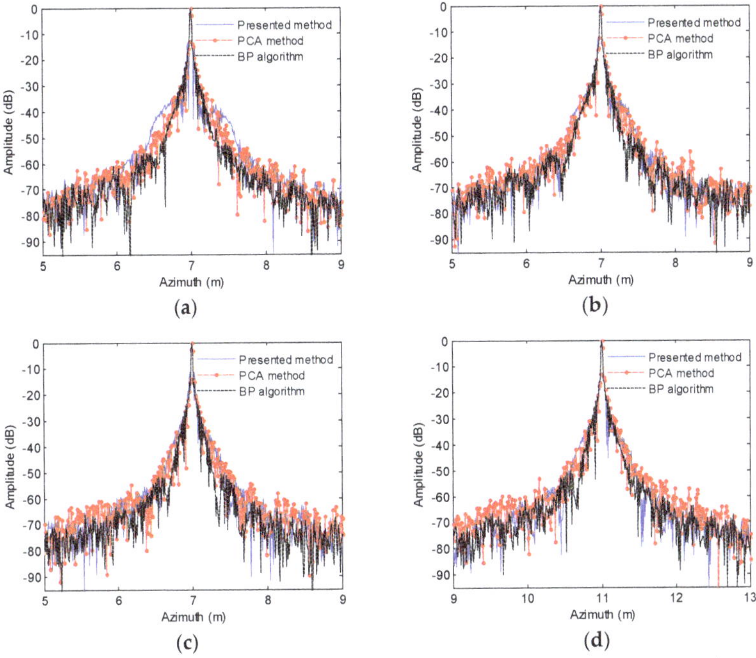

Figure 8. *Cont.*

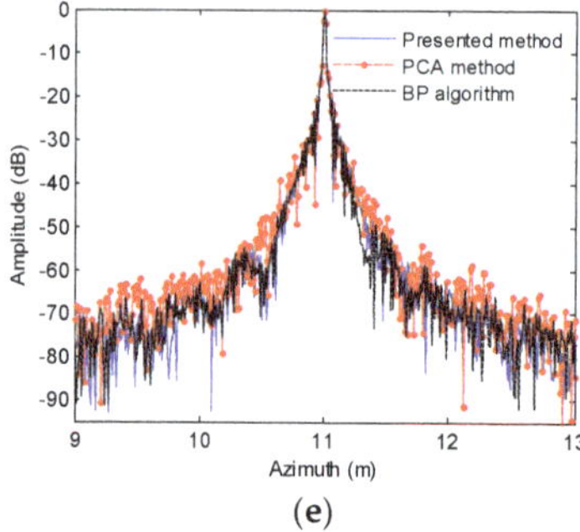

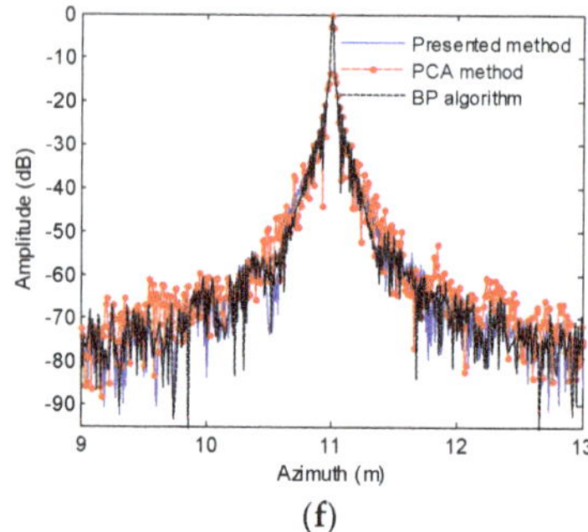

Figure 8. Azimuth slices of close targets. (**a**) Sub-block width with 20 m; (**b**) sub-block width with 14 m; (**c**) sub-block width with 12 m; (**d**) sub-block width with 10 m; (**e**) sub-block width with 8 m; and (**f**) sub-block width with 2 m.

From Figure 8, the performance of the presented method is improved by decreasing the sub-block width. The large width of the sub-block generates great residual error, which noticeably degrades the imaging performance. When the sub-block widths are 14, 12 and 10 m, the performance of the presented method is inferior to that of the PCA method. The performance of the presented method is nearly close to that of the PCA method when the sub-block width is decreased to 8 m. In other words, this sub-block width can satisfy the imagery with high performance at close range. The improvement is still obtained when we choose much narrower sub-block width. However, the improvement is not noticeable. Figure 8f enhances this conclusion.

Next, we use the peak sidelobe level ratio (PSLR) and integral sidelobe level ratio (ISLR) to evaluate the imaging performance. The quality parameters are shown in Table 2.

Table 2. Quality parameters for targets at close range. PCA: phase center approximation; BP: back projection; PSLR: peak sidelobe level ratio; ISLR: integral sidelobe level ratio.

	Presented Method		**PCA Method**		**BP Method**	
	PSLR/dB	**ISLR/dB**	**PSLR/dB**	**ISLR/dB**	**PSLR/dB**	**ISLR/dB**
T1	−11.61	−6.44	−14.38	−9.57	−14.83	−10.34
T2	−13.01	−8.36	−14.45	−9.49	−14.88	−10.25
T3	−10.82	−8.21	−14.54	−9.98	−14.90	−10.82
T4	−13.41	−9.39	−14.63	−9.68	−14.77	−10.44
T5	−14.61	−9.87	−14.44	−9.59	−14.81	−10.15
T6	−14.69	−10.12	−14.26	−9.60	−14.69	−10.06

From Table 2, we see that the presented method obtains a low resolution image with a large sub-block width. The PSLR and ISLR related to T1, T2 and T3 enhances this conclusion. For the target T4, the focusing performance of the presented method is mostly close to that of the PCA method. Considering the target T5, the focusing performance of the presented method is superior to that of the PCA method. In this case, the residual error introduced by the sub-block processing method can be negligible. Inspecting quality parameters of T6, the focusing performance is improved by decreasing the sub-block width. However, the improvement is slight, as the residual error does not dramatically influence the imaging performance. Therefore, we conclude that the sub-block width with 8 m can satisfy the high performance imagery at close range.

We now concentrate on the targets at medium range. In this case, the reference range used for the differential decoupling is 143 m. After the data processing, the azimuth slices are shown in Figure 9.

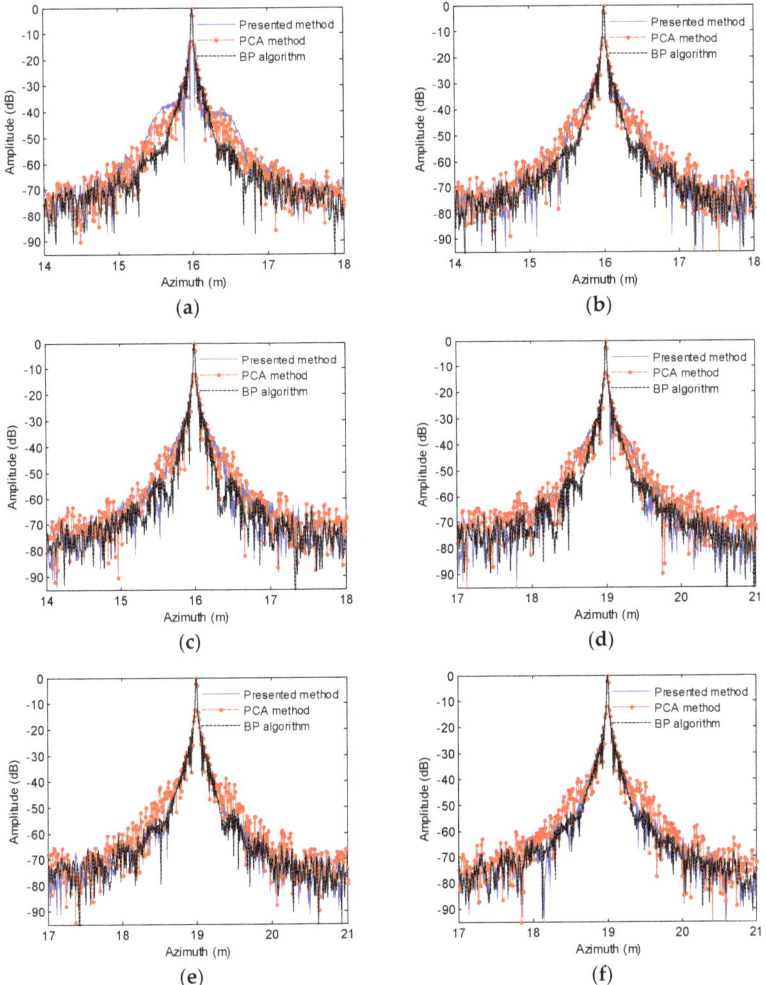

Figure 9. Azimuth slices of medium range targets. (**a**) Sub-block width with 20 m; (**b**) sub-block width with 14 m; (**c**) sub-block width with 12 m; (**d**) sub-block width with 10 m; (**e**) sub-block width with 8 m; and (**f**) sub-block width with 2 m.

Inspecting Figure 9, we nearly obtain the same conclusions drawn from Figure 8. When the targets are at medium range, the optimal width of the sub-block is still 8 m. Figure 10b also strengths this conclusion. Since the stop-and-hop error is not completely compensated, the performance of the PCA method is slightly degraded. Based on Figures 8 and 9, the optimal sub-block width of both cases is mostly identical. In other words, the optimal width of the sub-block is nearly range invariant. Table 3 lists quality parameters for targets at medium range.

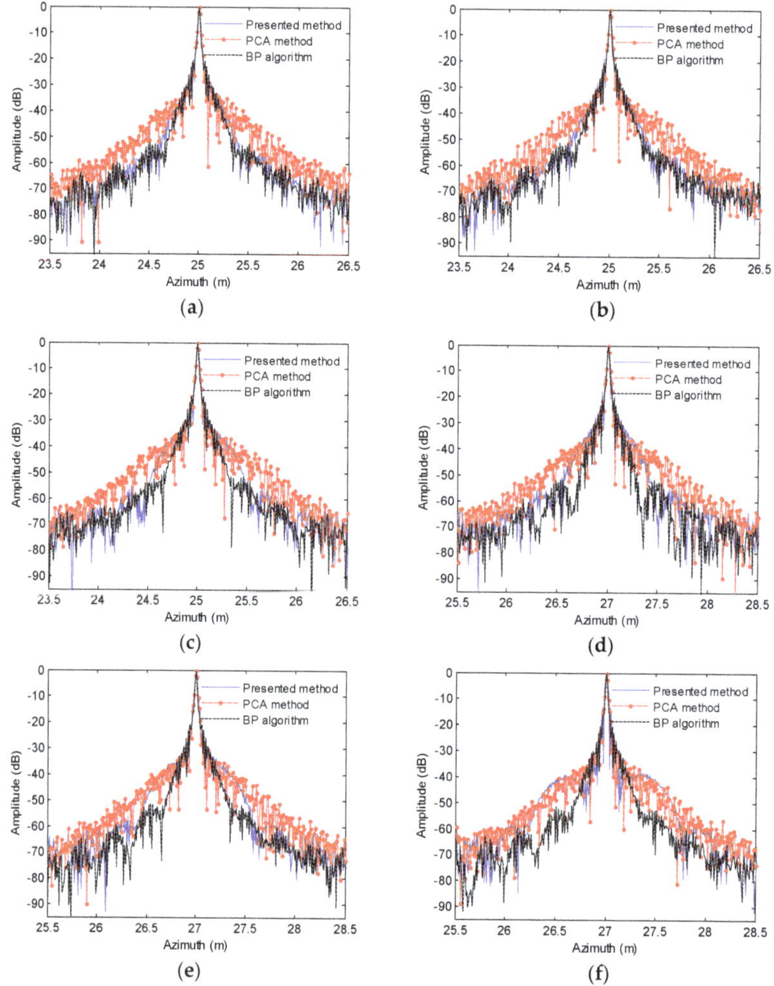

Figure 10. Azimuth slices of far targets. (**a**) Sub-block width with 2 m; (**b**) sub-block width with 8 m; (**c**) sub-block width with 10 m; (**d**) sub-block width with 12 m; (**e**) sub-block width with 14 m; and (**f**) sub-block width with 20 m.

Table 3. Quality parameters for targets at medium range.

	Presented Method		**PCA Method**		**BP Method**	
	PSLR/dB	**ISLR/dB**	**PSLR/dB**	**ISLR/dB**	**PSLR/dB**	**ISLR/dB**
T7	−11.61	−6.44	−13.92	−9.52	−14.81	−10.23
T8	−13.26	−8.67	−13.87	−9.37	−14.92	−10.25
T9	−10.36	−7.87	−13.86	−9.32	−14.77	−10.09
T10	−13.86	−9.38	−14.01	−9.71	−14.78	−10.28
T11	−14.02	−9.53	−13.73	−9.22	−14.87	−10.21
T12	−14.13	−10.13	−13.69	−9.17	−14.7	−10.12

Based on Table 3, the performance of the presented method is improved with the decreasing of the sub-block width. When the sub-block width is decreased to 8 m, we obtain the image which outperforms that of the PCA method. Due to the residual error of the stop-and-hop approximation, the quality parameters of the PCA method are slightly lowered.

The last experiment focuses on the imaging performance at far range. The reference range used for the differential decoupling in this sub-block is 194 m. The azimuth slices are shown in Figure 10. The residual error of the stop-and-hop assumption increases with the range. Consequently, the PCA method suffers from large residual error when the targets are located at far range. With the presented method, the imaging performance is nearly close to that of the PCA method when the sub-block width is 14 m. Figure 10e enhances this conclusion. By decreasing the sub-block width, the imaging performance of the presented method is improved. From Figure 10b, the sub-block width with 8 m satisfies the high performance imagery. In this case, we obtain the image which is mostly identical to that of the BP algorithm. However, the performance of the PCA method is inferior to that of the presented method and BP algorithm, because the residual error of the stop-and-hop approximation is not completely compensated. Inspecting Figures 8 and 9, the optimal sub-block width is about 8 m for three cases. In practice, the large sub-block width can be used at far range without loss of the imaging performance.

For targets at far range, the PSLR and ISLR are listed in Table 4. When the sub-block width is large, the error introduced by the sub-block processing method noticeably degrades the imaging performance of the presented method. The focusing performance of T15, T16, T17 and T18 also enhances this conclusion. The result of the presented method can be improved by decreasing the sub-block width. With the presented method, we obtain the high resolution results when the sub-block width is less than 8 m. The PSLR and ISLR related to T13 and T14 further strength this conclusion. However, the performance of the PCA method is greatly affected by the residual error of the stop-and-hop approximation at far range.

Table 4. Quality parameters for targets at far range.

	Presented Method		PCA Method		BP Method	
	PSLR/dB	**ISLR/dB**	**PSLR/dB**	**ISLR/dB**	**PSLR/dB**	**ISLR/dB**
T13	−14.49	−10.22	−13.23	−9.93	−14.76	−10.26
T14	−14.17	−9.97	−13.13	−9.8	−14.80	−10.22
T15	−13.23	−9.60	−12.87	−9.47	−14.72	−10.35
T16	−12.45	−7.83	−12.99	−9.39	−14.83	−10.18
T17	−11.96	−8.80	−13.15	−9.79	−14.86	−10.41
T18	−11.28	−6.51	−13.55	−9.67	−14.78	−10.21

Since the coupling term is space variant, the sub-block processing method is exploited to perform the decoupling operation. However, the sub-block method introduces the residual phase error, which is expressed as $|\varphi_{de\_i}(f_\tau, f_t; r_{\text{ref}\_n} \pm \Delta r/2) - \varphi_{de\_i}(f_\tau, f_t; r_{\text{ref}\_n})|$. Here, Δr denotes the sub-block width. Generally speaking, the imaging performance of the presented method highly depends on the sub-block width. In each sub-block, the residual phase error introduced by the decoupling operation should be limited within $\pi/4$ [31]. Under this condition, the influence of the phase error can be neglected.

Figures 8a, 9a and 10f are based on the sub-block width with 20 m. We find that the slices of the presented method are inferior to the slices of the PCA algorithm and BP method. The reason behind this is that the residual coupling error is not completely compensated. Hence, there is still the coupling between the range and azimuth dimensions. T1, T7 and T18 are far away from the sub-block center. The corresponding sub-blocks have a large width. Based on the decoupling phase of reference targets at the sub-block center, the decoupling operation across the whole sub-block is carried out. When the targets are at the sub-block edge, this operation introduces large residual coupling error, which is not limited within $\pi/4$. Due to this reason, the azimuth focusing performance such as the azimuth slice, PSLR and ISLR is seriously degraded. The focusing results of T1, T7 and T18 enhance this conclusion. When the targets are close to the sub-block center, the residual coupling error does not noticeably affect the SAS focusing performance. Under this condition, the residual coupling error limited within $\pi/4$ can be negligible. Considering the trade-off between the imaging efficiency and performance, the optimal width of the sub-block is often exploited. In this case, the imaging performance of the presented method is nearly close to that of the BP method. The processing results of T5, T11 and T14 enhance this conclusion. Generally, the presented method can obtain the high resolution image across the whole swath based on the optimal width of the sub-block. Besides, it is very suitable for the SAS imagery with wide swath.

6.1.3. Imaging Performance at Scenario Edge

Since the error of the stop-and-hop approximation is space variant, it is difficult to compensate this error using traditional methods. Fortunately, the presented method successfully solves this problem based on the accurate time delay of the signal. We now concentrate on the focusing performance at the scenario edge. Figure 11 shows the imaging scenario including three ideal targets.

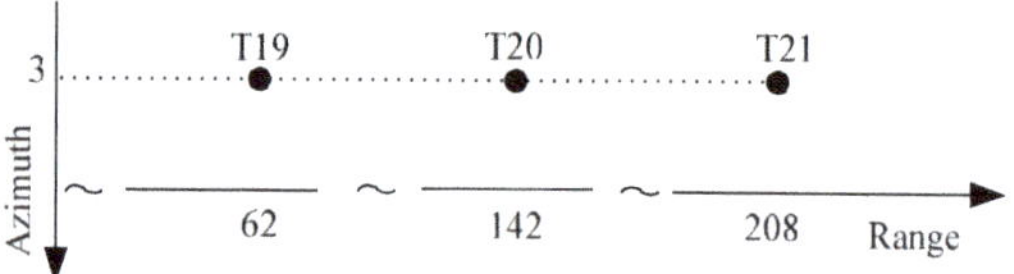

Figure 11. Simulated scenario with three point targets.

Based on the presented method, the PCA method and BP algorithm, the azimuth slices of focused targets are shown in Figure 12.

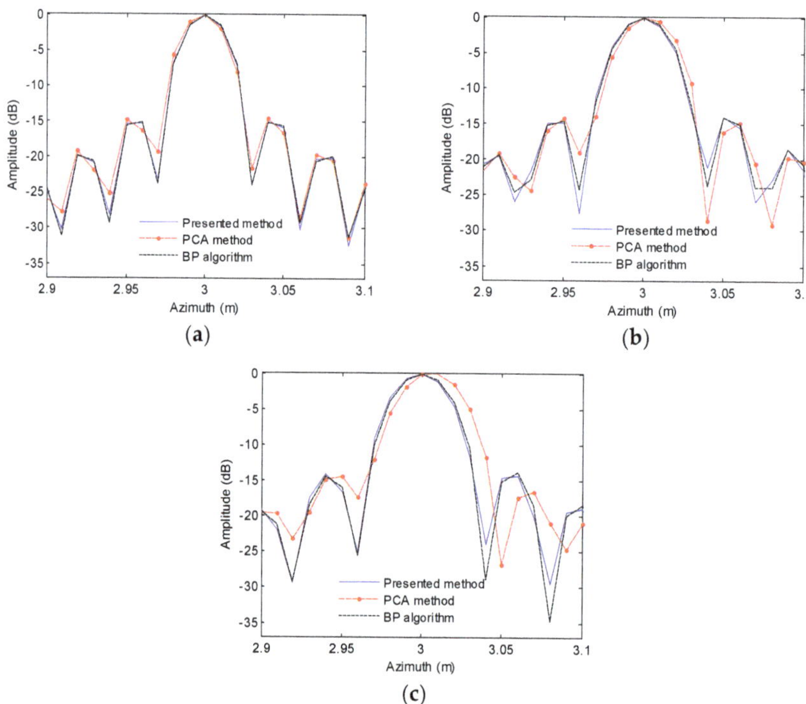

Figure 12. Azimuth slices of T19, T20 and T21. (**a**) T19; (**b**) T20; (**c**) T21.

With the presented method and PCA method, the target coordinates in Figure 12a are mostly accordant with coordinates shown in Figure 11. Considering the PCA method, there is a slight deviation of the azimuth coordinate in Figure 12b. Generally, this deviation can be negligible at close range. Inspecting Figure 12c, the PCA method suffers from noticeable deviation of the azimuth coordinate, and the deviation is about 0.01 m. Since the PCA method does not completely compensate the error of the stop-and-hop approximation, the residual error is introduced. It increases with the range and slow time. Due to this reason, the deviation can be negligible at close range. However, the deviation leads to the distortion when there are distributed targets at far range. Using the presented method, the targets across the whole swath are well focused.

6.2. Real Data Processing

We tested the presented method based on the real data. The data has 4800 sampling points in the range dimension and 3200 spatial sampling points in the azimuth dimension. For the transmitted signal, the center frequency and bandwidth are 150 kHz and 20 kHz, respectively. The receiver array, including 40 uniformly spaced receivers, is 1.6 m in azimuth. The velocity of the sonar platform is 2.5 m/s. When it comes to the operation of the differential decoupling with the presented method, two cases including four sub-blocks and eight sub-blocks in the range dimension are considered. The corresponding results are shown in Figure 13a,b, respectively. From Figure 13, it can be seen that the processing results of both cases are mostly identical. Therefore, we can draw a conclusion that the requirement of the sub-block segmentation in the range dimension can be relaxed. In practice, we can use large sub-block width without loss of performance when the real data is processed based on the presented method. For comparison, the real data is still processed by the BP algorithm. Figure 14 shows the processing result of the BP algorithm. Inspecting Figures 13 and 14, the presented method provide the high resolution result, which is mostly identical to that of the BP algorithm.

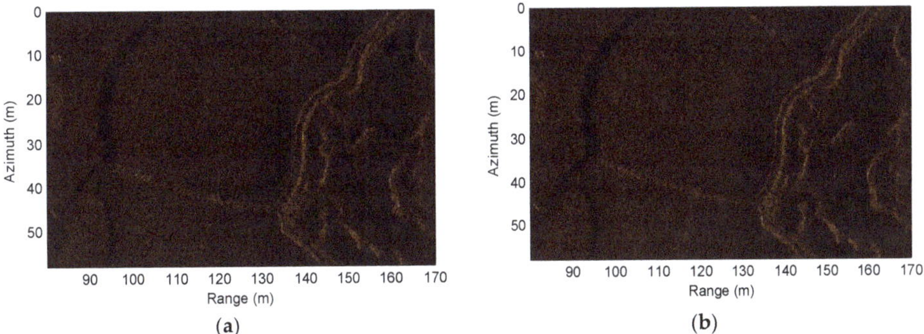

Figure 13. Processing results of the presented method. (**a**) Four sub-blocks; (**b**) eight sub-blocks.

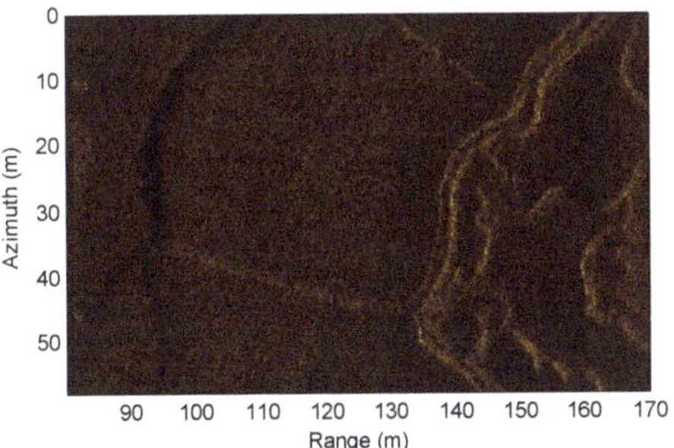

Figure 14. Processing results of the BP algorithm.

The processing time with both methods are listed in Table 5.

Table 5. Processing time of imaging methods.

	Presented Method		BP Algorithm
	Four Sub-Blocks	**Eight Sub-Blocks**	
Processing time/s	608	1149	35234

Both algorithms are developed based on Matlab 2012a. The processing time of the BP algorithm based on the sinc interpolation is 35,234 s. There are two schemes to develop the program of the presented method. With the first scheme, the calculation of the PSP and azimuth PSP are integrated with the imaging algorithm. It is time consuming due to the numerical evaluation of PSPs. In practice, the numerical calculation of PSPs can be carried out ahead of the focusing. With stored PSPs, we run the imaging algorithm. This is the second scheme. With this scheme, the processing time with four sub-blocks is decreased to 608 s. With the second scheme, the efficiency is dramatically improved compared with the first scheme. At this point, the second scheme is used with the real data processing. From Table 5, the presented algorithm with eight sub-blocks costs 1149 s. Overall, the processing time of the presented method increases with sub-blocks in range, because more time is needed to perform the differential decoupling. In comparison with the BP algorithm, the efficiency of presented method has been improved 30.7 times at least. Nowadays, the parallel algorithm and tools such as graphics processing unit (GPU), faster Fourier transform in the west (FFTW) and intel@ math kernel library (MKL) can be used to improve the efficiency of the presented method dramatically. Our future work is to optimize our method using parallel algorithms.

7. Conclusions

In this paper, we present a novel imaging algorithm for the multireceiver SAS based on the accurate time delay and numerical evaluation method. The presented method first deduces the PTRS and azimuth modulation using the numerical evaluation method. Then the difference between the PTRS and azimuth modulation denotes the coupling term. The key issue of the SAS imagery is the decoupling operation, which consists of two parts: the bulk and differential decoupling. The bulk decoupling mainly deals with the spatial invariance of the coupling term. Considering targets at reference range, the coupling is completely removed after this step. However, other targets suffer from the residual coupling error. The differential decoupling is carried out to solve this problem. Considering the spatial variance of the residual coupling error, the sub-block processing in range is exploited.

Based on the simulations, the focusing performance of the traditional method is greatly degraded at far range, as the residual error introduced by the stop-and-hop approximation increases with the range. Using the optimal width of the sub-block, the presented method achieves high performance results compared with traditional method.

Author Contributions: X.Z. and W.Y. carried out all of the analysis and algorithms and wrote the paper. C.T. designed part of experiments. All authors read and approved the final manuscript.

Funding: This research was funded by the National Natural Science Foundation of China (61601473) and National Key Laboratory Foundation (9140C290401150C29132).

Acknowledgments: The authors would like to thank the anonymous reviewers and editors for their constructive comments and suggestions. Wei Chen offered advice on the language and revision.

Conflicts of Interest: The authors declare no conflict of interest.

References

1. Williams, D.P. The Mondrian detection algorithm for sonar imagery. *IEEE Trans. Geosci. Remote Sens.* **2018**, *56*, 1091–1102. [CrossRef]

2. Larsen, L.J.; Wilby, A.; Stewart, C. Deepwater ocean survey and search using synthetic aperture sonar. In Proceedings of the 2010 MTS/IEEE Oceans Conference, Seattle, WA, USA, 20–23 September 2010; pp. 1–4.
3. LeHardy, P.K.; Larsen, L.J. Deepwater synthetic aperture sonar and the search for MH 370. In Proceedings of the 2015 MTS/IEEE Oceans Conference, Washington, DC, USA, 19–22 October 2015; pp. 1–4.
4. Odegard, O.; Ludvigsen, M.; Lagstad, A. Using synthetic aperture sonar in marine archaeological surveys—Some first experiences. In Proceedings of the 2013 MTS/IEEE Oceans Conference, Bergen, Norway, 10–14 June 2013; pp. 1–7.
5. Carballini, J.; Viana, F. Using synthetic aperture sonar as an effective tool for pipeline inspection survey projects. In Proceedings of the 2015 MTS/OES RIO Acoustics, Rio de Janeiro, Brazil, 29–31 July 2015; pp. 1–5.
6. Groen, J.; Coiras, E.; Vera, J.D.R.; Evans, B. Model-based sea mine classification with synthetic aperture sonar. *IET Radar Sonar Navig.* **2010**, *4*, 62–73. [CrossRef]
7. Wu, H.; Tang, J.; Zhong, H. A correction approach for the inclined array of hydrophones in synthetic aperture sonar. *Sensors* **2018**, *18*, 2000. [CrossRef] [PubMed]
8. Tang, S.; Guo, P.; Zhang, L.; Lin, C. Modeling and precise processing for spaceborne transmitter/missile-borne receiver SAR signals. *Remote Sens.* **2019**, *11*, 346. [CrossRef]
9. Cumming, I.G.; Wong, F.H. *Digital Processing of Synthetic Aperture Radar Data: Algorithms and Implementation*; Artech House: Norwood, MA, USA, 2005.
10. Wilkinson, D.R. Efficient Image Reconstruction Techniques for a Multiple-Receiver Synthetic Aperture Sonar. Master's Thesis, Department of Electrical and Computer Engineering, University of Canterbury, Christchurch, New Zealand, July 1999.
11. Bonifant, W.W. Interferometric Synthetic Aperture Sonar Processing. Master's Thesis, Georgia Institute of Technology, Atlanta, GA, USA, July 1999.
12. Zhang, X.; Tang, J.; Zhong, H. Multireceiver correction for the chirp scaling algorithm in synthetic aperture sonar. *IEEE J. Ocean. Eng.* **2014**, *39*, 472–481. [CrossRef]
13. Gough, P.T.; Hayes, M.P. Fast Fourier techniques for SAS imagery. In Proceedings of the 2005 MTS/IEEE Oceans Conference, Brest, France, 20–23 June 2005; pp. 563–568.
14. Loffeld, O.; Nies, H.; Peters, V.; Knedlik, S. Models and useful relations for bistatic SAR processing. *IEEE Trans. Geosci. Remote Sens.* **2004**, *42*, 2031–2038. [CrossRef]
15. Zhang, X.; Yang, P. Imaging algorithm for multireceiver synthetic aperture sonar. *J. Electr. Eng. Technol.* **2019**, *14*, 471–478. [CrossRef]
16. Yang, H.; Tang, J.; Chen, M.; Chen, X. A multiple-receiver synthetic aperture sonar wavenumber imaging algorithm with non-uniform sampling in azimuth. *J. Wuhan Univ. Technol. (Transp. Sci. Eng.)* **2011**, *35*, 993–996.
17. Zhang, X.; Chen, X.; Qu, W. Influence of the stop-and-hop assumption on synthetic aperture sonar imagery. In Proceedings of the 2017 International Conference on Communication Technology, Chengdu, China, 27–30 October 2017; pp. 1601–1607.
18. Wang, G.; Zhang, L.; Li, J.; Hu, Q. Precise aperture-dependent motion compensation for high-resolution synthetic aperture radar imaging. *IET Radar Sonar Navig.* **2017**, *11*, 204–211. [CrossRef]
19. Zhang, X.; Tang, J.; Zhang, S.; Bai, S.; Zhong, H. Four order polynomial based range-Doppler algorithm for multireceiver synthetic aperture sonar. *J. Electron. Inf. Technol.* **2014**, *36*, 1592–1598.
20. Wu, Q.; Xing, M.; Shi, H.; Hu, X.; Bao, Z. Exact analytical two-dimensional spectrum for bistatic synthetic aperture radar in tandem configuration. *IET Radar Sonar Navig.* **2011**, *5*, 349–360. [CrossRef]
21. Tian, Z.; Tang, J.; Zhong, H.; Zhang, S. Extended range Doppler algorithm for multiple-receiver synthetic aperture sonar based on exact analytical two-dimensional spectrum. *IEEE J. Ocean. Eng.* **2016**, *41*, 3350–3358.
22. Xu, J.; Tang, J.; Zhang, C. Multi-aperture synthetic aperture sonar imaging algorithm. *Signal Process.* **2003**, *19*, 157–160.
23. Wu, J.; Xu, Y.; Zhong, X.; Sun, Z.; Yang, J. A three-dimensional localization method for multistatic SAR based on numerical range-Doppler algorithm and entropy minimization. *Remote Sens.* **2017**, *9*, 470. [CrossRef]
24. Jin, M.Y.; Wu, C. A SAR correlation algorithm which accommodates large-range migration. *IEEE Trans. Geosci. Remote Sens.* **1984**, *GE-22*, 592–597. [CrossRef]
25. Chen, P.; Kang, J. Improved chirp scaling algorithms for SAR imaging under high squint angles. *IET Radar Sonar Navig.* **2017**, *11*, 1629–1636. [CrossRef]

26. Wong, F.H.; Yeo, T.S. New applications of nonlinear chirp scaling in SAR data processing. *IEEE Trans. Geosci. Remote Sens.* **2001**, *39*, 946–953. [CrossRef]
27. Li, Y.; Huang, P.; Lin, C. Focus improvement of highly squint bistatic synthetic aperture radar based on non-linear chirp scaling. *IET Radar Sonar Navig.* **2017**, *11*, 171–176. [CrossRef]
28. Wang, K.; Liu, X. Quartic-phase algorithm for highly squinted SAR data processing. *IEEE Geosci. Remote Sens. Lett.* **2007**, *4*, 246–250. [CrossRef]
29. Guo, P.; Tang, S.; Zhang, L.; Sun, G. Improved focusing approach for highly squinted beam steering SAR. *IET Radar Sonar Navig.* **2016**, *10*, 1394–1399. [CrossRef]
30. Ku, C.S.; Chen, K.S.; Chang, P.C.; Chang, Y.L. Imaging simulation for synthetic aperture radar: A full-wave approach. *Remote Sens.* **2018**, *10*, 1404. [CrossRef]
31. Wu, J.; An, H.; Zhang, Q.; Sun, Z.; Li, Z.; Du, K.; Huang, Y.; Yang, J. Two-dimensional frequency decoupling method for curved trajectory synthetic aperture radar imaging. *IET Radar Sonar Navig.* **2018**, *12*, 766–773. [CrossRef]

Article

An Adaptive Denoising and Detection Approach for Underwater Sonar Image

Xingmei Wang [1,*], Qiming Li [1], Jingwei Yin [2,3,4], Xiao Han [2,3,4] and Wenqian Hao [5]

1 College of Computer Science and Technology, Harbin Engineering University, Harbin 150001, China; liqiming@hrbeu.edu.cn
2 Acoustic science and Technology laboratory, Harbin Engineering University, Harbin 150001, China; yinjingwei@hrbeu.edu.cn (J.Y.); hanxiao1322@hrbeu.edu.cn (X.H.)
3 College of Underwater Acoustic Engineering, Harbin Engineering University, Harbin 150001, China
4 Key Laboratory of Marine Information Acquisition and Security, Harbin Engineering University, Harbin 150001, China
5 Institute of Acoustics, Chinese Academy of Science, Beijing 100190, China; haowenqian@mail.ioa.ac.cn
* Correspondence: wxm_hrbeu@sohu.com; Tel.: +86-189-4505-5955

Received: 14 January 2019; Accepted: 12 February 2019; Published: 15 February 2019

Abstract: An adaptive approach is proposed to denoise and detect the underwater sonar image in this paper. Firstly, to improve the denoising performance of non-local spatial information in the underwater sonar image, an adaptive non-local spatial information denoising method based on the golden ratio is proposed. Then, a new adaptive cultural algorithm (NACA) is proposed to accurately and quickly complete the underwater sonar image detection in this paper. Concretely, NACA has two improvements. In the first place, to obtain better initial clustering centres, an adaptive initialization algorithm based on data field (AIA-DF) is proposed in this paper. Secondly, in the belief space of NACA, a new update strategy is adopted to update cultural individuals in terms of the quantum-inspired shuffled frog leaping algorithm (QSFLA). The experimental results show that the proposed denoising method in this paper can effectively remove relatively large and small filtering degree parameters and improve the denoising performance to some extent. Compared with other comparison algorithms, the proposed NACA can converge to the global optimal solution within small epochs and accurately complete the object detection, having better effectiveness and adaptability.

Keywords: underwater sonar image; adaptive denoising; detection; adaptive initialization

1. Introduction

With the development of sonar imaging technology, the underwater sonar image detection technology has been extensively used in marine exploration, research, and investigation [1]. The underwater sonar image must be detected in different regions including object-highlight, sea-bottom-reverberation, and shadow before the underwater object can be recognized. [2]. With the help of the underwater sonar image detection technology, we can accurately detect the above regions and effectively retain the underwater sonar image details [3]. However, because of the complexity of the underwater environment, the underwater sonar image is easily affected by the reverberation effect, strong speckle noise, fuzzy edge, and weak texture information [4]. Therefore, before the underwater sonar image detection, the denoising algorithm can be adopted to maximize noise removal.

The spatial-based method has been widely used in image processing. For cartoon components in the image fusion method, a proper spatial-based method is presented for morphological structure preservation [5]. To mitigate the boundary seams produced by spatial domain methods, a non-local means filtering is introduced on the final decision maps to generate the fusion weight maps for each of the source images [6]. In the image denoising method, the spatial information mainly includes local spatial information and non-local spatial information. Compared with the disadvantage that local spatial information only uses a single neighborhood information, non-local spatial information can utilize multiple similar neighborhood information in subregions. Therefore, the non-local spatial information makes full use of the image, which is more conducive to image denoising. The filtering degree parameter is a significant parameter. It is intensively related to denoising results [7]. The relatively large filtering degree parameter will cause the loss of some image detail information, especially the image edge information. On the contrary, the relatively small filtering degree parameter will mean that the image noise cannot be effectively eliminated. Therefore, how to select the appropriate filtering degree parameter is a challenging problem. To select the appropriate filtering degree parameter, a selection method based on the image noise level was proposed. In this method, filtering degree parameters were a single value [8]. In order to get over the disadvantage that the single filtering degree parameters are sensitive to the noise, another selection method on the basis of the statistical characteristic in the search window was proposed [9]. However, when the search window is small, the denoising ability of the non-local spatial information will be greatly reduced. Furthermore, the idea of threshold was introduced to remove the abnormal parameters and select the appropriate filtering degree parameters [7]. Nevertheless, the two thresholds are fixed values and only refer to one underwater sonar image. Different underwater sonar images have different filtering parameter distribution characteristics. Therefore, in some cases, the relatively large and small filtering degree parameters cannot be effectively removed.

Various techniques are widely used in the underwater sonar image detection. The Fuzzy clustering algorithm is used for underwater sonar image detection [10]. However, this detection algorithm is susceptible to noise. The Markov algorithm is proposed for underwater sonar image detection [11,12]. Although the detection results were satisfactory, the processing procedure was quite complicated and computationally costly. Later, a promising and multiphase detection framework related to the level set was developed [13], the key idea of which was to minimize the energy with the local mean. However, this framework was not suitable for the detection of underwater sonar images in the presence of speckle noise. To obtain better detection results, a large number of improved algorithms were proposed. Active contours were introduced in the level set method [14,15]. In this case, Markov random field and the level set are combined to extract the features from the underwater sonar image [2]. The morphological top-hat and bottom-hat transformation were used in the level set [16]. It is conducive to the extraction of underwater objects. Subsequently, an adaptive initialization narrowband Chan-Vese model is proposed and applied to underwater sonar image detection [1].

Over the years, as a framework to solve complex problems, cultural algorithm (CA) is attracting more and more attention [17]. Meanwhile, the intelligent optimization algorithm is also widely used in image processing [18,19] and has achieved good results. In order to further enhance the search ability, many intelligent optimization algorithms are used to improve the cultural algorithm. For accurate selection of business partners, a cultural particle swarm optimization algorithm (CPSO) was proposed can get better options [20]. Quantum-behaved particle swarm optimization (QPSO) was introduced into CA to figure out multi-objective optimization problems [21], which has high efficiency. Then, an adaptive cultural algorithm with improved quantum-behaved particle swarm optimization (ACA-IQPSO) was proposed to solve the problem of the underwater sonar image detection [22], which accurately completes underwater object detection by searching clustering centres. However, because the clustering centres are randomly initialized in the solution space, the situation when the clustering centres are very close can occur to some extent. Therefore, the convergence speed of the algorithm is too slow to complete the underwater sonar image detection accurately within small epochs.

A clustering algorithm is proposed to calculate the clustering centres by the potential entropy of the data field [23]. The comparative experiments have shown that this algorithm can get better clustering results. Nevertheless, it does not take the characteristics of underwater sonar images into account when calculating potential entropy, which is not conducive to the initialization of the clustering centres in the underwater sonar images.

In order to reduce the complexity of the algorithm and enhance the search ability in the process of detection, the idea of the shuffled frog leaping algorithm (SFLA) is used as the belief space update strategy in ACA-IQPSO. However, SFLA is easy to get into the local optimal solution. To make the population more diverse and improve the search ability, Quantum computing theory is used to improve SFLA [24]. A quantum-inspired shuffled frog leaping algorithm combining the new search mechanism (QSFLA-NSM) is proposed to the underwater sonar image detection [7]. The experimental results of QSFLA-NSM show that it can further enhance population diversity and search ability.

To overcome the drawback of two thresholds in work [7], in this work we proposed an adaptive non-local spatial information denoising method based on the golden ratio. Two thresholds are set adaptively according to the golden radio. Then, NACA is proposed in this paper to accurately detect underwater sonar image. The improvement of the detection algorithm focuses on two aspects: Firstly, inspired by the idea of data field and entropy [23], the AIA-DF is adopted to more accurately calculate the initialization of the clustering centres in this paper to enhance convergence efficiency. The AIA-DF can automatically extract the optimal value of threshold by using the potential entropy of data field from the underwater sonar images dataset. Subsequently, the threshold from the data field can adaptively initialize the clustering centres. Secondly, in the belief space, to improve the limitations of the update strategy, a new update strategy is used to update the cultural individuals in terms of QSFLA.

2. Non-Local Spatial Information Denoising Method

2.1. Non-Local Spatial Information

The underwater sonar image is represented as $X = \{x_1, x_2 \cdots, x_n\}$, n is the total number of pixels. x_i is the ith pixel. $\overline{x_i}$ is the non-local spatial information of x_i and can be expressed as:

$$\overline{x_i} = \sum_{p \in W_i^r} w_{ip} x_p \tag{1}$$

where W_i^r represents the search window, its centre is x_i and the radius is r. w_{ip} and z_i can be calculated by:

$$w_{ip} = \frac{1}{Z_i} \exp\left(-\|x(N_i) - x(N_p)\|_{2,\rho}^2 / h\right) \tag{2}$$

$$Z_i = \sum_{p \in W_i^r} \exp\left(-\|x(N_i) - x(N_p)\|_{2,\rho}^2 / h\right) \tag{3}$$

where N_i is the neighborhood window, its centre is the x_i and the radius is s. $x(N_i)$ is the vector and represents all the pixels in N_i. h is the filtering degree parameter. $\|x(N_i) - x(N_p)\|_{2,\rho}^2$ is defined as:

$$\|x(N_i) - x(N_p)\|_{2,\rho}^2 = \sum_{q=1}^{(2s+1)^2} \rho^{(q)} \left(x^{(q)}(N_i) - x^{(q)}(N_p)\right)^2 \tag{4}$$

where $x^{(q)}(N_i)$ is the qth pixel in $x(N_i)$.

$\rho^{(q)}$ can be presented as:

$$\rho^{(q)} = \sum_{t=\max(d,1)}^{s} \frac{1}{(2t+1)^2 s} \tag{5}$$

$$d = \max\left(|y_q - s - 1|, |z_q - s - 1|\right) \tag{6}$$

where $y_q = \mathrm{mod}(q, (2s+1))$, $z_q = floor(q, (2s+1)) + 1$. (y_q, z_q) represents the coordinates of the qth dimension in N_i.

2.2. Our Proposed Denoising Method—An Adaptive Non-Local Spatial Information Denoising Algorithm Based on the Golden Ratio

h is a significant parameter. Its value can powerfully impact the weight of neighborhood configuration in Equation (2) and Equation (3). Therefore, h is closely related to the denoising result in the non-local spatial information denoising method. In the process of denoising, too large h will cause missing the image detail information. Meanwhile, too small h will reduce the denoising performance of non-local spatial information.

In our previous work [7], to select the appropriate h, two thresholds $hmin = 0.01$ and $hmax = 0.05$ were set to remove the inappropriate h. Experimental results showed that this method can effectively remove noise and had certain adaptability. However, it is found that different underwater sonar images have different h distribution characteristics. The two thresholds $hmin = 0.01$ and $hmax = 0.05$ are determined only based on one sonar image. In other words, we use the same thresholds for different underwater sonar images denoising. It obviously has limitations.

Although the distribution characteristics of h in the different underwater sonar images are different, further experiments show that the proportion of appropriate h in the total h is approximately the same, which satisfies the golden ratio within a certain error. Based on this discovery, in this paper, we propose an adaptive non-local spatial information denoising method based on the golden ratio which realizes two threshold adaptive settings and removes the underwater sonar image noise more effectively. The statistical law that the proportion satisfies is investigated on 20 different underwater sonar images by calculating the proportion of the h within two thresholds in the total h. Owing to space constraints, we only show partial results from 20 sonar images. Figures 1–6 show the estimated values of the optimal thresholds based on different underwater sonar images.

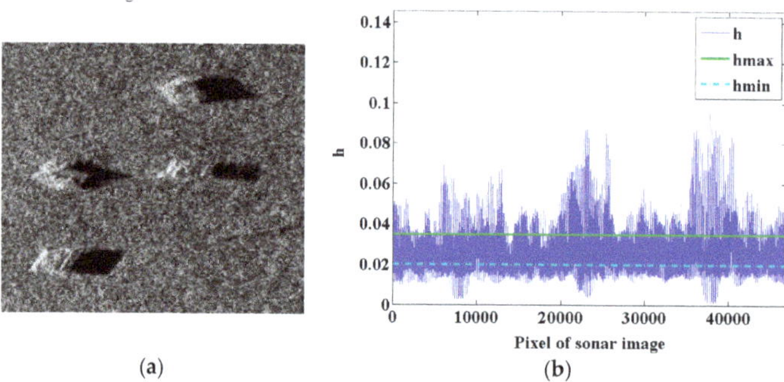

Figure 1. The estimated values of the optimal threshold (image size: 277 × 325): (**a**) Original sonar image; (**b**) The estimated values of the optimal threshold.

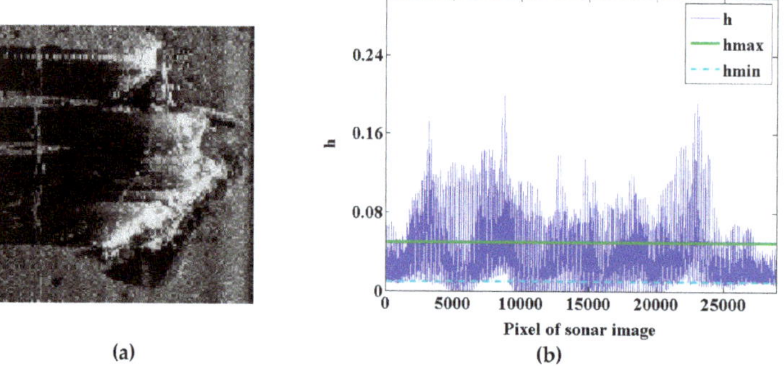

Figure 2. The estimated values of the optimal threshold (image size: 173 × 167): (**a**) Original sonar image; (**b**) The estimated values of the optimal threshold.

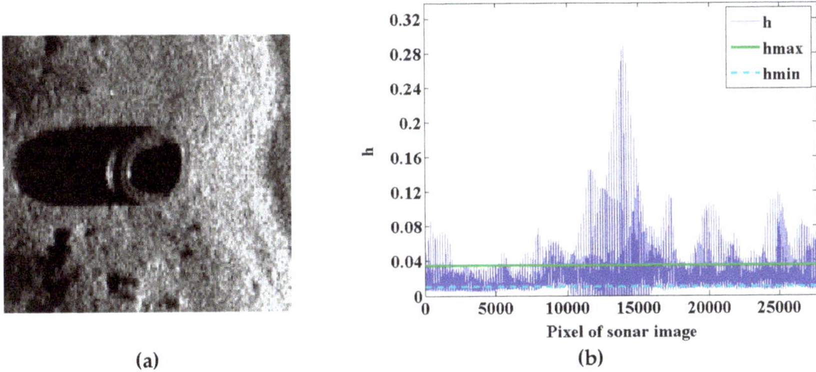

Figure 3. The estimated values of the optimal threshold (image size: 197 × 211): (**a**) Original sonar image; (**b**) The estimated values of the optimal threshold.

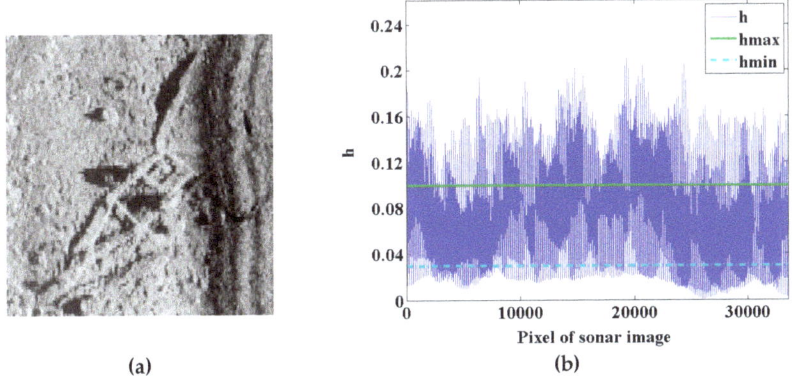

Figure 4. The estimated values of the optimal threshold (image size: 205 × 201): (**a**) Original sonar image; (**b**) The estimated values of the optimal threshold.

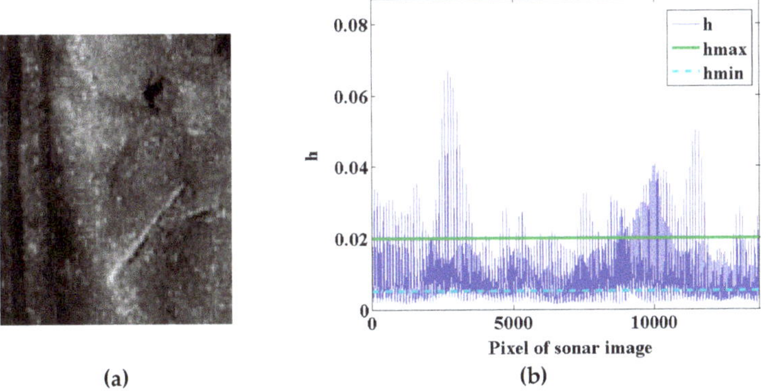

Figure 5. The estimated values of the optimal threshold (image size: 130 × 106): (**a**) Original sonar image; (**b**) The estimated values of the optimal threshold.

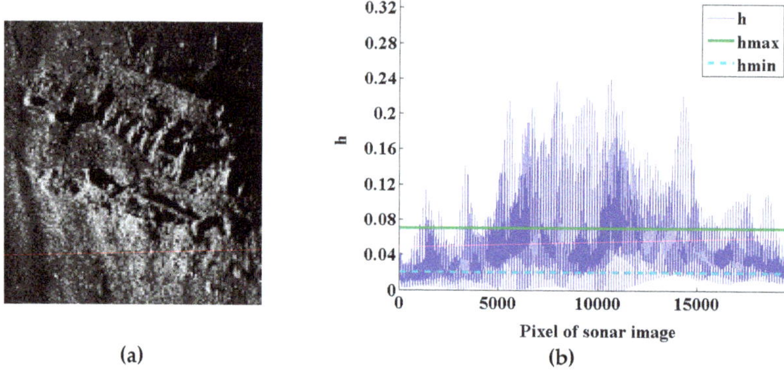

Figure 6. The estimated values of the optimal threshold (image size: 146 × 135): (**a**) Original sonar image; (**b**) The estimated values of the optimal threshold.

As seen from Figure 1b, Figure 2b, Figure 3b, Figure 4b, Figure 5b, and Figure 6b, different underwater sonar images have different distribution characteristics of h. This means that the estimated optimal thresholds are different. A linear fitting method is used to calculate the proportion satisfies based on different underwater sonar images.

For each underwater sonar image, the set of total h is defined as $H = \{h_i | i \in NU\}$, h_i is the filtering degree parameter corresponding to the ith pixel, NU is the number of the total pixels on the underwater sonar image. The set of h within two thresholds is defined as $H\prime = \{h_i | i \in NU \wedge h_i \geq hmin \wedge h_i \leq hmax\}$. The ratio of the cardinality of H to the cardinality of $H\prime$ is defined as:

$$\eta = \frac{card\ H\prime}{card\ H} \tag{7}$$

Table 1 shows the estimated optimal threshold and η of 20 different underwater sonar images.

Table 1. The estimated optimal threshold and η of 20 different underwater sonar images.

-	*hmin*	*hmax*	η
1	0.030	0.050	0.539
2	0.040	0.200	0.609
3	0.015	0.030	0.689
4	0.020	0.035	0.624
5	0.010	0.050	0.634
6	0.010	0.040	0.584
7	0.005	0.040	0.617
8	0.010	0.035	0.681
9	0.005	0.030	0.532
10	0.010	0.040	0.575
11	0.002	0.005	0.567
12	0.030	0.100	0.604
13	0.005	0.040	0.676
14	0.010	0.060	0.636
15	0.005	0.015	0.650
16	0.005	0.020	0.649
17	0.010	0.025	0.605
18	0.020	0.035	0.571
19	0.005	0.070	0.556
20	0.030	0.050	0.539

Figure 7 shows the result of the linear fitting based on Table 1.

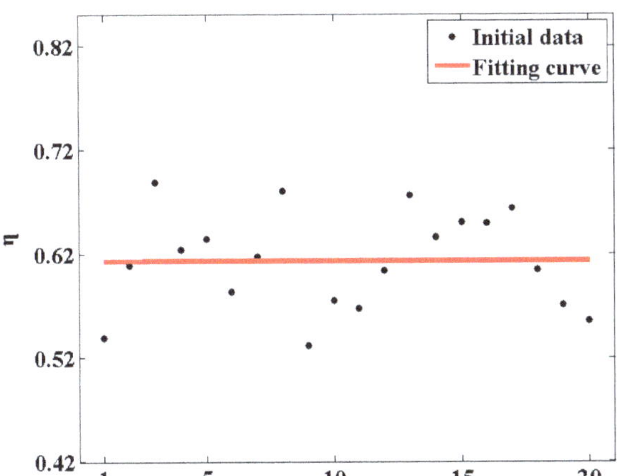

Figure 7. The result of the linear fitting based on Table 1.

The fitting curve is $y = 7.14e - 06x + 0.613$ in Figure 7. It demonstrates that η is approximately distributed in the vicinity of 0.613. The statistical law that the proportion satisfies is the golden radio within the range of error permitting. Two thresholds are set adaptively according to the golden ratio to remove too large and small h. On this basis, an adaptive non-local spatial information denoising method based on the golden ratio is proposed in this paper to effectively improve the denoising performance.

This is set that h less than $hmin$ is defined as $H_1 = \{h_i | i \in NU \wedge h_i \leq hmin\}$ and that h larger than $hmax$ is defined as $H_1 = \{h_i | i \in NU \wedge h_i \geq hmax\}$. On this basis, the ratio of the cardinality of H_1 to the cardinality of H is defined as $\eta_1 = \frac{card\ H_1}{card\ H}$. The ratio of the cardinality of H_2 to the cardinality of H is defined as $\eta_2 = \frac{card\ H_2}{card\ H}$. Comparing η_1 and η_2 with $\frac{1-\eta}{2}$, when the value exceeds the range of error permitting, the two thresholds are constantly updated by the dichotomy. The update process is not finished until the experiment result simultaneously satisfies $\left|\eta_2 - \frac{1-\eta}{2}\right| \leq \varepsilon$ and $\left|\eta_1 - \frac{1-\eta}{2}\right| \leq \varepsilon$. ε is the range of error permitting. The result of two adaptive thresholds is respectively defined as $Ahmin$ and $Ahmin$.

For the ith pixel in the sonar image, h_i is defined as:

$$h_i = \begin{cases} Ahmin & hm_i \leq Ahmin \\ Ahmax & hm_i \geq Ahmax \\ hm_i & otherwise \end{cases} \tag{8}$$

$$hm_i = \underset{p \in W_i^r}{\text{mean}} \left\{ \|x(N_j) - x(N_p)\|_{2,\rho}^2 \right\} \tag{9}$$

The adaptive non-local spatial information denoising method based on the golden ratio is described in Algorithm 1.

Algorithm 1: the proposed denoising method

Input: Original sonar image;
Output: Denoising image;
Initialization: r (the radius of search window), s (the radius of neighbourhood window);
Procedure:
1: For $q = 1$ to s^2 do:
2: Calculate d using Equation (6);
3: Calculate $\rho^{(q)}$ using Equation (5);
4: For every search window in the input image, do:
5: For any two different neighbourhood windows in the search window, do:
6: Calculate the weighted Euclidean distance using Equation (4);
7: Calculate the filtering degree parameter h using Equation (9);
8: Calculate $Ahmin$ and $Ahmax$ when the result of Equation (8) satisfies the golden ratio;
9: For every search window in the input image, do:
10: Calculate the filtering degree parameter h using Equation (8);
11: Calculate Z using Equation (3);
12: Calculate w using Equation (2);
13: Calculate non-local spatial information $\overline{x}$ using Equation (1).

3. Culture Algorithm

3.1. Cultural Algorithm Framework

The bilevel evolutionary mechanism in the cultural algorithm mainly includes three elements: the population space, the belief space, and the communication protocol. The communication protocol consists of influence function and accept function. The population space and the belief space are two relatively independent evolutionary and update processes. However, they rely on the influence function and accept function to communicate with each other, manage relevant information, and guide the evolution and update of space. The important feature of the cultural algorithm is the introduction of belief space. The individual evolutionary experience in the population space can be transferred to the belief space by the accept function. The belief space transforms advanced experience into knowledge through certain rules. The knowledge in the belief space guides the evolution of the population space towards a more accurate direction through the influence function. Figure 8 shows the schematic diagram of CA.

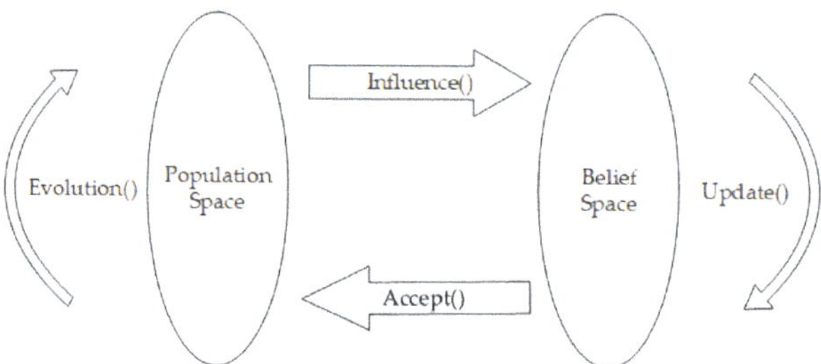

Figure 8. Schematic diagram of CA (cultural algorithm).

3.2. Our Proposed Cultural Algorithm—A New Adaptive Culture Algorithm

In our previous work [22], we proposed an adaptive cultural algorithm with improved quantum-behaved particle swarm optimization (ACA-IQPSO) and applied it to the underwater sonar image detection. In the population space, to enhance the search ability of the ACA-IQPSO, we introduced the IQPSO as the evolution strategy. In the belief space, we adopted a new update strategy based on the idea of SFLA. Situational knowledge, normative knowledge, and domain knowledge were selected to form the knowledge structure. A new communication mechanism was designed to make better use of the information in the population space and belief space. Experimental results show that ACA-IQPSO has better searching ability and convergence efficiency.

However, we found that ACA-IQPSO could not obtain good initial clustering centres, and the update strategy of the belief space easily gets into the local optimal solution. Therefore, the convergence ability of ACA-IQPSO needs to be improved and the efficiency of underwater sonar image detection is reduced.

Therefore, in order to complete the underwater sonar image detection more accurately in shorter iterations, NACA is proposed for the underwater sonar image detection. NACA has two improvements over ACA-IQPSO. To begin with, inspired by the idea of data field and maximum entropy [23], we propose an adaptive initialization algorithm based on data field (AIA-DF) for underwater sonar image detection. AIA-DF is used in the population space to more accurately find the initial clustering centres. Secondly, in our previous work [7], we proposed QSFLA-NSM for the underwater sonar image detection. Experimental results show that compared with SFLA, QSFLA-NSM has stronger search ability. Therefore, we propose a new update strategy in the belief space based on the idea of QSFLA-NSM.

3.2.1. Adaptive Initialization Algorithm Based on Data Field

The dataset R is defined as $R = \{R_1, R_2, \cdots, R_N\}$ in the data space Ω. The ith data point is defined as $R_i = (R_{i1}, R_{i2}, \cdots, R_{iD})$, N is the size of R. D is the dimension of a data point. Every data point in R interacts with the data points around it. In this way, a data field is formed according to the interaction of different data points. The interaction between different data points is measured by potential functions in the data field. Generally, the closer data points have stronger interaction and greater potential. In contrast, the further data points have weaker interaction and lower potential. The potential function is defined as:

$$\varphi_{R_{i1}R_{i2}} = m_{R_i} e^{\frac{-\|R_i - R_*\|^2}{\sigma^2}} \tag{10}$$

where $\varphi_{R_{i1}R_{i2}}$ is the potential of the data point $R_i = (R_{i1}, R_{i2})$ at the threshold R_*, R_{i1} is the grey value of the ith pixel in the underwater sonar image. R_{i2} is the average of the grey values of pixels in the neighborhood window. The neighborhood window centre is the ith pixel. m_{R_i} is the frequency corresponding to the ith pixel in the two-dimensional histogram of the underwater sonar image. $\| R_i - R_* \|$ is the Euclidean distance and σ is the impact factor.

Furthermore, in the data field, the sum of the potential of all data points at the threshold R_* is defined as:

$$\varphi = \sum_{i=1}^{N} m_{R_i} e^{\frac{-\|R_i - R_*\|^2}{\sigma^2}} \tag{11}$$

where N is the number of total pixels in the underwater sonar images.

The influence radius of every data point is $\frac{3\sigma}{\sqrt{2}}$ in the data field [23]. Therefore, when calculating φ, the data points at work are in the region O, of which its centre is R_* and the radius is $\frac{3\sigma}{\sqrt{2}}$. On this basis, φ is redefined as

$$\varphi = \sum_{R_i \in O} m_{R_i} e^{\frac{-\|R_i - R_*\|^2}{\sigma^2}} \tag{12}$$

The impact factor σ is a significant parameter. It is closely related to φ. The relatively large σ means that φ is relatively close, which makes the initialization algorithm easy to fall into the local optimal solution during the iteration. In contrast, the relatively small σ means that φ R_* is relatively close to 0, which also affects the subsequent solution of optimal threshold. To accurately calculate the optimal threshold, σ is redesigned.

According to the criterion:

$$P(\mu - 3\sigma \leq K \leq \mu + 3\sigma) = 0.997 \tag{13}$$

where K is the possible grey levels in the underwater sonar image and $\mu = \frac{L}{2}$. L is grey levels in the underwater sonar image. σ is defined as:

$$\sigma = \frac{6}{L} \tag{14}$$

According to the principle of the data field [23], when the uncertainty is the smallest, the potential can perfectly reflect the distribution of data points in Ω. The uncertainty is usually described as the entropy. The optimal threshold is calculated by entropy. The size of the data field is $L \times L$, the scope of R_* is $0 \leq R_{*1} \leq L-1, 0 \leq R_{*2} \leq L-1$. Data points in the data field are divided into two classes: $A = \{(0,0), \cdots, (R_{*1}, R_{*2})\}$ and $B = \{(R_{*1}+1, R_{*2}+1), \cdots, (L-1, L-1)\}$. The entropy of A and B are respectively expressed as follows:

$$H(A) = -\sum_{i1=0}^{R_{*1}} \sum_{i2=0}^{R_{*2}} \frac{\varphi_{R_{i1}R_{i2}}}{\Psi_A} \log \frac{\varphi_{R_{i1}R_{i2}}}{\Psi_A} \tag{15}$$

$$H(B) = -\sum_{i1=R_{*1}+1}^{L-1} \sum_{i2=R_{*2}+1}^{L-1} \frac{\varphi_{R_{i1}R_{i2}}}{\Psi_B} \log \frac{\varphi_{R_{i1}R_{i2}}}{\Psi_B} \tag{16}$$

where $\Psi_A = \sum_{i1=0}^{R_{*1}} \sum_{i2=0}^{R_{*2}} \varphi_{R_{i1}R_{i2}}$, $\Psi_B = \sum_{i1=R_{*1}+1}^{L-1} \sum_{i2=R_{*2}+1}^{L-1} \varphi_{R_{i1}R_{i2}}$.

The sum of posterior entropy in the data field is defined as:

$$H(R_*) = H(A) + H(B) \tag{17}$$

When the maximum of $H(R_*)$ is obtained, R_* is the optimal threshold.

AIA-DF is proposed in this paper according to Equations (12)–(17). The initialization algorithm adaptively calculates initial clustering centres in population space on the basis of the characteristics of the underwater sonar images. It improves the disadvantages that the initial cluster centers are randomly generated in the solution space and effectively avoids the prematurity of detection algorithm.

The influence region O of the optimal threshold is limited because of the influence radius. Besides, in the data field, the distribution of potentials has different characteristics in different underwater sonar images. Therefore, the optimal threshold is also different to some extent.

$U = \{(R_{*1}, R_{*2}) | 0 \leq R_{*1} \leq L-1, 0 \leq R_{*2} \leq L-1\}$ is the initial scope of threshold R_*. When the optimal threshold is obtained, the grey value is randomly selected in region O for initializing the clustering centres. The update formula for U is given as:

$$U(num+1) = U(num) - U* \tag{18}$$

where num is the number of optimal thresholds currently obtained.

U^* is defined as::

$$U* = \left\{ \begin{array}{l} (R(num+1)_{*1}, (R(num+1)_{*2}| \\ Rnum_{*1} - \frac{3\sigma}{\sqrt{2}} \leq R(num+1)_{*1} \leq Rnum_{*1} + \frac{3\sigma}{\sqrt{2}}, \\ Rnum_{*2} - \frac{3\sigma}{\sqrt{2}} \leq R(num+1)_{*2} \leq Rnum_{*2} + \frac{3\sigma}{\sqrt{2}} \end{array} \right\} \tag{19}$$

When the value of $R(num+1)_{*1}$ and $R(num+1)_{*2}$ are less than 0 or larger than $L-1$, the update process of U is over. AIA-DF is described in Algorithm 2.

Algorithm 2: the proposed AIA-DF

Input: Denoising image;
Output: Initialized image;
Initialization: s (the radius of the neighbourhood window), num (the number of the optimal threshold), U (the set of optimal threshold potential solutions);
Procedure:
For every pixel in the input image, do:
    Calculate the mean grey of its neighbourhood;
  Establish the two-dimensional grey histogram of the input image;
Establish the data field according to the grey histogram;
repeat:
    Calculate the entropy of $H(A)$ using Equation (15);
    Calculate the entropy of $H(B)$ using Equation (16);
    Calculate the optimal threshold when the maximum of Equation (17) is obtained;
  Calculate U^* using Equation (19);
  Update U using Equation (18):
    Update num using $num = num + 1$:
Until: the stopping criteria are met.

3.2.2. New Update Strategy

In the belief space, to complete the update process by using the new update strategy in terms of QSFLA-NSM. In the first place, cultural individuals are divided into the multiple subpopulations and sorted from large to small according to the fitness function value [25]. Each subpopulation executes the local search strategy. In the second place, with the subpopulation complete local search strategy, all subpopulations are mixed for global information exchange. Then, the belief space is divided into subpopulations again. Local search and global information exchange will alternate until the number of iterations is satisfied

Local search completes the update of the worst cultural individual in each subpopulation. The first step of the local search [5] is expressed as:

$$newY_w(t+1) = \begin{cases} P(t) + \beta \cdot |S_b(t) - Y_w(t)| \cdot \ln\left(\frac{1}{\mu_1}\right) & \mu_1 < 0.5 \\ P(t) - \beta \cdot |S_b(t) - Y_w(t)| \cdot \ln\left(\frac{1}{\mu_1}\right) & \text{otherwise} \end{cases} \quad (20)$$

where t is the current local iterative times, μ_1 is a random number in [0,1], $P(t)$ is the local attractor, and β is the contraction–expansion coefficient. S_b is the situational knowledge of the subpopulation on behalf of the local best cultural individual. Y_w is the local worst cultural individual.

Comparing the fitness function values of $newY_w$ and Y_w, if the fitness function value of $newY_w$ is larger, Y_w is replaced by $newY_w$. Whereas, S_g which represents the global best individual, will replace S_b in Equation (20). $newY_w$ is calculated by Equation (20) again. If the fitness function value of $newY_w$ is still less than Y_w, it will randomly generate an individual from the solution space and replace Y_w. When the upper limit of the local iteration number is reached, the local search is over.

4. Experiments and Discussion

4.1. The Characteristics of the Underwater Sonar Images

The underwater sonar images include the object-highlight region, sea-bottom-reverberation region, and shadow region. However, because of the complexity of the underwater environment, the underwater sonar image is easily affected by the reverberation effect, strong speckle noise, fuzzy edge, and weak texture information. This means that the underwater sonar images contain a lot of noise.

4.2. The Effectiveness Verification of the Proposed Denoising Method

To prove the superiority of the proposed denoising method, Figure 9 shows the denoising results based on Figure 1a of the proposed denoising method and the previous denoising method [7]. The experiment environment is using Matlab R2012b with a 2.7 GHz Core processor and 8 GB of RAM.

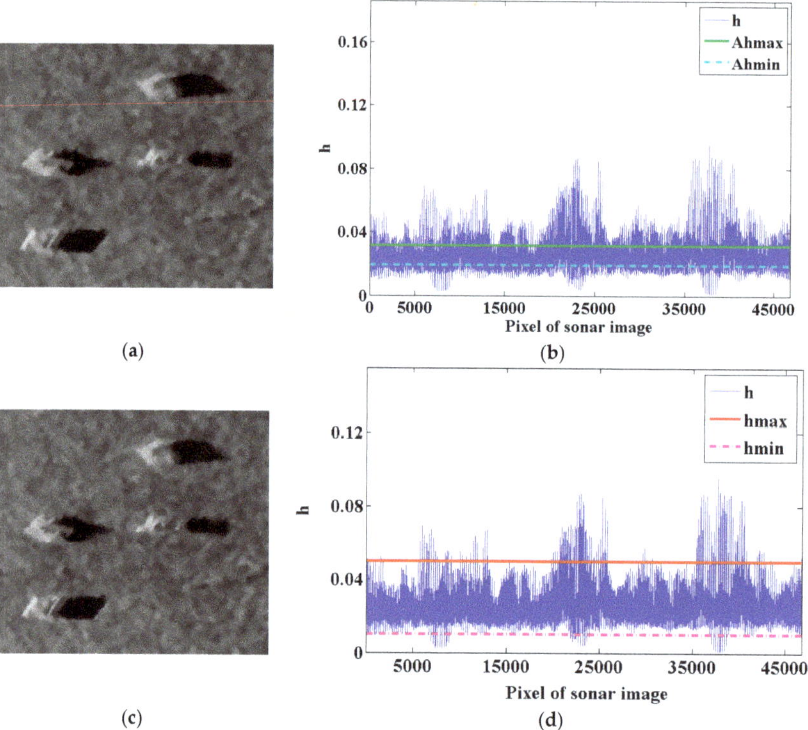

Figure 9. The denoising results based on Figure 1a of the proposed denoising method in this paper and the previous denoising method: (**a**) The denoising result of the denoising method proposed in this paper; (**b**) The result of filtering the degree parameter is selected by two adaptive thresholds in the proposed denoising method; (**c**) The denoising result of the previous denoising method; (**d**) The result of the filtering degree parameter is selected by two previous thresholds in the previous denoising method.

As can be seen from Figure 9b,d, compared with the previous thresholds $hmin = 0.01$, the adaptive thresholds $Ahmin = 0.019$ can remove more fairly small filtering degree parameters and effectively remove image noise points. Meanwhile, the threshold $Ahmax = 0.032$ is smaller than $hmax = 0.05$ which can remove more fairly large filtering degree parameters. It contributes to keeping the underwater sonar image details better when the image noise points are removed.

A relatively simple and effective the fuzzy c-means (FCM) [10] is used to further qualitatively verify the effectiveness of the denoising method proposed in this paper. Figure 10 shows the detection results of FCM based on the thresholds in Figure 9b,d.

As can be seen from Figure 10a,b, compared with the previous threshold $hmin$, $Ahmin$ removes smaller parameters in a larger range. Therefore, it can effectively improve the performance of sonar images. Meanwhile, as shown in Figure 10c,d, compared with the previous threshold $hmax$, $Ahmax$ can more accurately define the boundaries of relatively large h, which benefits preserving detail information in sonar images. The noise can be effectively removed on the basis of preserving image details by the proposed denoising method which contributes to the remainder of the image processing.

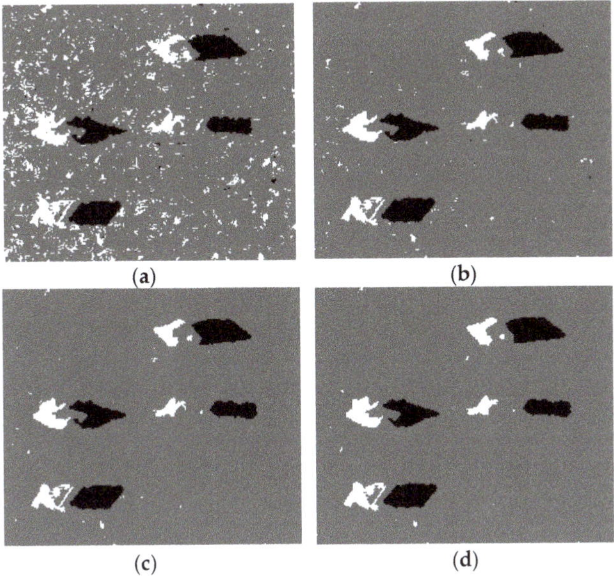

Figure 10. The detection results of the fuzzy c-means (FCM) based on Figure 9b,d: (**a**) FCM detection result when $h = 0.01$; (**b**) FCM detection result when $h = 0.019$; (**c**) FCM detection result when $h = 0.032$; (**d**) FCM detection result when $h = 0.05$.

4.3. The Effectiveness Demonstration of the Proposed Detection Method

To testify the effectiveness of NACA and AIA-DF, the detection experiments based on different underwater sonar images are presented. The proposed non-local spatial information denoising method based on the golden ratio in this paper is used for removing noise in the underwater sonar images. On that basis, the proposed NACA is used for underwater sonar image detection. Meanwhile, the proposed NACA is compared with QSFLA-NSM [7], ACA-IQPSO [22], QSFLA [26,27], CPSO [20], and QPSO [28]. Except that the number of clustering centres is adaptively set according to AIA-DF in NACA, the number of clustering centres is 3 in other intelligent optimization algorithms. The population size is 20, the sub-populations size is 5 in proposed NACA, QSFLA-NSM, and QSFLA, the size of the belief space is 8 in NACA, ACA-IQPSO, and CPSO, the global maximum number of iterations is 10, the local maximum number of iterations is 2 in NACA, QSFLA-NSM, and QSFLA. The experiment environment is using Matlab R2012b with a 2.7 GHz Core processor and 8 GB of RAM.

To demonstrate the effectiveness of NACA, Figure 11 shows the detection results of the underwater sonar image shown in Figure 9a.

As is described in Figure 11, the regions of object-highlight and shadow are more accurately detected by NACA in Figure 11a. It is found that many noise points cannot be removed in Figure 11b,d. In Figure 11e,f, the detection results have poor integrity in the regions of object-highlight and shadow, respectively. It will cause serious difficulty to the subsequent underwater sonar image recognition. Therefore, compared with the detection results of QSFLA-NSM, ACA-IQPSO, QSFLA-NSM, QSFLA, CPSO, and QPSO, the proposed NACA can effectively remove the noise and accurately complete underwater sonar image detection.

Subsequently, the quantitative analysis of the detection results in Figure 12 is performed. The values of fitness function [25] after each iteration in the detection process of NACA and contrast algorithms is shown in Table 2.

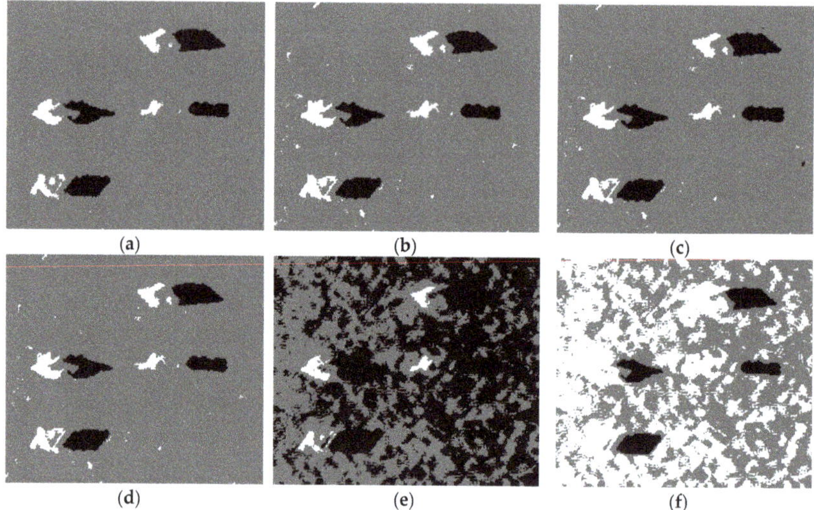

Figure 11. Detection results of the image shown in Figure 9a: (**a**) Detection result of NACA (new adaptive cultural algorithm); (**b**) Detection result of QSFLA-NSM (quantum-inspired shuffled frog leaping algorithm combining the new search mechanism); (**c**) Detection result of ACA-IQPSO (adaptive cultural algorithm with improved quantum-behaved particle swarm optimization); (**d**) Detection result of QSFLA (quantum-inspired shuffled frog leaping algorithm); (**e**) Detection result of CPSO (cultural particle swarm optimization algorithm); (**f**) Detection result of QPSO (quantum-behaved particle swarm optimization).

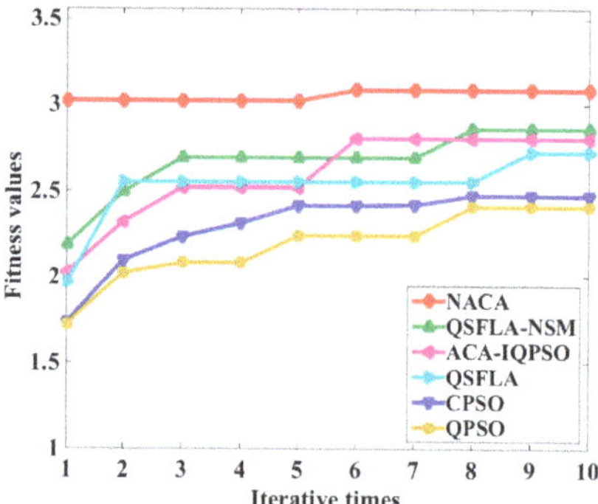

Figure 12. The chart of the best fitness function values in Table 2.

Table 2. The values of fitness function after each iteration in the detection process.

Iterative Times	NACA	QSFLA-NSM	ACA-IQPSO	QSFLA	CPSO	QPSO
1	3.024	2.188	2.028	1.969	1.737	1.724
2	3.024	2.495	2.320	2.554	2.100	2.024
3	3.024	2.695	2.520	2.554	2.236	2.083
4	3.024	2.695	2.520	2.554	2.315	2.083
5	3.024	2.695	2.520	2.554	2.417	2.241
6	3.087	2.695	2.806	2.554	2.417	2.241
7	3.087	2.695	2.806	2.554	2.421	2.241
8	3.087	2.862	2.806	2.727	2.475	2.410
9	3.087	2.862	2.806	2.727	2.475	2.410
10	3.087	2.862	2.806	2.727	2.475	2.410

Figure 12 shows the chart of the best fitness function values in Table 2.

As can be seen from Table 2 and Figure 12, the best fitness values of the proposed NACA is the largest after each iteration. Only NACA converges after four iterations. The results show that the proposed NACA has a better search ability and a faster convergence speed (the speed corresponds to the number of iterations). Meanwhile, the fitness function values after the first iteration are close to the fitness function values after 10 iterations in Figure 12. This means that AIA-DF can obtain better initial clustering centres in the proposed NACA. Through qualitative analysis and quantitative analysis of the underwater sonar detection results, the effectiveness of the proposed NACA is verified.

To further verify the effectiveness of the proposed NACA, Figure 13 shows the detection results of the underwater sonar image.

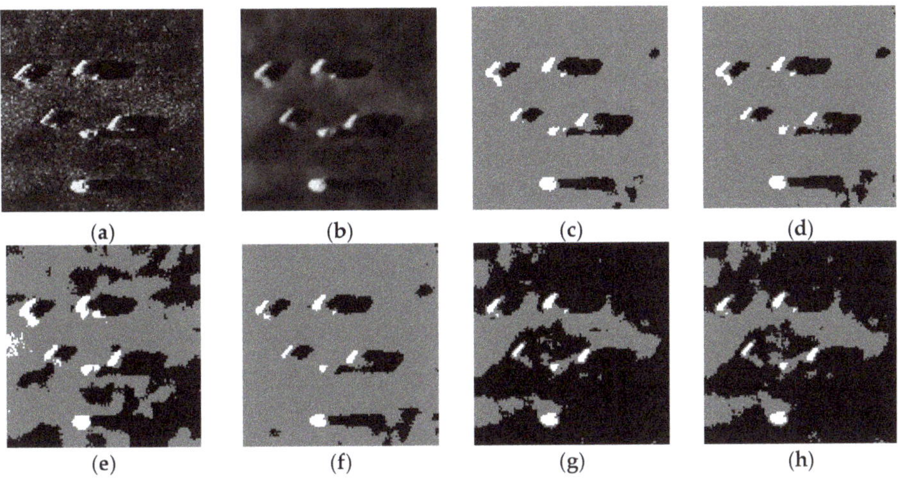

Figure 13. Detection results of the underwater sonar image (image size: 112×117): (**a**) Original sonar image; (**b**) The denoising result of the denoising method proposed in this paper; (**c**) Detection result of NACA; (**d**) Detection result of QSFLA-NSM; (**e**) Detection result of ACA-IQPSO; (**f**) Detection result of QSFLA; (**g**) Detection result of CPSO; (**h**) Detection result of QPSO.

The values of fitness function [25] after each iteration in the detection process of NACA and contrast algorithms are shown Table 3. Figure 14 shows the chart of the best fitness function values in Table 3.

Table 3. The values of fitness function after each iteration in the detection process.

Iterative Times	NACA	QSFLA-NSM	ACA-IQPSO	QSFLA	CPSO	QPSO
1	2.365	2.184	2.003	1.956	1.794	2.066
2	2.477	2.184	2.003	2.080	1.951	2.066
3	2.477	2.212	2.003	2.080	2.103	2.170
4	2.477	2.212	2.133	2.312	2.103	2.170
5	2.532	2.212	2.133	2.337	2.162	2.170
6	2.532	2.426	2.133	2.337	2.226	2.170
7	2.532	2.426	2.273	2.337	2.226	2.184
8	2.532	2.426	2.273	2.337	2.253	2.209
9	2.532	2.426	2.273	2.337	2.253	2.209
10	2.532	2.426	2.273	2.337	2.253	2.209

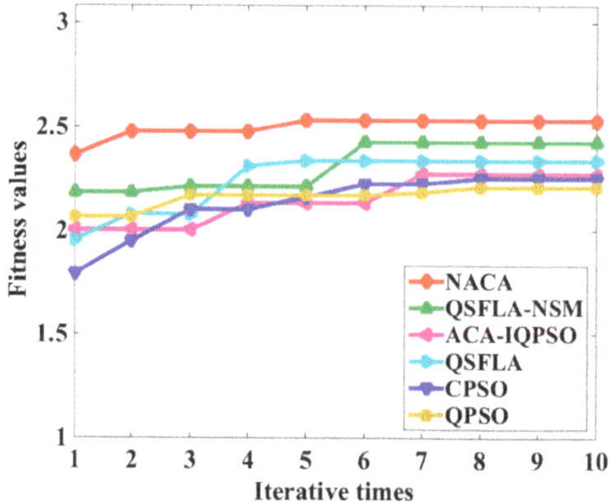

Figure 14. The chart of the best fitness function values in Table 3.

As can be seen in Figure 13, NACA more accurately completes the detection of different regions in Figure 13c. Figure 13d,f show over detection in the shadow region. The detection result fails to effectively remove the noise in Figure 13e,g,h. At the same time, the best fitness values of NACA are also the largest after each iteration in Table 3 and Figure 14. Comparing with other algorithms, the detection results demonstrate that NACA has the better search ability and a faster convergence speed (the speed corresponds to the number of iterations) based on the better initialization results that were obtained by AIA-DF in the paper. Therefore, through qualitative and quantitative analysis, the proposed NACA can effectively remove the noise and better complete the process of underwater sonar image detection.

To demonstrate the effectiveness of AIA-DF, Figure 15 shows the detection result after the first iteration in the detection process of Figure 11. The fitness function value after iteration is shown in the first line of Table 2.

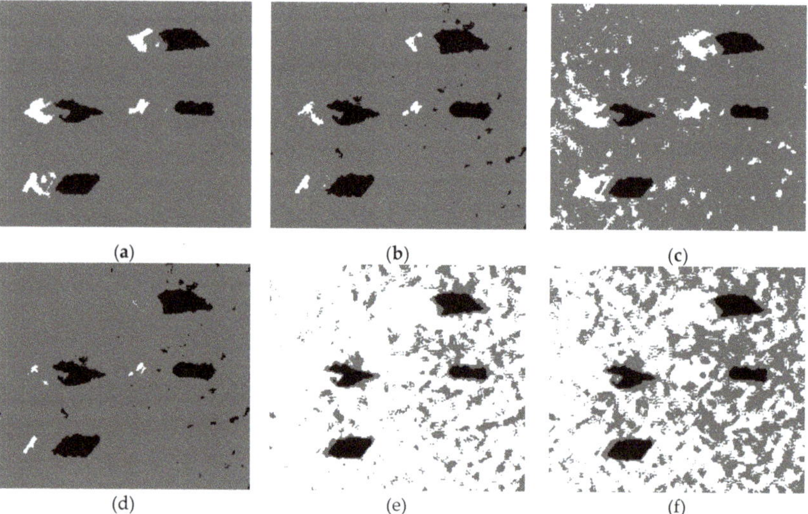

Figure 15. Detection result after the first iteration in the detection process of Figure 11: (**a**) Detection result of NACA; (**b**) Detection result of QSFLA-NSM; (**c**) Detection result of ACA-IQPSO; (**d**) Detection result of QSFLA; (**e**) Detection result of CPSO; (**f**) Detection result of QPSO.

As can be seen in Figure 15, only NACA can accurately detect the region of the object highlight region and shadow. It also effectively removes the noise in the underwater sonar image in Figure 15a. These results show that AIA-DF can find more accurate clustering centres. In addition, it can be seen from Figures 15a and 11a that the detection result after the first iteration is almost identical with the final detection result. The experimental results demonstrate that AIA-DF contributes to improving convergence speed (the speed corresponds to the number of iterations), accuracy, and search ability in the proposed NACA. Through qualitative analysis and quantitative analysis of the underwater sonar image detection results after the first iteration, the effectiveness of AIA-DF is proved.

4.4. The Performance Analysis of the New Update Strategy in NACA

To demonstrate the superiority in the search ability of the new update strategy in NACA, in the benchmark functions, Sphere function and Griewank function are used to test the position distribution of particles in this paper. Sphere function is unimodal and only has one global optimal solution. Griewank function is multimodal and has many local optimal solutions, however only has one global optimal solution. Figure 16 shows the distribution of particle positions in the new update strategy and old update strategy [22]. The relevant parameters are as follows: the dimension of the solution space is 2, the size of the belief space is 30, the size of the subspaces is 5, the global maximum number of iterations is 10, and the local maximum number of iterations is 2.

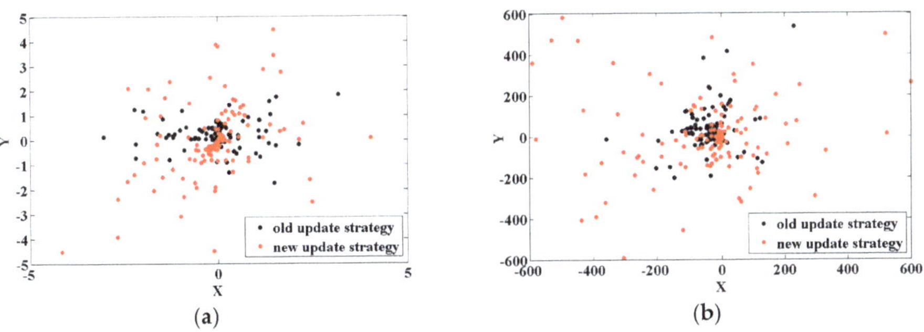

Figure 16. The distribution of particle positions in the new update strategy and old update strategy: (**a**) Position distribution of particles on Sphere function; (**b**) Position distribution of particles on Griewank function.

It can be seen in Figure 16 that compared with the old update strategy, the distribution of particle positions in the new update strategy are more dispersed. The new update strategy is easier to obtain the global optimal solution and has a stronger search ability in this paper.

In order to further verify the effectiveness of the new update strategy in its search ability, the fitness function values are calculated by Sphere function and Griewank function in the new update strategy and old update strategy, and the result of the fitness function values are shown in Figure 17. The relevant parameters are as follows: the dimension of the solution space is 30, the size of the belief space is 10, the size of the subspaces is 2, the global maximum number of iterations is 100, and the local maximum number of iterations is 10.

As can be seen in Figure 17, the fitness function values of the old update strategy are always greater than that of the new update strategy in each iteration. Compared with the old update strategy, the new update strategy is more likely to jump out of the local optimization. The experiment results further demonstrate that the search ability of the new update strategy is better than the old update strategy.

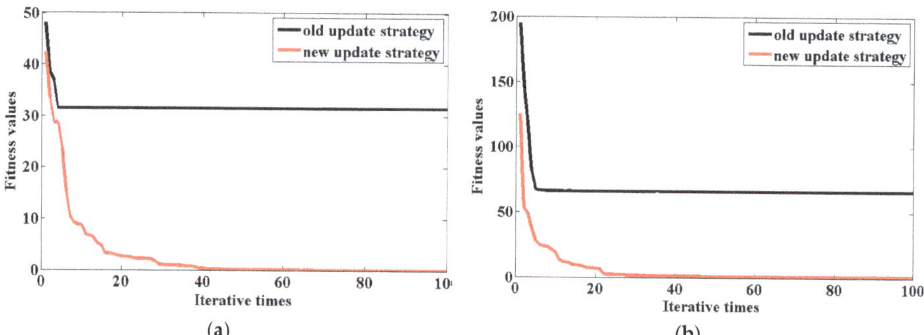

Figure 17. The result of the fitness value calculation in the new update strategy and old update strategy: (**a**) The optimization results of the Sphere function; (**b**) The optimization results of the Griewank function.

4.5. The Adaptability Demonstration of the Proposed Denoising Method and Detection Method

To prove the adaptability of the proposed denoising method and NACA, Figure 18 shows the detection results of the structured seabed which is an object in sand ripples. Figure 19 shows the detection results of the starboard original sonar image with a ship. Figure 20 shows the detection results of a floating object. The experiment environment uses Matlab R2012b with a 2.7 GHz Core processor and 8 GB of RAM.

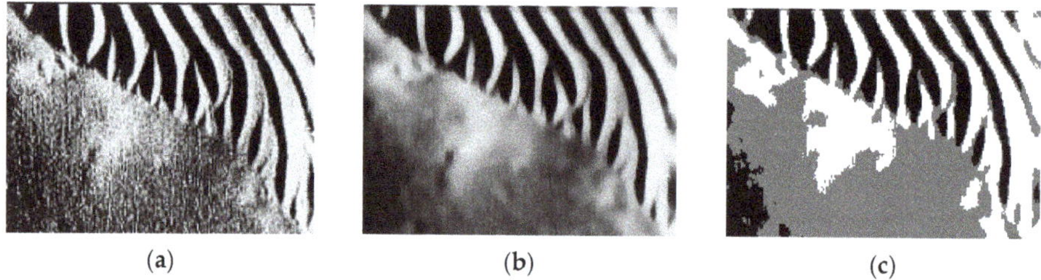

Figure 18. Detection results of the underwater sonar image (image size: 259×368): (**a**) Original sonar image; (**b**) The denoising result of the denoising method proposed in this paper; (**c**) Detection result of NACA.

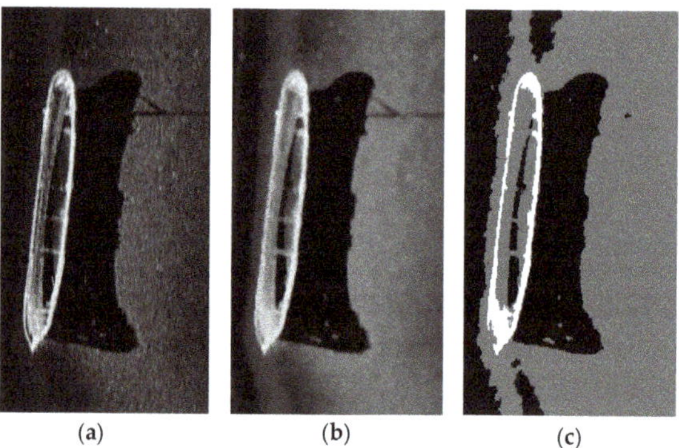

Figure 19. Detection results of the underwater sonar image (image size: 259×368): (**a**) Original sonar image; (**b**) The denoising result of the denoising method proposed in this paper; (**c**) Detection result of NACA.

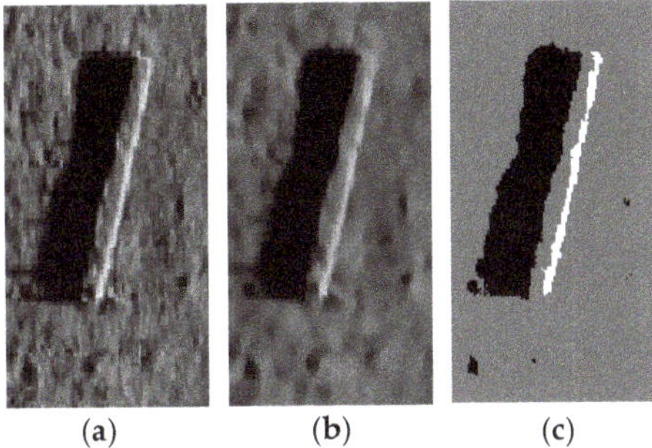

Figure 20. Detection results of the underwater sonar image (image size: 203×101): (**a**) Original sonar image; (**b**) The denoising result of the denoising method proposed in this paper; (**c**) Detection result of NACA.

It can be seen from the detection results in Figure 18b, Figure 19b, and Figure 20b that the proposed denoising method can remove noise effectively in polytype underwater sonar images. The boundary information of the different regions is accurately detected in Figure 18c, Figure 19c, and Figure 20c. Moreover, the iteration times of the algorithm are only two times, which further proves the adaptability of the AIA-DF in this paper. Therefore, the adaptability of the proposed method has been effectively proved.

5. Conclusions

This paper proposed an adaptive approach to denoise and detect the underwater sonar image. The problem of the inappropriate filtering degree parameter means that it seriously affects the denoising performance in underwater sonar images, which is solved by the proposed adaptive non-local spatial information denoising method based on the golden ratio in the paper. NACA is proposed in this paper. Firstly, in the population space, AIA-DF is adopted to obtain good initial clustering centres by calculating the potential entropy of the data field from the underwater sonar images dataset. In the belief space, a new update strategy is adopted to update cultural individuals according to the idea of QSFLA. The new update strategy further improves the search ability of NACA.

We apply the proposed method to the different underwater sonar images. The experimental results demonstrate that the proposed denoising method can effectively remove noise and reduce the difficulty of the following underwater sonar image recognition. Compared to the comparison algorithms, the proposed NACA has more advantages in its search ability and convergence speed (the speed corresponds to the number of iterations). AIA-DF can better locate initial clustering centres and enhance convergence efficiency of NACA which contributes to accurately complete underwater sonar image detection and convergence to the global optimal solution within small epochs. This paper provides an effective and important method for underwater sonar image detection. It has much academic and practical importance.

Author Contributions: Conceptualization, X.W., J.Y., and Q.L.; Methodology, X.H. and W.H.; Software, W.H.; Investigation, X.H. and W.H.; Resources, X.W.; Writing—original draft preparation, Q.L. and W.H.; Writing—review and editing, X.W., Q.L., J.Y., and X.H.; Visualization, X.H.; Supervision, X.W. and J.Y.; Project administration, X.W.; Funding acquisition, X.W.

Funding: This research was supported in part by the National Natural Science Foundation of China grant number 41876110, the Fundamental Research Funds for the Central Universities of China grant number HEUCF180601, the Heilongjiang Province Outstanding Youth Science Fund of China grant number JC2017017, and in part by the Fok Ying-Tong Education Foundation of China grant number 151007.

Conflicts of Interest: The authors declare no conflict of interest.

References

1. Wang, X.; Guo, L.; Yin, J.; Liu, Z.; Han, X. Narrowband Chan-Vese model of sonar image segmentation: An adaptive ladder initialization approach. *Appl. Acoust.* **2016**, *113*, 238–254. [CrossRef]
2. Ye, X.; Zhang, Z.; Liu, P.; Guan, H. Sonar image segmentation based on GMRF and Level-set models. *Ocean Eng.* **2010**, *37*, 891–901. [CrossRef]
3. Wang, X.; Liu, S.; Teng, X.; Sun, J.; Jiao, J. SFLA with PSO local search for detection sonar image. In Proceedings of the 2016 35th Chinese Control Conference, Chengdu, China, 27–29 July 2016; pp. 3852–3857.
4. Wu, J.P.; Guo, H. A method for sonar image segmentation based on combination of MRF and region growing. In Proceedings of the 2015 5th International Conference on Communication Systems and Network Technologies, Gwalior, India, 4–6 April 2015; pp. 457–460.
5. Zhu, Z.; Yin, H.; Chai, Y.; Li, Y.; Qi, G. A novel multi-modality image fusion method based on image decomposition and sparse representation. *Inf. Sci.* **2018**, *432*, 516–529. [CrossRef]
6. Li, H.; Qiu, H.; Yu, Z.; Li, B. Multifocus image fusion via fixed window technique of multiscale images and non-local means filtering. *Signal Process.* **2017**, *138*, 71–85. [CrossRef]
7. Wang, X.; Liu, S.; Liu, Z. Underwater sonar image detection: A combination of non-local spatial information and quantum-inspired shuffled frog leaping algorithm. *PLoS ONE* **2017**, *12*, e0177666. [CrossRef]
8. Zhao, F.; Jiao, L.; Liu, H.; Gao, X. A novel fuzzy clustering algorithm with non-local adaptive spatial constraint for image segmentation. *Signal Process.* **2011**, *91*, 988–999. [CrossRef]
9. Zhao, F. Fuzzy clustering algorithms with self-tuning non-local spatial information for image segmentation. *Neurocomputing* **2013**, *106*, 115–125. [CrossRef]
10. Wang, L. Segmentation algorithm of fuzzy clustering on side scan sonar image. *J. Huazhong Univ. Sci. Technol.* **2012**, *40*, 25–29.
11. Mignotte, M.; Collect, C.; Perez, P.; Bouthemy, P. Three-class markovian segmentation of high-resolution sonar image. *Comput. Vis. Image Underst.* **1999**, *76*, 191–204. [CrossRef]
12. Mignotte, M.; Collect, C.; Perez, P.; Bouthemy, P. Sonar image segmentation using an unsupervised hierarchical MRF model. *IEEE Trans. Signal Process.* **2000**, *9*, 1216–1231. [CrossRef]
13. Vese, L.; Chan, T. A multiphase level set framework for image segmentation using the mumford and shah model. *Int. J. Comput. Vis.* **2002**, *50*, 271–293. [CrossRef]
14. Lianantonakis, M.; Petillot, Y. Sidescan sonar segmentation using active contours and level set methods. In Proceedings of the Oceans Europe 2005, Brest, France, France, 20–23 June 2005; pp. 719–724.
15. Lianantonakis, M.; Petillot, Y. Sidescan sonar segmentation using texture descriptors and active contours. *IEEE J. Ocean. Eng.* **2007**, *32*, 744–752. [CrossRef]
16. Liu, G.; Bian, H.; Shi, H. Sonar image segmentation based on an improved level set method. *Phys. Procedia* **2012**, *33*, 1168–1175. [CrossRef]
17. Awad, N.; Ali, M.; Suganthan, P.; Reynolds, R. CADE: A Hybridization of cultural algorithm and differential evolution for numerical optimization. *Inf. Sci.* **2017**, *378*, 215–241. [CrossRef]
18. Khatami, A.; Mirghasemi, S.; Khosravi, A.; Lim, C.P.; Nahavandi, S. A new PSO-based approach to fire flame detection using K-Medoids clustering. *Expert Syst. Appl.* **2017**, *68*, 69–80. [CrossRef]
19. Morra, L.; Coccia, N.; Cerquitelli, T. Optimization of computer aided detection systems: An evolutionary approach. *Expert Syst. Appl.* **2018**, *100*, 45–156. [CrossRef]
20. Wei, Z.; Bu, Y. Cultural particle swarm optimization algorithm and its application. In Proceedings of the 2012 24th Chinese Control and Decision Conference, Taiyuan, China, 23–25 May 2012; pp. 740–744.
21. Liu, T.; Jiao, L.; Ma, W.; Ma, J.; Shang, R. A new quantum-behaved particle swarm optimization based on cultural evolution mechanism for multiobjective problems. *Knowl. Based Syst.* **2016**, *101*, 90–99. [CrossRef]
22. Wang, X.; Hao, W.; Li, Q. An adaptive cultural algorithm with improved quantum-behaved particle swarm optimization for sonar image detection. *Sci. Rep.* **2017**, *7*, 17733. [CrossRef]
23. Wang, S.; Wang, D.; Li, C.; Li, Y.; Ding, G. Clustering by fast search and find of density peaks with data field. *Chin. J. Electron.* **2016**, *25*, 397–402. [CrossRef]
24. Zhang, Y.; Lu, K.; Gao, Y. Quantum algorithms and quantum-inspired algorithms. *Chin. J. Comput.* **2013**, *36*, 1835–1842. [CrossRef]
25. Wang, X.; Liu, S.; Li, Q.; Liu, Z. Underwater sonar image detection: A novel quantum-inspired shuffled frog leaping algorithm. *Chin. J. Electron.* **2018**, *27*, 588–594. [CrossRef]

26. Ding, W.; Wang, J.; Guan, Z.; Shi, Q. Enhanced minimum attribute reduction based on quantum-inspired shuffled frog leaping algorithm. *J. Syst. Eng. Electron.* **2013**, *24*, 426–434. [CrossRef]
27. Ding, W.; Wang, J.; Guan, Z. A minimum attribute self-adaptive cooperative co-evolutionary reduction algorithm based on quantum elitist frogs. *J. Comput. Res. Dev.* **2014**, *51*, 743–753.
28. Zhang, B.; Qi, H.; Sun, S.; Ruan, L.; Tan, H. Solving inverse problems of radiative heat transfer and phase change in semitransparent medium by using improved quantum particle swarm optimization. *Int. J. Heat Mass Transf.* **2015**, *85*, 300–310. [CrossRef]

Article

Comparison of Computational Intelligence Methods Based on Fuzzy Sets and Game Theory in the Synthesis of Safe Ship Control Based on Information from a Radar ARPA System

Józef Lisowski * and Mostefa Mohamed-Seghir

Department of Ship Automation, Gdynia Maritime University, 81-225 Gdynia, Poland; m.mohamed-seghir@we.umg.edu.pl

* Correspondence: j.lisowski@we.umg.edu.pl; Tel.: +48-694-458-333

Received: 28 November 2018; Accepted: 26 December 2018; Published: 4 January 2019

Abstract: This article presents safe ship control optimization design for navigator advisory system. Optimal safe ship control is presented as multistage decision-making in a fuzzy environment and as multistep decision-making in a game environment. The navigator's subjective and the maneuvering parameters are taken under consideration in the model process. A computer simulation of fuzzy neural anticollision (FNAC) and matrix game anticollision (MGAC) algorithms was carried out on MATLAB software on an example of the real navigational situation of passing three encountered ships in the Skagerrak Strait, in good and restricted visibility at sea. The developed solution can be applied in decision-support systems on board a ship.

Keywords: radar; fuzzy sets theory; artificial neural network; game theory; safe ship trajectory; computer simulation; computer decision support

1. Introduction

One of the most important transport issues is the safe control of movement, measured by the probability of collision risk when passing ships on the route, using information from the radar anticollision system [1,2]. In practice, there are many safe trajectories for the ship, from which an optimal trajectory can be chosen that ensures minimum collision risk and the smallest path loss on passing encountered ships [3–5]. The development of modern information technologies creates appropriate opportunities for the automation of navigation and construction of decision support systems to safety control the movement of a ship (Figure 1).

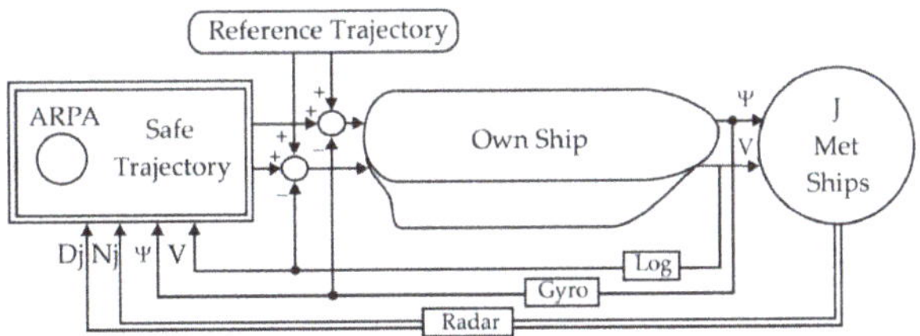

Figure 1. Safe ship-control system: D_j—distance to j-th met ship, N_j—bearing to j-th met ship, ψ—course of the own ship, V—speed of the own ship, J—number of met ships.

According to Lloyd's statistics, in about 87% of marine accidents, the cause of ship collisions is the navigator's subjectivity in maneuvering decisions, often under conditions of ambiguity and conflict.

Therefore, among the many possibilities of describing this process, models of fuzzy control and game control become useful [6–8].

The methods of static and dynamic optimization used so far, evolutionary algorithms or particle swarm methods, do not include the fuzzy and game properties of the real anticollision problems of the ship [9,10].

Therefore, the aim of this paper is to determine an optimal and safe ship trajectory using the theory of fuzzy sets and game theory (Figures 2 and 3).

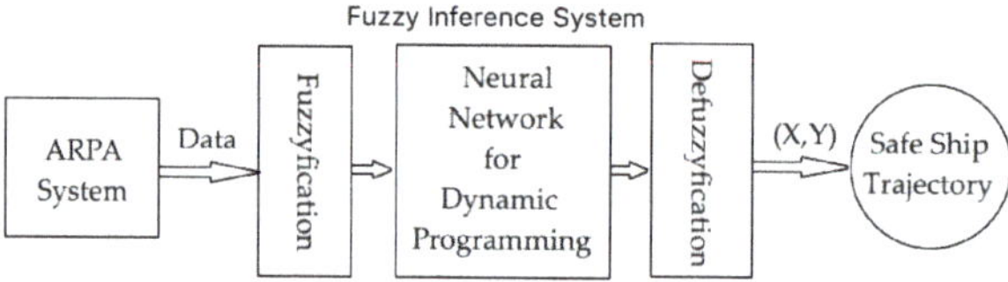

Figure 2. Fuzzy control system.

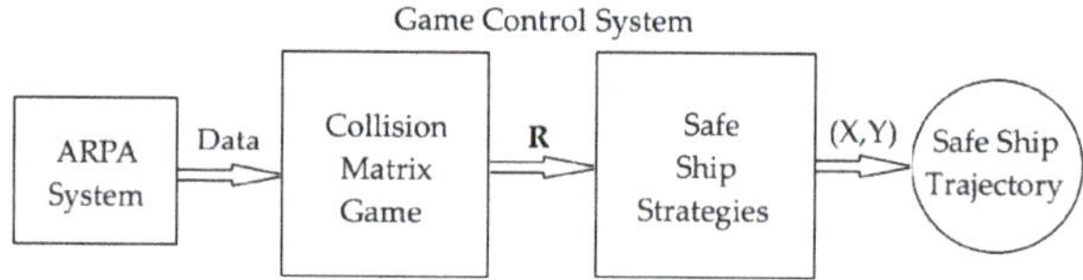

Figure 3. Game control system.

2. Kinematic Model of the Ship

In practice, the kinematics parameters of passing ships at sea in the form of:

- speed V_j,
- course ψ_j,
- distance of the closest point of approach $DCPA_j = D^j_{min}$,
- time to the closest point of approach $TCPA_j = T^j_{min}$.

These are identified by the automatic radar plotting aids (ARPA) anticollision system (Figure 4).

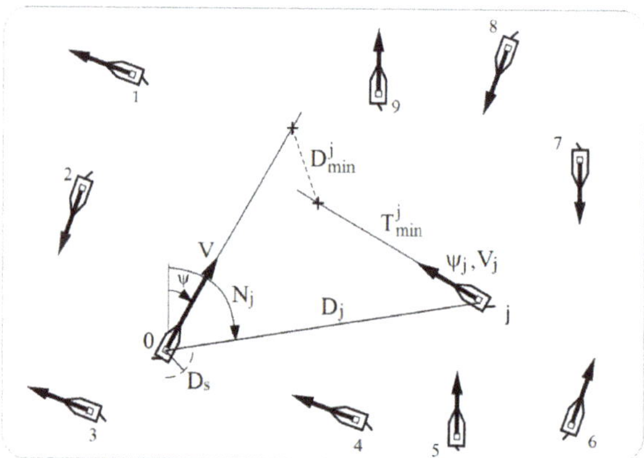

Figure 4. Situation of passing own ship with other ships.

The kinematic model of the control can be presented in the form of state equation

$$X_{k+1} = f\,(X_k, U_k),\ k = 1, 2, \ldots, n \tag{1}$$

where: (X_{k+1}, X_k) is the set of real ship position co-ordinates, and U_k is the control set [11].

The process ends when the ship reaches the back points called the final states

$$W = \{a_{p+1}, a_{p+2}, a_n\} \tag{2}$$

The set of final states must meet this condition

$$\left.\begin{array}{l} \psi_{opt} = \psi_S \\ V_{opt} = V_S \\ \mu_R \leq \mu_{Rsafe} \end{array}\right\} \tag{3}$$

where: (ψ_{opt}, V_{opt}) are the optimal own-ship course and speed, (ψ_S, V_S) are set point own-ship course and speed, and μ_R is the membership function of collision risk [12].

3. Fuzzy Control Model of the Process

3.1. Membership Function of Fuzzy Goal

Different security assessments made by navigators can be described as the membership function of the fuzzy goal, allowing a subjective assessment of

$$\mu_G(k,j) = 1 - \frac{1}{\exp(\lambda_d(k,j)DCPA_j^2)} \tag{4}$$

where λ_d is the navigator's subjective parameter.

3.2. Membership Function of Fuzzy Constraints

The membership function of the fuzzy constraints can be defined as

$$\mu_C(k) = \frac{1}{\exp(\lambda_c(k)(V\cos\psi(k) - V\cos\psi(k-1))t_k^2)} \tag{5}$$

where λ_c is the navigator's subjective parameter.

3.3. Membership Function of Fuzzy Collision Risk

Ships that can take part in a collision should be sorted according to the degree of threat by the collision-risk rating index. In many works, the ship's domain is treated as an assessment of collision risk [8]. In this article, the collision-risk transfer function was used as collision-risk assessment

$$\mu_R(k,j) = \frac{1}{\exp(\lambda_{rd}(k,j)DCPA_j^2 + \lambda_{rt}(k,j)TCPA_j^2)} \tag{6}$$

where λ_{rd} and λ_{rd} are the navigator's subjective parameters.

The fuzzy-set decision is determined as result of an operation of the fuzzy set of a goal and fuzzy set of constraints

$$\mu_D(.,.) = \mu_C(.,.) * \mu_G(.,.) \tag{7}$$

3.4. Fuzzy Neural Anticollision (FNAC) Algorithm

To present the idea of dynamic programming in this case, we first have to present the task in a slightly fuller form

$$\begin{aligned}&\mu_D(u_0,\ldots,u_{N-1}|X_0) = \\ &\max_{u_0,\ldots,u_{N-1}}[\mu_C^0(u_0) \wedge \mu_G^1(X_1) \wedge \mu_C^1(u_1) \wedge \mu_G^2(X_2) \wedge \ldots \\ &\ldots \wedge \mu_C^{N-1}(u_{N-1}) \wedge \mu_G^N(f(X_{N-1}, u_{N-1}))]\end{aligned} \quad (8)$$

After transformation and maximization in relation to controls $(u_0, u_1, \ldots, u_{N-1})$, we obtain the following system of recursive equations

$$\begin{cases} \mu_G^{N-i}(X_{N-i}) = \max\limits_{u_{N-i}}[\mu_C^{N-i}(u_{N-i}) \wedge \mu_G^{N-i+1}(f(X_{N-i+1})] \\ X_{N-i+1} = f(X_{N-i}, u_{N-i}) \; i = 0,1,\ldots,N \end{cases} \quad (9)$$

Referring to Formula (9), going back from stage t = N to t = 0, at each stage there are two phases: minimization and maximization. Such operations can be implemented using the special neural network proposed in [13]. The traditional artificial neural network does not perform the minimum and maximum operations of a finite set. Appropriate neurons were proposed by Rocha, which will be presented in the next sections [14–18].

3.4.1. Neural Network

The operations presented above require special neurons that can be used in an artificial neural network. Such neurons were proposed by Rocha [19], neurons of maximum type and minimum type. We assume that the neuron has n inputs $(b_1, b_2 \ldots, b_n)$ and the weighted sum of these n inputs is defined by the following formula

$$y = \sum_{k=1}^{n} w_k b_k \quad (10)$$

where w_k are the synapse weights connecting the input neurons.

The following pattern shows that the resulting obtained value u is encoded as the axonal activation b_p of the postsynaptic neuron.

$$b_p = \begin{cases} 1 & \text{if} \;\; u \geq \alpha_2 \\ f(u) & \text{if} \;\; \alpha_1 \leq u \leq \alpha_2 \\ 0 & \text{otherwise} \end{cases} \quad (11)$$

We have introduced two axonal thresholds α_1, α_2, which are defined by polarized neurons, where f transition function. We can now define the maximum-type neuron and the minimum-type neuron.

Maximum-Type Neuron

Defined as such whose axonal threshold α_t in stage t is described as

$$\alpha(t) = \begin{cases} 1 & \text{if} \;\; t = 0 \\ b_p(t-1) & \text{otherwise} \end{cases} \quad (12)$$

However, axionic activation b_p is presented as

$$b_p(t) = \begin{cases} \alpha(t) & \text{if}\, u(t) \leq \alpha(t) \\ u(t) & \text{otherwise} \end{cases} \quad (13)$$

where u(t) is the postsynaptic activation at stage t [20].

Combining the two functions, at the output of the max neuron, we obtain a maximum value of the inputs if weight $w_k = 1$

$$b_p(t) = \max_{k=1,2,\ldots,t} [w_k b_k] \tag{14}$$

Minimum-Type Neuron

Defined as a neuron whose axionic threshold α_t at stage t is

$$\alpha(t) = \begin{cases} 1 & \text{if } t = 0 \\ b_p(t-1) & \text{otherwise} \end{cases} \tag{15}$$

However, axionic activation

$$b_p(t) = \begin{cases} \alpha(t) & \text{if } u(t) \geq \alpha(t) \\ u(t) & \text{otherwise} \end{cases} \tag{16}$$

The equations show that the output of the min neuron encodes at least the minimum value of the inputs if weight $w_k = 1$

$$b_p(t) = \min_{k=1,2,\ldots,t} [w_k b_k] \tag{17}$$

In this way, defined neurons allow to build the neural network to solve the task of the optimal safe ship trajectory.

3.4.2. Structure of Neural Networks in Relation to Multistage Control

The neural network proposed in Reference [13] allows solving the task presented above. Its structure consists of alternating layers of minimum and maximum neurons. Weights values of neuron entrances are not given by learning in the ordinary sense, but result from the description the task, i.e., state transitions, fuzzy constraints, and fuzzy goals. Therefore, in order for the structure of the neural network to properly work, it is necessary to determine the connections between minimum and maximum neurons on the same layer, the maximum neurons of the preceding layer, and the minimum neurons on a current layer.

In the following, the neurons are described as

- $M^i{}_k$—max neuron at stage k,
- $m^i{}_k$—min neuron at stage k.

3.4.3. Generating Interconnections between Max and Min Neurons at the Same Layer

The connection of the two types of neurons from the same layer is done using the state-transitions equations, the connection function is as

$$W(m^i_k, M^j_k) = \begin{cases} 1 & \text{if } f_{N-k}(q_R(m^i_k), q_T(M^j_k)) \neq 0 \\ 0 & \text{otherwise} \end{cases} \tag{18}$$

where $q_R(m^i{}_k)$ is the number of receptor control, $q_T(M^i{}_k)$ is the number of relay states, $f_{N\text{-}k}(.,.)$ is the state-transition equation. A value of 1 means there is a connection, while 0 means no connection.

3.4.4. Generating Interconnections between Max Neurons and Min Neurons at the Given Layer

The combination of max neuron $M^j{}_{k-1}$ (layer k − 1) and min neuron $m^i{}_k$ (layer k) is executed by the number of receptors $q_R(m^i{}_k)$ and the number of the relay. It allows to obtain state x_{N-k} using the state equation, running neuron driver $q_C(m^i{}_k)$. That can be presented as the equation

$$q_C(m^i_k) = x_{N-k} = f_{N-k}(q_R(m^i_k), q_T(M^j_k)) \tag{19}$$

This driver is designed to send to all max neurons in layer (k − 1) the number of receptors $q_C(m^i{}_k)$. Neurons, which, due to the sent value, are activated, have the same value as receptor $q_R(M^l{}_{k-1})$. Just like in computer networks, between neuron $m^i{}_k$ and $M^l{}_{k-1}$, a connection is established, which can be defined as

$$W(M^l_{k-1}, m^j_k) = \begin{cases} 1 \text{ if } q_R(M^l_{k-1}) = q_T(m^i_k) = q_C(m^i_k) \\ 0 \text{ otherwise} \end{cases} \tag{20}$$

The connections presented in this fashion give the possibility to form an algorithm based on an artificial neural network, which will emulate solving the problem of dynamic programming in fuzzy environment, that is, the fuzzy neural anticollision (FNAC) algorithm [21].

The structure of the neural network to determine the safe ship trajectory is atypical, the network consists of six stages, in the first stage there are two layers of neurons, one max neuron and nine min neurons. The next has 9 max neurons and 32 min neurons. However, in third stage there are 25 max neurons and 38 min neurons. The penultimate stage has nine max neurons and nine min neurons. The last has one max neurons. The neurons weight result from the function state transitions and the membership function of fuzzy constraints and the membership function fuzzy goals.

Output step is to find a series of connections of maximum neurons, whose outputs have the highest value fuzzy decision μ_{Dmax}.

The initialization step it is to create a neural network. As in the dynamic programming steps proceed from the latter to zero (initial).

Thus, the steps of numbering 0 (the last stage with respect to X_0) to N (last step k, the first with respect to X_0) initialize the first layer of neurons maximum at stage k = 0, and increase by 1 (k = 1). From this step on, in each subsequent step we first initialize the layer of minimum neurons and then the layer of maximum neurons (Figure 5).

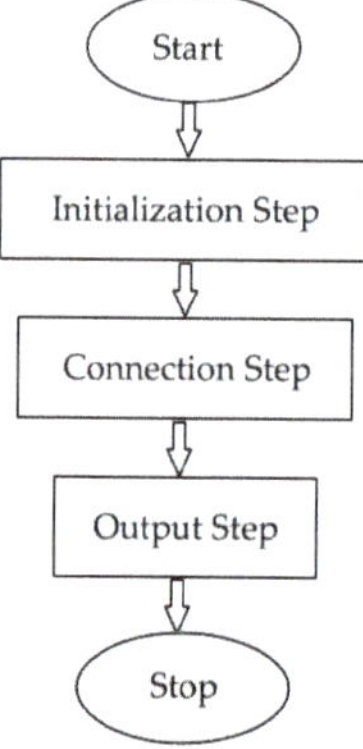

Figure 5. Fuzzy neural anticollision (FNAC) algorithm block diagram.

The connection step, as the name suggests, is to combine the maximum and minimum neurons of the same stage k (Phase 1), the maximum neurons at stage k − 1, and the minimum neurons at stage k (Phase 2). The output step is to find a series of connections of maximum neurons whose outputs have the highest value, μ_{Dmax}.

Of course, this is also true for control, which made it possible to obtain a value $\mu_D(u_0^*, \ldots, u_{N-1}^* | X_0)$ so there is also a connection with the minimum neurons. This time, these values are being sought in an order according to the initial state, that is x_0 to x_N (final state).

The initial state and the final state are single (this is related to the maneuvers of the ship). In some structures of the neural network, this may occur more than once, in the initial and final states, and, in this case, the connection may vary depending on the selection state at the initial stage.

Using the fuzzy toolbox and neural toolbox contained in the MATLAB_R2016a software, the fuzzy neural anti-collision (FNAC) computer program was designed for the determination of the safe own-ship trajectory in a collision situation [11].

4. Game Control Model of the Process

4.1. Base-Differential Game Model

The most general description of the own ship passing j other encountered ships is the model of a differential game of moving control objects (Figure 6).

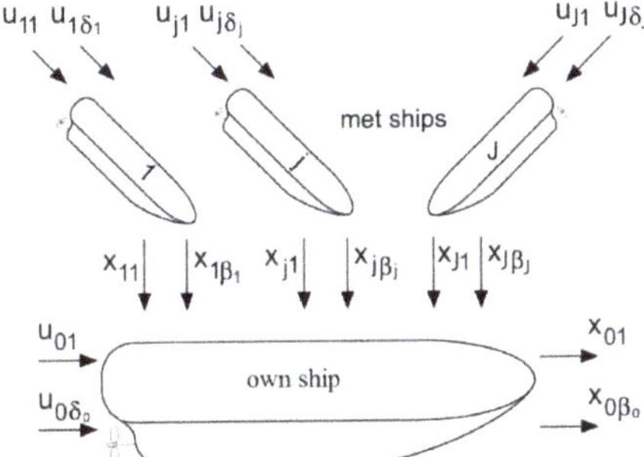

Figure 6. Block diagram of the basic model of differential game.

The properties of the process are described by the state equation

$$\begin{gathered}\dot{x}_i = f_i[(x_{0,\beta_0}, x_{1,\beta_1}, \ldots, x_{j,\beta_j}, \ldots, x_{J,\beta_J}), \\ (u_{0,\delta_0}, u_{1,\delta_1}, \ldots, u_{j,\delta_j}, \ldots, u_{J,\delta_J}), t] \\ j = 1, 2, \ldots, J\end{gathered} \tag{21}$$

where $\vec{x}_{0,\beta_0}(t)$ is the β_0 dimensional vector of the process state of the own-ship determined in a time span $t \in [t_0, t_k]$; $\vec{x}_{j,\beta_j}(t)$ is the β_j dimensional vector of the process state for the j-th met ship; $\vec{u}_{0,\delta_0}(t)$ is the δ_0 dimensional control vector of the own-ship; $\vec{u}_{j,\delta_j}(t)$ is the δ_j dimensional control vector of j-th met ship [22].

Control constraints and the state of the process are connected with the basic condition for the safe passing of the ships at a safe distance D_s in compliance with the International Regulations for Preventing Collisions at Sea (COLREGs Rules)

$$g_j(x_{j,\beta_j}, u_{j,\delta_j}) \leq 0 \tag{22}$$

Goal function has the form of the payments, the integral payment and the final one

$$I_{0,j} = \int_{t_0}^{t_k} [x_{0,\beta_0}(t)]^2 dt + r_j(t_k) + d(t_k) \rightarrow \min \tag{23}$$

The integral payment represents the additional distance traveled by the own-ship while passing the encountered ships and the final payment determines the final collision-risk $r_j(t_k)$ relative to the j ship and the final deflection of the own-ship $d(t_k)$ from the reference trajectory [23].

Two types of control goals were taken into consideration, programmed control $u_0(t)$ and positional control $u_0[x_0(t),t]$. The basis for the decision-making control are the decision-making patterns of the positional control processes, the patterns with the feedback arrangement representing the differential games.

While formulating the model of the control process, it is essential to take into consideration both the kinematics and the dynamics of the own-ship movement, the disturbances, the strategy of the encountered ships, and the assumed formula as the goal of the own-ship handling.

The diversity of selection of the possible models directly affects the synthesis of the own-ship control algorithms, which are afterwards affected by the ship-handling device, directly linked to the ARPA system and, consequently, determine the effects of safe and optimal control.

The application of reductions in the description of own-ship dynamics and the dynamics of the j-th encountered ship, and their movement kinematics, leads to the approximated model-positional and matrix.

4.2. Approximate Matrix Game Model

When leaving aside the own-ship dynamics equations, the general model of a differential game for the process of preventing collisions is reduced to the matrix game of J participants non-co-operating or co-operating among them. The state and control variables are represented by the values

$$\begin{gathered} x_{j1} = D_j;\ x_{j2} = N_j;\ u_{01} = \psi;\ u_{02} = V;\ u_{j1} = \psi_j;\ u_{j2} = V_j \\ j = 1,\ 2,\ \ldots,\ J \end{gathered} \tag{24}$$

4.3. Matrix Game Anticollision (MGAC) Algorithm

Collision matrix risk R includes the values previously determined on the basis of data taken from anticollision system ARPA; the value of collision-risk r_j with regard to the determined strategies of the own-ship and those of j-th encountered ships (Figure 7).

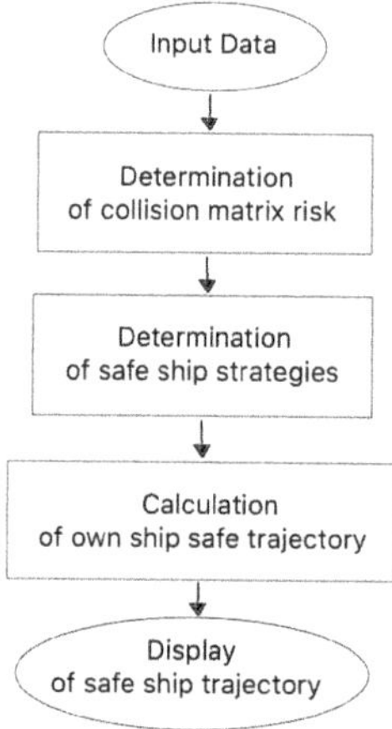

Figure 7. Matrix game anticollision (MGAC) algorithm block diagram.

The form of such a game is represented by the following risk matrix

$$R = [r_j(\delta_0, \delta_j)] \tag{25}$$

containing the same number of columns as the number of participant O (own-ship) strategies. It has; e.g., a constant course and speed, course alteration 20° to starboard, 20° to port, etc., and contains a number of lines that correspond to a joint number of participants J (j-th encountered ships) strategies

$$R = [r_j(\delta_0,\ \delta_j)] = \left|\begin{array}{ccccc} r_{11} & r_{12} & \cdots & r_{1,\nu_0-1} & r_{1\nu_0} \\ r_{21} & r_{22} & \cdots & r_{2,\delta_0-1} & r_{2\delta_0} \\ \cdots & \cdots & \cdots & \cdots & \cdots \\ r_{\delta_1 1} & r_{\delta_1 2} & \cdots & r_{\delta_1,\delta_0-1} & r_{\delta_1\delta_0} \\ \cdots & \cdots & \cdots & \cdots & \cdots \\ r_{\delta_j 1} & r_{\delta_j 2} & \cdots & r_{\delta_j,\delta_0-1} & r_{\delta_j\delta_0} \\ \cdots & \cdots & \cdots & \cdots & \cdots \\ r_{\delta_J 1} & r_{\delta_J 2} & \cdots & r_{\delta_J,\delta_0-1} & r_{\delta_J\delta_0} \end{array}\right| \quad (26)$$

The value of the collision-risk r_j is defined as the reference of the current situation of the approach described by parameters D^j_{min} and T^j_{min} to the assumed assessment of the situation defined as safe and determined by the safe distance of approach D_s and the safe time T_s—which are necessary to execute a maneuver avoiding a collision with consideration of actual distance D_j between the own-ship and the encountered j-th ship (Figure 8)

$$r_j = \left[\varepsilon_1\left(\frac{D^j_{min}}{D_s}\right)^2 + \varepsilon_2\left(\frac{T^j_{min}}{T_s}\right)^2 + \varepsilon_3\left(\frac{D_j}{D_s}\right)^2\right]^{-\frac{1}{2}} \quad (27)$$

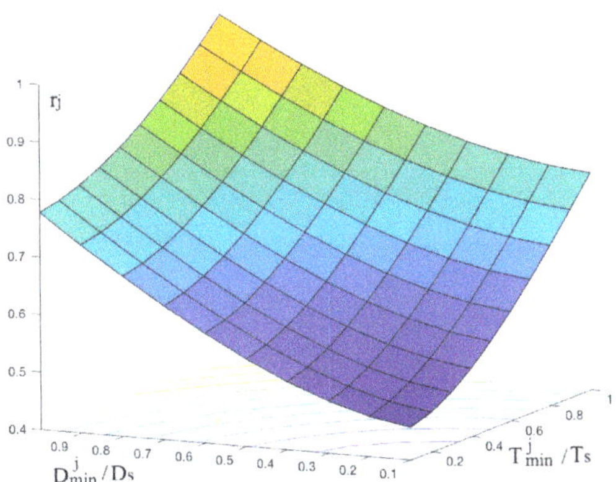

Figure 8. Example of the dependence of collision risk on the relative values of distance and time of approaching ships.

Weight coefficients ε_1, ε_2, ε_3 are dependent on the state of visibility at sea (good or restricted), the kind of water region (open or restricted), speed V of the ship, static L, and dynamic L_d length of ship, and static B and dynamic B_d beam of ship [10]

$$\left.\begin{array}{l} 0.1 \le \varepsilon_1 \le 10 \\ 0.1 \le \varepsilon_2 \le 10 \\ 0.1 \le \varepsilon_3 \le 10 \end{array}\right\} \quad (28)$$

$$L_d = 1.1\ (1 + 0.345\ V^{1.6}) \quad (29)$$

$$B_d = 1.1\ (B + 0.767\ LV^{0.4}) \quad (30)$$

Assuming higher values of particular weight coefficients ε_1, ε_2 and ε_3; the share of the risk of collision r_j depending on the distance of ships D_j and $D^j{}_{min}$ or the time of excessive approach of ships $T^j{}_{min}$ increases.

The constraints affecting the choice of strategies are a result of the recommendations of the way of priority at sea. Player O (own-ship) may use δ_0 of various pure strategies in a matrix game and player J (encountered ships) has δ_j of various pure strategies.

As the game, most frequently does not have a saddle point, the state of balance is not guaranteed, and there is a lack of pure strategies for both players in the game. In order to solve this problem, dual linear programming may be used.

In a dual problem, player O, having δ_0 various strategies to be chosen, tries to minimize the risk of collision

$$I_0 = \min_{\delta_0} r_j \tag{31}$$

while player J, having δ_j strategies to be chosen, tries to maximize the collision-risk

$$I^j = \max_{\delta_j} r_j \tag{32}$$

For a non-co-operative matrix game, the problem of determining an optimal strategy may be reduced to the task of solving a dual linear programming problem [24]

$$\left(I_0^j\right)^* = \min_{\delta_0} \max_{\delta_j} r_j \tag{33}$$

For a co-operative matrix game, the problem of determining an optimal strategy may be reduced to the task of solving a dual linear programming problem

$$\left(I_0^j\right)^* = \min_{\delta_0} \min_{\delta_j} r_j \tag{34}$$

Mixed strategy components express probability distribution $P = [p_j(\delta_0,\delta_j)]$ of players using pure strategies

$$P = [p_j(\delta_0,\ \delta_j)] = \left|\begin{array}{ccccc} p_{11} & p_{12} & \cdots & p_{1,\delta_0-1} & p_{1\delta_0} \\ p_{21} & p_{22} & \cdots & p_{2,\delta_0-1} & p_{2\delta_0} \\ \cdots & \cdots & \cdots & \cdots & \cdots \\ p_{\delta_1 1} & p_{\delta_1 2} & \cdots & p_{\delta_1,\delta_0-1} & p_{\delta_1\delta_0} \\ \cdots & \cdots & \cdots & \cdots & \cdots \\ p_{\delta_j 1} & p_{\delta_j 2} & \cdots & p_{\delta_j,\delta_0-1} & p_{\delta_j\delta_0} \\ \cdots & \cdots & \cdots & \cdots & \cdots \\ p_{\delta_J 1} & p_{\delta_J 2} & \cdots & p_{\delta_J,\delta_0-1} & p_{\delta_J\delta_0} \end{array}\right| \tag{35}$$

The solution for the steering goal is the strategy of the highest probability and will also be the optimal value approximated to the pure strategy

$$\left(u_0^{\delta_0}\right)^{\bullet} = u_0^{\delta_0}\left\{\left[p_j(\delta_0, \delta_j)\right]_{max}\right\} \tag{36}$$

The safe trajectory of the own-ship was treated here as a sequence of changes to the course and speed. The established values are as follows: safe passing distances among the ships under given visibility conditions at sea D_s, time delay of manoeuvring and the duration of one stage of the trajectory as one calculation step. At each step, the most dangerous ship is determined with regard to the value of collision risk r_j.

Consequently, on the basis of the semantic interpretation of the COLREGs, the direction of a turn of the own-ship is selected to the most dangerous encountered ship [25–27].

Collision matrix risk R is determined for the admissible strategies of the own-ship δ_0 and those δ_j for j-th ships encountered. By applying dual linear programming in order to solve the matrix game, we obtain the optimal values of the own-ship course and those of the j-th ship at the smallest deviation from their initial values.

If, at a given step, no solution can be found at the speed of the own-ship V, the calculations are repeated at reduced speed by 25% until the game is solved.

The calculations are repeated step by step until the moment when all elements of matrix R become equal to zero, and the own-ship, after having passed the encountered ships, returns to its initial course and speed.

Using the *linprog* function, that is, linear programming from the *optimtool* optimization toolbox contained in the MATLAB_R2017a software, the matrix game anti-collision (MGAC) computer program was designed for the determination of the safe own-ship trajectory in a collision situation [10].

5. Research Results

The aim of the computer simulation research of the FNAC and MGAC algorithms to determine the optimal safe-ship trajectory in collision situations was to evaluate methods to solve the problem formulated in this work by using fuzzy-set theory as a multistage and matrix game theory as a multistep decision-making process.

The computer simulation of the FNAC and MGAC algorithms was carried out in MATLAB software on an example of the real navigational situation of passing J = 3 encountered ships in the Skagerrak Strait in good visibility D_s = 0.2 nm and the restricted visibility D_s = 2.0 nm (nautical miles) (Table 1).

Table 1. Data of own-ship and met ships: 1, 2 and 3.

	Bearing N_j (°)	Distance D_j (nm)	Speed V_j (kn)	Course ψ_j (°)
Own-ship	-	-	20	0
Ship 1	326	8.8	13.5	90
Ship 2	6	14.3	16.2	180
Ship 3	11	7.5	16.0	200

The situation was registered on board r/v HORYZONT II, a research and training vessel of the Gdynia Maritime University, on the radar screen of the ARPA anticollision system Raytheon. The sample results of the performed computer simulations for the navigational situation when the own-ship passed three met ships in a good and restricted visibility at sea are presented in Figures 9 and 10, respectively.

The trajectories of the own-ship, shown in Figures 9 and 10, are optimal, but in different ways.

The FNAC trajectory ensures minimum risk of collision of own-ship, taking into account the uncertainty of the control process, described by the fuzzy set membership functions of state and control constraints, and collision risk, not including maneuvers of encountered ships.

On the other hand, the trajectory of MGAC ensures a minimum collision-risk of the own-ship taking into account the maneuvering of encountered ships in a co-operative or non-co-operative way.

Designated safe trajectories FNAC and MGAC of the own-ship are the reference trajectories for automatic ship's control systems using the autopilot and main engine.

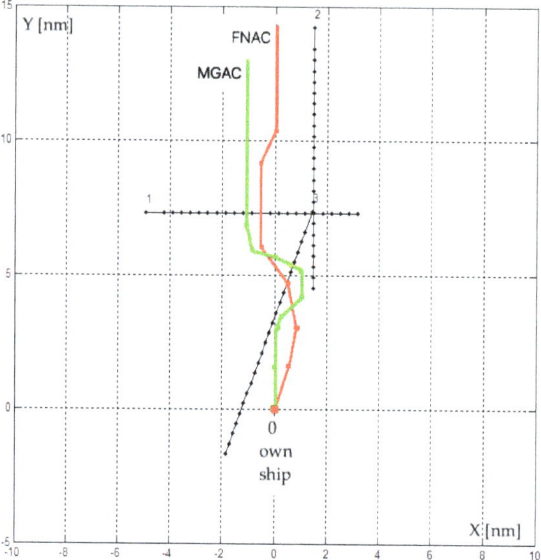

Figure 9. Comparison of safe and optimal trajectories of own-ship in the situation of passing three ships, in conditions of good visibility at sea at D_s = 0.2 nm, determined using the following algorithms: FNAC decision chosen trajectory = 0.5766; MGAC final payment = 1.08 nm.

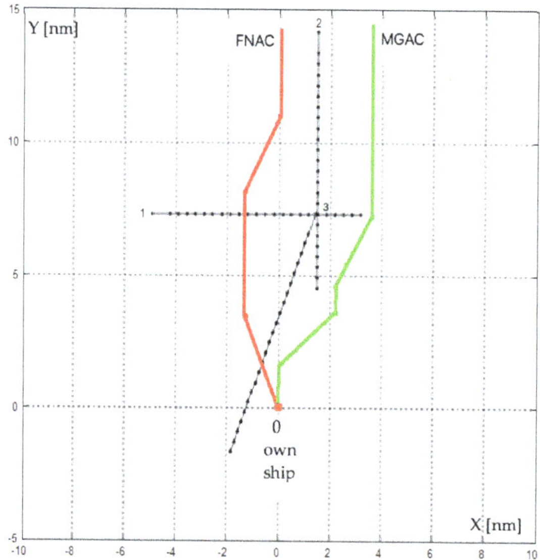

Figure 10. Comparison of safe and optimal trajectories of own-ship in the situation of passing three ships, in conditions of restricted visibility at sea at D_s = 2.0 nm, determined using the following algorithms: FNAC decision chosen trajectory = 0.6452; MGAC final payment = 3.83 nm.

6. Conclusions

This work showed that the proposed concept of applying the fuzzy neural anticollision and matrix game anticollision algorithms is a promising way to solve the considered problem and design a novel anticollision system, allowing to increase the safety of maritime transport.

The FNAC algorithm achieves a short computation time of about one second, while the MGAC algorithm scope of the calculation time is from about two to five seconds.

The obtained solutions obtained using both algorithms meet the requirements of the COLREGs.

The MGAC computer program, designed in MATLAB software, takes into consideration the following: degree of co-operation with the own-ship and encountered ships, COLREGs Rules, advance time for a maneuver calculated with regard to the own-ship dynamic features, and the assessment of the final deviation between the real and reference trajectories.

In summary of these results, the created algorithms can be used as a tool to assist navigator in making maneuver decision, in complex collision situations, when passing through more ships, especially in restricted visibility at sea.

In future works, the sensitivity analysis of safe ship control should be performed to change the parameters of the process model and the inaccuracy of information from the ARPA radar system; moreover, the design of the ARPA system may be considered, extended with the function of a computer-aided maneuvering navigator decision, using the FNAC and MGAC algorithms.

Author Contributions: Conceptualization, J.L.; Methodology, M.M.-S.; Software, M.M.-S.; Validation, J.L. and M.M.-S.; Formal analysis, J.L. and M.M.-S.; Investigation, J.L. and M.M.-S.; Resources, J.L. and M.M.-S.; Data curation, M.M.-S.; Writing—original draft preparation, M.M.-S.; Writing—review and editing, J.L.; Visualization, J.L.; Supervision, J.L.; Project administration, J.L.; Funding acquisition, M.M.-S.

Funding: This research was funded by a statute research project of Gdynia Maritime University in Poland, No. 446/DS/2018: "Design and simulation tests of marine automation systems in MATLAB/Simulink and LabVIEW software".

Conflicts of Interest: The authors declare no conflict of interest regarding the publication of this paper. The funders had no role in the design of the study; in the collection, analyses, or interpretation of data; in the writing of the manuscript; or in the decision to publish the results.

Nomenclature

A_p	axonic activation
C	fuzzy-set goal
D	fuzzy-set decision
D_s	safe distance of approach
D_j	distance between own-ship and the j-th met ship
DCPA	distance to closest point of approach
G	fuzzy-set contraints
P	probability distribution
R	collision-risk matrix
r_j	value of the collision-risk
u_t	controls
TCPA	time to closest point of approach
T_s	safe time of approach
U	control-set
$u_k(t)$	postsynaptic activation at stage t
V	ship speed
V_{opt}	optimal ship speed
W	set of final states
X_{t+1}, X_t	ship position co-ordinates

X	set of real ship position co-ordinates
α_t	axonic threshold at stage t
μ_R	membership function of fuzzy-set collision-risk
μ_{Rsafe}	value of μ_R at which the process is assumed safe
$\lambda_c, \lambda_d, \lambda_{rd}, \lambda_{rt}$	navigator's subjective parameters
ψ	ship course
ψ_{opt}	optimal ship course
$\wedge$	minimum operator

References

1. Bist, D.S. *Safety and Security at Sea*; Butterworth Heinemann: Oxford, UK; New Delhi, India, 2000; ISBN 0-75064-774-4.
2. Kazimierski, W.; Stateczny, A. Fusion of Data from AIS and Tracking Radar for the Needs of ECDIS. In Proceedings of the Signal Processing Symposium, Jachranka, Poland, 5–7 June 2013.
3. Borkowski, P. Inference engine in an intelligent ship course-keeping system. *Comput. Intell. Neurosci.* **2017**, *2017*, 1–9. [CrossRef] [PubMed]
4. Tomera, M. Nonlinear observers design for multivariable ship motion control. *Pol. Marit. Res.* **2012**, *19*, 50–56. [CrossRef]
5. Lebkowski, A. Design of an autonomous transport system for coastal areas. *TransNav Int. J. Mar. Navig. Saf. Sea* **2018**, *12*, 117–124. [CrossRef]
6. Kazimierski, W.; Lubczonek, J. Verification of Marine Multiple Model Neural Tracking Filter for the Needs of Shore Radar Stations. In Proceedings of the 13 International Radar Symposium, Warsaw, Poland, 23–25 May 2012; pp. 534–539.
7. Bellman, R.E.; Zadeh, L.A. Decision making in a fuzzy environment. *Manag. Sci.* **1970**, *17*, 12–19. [CrossRef]
8. Isaacs, R. *Differential Games*; John Wiley and Sons: New York, NY, USA; London, UK, 1965; ISBN 0-48640-682-2.
9. Szlapczynski, R.; Szlapczynska, J. An analysis of domain-based ship collision risk parameters. *Ocean Eng.* **2016**, *126*, 47–56. [CrossRef]
10. Lisowski, J. Game control methods in avoidance of ships collision. *Pol. Marit. Res.* **2012**, *19*, 3–10. [CrossRef]
11. Mohamed-Seghir, M. The branch-and-bound method, genetic algorithm, and dynamic programming to determine a safe ship trajectory in fuzzy, knowledge-based and intelligent information and engineering systems. *Procedia Comput. Sci.* **2014**, *35*, 348–357.
12. Modarres, M. *Risk Analysis in Engineering*; Taylor and Francis Group: Boca Raton, FL, USA, 2006; ISBN 1-57444-794-7.
13. Francelin, R.; Kacprzyk, J.; Gomide, F. Neural network based algorithm for dynamic system optimization. *Asian J. Contr.* **2001**, *3*, 131–142. [CrossRef]
14. Lyu, H.; Yin, Y. Fast path planning for autonomous ships in restricted waters. *Appl. Sci.* **2018**, *12*, 2592. [CrossRef]
15. Deng, W.; Gan, L.; Zhou, C.; Zheng, Y.; Liu, M.; Zhang, L. Study on Path Planning of Ship Collision Avoidance in Restricted Water base on AFS Algorithm. In Proceedings of the 27th International Ocean and Polar Engineering Conference, San Francisco, CA, USA, 25–30 June 2017; pp. 1–7.
16. Gia, H.D.; Nam-Kyun, I. Study on the construction of stage discrimination model and consecutive waypoints generation method for ship's automatic avoiding action. *Int. J. Fuzzy Log. Intell. Syst.* **2017**, *17*, 294–306.
17. Lyu, H.; Yin, Y. COLREGS-constrained real-time path planning for autonomous ships using modified artificial potential fields. *J. Navig.* **2018**, 1–21. [CrossRef]
18. Lazarowska, A. A new deterministic approach in a decision support system for ship's trajectory planning. *Expert Syst. Appl.* **2017**, *71*, 469–478. [CrossRef]
19. Rocha, A.F. *Neural Nets—A Theory of Brain an Machines*; Springer: Berlin, Germany; New York, NY, USA, 1992; ISBN 0-8493-2643-5.
20. Kacprzyk, J.; Romeo, R.A.; Gomide, F.A.C. Involving objective and subjective aspects in multistage decision making and control under fuzziness: dynamic programming and neural network. *Int. J. Intell. Syst.* **1999**, *14*, 79–104. [CrossRef]

21. Pedrycz, W.; Gomide, E. *Fuzzy Systems Engineering Toward Human Centric Computing*; Wiley: Hoboken, NJ, USA, 2007; ISBN 978-0-471-78857-7.
22. Osborne, M.J. *An Introduction to Game Theory*; Oxford University Press: New York, NY, USA, 2004; ISBN 978-0-19-512895-6.
23. Lisowski, J. Optimization-supported decision-making in the marine game environment. *Solid State Phenom.* **2014**, *210*, 215–222. [CrossRef]
24. Basar, T.; Olsder, G.J. *Dynamic Non-Cooperative Game Theory*; SIAM: Philadelphia, PA, USA, 2013; ISBN 978-0-89871-429-6.
25. Nisan, N.; Roughgarden, T.; Tardos, E.; Vazirani, V.V. *Algorithmic Game Theory*; Cambridge University Press: New York, NY, USA, 2007; ISBN 978-0-521-87282-9.
26. Millington, I.; Funge, J. *Artificial Intelligence for Games*; Elsevier: Amsterdam, The Netherlands; Tokyo, Japan, 2009; ISBN 978-0-12-374731-0.
27. Engwerda, J.C. *LQ Dynamic Optimization and Differential Games*; John Wiley and Sons: West Sussex, UK, 2005; ISBN 978-0-470-01524-7.

MDPI
St. Alban-Anlage 66
4052 Basel
Switzerland
Tel. +41 61 683 77 34
Fax +41 61 302 89 18
www.mdpi.com

Remote Sensing Editorial Office
E-mail: remotesensing@mdpi.com
www.mdpi.com/journal/remotesensing

www.ingramcontent.com/pod-product-compliance
Lightning Source LLC
LaVergne TN
LVHW071622170726
843515LV00009B/2460